MW00768456

WIN the complete library of RSMeans Online data!

Register your book below to receive your FREE quarterly updates, PLUS you will be entered into a quarterly drawing to win the COMPLETE RSMeans Online library of 2015 data!

Be sure to keep up-to-date in 2015!

Fill out the card below for RSMeans' free quarterly updates,
as well as a chance to win the complete RSMeans Online library.
Please provide your name, address, and email below and return
this card by mail, or to register online:

http://info.thegordiangroup.com/RSMeans.html

Name _____

email_____ Title_____

Company _____

Street _____

City/Town _____ State/Prov. _____ Zip/Postal Code _____

RSMeans

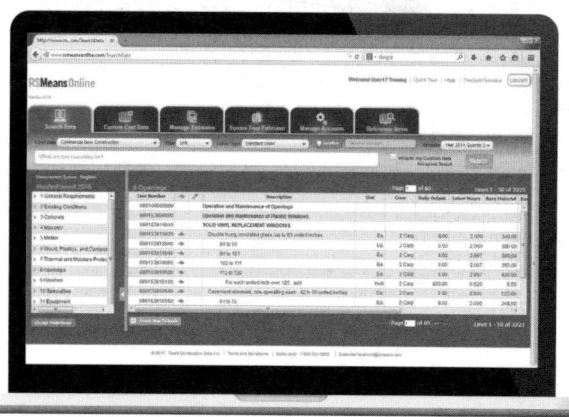

WIN the complete library of RSMeans Online 2015 data!

See other side for details.

RSMeans Facilities Maintenance & Repair Cost Data

2015 22nd annual edition

Stephen C. Plotner, Senior Editor

Engineering Director
Bob Mewis, CCC (*1, 14, 41*)

Contributing Editors
Christopher Babbitt
Adrian C. Charest, PE
(*26, 27, 28, 48*)
Gary W. Christensen
Cheryl Elsmore
David Fiske
Robert Fortier, PE
(*2, 31, 32, 33, 34, 35, 44, 46*)
Robert J. Kuchta (*8*)

Thomas Lane (*6, 7*)
Robert C. McNichols
Genevieve Medeiros
Melville J. Mossman, PE (*21, 22, 23*)
Jeannene D. Murphy
Marilyn Phelan, AIA (*9, 10, 11, 12*)
Stephen C. Plotner (*3, 5*)
Siobhan Soucie (*4, 13*)

Vice President Data & Engineering
Chris Anderson

Cover Design
Sara Rutan

Product Manager
Andrea Sillah

Production Manager
Debbie Panarelli

Production
Jill Goodman
Jonathan Forgit
Sara Rutan
Mary Lou Geary

Technical Support
Gary L. Hoitt
Kathryn S. Rodriguez

Numbers in italics are the divisional responsibilities for each editor.
Please contact the designated editor directly with any questions.

RSMeans
Construction Publishers & Consultants
700 Longwater Drive
Norwell, MA 02061
United States of America
1-877-801-5724
www.rsmeans.com

Copyright 2014 by RSMeans
All rights reserved.
Cover photo ©
jongjet303/iStock/Thinkstock

Printed in the United States of America
ISSN 1074-0953
ISBN 978-1-940238-56-2

| 0305 | $461.95 per copy (in United States) |
| | Price is subject to change without prior notice |

Related RSMeans Products and Services

The engineers at RSMeans suggest the following products and services as companion information resources to *RSMeans Facilities Maintenance & Repair Cost Data:*

Construction Cost Data Books
Facilities Construction Cost Data 2015
Electrical Cost Data 2015
Mechanical Cost Data 2015
Plumbing Cost Data 2015

Reference Books
Facilities Planning and Relocation
Life Cycle Costing for Facilities
Cost Planning & Estimating for Facilities Maintenance
Estimating Building Costs
RSMeans Estimating Handbook
Building Security: Strategies & Costs

Seminars and In-House Training
Facilities Maintenance & Repair Estimating
Unit Price Estimating
RSMeans Online® Training
Mechanical & Electrical Estimating
Scheduling with MSProject for Construction Professionals

RSMeans Online Bookstore
Visit RSMeans at **www.rsmeans.com** for a one-stop portal for the most reliable and current resources available on the market. Here you can learn about our more than 20 annual cost data books and eBooks, also available on CD and online versions, as well as our libary of reference books and RSMeans seminars.

RSMeans Electronic Data
Get the information found in RSMeans cost books electronically at **RSMeansOnline.com**.

RSMeans Business Solutions
Cost engineers and analysts conduct facility life cycle and benchmark research studies and predictive cost modeling, as well as offer consultation services for conceptual estimating and real property management. Research studies are designed with the application of proprietary cost and project data from RSMeans' extensive North American databases. Analysts offer building product manufacturers qualitative and quantitative market research, as well as time/motion studies for new products, and Web-based dashboards for market opportunity analysis. Clients are from both the public and private sectors, including federal, state, and municipal agencies; corporations; institutions; construction management firms; hospitals; and associations.

Construction Costs for Software Applications
More than 25 unit price and assemblies cost databases are available through a number of leading estimating and facilities management software partners. For more information see the "Other RSMeans Products and Services" pages at the back of this publication.

RSMeans data is also available to federal, state, and local government agencies as multi-year, multi-seat licenses.

For information on our current partners, visit: **www.rsmeans.com/partners** or call 1-877-807-5724.

Table of Contents

Foreword

Our Mission

Since 1942, RSMeans has been actively engaged in construction cost publishing and consulting throughout North America.

Today, more than 70 years after RSMeans began, our primary objective remains the same: to provide you, the construction and facilities professional, with the most current and comprehensive construction cost data possible.

Whether you are a contractor, owner, architect, engineer, facilities manager, or anyone else who needs a reliable construction cost estimate, you'll find this publication to be a highly useful and necessary tool.

With the constant flow of new construction methods and materials today, it's difficult to find the time to look at and evaluate all the different construction cost possibilities. In addition, because labor and material costs keep changing, last year's cost information is not a reliable basis for today's estimate or budget.

That's why so many construction professionals turn to RSMeans. We keep track of the costs for you, along with a wide range of other key information, from city cost indexes . . . to productivity rates . . . to crew composition . . . to contractor's overhead and profit rates.

RSMeans performs these functions by collecting data from all facets of the industry and organizing it in a format that is instantly accessible to you. From the preliminary budget to the detailed unit price estimate, you'll find the data in this book useful for all phases of construction cost determination.

The Staff, the Organization, and Our Services

When you purchase one of RSMeans' publications, you are, in effect, hiring the services of a full-time staff of construction and engineering professionals.

Our thoroughly experienced and highly qualified staff works daily at collecting, analyzing, and disseminating comprehensive cost information for your needs. These staff members have years of practical construction experience and engineering training prior to joining the firm. As a result, you can count on them not only for accurate cost figures, but also for additional background reference information that will help you create a realistic estimate.

The RSMeans organization is always prepared to help you solve construction problems through its variety of data solutions, including online, CD, and print book formats, as well as cost estimating expertise available via our business solutions, training, and seminars.

Besides a full array of construction cost estimating books, RSMeans also publishes a number of other reference works for the construction industry. Subjects include construction estimating and project and business management, green building, and a library of facility management references.

In addition, you can access all of our construction cost data electronically in convenient CD format or on the web. Visit **RSMeansOnline.com** for more information on our 24/7 online cost data.

What's more, you can increase your knowledge and improve your construction estimating and management performance with an RSMeans construction seminar or in-house training program. These two-day seminar programs offer unparalleled opportunities for everyone in your organization to become updated on a wide variety of construction-related issues.

RSMeans is also a worldwide provider of construction cost management and analysis services for commercial and government owners.

In short, RSMeans can provide you with the tools and expertise for constructing accurate and dependable construction estimates and budgets in a variety of ways.

Robert Snow Means Established a Tradition of Quality That Continues Today

Robert Snow Means spent years building RSMeans, making certain he always delivered a quality product.

Today, at RSMeans, we do more than talk about the quality of our data and the usefulness of our books. We stand behind all of our data, from historical cost indexes to construction materials and techniques to current costs.

If you have any questions about our products or services, please call us toll-free at 1-877-807-5724. Our customer service representatives will be happy to assist you. You can also visit our website at **www.rsmeans.com**

How the Book Is Built: An Overview

A Complete Reference for the Facility Manager

This book was designed as a complete reference and cost data source for facility managers, owners, engineers, and contractors, or anyone who manages real estate. A complete facilities management program starts with a building audit. The audit describes the structures, their characteristics and major equipment, and apparent deficiencies. An audit is not a prerequisite to using any data contained in this book, but it does provide an organizational framework. A list of equipment and deficiencies forms the basis for a repair budget and preventive maintenance program. An audit also provides facility size and usage criteria on which the general maintenance (cleaning) program will be based. Together, the five sections of this book provide a framework and definitive data for developing a complete program for all facets of facilities maintenance.

Even if you have a regular building audit and preventive maintenance program in place, the information in this book will be useful.

For the occasional user, the book provides:

- a reference for time and material requirements for Maintenance & Repair tasks to fill in where your organization lacks its own historical data;
- Preventive Maintenance (PM) checklists with labor-hour standards that can be used to benchmark your program;
- a quick reference for General Maintenance (cleaning) productivity rates to analyze proposals from outside contractors;
- an explanation of Life Cycle Cost Analysis as a tool to assist in making the correct budget decisions; and
- data to validate line items in budgets.

For the frequent user, the book supplies all of the above, plus:

- guidance for getting the audit under way;
- a foundation for zero-based budgeting of maintenance & repair, preventive maintenance, and general maintenance;
- cost data that can be used to estimate preventive maintenance, deferred maintenance (repairs), and general maintenance (cleaning) programs; and
- detailed task descriptions for establishing a complete PM program.

How to Use the Book: The Details

This section contains an in-depth explanation of how the book is arranged . . . and how you can use it to determine a reliable construction cost estimate. It includes information about how we develop our cost figures and how to completely prepare your estimate.

Section 1: Maintenance & Repair (M&R)

The Maintenance & Repair section is a listing of common maintenance tasks performed at facilities. The tasks include removal and replacement, repair, and refinishing. This section is organized according to the UNIFORMAT II system. (UNIFORMAT was originally developed for the General Services Administration as a systems approach to organizing building construction and has since become widely used in the building industry. It is now maintained and published by CSI, the Construction Specifications Institute.)

The purpose of Section 1 is to provide cost data and an approximate frequency of occurrence for each maintenance and repair task. This data is used by facility managers, owners, plant engineers, and contractors to prepare estimates for deferred maintenance projects. The data is also used to estimate labor-hours and material costs for in-house staff. The frequency listing facilitates preparation of a zero-based budget.

Under the System Description column, each task, including its components, is described. In some cases additional information (such as "2% of total roof area") appears in parentheses. The next column, Frequency, refers to how often one should expect, and therefore estimate, that this task will have to be performed. Labor hours have been enhanced, where appropriate, by 30% to include setup and delay time. In some cases, the list of steps comprising the task includes most of the potential repairs that could be made on a system. For those tasks, individual items, as appropriate, can be selected from the all-encompassing list.

Section 2: Preventive Maintenance (PM)

The Preventive Maintenance section provides the framework for a complete PM program. The establishment of a program is frequently

Maintenance & Repair

B30	ROOFING			B3013		Roof Covering						

B3013	105			Built-Up Roofing								

	System Description	Freq. (Years)	Crew	Unit	Labor Hours	2015 Bare Costs				Total In-House	Total w/O&P
						Material	Labor	Equipment	Total		
0600	Place new BUR membrane over existing	20	G-5	Sq.						1.20	1.45
	Set up, secure and take down ladder				.020		.80		.80	1.20	1.45
	Sweep / spud ballast clean				.500		20		20	30	36
	Vent existing membrane				.130		5.20		5.20	7.80	9.40
	Cut out buckled flashing				.024		.96		.96	1.44	1.74
	Install 2 ply membrane flashing				.027	.39	1.06		1.45	2.02	2.42
	Install 4 ply bituminous roofing				3.733	136	140	38	314	395	465
	Reinstall ballast				.381		15.30		15.30	23	27.50
	Clean up				.390		15.65		15.65	23.50	28.50
	Total				5.205	136.39	198.97	38	373.36	483.96	572.01
0700	Total BUR roof replacement	28	G-1	Sq.							
	Set up, secure and take down ladder				.020		.80		.80	1.20	1.45
	Sweep / spud ballast clean				.500		20		20	30	36
	Remove built-up roofing				2.500		95		95	126	156
	Remove insulation board				1.026		39		39	52	64
	Remove flashing				.024		.96		.96	1.44	1.74
	Install 2" perlite insulation				.879	100	35		135	163	189
	Install 2 ply membrane flashing				.027	.39	1.06		1.45	2.02	2.42
	Install 4 ply bituminous membrane				2.800	136	105	28.50	269.50	335	390
	Clean up				1.000		40		40	60	72.50
	Total				8.776	236.39	336.82	28.50	601.71	770.66	913.11

B3013	120			Modified Bituminous / Thermoplastic								

	System Description	Freq. (Years)	Crew	Unit	Labor Hours	2015 Bare Costs				Total In-House	Total w/O&P
						Material	Labor	Equipment	Total		
0100	Debris removal by hand & visual inspection	1	2 ROFC	M.S.F.							
	Set up, secure and take down ladder				.052		2.08		2.08	3.12	3.76
	Pick up trash / debris & clean up				.327		13.10		13.10	19.60	23.50
	Visual inspection				.327		13.10		13.10	19.60	23.50
	Total				.705		28.28		28.28	42.32	50.76

hindered by the lack of a comprehensive list of equipment, actual PM steps, and budget documentation. This section fulfills all these needs. The facility manager, plant engineer, or owner can use these schedules to establish labor-hours and a budget. The maintenance contractor can use these PM checklists as the basis for a comprehensive maintenance proposal or as an estimating aid when bidding PM contracts. Selected pages, with preventive maintenance steps, can be photocopied and assembled into a customized PM program. Copies of individual sheets can also be distributed to maintenance personnel to identify the procedures required. The hours listed are predicated on work being performed by experienced technicians familiar with the

PM tasks and equipped with the proper tools and materials.

The PM section lists the tasks and their frequency, whether weekly, monthly, quarterly, semi-annually, or annually. The frequency of those procedures is based on non-critical usage (e.g., in "normal use" situations, versus in facilities such as surgical suites or computer rooms that demand absolute adherence to a limited range of environmental conditions). The labor-hours to perform each item on the checklist are listed in the next column. Beneath the table is cost data for the PM schedule. The data is shown both annually and annualized, to provide the facility manager with an estimating range. If all tasks on the schedule are performed

once a year, the *annually* line should be used in the PM estimate. The *annualized* data is used when all items on the schedule are performed at the frequency shown.

Section 3: General Maintenance (GM)

This section provides labor-hour estimates and costs to perform day-to-day cleaning. The data is used to estimate cleaning times, compare and assess estimates submitted by cleaning companies, or to budget in-house staff. The information is divided into *Interior* and *Exterior* maintenance. A common maintenance laborer (*Clam*) is used to perform these tasks.

Section 4: Facilities Audits

The Facilities Audit is generally the basis from which the Maintenance & Repair, Preventive Maintenance, and General Maintenance estimates are prepared. The audit provides the following:

1. A list of deficiencies for the Maintenance & Repair estimate
2. A list of equipment and other items from which to prepare a Preventive Maintenance estimate
3. A list of facilities' usage and size requirements to prepare a General Maintenance estimate

The Introduction to Facilities Audits in this section explains the rationale for the audit and the steps required to complete it successfully.

This section also provides forms for listing all of the organization's facilities, and a separate detailed form for the specifics of each facility Checklists are provided for major building components. Each checklist should be accompanied by a blank audit form. This form is used to record the audit findings, prioritize the deficiencies, and estimate the cost to remedy these deficiencies.

Preventive Maintenance

General Maintenance

Reference Section

This section includes information on Equipment Rental Costs, Crew Listings, Travel Costs, Reference Tables, Life Cycle Costing, and a listing of Abbreviations.

Equipment Rental Costs: This section contains the average costs to rent and operate hundreds of pieces of construction equipment.

Crew Listings: This section lists all of the crews referenced in the book. For the purposes of this book, a crew is composed of more than one trade classification and/or the addition of power equipment to any trade classification. Power equipment is included in the cost of the crew. Costs are shown both with bare labor rates and with the installing contractor's overhead and profit added. For each, the total crew cost per eight-hour day and the composite cost per labor-hour are listed.

Travel Costs: This chart provides labor-hour costs for round-trip travel between a base of operation and the project site.

Reference Tables: In this section, you'll find reference tables, explanations, estimating information that support how we develop the unit price data, technical data, and estimating procedures.

Life Cycle Costing: This section provides definitions and basic equations for performing life cycle analysis. A sample problem is solved to demonstrate the methodology.

Abbreviations: A listing of the abbreviations used throughout this book, along with the terms they represent, is included in this section.

The Design Assumptions of this Book

This book is designed to be as easy to use as possible. To that end, we have made certain assumptions and limited its scope.

1. Unlike any other RSMeans publication, *Facilities Maintenance & Repair Cost Data* was designed for estimating a wide range of maintenance tasks in diverse environments. Due to this diversity, the level of accuracy of the data is +/− 20%.

2. We have established material prices based on a national average.

3. We have computed labor costs based on a 30-city national average of union wage rates.

4. Except where major equipment or component replacement is described, the projects in this book are small; therefore, material prices and labor-hours have been enhanced to reflect the increased costs of small-scale work.

5. The PM frequencies are based on non-critical applications. Increased frequencies are required for critical environments.

How to Use the Book: The Details

What's Behind the Numbers? The Development of Cost Data

The staff at RSMeans continually monitors developments in the construction industry in order to ensure reliable, thorough, and up-to-date cost information. While overall construction costs may vary relative to general economic conditions, price fluctuations within the industry are dependent upon many factors. Individual price variations may, in fact, be opposite to overall economic trends. Therefore, costs are constantly tracked and complete updates are published yearly. Also, new items are frequently added in response to changes in materials and methods.

Costs—$ (U.S.)

All costs represent U.S. national averages and are given in U.S. dollars. The RSMeans City Cost Indexes can be used to adjust costs to a particular location. The City Cost Indexes for Canada can be used to adjust U.S. national averages to local costs in Canadian dollars. No exchange rate conversion is necessary.

G The processes or products identified by the green symbol in our publications have been determined to be environmentally responsible and/or resource-efficient solely by the RSMeans engineering staff. The inclusion of the green symbol does not represent compliance with any specific industry association or standard.

Material Costs

The RSMeans staff contacts manufacturers, dealers, distributors, and contractors all across the U.S. and Canada to determine national average material costs. If you have access to current material costs for your specific location, you may wish to make adjustments to reflect differences from the national average. Included within material costs are fasteners for a normal installation. RSMeans engineers use manufacturers' recommendations, written specifications, and/or standard construction practice for size and spacing of fasteners. Adjustments to material costs may be required for your specific application or location. The manufacturer's warranty is assumed. Extended warranties are not included in the material costs. Material costs do not include sales tax.

Labor Costs

Labor costs are based on the average of wage rates from 30 major U.S. cities. Rates are determined from labor union agreements or prevailing wages for construction trades for the current year. Rates, along with overhead and profit markups, are listed on the inside back cover of this book.

- If wage rates in your area vary from those used in this book, or if rate increases are expected within a given year, labor costs should be adjusted accordingly.

Labor costs reflect productivity based on actual working conditions. In addition to actual installation, these figures include time spent during a normal weekday on tasks such as, material receiving and handling, mobilization at site, site movement, breaks, and cleanup.

Productivity data is developed over an extended period so as not to be influenced by abnormal variations and reflects a typical average.

Equipment Costs

Equipment costs include not only rental, but also operating costs for equipment under normal use. The operating costs include parts and labor for routine servicing such as repair and replacement of pumps, filters, and worn lines. Normal operating expendables, such as fuel, lubricants, tires, and electricity (where applicable), are also included. Extraordinary operating expendables with highly variable wear patterns, such as diamond bits and blades, are excluded. These costs are included under materials. Equipment rental rates are obtained from industry sources throughout North America—contractors, suppliers, dealers, manufacturers, and distributors.

Rental rates can also be treated as reimbursement costs for contractor-owned equipment. Owned equipment costs include depreciation, loan payments, interest, taxes, insurance, storage and major repairs.

Equipment costs do not include operators' wages; nor do they include the cost to move equipment to a job site (mobilization) or from a job site (demobilization).

Equipment Cost/Day—The cost of power equipment required for each crew is included in the Crew Listings in the Reference Section (small tools that are considered as essential everyday tools are not listed out separately). The Crew Listings itemize specialized tools and heavy equipment along with labor trades. The daily cost of itemized equipment included in a crew is based on dividing the weekly bare rental rate by 5 (number of working days per week) and then adding the hourly operating cost times 8 (the number of hours per day). This Equipment Cost/Day is shown in the last column of the Equipment Rental Cost pages in the Reference Section.

Mobilization/Demobilization—The cost to move construction equipment from an equipment yard or rental company to the job site and back again is not included in equipment costs. Mobilization (to the site) and demobilization (from the site) costs can be found in the Unit Price Section. If a piece of equipment is already at the job site, it is not appropriate to utilize mob./demob. costs again in an estimate.

Overhead and Profit

Total Cost including O&P for the *Installing Contractor* is shown in the last column on the Unit Price and/or the Assemblies pages of this book. This figure is the sum of the bare material cost plus 10% for profit, the bare labor cost plus total overhead and profit, and the bare equipment cost plus 10% for profit. Details for the calculation of Overhead and Profit on labor are shown on the inside back cover and in the Reference Section of this book. (See "How RSMeans Unit Price Data Works" for an example of this calculation.)

General Conditions

Cost data in this book is presented in two ways: Bare Costs and Total Cost including O&P (Overhead and Profit). General Conditions, or General Requirements, of the contract should also be added to the Total Cost including O&P when applicable. Costs for General Conditions are listed in Division 1 of the Unit Price Section and the Reference

Section of this book. General Conditions for the *Installing Contractor* may range from 0% to 10% of the Total Cost including O&P. For the *General* or *Prime Contractor*, costs for General Conditions may range from 5% to 15% of the Total Cost including O&P, with a figure of 10% as the most typical allowance. If applicable, the Assemblies and Models sections of this book use costs that include the installing contractor's overhead and profit (O&P).

Factors Affecting Costs

Costs can vary depending upon a number of variables. Here's how we have handled the main factors affecting costs.

Quality—The prices for materials and the workmanship upon which productivity is based represent sound construction work. They are also in line with U.S. government specifications.

Overtime—We have made no allowance for overtime. If you anticipate premium time or work beyond normal working hours, be sure to make an appropriate adjustment to your labor costs.

Productivity—The productivity, daily output, and labor-hour figures for each line item are based on working an eight-hour day in daylight hours in moderate temperatures. For work that extends beyond normal work hours or is performed under adverse conditions, productivity may decrease. (See "How RSMeans Unit Price Data Works" for more on productivity.)

Size of Project—The size, scope of work, and type of construction project will have a significant impact on cost. Economies of scale can reduce costs for large projects. Unit costs can often run higher for small projects.

Location—Material prices in this book are for metropolitan areas. However, in dense urban areas, traffic and site storage limitations may increase costs. Beyond a 20-mile radius of large cities, extra trucking or transportation charges may also increase the material costs slightly. On the other hand, lower wage rates may be in effect. Be sure to consider both of these factors when preparing an estimate, particularly if the job site is located in a central city or remote rural location. In addition, highly specialized subcontract items may require travel and per-diem expenses for mechanics.

Other Factors——

- season of year
- contractor management
- weather conditions
- local union restrictions
- building code requirements
- availability of:
  - adequate energy
  - skilled labor
  - building materials
- owner's special requirements/restrictions
- safety requirements
- environmental considerations

Unpredictable Factors—General business conditions influence "in-place" costs of all items. Substitute materials and construction methods may have to be employed. These may affect the installed cost and/or life cycle costs. Such factors may be difficult to evaluate and cannot necessarily be predicted on the basis of the job's location in a particular section of the country. Thus, where these factors apply, you may find significant but unavoidable cost variations for which you will have to apply a measure of judgment to your estimate.

Rounding of Costs

In general, all unit prices in excess of $5.00 have been rounded to make them easier to use and still maintain adequate precision of the results. The rounding rules we have chosen are in the following table.

Prices from . . .	Rounded to the nearest . . .
$.01 to $5.00	$.01
$5.01 to $20.00	$.05
$20.01 to $100.00	$.50
$100.01 to $300.00	$1.00
$300.01 to $1,000.00	$5.00
$1,000.01 to $10,000.00	$25.00
$10,000.01 to $50,000.00	$100.00
$50,000.01 and above	$500.00

How Subcontracted Items Affect Costs

A considerable portion of all large construction jobs is usually subcontracted. In fact, the percentage done by subcontractors is constantly increasing and may run over 90%. Since the workers employed by these companies do nothing else but install their particular product, they soon become expert in that line. The result is, installation by these firms is accomplished so efficiently that the total in-place cost, even adding the general contractor's overhead and profit, is no more, and often less, than if the principal contractor had handled the installation himself/herself. Companies that deal with construction specialties are anxious to have their product perform well and, consequently, the installation will be the best possible.

Contingencies

The allowance for contingencies generally provides for unforeseen construction difficulties. On alterations or repair jobs, 20% is not too much. If drawings are final and only field contingencies are being considered, 2% or 3% is probably sufficient, and often nothing need be added. Contractually, changes in plans will be covered by extras. The contractor should consider inflationary price trends and possible material shortages during the course of the job. These escalation factors are dependent upon both economic conditions and the anticipated time between the estimate and actual construction. If drawings are not complete or approved, or a budget cost is wanted, it is wise to add 5% to 10%. Contingencies, then, are a matter of judgment. Additional allowances for contingencies are shown in Division 1.

Important Estimating Considerations

The productivity or daily output of each craftsman or crew assumes a well-managed job where tradesmen with the proper tools and equipment, along with the appropriate construction materials, are present. Included are daily mobilization and cleanup time, break time and plan layout time. Unless otherwise indicated, time for material movement on site (for items that can be transported by hand) of up to 200' into the building and to the first or second floor is also included. If material has to be transported by other means, over greater distances, or to higher floors, an additional allowance should be considered by the estimator.

While horizontal movement is typically a sole function of distances, vertical transport introduces other variables that can significantly impact productivity. In an occupied building, the use of elevators (assuming access, size, and required protective measures are acceptable) must be understood at the time of the estimate. For new construction, hoist wait and cycle times can easily be 15 minutes and may result in scheduled access extending beyond the normal work day. Finally, all vertical transport will impose strict weight limits likely to preclude the use of any motorized material handling.

The productivity, or daily output, also assumes installation that meets manufacturer/designer/ standard specifications. A time allowance for quality control checks, minor adjustments and any task required to ensure the proper function or operation is also included. For items that require connections to services, time is included for positioning, leveling, securing the unit and for making all the necessary connections (and start up where applicable) ensuring a complete installation. Estimating of the services themselves (electrical, plumbing, water, steam, hydraulics, dust collection, etc.) is separate.

In some cases, the estimator must consider the use of a crane and an appropriate crew for the installation of large or heavy items. For those situations where a crane is not included in the assigned crew and as part of the line item cost, equipment rental costs, mobilization and demobilization costs, and operator and support personnel costs must be considered.

Labor-Hours

The labor-hours expressed in this publication are derived by dividing the total daily labor hours for the crew by the daily output. Based on average installation time and the assumptions listed above, the labor-hours include: direct labor, indirect labor, and nonproductive time. A typical day for a craftsman might include, but not be limited to:

- Direct Work
  - Measuring and layout
  - Preparing materials
  - Actual installation
  - Quality assurance/quality control
- Indirect Work
  - Reading plans or specifications
  - Preparing space
  - Receiving materials
  - Material movement
  - Giving or receiving instruction
  - Miscellaneous
- Non-Work
  - Chatting
  - Personal issues
  - Breaks
  - Interruptions (i.e., sickness, weather, material or equipment shortages, etc.)

If any of the items for a typical day do not apply to the particular work or project situation, the estimator should make any necessary adjustments.

Final Checklist

Estimating can be a straightforward process provided you remember the basics. Here's a checklist of some of the steps you should remember to complete before finalizing your estimate.

Did you remember to . . .

- factor in the City Cost Index for your locale?
- take into consideration which items have been marked up and by how much?
- mark up the entire estimate sufficiently for your purposes?
- read the background information on techniques and technical matters that could impact your project time span and cost?
- include all components of your project in the final estimate?
- double check your figures for accuracy?
- call RSMeans if you have any questions about your estimate or the data you've found in our publications? Remember, RSMeans stands behind its publications. If you have any questions about your estimate . . . about the costs you've used from our books . . . or even about the technical aspects of the job that may affect your estimate, feel free to call the RSMeans editors at 1-877-807-5724.

Free Quarterly Updates

Stay up-to-date throughout 2015 with RSMeans' free cost data updates four times a year. Sign up online to make sure you have access to the newest data. Every quarter we provide the city cost adjustment factors for hundreds of cities and key materials. Register at: http://info.thegordiangroup.com/RSMeans.html.

Maintenace & Repair

Table of Contents

Table of Contents (cont.)

Table of Contents (cont.)

3

How to Use the Maintenance & Repair Assemblies Cost Tables

The following is a detailed explanation of a sample Maintenance & Repair Assemblies Cost Table. Next to each bold number that follows is the described item with the appropriate component of the sample entry following in parantheses.

Total system costs, as well as the individual component costs, are shown. In most cases, the intent is for the user to apply the total system costs. However, changes and adjustments to the components or partial use of selected components is also appropriate. In particular, selected equipment system tables in the mechanical section include complete listings of operations that are meant to be chosen from rather than used in total.

1

System/Line Numbers (B3013 105 0700)

Each Maintenance & Repair Assembly has been assigned a unique identification number based on the UNIFORMAT II classification system.

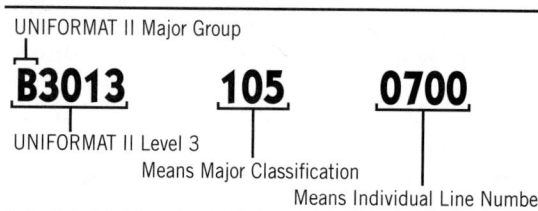

UNIFORMAT II Major Group

B3013 **105** **0700**

UNIFORMAT II Level 3

Means Major Classification

Means Individual Line Number

B30 ROOFING		B3013	Roof Covering									
B3013 105			**Built-Up Roofing**									

	System Description	Freq. (Years)	Crew	Unit	Labor Hours	2015 Bare Costs				Total In-House	Total w/O&P
						Material	Labor	Equipment	Total		
0600	Place new BUR membrane over existing	20	G-5	Sq.							
	Set up, secure and take down ladder				.020		.80		.80	1.20	1.45
	Sweep / spud ballast clean				.500		20		20	30	36
	Vent existing membrane				.130		5.20		5.20	7.80	9.40
	Cut out buckled flashing				.024		.96		.96	1.44	1.74
	Install 2 ply membrane flashing				.027	.39	1.06		1.45	2.02	2.42
	Install 4 ply bituminous roofing				3.733	136	140	38	314	395	465
	Reinstall ballast				.381		15.30		15.30	23	27.50
	Clean up				.390		15.65		15.65	23.50	28.50
	Total				5.205	136.39	198.97	38	373.36	**483.96**	**572.01**
0700	Total BUR roof replacement	28	G-1	Sq.							
	Set up, secure and take down ladder				.020		.80		.80	1.20	1.45
	Sweep / spud ballast clean				.500		20		20	30	36
	Remove built-up roofing				2.500		95		95	126	156
	Remove insulation board				1.026		39		39	52	64
	Remove flashing				.024		.96		.96	1.44	1.74
	Install 2" perlite insulation				.879	100	35		135	163	189
	Install 2 ply membrane flashing				.027	.39	1.06		1.45	2.02	2.42
	Install 4 ply bituminous membrane				2.800	136	105	28.50	269.50	335	390
	Clean up				1.000		40		40	60	72.50
	Total				8.776	236.39	336.82	28.50	601.71	**770.66**	**913.11**

B3013 120			**Modified Bituminous / Thermoplastic**								

	System Description	Freq. (Years)	Crew	Unit	Labor Hours	2015 Bare Costs				Total In-House	Total w/O&P
						Material	Labor	Equipment	Total		
0100	Debris removal by hand & visual inspection	1	2 ROFC	M.S.F.							
	Set up, secure and take down ladder				.052		2.08		2.08	3.12	3.76
	Pick up trash / debris & clean up				.327		13.10		13.10	19.60	23.50
	Visual inspection				.327		13.10		13.10	19.60	23.50
	Total				.705		28.28		28.28	**42.32**	**50.76**
0200	Non - destructive moisture inspection	5	2 ROFC	M.S.F.							
	Set up, secure and take down ladder				.052		2.08		2.08	3.12	3.76
	Infrared inspection of roof membrane				2.133		85.50		85.50	128	155
	Total				2.185		87.58		87.58	**131.12**	**158.76**

The table above contains callout numbers overlaid on it: **1** (System/Line Numbers), **2** (System Description), **3** (Freq. (Years)), **4** (Crew), **5** (Unit), **6** (Labor Hours), **7** (2015 Bare Costs), **8** (Total In-House), **9** (Total w/O&P).

2 System Description

A one-line description of each system is given, followed by a description of each component that makes up the system. In selected items, a percentage factor for planning annual costs is also given.

3 Frequency (28)

The projected frequency of occurrence in years for each system is listed. For example, **28** indicates that this operation should be budgeted for once every 28 years. If the frequency is less than 1, then the operation should be planned for more than once a year. For example, a value of **.5** indicates a planned occurrence of 2 times a year.

4 Crew (G-1)

The "Crew" column designates the typical trade or crew used for the task, although other trades may be used for minor portions of the work. If the task can be accomplished by one trade and requires no power equipment, that trade and the number of workers are listed (for example, "4 Rofc"). If the task requires a composite crew, a crew code designation is listed (for example, "G-1"). You'll find full details on all composite crews in the Crew Listings.

- For a complete list of all trades utilized in this book and their abbreviations, see the inside back cover of the book.

Crews										
Crew No.	Bare Costs		In-House Costs		Incl. Subs O&P				Cost Per Labor-Hour	
Crew B-80	Hr.	Daily	Hr.	Daily	Hr.	Daily	Bare Costs	In House	Incl. O&P
1 Labor Foreman (outside)	$39.60	$ 316.80	$52.65	$ 421.20	$64.90	$ 519.20	$41.23	$53.89	$66.69
1 Laborer	37.60	300.80	49.95	399.60	61.65	493.20			
1 Truck Driver (light)	39.10	312.80	50.90	407.20	63.05	504.40			
1 Equip. Oper. (light)	48.60	388.80	62.05	496.40	77.15	617.20			
1 Flatbed Truck, Gas, 3 Ton		303.20		303.20		333.52			
1 Earth Auger, Truck-Mtd.		417.20		417.20		458.92	22.51	22.51	24.76
32 L.H., Daily Totals		$2039.60		$2444.80		$2926.44	$63.74	$76.40	$91.45

5 Unit of Measure for Each System (Sq.)

The abbreviated designation indicates the unit of measure, as defined by industry standards, upon which the price of the component is base. In this example, "Cost per Square" is the unit of measure for this system or "assembly."

6 Labor-Hours (8.776)

The "Labor-Hours" figure represents the number of labor-hours required to perform one unit of work. To calculate the number of labor-hours required for a particular task, multiply the quantity of the item times the number of labor-hours shown. For example:

Quantity	X	Productivity Rate	=	Duration
30 Sq.	X	8.776 Labor-Hours/ Sq.	=	263 Labor-Hours

7 Bare Costs:

Material (236.39)

This figure for the unit material cost for the line item is the "bare" material cost with no overhead and profit allowances included. *Costs shown reflect national average material prices for the current year and generally include delivery to the facility. Small purchases may require an additional deliver charge.*
No sales taxes are included.

Labor (336.82)

The unit labor costs is derived by multiplying bare labor-hour costs for Crew G-1 and other trades by labor-hour units. In this case, the bare labor-hour cost for the G-1 crew is found in the Crew Section. (If a trade is listed or used for minor portions of the work, the hourly labor cost—the wage rate—is found on the inside back cover of this book).

Equipment (28.50)

Equipment costs for each crew are listed in the description of each crew. The unit equipment cost is derived by multiplying the bare equipment hourly cost by the labor-hour units.

Total (601.71)

The total of the bare costs is the arithmetic total of the three previous columns: mat., labor, and equip.

Material	+	Labor	+	Equip.	=	Total
$236.39	+	$336.82	+	$28.50	=	$601.71

8 **Total In-House Costs (770.66)**

"Total In-House Costs" include suggested markups to bare costs for the direct overhead requirements of in-house maintenance staff. The figure in this column is the sum of three components: the bare material cost plus 10% (for purchasing and handling small quantities); the bare labor cost plus workers' compensation and average fixed overhead (per the labor rate table on the inside back cover of this books or, if a crew is listed, from the Crew Listings); and the bare equipment cost.

9 **Total Costs Including O&P (913.11)**

"Total Costs Including O&P" include suggested markups to bare costs for an outside subcontractor's overhead and profit requirements. The figure in this column is the sum of three components: the bare material cost plus 25% for profit; the bare labor cost plus total overhead and profit (per the labor rate table on the inside back cover of the book or, if a crew is listed, from the Crew Listings); and the bare equipment cost plus 10% for profit.

A1033 110 | Concrete, Unfinished

System Description	Freq. (Years)	Crew	Unit	Labor Hours	2015 Bare Costs				Total In-House	Total w/O&P
					Material	Labor	Equipment	Total		
0010 Minor repairs to concrete floor	15	1 CEFI	S.F.							
Repair spall				.612		23.50	3.56	27.06	34.50	42
Repair concrete flooring				.061	2.92	2.75		5.67	6.70	8
Total				.673	2.92	26.25	3.56	32.73	41.20	50
0020 Replace unfinished concrete floor	75	2 CEFI	C.S.F.							
Break up concrete slab				1.664		68.80	25.28	94.08	115	139
Load into truck				1.950		75.75	42	117.75	142	170
Rebar #3 to #7				.139	10	7.30		17.30	21	24.50
Edge form				.693	3.50	30.90		34.40	45	55
Slab on grade 4" in place with finish				1.577	99	67.50	.75	167.25	197	233
Expansion joint				.003	.05	.12		.17	.21	.26
Cure concrete				.109	5.53	4.13		9.66	11.50	13.65
Total				6.135	118.08	254.50	68.03	440.61	531.71	635.41

For customer support on your Facilities Maintenance & Repair Cost Data, call 877.801.5724.

7

B1013 510 Exterior Concrete Stairs

	System Description	Freq. (Years)	Crew	Unit	2015 Bare Costs					Total In-House	Total w/O&P
					Labor Hours	Material	Labor	Equipment	Total		
1010	**Repair exterior concrete steps**	30	1 CEFI	S.F.							
	Repair spalls in steps				.104	12.10	4.68		16.78	19.25	22.50
	Total				.104	12.10	4.68		16.78	**19.25**	**22.50**
1020	**Replace exterior concrete steps**	75	1 CEFI	S.F.							
	Remove concrete				.520		19.75	3.03	22.78	29.50	36
	Install forms				.328	1.40	15		16.40	21.50	26.50
	Pour concrete				.567	2.60	22	.74	25.34	33	40
	Total				1.416	4	56.75	3.77	64.52	**84**	**102.50**
1030	**Refinish exterior concrete steps**	3	1 PORD	S.F.							
	Prepare surface				.016		.65		.65	.84	1.04
	Refinish surface, brushwork, 1 coat				.013	.08	.50		.58	.74	.91
	Total				.029	.08	1.15		1.23	**1.58**	**1.95**
2030	**Refinish exterior concrete landing**	3	1 PORD	S.F.							
	Prepare surface				.016		.65		.65	.84	1.04
	Refinish surface, brushwork, 1 coat				.013	.08	.50		.58	.74	.91
	Total				.029	.08	1.15		1.23	**1.58**	**1.95**

B1013 520 Exterior Masonry Steps

	System Description	Freq. (Years)	Crew	Unit	2015 Bare Costs					Total In-House	Total w/O&P
					Labor Hours	Material	Labor	Equipment	Total		
1010	**Repair exterior masonry steps**	4	1 BRIC	S.F.							
	Remove damaged masonry				.074		2.79		2.79	3.71	4.58
	Clean area				.019	.04	.86		.90	1.17	1.44
	Replace with new masonry				.236	3.04	10.15		13.19	16.70	20.50
	Total				.329	3.08	13.80		16.88	**21.58**	**26.52**

B10 SUPERSTRUCTURE | B1013 | Floor Construction

B1013 530 | Exterior Wood Stairs

	System Description	Freq. (Years)	Crew	Unit	Labor Hours	2015 Bare Costs				Total In-House	Total w/O&P
						Material	Labor	Equipment	Total		
1030	**Refinish exterior wood steps**	3	1 PORD	S.F.							
	Prepare surface				.016		.65		.65	.84	1.04
	Wash / dry surface				.001		.05		.05	.06	.08
	Refinish surface, brushwork, 1 coat				.013	.08	.50		.58	.74	.91
	Total				.030	.08	1.20		1.28	**1.64**	**2.03**
1040	**Replace exterior wood steps**	30	2 CARP	S.F.							
	Remove stairs (to 6' wide)				.100		3.76		3.76	5	6.15
	New stringers (to 5 risers)				.074	1.17	3.48		4.65	5.90	7.15
	Risers				.083	.78	3.91		4.69	6.05	7.40
	Treads				.123	1.95	5.80		7.75	9.85	11.90
	Total				.380	3.90	16.95		20.85	**26.80**	**32.60**
2030	**Refinish exterior wood landing**	3	1 PORD	S.F.							
	Prepare surface				.016		.65		.65	.84	1.04
	Wash / dry surface				.001		.05		.05	.06	.08
	Refinish surface, brushwork, 1 coat				.013	.08	.50		.58	.74	.91
	Total				.030	.08	1.20		1.28	**1.64**	**2.03**

B1013 540 | Exterior Wood Railing

	System Description	Freq. (Years)	Crew	Unit	Labor Hours	2015 Bare Costs				Total In-House	Total w/O&P
						Material	Labor	Equipment	Total		
1010	**Replace exterior wood balustrade**	20	1 CARP	L.F.							
	Remove damaged balusters and handrail				.087		3.26		3.26	4.33	5.35
	Install new balusters and handrail				.473	24.50	22		46.50	56.50	67
	Total				.559	24.50	25.26		49.76	**60.83**	**72.35**
1030	**Refinish exterior wood balustrade**	7	1 PORD	L.F.							
	Prepare balusters and handrails for painting				.052		2.10		2.10	2.72	3.37
	Wash / dry surface				.001		.05		.05	.06	.08
	Paint balusters and handrails, brushwork, primer + 1 coat				.089	.81	3.59		4.40	5.55	6.75
	Total				.142	.81	5.74		6.55	**8.33**	**10.20**

For customer support on your Facilities Maintenance & Repair Cost Data, call 877.801.5724.

9

B1013 550   Exterior Metal Stairs

System Description	Freq. (Years)	Crew	Unit	Labor Hours	2015 Bare Costs				Total In-House	Total w/O&P
					Material	Labor	Equipment	Total		
1010 Repair exterior metal steps										
Remove damaged stair tread	15	1 SSWK	S.F.	.099		5.22		5.22	7.85	9.45
New stair tread				.104	42.61	5.48		48.09	55	63
Total				.203	42.61	10.70		53.31	**62.85**	**72.45**
1040 Replace exterior metal steps										
Remove stairs	40	2 SSWK	S.F.	.168		8.85		8.85	13.25	16.05
New stairs				.239	157.30	12.44	6.44	176.18	197	225
Total				.408	157.30	21.29	6.44	185.03	**210.25**	**241.05**
1050 Refinish exterior metal steps										
Prepare metal railing for painting	3	1 PORD	S.F.	.019		.77		.77	1	1.24
Wash / dry surface				.001		.05		.05	.06	.08
Paint metal steps, brushwork, primer + 1 coat				.015	.11	.62		.73	.93	1.14
Total				.036	.11	1.44		1.55	**1.99**	**2.46**
2030 Refinish exterior metal landing										
Prepare metal railing for painting	3	1 PORD	S.F.	.019		.77		.77	1	1.24
Wash / dry surface				.001		.05		.05	.06	.08
Paint metal landing, brushwork, primer + 1 coat				.015	.11	.62		.73	.93	1.14
Total				.036	.11	1.44		1.55	**1.99**	**2.46**

B1013 560   Exterior Metal Railing

System Description	Freq. (Years)	Crew	Unit	Labor Hours	2015 Bare Costs				Total In-House	Total w/O&P
					Material	Labor	Equipment	Total		
1010 Replace exterior metal hand rail										
Remove damaged railings	30	1 SSWK	L.F.	.087		3.26		3.26	4.33	5.35
Install new metal railings				.163	25	8.65	.74	34.39	41	47.50
Total				.250	25	11.91	.74	37.65	**45.33**	**52.85**
1030 Refinish exterior metal hand rail										
Prepare metal railing for painting	7	1 PORD	L.F.	.019		.77		.77	1	1.24
Wash / dry surface				.001		.05		.05	.06	.08
Paint metal railing, brushwork, primer + 1 coat				.015	.11	.62		.73	.93	1.14
Total				.036	.11	1.44		1.55	**1.99**	**2.46**

For customer support on your Facilities Maintenance & Repair Cost Data, call 877.801.5724.

B1013 570   Exterior Wrought Iron Balustrade

	System Description	Freq. (Years)	Crew	Unit	Labor Hours	2015 Bare Costs				Total In-House	Total w/O&P
						Material	Labor	Equipment	Total		
1010	**Repair exterior wrought iron balustrade**	20	1 SSWK	L.F.							
	Cut / remove damaged section				.325		12.20		12.20	16.25	20
	Reset new section				.867	81	45.50		126.50	157	184
	Total				1.192	81	57.70		138.70	**173.25**	**204**
1030	**Refinish exterior wrought iron balustrade**	7	1 PORD	L.F.							
	Prepare metal balustrade for painting				.019		.77		.77	1	1.24
	Wash / dry surface				.001		.05		.05	.06	.08
	Paint metal balustrade, brushwork, primer + 1 coat				.015	.11	.62		.73	.93	1.14
	Total				.036	.11	1.44		1.55	**1.99**	**2.46**

B1013 580   Exterior Metal Fire Escapes

	System Description	Freq. (Years)	Crew	Unit	Labor Hours	2015 Bare Costs				Total In-House	Total w/O&P
						Material	Labor	Equipment	Total		
1010	**Refinish exterior fire escape balcony, 2' wide**	7	1 PORD	L.F.							
	Prepare metal fire escape balcony for painting				.100		4.04		4.04	5.25	6.50
	Brush on primer coat				.100	2.60	4.04		6.64	8.10	9.70
	Brush on finish coat				.100	2.27	4.04		6.31	7.70	9.30
	Total				.300	4.87	12.12		16.99	**21.05**	**25.50**
1020	**Replace exterior fire escape balcony, 2' wide**	25	2 SSWK	L.F.							
	Remove metal fire escape balcony				.500		26.50		26.50	39.50	47.50
	Install new metal fire escape balcony				2.000	58.50	105		163.50	222	263
	Total				2.500	58.50	131.50		190	**261.50**	**310.50**
2010	**Refinish exterior fire escape stair and platform**	7	1 PORD	Flight							
	Prepare fire escape stair and platform for painting				2.400		97		97	126	156
	Brush on primer coat				2.400	62.28	96.84		159.12	194	233
	Brush on finish coat				2.400	54.36	96.84		151.20	185	223
	Total				7.200	116.64	290.68		407.32	**505**	**612**
2020	**Replace exterior fire escape stair and platform**	25	2 SSWK	Flight							
	Remove metal fire escape stair and platform				14.995		790		790	1,175	1,425
	Install new metal fire escape stair and platform				59.925	1,075	3,150		4,225	5,900	7,050
	Total				74.920	1,075	3,940		5,015	7,075	8,475

For customer support on your Facilities Maintenance & Repair Cost Data, call 877.801.5724.

11

B1013 580   Exterior Metal Fire Escapes

	System Description	Freq. (Years)	Crew	Unit	Labor Hours	2015 Bare Costs				Total In-House	Total w/O&P
						Material	Labor	Equipment	Total		
3010	**Refinish exterior fire escape ladder**	7	1 PORD	V.L.F.							
	Prepare metal fire escape ladder for painting				.020		.81		.81	1.05	1.30
	Brush on primer coat				.020	.35	.81		1.16	1.43	1.73
	Brush on finish coat				.020	.30	.81		1.11	1.38	1.68
	Total				.060	.65	2.43		3.08	3.86	4.71
3020	**Replace exterior fire escape ladder**	25	2 SSWK	V.L.F.							
	Remove metal fire escape ladder				.250		13.15		13.15	19.70	24
	Install new metal fire escape ladder				.500	38	26.50		64.50	81	95
	Total				.750	38	39.65		77.65	100.70	119

B1013 910   Exterior Steel Decking

	System Description	Freq. (Years)	Crew	Unit	Labor Hours	2015 Bare Costs				Total In-House	Total w/O&P
						Material	Labor	Equipment	Total		
0020	**Replace exterior steel decking**	30	2 SSWK	S.F.							
	Remove decking				.011	2.60	.62	.21	3.43	4	4.61
	Assemble and weld decking				.011	2.60	.62	.21	3.43	4	4.61
	Total				.023	5.20	1.24	.42	6.86	8	9.22

B1013 920   Exterior Metal Grating

	System Description	Freq. (Years)	Crew	Unit	Labor Hours	2015 Bare Costs				Total In-House	Total w/O&P
						Material	Labor	Equipment	Total		
0010	**Repair exterior metal floor grating - (2% of grating)**	10	1 SSWK	S.F.							
	Misc. welding of floor grating				.348	1.08	19	6.35	26.43	36	43
	Total				.348	1.08	19	6.35	26.43	36	43
0020	**Replace exterior metal floor grating**	30	2 SSWK	S.F.							
	Remove grating				.011	2.60	.62	.21	3.43	4	4.61
	Assemble and weld grating				.059	19.65	3.16	.27	23.08	26.50	30.50
	Total				.071	22.25	3.78	.48	26.51	30.50	35.11

B2013 104 Copings

System Description	Freq. (Years)	Crew	Unit	Labor Hours	2015 Bare Costs				Total In-House	Total w/O&P
					Material	Labor	Equipment	Total		
2010 Replace precast concrete coping, 12" wide	50	2 BRIC	L.F.							
Set up and secure scaffold, 3 floors to roof				.429		20.16		20.16	26.50	33
Remove coping				.050		1.88		1.88	2.50	3.08
Install new precast concrete coping				.229	18.45	9.60		28.05	33	38.50
Remove scaffold				.429		20.16		20.16	26.50	33
Total				1.137	18.45	51.80		70.25	88.50	107.58
2020 Replace limestone coping, 12" wide	50	2 BRIC	L.F.							
Set up and secure scaffold, 3 floors to roof				.429		20.16		20.16	26.50	33
Remove coping				.050		1.88		1.88	2.50	3.08
Install new limestone coping				.178	14.70	7.50		22.20	26	30.50
Remove scaffold				.429		20.16		20.16	26.50	33
Total				1.086	14.70	49.70		64.40	81.50	99.58
2030 Replace marble coping, 12" wide	50	2 BRIC	L.F.							
Set up and secure scaffold, 3 floors to roof				.429		20.16		20.16	26.50	33
Remove coping				.050		1.88		1.88	2.50	3.08
Install new marble coping				.200	16.15	8.40		24.55	29	34
Remove scaffold				.429		20.16		20.16	26.50	33
Total				1.108	16.15	50.60		66.75	84.50	103.08
2040 Replace terra cotta coping, 12" wide	50	2 BRIC	L.F.							
Set up and secure scaffold, 3 floors to roof				.429		20.16		20.16	26.50	33
Remove coping				.050		1.88		1.88	2.50	3.08
Install new terra cotta coping				.200	8.65	8.40		17.05	20.50	24.50
Remove scaffold				.429		20.16		20.16	26.50	33
Total				1.108	8.65	50.60		59.25	76	93.58

For customer support on your Facilities Maintenance & Repair Cost Data, call 877.801.5724.

13

B2013 109 — Concrete Block

	System Description	Freq. (Years)	Crew	Unit	Labor Hours	2015 Bare Costs				Total In-House	Total w/O&P
						Material	Labor	Equipment	Total		
1010	**Repair 8" concrete block wall - 1st floor**	25	1 BRIC	S.F.							
	Set up, secure and take down ladder				.130		6		6	7.90	9.75
	Remove damaged block				.183		6.88		6.88	9.10	11.30
	Wire truss reinforcing				.005	.26	.24		.50	.60	.72
	Install block				.111	2.86	4.77		7.63	9.45	11.35
	Total				.429	3.12	17.89		21.01	27.05	33.12
1020	**Replace 8" concrete block wall - 1st floor**	60	2 BRIC	C.S.F.							
	Set up and secure scaffold				.750		35.25		35.25	47	57.50
	Remove damaged block				4.571		172		172	228	282
	Wire truss reinforcing				.260	13	12		25	30	36
	Install block				11.111	286	477		763	945	1,125
	Remove scaffold				.750		35.25		35.25	47	57.50
	Total				17.442	299	731.50		1,030.50	1,297	1,558
1030	**Waterproof concrete block wall - 1st floor**	10	1 ROFC	C.S.F.							
	Steam clean				2.971		113	17	130	167	204
	Spray surface, silicone				.260	34	10		44	53	61.50
	Total				3.231	34	123	17	174	220	265.50
2020	**Replace 8" concrete block wall - 2nd floor**	60	2 BRIC	C.S.F.							
	Set up and secure scaffold				1.500		70.50		70.50	93.50	115
	Remove block				4.571		172		172	228	282
	Wire truss reinforcing				.260	13	12		25	30	36
	Install block				11.111	286	477		763	945	1,125
	Remove scaffold				1.500		70.50		70.50	93.50	115
	Total				18.942	299	802		1,101	1,390	1,673
2030	**Waterproof concrete block wall - 2nd floor**	10	G-8	C.S.F.							
	Position aerial lift truck, raise & lower platform				.250		10.05	19.85	29.90	35	40
	Steam clean				11.765		442	111	553	700	845
	Spray surface, silicone				.260	34	10	21	65	74	84.50
	Total				12.275	34	462.05	151.85	647.90	809	969.50

For customer support on your Facilities Maintenance & Repair Cost Data, call 877.801.5724.

B20 EXTERIOR CLOSURE B2013 Exterior Walls

B2013 109 Concrete Block

System Description	Freq. (Years)	Crew	Unit	Labor Hours	2015 Bare Costs				Total In-House	Total w/O&P
					Material	Labor	Equipment	Total		
3020 Replace 8" concrete block wall - 3rd floor	60	2 BRIC	C.S.F.							
Set up and secure scaffold				1.500		70.50		70.50	94	116
Remove damaged block				4.571		172		172	228	282
Wire truss reinforcing				.260	13	12		25	30	36
Install block				11.111	286	477		763	945	1,125
Remove scaffold				1.500		70.50		70.50	94	116
Total				18.942	299	802		1,101	1,391	1,675
3030 Waterproof concrete block wall - 3rd floor	10	G-8	C.S.F.							
Position aerial lift truck, raise & lower platform				.300		12.05	24	36.05	42	47.50
Steam clean				11.765		442	111	553	700	845
Spray surface, silicone				.260	34	10	21	65	74	84.50
Total				12.325	34	464.05	156	654.05	816	977

B2013 114 Concrete Block Wall, Painted

System Description	Freq. (Years)	Crew	Unit	Labor Hours	2015 Bare Costs				Total In-House	Total w/O&P
					Material	Labor	Equipment	Total		
1030 Point & refinish block wall - 1st floor	25	1 BRIC	C.S.F.							
Set up and secure scaffold				.750		35.25		35.25	47	57.50
Steam clean				2.600		98		98	130	160
Clean & point block				3.924		181		181	239	295
Paint surface, brushwork, 1 coat				1.040	14.30	41.60		55.90	70	85.50
Remove scaffold				.750		35.25		35.25	47	57.50
Total				9.064	14.30	391.10		405.40	533	655.50
2030 Point & refinish block wall - 2nd floor	25	1 BRIC	C.S.F.							
Set up and secure scaffold				1.500		70.50		70.50	93.50	115
Steam clean				2.600		98		98	130	160
Clean & point block				3.924		181		181	239	295
Paint surface, brushwork, 1 coat				1.040	14.30	41.60		55.90	70	85.50
Remove scaffold				1.500		70.50		70.50	93.50	115
Total				10.564	14.30	461.60		475.90	626	770.50

B20 EXTERIOR CLOSURE — B2013 Exterior Walls

B2013 114 Concrete Block Wall, Painted

	System Description	Freq. (Years)	Crew	Unit	Labor Hours	2015 Bare Costs Material	Labor	Equipment	Total	Total In-House	Total w/O&P
3020	**Point & refinish block wall - 3rd floor**	25	1 BRIC	C.S.F.							
	Set up and secure scaffold				2.250		105.75		105.75	140	173
	Steam clean				2.600		98		98	130	160
	Clean & point block				3.924		181		181	239	295
	Paint surface, brushwork, 1 coat				1.040	14.30	41.60		55.90	70	85.50
	Remove scaffold				2.250		105.75		105.75	140	173
	Total				12.064	14.30	532.10		546.40	719	886.50

B2013 118 Terra Cotta

	System Description	Freq. (Years)	Crew	Unit	Labor Hours	2015 Bare Costs Material	Labor	Equipment	Total	Total In-House	Total w/O&P
1020	**Replace terra cotta wall - 1st floor**	50	2 BRIC	C.S.F.							
	Set up and secure scaffold				.750		35.25		35.25	47	57.50
	Remove terra cotta				1.040		39.50	6.05	45.55	58.50	71.50
	Replace terra cotta				10.400	545	446		991	1,200	1,400
	Remove scaffold				.750		35.25		35.25	47	57.50
	Total				12.940	545	556	6.05	1,107.05	1,352.50	1,586.50
2010	**Replace terra cotta wall - 2nd floor**	50	2 BRIC	C.S.F.							
	Set up and secure scaffold				1.500		70.50		70.50	93.50	115
	Remove terra cotta				1.040		39.50	6.05	45.55	58.50	71.50
	Replace terra cotta				10.400	545	446		991	1,200	1,400
	Remove scaffold				1.500		70.50		70.50	93.50	115
	Total				14.440	545	626.50	6.05	1,177.55	1,445.50	1,701.50
2020	**Replace terra cotta wall - 3rd floor**	50	2 BRIC	C.S.F.							
	Set up and secure scaffold				2.250		105.75		105.75	140	173
	Remove terra cotta				1.040		39.50	6.05	45.55	58.50	71.50
	Replace terra cotta				10.400	545	446		991	1,200	1,400
	Remove scaffold				2.250		105.75		105.75	140	173
	Total				15.940	545	697	6.05	1,248.05	1,538.50	1,817.50

B2013 119 | **Clay Brick**

	System Description	Freq. (Years)	Crew	Unit	Labor Hours	2015 Bare Costs				Total In-House	Total w/O&P
						Material	Labor	Equipment	Total		
1010	**Repair brick wall - 1st floor**	25	1 BRIC	S.F.							
	Set up, secure and take down ladder				.130		6		6	7.90	9.75
	Remove damage				.451		16.90		16.90	22.50	28
	Install brick				.236	3.04	10.15		13.19	16.70	20.50
	Total				.817	3.04	33.05		36.09	**47.10**	**58.25**
1020	**Point brick wall - 1st floor**	25	2 BRIC	C.S.F.							
	Set up and secure scaffold				1.333		54		54	70	86.50
	Steam clean				2.971		113	17	130	167	204
	Clean & point brick				10.000	3	462		465	610	755
	Remove scaffold				1.333		54		54	70	86.50
	Total				15.638	3	683	17	703	**917**	**1,132**
1030	**Waterproof brick wall - 1st floor**	10	1 ROFC	C.S.F.							
	Steam clean				2.971		113	17	130	167	204
	Spray surface, silicone				.260	34	10		44	53	61.50
	Total				3.231	34	123	17	174	**220**	**265.50**
2010	**Replace brick wall - 2nd floor**	75	2 BRIC	C.S.F.							
	Set up and secure scaffold				1.500		70.50		70.50	93.50	115
	Remove damage				10.401		390		390	520	640
	Install brick				23.636	304	1,015		1,319	1,675	2,025
	Remove scaffold				1.500		70.50		70.50	93.50	115
	Total				37.037	304	1,546		1,850	**2,382**	**2,895**
2020	**Point brick wall - 2nd floor**	25	2 BRIC	C.S.F.							
	Set up and secure scaffold				1.500		70.50		70.50	93.50	115
	Steam clean				2.971		113	17	130	167	204
	Clean & point brick				10.000	3	462		465	610	755
	Remove scaffold				1.500		70.50		70.50	93.50	115
	Total				15.971	3	716	17	736	**964**	**1,189**
2030	**Waterproof brick wall - 2nd floor**	10	G-8	C.S.F.							
	Position aerial lift truck, raise & lower platform				.250		10.05	19.85	29.90	35	40
	Steam clean				11.765		442	111	553	700	845
	Spray surface, silicone				.260	34	10	21	65	74	84.50
	Total				12.275	34	462.05	151.85	647.90	**809**	**969.50**

B2013 119 — Clay Brick

	System Description	Freq. (Years)	Crew	Unit	Labor Hours	2015 Bare Costs				Total In-House	Total w/O&P
						Material	Labor	Equipment	Total		
3010	**Replace brick wall - 3rd floor**	75	2 BRIC	C.S.F.							
	Set up and secure scaffold				1.500		70.50		70.50	94	116
	Remove damage				10.401		390		390	520	640
	Install brick				23.636	304	1,015		1,319	1,675	2,025
	Remove scaffold				1.500		70.50		70.50	94	116
	Total				37.037	304	1,546		1,850	2,383	2,897
3020	**Point brick wall - 3rd floor**	25	2 BRIC	C.S.F.							
	Set up and secure scaffold				1.500		70.50		70.50	94	116
	Steam clean				2.971		113	17	130	167	204
	Clean & point brick				10.000	3	462		465	610	755
	Remove scaffold				1.500		70.50		70.50	94	116
	Total				15.971	3	716	17	736	965	1,191
3030	**Waterproof brick wall - 3rd floor**	10	G-8	C.S.F.							
	Position aerial lift truck, raise & lower platform				.300		12.05	24	36.05	42	47.50
	Steam clean				11.765		442	111	553	700	845
	Spray surface, silicone				.260	34	10	21	65	74	84.50
	Total				12.325	34	464.05	156	654.05	816	977

B2013 121 — Clay Brick, Painted

	System Description	Freq. (Years)	Crew	Unit	Labor Hours	2015 Bare Costs				Total In-House	Total w/O&P
						Material	Labor	Equipment	Total		
1010	**Repair painted brick wall - 1st floor**	25	1 BRIC	S.F.							
	Set up, secure and take down ladder				.130		6		6	7.90	9.75
	Remove damage				.451		16.90		16.90	22.50	28
	Install brick				.236	3.04	10.15		13.19	16.70	20.50
	Paint surface, brushwork, 1 coat				.010	.14	.42		.56	.71	.86
	Total				.827	3.18	33.47		36.65	47.81	59.11

B2013 121   Clay Brick, Painted

System Description	Freq. (Years)	Crew	Unit	Labor Hours	2015 Bare Costs				Total In-House	Total w/O&P
					Material	Labor	Equipment	Total		
1030 Point brick wall - 1st floor	25	1 BRIC	C.S.F.							
Set up and secure scaffold				.500		23.50		23.50	31.50	38.50
Steam clean				2.971		113	17	130	167	204
Clean & point brick				10.000	3	462		465	610	755
Paint surface, brushwork, 1 coat				1.040	14.30	41.60		55.90	70	85.50
Remove scaffold				.500		23.50		23.50	31.50	38.50
Total				15.011	17.30	663.60	17	697.90	910	1,121.50
2010 Replace brick wall - 2nd floor	75	1 BRIC	C.S.F.							
Set up and secure scaffold				1.500		70.50		70.50	93.50	115
Remove damage				10.401		390		390	520	640
Install brick				23.636	304	1,015		1,319	1,675	2,025
Paint surface, brushwork, 1 coat				1.040	14.30	41.60		55.90	70	85.50
Remove scaffold				1.500		70.50		70.50	93.50	115
Total				38.077	318.30	1,587.60		1,905.90	2,452	2,980.50
2020 Point and refinish brick wall - 2nd floor	25	1 BRIC	C.S.F.							
Set up and secure scaffold				1.500		70.50		70.50	93.50	115
Steam clean				2.971		113	17	130	167	204
Clean & point brick				10.000	3	462		465	610	755
Paint surface, brushwork, 1 coat				1.040	14.30	41.60		55.90	70	85.50
Remove scaffold				1.500		70.50		70.50	93.50	115
Total				17.011	17.30	757.60	17	791.90	1,034	1,274.50
3010 Replace brick wall - 3rd floor	75	1 BRIC	C.S.F.							
Set up and secure scaffold				1.500		70.50		70.50	94	116
Remove damage				10.401		390		390	520	640
Install brick				23.636	304	1,015		1,319	1,675	2,025
Paint surface, brushwork, 1 coat				.010	.14	.42		.56	.71	.86
Remove scaffold				1.500		70.50		70.50	94	116
Total				37.047	304.14	1,546.42		1,850.56	2,383.71	2,897.86

For customer support on your Facilities Maintenance & Repair Cost Data, call 877.801.5724.

19

B2013 121 Clay Brick, Painted

	System Description	Freq. (Years)	Crew	Unit	Labor Hours	2015 Bare Costs				Total In-House	Total w/O&P
						Material	Labor	Equipment	Total		
3020	**Point and refinish brick wall - 3rd floor**	25	1 BRIC	C.S.F.							
	Set up and secure scaffold				1.500		70.50		70.50	94	116
	Steam clean				2.971		113	17	130	167	204
	Clean & point brick				10.000	3	462		465	610	755
	Paint surface, brushwork, 1 coat				1.040	14.30	41.60		55.90	70	85.50
	Remove scaffold				1.500		70.50		70.50	94	116
	Total				17.011	17.30	757.60	17	791.90	1,035	1,276.50

B2013 123 Field Stone

	System Description	Freq. (Years)	Crew	Unit	Labor Hours	2015 Bare Costs				Total In-House	Total w/O&P
						Material	Labor	Equipment	Total		
1010	**Repair field stone wall - 1st floor**	25	1 BRIC	S.F.							
	Set up, secure and take down ladder				.390		14.90		14.90	19.85	24.50
	Remove damaged stones				.173		6.60	1.01	7.61	9.75	11.90
	Reset stones				.416	19	19.15	6.15	44.30	52	61.50
	Total				.979	19	40.65	7.16	66.81	81.60	97.90
1020	**Point stone wall, 8" x 8" stones - 1st floor**	25	2 BRIC	C.S.F.							
	Set up and secure scaffold				1.333		54		54	70	86.50
	Steam clean				2.971		113	17	130	167	204
	Clean & point stonework joints				5.714	74	264		338	430	520
	Remove scaffold				1.333		54		54	70	86.50
	Total				11.352	74	485	17	576	737	897
2020	**Point stone wall, 8" x 8" stones - 2nd floor**	25	2 BRIC	C.S.F.							
	Set up and secure scaffold				1.500		70.50		70.50	93.50	115
	Steam clean				2.971		113	17	130	167	204
	Clean & point stonework joints				5.714	74	264		338	430	520
	Remove scaffold				1.500		70.50		70.50	93.50	115
	Total				11.685	74	518	17	609	784	954

B2013 123 Field Stone

	System Description	Freq. (Years)	Crew	Unit	Labor Hours	2015 Bare Costs				Total In-House	Total w/O&P
						Material	Labor	Equipment	Total		
3020	Point stone wall, 8" x 8" stones - 3rd floor	25	2 BRIC	C.S.F.							
	Set up and secure scaffold				1.500		70.50		70.50	94	116
	Steam clean				2.971		113	17	130	167	204
	Clean & point stonework joints				5.714	74	264		338	430	520
	Remove scaffold				1.500		70.50		70.50	94	116
	Total				11.685	74	518	17	609	785	956

B2013 140 Glass Block

	System Description	Freq. (Years)	Crew	Unit	Labor Hours	2015 Bare Costs				Total In-House	Total w/O&P
						Material	Labor	Equipment	Total		
1020	Replace glass block wall - 1st floor	75	2 BRIC	C.S.F.							
	Set up and secure scaffold				.750		35.25		35.25	47	57.50
	Remove blocks				5.200		198	30	228	293	355
	Replace blocks				45.217	2,650	1,940		4,590	5,450	6,475
	Remove scaffold				.750		35.25		35.25	47	57.50
	Total				51.917	2,650	2,208.50	30	4,888.50	5,837	6,945
2020	Replace glass block wall - 2nd floor	75	2 BRIC	C.S.F.							
	Set up and secure scaffold				1.500		70.50		70.50	93.50	115
	Remove blocks				5.200		198	30	228	293	355
	Replace blocks				45.217	2,650	1,940		4,590	5,450	6,475
	Remove scaffold				1.500		70.50		70.50	93.50	115
	Total				53.417	2,650	2,279	30	4,959	5,930	7,060
3020	Replace glass block wall - 3rd floor	75	2 BRIC	C.S.F.							
	Set up and secure scaffold				2.250		105.75		105.75	140	173
	Remove blocks				5.200		198	30	228	293	355
	Replace blocks				45.217	2,650	1,940		4,590	5,450	6,475
	Remove scaffold				2.250		105.75		105.75	140	173
	Total				54.917	2,650	2,349.50	30	5,029.50	6,023	7,176

For customer support on your Facilities Maintenance & Repair Cost Data, call 877.801.5724.

21

B2013 141   Aluminum Siding

	System Description	Freq. (Years)	Crew	Unit	Labor Hours	Material	Labor	Equipment	Total	Total In-House	Total w/O&P
							2015 Bare Costs				
1010	**Replace aluminum siding - 1st floor**	35	1 CARP	C.S.F.							
	Set up, secure and take down ladder				.520		24.40		24.40	32.50	40
	Remove damaged siding				1.434		54		54	72	88
	Install siding				5.368	166	251		417	505	610
	Total				7.322	166	329.40		495.40	**609.50**	**738**
1020	**Refinish aluminum siding - 1st floor**	20	1 PORD	C.S.F.							
	Set up, secure and take down ladder				1.300		52.50		52.50	68	84.50
	Prepare surface				1.159		47		47	61	75
	Refinish surface, brushwork, primer + 1 coat				1.231	13	50		63	78	96.50
	Total				3.690	13	149.50		162.50	**207**	**256**
2010	**Replace aluminum siding - 2nd floor**	35	1 CARP	C.S.F.							
	Set up and secure scaffold				1.500		70.50		70.50	93.50	115
	Remove damaged siding				1.434		54		54	72	88
	Install siding				5.368	166	251		417	505	610
	Remove scaffold				1.500		70.50		70.50	93.50	115
	Total				9.802	166	446		612	**764**	**928**
2020	**Refinish aluminum siding - 2nd floor**	20	1 PORD	C.S.F.							
	Set up and secure scaffold				1.500		70.50		70.50	93.50	115
	Prepare surface				1.159		47		47	61	75
	Refinish surface, brushwork, primer + 1 coat				1.231	13	50		63	78	96.50
	Remove scaffold				1.500		70.50		70.50	93.50	115
	Total				5.390	13	238		251	**326**	**401.50**
3010	**Replace aluminum siding - 3rd floor**	35	1 CARP	C.S.F.							
	Set up and secure scaffold				2.250		105.75		105.75	140	173
	Remove damaged siding				1.434		54		54	72	88
	Install siding				5.368	166	251		417	505	610
	Remove scaffold				2.250		105.75		105.75	140	173
	Total				11.302	166	516.50		682.50	**857**	**1,044**

For customer support on your Facilities Maintenance & Repair Cost Data, call 877.801.5724.

System Description	Freq. (Years)	Crew	Unit	Labor Hours	2015 Bare Costs				Total In-House	Total w/O&P
					Material	Labor	Equipment	Total		
3020 Refinish aluminum siding - 3rd floor	20	1 PORD	C.S.F.							
Set up and secure scaffold				2.250		105.75		105.75	140	173
Prepare surface				1.159		47		47	61	75
Refinish surface, brushwork, primer + 1 coat				1.231	13	50		63	78	96.50
Remove scaffold				2.250		105.75		105.75	140	173
Total				6.890	13	308.50		321.50	419	517.50
4120 Spray refinish aluminum siding - 1st floor	20	1 PORD	C.S.F.							
Set up, secure and take down ladder				1.300		52.50		52.50	68	84.50
Prepare surface				1.159		47		47	61	75
Protect/mask unpainted surfaces				.031	.30	1.20		1.50	1.90	2.38
Spray full primer coat				.444	9	18		27	39	45.50
Spray finish coat				.500	8	21		29	41	49
Total				3.434	17.30	139.70		157	210.90	256.38
4220 Spray refinish aluminum siding - 2nd floor	20	2 PORD	C.S.F.							
Set up and secure scaffold				1.500		70.50		70.50	93.50	115
Prepare surface				1.159		47		47	61	75
Ptotect/mask unpainted surfaces				.031	.30	1.20		1.50	1.90	2.38
Spray full primer coat				.444	9	18		27	39	45.50
Spray finish coat				.500	8	21		29	41	49
Remove scaffold				1.500		70.50		70.50	93.50	115
Total				5.134	17.30	228.20		245.50	329.90	401.88
4320 Spray refinish aluminum siding - 3rd floor	20	2 PORD	C.S.F.							
Set up and secure scaffold				2.250		105.75		105.75	140	173
Prepare surface				1.159		47		47	61	75
Protect/mask unpainted surfaces				.031	.30	1.20		1.50	1.90	2.38
Spray full primer coat				.444	9	18		27	39	45.50
Spray finish coat				.500	8	21		29	41	49
Remove scaffold				2.250		105.75		105.75	140	173
Total				6.634	17.30	298.70		316	422.90	517.88

	System Description	Freq. (Years)	Crew	Unit	Labor Hours	2015 Bare Costs				Total In-House	Total w/O&P
						Material	Labor	Equipment	Total		
1010	**Replace steel siding - 1st floor**	35	1 CARP	C.S.F.							
	Set up, secure and take down ladder				.520		24.40		24.40	32.50	40
	Remove damaged siding				1.434		54		54	72	88
	Install siding				5.266	180	246		426	515	620
	Total				7.220	180	324.40		504.40	619.50	748
1020	**Refinish steel siding - 1st floor**	20	1 PORD	C.S.F.							
	Set up, secure and take down ladder				1.300		52.50		52.50	68	84.50
	Prepare surface				1.159		47		47	61	75
	Refinish surface, brushwork, primer + 1 coat				1.231	13	50		63	78	96.50
	Total				3.690	13	149.50		162.50	207	256
2010	**Replace steel siding - 2nd floor**	35	1 CARP	C.S.F.							
	Set up and secure scaffold				1.500		70.50		70.50	93.50	115
	Remove damaged siding				1.434		54		54	72	88
	Install siding				5.266	180	246		426	515	620
	Remove scaffold				1.500		70.50		70.50	93.50	115
	Total				9.700	180	441		621	774	938
2020	**Refinish steel siding - 2nd floor**	20	1 PORD	C.S.F.							
	Set up and secure scaffold				1.500		70.50		70.50	93.50	115
	Prepare surface				1.159		47		47	61	75
	Refinish surface, brushwork, primer + 1 coat				1.231	13	50		63	78	96.50
	Remove scaffold				1.500		70.50		70.50	93.50	115
	Total				5.390	13	238		251	326	401.50
3010	**Replace steel siding - 3rd floor**	35	1 CARP	C.S.F.							
	Set up and secure scaffold				2.250		105.75		105.75	140	173
	Remove damaged siding				1.434		54		54	72	88
	Install siding				5.266	180	246		426	515	620
	Remove scaffold				2.250		105.75		105.75	140	173
	Total				11.200	180	511.50		691.50	867	1,054

For customer support on your Facilities Maintenance & Repair Cost Data, call 877.801.5724.

B2013 142 Steel Siding

System Description	Freq. (Years)	Crew	Unit	Labor Hours	2015 Bare Costs				Total In-House	Total w/O&P
					Material	Labor	Equipment	Total		
3020 Refinish steel siding - 3rd floor	20	1 PORD	C.S.F.							
Set up and secure scaffold				2.250		105.75		105.75	140	173
Prepare surface				1.159		47		47	61	75
Refinish surface, brushwork, primer + 1 coat				1.231	13	50		63	78	96.50
Remove scaffold				2.250		105.75		105.75	140	173
Total				6.890	13	308.50		321.50	419	517.50
4120 Spray refinish steel siding - 1st floor	20	1 PORD	C.S.F.							
Set up, secure and take down ladder				1.300		52.50		52.50	68	84.50
Prepare surface				1.159		47		47	61	75
Protect/mask unpainted surfaces				.031	.30	1.20		1.50	1.90	2.38
Spray full primer coat				.444	9	18		27	39	45.50
Spray finish coat				.500	8	21		29	41	49
Total				3.434	17.30	139.70		157	210.90	256.38
4220 Spray refinish steel siding - 2nd floor	20	2 PORD	C.S.F.							
Set up and secure scaffold				1.500		70.50		70.50	93.50	115
Prepare surface				1.159		47		47	61	75
Protect/mask unpainted surfaces				.031	.30	1.20		1.50	1.90	2.38
Spray full primer coat				.444	9	18		27	39	45.50
Spray finish coat				.500	8	21		29	41	49
Remove scaffold				1.500		70.50		70.50	93.50	115
Total				5.134	17.30	228.20		245.50	329.90	401.88
4320 Spray refinish steel siding - 3rd floor	20	2 PORD	C.S.F.							
Set up and secure scaffold				2.250		105.75		105.75	140	173
Prepare surface				1.159		47		47	61	75
Protect/mask unpainted surfaces				.031	.30	1.20		1.50	1.90	2.38
Spray full primer coat				.444	9	18		27	39	45.50
Spray finish coat				.500	8	21		29	41	49
Remove scaffold				2.250		105.75		105.75	140	173
Total				6.634	17.30	298.70		316	422.90	517.88

B2013 143 Masonite Panel, Sealed

	System Description	Freq. (Years)	Crew	Unit	Labor Hours	2015 Bare Costs				Total In-House	Total w/O&P
						Material	Labor	Equipment	Total		
1010	**Replace hardboard panels - 1st floor**	12	1 CARP	C.S.F.							
	Set up, secure and take down ladder				.520		24.40		24.40	32.50	40
	Remove damaged panel				1.434		54		54	72	88
	Cut new hardboard panel to fit				2.773	242	130		372	440	515
	Total				4.727	242	208.40		450.40	**544.50**	**643**
2010	**Replace hardboard panels - 2nd floor**	12	1 CARP	C.S.F.							
	Set up and secure scaffold				1.500		70.50		70.50	93.50	115
	Remove damaged panel				1.434		54		54	72	88
	Cut new hardboard panel to fit				2.773	242	130		372	440	515
	Remove scaffold				1.500		70.50		70.50	93.50	115
	Total				7.207	242	325		567	**699**	**833**
3010	**Replace hardboard panels - 3rd floor**	12	1 CARP	C.S.F.							
	Set up and secure scaffold				2.250		105.75		105.75	140	173
	Remove damaged panel				1.434		54		54	72	88
	Cut new hardboard panel to fit				2.773	242	130		372	440	515
	Remove scaffold				2.250		105.75		105.75	140	173
	Total				8.707	242	395.50		637.50	**792**	**949**

B2013 145 Wood, Clapboard Finished, 1 Coat

	System Description	Freq. (Years)	Crew	Unit	Labor Hours	2015 Bare Costs				Total In-House	Total w/O&P
						Material	Labor	Equipment	Total		
1010	**Replace & finish wood clapboards - 1st floor**	25	1 CARP	C.S.F.							
	Set up, secure and take down ladder				.520		24.40		24.40	32.50	40
	Remove damage				2.600		122		122	162	200
	Replace wood clapboards				2.711	318.46	126.92		445.38	520	605
	Refinish surface, brushwork, primer + 1 coat				2.000	20	81		101	127	155
	Total				7.831	338.46	354.32		692.78	**841.50**	**1,000**

B2013 145 Wood, Clapboard Finished, 1 Coat

	System Description	Freq. (Years)	Crew	Unit	Labor Hours	2015 Bare Costs				Total In-House	Total w/O&P
						Material	Labor	Equipment	Total		
2010	**Replace & finish wood clapboards - 2nd floor**	25	1 CARP	C.S.F.							
	Set up and secure scaffold				1.500		70.50		70.50	93.50	115
	Remove damage				2.600		122		122	162	200
	Replace wood clapboards				3.524	414	165		579	675	790
	Refinish surface, brushwork, primer + 1 coat				2.000	20	81		101	127	155
	Remove scaffold				1.500		70.50		70.50	93.50	115
	Total				11.124	434	509		943	1,151	1,375
3010	**Replace & finish wood clapboards - 3rd floor**	25	1 CARP	C.S.F.							
	Set up and secure scaffold				2.250		105.75		105.75	140	173
	Remove damage				2.600		122		122	162	200
	Replace wood clapboards				3.489	409.90	163.37		573.27	670	780
	Refinish surface, brushwork, primer + 1 coat				2.000	20	81		101	127	155
	Remove scaffold				2.250		105.75		105.75	140	173
	Total				12.589	429.90	577.87		1,007.77	1,239	1,481

B2013 147 Wood Shingles, Unfinished

	System Description	Freq. (Years)	Crew	Unit	Labor Hours	2015 Bare Costs				Total In-House	Total w/O&P
						Material	Labor	Equipment	Total		
1010	**Repair wood shingles - 1st floor**	12	1 CARP	S.F.							
	Set up, secure and take down ladder				.052		2.44		2.44	3.24	4
	Remove damaged shingles				.039		1.47		1.47	1.95	2.40
	Replace shingles				.056	2.50	2.64		5.14	6.25	7.50
	Total				.147	2.50	6.55		9.05	11.44	13.90
1020	**Replace wood shingles - 1st floor**	40	2 CARP	C.S.F.							
	Set up and secure scaffold				.750		35.25		35.25	47	57.50
	Remove damaged shingles				2.600		98		98	130	160
	Replace shingles				4.324	192	203		395	480	575
	Remove scaffold				.750		35.25		35.25	47	57.50
	Total				8.424	192	371.50		563.50	704	850

For customer support on your Facilities Maintenance & Repair Cost Data, call 877.801.5724.

27

B2013 147 Wood Shingles, Unfinished

System Description	Freq. (Years)	Crew	Unit	Labor Hours	2015 Bare Costs				Total In-House	Total w/O&P
					Material	Labor	Equipment	Total		
2010 Replace wood shingles - 2nd floor	40	1 CARP	C.S.F.							
Set up and secure scaffold				1.000		47		47	62.50	77
Remove damaged shingles				2.600		98		98	130	160
Replace shingles				4.324	192	203		395	480	575
Remove scaffold				1.000		47		47	62.50	77
Total				8.924	192	395		587	735	889
3010 Replace wood shingles - 3rd floor	40	1 CARP	C.S.F.							
Set up and secure scaffold				1.500		70.50		70.50	94	116
Remove damaged shingles				2.600		98		98	130	160
Replace shingles				4.324	192	203		395	480	575
Remove scaffold				1.500		70.50		70.50	94	116
Total				9.924	192	442		634	798	967

B2013 148 Wood Shingles, Finished

System Description	Freq. (Years)	Crew	Unit	Labor Hours	2015 Bare Costs				Total In-House	Total w/O&P
					Material	Labor	Equipment	Total		
1020 Replace & finish wood shingles - 1st floor	45	1 CARP	C.S.F.							
Set up, secure and take down ladder				1.500		70.50		70.50	93.50	115
Remove damage				2.600		98		98	130	160
Replace shingles				4.324	192	203		395	480	575
Refinish surface, brushwork, primer + 1 coat				2.000	20	81		101	127	155
Total				10.424	212	452.50		664.50	830.50	1,005
1030 Refinish wood shingles - 1st floor	5	1 PORD	C.S.F.							
Set up and secure scaffold				.750		35.25		35.25	47	57.50
Prepare surface				1.667		67		67	87	108
Refinish surface, brushwork, primer + 1 coat				2.000	20	81		101	127	155
Remove scaffold				.750		35.25		35.25	47	57.50
Total				5.167	20	218.50		238.50	308	378

For customer support on your Facilities Maintenance & Repair Cost Data, call 877.801.5724.

B2013 148 Wood Shingles, Finished

	System Description	Freq. (Years)	Crew	Unit	Labor Hours	2015 Bare Costs				Total In-House	Total w/O&P
						Material	Labor	Equipment	Total		
2020	**Refinish wood shingles - 2nd floor**	5	1 PORD	C.S.F.							
	Set up and secure scaffold				1.500		70.50		70.50	93.50	115
	Prepare surface				1.667		67		67	87	108
	Refinish surface, brushwork, primer + 1 coat				2.000	20	81		101	127	155
	Remove scaffold				1.500		70.50		70.50	93.50	115
	Total				6.667	20	289		309	401	493
3020	**Refinish wood shingles - 3rd floor**	5	1 PORD	C.S.F.							
	Set up and secure scaffold				2.250		105.75		105.75	140	173
	Prepare surface				1.667		67		67	87	108
	Refinish surface, brushwork, primer + 1 coat				2.000	20	81		101	127	155
	Remove scaffold				2.250		105.75		105.75	140	173
	Total				8.167	20	359.50		379.50	494	609
4120	**Spray refinish wood siding - 1st floor**	5	1 PORD	C.S.F.							
	Set up, secure and take down ladder				1.300		52.50		52.50	68	84.50
	Prepare surface				1.159		47		47	61	75
	Protect/mask unpainted surfaces				.031	.30	1.20		1.50	1.90	2.38
	Spray full primer coat				.410	13	17		30	35	43.50
	Spray finish coat				.410	15	17		32	38	46
	Total				3.310	28.30	134.70		163	203.90	251.38
4220	**Spray refinish wood siding - 2nd floor**	5	2 PORD	C.S.F.							
	Set up and secure scaffold				1.500		70.50		70.50	93.50	115
	Prepare surface				1.159		47		47	61	75
	Protect/mask unpainted surfaces				.031	.30	1.20		1.50	1.90	2.38
	Spray full primer coat				.410	13	17		30	35	43.50
	Spray finish coat				.410	15	17		32	38	46
	Remove scaffold				1.500		70.50		70.50	93.50	115
	Total				5.010	28.30	223.20		251.50	322.90	396.88

For customer support on your Facilities Maintenance & Repair Cost Data, call 877.801.5724.

29

B2013 148 Wood Shingles, Finished

	System Description	Freq. (Years)	Crew	Unit	Labor Hours	2015 Bare Costs				Total In-House	Total w/O&P
						Material	Labor	Equipment	Total		
4320	**Spray refinish wood siding - 3rd floor**	5	2 PORD	C.S.F.							
	Set up and secure scaffold				2.250		105.75		105.75	140	173
	Prepare surface				1.159		47		47	61	75
	Protect/mask unpainted surfaces				.031	.30	1.20		1.50	1.90	2.38
	Spray full primer coat				.410	13	17		30	35	43.50
	Spray finish coat				.410	15	17		32	38	46
	Remove scaffold				2.250		105.75		105.75	140	173
	Total				6.510	28.30	293.70		322	415.90	512.88

B2013 150 Fiberglass Panel, Rigid

	System Description	Freq. (Years)	Crew	Unit	Labor Hours	2015 Bare Costs				Total In-House	Total w/O&P
						Material	Labor	Equipment	Total		
1010	**Replace fiberglass panels - 1st floor**	9	1 CARP	C.S.F.							
	Set up, secure and take down ladder				.520		24.40		24.40	32.50	40
	Remove damaged fiberglass panel				1.434		54		54	72	88
	Install new fiberglass panel				4.727	375	221		596	700	825
	Total				6.681	375	299.40		674.40	804.50	953
2010	**Replace fiberglass panels - 2nd floor**	9	1 CARP	C.S.F.							
	Set up and secure scaffold				1.500		70.50		70.50	93.50	115
	Remove damaged fiberglass panel				1.434		54		54	72	88
	Install new fiberglass panel				4.727	375	221		596	700	825
	Remove scaffold				1.500		70.50		70.50	93.50	115
	Total				9.161	375	416		791	959	1,143
3010	**Replace fiberglass panels - 3rd floor**	9	1 CARP	C.S.F.							
	Set up and secure scaffold				2.250		105.75		105.75	140	173
	Remove damaged fiberglass panel				1.434		54		54	72	88
	Install new fiberglass panel				4.727	375	221		596	700	825
	Remove scaffold				2.250		105.75		105.75	140	173
	Total				10.661	375	486.50		861.50	1,052	1,259

B2013 152 Synthetic Veneer Plaster

System Description	Freq. (Years)	Crew	Unit	Labor Hours	2015 Bare Costs				Total In-House	Total w/O&P
					Material	Labor	Equipment	Total		
1030 Refinish veneer plaster - 1st floor	10	1 BRIC	C.S.F.							
Set up and secure scaffold				.750		35.25		35.25	47	57.50
Pressure wash surface				.260		10	2	12	15	19
Refinish surface				1.270	154	51		205	235	275
Remove scaffold				.750		35.25		35.25	47	57.50
Total				3.030	154	131.50	2	287.50	344	409
2020 Refinish veneer plaster - 2nd floor	10	1 BRIC	C.S.F.							
Set up and secure scaffold				1.500		70.50		70.50	93.50	115
Pressure wash surface				.260		10	2	12	15	19
Refinish surface				1.270	154	51		205	235	275
Remove scaffold				1.500		70.50		70.50	93.50	115
Total				4.530	154	202	2	358	437	524
3020 Refinish veneer plaster - 3rd floor	10	1 BRIC	C.S.F.							
Set up and secure scaffold				2.250		105.75		105.75	140	173
Pressure wash surface				.260		10	2	12	15	19
Refinish surface				1.270	154	51		205	235	275
Remove scaffold				2.250		105.75		105.75	140	173
Total				6.030	154	272.50	2	428.50	530	640

B2013 156 Exterior Soffit Gypsum Board, Painted

System Description	Freq. (Years)	Crew	Unit	Labor Hours	2015 Bare Costs				Total In-House	Total w/O&P
					Material	Labor	Equipment	Total		
1030 Refinish exterior painted gypsum board - 1st floor soffit	5	1 PORD	C.S.F.							
Set up and secure scaffold				.750		35.25		35.25	47	57.50
Prepare surface				1.159		47		47	61	75
Paint surface, roller, primer + 1 coat				1.231	10	50		60	75	92.50
Remove scaffold				.750		35.25		35.25	47	57.50
Total				3.890	10	167.50		177.50	230	282.50

For customer support on your Facilities Maintenance & Repair Cost Data, call 877.801.5724.

31

B2013 156 Exterior Soffit Gypsum Board, Painted

	System Description	Freq. (Years)	Crew	Unit	Labor Hours	2015 Bare Costs				Total In-House	Total w/O&P
						Material	Labor	Equipment	Total		
1040	**Replace exterior painted gypsum board - 1st floor soffit**	25	2 CARP	C.S.F.							
	Set up and secure scaffold				.750		35.25		35.25	47	57.50
	Remove old exterior gypsum board				2.737		103		103	137	169
	Install new exterior gypsum board				3.467	72	163		235	295	355
	Paint surface, roller, primer + 1 coat				1.231	10	50		60	75	92.50
	Remove scaffold				.750		35.25		35.25	47	57.50
	Total				8.935	82	386.50		468.50	601	731.50
2010	**Replace exterior painted gypsum board - 2nd floor soffit**	25	1 CARP	C.S.F.							
	Set up and secure scaffold				1.500		70.50		70.50	93.50	115
	Remove old exterior gypsum board				2.737		103		103	137	169
	Install new exterior gypsum board				3.467	72	163		235	295	355
	Paint surface, roller, primer + 1 coat				2.462	20	100		120	150	185
	Remove scaffold				1.500		70.50		70.50	93.50	115
	Total				11.666	92	507		599	769	939
2020	**Refinish exterior painted gypsum board - 2nd floor soffit**	5	1 PORD	C.S.F.							
	Set up and secure scaffold				1.500		70.50		70.50	93.50	115
	Prepare surface				1.159		47		47	61	75
	Paint surface, roller, primer + 1 coat				1.231	10	50		60	75	92.50
	Remove scaffold				1.500		70.50		70.50	93.50	115
	Total				5.390	10	238		248	323	397.50
3010	**Replace exterior painted gypsum board - 3rd floor soffit**	25	1 CARP	C.S.F.							
	Set up and secure scaffold				2.250		105.75		105.75	140	173
	Remove old exterior gypsum board				2.737		103		103	137	169
	Install new exterior gypsum board				3.467	72	163		235	295	355
	Paint surface, roller, primer + 1 coat				3.693	30	150		180	225	278
	Remove scaffold				2.250		105.75		105.75	140	173
	Total				14.397	102	627.50		729.50	937	1,148
3020	**Refinish exterior painted gypsum board - 3rd floor soffit**	5	1 PORD	C.S.F.							
	Set up and secure scaffold				2.250		105.75		105.75	140	173
	Prepare surface				1.159		47		47	61	75
	Paint surface, roller, primer + 1 coat				1.231	10	50		60	75	92.50
	Remove scaffold				2.250		105.75		105.75	140	173
	Total				6.890	10	308.50		318.50	416	513.50

For customer support on your Facilities Maintenance & Repair Cost Data, call 877.801.5724.

B2013 157 Overhang, Exterior Entry

	System Description	Freq. (Years)	Crew	Unit	Labor Hours	2015 Bare Costs				Total In-House	Total w/O&P
						Material	Labor	Equipment	Total		
1030	**Refinish wood overhang**	5	1 PORD	S.F.							
	Prepare surface				.016		.65		.65	.84	1.04
	Wash / dry surface				.001		.05		.05	.06	.08
	Refinish surface, brushwork, primer + 1 coat				.020	.20	.81		1.01	1.27	1.55
	Total				.037	.20	1.51		1.71	2.17	2.67

B2013 158 Recaulking

	System Description	Freq. (Years)	Crew	Unit	Labor Hours	2015 Bare Costs				Total In-House	Total w/O&P
						Material	Labor	Equipment	Total		
1010	**Recaulk expansion & control joints**	20	1 BRIC	L.F.							
	Set up, secure and take down ladder				.130		6		6	7.90	9.75
	Cut out and recaulk, silicone				.067	4.25	3.08		7.33	8.75	10.30
	Total				.197	4.25	9.08		13.33	16.65	20.05
2010	**Recaulk door**	20	1 BRIC	L.F.							
	Set up, secure and take down ladder				.130		6		6	7.90	9.75
	Cut out and recaulk, silicone				.067	4.25	3.08		7.33	8.75	10.30
	Total				.197	4.25	9.08		13.33	16.65	20.05
3010	**Recaulk window, 4' x 6' - 1st floor**	20	2 BRIC	Ea.							
	Set up and secure scaffold				1.333		61.50		61.50	81	100
	Cut out and recaulk, silicone				1.600	102	73.92		175.92	210	248
	Remove scaffold				1.333		61.50		61.50	81	100
	Total				4.267	102	196.92		298.92	372	448
3020	**Recaulk window, 4' x 6' - 2nd floor**	20	2 BRIC	Ea.							
	Set up and secure scaffold				1.500		70.50		70.50	93.50	115
	Cut out and recaulk, silicone				1.600	102	73.92		175.92	210	248
	Remove scaffold				1.500		70.50		70.50	93.50	115
	Total				4.600	102	214.92		316.92	397	478
3030	**Recaulk window, 4' x 6' - 3rd floor**	20	2 BRIC	Ea.							
	Set up and secure scaffold				1.500		70.50		70.50	94	116
	Cut out and recaulk, silicone				1.600	102	73.92		175.92	210	248
	Remove scaffold				1.500		70.50		70.50	94	116
	Total				4.600	102	214.92		316.92	398	480

For customer support on your Facilities Maintenance & Repair Cost Data, call 877.801.5724.

33

B2013 310 Wood Louvers and Shutters

System Description	Freq. (Years)	Crew	Unit	Labor Hours	2015 Bare Costs				Total In-House	Total w/O&P
					Material	Labor	Equipment	Total		
1010 Repair wood louver in frame 1st floor	6	1 CARP	Ea.							
Set up, secure and take down ladder				.001		.05		.05	.07	.09
Remove louver				.473		17.75		17.75	23.50	29
Remove damaged stile				.078		3.69		3.69	4.89	6.05
Remove damaged slat				.039		1.82		1.82	2.42	2.98
Install new slat				.077	.70	3.62		4.32	5.60	6.80
Install new stile				.157	2.37	7.38		9.75	12.40	15
Reinstall louver				.075		3.51		3.51	4.67	5.75
Total				.900	3.07	37.82		40.89	53.55	65.67
1030 Refinish wood louver - 1st floor	5	1 PORD	Ea.							
Set up and secure scaffold				.444		17.95		17.95	23.50	29
Prepare louver for painting				.434		17.44		17.44	22.50	28
Paint louvered surface (to 16 S.F.), brushwork, primer + 1 coat				.800	2.29	32.50		34.79	44.50	55
Remove scaffold				.444		17.95		17.95	23.50	29
Total				2.123	2.29	85.84		88.13	114	141
1040 Replace wood louver - 1st floor	45	1 CARP	Ea.							
Set up and secure scaffold				.444		21		21	27.50	34
Remove louver				.473		17.75		17.75	23.50	29
Install new wood louver				.520	135	24.50		159.50	180	209
Remove scaffold				.444		21		21	27.50	34
Total				1.882	135	84.25		219.25	258.50	306
2010 Repair wood louver - 2nd floor	6	1 CARP	Ea.							
Set up and secure scaffold				.889		41.50		41.50	55.50	68.50
Remove louver				.473		17.75		17.75	23.50	29
Remove damaged stile				.078		3.69		3.69	4.89	6.05
Remove damaged slat				.039		1.82		1.82	2.42	2.98
Install new slat				.077	.70	3.62		4.32	5.60	6.80
Install new stile				.157	2.37	7.38		9.75	12.40	15
Reinstall louver				.075		3.51		3.51	4.67	5.75
Remove scaffold				.889		41.50		41.50	55.50	68.50
Total				2.677	3.07	120.77		123.84	164.48	202.58

System Description	Freq. (Years)	Crew	Unit	Labor Hours	2015 Bare Costs				Total In-House	Total w/O&P
					Material	Labor	Equipment	Total		
2030 Refinish wood louver - 2nd floor	4	1 PORD	Ea.							
Set up and secure scaffold				.889		36		36	46.50	57.50
Prepare louver for painting				.434		17.44		17.44	22.50	28
Paint louvered surface (to 16 S.F.), brushwork, primer + 1 coat				.800	2.29	32.50		34.79	44.50	55
Remove scaffold				.889		36		36	46.50	57.50
Total				3.012	2.29	121.94		124.23	160	198
2040 Replace wood louver - 2nd floor	45	1 CARP	Ea.							
Set up and secure scaffold				.889		41.50		41.50	55.50	68.50
Remove louver				.473		17.75		17.75	23.50	29
Install new wood louver				.520	135	24.50		159.50	180	209
Remove scaffold				.889		41.50		41.50	55.50	68.50
Total				2.771	135	125.25		260.25	314.50	375
3010 Repair wood louver - 3rd floor	6	1 CARP	Ea.							
Set up and secure scaffold				1.333		62.50		62.50	83	103
Remove louver				.473		17.75		17.75	23.50	29
Remove damaged stile				.078		3.69		3.69	4.89	6.05
Remove damaged slat				.039		1.82		1.82	2.42	2.98
Install new slat				.077	.70	3.62		4.32	5.60	6.80
Install new stile				.157	2.37	7.38		9.75	12.40	15
Reinstall louver				.075		3.51		3.51	4.67	5.75
Remove scaffold				1.333		62.50		62.50	83	103
Total				3.565	3.07	162.77		165.84	219.48	271.58
3030 Refinish wood louver - 3rd floor	4	1 PORD	Ea.							
Set up and secure scaffold				1.333		54		54	70	86.50
Prepare louver for painting				.434		17.44		17.44	22.50	28
Paint louvered surface (to 16 S.F.), brushwork, primer + 1 coat				.800	2.29	32.50		34.79	44.50	55
Remove scaffold				1.333		54		54	70	86.50
Total				3.901	2.29	157.94		160.23	207	256
3040 Replace wood louver - 3rd floor	45	1 CARP	Ea.							
Set up and secure scaffold				1.333		62.50		62.50	83	103
Remove louver				.473		17.75		17.75	23.50	29
Install new wood louver				.520	135	24.50		159.50	180	209
Remove scaffold				1.333		62.50		62.50	83	103
Total				3.660	135	167.25		302.25	369.50	444

For customer support on your Facilities Maintenance & Repair Cost Data, call 877.801.5724.

35

B2013 320 Aluminum Louver

System Description	Freq. (Years)	Crew	Unit	Labor Hours	2015 Bare Costs				Total In-House	Total w/O&P
					Material	Labor	Equipment	Total		
1030 Refinish aluminum louver - 1st floor	5	1 PORD	Ea.							
Set up and secure scaffold				.444		17.95		17.95	23.50	29
Prepare louver for painting				.434		17.44		17.44	22.50	28
Paint louvered surface (to 16 S.F.), brushwork, primer + 1 coat				.800	2.29	32.50		34.79	44.50	55
Remove scaffold				.444		17.95		17.95	23.50	29
Total				2.123	2.29	85.84		88.13	114	141
1040 Replace aluminum louver - 1st floor	60	1 CARP	Ea.							
Set up and secure scaffold				.444		21		21	27.50	34
Remove louver				.473		17.75		17.75	23.50	29
Install new louver				1.783	246	99.90		345.90	395	465
Remove scaffold				.444		21		21	27.50	34
Total				3.145	246	159.65		405.65	473.50	562
2020 Refinish aluminum louver - 2nd floor	12	1 PORD	Ea.							
Set up and secure scaffold				.889		41.50		41.50	55.50	68.50
Prepare louver for painting				.434		17.44		17.44	22.50	28
Paint louvered surface (to 16 S.F.), brushwork, primer + 1 coat				.800	2.29	32.50		34.79	44.50	55
Remove scaffold				.889		41.50		41.50	55.50	68.50
Total				3.012	2.29	132.94		135.23	178	220
2040 Replace aluminum louver - 2nd floor	60	1 CARP	Ea.							
Set up and secure scaffold				.889		41.50		41.50	55.50	68.50
Remove louver				.473		17.75		17.75	23.50	29
Install new louver				1.783	246	99.90		345.90	395	465
Remove scaffold				.889		41.50		41.50	55.50	68.50
Total				4.034	246	200.65		446.65	529.50	631
3030 Refinish aluminum louver - 3rd floor	5	1 PORD	Ea.							
Set up and secure scaffold				1.333		54		54	70	86.50
Prepare louver for painting				.434		17.44		17.44	22.50	28
Paint louvered surface (to 16 S.F.), brushwork, primer + 1 coat				.800	2.29	32.50		34.79	44.50	55
Remove scaffold				1.333		54		54	70	86.50
Total				3.901	2.29	157.94		160.23	207	256

For customer support on your Facilities Maintenance & Repair Cost Data, call 877.801.5724.

B2013 320   Aluminum Louver

	System Description	Freq. (Years)	Crew	Unit	Labor Hours	2015 Bare Costs				Total In-House	Total w/O&P
						Material	Labor	Equipment	Total		
3040	**Replace aluminum louver - 3rd floor**	60	1 CARP	Ea.							
	Set up and secure scaffold				1.333		62.50		62.50	83	103
	Remove louver				.473		17.75		17.75	23.50	29
	Install new louver				1.783	246	99.90		345.90	395	465
	Remove scaffold				1.333		62.50		62.50	83	103
	Total				4.922	246	242.65		488.65	584.50	700

B2013 330   Steel Louver

	System Description	Freq. (Years)	Crew	Unit	Labor Hours	2015 Bare Costs				Total In-House	Total w/O&P
						Material	Labor	Equipment	Total		
1030	**Refinish steel louver - 1st floor**	5	1 PORD	Ea.							
	Set up and secure scaffold				.444		17.95		17.95	23.50	29
	Prepare louver for painting				.434		17.44		17.44	22.50	28
	Paint louvered surface (to 16 S.F.), brushwork, primer + 1 coat				.800	2.29	32.50		34.79	44.50	55
	Remove scaffold				.444		17.95		17.95	23.50	29
	Total				2.123	2.29	85.84		88.13	114	141
1040	**Replace steel louver - 1st floor**	40	1 CARP	Ea.							
	Set up and secure scaffold				.444		21		21	27.50	34
	Remove louver				.473		17.75		17.75	23.50	29
	Install new louver				1.783	246	99.90		345.90	395	465
	Remove scaffold				.444		21		21	27.50	34
	Total				3.145	246	159.65		405.65	473.50	562
2030	**Refinish steel louver - 2nd floor**	5	1 PORD	Ea.							
	Set up and secure scaffold				.889		36		36	46.50	57.50
	Prepare louver for painting				.434		17.44		17.44	22.50	28
	Paint louvered surface (to 16 S.F.), brushwork, primer + 1 coat				.800	2.29	32.50		34.79	44.50	55
	Remove scaffold				.889		36		36	46.50	57.50
	Total				3.012	2.29	121.94		124.23	160	198

For customer support on your Facilities Maintenance & Repair Cost Data, call 877.801.5724.

37

B2013 330 | **Steel Louver**

	System Description	Freq. (Years)	Crew	Unit	Labor Hours	2015 Bare Costs				Total In-House	Total w/O&P
						Material	Labor	Equipment	Total		
2040	**Replace steel louver - 2nd floor**	40	1 CARP	Ea.							
	Set up and secure scaffold				.889		41.50		41.50	55.50	68.50
	Remove louver				.473		17.75		17.75	23.50	29
	Install new louver				1.783	246	99.90		345.90	395	465
	Remove scaffold				.889		41.50		41.50	55.50	68.50
	Total				4.034	246	200.65		446.65	529.50	631
3030	**Refinish steel louver - 3rd floor**	5	1 PORD	Ea.							
	Set up and secure scaffold				1.333		54		54	70	86.50
	Prepare louver for painting				.434		17.44		17.44	22.50	28
	Paint louvered surface (to 16 S.F.), brushwork, primer + 1 coat				.800	2.29	32.50		34.79	44.50	55
	Remove scaffold				1.333		54		54	70	86.50
	Total				3.901	2.29	157.94		160.23	207	256
3040	**Replace steel louver - 3rd floor**	40	1 CARP	Ea.							
	Set up and secure scaffold				1.333		62.50		62.50	83	103
	Remove louver				.473		17.75		17.75	23.50	29
	Install new louver				1.783	246	99.90		345.90	395	465
	Remove scaffold				1.333		62.50		62.50	83	103
	Total				4.922	246	242.65		488.65	584.50	700

B20 EXTERIOR CLOSURE | **B2023** | **Exterior Windows**

B2023 102 | **Steel Frame, Operating**

System Description	Freq. (Years)	Crew	Unit	Labor Hours	2015 Bare Costs				Total In-House	Total w/O&P
					Material	Labor	Equipment	Total		
1010 Replace glass - 1st floor (1% of glass)	1	1 CARP	S.F.							
Remove damaged glass				.052		1.95		1.95	2.59	3.20
Install new glass				.043	5.90	1.95		7.85	9.05	10.55
Total				.095	5.90	3.90		9.80	11.64	13.75
1020 Repair 3'-9" x 5'-5" steel frame window - 1st floor	20	1 CARP	Ea.							
Set up and secure scaffold				.444		21		21	27.50	34
Remove window gaskets				.250		11.75		11.75	15.60	19.25
Install new gaskets				.564	105.46	26.41		131.87	151	175
Misc. hardware replacement				.333	86	15.65		101.65	115	133
Remove scaffold				.444		21		21	27.50	34
Total				2.037	191.46	95.81		287.27	336.60	395.25
1030 Refinish 3'-9" x 5'-5" steel frame window - 1st floor	5	1 PORD	Ea.							
Set up, secure and take down ladder				.267		10.77		10.77	14.10	17.40
Prepare window for painting				.473		19.10		19.10	25	30.50
Paint window, brushwork, primer + 1 coat				1.000	2.07	40.50		42.57	54.50	67.50
Total				1.739	2.07	70.37		72.44	93.60	115.40
1040 Replace 3'-9" x 5'-5" steel frame window - 1st floor	45	1 CARP	Ea.							
Set up and secure scaffold				.444		21		21	27.50	34
Remove window				.650		24.50		24.50	32.50	40
Install new window with frame & glazing (to 21 S.F.)				2.078	1,225	109		1,334	1,500	1,725
Install sealant				.764	8.25	35.21		43.46	55.50	67.50
Remove scaffold				.444		21		21	27.50	34
Total				4.381	1,233.25	210.71		1,443.96	1,643	1,900.50
1050 Finish new 3'-9" x 5'-5" steel frame window - 1st floor	45	1 PORD	Ea.							
Set up, secure and take down ladder				.267		10.77		10.77	14.10	17.40
Prepare window for painting				.236		9.55		9.55	12.50	15.25
Paint window, brushwork, primer + 1 coat				1.000	2.07	40.50		42.57	54.50	67.50
Total				1.503	2.07	60.82		62.89	81.10	100.15

For customer support on your Facilities Maintenance & Repair Cost Data, call 877.801.5724.

39

Freq. (Years)	System Description	Crew	Unit	Labor Hours	2015 Bare Costs				Total In-House	Total w/O&P
					Material	Labor	Equipment	Total		
2010	**Replace glass - 2nd floor (1% of glass)**	1 CARP	S.F.							
1	Set up and secure scaffold			.889		41.50		41.50	55.50	68.50
	Remove damaged glass			.052		1.95		1.95	2.59	3.20
	Install new glass			.043	5.90	1.95		7.85	9.05	10.55
	Remove scaffold			.889		41.50		41.50	55.50	68.50
	Total			1.873	5.90	86.90		92.80	122.64	150.75
2020	**Repair 3'-9" x 5'-5" steel frame window - 2nd floor**	1 CARP	Ea.							
20	Set up and secure scaffold			.889		41.50		41.50	55.50	68.50
	Remove window gaskets			.250		11.75		11.75	15.60	19.25
	Install new gaskets			.564	105.46	26.41		131.87	151	175
	Misc. hardware replacement			.333	86	15.65		101.65	115	133
	Remove scaffold			.889		41.50		41.50	55.50	68.50
	Total			2.925	191.46	136.81		328.27	392.60	464.25
2030	**Refinish 3'-9" x 5'-5" steel frame window - 2nd floor**	1 PORD	Ea.							
5	Set up and secure scaffold			.889		36		36	46.50	57.50
	Prepare window for painting			.473		19.10		19.10	25	30.50
	Paint window, brushwork, primer + 1 coat			1.000	2.07	40.50		42.57	54.50	67.50
	Remove scaffold			.889		36		36	46.50	57.50
	Total			3.251	2.07	131.60		133.67	172.50	213
2040	**Replace 3'-9" x 5'-5" steel frame window - 2nd floor**	1 CARP	Ea.							
45	Set up and secure scaffold			.889		41.50		41.50	55.50	68.50
	Remove window			.650		24.50		24.50	32.50	40
	Install new window with frame & glazing (to 21 S.F.)			2.078	1,225	109		1,334	1,500	1,725
	Install sealant			.764	8.25	35.21		43.46	55.50	67.50
	Remove scaffold			.889		41.50		41.50	55.50	68.50
	Total			5.270	1,233.25	251.71		1,484.96	1,699	1,969.50
2050	**Finish new 3'-9" x 5'-5" steel frame window - 2nd floor**	1 PORD	Ea.							
45	Set up and secure scaffold			.889		36		36	46.50	57.50
	Prepare window for painting			.236		9.55		9.55	12.50	15.25
	Paint window, brushwork, primer + 1 coat			1.000	2.07	40.50		42.57	54.50	67.50
	Remove scaffold			.889		36		36	46.50	57.50
	Total			3.014	2.07	122.05		124.12	160	197.75

For customer support on your Facilities Maintenance & Repair Cost Data, call 877.801.5724.

B2023 102 | Steel Frame, Operating

	System Description	Freq. (Years)	Crew	Unit	Labor Hours	2015 Bare Costs				Total In-House	Total w/O&P
						Material	Labor	Equipment	Total		
3010	**Replace glass - 3rd floor (1% of glass)**	1	1 CARP	S.F.							
	Set up and secure scaffold				.123		5.80		5.80	7.70	9.45
	Remove damaged glass				.052		1.95		1.95	2.59	3.20
	Install new glass				.043	5.90	1.95		7.85	9.05	10.55
	Remove scaffold				.123		5.80		5.80	7.70	9.45
	Total				.341	5.90	15.50		21.40	27.04	32.65
3020	**Repair 3'-9" x 5'-5" steel frame window - 3rd floor**	20	1 CARP	Ea.							
	Set up and secure scaffold				1.333		62.50		62.50	83	103
	Remove window gaskets				.250		11.75		11.75	15.60	19.25
	Install new gaskets				.564	105.46	26.41		131.87	151	175
	Misc. hardware replacement				.333	86	15.65		101.65	115	133
	Remove scaffold				1.333		62.50		62.50	83	103
	Total				3.814	191.46	178.81		370.27	447.60	533.25
3030	**Refinish 3'-9" x 5'-5" steel frame window - 3rd floor**	5	1 PORD	Ea.							
	Set up and secure scaffold				1.333		54		54	70	86.50
	Prepare window for painting				.473		19.10		19.10	25	30.50
	Paint window, brushwork, primer + 1 coat				1.000	2.07	40.50		42.57	54.50	67.50
	Remove scaffold				1.333		54		54	70	86.50
	Total				4.139	2.07	167.60		169.67	219.50	271
3040	**Replace 3'-9" x 5'-5" steel frame window - 3rd floor**	45	1 CARP	Ea.							
	Set up and secure scaffold				1.333		62.50		62.50	83	103
	Remove window				.650		24.50		24.50	32.50	40
	Install new window with frame & glazing (to 21 S.F.)				2.078	1,225	109		1,334	1,500	1,725
	Install sealant				.764	8.25	35.21		43.46	55.50	67.50
	Remove scaffold				1.333		62.50		62.50	83	103
	Total				6.159	1,233.25	293.71		1,526.96	1,754	2,038.50
3050	**Finish new 3'-9" x 5'-5" steel frame window - 3rd floor**	45	1 PORD	Ea.							
	Set up and secure scaffold				1.333		54		54	70	86.50
	Prepare window for painting				.236		9.55		9.55	12.50	15.25
	Paint window, brushwork, primer + 1 coat				1.000	2.07	40.50		42.57	54.50	67.50
	Remove scaffold				1.333		54		54	70	86.50
	Total				3.903	2.07	158.05		160.12	207	255.75

For customer support on your Facilities Maintenance & Repair Cost Data, call 877.801.5724.

41

B20 EXTERIOR CLOSURE | **B2023** | **Exterior Windows**

B2023 103 | **Steel Frame, Fixed**

	System Description	Freq. (Years)	Crew	Unit	Labor Hours	2015 Bare Costs Material	Labor	Equipment	Total	Total In-House	Total w/O&P
1010	**Replace glass - 1st floor (1% of glass)**	1	1 CARP	S.F.							
	Remove damaged glass				.052		1.95		1.95	2.59	3.20
	Install new glass				.043	5.90	1.95		7.85	9.05	10.55
	Total				.095	5.90	3.90		9.80	11.64	13.75
1020	**Repair 2'-0" x 3'-0" steel frame window - 1st flr.**	20	1 CARP	Ea.							
	Set up and secure scaffold				.444		21		21	27.50	34
	Remove window gaskets				.250		11.75		11.75	15.60	19.25
	Install new gaskets				.308	57.50	14.40		71.90	82.50	95.50
	Remove scaffold				.444		21		21	27.50	34
	Total				1.447	57.50	68.15		125.65	153.10	182.75
1030	**Refinish 2'-0" x 3'-0" st. frame window - 1st flr.**	5	1 PORD	Ea.							
	Set up, secure and take down ladder				.267		10.77		10.77	14.10	17.40
	Prepare window for painting				.473		19.10		19.10	25	30.50
	Paint window, brushwork, primer + 1 coat				1.000	2.07	40.50		42.57	54.50	67.50
	Total				1.739	2.07	70.37		72.44	93.60	115.40
1040	**Replace 2'-0" x 3'-0" steel frame window - 1st flr.**	45	1 CARP	Ea.							
	Set up and secure scaffold				.444		21		21	27.50	34
	Remove window				.650		24.50		24.50	32.50	40
	Install new window with frame & glazing (2'-0" x 3'-0")				2.078	208	109		317	395	460
	Install sealant				.764	8.25	35.21		43.46	55.50	67.50
	Remove scaffold				.444		21		21	27.50	34
	Total				4.381	216.25	210.71		426.96	538	635.50
2010	**Replace glass - 2nd floor (1% of glass)**	1	1 CARP	S.F.							
	Set up and secure scaffold				.044		2.08		2.08	2.78	3.43
	Remove damaged glass				.052		1.95		1.95	2.59	3.20
	Install new glass				.043	5.90	1.95		7.85	9.05	10.55
	Remove scaffold				.044		2.08		2.08	2.78	3.43
	Total				.184	5.90	8.06		13.96	17.20	20.61

	System Description	Freq. (Years)	Crew	Unit	Labor Hours	2015 Bare Costs				Total In-House	Total w/O&P
						Material	Labor	Equipment	Total		
2020	**Repair 2'-0" x 3'-0" steel frame window - 2nd flr.**	20	1 CARP	Ea.							
	Set up and secure scaffold				.889		41.50		41.50	55.50	68.50
	Remove window gaskets				.250		11.75		11.75	15.60	19.25
	Install new gaskets				.308	57.50	14.40		71.90	82.50	95.50
	Remove scaffold				.889		41.50		41.50	55.50	68.50
	Total				2.335	57.50	109.15		166.65	209.10	251.75
2030	**Refinish 2'-0" x 3'-0" steel frame window - 2nd flr.**	5	1 PORD	Ea.							
	Set up and secure scaffold				.889		36		36	46.50	57.50
	Prepare window for painting				.473		19.10		19.10	25	30.50
	Paint window, brushwork, primer + 1 coat				1.000	2.07	40.50		42.57	54.50	67.50
	Remove scaffold				.889		36		36	46.50	57.50
	Total				3.251	2.07	131.60		133.67	172.50	213
2040	**Replace 2'-0" x 3'-0" steel frame window - 2nd flr.**	45	1 CARP	Ea.							
	Set up and secure scaffold				.889		41.50		41.50	55.50	68.50
	Remove window				.650		24.50		24.50	32.50	40
	Install new window with frame & glazing (2'-0" x 3'-0")				2.078	208	109		317	395	460
	Install sealant				.764	8.25	35.21		43.46	55.50	67.50
	Remove scaffold				.889		41.50		41.50	55.50	68.50
	Total				5.270	216.25	251.71		467.96	594	704.50
3010	**Replace glass - 3rd floor (1% of glass)**	1	1 CARP	S.F.							
	Set up and secure scaffold				.123		5.80		5.80	7.70	9.45
	Remove damaged glass				.052		1.95		1.95	2.59	3.20
	Install new glass				.043	5.90	1.95		7.85	9.05	10.55
	Remove scaffold				.123		5.80		5.80	7.70	9.45
	Total				.341	5.90	15.50		21.40	27.04	32.65
3020	**Repair 2'-0" x 3'-0" steel frame window - 3rd flr.**	20	1 CARP	Ea.							
	Set up and secure scaffold				1.333		62.50		62.50	83	103
	Remove window gaskets				.250		11.75		11.75	15.60	19.25
	Install new gaskets				.308	57.50	14.40		71.90	82.50	95.50
	Remove scaffold				1.333		62.50		62.50	83	103
	Total				3.224	57.50	151.15		208.65	264.10	320.75

B2023 103 | Steel Frame, Fixed

	System Description	Freq. (Years)	Crew	Unit	Labor Hours	2015 Bare Costs				Total In-House	Total w/O&P
						Material	Labor	Equipment	Total		
3030	Refinish 2'-0" x 3'-0" steel frame window - 3rd floor	5	1 PORD	Ea.							
	Set up and secure scaffold				1.333		54		54	70	86.50
	Prepare window for painting				.473		19.10		19.10	25	30.50
	Paint window, brushwork, primer + 1 coat				1.000	2.07	40.50		42.57	54.50	67.50
	Remove scaffold				1.333		54		54	70	86.50
	Total				4.139	2.07	167.60		169.67	219.50	271
3040	Replace 2'-0" x 3'-0" steel frame window - 3rd floor	45	1 CARP	Ea.							
	Set up and secure scaffold				1.333		62.50		62.50	83	103
	Remove window				.650		24.50		24.50	32.50	40
	Install new window with frame & glazing (2'-0" x 3'-0")				2.078	208	109		317	395	460
	Install sealant				.764	8.25	35.21		43.46	55.50	67.50
	Remove scaffold				1.333		62.50		62.50	83	103
	Total				6.159	216.25	293.71		509.96	649	773.50

B2023 110 | Aluminum Window, Operating

	System Description	Freq. (Years)	Crew	Unit	Labor Hours	2015 Bare Costs				Total In-House	Total w/O&P
						Material	Labor	Equipment	Total		
1010	Replace glass - 1st flr. (1% of glass)	1	1 CARP	S.F.							
	Remove damaged glass				.052		1.95		1.95	2.59	3.20
	Install new glass				.043	5.90	1.95		7.85	9.05	10.55
	Total				.095	5.90	3.90		9.80	11.64	13.75
1020	Repair 3' x 4' aluminum window - 1st floor	20	1 CARP	Ea.							
	Set up and secure scaffold				.444		21		21	27.50	34
	Remove window gaskets				.250		11.75		11.75	15.60	19.25
	Install new gaskets				.431	80.50	20.16		100.66	116	134
	Misc. hardware replacement				.333	86	15.65		101.65	115	133
	Remove scaffold				.444		21		21	27.50	34
	Total				1.903	166.50	89.56		256.06	301.60	354.25

For customer support on your Facilities Maintenance & Repair Cost Data, call 877.801.5724.

B20 EXTERIOR CLOSURE | **B2023** | **Exterior Windows**

B2023 110 | **Aluminum Window, Operating**

System Description	Freq. (Years)	Crew	Unit	Labor Hours	2015 Bare Costs				Total In-House	Total w/O&P
					Material	Labor	Equipment	Total		
1030 Replace 3' x 4' aluminum window - 1st floor	50	1 CARP	Ea.							
Set up and secure scaffold				.444		21		21	27.50	34
Remove window				.650		24.50		24.50	32.50	40
Install new window with frame & glazing (to 12 S.F.)				2.078	490	109		599	705	810
Install sealant				.555	6	25.59		31.59	40.50	49
Remove scaffold				.444		21		21	27.50	34
Total				4.173	496	201.09		697.09	833	967
2010 Replace glass - 2nd flr. (1% of glass)	1	1 CARP	S.F.							
Set up and secure scaffold				.044		2.08		2.08	2.78	3.43
Remove damaged glass				.052		1.95		1.95	2.59	3.20
Install new glass				.043	5.90	1.95		7.85	9.05	10.55
Remove scaffold				.044		2.08		2.08	2.78	3.43
Total				.184	5.90	8.06		13.96	17.20	20.61
2020 Repair 3' x 4' aluminum window - 2nd floor	20	1 CARP	Ea.							
Set up and secure scaffold				.889		41.50		41.50	55.50	68.50
Remove window gaskets				.250		11.75		11.75	15.60	19.25
Install new gaskets				.431	80.50	20.16		100.66	116	134
Misc. hardware replacement				.333	86	15.65		101.65	115	133
Remove scaffold				.889		41.50		41.50	55.50	68.50
Total				2.792	166.50	130.56		297.06	357.60	423.25
2030 Replace 3' x 4' aluminum window - 2nd floor	50	1 CARP	Ea.							
Set up and secure scaffold				.889		41.50		41.50	55.50	68.50
Remove window				.650		24.50		24.50	32.50	40
Install new window with frame & glazing (to 12 S.F.)				2.078	490	109		599	705	810
Install sealant				.555	6	25.59		31.59	40.50	49
Remove scaffold				.889		41.50		41.50	55.50	68.50
Total				5.062	496	242.09		738.09	889	1,036
3010 Replace glass - 3rd floor (1% of glass)	1	1 CARP	S.F.							
Set up and secure scaffold				.123		5.80		5.80	7.70	9.45
Remove damaged glass				.052		1.95		1.95	2.59	3.20
Install new glass				.043	5.90	1.95		7.85	9.05	10.55
Remove scaffold				.123		5.80		5.80	7.70	9.45
Total				.341	5.90	15.50		21.40	27.04	32.65

For customer support on your Facilities Maintenance & Repair Cost Data, call 877.801.5724.

45

B2023 110 Aluminum Window, Operating

	System Description	Freq. (Years)	Crew	Unit	Labor Hours	2015 Bare Costs				Total In-House	Total w/O&P
						Material	Labor	Equipment	Total		
3020	**Repair 3' x 4' aluminum window - 3rd floor**	20	1 CARP	Ea.							
	Set up and secure scaffold				1.333		62.50		62.50	83	103
	Remove window gaskets				.250		11.75		11.75	15.60	19.25
	Install new gaskets				.431	80.50	20.16		100.66	116	134
	Misc. hardware replacement				.333	86	15.65		101.65	115	133
	Remove scaffold				1.333		62.50		62.50	83	103
	Total				3.681	166.50	172.56		339.06	412.60	492.25
3030	**Replace 3' x 4' aluminum window - 3rd floor**	50	1 CARP	Ea.							
	Set up and secure scaffold				1.333		62.50		62.50	83	103
	Remove window				.650		24.50		24.50	32.50	40
	Install new window with frame & glazing (to 12 S.F.)				2.078	490	109		599	705	810
	Install sealant				.555	6	25.59		31.59	40.50	49
	Remove scaffold				1.333		62.50		62.50	83	103
	Total				5.950	496	284.09		780.09	944	1,105

B2023 111 Aluminum Window, Fixed

	System Description	Freq. (Years)	Crew	Unit	Labor Hours	2015 Bare Costs				Total In-House	Total w/O&P
						Material	Labor	Equipment	Total		
1010	**Replace glass - 1st floor (1% of glass)**	1	1 CARP	S.F.							
	Remove damaged glass				.052		1.95		1.95	2.59	3.20
	Install new glass				.043	5.90	1.95		7.85	9.05	10.55
	Total				.095	5.90	3.90		9.80	11.64	13.75
1015	**Replace glass - 1st floor, 1" insulating panel with heat reflective glass**	30	2 GLAZ	S.F.							
	Remove damaged glass				.052		1.95		1.95	2.59	3.20
	Tinted insulated glass				.213	23.50	9.60		33.10	38.50	45
	Total				.265	23.50	11.55		35.05	41.09	48.20
1020	**Repair 2'-0" x 3'-0" aluminum window - 1st floor**	20	1 CARP	Ea.							
	Set up and secure scaffold				.444		21		21	27.50	34
	Remove window gaskets				.250		11.75		11.75	15.60	19.25
	Install new gaskets				.308	57.50	14.40		71.90	82.50	95.50
	Remove scaffold				.444		21		21	27.50	34
	Total				1.447	57.50	68.15		125.65	153.10	182.75

B20 EXTERIOR CLOSURE | **B2023** | **Exterior Windows**

B2023 111 | **Aluminum Window, Fixed**

	System Description	Freq. (Years)	Crew	Unit	Labor Hours	2015 Bare Costs				Total In-House	Total w/O&P
						Material	Labor	Equipment	Total		
1030	**Replace 2'-0" x 3'-0" aluminum window - 1st floor**	50	1 CARP	Ea.							
	Set up and secure scaffold				.444		21		21	27.50	34
	Remove window				.650		24.50		24.50	32.50	40
	Install new window with frame & glazing (2'-0" x 3'-0")				2.078	208	109		317	395	460
	Install sealant				.555	6	25.59		31.59	40.50	49
	Remove scaffold				.444		21		21	27.50	34
	Total				4.173	214	201.09		415.09	523	617
2010	**Replace glass - 2nd floor (1% of glass)**	1	1 CARP	S.F.							
	Set up and secure scaffold				.889		41.50		41.50	55.50	68.50
	Remove damaged glass				.052		1.95		1.95	2.59	3.20
	Install new glass				.043	5.90	1.95		7.85	9.05	10.55
	Remove scaffold				.889		41.50		41.50	55.50	68.50
	Total				1.873	5.90	86.90		92.80	122.64	150.75
2020	**Repair 2'-0" x 3'-0" aluminum window - 2nd floor**	20	1 CARP	Ea.							
	Set up and secure scaffold				.889		41.50		41.50	55.50	68.50
	Remove window gaskets				.250		11.75		11.75	15.60	19.25
	Install new gaskets				.308	57.50	14.40		71.90	82.50	95.50
	Remove scaffold				.889		41.50		41.50	55.50	68.50
	Total				2.335	57.50	109.15		166.65	209.10	251.75
2030	**Replace 2'-0" x 3'-0" aluminum window - 2nd floor**	50	1 CARP	Ea.							
	Set up and secure scaffold				.889		41.50		41.50	55.50	68.50
	Remove window				.650		24.50		24.50	32.50	40
	Install new window with frame & glazing (2'-0" x 3'-0")				2.078	208	109		317	395	460
	Install sealant				.555	6	25.59		31.59	40.50	49
	Remove scaffold				.889		41.50		41.50	55.50	68.50
	Total				5.062	214	242.09		456.09	579	686
3010	**Replace glass - 3rd floor (1% of glass)**	1	1 CARP	S.F.							
	Set up and secure scaffold				.123		5.80		5.80	7.70	9.45
	Remove damaged glass				.052		1.95		1.95	2.59	3.20
	Install new glass				.043	5.90	1.95		7.85	9.05	10.55
	Remove scaffold				.123		5.80		5.80	7.70	9.45
	Total				.341	5.90	15.50		21.40	27.04	32.65

For customer support on your Facilities Maintenance & Repair Cost Data, call 877.801.5724.

47

B2023 111 — Aluminum Window, Fixed

	System Description	Freq. (Years)	Crew	Unit	Labor Hours	2015 Bare Costs				Total In-House	Total w/O&P
						Material	Labor	Equipment	Total		
3020	**Repair 2'-0" x 3'-0" aluminum window - 3rd floor**	20	1 CARP	Ea.							
	Set up and secure scaffold				1.333		62.50		62.50	83	103
	Remove window gaskets				.250		11.75		11.75	15.60	19.25
	Install new gaskets				.308	57.50	14.40		71.90	82.50	95.50
	Remove scaffold				1.333		62.50		62.50	83	103
	Total				3.224	57.50	151.15		208.65	**264.10**	**320.75**
3030	**Replace 2'-0" x 3'-0" aluminum window - 3rd floor**	50	1 CARP	Ea.							
	Set up and secure scaffold				1.333		62.50		62.50	83	103
	Remove window				.650		24.50		24.50	32.50	40
	Install new window with frame & glazing (2'-0" x 3'-0")				2.078	208	109		317	395	460
	Install sealant				.555	6	25.59		31.59	40.50	49
	Remove scaffold				1.333		62.50		62.50	83	103
	Total				5.950	214	284.09		498.09	**634**	**755**

B2023 112 — Wood Frame, Operating

	System Description	Freq. (Years)	Crew	Unit	Labor Hours	2015 Bare Costs				Total In-House	Total w/O&P
						Material	Labor	Equipment	Total		
1010	**Replace glass - 1st floor (1% of glass)**	1	1 CARP	S.F.							
	Remove damaged glass				.052		1.95		1.95	2.59	3.20
	Install new glass				.043	5.90	1.95		7.85	9.05	10.55
	Total				.095	5.90	3.90		9.80	**11.64**	**13.75**
1020	**Repair 2'-3" x 6'-0" wood frame window - 1st flr.**	15	1 CARP	Ea.							
	Set up and secure scaffold				.444		21		21	27.50	34
	Remove putty				.667		31.50		31.50	41.50	51.50
	Install new putty				.333	5.50	15.65		21.15	27	32.50
	Misc. hardware replacement				.333	86	15.65		101.65	115	133
	Remove scaffold				.444		21		21	27.50	34
	Total				2.222	91.50	104.80		196.30	**238.50**	**285**
1030	**Refinish 2'-3" x 6'-0" wood frame window - 1st flr.**	4	1 PORD	Ea.							
	Set up, secure and take down ladder				.267		10.75		10.75	13.95	17.30
	Prepare window for painting				.473		19.10		19.10	25	30.50
	Paint window, brushwork, primer + 1 coat				1.000	2.07	40.50		42.57	54.50	67.50
	Total				1.739	2.07	70.35		72.42	**93.45**	**115.30**

For customer support on your Facilities Maintenance & Repair Cost Data, call 877.801.5724.

B2023 112 Wood Frame, Operating

	System Description	Freq. (Years)	Crew	Unit	Labor Hours	2015 Bare Costs				Total In-House	Total w/O&P
						Material	Labor	Equipment	Total		
1040	**Replace 2'-3" x 6'-0" wood frame window - 1st flr.**	40	1 CARP	Ea.							
	Set up and secure scaffold				.444		21		21	27.50	34
	Remove window				.473		17.75		17.75	23.50	29
	Install new window with frame & glazing (to 14 S.F.)				1.143	330	53.50		383.50	435	500
	Install sealant				.625	6.75	28.80		35.55	45.50	55.50
	Remove scaffold				.444		21		21	27.50	34
	Total				3.130	336.75	142.05		478.80	559	652.50
2010	**Replace glass - 2nd floor (1% of glass)**	1	1 CARP	S.F.							
	Set up and secure scaffold				.044		2.08		2.08	2.78	3.43
	Remove damaged glass				.052		1.95		1.95	2.59	3.20
	Install new glass				.043	5.90	1.95		7.85	9.05	10.55
	Remove scaffold				.044		2.08		2.08	2.78	3.43
	Total				.184	5.90	8.06		13.96	17.20	20.61
2020	**Repair 2'-3" x 6'-0" wood frame window - 2nd flr.**	15	1 CARP	Ea.							
	Set up and secure scaffold				.889		41.50		41.50	55.50	68.50
	Remove putty				.667		31.50		31.50	41.50	51.50
	New putty				.333	5.50	15.65		21.15	27	32.50
	Misc. hardware replacement				.333	86	15.65		101.65	115	133
	Remove scaffold				.889		41.50		41.50	55.50	68.50
	Total				3.111	91.50	145.80		237.30	294.50	354
2030	**Refinish 2'-3" x 6'-0" wood frame window - 2nd flr.**	5	1 PORD	Ea.							
	Set up and secure scaffold				.889		36		36	46.50	57.50
	Prepare window for painting				.473		19.10		19.10	25	30.50
	Paint window, brushwork, primer + 1 coat				1.000	2.07	40.50		42.57	54.50	67.50
	Remove scaffold				.889		36		36	46.50	57.50
	Total				3.251	2.07	131.60		133.67	172.50	213
2040	**Replace 2'-3" x 6'-0" wood frame window - 2nd flr.**	40	1 CARP	Ea.							
	Set up and secure scaffold				.889		41.50		41.50	55.50	68.50
	Remove window				.473		17.75		17.75	23.50	29
	Install new window with frame & glazing (to 14 S.F.)				1.143	330	53.50		383.50	435	500
	Install sealant				.625	6.75	28.80		35.55	45.50	55.50
	Remove scaffold				.889		41.50		41.50	55.50	68.50
	Total				4.018	336.75	183.05		519.80	615	721.50

For customer support on your Facilities Maintenance & Repair Cost Data, call 877.801.5724.

49

B2023 112 Wood Frame, Operating

System Description	Freq. (Years)	Crew	Unit	Labor Hours	2015 Bare Costs				Total In-House	Total w/O&P
					Material	Labor	Equipment	Total		
3010 Replace glass - 3rd floor (1% of glass)	1	1 CARP	Ea.							
Set up and secure scaffold				.123		5.80		5.80	7.70	9.45
Remove damaged glass				.052		1.95		1.95	2.59	3.20
Install new glass				.043	5.90	1.95		7.85	9.05	10.55
Remove scaffold				.123		5.80		5.80	7.70	9.45
Total				.341	5.90	15.50		21.40	27.04	32.65
3020 Repair 2'-3" x 6'-0" wood frame window - 3rd flr.	15	1 CARP	Ea.							
Set up and secure scaffold				1.333		62.50		62.50	83	103
Remove putty				.667		31.50		31.50	41.50	51.50
Install new putty				.333	5.50	15.65		21.15	27	32.50
Misc. hardware replacement				.333	86	15.65		101.65	115	133
Remove scaffold				1.333		62.50		62.50	83	103
Total				4.000	91.50	187.80		279.30	349.50	423
3030 Refinish 2'-3" x 6'-0" wood frame window - 3rd flr.	5	1 PORD	Ea.							
Set up and secure scaffold				1.333		54		54	70	86.50
Prepare window for painting				.473		19.10		19.10	25	30.50
Paint window, brushwork, primer + 1 coat				1.000	2.07	40.50		42.57	54.50	67.50
Remove scaffold				1.333		54		54	70	86.50
Total				4.139	2.07	167.60		169.67	219.50	271
3040 Replace 2'-3" x 6'-0" wood frame window - 3rd flr.	40	1 CARP	Ea.							
Set up and secure scaffold				1.333		62.50		62.50	83	103
Remove window				.473		17.75		17.75	23.50	29
Install new window with frame & glazing (to 14 S.F.)				1.143	330	53.50		383.50	435	500
Install sealant				.625	6.75	28.80		35.55	45.50	55.50
Remove scaffold				1.333		62.50		62.50	83	103
Total				4.907	336.75	225.05		561.80	670	790.50

For customer support on your Facilities Maintenance & Repair Cost Data, call 877.801.5724.

B20 EXTERIOR CLOSURE — B2023 — Exterior Windows

B2023 113 — Wood Frame, Fixed

System Description	Freq. (Years)	Crew	Unit	Labor Hours	2015 Bare Costs Material	Labor	Equipment	Total	Total In-House	Total w/O&P
1010 Replace glass - 1st floor	1	1 CARP	S.F.							
Remove damaged glass				.052		1.95		1.95	2.59	3.20
Install new glass				.043	5.90	1.95		7.85	9.05	10.55
Total				.095	5.90	3.90		9.80	**11.64**	**13.75**
1020 Repair 3'-6" x 4'-0" wood frame window - 1st flr.	20	1 CARP	Ea.							
Set up and secure scaffold				.444		21		21	27.50	34
Remove putty				.667		31.50		31.50	41.50	51.50
New putty				.333	5.50	15.65		21.15	27	32.50
Remove scaffold				.444		21		21	27.50	34
Total				1.889	5.50	89.15		94.65	**123.50**	**152**
1030 Refinish 3'-6" x 4'-0" wood frame window - 1st flr.	5	1 PORD	Ea.							
Set up, secure and take down ladder				.444		17.95		17.95	23.50	29
Prepare window for painting				.473		19.10		19.10	25	30.50
Paint window, brushwork, primer + 1 coat				1.000	2.07	40.50		42.57	54.50	67.50
Total				1.917	2.07	77.55		79.62	**103**	**127**
1040 Replace 3'-6" x 4'-0" wood frame window - 1st flr.	40	1 CARP	Ea.							
Set up and secure scaffold				.444		21		21	27.50	34
Remove window				.473		17.75		17.75	23.50	29
Install new window with frame & glazing (to 14 S.F.)				1.733	420	81.50		501.50	570	660
Install sealant				.625	6.75	28.80		35.55	45.50	55.50
Remove scaffold				.444		21		21	27.50	34
Total				3.720	426.75	170.05		596.80	**694**	**812.50**
2010 Replace glass - 2nd floor (1% of glass)	1	1 CARP	S.F.							
Set up and secure scaffold				.044		2.08		2.08	2.78	3.43
Remove damaged glass				.052		1.95		1.95	2.59	3.20
Install new glass				.043	5.90	1.95		7.85	9.05	10.55
Remove scaffold				.044		2.08		2.08	2.78	3.43
Total				.184	5.90	8.06		13.96	**17.20**	**20.61**

For customer support on your Facilities Maintenance & Repair Cost Data, call 877.801.5724.

51

B2023 113   Wood Frame, Fixed

System Description	Freq. (Years)	Crew	Unit	Labor Hours	2015 Bare Costs				Total In-House	Total w/O&P
					Material	Labor	Equipment	Total		
2020 Repair 3'-6" x 4'-0" wood frame window - 2nd flr.	20	1 CARP	Ea.							
Set up and secure scaffold				.889		41.50		41.50	55.50	68.50
Remove putty				.667		31.50		31.50	41.50	51.50
Install new putty				.333	5.50	15.65		21.15	27	32.50
Remove scaffold				.889		41.50		41.50	55.50	68.50
Total				2.778	5.50	130.15		135.65	**179.50**	**221**
2030 Refinish 3'-6" x 4'-0" wood frame window - 2nd flr.	5	1 PORD	Ea.							
Set up and secure scaffold				.889		36		36	46.50	57.50
Prepare window for painting				.473		19.10		19.10	25	30.50
Paint window, brushwork, primer + 1 coat				1.000	2.07	40.50		42.57	54.50	67.50
Remove scaffold				.889		36		36	46.50	57.50
Total				3.251	2.07	131.60		133.67	**172.50**	**213**
2040 Replace 3'-6" x 4'-0" wood frame window - 2nd flr.	40	1 CARP	Ea.							
Set up and secure scaffold				.889		41.50		41.50	55.50	68.50
Remove window				.473		17.75		17.75	23.50	29
Install new window with frame & glazing (to 14 S.F.)				1.733	420	81.50		501.50	570	660
Install sealant				.625	6.75	28.80		35.55	45.50	55.50
Remove scaffold				.889		41.50		41.50	55.50	68.50
Total				4.609	426.75	211.05		637.80	**750**	**881.50**
3010 Replace glass - 3rd floor (1% of glass)	1	1 CARP	S.F.							
Set up and secure scaffold				.123		5.80		5.80	7.70	9.45
Remove damaged glass				.052		1.95		1.95	2.59	3.20
Install new glass				.043	5.90	1.95		7.85	9.05	10.55
Remove scaffold				.123		5.80		5.80	7.70	9.45
Total				.341	5.90	15.50		21.40	**27.04**	**32.65**
3020 Repair 3'-6" x 4'-0" wood frame window - 3rd flr.	20	1 CARP	Ea.							
Set up and secure scaffold				1.333		62.50		62.50	83	103
Remove putty				.667		31.50		31.50	41.50	51.50
Install new putty				.333	5.50	15.65		21.15	27	32.50
Remove scaffold				1.333		62.50		62.50	83	103
Total				3.667	5.50	172.15		177.65	**234.50**	**290**

B20 EXTERIOR CLOSURE — B2023 — Exterior Windows

B2023 113 — Wood Frame, Fixed

System Description	Freq. (Years)	Crew	Unit	Labor Hours	2015 Bare Costs				Total In-House	Total w/O&P
					Material	Labor	Equipment	Total		
3030 **Refinish 3'-6" x 4'-0" wood frame window - 3rd flr.**	5	1 PORD	Ea.							
Set up and secure scaffold				1.333		54		54	70	86.50
Prepare window for painting				.473		19.10		19.10	25	30.50
Paint window, brushwork, primer + 1 coat				1.000	2.07	40.50		42.57	54.50	67.50
Remove scaffold				1.333		54		54	70	86.50
Total				4.139	2.07	167.60		169.67	219.50	271
3040 **Replace 3'-6" x 4'-0" wood frame window - 3rd flr.**	40	1 CARP	Ea.							
Set up and secure scaffold				1.333		62.50		62.50	83	103
Remove window				.473		17.75		17.75	23.50	29
Install new window with frame & glazing (to 14 S.F.)				1.733	420	81.50		501.50	570	660
Install sealant				.625	6.75	28.80		35.55	45.50	55.50
Remove scaffold				1.333		62.50		62.50	83	103
Total				5.498	426.75	253.05		679.80	805	950.50

B2023 116 — Aluminum Frame Storm Window

System Description	Freq. (Years)	Crew	Unit	Labor Hours	2015 Bare Costs				Total In-House	Total w/O&P
					Material	Labor	Equipment	Total		
1010 **Replace glass - (1% of glass)**	1	1 CARP	S.F.							
Remove glass				.052		1.95		1.95	2.59	3.20
Install glass / storm window				.173	5.95	7.80		13.75	16.80	20
Total				.225	5.95	9.75		15.70	19.39	23.20
1030 **Refinish 3'-0" x 4'-0" aluminum frame storm window - 1st floor**	10	1 PORD	Ea.							
Set up and secure scaffold				.444		17.95		17.95	23.50	29
Prepare window frame surface				.473		19.05		19.05	25	30.50
Refinish window frame surface, brushwork, primer + 1 coat				.615	1.17	25		26.17	33.50	41.50
Remove scaffold				.444		17.95		17.95	23.50	29
Total				1.977	1.17	79.95		81.12	105.50	130
1040 **Replace 3'-0" x 4'-0" aluminum frame storm window - 1st floor**	50	1 CARP	Ea.							
Set up and secure scaffold				.444		21		21	27.50	34
Remove old window				.385		14.50		14.50	19.25	24
Install new window (to 12 S.F.)				.743	117	35		152	175	203
Remove scaffold				.444		21		21	27.50	34
Total				2.017	117	91.50		208.50	249.25	295

B2023 116 Aluminum Frame Storm Window

System Description	Freq. (Years)	Crew	Unit	Labor Hours	2015 Bare Costs				Total In-House	Total w/O&P
					Material	Labor	Equipment	Total		
2030 Refinish 3'-0" x 4'-0" aluminum frame storm window - 2nd floor	10	1 PORD	Ea.							
Set up and secure scaffold				.889		36		36	46.50	57.50
Prepare window frame surface				.473		19.05		19.05	25	30.50
Refinish window frame surface, brushwork, primer + 1 coat				.615	1.17	25		26.17	33.50	41.50
Remove scaffold				.889		36		36	46.50	57.50
Total				2.866	1.17	116.05		117.22	**151.50**	187
2040 Replace 3'-0" x 4'-0" aluminum frame storm window - 2nd floor	50	1 CARP	Ea.							
Set up and secure scaffold				.889		41.50		41.50	55.50	68.50
Remove old window				.385		14.50		14.50	19.25	24
Install new window (to 12 S.F.)				.743	117	35		152	175	203
Remove scaffold				.889		41.50		41.50	55.50	68.50
Total				2.906	117	132.50		249.50	**305.25**	364
3030 Refinish 3'-0" x 4'-0" aluminum frame storm window - 3rd floor	10	1 PORD	Ea.							
Set up and secure scaffold				1.333		54		54	70	86.50
Prepare window frame surface				.473		19.05		19.05	25	30.50
Refinish window frame surface, brushwork, primer + 1 coat				.615	1.17	25		26.17	33.50	41.50
Remove scaffold				1.333		54		54	70	86.50
Total				3.755	1.17	152.05		153.22	**198.50**	245
3040 Replace 3'-0" x 4'-0" aluminum frame storm window - 3rd floor	50	1 CARP	Ea.							
Set up and secure scaffold				1.333		62.50		62.50	83	103
Remove old window				.385		14.50		14.50	19.25	24
Install new window (to 12 S.F.)				.743	117	35		152	175	203
Remove scaffold				1.333		62.50		62.50	83	103
Total				3.795	117	174.50		291.50	**360.25**	433

B2023 117 Steel Frame Storm Window

System Description	Freq. (Years)	Crew	Unit	Labor Hours	2015 Bare Costs				Total In-House	Total w/O&P
					Material	Labor	Equipment	Total		
1010 Replace glass - (1% of glass)	1	1 CARP	S.F.							
Remove glass				.052		1.95		1.95	2.59	3.20
Install glass / storm window				.173	5.95	7.80		13.75	16.80	20
Total				.225	5.95	9.75		15.70	**19.39**	23.20

For customer support on your Facilities Maintenance & Repair Cost Data, call 877.801.5724.

B2023 117   Steel Frame Storm Window

	System Description	Freq. (Years)	Crew	Unit	Labor Hours	2015 Bare Costs				Total In-House	Total w/O&P
						Material	Labor	Equipment	Total		
1030	**Refinish 3'-0" x 4'-0" steel frame storm window - 1st floor**	5	1 PORD	Ea.							
	Set up and secure scaffold				.444		17.95		17.95	23.50	29
	Prepare window frame surface				.473		19.05		19.05	25	30.50
	Refinish window frame surface, brushwork, primer + 1 coat				.615	1.17	25		26.17	33.50	41.50
	Remove scaffold				.444		17.95		17.95	23.50	29
	Total				1.977	1.17	79.95		81.12	105.50	130
1040	**Replace 3'-0" x 4'-0" steel frame storm window - 1st floor**	25	1 CARP	Ea.							
	Set up and secure scaffold				.444		21		21	27.50	34
	Remove old window				.385		14.50		14.50	19.25	24
	Install new window (to 12 S.F.)				.743	117	35		152	175	203
	Remove scaffold				.444		21		21	27.50	34
	Total				2.017	117	91.50		208.50	249.25	295
2030	**Refinish 3'-0" x 4'-0" steel frame storm window - 2nd floor**	5	1 PORD	Ea.							
	Set up and secure scaffold				.889		36		36	46.50	57.50
	Prepare window frame surface				.473		19.05		19.05	25	30.50
	Refinish window frame surface, brushwork, primer + 1 coat				.615	1.17	25		26.17	33.50	41.50
	Remove scaffold				.889		36		36	46.50	57.50
	Total				2.866	1.17	116.05		117.22	151.50	187
2040	**Replace 3'-0" x 4'-0" steel frame storm window - 2nd floor**	25	1 CARP	Ea.							
	Set up and secure scaffold				.889		41.50		41.50	55.50	68.50
	Remove old window				.385		14.50		14.50	19.25	24
	Install new window (to 12 S.F.)				.743	117	35		152	175	203
	Remove scaffold				.889		41.50		41.50	55.50	68.50
	Total				2.906	117	132.50		249.50	305.25	364
3030	**Refinish 3'-0" x 4'-0" steel frame storm window - 3rd floor**	5	1 PORD	Ea.							
	Set up and secure scaffold				1.333		54		54	70	86.50
	Prepare window frame surface				.473		19.05		19.05	25	30.50
	Refinish window frame surface, brushwork, primer + 1 coat				.615	1.17	25		26.17	33.50	41.50
	Remove scaffold				1.333		54		54	70	86.50
	Total				3.755	1.17	152.05		153.22	198.50	245

B2023 117 | Steel Frame Storm Window

System Description	Freq. (Years)	Crew	Unit	Labor Hours	2015 Bare Costs				Total In-House	Total w/O&P
					Material	Labor	Equipment	Total		
3040 **Replace 3'-0" x 4'-0" steel frame storm window - 3rd floor**	25	1 CARP	Ea.							
Set up and secure scaffold				1.333		62.50		62.50	83	103
Remove old window				.385		14.50		14.50	19.25	24
Install new window (to 12 S.F.)				.743	117	35		152	175	203
Remove scaffold				1.333		62.50		62.50	83	103
Total				3.795	117	174.50		291.50	360.25	433

B2023 119 | Wood Frame Storm Window

System Description	Freq. (Years)	Crew	Unit	Labor Hours	2015 Bare Costs				Total In-House	Total w/O&P
					Material	Labor	Equipment	Total		
1010 **Replace glass - (1% of glass)**	1	1 CARP	S.F.							
Remove glass				.052		1.95		1.95	2.59	3.20
Install new float glass				.043	5.90	1.95		7.85	9.05	10.55
Total				.095	5.90	3.90		9.80	11.64	13.75
1020 **Repair 3'-0" x 5'-0" wood frame storm window**	10	1 CARP	Ea.							
Remove putty				.667		31.50		31.50	41.50	51.50
Install new putty				.333	5.50	15.65		21.15	27	32.50
Total				1.000	5.50	47.15		52.65	68.50	84
1030 **Refinish 3'-0" x 5'-0" wood frame storm window**	5	1 PORD	Ea.							
Prepare window frame surface				.473		19.05		19.05	25	30.50
Refinish window frame surface, brushwork, primer + 1 coat				.615	1.17	25		26.17	33.50	41.50
Total				1.088	1.17	44.05		45.22	58.50	72
1040 **Replace 3'-0" x 5'-0" wood frame storm window - 1st floor**	40	1 CARP	Ea.							
Set up and secure scaffold				.444		21		21	27.50	34
Remove old window				.473		17.75		17.75	23.50	29
Install new window (to 15 S.F.)				.693	168	32.60		200.60	228	263
Remove scaffold				.444		21		21	27.50	34
Total				2.055	168	92.35		260.35	306.50	360

B2023 119 Wood Frame Storm Window

	System Description	Freq. (Years)	Crew	Unit	Labor Hours	2015 Bare Costs				Total In-House	Total w/O&P
						Material	Labor	Equipment	Total		
2040	**Replace 3'-0" x 5'-0" wood frame storm window - 2nd floor**	40	1 CARP	Ea.							
	Set up and secure scaffold				.889		41.50		41.50	55.50	68.50
	Remove old window				.473		17.75		17.75	23.50	29
	Install new window (to 15 S.F.)				.693	168	32.60		200.60	228	263
	Remove scaffold				.889		41.50		41.50	55.50	68.50
	Total				2.944	168	133.35		301.35	362.50	429
3040	**Replace 3'-0" x 5'-0" wood frame storm window - 3rd floor**	40	1 CARP	Ea.							
	Set up and secure scaffold				1.333		62.50		62.50	83	103
	Remove old window				.473		17.75		17.75	23.50	29
	Install new window (to 15 S.F.)				.693	168	32.60		200.60	228	263
	Remove scaffold				1.333		62.50		62.50	83	103
	Total				3.833	168	175.35		343.35	417.50	498

B2023 122 Vinyl Clad Wood Window, Operating

	System Description	Freq. (Years)	Crew	Unit	Labor Hours	2015 Bare Costs				Total In-House	Total w/O&P
						Material	Labor	Equipment	Total		
1030	**Repl. 3'-0" x 4'-0" vinyl clad window - 1st flr.**	60	1 CARP	Ea.							
	Set up and secure scaffold				.444		21		21	27.50	34
	Remove window				.473		17.75		17.75	23.50	29
	Install new window (3'-0" x 4'-0")				1.156	395	54.50		449.50	505	585
	Remove scaffold				.444		21		21	27.50	34
	Total				2.518	395	114.25		509.25	583.50	682
2030	**Repl. 3'-0" x 4'-0" vinyl clad window - 2nd flr.**	60	1 CARP	Ea.							
	Set up and secure scaffold				.889		41.50		41.50	55.50	68.50
	Remove window				.473		17.75		17.75	23.50	29
	Install new window (3'-0" x 4'-0")				1.156	395	54.50		449.50	505	585
	Remove scaffold				.889		41.50		41.50	55.50	68.50
	Total				3.407	395	155.25		550.25	639.50	751

B2023 122 Vinyl Clad Wood Window, Operating

	System Description	Freq. (Years)	Crew	Unit	Labor Hours	Material	2015 Bare Costs Labor	Equipment	Total	Total In-House	Total w/O&P
3030	**Repl. 3'-0" x 4'-0" vinyl clad window - 3rd flr.**	60	1 CARP	Ea.							
	Set up and secure scaffold				1.333		62.50		62.50	83	103
	Remove window				.473		17.75		17.75	23.50	29
	Install new window (3'0" x 4'0")				1.156	395	54.50		449.50	505	585
	Remove scaffold				1.333		62.50		62.50	83	103
	Total				4.295	395	197.25		592.25	694.50	820

B2023 123 Vinyl Clad Wood Window, Fixed

	System Description	Freq. (Years)	Crew	Unit	Labor Hours	Material	2015 Bare Costs Labor	Equipment	Total	Total In-House	Total w/O&P
1030	**Repl. 4'-0" x 4'-0" vinyl clad window - 1st flr.**	60	1 CARP	Ea.							
	Set up and secure scaffold				.444		21		21	27.50	34
	Remove window				.473		17.75		17.75	23.50	29
	Install new window (4'-0" x 4'-0")				1.733	510	81.50		591.50	670	770
	Remove scaffold				.444		21		21	27.50	34
	Total				3.095	510	141.25		651.25	748.50	867
2030	**Repl. 4'-0" x 4'-0" vinyl clad window - 2nd flr.**	60	1 CARP	Ea.							
	Set up and secure scaffold				.889		41.50		41.50	55.50	68.50
	Remove window				.473		17.75		17.75	23.50	29
	Install new window (4'-0" x 4'-0")				1.733	510	81.50		591.50	670	770
	Remove scaffold				.889		41.50		41.50	55.50	68.50
	Total				3.984	510	182.25		692.25	804.50	936
3030	**Repl. 4'-0" x 4'-0" vinyl clad window - 3rd flr.**	60	1 CARP	Ea.							
	Set up and secure scaffold				1.333		62.50		62.50	83	103
	Remove window				.473		17.75		17.75	23.50	29
	Install new window (4'-0" x 4'-0")				1.733	510	81.50		591.50	670	770
	Remove scaffold				1.333		62.50		62.50	83	103
	Total				4.873	510	224.25		734.25	859.50	1,005

For customer support on your Facilities Maintenance & Repair Cost Data, call 877.801.5724.

B2023 150 Glass Block, Fixed

System Description	Freq. (Years)	Crew	Unit	Labor Hours	Material	Labor	Equipment	Total	Total In-House	Total w/O&P
1020 Repair window - 1st floor (2% of glass)	8	1 BRIC	S.F.							
Cut out damaged block				.433		16.50	2.53	19.03	24.50	30
Replace glass block				.452	26.50	19.40		45.90	54.50	64.50
Total				.886	26.50	35.90	2.53	64.93	79	94.50
1030 Replace glass block window - 1st floor	75	1 BRIC	S.F.							
Set up, secure and take down ladder				.267		12.30		12.30	16.20	20
Demolish glass block				.074		2.79		2.79	3.71	4.58
Replace glass block				.452	26.50	19.40		45.90	54.50	64.50
Total				.793	26.50	34.49		60.99	74.41	89.08
2020 Repair window - 2nd floor (2% of glass)	8	1 BRIC	S.F.							
Set up and secure scaffold				.889		41		41	54	67
Cut out damaged block				.433		16.50	2.53	19.03	24.50	30
Replace glass block				.452	26.50	19.40		45.90	54.50	64.50
Remove scaffold				.889		41		41	54	67
Total				2.663	26.50	117.90	2.53	146.93	187	228.50
2030 Replace glass block window - 2nd floor	75	1 BRIC	S.F.							
Set up and secure scaffold				.889		41		41	54	67
Demolish glass block				.074		2.79		2.79	3.71	4.58
Replace glass block				.452	26.50	19.40		45.90	54.50	64.50
Remove scaffold				.889		41		41	54	67
Total				2.304	26.50	104.19		130.69	166.21	203.08
3020 Repair window - 3rd floor (2% of glass)	8	1 BRIC	S.F.							
Set up and secure scaffold				1.333		61.50		61.50	81	100
Cut out damaged block				.433		16.50	2.53	19.03	24.50	30
Replace glass block				.452	26.50	19.40		45.90	54.50	64.50
Remove scaffold				1.333		61.50		61.50	81	100
Total				3.552	26.50	158.90	2.53	187.93	241	294.50
3030 Replace glass block window - 3rd floor	75	1 BRIC	S.F.							
Set up and secure scaffold				1.333		61.50		61.50	81	100
Demolish glass block				.074		2.79		2.79	3.71	4.58
Replace glass block				.452	26.50	19.40		45.90	54.50	64.50
Remove scaffold				1.333		61.50		61.50	81	100
Total				3.193	26.50	145.19		171.69	220.21	269.08

For customer support on your Facilities Maintenance & Repair Cost Data, call 877.801.5724.

59

System Description	Freq. (Years)	Crew	Unit	Labor Hours	2015 Bare Costs				Total In-House	Total w/O&P
					Material	Labor	Equipment	Total		
1030 Refinish aluminum shutter - 1st floor	5	1 PORD	Ea.							
Set up and secure scaffold				.444		17.95		17.95	23.50	29
Prepare shutter for painting				.434		17.44		17.44	22.50	28
Paint shutter surface (to 16 S.F.), brushwork, primer + 1 coat				.800	2.29	32.50		34.79	44.50	55
Remove scaffold				.444		17.95		17.95	23.50	29
Total				2.123	2.29	85.84		88.13	114	141
1040 Replace aluminum shutter - 1st floor	60	1 CARP	Ea.							
Set up and secure scaffold				.444		21		21	27.50	34
Remove shutter				.260		12.25		12.25	16.25	20
Install new shutter				.520	100	24.50		124.50	143	165
Remove scaffold				.444		21		21	27.50	34
Total				1.669	100	78.75		178.75	214.25	253
2030 Refinish aluminum shutter - 2nd floor	5	1 PORD	Ea.							
Set up and secure scaffold				.889		36		36	46.50	57.50
Prepare shutter for painting				.434		17.44		17.44	22.50	28
Paint shutter surface (to 16 S.F.), brushwork, primer + 1 coat				.800	2.29	32.50		34.79	44.50	55
Remove scaffold				.889		36		36	46.50	57.50
Total				3.012	2.29	121.94		124.23	160	198
2040 Replace aluminum shutter - 2nd floor	60	1 CARP	Ea.							
Set up and secure scaffold				.889		41.50		41.50	55.50	68.50
Remove shutter				.260		12.25		12.25	16.25	20
Install new shutter				.520	100	24.50		124.50	143	165
Remove scaffold				.889		41.50		41.50	55.50	68.50
Total				2.558	100	119.75		219.75	270.25	322
3030 Refinish aluminum shutter - 3rd floor	5	1 PORD	Ea.							
Set up and secure scaffold				1.333		54		54	70	86.50
Prepare shutter for painting				.434		17.44		17.44	22.50	28
Paint shutter surface (to 16 S.F.), brushwork, primer + 1 coat				.800	2.29	32.50		34.79	44.50	55
Remove scaffold				1.333		54		54	70	86.50
Total				3.901	2.29	157.94		160.23	207	256

For customer support on your Facilities Maintenance & Repair Cost Data, call 877.801.5724.

B2023 160 Aluminum Shutter

System Description	Freq. (Years)	Crew	Unit	Labor Hours	2015 Bare Costs				Total In-House	Total w/O&P
					Material	Labor	Equipment	Total		
3040 **Replace aluminum shutter - 3rd floor**	60	1 CARP	Ea.							
Set up and secure scaffold				1.333		62.50		62.50	83	103
Remove shutter				.260		12.25		12.25	16.25	20
Install new shutter				.520	100	24.50		124.50	143	165
Remove scaffold				1.333		62.50		62.50	83	103
Total				3.447	100	161.75		261.75	325.25	391

B2023 162 Steel Shutter

System Description	Freq. (Years)	Crew	Unit	Labor Hours	2015 Bare Costs				Total In-House	Total w/O&P
					Material	Labor	Equipment	Total		
1030 **Refinish steel shutter - 1st floor**	5	1 PORD	Ea.							
Set up and secure scaffold				.444		17.95		17.95	23.50	29
Prepare shutter for painting				.434		17.44		17.44	22.50	28
Paint shutter surface (to 16 S.F.), brushwork, primer + 1 coat				.800	2.29	32.50		34.79	44.50	55
Remove scaffold				.444		17.95		17.95	23.50	29
Total				2.123	2.29	85.84		88.13	114	141
1040 **Replace steel shutter - 1st floor**	40	1 CARP	Ea.							
Set up and secure scaffold				.444		21		21	27.50	34
Remove old shutter				.308		16.38	1.38	17.76	26	31
Install new steel shutter				.617	37.50	32.70	2.82	73.02	93.50	109
Remove scaffold				.444		21		21	27.50	34
Total				1.814	37.50	91.08	4.20	132.78	174.50	208
2030 **Refinish steel shutter - 2nd floor**	5	1 PORD	Ea.							
Set up and secure scaffold				.889		36		36	46.50	57.50
Prepare shutter for painting				.434		17.44		17.44	22.50	28
Paint shutter surface (to 16 S.F.), brushwork, primer + 1 coat				.800	2.29	32.50		34.79	44.50	55
Remove scaffold				.889		36		36	46.50	57.50
Total				3.012	2.29	121.94		124.23	160	198

B2023 162 — Steel Shutter

	System Description	Freq. (Years)	Crew	Unit	Labor Hours	2015 Bare Costs				Total In-House	Total w/O&P
						Material	Labor	Equipment	Total		
2040	**Replace steel shutter - 2nd floor**	40	1 CARP	Ea.							
	Set up and secure scaffold				.889		41.50		41.50	55.50	68.50
	Remove old shutter				.308		16.38	1.38	17.76	26	31
	Install new steel shutter				.617	37.50	32.70	2.82	73.02	93.50	109
	Remove scaffold				.889		41.50		41.50	55.50	68.50
	Total				2.703	37.50	132.08	4.20	173.78	230.50	277
3030	**Refinish steel shutter - 3rd floor**	5	1 PORD	Ea.							
	Set up and secure scaffold				1.333		54		54	70	86.50
	Prepare shutter for painting				.434		17.44		17.44	22.50	28
	Paint shutter surface (to 16 S.F.), brushwork, primer + 1 coat				.800	2.29	32.50		34.79	44.50	55
	Remove scaffold				1.333		54		54	70	86.50
	Total				3.901	2.29	157.94		160.23	207	256
3040	**Replace steel shutter - 3rd floor**	40	1 CARP	Ea.							
	Set up and secure scaffold				1.333		62.50		62.50	83	103
	Remove old shutter				.308		16.38	1.38	17.76	26	31
	Install new steel shutter				.617	37.50	32.70	2.82	73.02	93.50	109
	Remove scaffold				1.333		62.50		62.50	83	103
	Total				3.592	37.50	174.08	4.20	215.78	285.50	346

B2023 164 — Wood Shutter

	System Description	Freq. (Years)	Crew	Unit	Labor Hours	2015 Bare Costs				Total In-House	Total w/O&P
						Material	Labor	Equipment	Total		
1010	**Repair wood shutter - 1st floor**	6	1 CARP	Ea.							
	Remove shutter				.260		12.25		12.25	16.25	20
	Remove damaged panel				.157		7.38		7.38	9.80	12.05
	Install new panel				.314	4.74	14.76		19.50	25	30
	Install shutter				.520		24.50		24.50	32.50	40
	Total				1.251	4.74	58.89		63.63	83.55	102.05

For customer support on your Facilities Maintenance & Repair Cost Data, call 877.801.5724.

ID	System Description	Freq. (Years)	Crew	Unit	Labor Hours	2015 Bare Costs				Total In-House	Total w/O&P
						Material	Labor	Equipment	Total		
1030	**Refinish wood shutter - 1st floor**	5	1 PORD	Ea.							
	Set up and secure scaffold				.444		17.95		17.95	23.50	29
	Prepare shutter for painting				.434		17.44		17.44	22.50	28
	Paint shutter surface (to 16 S.F.), brushwork, primer + 1 coat				.800	2.29	32.50		34.79	44.50	55
	Remove scaffold				.444		17.95		17.95	23.50	29
	Total				2.123	2.29	85.84		88.13	114	141
1040	**Replace wood shutter - 1st floor**	40	1 CARP	Ea.							
	Set up and secure scaffold				.444		21		21	27.50	34
	Remove shutter				.260		12.25		12.25	16.25	20
	Install new wood shutter				.520	135	24.50		159.50	180	209
	Remove scaffold				.444		21		21	27.50	34
	Total				1.669	135	78.75		213.75	251.25	297
2010	**Repair wood shutter - 2nd floor**	6	1 CARP	Ea.							
	Set up and secure scaffold				.889		41.50		41.50	55.50	68.50
	Remove shutter				.260		12.25		12.25	16.25	20
	Remove damaged panel				.157		7.38		7.38	9.80	12.05
	Install new panel				.314	4.74	14.76		19.50	25	30
	Install shutter				.520		24.50		24.50	32.50	40
	Remove scaffold				.889		41.50		41.50	55.50	68.50
	Total				3.029	4.74	141.89		146.63	194.55	239.05
2030	**Refinish wood shutter - 2nd floor**	5	1 PORD	Ea.							
	Set up and secure scaffold				.889		36		36	46.50	57.50
	Prepare shutter for painting				.434		17.44		17.44	22.50	28
	Paint shutter surface (to 16 S.F.), brushwork, primer + 1 coat				.800	2.29	32.50		34.79	44.50	55
	Remove scaffold				.889		36		36	46.50	57.50
	Total				3.012	2.29	121.94		124.23	160	198
2040	**Replace wood shutter - 2nd floor**	40	1 CARP	Ea.							
	Set up and secure scaffold				.889		41.50		41.50	55.50	68.50
	Remove shutter				.260		12.25		12.25	16.25	20
	Install new wood shutter				.520	135	24.50		159.50	180	209
	Remove scaffold				.889		41.50		41.50	55.50	68.50
	Total				2.558	135	119.75		254.75	307.25	366

For customer support on your Facilities Maintenance & Repair Cost Data, call 877.801.5724.

63

B2023 164 Wood Shutter

	System Description	Freq. (Years)	Crew	Unit	Labor Hours	2015 Bare Costs				Total In-House	Total w/O&P
						Material	Labor	Equipment	Total		
3010	**Repair wood shutter - 3rd floor**	6	1 CARP	Ea.							
	Set up and secure scaffold				1.333		62.50		62.50	83	103
	Remove shutter				.260		12.25		12.25	16.25	20
	Remove damaged panel				.157		7.38		7.38	9.80	12.05
	Install new panel				.314	4.74	14.76		19.50	25	30
	Install shutter				.520		24.50		24.50	32.50	40
	Remove scaffold				1.333		62.50		62.50	83	103
	Total				3.917	4.74	183.89		188.63	249.55	308.05
3030	**Refinish wood shutter - 3rd floor**	5	1 PORD	Ea.							
	Set up and secure scaffold				1.333		54		54	70	86.50
	Prepare shutter for painting				.434		17.44		17.44	22.50	28
	Paint shutter surface (to 16 S.F.), brushwork, primer + 1 coat				.800	2.29	32.50		34.79	44.50	55
	Remove scaffold				1.333		54		54	70	86.50
	Total				3.901	2.29	157.94		160.23	207	256
3040	**Replace wood shutter - 3rd floor**	40	1 CARP	Ea.							
	Set up and secure scaffold				1.333		62.50		62.50	83	103
	Remove shutter				.260		12.25		12.25	16.25	20
	Install new wood shutter				.520	135	24.50		159.50	180	209
	Remove scaffold				1.333		62.50		62.50	83	103
	Total				3.447	135	161.75		296.75	362.25	435

B2023 166 Metal Window Grating

	System Description	Freq. (Years)	Crew	Unit	Labor Hours	2015 Bare Costs				Total In-House	Total w/O&P
						Material	Labor	Equipment	Total		
1020	**Refinish 3'-0" x 4'-0" metal window grating**	5	1 PORD	Ea.							
	Prepare window grating surface				.308		12.40		12.40	16.10	19.95
	Paint grated surface, brushwork, primer + 1 coat				.167	1.92	6.72		8.64	10.80	13.20
	Total				.474	1.92	19.12		21.04	26.90	33.15
1030	**Replace 3'-0" x 4'-0" metal window grating - 1st floor**	30	1 SSWK	Ea.							
	Remove old grating				.800		30		30	40	49.50
	Fabricate and install new window grating				6.275	314	330		644	840	995
	Total				7.075	314	360		674	880	1,044.50

For customer support on your Facilities Maintenance & Repair Cost Data, call 877.801.5724.

B2023 166   Metal Window Grating

	System Description	Freq. (Years)	Crew	Unit	Labor Hours	2015 Bare Costs				Total In-House	Total w/O&P
						Material	Labor	Equipment	Total		
2030	**Replace 3'-0" x 4'-0" metal window grating - 2nd floor**	30	1 SSWK	Ea.							
	Set up and secure scaffold				.889		47		47	70	84.50
	Remove old grating				.800		30		30	40	49.50
	Fabricate and install new window grating				6.275	314	330		644	840	995
	Remove scaffold				.889		47		47	70	84.50
	Total				8.852	314	454		768	1,020	**1,213.50**
3030	**Replace 3'-0" x 4'-0" metal window grating - 3rd floor**	30	1 SSWK	Ea.							
	Set up and secure scaffold				1.333		70		70	105	127
	Remove old window grating				.800		30		30	40	49.50
	Fabricate and install new window grating				6.275	314	330		644	840	995
	Remove scaffold				1.333		70		70	105	127
	Total				9.741	314	500		814	1,090	**1,298.50**

B2023 168   Metal Wire Mesh Cover

	System Description	Freq. (Years)	Crew	Unit	Labor Hours	2015 Bare Costs				Total In-House	Total w/O&P
						Material	Labor	Equipment	Total		
1020	**Refinish 3'-0" x 4'-0" mtl. wire mesh window cover**	5	1 PORD	Ea.							
	Prepare metal mesh surface				.103		4.14		4.14	5.35	6.65
	Paint metal mesh surface, brushwork, primer + 1 coat				.167	1.92	6.72		8.64	10.80	13.20
	Total				.269	1.92	10.86		12.78	16.15	**19.85**
1030	**Repl. 3'-0" x 4'-0" metal wire mesh window cover - 1st floor**	30	1 SSWK	Ea.							
	Set up and secure scaffold				.444		23.50		23.50	35	42.50
	Remove old mesh				.800		30		30	40	49.50
	Install new wire mesh & frame (to 12 S.F.)				.452	60	21.25		81.25	94	110
	Remove scaffold				.444		23.50		23.50	35	42.50
	Total				2.141	60	98.25		158.25	204	**244.50**
2030	**Repl. 3'-0" x 4'-0" metal wire mesh window cover - 2nd floor**	30	1 SSWK	Ea.							
	Set up and secure scaffold				.889		47		47	70	84.50
	Remove old mesh				.800		30		30	40	49.50
	Install new wire mesh & frame (to 12 S.F.)				.452	60	21.25		81.25	94	110
	Remove scaffold				.889		47		47	70	84.50
	Total				3.030	60	145.25		205.25	274	**328.50**

For customer support on your Facilities Maintenance & Repair Cost Data, call 877.801.5724.

65

B2023 168 Metal Wire Mesh Cover

	System Description	Freq. (Years)	Crew	Unit	Labor Hours	Material	Labor	Equipment	Total	Total In-House	Total w/O&P
						2015 Bare Costs					
3030	Repl. 3'-0" x 4'-0" metal wire mesh window cover - 3rd floor	30	1 SSWK	Ea.							
	Set up and secure scaffold				1.333		70		70	105	127
	Remove old mesh				.800		30		30	40	49.50
	Install new wire mesh & frame (to 12 S.F.)				.452	60	21.25		81.25	94	110
	Remove scaffold				1.333		70		70	105	127
	Total				3.919	60	191.25		251.25	344	413.50

B2023 170 Aluminum Frame Window Screen

	System Description	Freq. (Years)	Crew	Unit	Labor Hours	Material	Labor	Equipment	Total	Total In-House	Total w/O&P
						2015 Bare Costs					
1010	Repair aluminum frame window screen	20	1 CARP	S.F.							
	Remove screen from mountings				.052		2.74		2.74	4.10	4.95
	Repair damaged screen				.017	3.15	.91		4.06	4.84	5.60
	Remount screen				.104		5.50		5.50	8.20	9.90
	Total				.173	3.15	9.15		12.30	17.14	20.45
1020	Refinish 3'-0" x 4'-0" aluminum frame window screen	5	1 PORD	Ea.							
	Prepare screen frame surface				.473		19.05		19.05	25	30.50
	Refinish screen frame surface, brushwork, primer + 1 coat				.615	1.17	25		26.17	33.50	41.50
	Total				1.088	1.17	44.05		45.22	58.50	72
1030	Replace 3'-0" x 4'-0" aluminum frame window screen - 1st floor	50	1 CARP	Ea.							
	Set up and secure scaffold				.444		21		21	27.50	34
	Remove screen from mountings				.624		32.88		32.88	49	59.50
	Install new screen				1.248	258	66		324	385	440
	Remove scaffold				.444		21		21	27.50	34
	Total				2.761	258	140.88		398.88	489	567.50
2030	Replace 3'-0" x 4'-0" aluminum frame window screen - 2nd floor	50	1 CARP	Ea.							
	Set up and secure scaffold				.889		41.50		41.50	55.50	68.50
	Remove screen from mountings				.624		32.88		32.88	49	59.50
	Install new screen				1.248	258	66		324	385	440
	Remove scaffold				.889		41.50		41.50	55.50	68.50
	Total				3.650	258	181.88		439.88	545	636.50

B2023 170 — Aluminum Frame Window Screen

	System Description	Freq. (Years)	Crew	Unit	Labor Hours	2015 Bare Costs				Total In-House	Total w/O&P
						Material	Labor	Equipment	Total		
3030	**Replace 3'-0" x 4'-0" aluminum frame window screen - 3rd floor**	50	1 CARP	Ea.							
	Set up and secure scaffold				1.333		62.50		62.50	83	103
	Remove screen from mountings				.624		32.88		32.88	49	59.50
	Install new screen				1.248	258	66		324	385	440
	Remove scaffold				1.333		62.50		62.50	83	103
	Total				4.539	258	223.88		481.88	600	705.50

B2023 172 — Steel Frame Window Screen

	System Description	Freq. (Years)	Crew	Unit	Labor Hours	2015 Bare Costs				Total In-House	Total w/O&P
						Material	Labor	Equipment	Total		
1010	**Repair steel frame window screen**	10	1 CARP	S.F.							
	Remove screen from mountings				.052		2.74		2.74	4.10	4.95
	Repair damaged screen				.017	3.15	.91		4.06	4.84	5.60
	Remount screen				.104		5.50		5.50	8.20	9.90
	Total				.173	3.15	9.15		12.30	17.14	20.45
1020	**Refinish 3'-0" x 4'-0" steel frame window screen**	5	1 PORD	Ea.							
	Prepare screen frame surface				.473		19.05		19.05	25	30.50
	Refinish screen frame surface, brushwork, primer + 1 coat				.615	1.17	25		26.17	33.50	41.50
	Total				1.088	1.17	44.05		45.22	58.50	72
1030	**Replace 3'-0" x 4'-0" steel frame window screen - 1st floor**	40	1 CARP	Ea.							
	Set up and secure scaffold				.444		21		21	27.50	34
	Remove screen from mountings				.624		32.88		32.88	49	59.50
	Install new screen				1.248	51.60	66		117.60	155	183
	Remove scaffold				.444		21		21	27.50	34
	Total				2.761	51.60	140.88		192.48	259	310.50
2030	**Replace 3'-0" x 4'-0" steel frame window screen - 2nd floor**	40	1 CARP	Ea.							
	Set up and secure scaffold				.889		41.50		41.50	55.50	68.50
	Remove screen from mountings				.624		32.88		32.88	49	59.50
	Install new screen				1.248	51.60	66		117.60	155	183
	Remove scaffold				.889		41.50		41.50	55.50	68.50
	Total				3.650	51.60	181.88		233.48	315	379.50

B2023 172 — Steel Frame Window Screen

	System Description	Freq. (Years)	Crew	Unit	Labor Hours	2015 Bare Costs				Total In-House	Total w/O&P
						Material	Labor	Equipment	Total		
3030	**Replace 3'-0" x 4'-0" steel frame window screen - 3rd floor**	40	1 CARP	Ea.							
	Set up and secure scaffold				1.333		62.50		62.50	83	103
	Remove screen from mountings				.624		32.88		32.88	49	59.50
	Install new screen				1.248	51.60	66		117.60	155	183
	Remove scaffold				1.333		62.50		62.50	83	103
	Total				4.539	51.60	223.88		275.48	370	448.50

B2023 174 — Wood Frame Window Screen

	System Description	Freq. (Years)	Crew	Unit	Labor Hours	2015 Bare Costs				Total In-House	Total w/O&P
						Material	Labor	Equipment	Total		
1010	**Repair wood frame window screen**	10	1 CARP	S.F.							
	Remove screen from mountings				.052		2.74		2.74	4.10	4.95
	Repair damaged screen				.017	3.15	.91		4.06	4.84	5.60
	Remount screen				.104		5.50		5.50	8.20	9.90
	Total				.173	3.15	9.15		12.30	17.14	20.45
1020	**Refinish 3'-0" x 4'-0" wood frame window screen**	5	1 PORD	Ea.							
	Prepare screen frame surface				.473		19.05		19.05	25	30.50
	Refinish screen frame surface, brushwork, primer + 1 coat				.615	1.17	25		26.17	33.50	41.50
	Total				1.088	1.17	44.05		45.22	58.50	72
1030	**Replace 3'-0" x 4'-0" wood frame window screen - 1st floor**	40	1 CARP	Ea.							
	Set up and secure scaffold				.444		21		21	27.50	34
	Remove screen from mountings				.624		32.88		32.88	49	59.50
	Install new screen				.666	98.40	31.20		129.60	150	174
	Remove scaffold				.444		21		21	27.50	34
	Total				2.179	98.40	106.08		204.48	254	301.50
2030	**Replace 3'-0" x 4'-0" wood frame window screen - 2nd floor**	40	1 CARP	Ea.							
	Set up and secure scaffold				.889		41.50		41.50	55.50	68.50
	Remove screen from mountings				.624		32.88		32.88	49	59.50
	Install new screen				.666	98.40	31.20		129.60	150	174
	Remove scaffold				.889		41.50		41.50	55.50	68.50
	Total				3.067	98.40	147.08		245.48	310	370.50

B20 EXTERIOR CLOSURE | B2023 | Exterior Windows

B2023 174 — Wood Frame Window Screen

System Description	Freq. (Years)	Crew	Unit	Labor Hours	2015 Bare Costs Material	2015 Bare Costs Labor	2015 Bare Costs Equipment	2015 Bare Costs Total	Total In-House	Total w/O&P
3030 Replace 3'-0" x 4'-0" wood frame window screen - 3rd floor	40	1 CARP	Ea.							
Set up and secure scaffold				1.333		62.50		62.50	83	103
Remove screen from mountings				.624		32.88		32.88	49	59.50
Install new screen				.666	98.40	31.20		129.60	150	174
Remove scaffold				1.333		62.50		62.50	83	103
Total				3.956	98.40	189.08		287.48	365	439.50

B2023 324 — Plate Glass

System Description	Freq. (Years)	Crew	Unit	Labor Hours	2015 Bare Costs Material	2015 Bare Costs Labor	2015 Bare Costs Equipment	2015 Bare Costs Total	Total In-House	Total w/O&P
1020 Replace plate glass storefront - 1st floor	50	2 CARP	C.S.F.							
Remove plate glass and frame				3.467		163		163	216	267
Replace plate glass and frame				16.000	2,800	720		3,520	4,050	4,675
Total				19.467	2,800	883		3,683	4,266	4,942
2020 Replace plate glass storefront - 2nd floor	50	2 CARP	C.S.F.							
Set up and secure scaffold				1.500		70.50		70.50	93.50	115
Remove plate glass and frame				3.467		163		163	216	267
Replace plate glass and frame				16.000	2,800	720		3,520	4,050	4,675
Remove scaffold				1.500		70.50		70.50	93.50	115
Total				22.467	2,800	1,024		3,824	4,453	5,172
3020 Replace plate glass storefront - 3rd floor	50	2 CARP	C.S.F.							
Set up and secure scaffold				2.250		105.75		105.75	140	173
Remove plate glass and frame				3.467		163		163	216	267
Replace plate glass and frame				16.000	2,800	720		3,520	4,050	4,675
Remove scaffold				2.250		105.75		105.75	140	173
Total				23.967	2,800	1,094.50		3,894.50	4,546	5,288

For customer support on your Facilities Maintenance & Repair Cost Data, call 877.801.5724.

69

System Description	Freq. (Years)	Crew	Unit	Labor Hours	2015 Bare Costs				Total In-House	Total w/O&P
					Material	Labor	Equipment	Total		
1020 Repair aluminum storefront door	12	1 CARP	Ea.							
Remove weatherstripping				.015		.69		.69	.92	1.13
Remove door closer				.167		7.85		7.85	10.40	12.85
Remove door hinge				.167		7.85		7.85	10.40	12.85
Install new weatherstrip				3.478	46.50	163		209.50	268	325
Install new door closer				.868	96	40.75		136.75	160	187
Install new hinge (50% of total hinges / 12 years)				.167		7.85		7.85	10.40	12.85
Hinge, 5 x 5, brass base					61			61	67	76.50
Oil / lubricate door closer				.065		3.03		3.03	4.03	4.96
Oil / lubricate hinges				.065		3.03		3.03	4.03	4.96
Total				4.990	203.50	234.05		437.55	535.18	638.10
1030 Replace 3'-0" x 7'-0" aluminum storefront doors	50	1 CARP	Ea.							
Remove door				.650		24.50		24.50	32.50	40
Remove door frame				1.300		61		61	81	100
Install new door, frame, and hardware				10.458	850	550		1,400	1,750	2,050
Install new glass				3.640	185.85	163.80		349.65	420	500
Total				16.048	1,035.85	799.30		1,835.15	2,283.50	2,690
2010 Replace insulating glass - (3% of glass)	1	1 CARP	S.F.							
Remove damaged glass				.052		1.95		1.95	2.59	3.20
Install new insulating glass				.277	31	12.50		43.50	50.50	59.50
Total				.329	31	14.45		45.45	53.09	62.70
2020 Repair aluminum storefront sliding door	12	1 CARP	Ea.							
Remove weatherstripping				.015		.69		.69	.92	1.13
Oil / lubricate door closer				.065		3.03		3.03	4.03	4.96
Oil / inspect wheel & track				.065		3.03		3.03	4.03	4.96
Install new door track / glide wheels (25% of total / 12 years)				.440	13.13	20.63		33.76	42	50
Install new weatherstrip				4.167	52	196		248	315	385
Total				4.750	65.13	223.38		288.51	365.98	446.05

For customer support on your Facilities Maintenance & Repair Cost Data, call 877.801.5724.

B2033 130 Glazed Wood

System Description	Freq. (Years)	Crew	Unit	Labor Hours	2015 Bare Costs				Total In-House	Total w/O&P
					Material	Labor	Equipment	Total		
1020 Repair glazed wood doors	12	1 CARP	Ea.							
Remove weatherstripping				.015		.69		.69	.92	1.13
Remove door closer				.167		7.85		7.85	10.40	12.85
Remove door hinge				.167		7.85		7.85	10.40	12.85
Install new weatherstrip				3.478	46.50	163		209.50	268	325
Install new door closer				.868	96	40.75		136.75	160	187
Install new hinge (50% of total hinges / 12 years)				.167		7.85		7.85	10.40	12.85
Hinge, 5 x 5, brass base					61			61	67	76.50
Oil / lubricate door closer				.065		3.03		3.03	4.03	4.96
Oil / lubricate hinges				.065		3.03		3.03	4.03	4.96
Total				4.990	203.50	234.05		437.55	535.18	638.10
1040 Replace 3'-0" x 7'-0" glazed wood door	40	1 CARP	Ea.							
Remove door				.650		24.50		24.50	32.50	40
Remove door frame				.743		35		35	46.50	57
Install new door frame				.943	134.30	44.20		178.50	207	240
Install new oak door sill				.832	82	39		121	142	167
Install exterior door				2.105	825	99		924	1,050	1,200
Total				5.274	1,041.30	241.70		1,283	1,478	1,704
1050 Refinish glazed wood doors	4	1 PORD	Ea.							
Paint door, door frame & trim, brushwork, primer + 2 coats				1.760	16.40	71		87.40	110	135
Total				1.760	16.40	71		87.40	110	135
2010 Replace insulating glass - (8% of glass)	1	1 CARP	S.F.							
Remove damaged glass				.052		1.95		1.95	2.59	3.20
Install new insulated glass				.277	31	12.50		43.50	50.50	59.50
Total				.329	31	14.45		45.45	53.09	62.70
2020 Repair glazed wood sliding door	14	1 CARP	Ea.							
Remove weatherstripping				.015		.69		.69	.92	1.13
Oil / lubricate door closer				.065		3.03		3.03	4.03	4.96
Oil / inspect wheel & track				.065		3.03		3.03	4.03	4.96
Install new door track / glide wheels (25% of total / 14 years)				.440	13.13	20.63		33.76	42	50
Install new weatherstrip				4.167	52	196		248	315	385
Total				4.750	65.13	223.38		288.51	365.98	446.05

For customer support on your Facilities Maintenance & Repair Cost Data, call 877.801.5724.

71

B2033 130 Glazed Wood

	System Description	Freq. (Years)	Crew	Unit	Labor Hours	2015 Bare Costs				Total In-House	Total w/O&P
						Material	Labor	Equipment	Total		
2030	**Prepare and refinish glazed wood sliding door**	4	1 PORD	Ea.							
	Prepare surface				.310		12.50		12.50	16.25	20
	Paint door surface, brushwork, primer + 2 coats				.310	7.54	12.48		20.02	24.50	29.50
	Total				.621	7.54	24.98		32.52	40.75	49.50
2040	**Replace 6'-0" x 7'-0" glazed wood sliding door**	40	2 CARP	Ea.							
	Remove sliding door				1.733		81.50		81.50	108	133
	Install new sliding door				5.200	1,425	244		1,669	1,900	2,175
	Total				6.933	1,425	325.50		1,750.50	2,008	2,308
2050	**Refinish glazed wood sliding door & frame**	4	1 PORD	Ea.							
	Prepare surface				.465		18.75		18.75	24.50	30
	Paint door, door frame & trim				1.760	16.40	71		87.40	110	135
	Total				2.225	16.40	89.75		106.15	134.50	165

B2033 206 Solid Core, Painted

	System Description	Freq. (Years)	Crew	Unit	Labor Hours	2015 Bare Costs				Total In-House	Total w/O&P
						Material	Labor	Equipment	Total		
1010	**Repair solid core door, painted**	12	1 CARP	Ea.							
	Remove weatherstripping				.015		.69		.69	.92	1.13
	Remove door hinge				.250		11.78		11.78	15.60	19.30
	Install new weatherstrip				3.478	46.50	163		209.50	268	325
	Install new hinge (50% of total hinges / 12 years)				.250		11.78		11.78	15.60	19.30
	Hinge, 5 x 5, brass base					91.50			91.50	101	114
	Oil / lubricate hinges				.065		3.03		3.03	4.03	4.96
	Total				4.058	138	190.28		328.28	405.15	483.69
1020	**Prepare and refinish solid core door, painted**	4	1 PORD	Ea.							
	Prepare door surface				.333		13.45		13.45	17.45	21.50
	Paint door surface, brushwork, primer + 2 coats				.549	13.34	22.08		35.42	43.50	52
	Total				.883	13.34	35.53		48.87	60.95	73.50

B2033 206 Solid Core, Painted

System Description	Freq. (Years)	Crew	Unit	Labor Hours	2015 Bare Costs				Total In-House	Total w/O&P
					Material	Labor	Equipment	Total		
1030 **Replace 3'-0" x 7'-0" solid core door, painted**	40	1 CARP	Ea.							
Remove door				.650		24.50		24.50	32.50	40
Remove door frame				.743		35		35	46.50	57
Install new door frame				.943	134.30	44.20		178.50	207	240
Install new oak door sill				.832	82	39		121	142	167
Install new solid core door				1.143	209	53.50		262.50	300	350
Install new weatherstrip				4.167	52	196		248	315	385
Install new lockset				.867	70	40.50		110.50	131	154
Total				9.345	547.30	432.70		980	1,174	1,393
2010 **Repair solid core sliding wood door**	14	1 CARP	Ea.							
Remove weatherstripping				.015		.69		.69	.92	1.13
Oil / lubricate door closer				.065		3.03		3.03	4.03	4.96
Oil / inspect wheel & track				.065		3.03		3.03	4.03	4.96
Install new door track / glide wheels (25% of total / 14 years)				.440	13.13	20.63		33.76	42	50
Install new weatherstrip				4.167	52	196		248	315	385
Total				4.750	65.13	223.38		288.51	365.98	446.05
2030 **Replace 3'-0" x 7'-0" solid core sliding wood door**	40	1 CARP	Ea.							
Remove sliding door				1.733		81.50		81.50	108	133
Install new sliding door frame				.998	118.80	46.80		165.60	193	225
Install new door track / glide wheels				1.760	52.50	82.50		135	168	201
Install new solid core door				1.143	209	53.50		262.50	300	350
Install new weatherstrip				3.478	46.50	163		209.50	268	325
Install new lockset				.867	70	40.50		110.50	131	154
Total				9.980	496.80	467.80		964.60	1,168	1,388
3010 **Replace sliding door glass - (3% of glass)**	1	1 CARP	S.F.							
Remove damaged glass				.052		1.95		1.95	2.59	3.20
Install new glass				.173	8.85	7.80		16.65	20	24
Total				.225	8.85	9.75		18.60	22.59	27.20

For customer support on your Facilities Maintenance & Repair Cost Data, call 877.801.5724.

73

B2033 206 Solid Core, Painted

	System Description	Freq. (Years)	Crew	Unit	Labor Hours	2015 Bare Costs				Total In-House	Total w/O&P
						Material	Labor	Equipment	Total		
3040	**Replace 3'-0" x 7'-0" solid core door, w/ safety glass, painted**	40	1 CARP	Ea.							
	Remove door				.650		24.50		24.50	32.50	40
	Remove door frame				.743		35		35	46.50	57
	Install new door frame				.943	134.30	44.20		178.50	207	240
	Install new oak door sill				.832	82	39		121	142	167
	Install new solid core door				1.143	209	53.50		262.50	300	350
	Install new glass (3% of total glass/year)				.201	10.27	9.05		19.32	23	27.50
	Install new weatherstrip				4.167	52	196		248	315	385
	Install new lockset				.867	70	40.50		110.50	131	154
	Total				9.546	557.57	441.75		999.32	1,197	1,420.50

B2033 210 Aluminum Doors

	System Description	Freq. (Years)	Crew	Unit	Labor Hours	2015 Bare Costs				Total In-House	Total w/O&P
						Material	Labor	Equipment	Total		
1010	**Repair aluminum door**	12	1 CARP	Ea.							
	Remove door closer				.167		7.85		7.85	10.40	12.85
	Remove door hinge				.250		11.78		11.78	15.60	19.30
	Install new door closer				1.735	192	81.50		273.50	320	375
	Install new hinge (50% of total hinges / 12 years)				.250		11.78		11.78	15.60	19.30
	Hinge, 5 x 5, brass base					91.50			91.50	101	114
	Oil / lubricate door closer				.065		3.03		3.03	4.03	4.96
	Oil / lubricate hinges				.065		3.03		3.03	4.03	4.96
	Total				2.531	283.50	118.97		402.47	470.66	550.37
1020	**Replace 3'-0" x 7'-0" flush aluminum door**	50	1 CARP	Ea.							
	Remove door				.650		24.50		24.50	32.50	40
	Remove cylinder lock				.400		18.80		18.80	25	31
	Reinstall cylinder lock				1.156		54.50		54.50	72	89
	Install new door				1.224	1,450	57.50		1,507.50	1,675	1,900
	Total				3.431	1,450	155.30		1,605.30	1,804.50	2,060
2010	**Replace wire glass - (3% of glass)**	1	1 CARP	S.F.							
	Remove damaged glass				.052		1.95		1.95	2.59	3.20
	Install new wire glass				.179	27.26	8.06		35.32	40.50	47
	Total				.231	27.26	10.01		37.27	43.09	50.20

For customer support on your Facilities Maintenance & Repair Cost Data, call 877.801.5724.

B2033 210   Aluminum Doors

System Description	Freq. (Years)	Crew	Unit	Labor Hours	2015 Bare Costs				Total In-House	Total w/O&P
					Material	Labor	Equipment	Total		
2020 Repair aluminum frame and door	12	1 CARP	Ea.							
Remove weatherstripping				.015		.69		.69	.92	1.13
Remove door closer				.167		7.85		7.85	10.40	12.85
Remove door hinge				.250		11.78		11.78	15.60	19.30
Install new weatherstrip				3.478	46.50	163		209.50	268	325
Install new door closer				1.735	192	81.50		273.50	320	375
Install new hinge (50% of total hinges / 12 years)				.250		11.78		11.78	15.60	19.30
Hinge, 5 x 5, brass base					91.50			91.50	101	114
Oil / lubricate door closer				.065		3.03		3.03	4.03	4.96
Oil / lubricate hinges				.065		3.03		3.03	4.03	4.96
Total				6.024	330	282.66		612.66	739.58	876.50
2030 Replace 3'-0" x 7'-0" aluminum door, incl. vision lite	50	1 CARP	Ea.							
Remove door closer				.167		7.85		7.85	10.40	12.85
Remove door				.650		24.50		24.50	32.50	40
Install new door				1.224	1,450	57.50		1,507.50	1,675	1,900
Reinstall door closer				1.735		81.50		81.50	108	134
Install new glass				.217	11.06	9.75		20.81	25	29.50
Total				3.993	1,461.06	181.10		1,642.16	1,850.90	2,116.35
3010 Replace float glass - (3% of glass)	1	1 CARP	S.F.							
Remove damaged glass				.052		1.95		1.95	2.59	3.20
Install new glass				.201	10.27	9.05		19.32	23	27.50
Total				.253	10.27	11		21.27	25.59	30.70
3020 Repair aluminum sliding door	12	1 CARP	Ea.							
Remove weatherstripping				.015		.69		.69	.92	1.13
Oil / lubricate door closer				.065		3.03		3.03	4.03	4.96
Install new door closer (25% of total / 12 years)				.434	48	20.38		68.38	80	93.50
Oil / inspect wheel & track				.065		3.03		3.03	4.03	4.96
Install new door track / glide wheels (25% of total / 12 years)				.440	13.13	20.63		33.76	42	50
Install new weatherstrip				4.167	52	196		248	315	385
Total				5.184	113.13	243.76		356.89	445.98	539.55

For customer support on your Facilities Maintenance & Repair Cost Data, call 877.801.5724.

75

B2033 210 Aluminum Doors

System Description	Freq. (Years)	Crew	Unit	Labor Hours	2015 Bare Costs				Total In-House	Total w/O&P
					Material	Labor	Equipment	Total		
3030 Replace 3'-0" x 7'-0" aluminum sliding door	50	1 CARP	Ea.							
Remove door				.650		24.50		24.50	32.50	40
Install new door track / glide wheels				1.760	52.50	82.50		135	168	201
Install new aluminum door				1.224	1,450	57.50		1,507.50	1,675	1,900
Total				3.634	1,502.50	164.50		1,667	**1,875.50**	**2,141**
4020 Repair flush aluminum door, wood core	12	1 CARP	Ea.							
Remove weatherstripping				.015		.69		.69	.92	1.13
Remove door closer				.167		7.85		7.85	10.40	12.85
Remove door hinge				.250		11.78		11.78	15.60	19.30
Install new weatherstrip				3.478	46.50	163		209.50	268	325
Install new door closer				1.735	192	81.50		273.50	320	375
Install new hinge (50% of total hinges / 12 years)				.250		11.78		11.78	15.60	19.30
Hinge, 5 x 5, brass base					91.50			91.50	101	114
Oil / lubricate door closer				.065		3.03		3.03	4.03	4.96
Oil / lubricate hinges				.065		3.03		3.03	4.03	4.96
Total				6.024	330	282.66		612.66	**739.58**	**876.50**
4030 Replace 3'-0" x 7'-0" aluminum door, wood core	50	1 CARP	Ea.							
Remove door				.650		24.50		24.50	32.50	40
Remove door closer				.167		7.85		7.85	10.40	12.85
Install new aluminum door				1.224	1,450	57.50		1,507.50	1,675	1,900
Install door closer				1.735		81.50		81.50	108	134
Total				3.777	1,450	171.35		1,621.35	**1,825.90**	**2,086.85**
5010 Replace tempered glass - (3% of glass)	1	1 CARP	S.F.							
Remove damaged glass				.052		1.95		1.95	2.59	3.20
Install new glass				.201	10.27	9.05		19.32	23	27.50
Total				.253	10.27	11		21.27	**25.59**	**30.70**

B2033 210 Aluminum Doors

	System Description	Freq. (Years)	Crew	Unit	Labor Hours	Material	Labor	Equipment	Total	Total In-House	Total w/O&P
								2015 Bare Costs			
5020	**Repair aluminum door, insulated**	12	1 CARP	Ea.							
	Remove weatherstripping				.015		.69		.69	.92	1.13
	Remove door closer				.167		7.85		7.85	10.40	12.85
	Remove door hinge				.250		11.78		11.78	15.60	19.30
	Install new weatherstrip				3.478	46.50	163		209.50	268	325
	Install new door closer				1.735	192	81.50		273.50	320	375
	Install new hinge (50% of total hinges / 12 years)				.250		11.78		11.78	15.60	19.30
	Hinge, 5 x 5, brass base					91.50			91.50	101	114
	Oil / lubricate door closer				.065		3.03		3.03	4.03	4.96
	Oil / lubricate hinges				.065		3.03		3.03	4.03	4.96
	Total				6.024	330	282.66		612.66	739.58	876.50
5030	**Replace 3'-0" x 7'-0" aluminum door, insulated**	50	1 CARP	Ea.							
	Remove door				.650		24.50		24.50	32.50	40
	Remove door closer				.167		7.85		7.85	10.40	12.85
	Install new aluminum door				1.224	1,450	57.50		1,507.50	1,675	1,900
	Install door closer				1.735		81.50		81.50	108	134
	Total				3.777	1,450	171.35		1,621.35	1,825.90	2,086.85

B2033 220 Steel, Painted

	System Description	Freq. (Years)	Crew	Unit	Labor Hours	Material	Labor	Equipment	Total	Total In-House	Total w/O&P
								2015 Bare Costs			
1010	**Repair steel door, painted**	14	1 CARP	Ea.							
	Remove weatherstripping				.015		.69		.69	.92	1.13
	Remove door closer				.167		7.85		7.85	10.40	12.85
	Remove door hinge				.250		11.78		11.78	15.60	19.30
	Install new weatherstrip				3.478	46.50	163		209.50	268	325
	Install new door closer				1.735	192	81.50		273.50	320	375
	Install new hinge (50% of total hinges / 14 years)				.250		11.78		11.78	15.60	19.30
	Hinge, 5 x 5, brass base					91.50			91.50	101	114
	Oil / lubricate door closer				.065		3.03		3.03	4.03	4.96
	Oil / lubricate hinges				.065		3.03		3.03	4.03	4.96
	Total				6.024	330	282.66		612.66	739.58	876.50

B20 EXTERIOR CLOSURE B2033 Exterior Doors

B2033 220 Steel, Painted

	System Description	Freq. (Years)	Crew	Unit	Labor Hours	Material	Labor	Equipment	Total	Total In-House	Total w/O&P
								2015 Bare Costs			
1020	**Refinish 3'-0" x 7'-0" steel door, painted**	4	1 PORD	Ea.							
	Paint on finished metal surface, brushwork, primer + 1 coat				.800	4.40	32.50		36.90	46.50	57.50
	Total				.800	4.40	32.50		36.90	46.50	57.50
1030	**Replace 3'-0" x 7'-0" steel door, painted**	45	1 CARP	Ea.							
	Remove door				.650		24.50		24.50	32.50	40
	Remove cylinder lock				.400		18.80		18.80	25	31
	Install new door				1.224	435	57.50		492.50	555	640
	Reinstall cylinder lock				1.156		54.50		54.50	72	89
	Total				3.431	435	155.30		590.30	684.50	800
2010	**Replace tempered glass - (3% of glass)**	1	1 CARP	S.F.							
	Remove damaged glass				.052		1.95		1.95	2.59	3.20
	Install new glass				.201	10.27	9.05		19.32	23	27.50
	Total				.253	10.27	11		21.27	25.59	30.70
2030	**Prepare and refinish 3'-0" x 7'-0" steel door, painted**	4	1 PORD	Ea.							
	Prepare surface				.333		13.45		13.45	17.45	21.50
	Refinish door surface, brushwork, primer + 1 coat				1.333	4.14	54		58.14	74.50	91.50
	Total				1.667	4.14	67.45		71.59	91.95	113
2040	**Replace 3'-0" x 7'-0" steel door w/ wire glass, painted**	45	1 CARP	Ea.							
	Remove door				.650		24.50		24.50	32.50	40
	Remove door closer				.167		7.85		7.85	10.40	12.85
	Remove door frame & trim				1.300		61		61	81	100
	Install new steel door				1.224	625	57.50		682.50	765	875
	Install new steel frame				1.301	151	61		212	247	289
	Install new wire glass				.179	32.48	8.06		40.54	46.50	53.50
	Reinstall door closer				1.735		81.50		81.50	108	134
	Total				6.556	808.48	301.41		1,109.89	1,290.40	1,504.35

For customer support on your Facilities Maintenance & Repair Cost Data, call 877.801.5724.

B2033 220 Steel, Painted

System Description	Freq. (Years)	Crew	Unit	Labor Hours	2015 Bare Costs Material	Labor	Equipment	Total	Total In-House	Total w/O&P
3010 Repair steel sliding door, painted	14	1 CARP	Ea.							
Remove weatherstripping				.015		.69		.69	.92	1.13
Oil / lubricate door closer				.065		3.03		3.03	4.03	4.96
Install new door closer (25% of total / 14 years)				.434	48	20.38		68.38	80	93.50
Oil / inspect wheel & track				.065		3.03		3.03	4.03	4.96
Install new door track / glide wheels (25% of total / 14 years)				.440	13.13	20.63		33.76	42	50
Install new weatherstrip				4.167	52	196		248	315	385
Total				5.184	113.13	243.76		356.89	445.98	539.55
3020 Refinish 3'-0" x 7'-0" steel sliding door, painted	4	1 PORD	Ea.							
Prepare surface				.333		13.45		13.45	17.45	21.50
Refinish door surface, brushwork, primer + 1 coat				1.333	4.14	54		58.14	74.50	91.50
Total				1.667	4.14	67.45		71.59	91.95	113
3030 Replace 3'-0" x 7'-0" steel sliding door, painted	45	1 CARP	Ea.							
Remove door				.650		24.50		24.50	32.50	40
Remove door closer				.167		7.85		7.85	10.40	12.85
Install new steel door				1.224	625	57.50		682.50	765	875
Install door closer				1.735		81.50		81.50	108	134
Total				3.777	625	171.35		796.35	915.90	1,061.85
4010 Replace steel sliding door glass - (3% of glass)	1	1 CARP	S.F.							
Remove damaged glass				.052		1.95		1.95	2.59	3.20
Install new glass				.201	10.27	9.05		19.32	23	27.50
Total				.253	10.27	11		21.27	25.59	30.70
4020 Repair steel door, insulated core, painted	14	1 CARP	Ea.							
Remove weatherstripping				.015		.69		.69	.92	1.13
Remove door closer				.167		7.85		7.85	10.40	12.85
Remove door hinge				.167		7.85		7.85	10.40	12.85
Install new weatherstrip				3.478	46.50	163		209.50	268	325
Install new door closer				1.735	192	81.50		273.50	320	375
Install new hinge (50% of total hinges / 14 years)				.167		7.85		7.85	10.40	12.85
Hinge, 5 x 5, brass base					61			61	67	76.50
Oil / lubricate door closer				.065		3.03		3.03	4.03	4.96
Oil / lubricate hinges				.065		3.03		3.03	4.03	4.96
Total				5.857	299.50	274.80		574.30	695.18	826.10

For customer support on your Facilities Maintenance & Repair Cost Data, call 877.801.5724.

79

B2033 220 — Steel, Painted

	System Description	Freq. (Years)	Crew	Unit	Labor Hours	Material	Labor	Equipment	Total	Total In-House	Total w/O&P
						2015 Bare Costs					
4030	**Refinish 3'-0" x 7'-0" steel door, insulated core, painted**	4	1 PORD	Ea.							
	Prepare surface				.333		13.45		13.45	17.45	21.50
	Refinish door surface, brushwork, primer + 1 coat				1.333	4.14	54		58.14	74.50	91.50
	Total				1.667	4.14	67.45		71.59	91.95	113
4040	**Replace 3'-0" x 7'-0" steel door, insulated core, painted**	45	1 CARP	Ea.							
	Remove door				.650		24.50		24.50	32.50	40
	Remove door closer				.167		7.85		7.85	10.40	12.85
	Remove door frame & trim				1.300		61		61	81	100
	Install new steel door				1.224	625	57.50		682.50	765	875
	Install new steel frame				1.301	151	61		212	247	289
	Install new insulated glass				.322	35.96	14.50		50.46	58.50	68.50
	Install door closer				1.735		81.50		81.50	108	134
	Total				6.699	811.96	307.85		1,119.81	1,302.40	1,519.35

B2033 221 — Steel, Unpainted

	System Description	Freq. (Years)	Crew	Unit	Labor Hours	Material	Labor	Equipment	Total	Total In-House	Total w/O&P
						2015 Bare Costs					
1010	**Repair steel door, unpainted**	14	1 CARP	Ea.							
	Remove weatherstripping				.015		.69		.69	.92	1.13
	Remove door closer				.167		7.85		7.85	10.40	12.85
	Remove door hinge				.167		7.85		7.85	10.40	12.85
	Install new weatherstrip				3.478	46.50	163		209.50	268	325
	Install new door closer				1.735	192	81.50		273.50	320	375
	Install new hinge (50% of total hinges / 14 years)				.167		7.85		7.85	10.40	12.85
	Hinge, 5 x 5, brass base					61			61	67	76.50
	Oil / lubricate door closer				.065		3.03		3.03	4.03	4.96
	Oil / lubricate hinges				.065		3.03		3.03	4.03	4.96
	Total				5.857	299.50	274.80		574.30	695.18	826.10
1020	**Replace 3'-0" x 7'-0" steel door, unpainted**	45	1 CARP	Ea.							
	Remove door				.650		24.50		24.50	32.50	40
	Remove door frame & trim				1.300		61		61	81	100
	Install new steel door				1.224	640	57.50		697.50	780	895
	Install new steel frame				1.301	151	61		212	247	289
	Total				4.475	791	204		995	1,140.50	1,324

B2033 221 Steel, Unpainted

	System Description	Freq. (Years)	Crew	Unit	Labor Hours	2015 Bare Costs				Total In-House	Total w/O&P
						Material	Labor	Equipment	Total		
2010	**Replace wire glass - (3% of glass)**	1	1 CARP	S.F.							
	Remove damaged glass				.052		1.95		1.95	2.59	3.20
	Install new wire glass				.179	27.26	8.06		35.32	40.50	47
	Total				.231	27.26	10.01		37.27	43.09	50.20
2020	**Repair steel door with safety glass, unpainted**	14	1 CARP	Ea.							
	Remove weatherstripping				.015		.69		.69	.92	1.13
	Remove door closer				.167		7.85		7.85	10.40	12.85
	Remove door hinge				.167		7.85		7.85	10.40	12.85
	Install new weatherstrip				3.478	46.50	163		209.50	268	325
	Install new door closer				1.735	192	81.50		273.50	320	375
	Install new hinge (50% of total hinges / 14 years)				.167		7.85		7.85	10.40	12.85
	Hinge, 5 x 5, brass base					61			61	67	76.50
	Oil / lubricate door closer				.065		3.03		3.03	4.03	4.96
	Oil / lubricate hinges				.065		3.03		3.03	4.03	4.96
	Total				5.857	299.50	274.80		574.30	695.18	826.10
2030	**Repl. 3'-0" x 7'-0" steel door with safety glass, unpainted**	45	1 CARP	Ea.							
	Remove door				.650		24.50		24.50	32.50	40
	Remove door closer				.167		7.85		7.85	10.40	12.85
	Remove door frame & trim				1.300		61		61	81	100
	Install new steel door				1.301	510	61		571	640	740
	Install new steel frame				1.301	151	61		212	247	289
	Install new wire glass				.308	56	13.90		69.90	80	92.50
	Install door closer				1.735		81.50		81.50	108	134
	Total				6.762	717	310.75		1,027.75	1,198.90	1,408.35
3010	**Repair steel sliding door, unpainted**	14	1 CARP	Ea.							
	Remove weatherstripping				.015		.69		.69	.92	1.13
	Oil / lubricate door closer				.065		3.03		3.03	4.03	4.96
	Install new door closer (25% of total / 14 years)				.434	48	20.38		68.38	80	93.50
	Oil / inspect wheel & track				.065		3.03		3.03	4.03	4.96
	New door track / glide wheels (25% of total / 14 years)				.440	13.13	20.63		33.76	42	50
	Total				1.018	61.13	47.76		108.89	130.98	154.55

For customer support on your Facilities Maintenance & Repair Cost Data, call 877.801.5724.

81

B2033 221 | **Steel, Unpainted**

System Description	Freq. (Years)	Crew	Unit	Labor Hours	2015 Bare Costs Material	Labor	Equipment	Total	Total In-House	Total w/O&P
3020 Replace 3'-0" x 7'-0" steel sliding door, unpainted	45	1 CARP	Ea.							
Remove door				.650		24.50		24.50	32.50	40
Remove door closer				.167		7.85		7.85	10.40	12.85
Install new steel door				1.224	640	57.50		697.50	780	895
Reinstall door closer				1.735		81.50		81.50	108	134
Total				3.776	640	171.35		811.35	930.90	1,081.85
4010 Replace tempered glass - (3% of glass)	1	1 CARP	S.F.							
Remove damaged glass				.052		1.95		1.95	2.59	3.20
Install new glass				.201	10.27	9.05		19.32	23	27.50
Total				.253	10.27	11		21.27	25.59	30.70
4020 Repair steel door, insulated core, unpainted	14	1 CARP	Ea.							
Remove weatherstripping				.015		.69		.69	.92	1.13
Remove door closer				.167		7.85		7.85	10.40	12.85
Remove door hinge				.167		7.85		7.85	10.40	12.85
Install new weatherstrip				3.478	46.50	163		209.50	268	325
Install new door closer (50% of total / 14 years)				.868	96	40.75		136.75	160	187
Install new hinge (50% of total hinges / 14 years)				.167		7.85		7.85	10.40	12.85
Hinge, 5 x 5, brass base					61			61	67	76.50
Oil / lubricate door closer				.065		3.03		3.03	4.03	4.96
Oil / lubricate hinges				.065		3.03		3.03	4.03	4.96
Total				4.990	203.50	234.05		437.55	535.18	638.10
4030 Repl. 3'-0" x 7'-0" steel door, insulated core, unpainted	45	1 CARP	Ea.							
Remove door				.650		24.50		24.50	32.50	40
Remove door closer				.167		7.85		7.85	10.40	12.85
Remove door frame & trim				1.300		61		61	81	100
Install new insulated steel door				1.301	685	61		746	835	955
Install new steel frame				1.301	151	61		212	247	289
Install new insulated glass				.322	35.96	14.50		50.46	58.50	68.50
Reinstall door closer				1.735		81.50		81.50	108	134
Total				6.776	871.96	311.35		1,183.31	1,372.40	1,599.35

For customer support on your Facilities Maintenance & Repair Cost Data, call 877.801.5724.

B2033 410 — Steel Roll-Up, Single

	System Description	Freq. (Years)	Crew	Unit	Labor Hours	2015 Bare Costs				Total In-House	Total w/O&P
						Material	Labor	Equipment	Total		
1010	**Repair 12' x 12' steel single roll-up door**	10	1 CARP	Ea.							
	Remove overhead type door section				1.303		61.25		61.25	81.50	100
	Install new roll-up door section				4.348	487.50	228.75		716.25	880	1,025
	Total				5.651	487.50	290		777.50	961.50	1,125
1020	**Refinish 12' x 12' steel single roll-up door**	5	1 PORD	Ea.							
	Set up, secure and take down ladder				.258		10.40		10.40	13.50	16.75
	Prepare door surface				2.000		80.70		80.70	105	129
	Paint overhead door, brushwork, primer + 1 coat, 1 side				1.493	34.02	60.48		94.50	116	140
	Total				3.751	34.02	151.58		185.60	234.50	285.75
1030	**Replace 12' x 12' steel single roll-up door**	35	2 CARP	Ea.							
	Remove overhead type door				5.212		245		245	325	400
	Install new roll-up door				17.391	1,950	915		2,865	3,525	4,100
	Total				22.603	1,950	1,160		3,110	3,850	4,500

B2033 412 — Steel Roll-Up, Double

	System Description	Freq. (Years)	Crew	Unit	Labor Hours	2015 Bare Costs				Total In-House	Total w/O&P
						Material	Labor	Equipment	Total		
1020	**Refinish 12' x 24' steel double roll-up door**	5	1 PORD	Ea.							
	Set up, secure and take down ladder				.533		21.50		21.50	28	34.50
	Prepare door surface				4.000		161.40		161.40	209	258
	Paint overhead door, brushwork, primer + 1 coat, 1 side				2.986	68.04	120.96		189	232	279
	Total				7.519	68.04	303.86		371.90	469	571.50
1040	**Replace 12' x 24' steel double roll-up door**	35	2 CARP	Ea.							
	Remove overhead type door				10.423		490		490	650	800
	Install new door				34.783	3,900	1,830		5,730	7,050	8,175
	Total				45.206	3,900	2,320		6,220	7,700	8,975

For customer support on your Facilities Maintenance & Repair Cost Data, call 877.801.5724.

83

B2033 413 | Steel Garage Door, Single Leaf, Spring

	System Description	Freq. (Years)	Crew	Unit	Labor Hours	2015 Bare Costs				Total In-House	Total w/O&P
						Material	Labor	Equipment	Total		
1010	**Repair 9' x 7' steel, single leaf, garage door**	10	1 CARP	Ea.							
	Repair damaged track				.068		3.21		3.21	4.27	5.25
	Remove cremone lockset				.400		18.80		18.80	25	31
	Install new cremone lockset				1.040	108	49		157	184	215
	Retension door coil spring				1.143		53.50		53.50	71.50	88
	Total				2.651	108	124.51		232.51	284.77	339.25
1020	**Refinish 9' x 7' steel, single leaf, garage door**	5	1 PORD	Ea.							
	Set up, secure and take down ladder				.258		10.40		10.40	13.50	16.75
	Prepare surface				2.000		80.70		80.70	105	129
	Paint garage door, brushwork, primer + 1 coat, 1 side				1.493	34.02	60.48		94.50	116	140
	Total				3.751	34.02	151.58		185.60	234.50	285.75
1030	**Replace 9' x 7' steel, single leaf, garage door**	35	2 CARP	Ea.							
	Remove door				2.602		122		122	162	200
	Install new door				2.602	965	122		1,087	1,225	1,400
	Total				5.203	965	244		1,209	1,387	1,600

B2033 420 | Aluminum Roll-Up, Single

	System Description	Freq. (Years)	Crew	Unit	Labor Hours	2015 Bare Costs				Total In-House	Total w/O&P
						Material	Labor	Equipment	Total		
1010	**Repair 12' x 12' aluminum single roll-up door**	10	1 CARP	Ea.							
	Remove door section				1.303		61.25		61.25	81.50	100
	Install new door section				3.478	675	163.75		838.75	965	1,125
	Total				4.781	675	225		900	1,046.50	1,225
1020	**Refinish 12' x 12' aluminum single roll-up door**	5	1 PORD	Ea.							
	Set up, secure and take down ladder				.258		10.40		10.40	13.50	16.75
	Prepare surface				2.000		80.70		80.70	105	129
	Paint overhead type door				1.493	34.02	60.48		94.50	116	140
	Total				3.751	34.02	151.58		185.60	234.50	285.75
1030	**Replace 12' x 12' aluminum single roll-up door**	35	2 CARP	Ea.							
	Paint overhead door, brushwork, primer + 1 coat, 1 side				5.212		245		245	325	400
	Install new door				13.913	2,700	655		3,355	3,850	4,450
	Total				19.125	2,700	900		3,600	4,175	4,850

B2033 422 Aluminum Roll-Up, Double

	System Description	Freq. (Years)	Crew	Unit	Labor Hours	2015 Bare Costs				Total In-House	Total w/O&P
						Material	Labor	Equipment	Total		
1010	**Repair 12' x 24' aluminum double roll-up door**	10	1 CARP	Ea.							
	Remove door section				2.606		122.50		122.50	163	200
	Install new door section				6.957	1,350	327.50		1,677.50	1,925	2,225
	Total				9.562	1,350	450		1,800	2,088	2,425
1020	**Refinish 12' x 24' aluminum double roll-up door**	5	1 PORD	Ea.							
	Set up, secure and take down ladder				.533		21.50		21.50	28	34.50
	Prepare surface				4.000		161.40		161.40	209	258
	Paint overhead door, brushwork, primer + 1 coat, 1 side				2.986	68.04	120.96		189	232	279
	Total				7.519	68.04	303.86		371.90	469	571.50
1030	**Replace 12' x 24' aluminum double roll-up door**	35	2 CARP	Ea.							
	Remove overhead type door				10.423		490		490	650	800
	Install new door				27.826	5,400	1,310		6,710	7,700	8,900
	Total				38.250	5,400	1,800		7,200	8,350	9,700

B2033 423 Aluminum Garage Door, Single Leaf, Spring

	System Description	Freq. (Years)	Crew	Unit	Labor Hours	2015 Bare Costs				Total In-House	Total w/O&P
						Material	Labor	Equipment	Total		
1010	**Repair 9' x 7' aluminum, single leaf, garage door**	10	1 CARP	Ea.							
	Repair damaged track				.684		32.10		32.10	42.50	52.50
	Remove cremone lockset				.400		18.80		18.80	25	31
	Install new cremone lockset				1.040	108	49		157	184	215
	Retension door coil spring				1.143		53.50		53.50	71.50	88
	Total				3.267	108	153.40		261.40	323	386.50
1020	**Refinish 9' x 7' aluminum, single leaf, garage dr.**	5	1 PORD	Ea.							
	Set up, secure and take down ladder				.258		10.40		10.40	13.50	16.75
	Prepare surface				2.000		80.70		80.70	105	129
	Paint garage door, brushwork, primer + 1 coat, 1 side				.747	17.01	30.24		47.25	58	70
	Total				3.005	17.01	121.34		138.35	176.50	215.75
1030	**Replace 9' x 7' aluminum, single leaf, garage dr.**	35	2 CARP	Ea.							
	Remove door				2.602		122		122	162	200
	Install new door				2.602	965	122		1,087	1,225	1,400
	Total				5.203	965	244		1,209	1,387	1,600

For customer support on your Facilities Maintenance & Repair Cost Data, call 877.801.5724.

85

B2033 430 Wood Roll-Up, Single

System Description	Freq. (Years)	Crew	Unit	Labor Hours	2015 Bare Costs				Total In-House	Total w/O&P
					Material	Labor	Equipment	Total		
1010 Repair 12' x 12' wood single roll-up door	8	1 CARP	Ea.							
Remove door section				1.303		61.25		61.25	81.50	100
Install new door section				3.478	512.50	163.75		676.25	780	910
Total				4.781	512.50	225		737.50	861.50	**1,010**
1020 Refinish 12' x 12' wood single roll-up door	5	1 PORD	Ea.							
Set up, secure and take down ladder				.258		10.40		10.40	13.50	16.75
Prepare door surface				2.000		80.70		80.70	105	129
Paint overhead door, brushwork, primer + 1 coat, 1 side				3.150	15.12	127.26		142.38	181	223
Total				5.408	15.12	218.36		233.48	**299.50**	**368.75**
1030 Replace 12' x 12' wood single roll-up door	16	2 CARP	Ea.							
Remove overhead type door				5.212		245		245	325	400
Install new door				13.913	2,050	655		2,705	3,125	3,650
Total				19.125	2,050	900		2,950	**3,450**	**4,050**

B2033 432 Wood Roll-Up, Double

System Description	Freq. (Years)	Crew	Unit	Labor Hours	2015 Bare Costs				Total In-House	Total w/O&P
					Material	Labor	Equipment	Total		
1010 Repair 12' x 24' wood double roll-up door	8	1 CARP	Ea.							
Remove door section				2.606		122.50		122.50	163	200
Install new door section				6.957	1,025	327.50		1,352.50	1,575	1,825
Total				9.562	1,025	450		1,475	**1,738**	**2,025**
1020 Refinish 12' x 24' wood double roll-up door	5	1 PORD	Ea.							
Set up, secure and take down ladder				.533		21.50		21.50	28	34.50
Prepare surface				4.000		161.40		161.40	209	258
Paint overhead door, brushwork, primer + 1 coat, 1 side				6.300	30.24	254.52		284.76	365	445
Total				10.833	30.24	437.42		467.66	**602**	**737.50**
1030 Replace 12' x 24' wood double roll-up door	16	2 CARP	Ea.							
Remove overhead type door				10.423		490		490	650	800
Install new door				27.826	4,100	1,310		5,410	6,250	7,275
Total				38.250	4,100	1,800		5,900	**6,900**	**8,075**

B2033 433 | Wood Garage Door, Single Leaf, Spring

System Description	Freq. (Years)	Crew	Unit	Labor Hours	2015 Bare Costs				Total In-House	Total w/O&P
					Material	Labor	Equipment	Total		
1010 Repair 9' x 7' wood, single leaf, garage door	8	1 CARP	Ea.							
Repair damaged track				.068		3.21		3.21	4.27	5.25
Remove cremone lockset				.400		18.80		18.80	25	31
Install new cremone lockset				1.040	108	49		157	184	215
Retension door coil spring				1.143		53.50		53.50	71.50	88
Total				2.651	108	124.51		232.51	284.77	339.25
1020 Refinish 9' x 7' wood, single leaf, garage door	5	1 PORD	Ea.							
Set up, secure and take down ladder				.258		10.40		10.40	13.50	16.75
Prepare surface				2.000		80.70		80.70	105	129
Paint garage door, brushwork, primer + 1 coat, 1 side				3.150	15.12	127.26		142.38	181	223
Total				5.408	15.12	218.36		233.48	299.50	368.75
1030 Replace 9' x 7' wood, single leaf, garage door	16	2 CARP	Ea.							
Remove door				2.602		122		122	162	200
Install new door				2.602	700	122		822	930	1,075
Total				5.203	700	244		944	1,092	1,275

B2033 513 | Electric Bifolding Hangar Door

System Description	Freq. (Years)	Crew	Unit	Labor Hours	2015 Bare Costs				Total In-House	Total w/O&P
					Material	Labor	Equipment	Total		
0120 Remove and replace electric bi-folding hangar door motor	15	2 SKWK	Ea.							
Hangar door,electric bi-folding door, replace motor				3.200	530	156		686	790	915
Total				3.200	530	156		686	790	915
0140 Remove and replace electric bi-folding hangar door cables	15	2 SKWK	Ea.							
Hangar door, elec bi-folding door, replace cables				8.000	93	390		483	615	745
Total				8.000	93	390		483	615	745
0160 Remove and replace electric bi-folding hangar door	20	2 SKWK	Ea.							
Hangar door demo				77.568		4,080		4,080	6,100	7,375
Hangar door, bi-fldg dr, remove and replace 20' x 80'				116.368	34,400	6,128		40,528	47,200	54,000
Total				193.936	34,400	10,208		44,608	53,300	61,375

For customer support on your Facilities Maintenance & Repair Cost Data, call 877.801.5724.

87

B2033 810 Hinges, Brass

	System Description	Freq. (Years)	Crew	Unit	Labor Hours	2015 Bare Costs				Total In-House	Total w/O&P
						Material	Labor	Equipment	Total		
1010	Replace brass door hinge	60	1 CARP	Ea.							
	Remove door				.650		24.50		24.50	32.50	40
	Remove door hinge				.167		7.85		7.85	10.40	12.85
	Install new hinge				.167		7.85		7.85	10.40	12.85
	Full mortise hinge, 5 x 5, brass base					61			61	67	76.50
	Reinstall exterior door				1.388		65		65	86.50	107
	Total				**2.371**	**61**	**105.20**		**166.20**	**206.80**	**249.20**

B2033 820 Lockset, Brass

	System Description	Freq. (Years)	Crew	Unit	Labor Hours	2015 Bare Costs				Total In-House	Total w/O&P
						Material	Labor	Equipment	Total		
1010	Replace brass door lockset	30	1 CARP	Ea.							
	Remove mortise lock set				.421		19.75		19.75	26.50	32.50
	Mortise lock				1.301	330	61		391	445	515
	Total				**1.722**	**330**	**80.75**		**410.75**	**471.50**	**547.50**

B2033 830 Door Closer, Brass

	System Description	Freq. (Years)	Crew	Unit	Labor Hours	2015 Bare Costs				Total In-House	Total w/O&P
						Material	Labor	Equipment	Total		
1010	Replace brass door closer	15	1 CARP	Ea.							
	Remove door closer				.167		7.85		7.85	10.40	12.85
	Install new door closer				1.735	192	81.50		273.50	320	375
	Total				**1.902**	**192**	**89.35**		**281.35**	**330.40**	**387.85**

88

For customer support on your Facilities Maintenance & Repair Cost Data, call 877.801.5724.

B2033 840 Door Opener

System Description	Freq. (Years)	Crew	Unit	Labor Hours	2015 Bare Costs Material	Labor	Equipment	Total	Total In-House	Total w/O&P
1010 Automatic door opener on existing door	7	1 ELEC	Ea.							
Cutout demolition of exterior wall for actuator button				1.739		66	10.10	76.10	98	119
Install 1/2" E.M.T. conduit				2.462	26.80	134.80		161.60	196	242
Install conductor, type THWN copper, #14				1.040	11.57	57.20		68.77	83	103
Install junction box				.533	12.05	29		41.05	49.50	60
Install outlet				.381	7.90	21		28.90	34.50	42.50
Install plate				.100	2.71	5.45		8.16	9.75	11.85
Automatic opener, button operation				20.000	3,600	975		4,575	5,250	6,075
Total				26.255	3,661.03	1,288.45	10.10	4,959.58	5,720.75	6,653.35

B2033 850 Deadbolt, Brass

System Description	Freq. (Years)	Crew	Unit	Labor Hours	2015 Bare Costs Material	Labor	Equipment	Total	Total In-House	Total w/O&P
1010 Replace brass deadbolt	20	1 CARP	Ea.							
Remove deadbolt				.154		7.20		7.20	9.60	11.85
Install new deadbolt				1.156	175	54.50		229.50	265	310
Total				1.309	175	61.70		236.70	274.60	321.85

B2033 860 Weatherstripping, Brass

System Description	Freq. (Years)	Crew	Unit	Labor Hours	2015 Bare Costs Material	Labor	Equipment	Total	Total In-House	Total w/O&P
1010 Replace brass door weatherstripping	20	1 CARP	Ea.							
Remove weatherstripping				.015		.69		.69	.92	1.13
Install new weatherstrip				3.478	46.50	163		209.50	268	325
Total				3.493	46.50	163.69		210.19	268.92	326.13

For customer support on your Facilities Maintenance & Repair Cost Data, call 877.801.5724.

89

B2033 870 Panic Device

	System Description	Freq. (Years)	Crew	Unit	Labor Hours	2015 Bare Costs				Total In-House	Total w/O&P
						Material	Labor	Equipment	Total		
1010	**Replace door panic device**	25	1 CARP	Ea.							
	Remove panic bar				.421		19.75		19.75	26.50	32.50
	Install new panic bar				2.080	865	97.50		962.50	1,075	1,250
	Total				2.501	865	117.25		982.25	1,101.50	1,282.50

B2033 905 Hollow Core, Painted

	System Description	Freq. (Years)	Crew	Unit	Labor Hours	2015 Bare Costs				Total In-House	Total w/O&P
						Material	Labor	Equipment	Total		
1010	**Repair hollow core door**	12	1 CARP	Ea.							
	Remove weatherstripping				.015		.69		.69	.92	1.13
	Remove door hinge				.167		7.85		7.85	10.40	12.85
	Install new weatherstrip				3.478	46.50	163		209.50	268	325
	Install new hinge (50% of total hinges / 12 years)				.167		7.85		7.85	10.40	12.85
	Hinge, 5 x 5, brass base					61			61	67	76.50
	Oil / lubricate hinges				.065		3.03		3.03	4.03	4.96
	Total				3.891	107.50	182.42		289.92	360.75	433.29
1020	**Refinish 3'-0" x 7'-0" hollow core door**	4	1 PORD	Ea.							
	Prepare door surface				.333		13.45		13.45	17.45	21.50
	Paint door surface, brushwork, primer + 1 coat				.549	13.34	22.08		35.42	43.50	52
	Total				.883	13.34	35.53		48.87	60.95	73.50
1030	**Replace 3'-0" x 7'-0" hollow core door**	30	1 CARP	Ea.							
	Remove door				.650		24.50		24.50	32.50	40
	Remove door frame				.743		35		35	46.50	57
	Install new door frame				.943	134.30	44.20		178.50	207	240
	Install new oak door sill				.832	82	39		121	142	167
	Install new hollow core door				1.000	110	47		157	183	215
	Install new weatherstrip				3.478	46.50	163		209.50	268	325
	Install new lockset				.867	70	40.50		110.50	131	154
	Total				8.514	442.80	393.20		836	1,010	1,198

For customer support on your Facilities Maintenance & Repair Cost Data, call 877.801.5724.

B20 EXTERIOR CLOSURE B2033 Exterior Doors

B2033 905 Hollow Core, Painted

	System Description	Freq. (Years)	Crew	Unit	Labor Hours	2015 Bare Costs				Total In-House	Total w/O&P
						Material	Labor	Equipment	Total		
2010	**Repair hollow core sliding wood door**	14	1 CARP	Ea.							
	Remove weatherstripping				.015		.69		.69	.92	1.13
	Oil / lubricate door closer				.065		3.03		3.03	4.03	4.96
	Oil / inspect wheel & track				.065		3.03		3.03	4.03	4.96
	Install new door track / glide wheels (25% of total / 14 years)				.440	13.13	20.63		33.76	42	50
	Install new weatherstrip				4.167	52	196		248	315	385
	Total				4.750	65.13	223.38		288.51	365.98	446.05
2030	**Replace 3'-0" x 7'-0" hollow core sliding wood door**	30	1 CARP	Ea.							
	Remove sliding door				1.733		81.50		81.50	108	133
	Install new sliding door frame				4.622	435	246	21	702	865	1,000
	Install new hollow core door				1.000	110	47		157	183	215
	Install new weatherstrip				3.478	46.50	163		209.50	268	325
	Lockset				.867	70	40.50		110.50	131	154
	Total				11.701	661.50	578	21	1,260.50	1,555	1,827

B2033 925 Metal Grated

	System Description	Freq. (Years)	Crew	Unit	Labor Hours	2015 Bare Costs				Total In-House	Total w/O&P
						Material	Labor	Equipment	Total		
1010	**Repair metal grated door**	15	1 SSWK	Ea.							
	Remove damaged metal bar				.836		44.16		44.16	66	79.50
	Oil / lubricate door closer				.065		3.40		3.40	5.10	6.15
	Oil / lubricate hinges				.065		3.03		3.03	4.03	4.96
	Metal grating repair (3% of door area / 15 years)				.320	1.30	17.50	5.84	24.64	33.50	39.50
	Replace damaged metal bar				1.600	13.30	86	14.60	113.90	158	188
	Total				2.885	14.60	154.09	20.44	189.13	266.63	318.11
1020	**Prepare and refinish metal grated painted door**	4	1 PORD	Ea.							
	Prepare grated door surface				.308		12.40		12.40	16.10	19.95
	Paint grated door surface, brushwork, 1 coat				.319	3.68	12.88		16.56	20.50	25.50
	Total				.627	3.68	25.28		28.96	36.60	45.45

B2033 925 Metal Grated

	System Description	Freq. (Years)	Crew	Unit	Labor Hours	Material	Labor	Equipment	Total	Total In-House	Total w/O&P
1030	**Replace 3'-0" x 7'-0" metal grated painted door**	30	2 SSWK	Ea.							
	Remove door				.650		24.50		24.50	32.50	40
	Remove door closer				.167		7.85		7.85	10.40	12.85
	Remove door hinge				.500		23.55		23.55	31	38.50
	Install new grated door				15.686	785	825		1,610	2,100	2,475
	Install new hinge (25% of total hinges / 30 years)				.500		23.55		23.55	31	38.50
	Hinge, 5 x 5					91.50			91.50	101	114
	Total				17.503	876.50	904.45		1,780.95	2,305.90	2,718.85

B2033 926 Metal Wire Mesh Door

	System Description	Freq. (Years)	Crew	Unit	Labor Hours	Material	Labor	Equipment	Total	Total In-House	Total w/O&P
1010	**Repair metal wire mesh door**	15	1 SSWK	Ea.							
	Remove wire mesh				1.750		92.40		92.40	138	166
	Wire mesh					4.94			4.94	5.45	6.15
	Install wire mesh				7.000		368.55		368.55	555	660
	Oil / lubricate door closer				.065		3.40		3.40	5.10	6.15
	Oil / lubricate hinges				.065		3.03		3.03	4.03	4.96
	Total				8.879	4.94	467.38		472.32	707.58	843.26
1020	**Prepare and refinish metal wire mesh door**	4	1 PORD	Ea.							
	Prepare metal mesh surface				.312		12.60		12.60	16.35	20
	Paint metal mesh door surface, brushwork, 1 coat				.319	3.68	12.88		16.56	20.50	25.50
	Total				.631	3.68	25.48		29.16	36.85	45.50
1030	**Replace metal wire mesh door**	30	2 SSWK	Ea.							
	Remove wire mesh door				.640		33.50		33.50	50.50	61
	Remove wire mesh door frame				2.602		137		137	205	248
	Install new woven wire swinging door				3.471	310	163		473	560	655
	Install new metal door frame				3.200	239	170	14.60	423.60	530	620
	Install new hinge (25% of total hinges / 30 years)				.500		23.55		23.55	31	38.50
	Hinge, 5 x 5					91.50			91.50	101	114
	Install new deadlock night latch				1.039	45	49		94	114	136
	Total				11.451	685.50	576.05	14.60	1,276.15	1,591.50	1,872.50

Note: 2015 Bare Costs

B2033 927 Louvered Door

System Description	Freq. (Years)	Crew	Unit	Labor Hours	2015 Bare Costs Material	Labor	Equipment	Total	Total In-House	Total w/O&P
1010 Repair aluminum louvered door	12	1 CARP	Ea.							
Remove door				.650		24.50		24.50	32.50	40
Remove weatherstripping				.015		.69		.69	.92	1.13
Remove damaged louver				.137		6.45		6.45	8.55	10.55
Install new louver				.274	16	12.85		28.85	34.50	41
Oil / lubricate door closer				.065		3.03		3.03	4.03	4.96
Install new weatherstrip				3.478	46.50	163		209.50	268	325
Reinstall exterior door				1.388		65		65	86.50	107
Total				6.006	62.50	275.52		338.02	435	529.64
1020 Prepare and refinish aluminum louvered door	4	1 PORD	Ea.							
Prepare door surface				.333		13.45		13.45	17.45	21.50
Paint door surface, brushwork, primer + 1 coat				.549	13.34	22.08		35.42	43.50	52
Total				.883	13.34	35.53		48.87	60.95	73.50
1040 Replace 3'-0" x 7'-0" aluminum louvered door	50	1 CARP	Ea.							
Remove door				.650		24.50		24.50	32.50	40
Remove door closer				.167		7.85		7.85	10.40	12.85
Install new aluminum door				1.224	1,450	57.50		1,507.50	1,675	1,900
Install new door closer				1.735	192	81.50		273.50	320	375
Total				3.777	1,642	171.35		1,813.35	2,037.90	2,327.85
2010 Repair steel louvered door	14	1 CARP	Ea.							
Remove door				.650		24.50		24.50	32.50	40
Remove weatherstripping				.015		.69		.69	.92	1.13
Remove damaged louver				.137		6.45		6.45	8.55	10.55
Install new louver				.274	16	12.85		28.85	34.50	41
Oil / lubricate door closer				.065		3.03		3.03	4.03	4.96
Install new weatherstrip				3.478	46.50	163		209.50	268	325
Reinstall exterior door				1.388		65		65	86.50	107
Total				6.006	62.50	275.52		338.02	435	529.64
2020 Prepare and refinish steel louvered door	4	1 PORD	Ea.							
Prepare door surfaces				.333		13.45		13.45	17.45	21.50
Paint door surface, brushwork, primer + 1 coat				.549	13.34	22.08		35.42	43.50	52
Total				.883	13.34	35.53		48.87	60.95	73.50

For customer support on your Facilities Maintenance & Repair Cost Data, call 877.801.5724.

93

B2033 927 **Louvered Door**

System Description	Freq. (Years)	Crew	Unit	Labor Hours	2015 Bare Costs				Total In-House	Total w/O&P
					Material	Labor	Equipment	Total		
2040 Replace 3'-0" x 7'-0" steel louvered door	45	1 CARP	Ea.							
Remove door				.650		24.50		24.50	32.50	40
Remove door closer				.167		7.85		7.85	10.40	12.85
Install new steel door, incl. louver				1.224	640	57.50		697.50	780	895
Install new door closer				1.735	192	81.50		273.50	320	375
Total				3.776	832	171.35		1,003.35	**1,142.90**	**1,322.85**
3010 Repair wood louvered door	7	1 CARP	Ea.							
Remove door				.650		24.50		24.50	32.50	40
Remove weatherstripping				.015		.69		.69	.92	1.13
Remove damaged louver				.137		6.45		6.45	8.55	10.55
Install new louver				.274	16	12.85		28.85	34.50	41
Oil / lubricate door closer				.065		3.03		3.03	4.03	4.96
Install new weatherstrip				3.478	46.50	163		209.50	268	325
Reinstall exterior door				1.388		65		65	86.50	107
Total				6.006	62.50	275.52		338.02	**435**	**529.64**
3020 Prepare and refinish wood louvered door	4	1 PORD	Ea.							
Prepare door surfaces				.333		13.45		13.45	17.45	21.50
Paint door surface, brushwork, primer + 1 coat				.549	13.34	22.08		35.42	43.50	52
Total				.883	13.34	35.53		48.87	**60.95**	**73.50**
3040 Replace 3'-0" x 7'-0" wood louvered door	40	1 CARP	Ea.							
Remove door				.650		24.50		24.50	32.50	40
Remove door closer				.167		7.85		7.85	10.40	12.85
Install new solid core door				1.143	209	53.50		262.50	300	350
Add for installed wood louver					282			282	310	355
Install new door closer				1.735	192	81.50		273.50	320	375
Total				3.695	683	167.35		850.35	**972.90**	**1,132.85**

B2033 935 Aluminum Storm/Screen Door

System Description	Freq. (Years)	Crew	Unit	Labor Hours	2015 Bare Costs				Total In-House	Total w/O&P
					Material	Labor	Equipment	Total		
1010 Repair aluminum storm/screen door	5	1 CARP	Ea.							
Remove weatherstripping				.015		.69		.69	.92	1.13
Remove screen / storm door threshold				.519		24.40		24.40	32.50	40
Remove door closer				.167		7.85		7.85	10.40	12.85
Install new screen / storm door closer				1.301	26.50	61		87.50	110	133
Install new weatherstrip				.870	11.63	40.75		52.38	67	81.50
Install new plain aluminum threshold				.520	9.40	24.50		33.90	43	52
Total				3.391	47.53	159.19		206.72	263.82	320.48
1020 Replace aluminum storm/screen door	20	1 CARP	Ea.							
Remove door				.650		24.50		24.50	32.50	40
Install new anodized aluminum screen / storm door				1.487	178	70		248	289	335
Total				2.137	178	94.50		272.50	321.50	375

B2033 960 Aluminized Steel Pedestrian Gate

System Description	Freq. (Years)	Crew	Unit	Labor Hours	2015 Bare Costs				Total In-House	Total w/O&P
					Material	Labor	Equipment	Total		
1010 Repair aluminized steel pedestrian gate	10	1 CARP	Ea.							
Remove gate from hinges				.286		13.40		13.40	17.85	22
Oil / lubricate hinges				.065		3.03		3.03	4.03	4.96
Install new 6' aluminized steel fencing				.360	70.50	13.71	3.81	88.02	99	115
Reinstall gate				.615		29		29	38.50	47.50
Total				1.326	70.50	59.14	3.81	133.45	159.38	189.46
1030 Prepare and refinish pedestrian gate	5	1 PORD	Ea.							
Lay drop cloth				.090		3.60		3.60	4.70	5.80
Prepare / brush surface				.042		1.68		1.68	2.16	2.70
Paint gate, brushwork, 1 coat				.150	1.44	6.12		7.56	9.55	11.50
Remove drop cloth				.090		3.60		3.60	4.70	5.80
Total				.371	1.44	15		16.44	21.11	25.80
1040 Replace pedestrian gate	40	1 CARP	Ea.							
Remove old gate from hinges				.286		13.40		13.40	17.85	22
Install new 6' x 3' aluminized steel gate				3.117	203	119	33	355	415	485
Total				3.403	203	132.40	33	368.40	432.85	507

B2033 961   Pedestrian Gate, Galvanized Steel

System Description	Freq. (Years)	Crew	Unit	Labor Hours	2015 Bare Costs				Total	Total In-House	Total w/O&P
					Material	Labor	Equipment	Total			
1010 Repair steel gate	10	1 CARP	Ea.								
Remove gate from hinges				.286		13.40		13.40	17.85	22
Oil / lubricate hinges				.065		3.03		3.03	4.03	4.96
Install new 6' galvanized steel fencing				.360	58.65	13.71	3.81	76.17	87	100
Reinstall gate				.615		29		29	38.50	47.50
Total				1.326	58.65	59.14	3.81	121.60	147.38	**174.46**
1030 Prepare and refinish steel gate	5	1 PORD	Ea.								
Lay drop cloth				.090		3.60		3.60	4.70	5.80
Prepare / brush surface				.042		1.68		1.68	2.16	2.70
Paint gate, brushwork, 1 coat				.150	1.44	6.12		7.56	9.55	11.50
Remove drop cloth				.090		3.60		3.60	4.70	5.80
Total				.371	1.44	15		16.44	21.11	**25.80**
1040 Replace steel gate	40	1 CARP	Ea.								
Remove old gate from hinges				.286		13.40		13.40	17.85	22
Install new 6' x 3' galvanized steel gate				3.117	202	119	33	354	410	485
Total				3.403	202	132.40	33	367.40	427.85	**507**

B2033 962   Pedestrian Gate, Wood

System Description	Freq. (Years)	Crew	Unit	Labor Hours	2015 Bare Costs				Total	Total In-House	Total w/O&P
					Material	Labor	Equipment	Total			
1010 Repair wood gate	7	1 CARP	Ea.								
Remove gate from hinges				.286		13.40		13.40	17.85	22
Oil / lubricate hinges				.065		3.03		3.03	4.03	4.96
Replace damaged fencing				1.171	123	44.70	12.36	180.06	207	240
Reinstall gate				.615		29		29	38.50	47.50
Total				2.136	123	90.13	12.36	225.49	267.38	**314.46**
1030 Prepare and refinish wood gate	4	1 PORD	Ea.								
Lay drop cloth				.090		3.60		3.60	4.70	5.80
Prepare / brush surface				.042		1.68		1.68	2.16	2.70
Paint gate, brushwork, primer + 1 coat				.138	1.44	5.58		7.02	8.80	10.80
Remove drop cloth				.090		3.60		3.60	4.70	5.80
Total				.359	1.44	14.46		15.90	20.36	**25.10**

For customer support on your Facilities Maintenance & Repair Cost Data, call 877.801.5724.

B2033 962 Pedestrian Gate, Wood

	System Description	Freq. (Years)	Crew	Unit	Labor Hours	2015 Bare Costs				Total In-House	Total w/O&P
						Material	Labor	Equipment	Total		
1040	**Replace wood gate**	25	1 CARP	Ea.							
	Remove old gate from hinges				.286		13.40		13.40	17.85	22
	New 6'-0" x 3'-0" wood gate				3.468	217	132	36.50	385.50	450	525
	Total				3.754	217	145.40	36.50	398.90	467.85	547

B2033 963 Pedestrian Gate, Wrought Iron

	System Description	Freq. (Years)	Crew	Unit	Labor Hours	2015 Bare Costs				Total In-House	Total w/O&P
						Material	Labor	Equipment	Total		
1010	**Repair wrought iron gate**	11	1 CARP	Ea.							
	Remove gate from hinges				.286		13.40		13.40	17.85	22
	Oil / lubricate hinges				.065		3.03		3.03	4.03	4.96
	Repair hand forged wrought iron				2.600	243	136.50		379.50	470	550
	Reinstall gate				.615		29		29	38.50	47.50
	Total				3.566	243	181.93		424.93	530.38	624.46
1030	**Prepare and refinish wrought iron gate**	5	1 PORD	Ea.							
	Lay drop cloth				.090		3.60		3.60	4.70	5.80
	Prepare / brush surface				.125		5.04		5.04	6.50	8.10
	Paint gate, brushwork, primer + 1 coat				.167	1.44	6.66		8.10	10.25	12.60
	Remove drop cloth				.090		3.60		3.60	4.70	5.80
	Total				.471	1.44	18.90		20.34	26.15	32.30
1040	**Replace wrought iron gate**	45	1 CARP	Ea.							
	Remove old gate from hinges				.286		13.40		13.40	17.85	22
	Install new hand forged wrought iron gate				3.467	324	182		506	630	735
	Total				3.753	324	195.40		519.40	647.85	757

For customer support on your Facilities Maintenance & Repair Cost Data, call 877.801.5724.

97

System Description	Freq. (Years)	Crew	Unit	2015 Bare Costs					Total In-House	Total w/O&P
				Labor Hours	Material	Labor	Equipment	Total		
0100 Debris removal and visual inspection	0.50	2 ROFC	M.S.F.							
Set up, secure and take down ladder				.052		2.08		2.08	3.12	3.76
Pick up trash / debris & clean-up				.327		13.10		13.10	19.60	23.50
Visual inspection				.327		13.10		13.10	19.60	23.50
Total				.705		28.28		28.28	42.32	50.76
0200 Non-destructive moisture inspection	5	2 ROFC	M.S.F.							
Set up, secure and take down ladder				.052		2.08		2.08	3.12	3.76
Infrared inspection of roof membrane				2.133		85.50		85.50	128	155
Total				2.185		87.58		87.58	131.12	158.76
0300 Minor BUR membrane repairs - (2% of roof area)	1	G-5	Sq.							
Set up, secure and take down ladder				1.000		40		40	60	72
Sweep / spud ballast clean				.640		25.50		25.50	38.50	46.50
Cross cut incision through bitumen				.258		10.35		10.35	15.50	18.70
Install base sheet and 2 plies of glass mopped				3.294	107	124	33.50	264.50	335	395
Reinstall ballast in bitumen				.390		15.65		15.65	23.50	28.50
Clean up				.390		15.65		15.65	23.50	28.50
Total				5.973	107	231.15	33.50	371.65	496	589.20
0400 BUR flashing repairs - (2 S.F. per sq. repaired)	1	2 ROFC	S.F.							
Set up, secure and take down ladder				.010		.40		.40	.60	.72
Sweep / spud ballast clean				.009		.36		.36	.54	.65
Cut out buckled flashing				.020		.80		.80	1.20	1.45
Install 2 ply membrane flashing				.013	.19	.53		.72	1.01	1.21
Reinstall ballast in bitumen				.005		.21		.21	.31	.38
Clean up				.004		.16		.16	.24	.29
Total				.062	.19	2.46		2.65	3.90	4.70
0500 Minor BUR membrane replacement - (25% of roof area)	15	G-5	Sq.							
Set up, secure and take down ladder				.078		3.13		3.13	4.69	5.65
Remove 5 ply built-up roof				3.252		124		124	164	203
Remove roof insulation board				1.333		51		51	67	83
Install 2" polystyrene insulation				.641	72	26		98	117	136
Install 4 ply bituminous roofing				3.733	136	140	38	314	395	465
Clean up				1.000		40		40	60	72.50
Total				10.037	208	384.13	38	630.13	807.69	965.15

B3013 105   Built-Up Roofing

No.	System Description	Freq. (Years)	Crew	Unit	Labor Hours	2015 Bare Costs				Total In-House	Total w/O&P
						Material	Labor	Equipment	Total		
0600	**Place new BUR membrane over existing**	20	G-5	Sq.							
	Set up, secure and take down ladder				.020		.80		.80	1.20	1.45
	Sweep / spud ballast clean				.500		20		20	30	36
	Vent existing membrane				.130		5.20		5.20	7.80	9.40
	Cut out buckled flashing				.024		.96		.96	1.44	1.74
	Install 2 ply membrane flashing				.027	.39	1.06		1.45	2.02	2.42
	Install 4 ply bituminous roofing				3.733	136	140	38	314	395	465
	Reinstall ballast				.381		15.30		15.30	23	27.50
	Clean up				.390		15.65		15.65	23.50	28.50
	Total				5.205	136.39	198.97	38	373.36	483.96	572.01
0700	**Total BUR roof replacement**	28	G-1	Sq.							
	Set up, secure and take down ladder				.020		.80		.80	1.20	1.45
	Sweep / spud ballast clean				.500		20		20	30	36
	Remove built-up roofing				2.500		95		95	126	156
	Remove insulation board				1.026		39		39	52	64
	Remove flashing				.024		.96		.96	1.44	1.74
	Install 2" perlite insulation				.879	100	35		135	163	189
	Install 2 ply membrane flashing				.027	.39	1.06		1.45	2.02	2.42
	Install 4 ply bituminous membrane				2.800	136	105	28.50	269.50	335	390
	Clean up				1.000		40		40	60	72.50
	Total				8.776	236.39	336.82	28.50	601.71	770.66	913.11

B3013 120   Modified Bituminous / Thermoplastic

No.	System Description	Freq. (Years)	Crew	Unit	Labor Hours	2015 Bare Costs				Total In-House	Total w/O&P
						Material	Labor	Equipment	Total		
0100	**Debris removal by hand & visual inspection**	1	2 ROFC	M.S.F.							
	Set up, secure and take down ladder				.052		2.08		2.08	3.12	3.76
	Pick up trash / debris & clean up				.327		13.10		13.10	19.60	23.50
	Visual inspection				.327		13.10		13.10	19.60	23.50
	Total				.705		28.28		28.28	42.32	50.76
0200	**Non - destructive moisture inspection**	5	2 ROFC	M.S.F.							
	Set up, secure and take down ladder				.052		2.08		2.08	3.12	3.76
	Infrared inspection of roof membrane				2.133		85.50		85.50	128	155
	Total				2.185		87.58		87.58	131.12	158.76

For customer support on your Facilities Maintenance & Repair Cost Data, call 877.801.5724.

99

B3013 120 Modified Bituminous / Thermoplastic

	System Description	Freq. (Years)	Crew	Unit	Labor Hours	2015 Bare Costs				Total In-House	Total w/O&P
						Material	Labor	Equipment	Total		
0300	**Minor thermoplastic membrane repairs - (2% of roof area)**	1	G-5	Sq.							
	Set up, secure and take down ladder				1.000		40		40	60	72
	Clean away loose surfacing				.258		10.35		10.35	15.50	18.70
	Install 150 mil mod. bit., fully adhered				2.597	80	95	13	188	243	285
	Clean up				.390		15.65		15.65	23.50	28.50
	Total				4.245	80	161	13	254	342	404.20
0500	**Thermoplastic flashing repairs - (2 S.F. per sq. repaired)**	1	2 ROFC	S.F.							
	Set up, secure and take down ladder				.010		.40		.40	.60	.72
	Clean away loose surfacing				.003		.10		.10	.16	.19
	Remove flashing				.020		.80		.80	1.20	1.45
	Install 120 mil mod. bit. flashing				.019	.76	.68	.09	1.53	1.95	2.28
	Clean up				.004		.16		.16	.24	.29
	Total				.055	.76	2.14	.09	2.99	4.15	4.93
0600	**Minor thermoplastic membrane replacement - (25% of roof area)**	20	G-5	Sq.							
	Set up, secure and take down ladder				.078		3.13		3.13	4.69	5.65
	Remove existing membrane / insulation				4.571		183		183	275	330
	Install 2" perlite insulation				1.140	100	46		146	178	208
	Clean adjacent membrane / patch				1.143		46		46	68.50	83
	Install fully adhered 150 mil membrane				2.597	80	95	13	188	243	285
	Clean up				1.000		40		40	60	72.50
	Total				10.529	180	413.13	13	606.13	829.19	984.15
0700	**Total thermoplastic roof replacement**	25	G-1	Sq.							
	Set up, secure and take down ladder				.020		.80		.80	1.20	1.45
	Remove existing membrane / insulation				3.501		140		140	210	254
	Remove flashing				.026		1.04		1.04	1.56	1.88
	Install 2" perlite insulation				.879	100	35		135	163	189
	Install flashing				.037	1.52	1.36		3.06	3.90	4.56
	Install fully adhered 180 mil membrane				2.000	89	73	10	172	217	254
	Clean up				1.000		40		40	60	72.50
	Total				7.463	190.52	291.20	10.18	491.90	656.66	777.39

B3013 125 | **Thermosetting**

	System Description	Freq. (Years)	Crew	Unit	Labor Hours	2015 Bare Costs				Total In-House	Total w/O&P
						Material	Labor	Equipment	Total		
0100	**Debris removal by hand & visual inspection**	1	2 ROFC	M.S.F.							
	Set up, secure and take down ladder				.052		2.08		2.08	3.12	3.76
	Pick up trash / debris & clean up				.327		13.10		13.10	19.60	23.50
	Visual inspection				.327		13.10		13.10	19.60	23.50
	Total				**.705**		**28.28**		**28.28**	**42.32**	**50.76**
0200	**Non - destructive moisture inspection**	5	2 ROFC	M.S.F.							
	Set up, secure and take down ladder				.052		2.08		2.08	3.12	3.76
	Infrared inspection of roof membrane				2.133		85.50		85.50	128	155
	Total				**2.185**		**87.58**		**87.58**	**131.12**	**158.76**
0300	**Minor thermoset membrane repairs - (2% of roof area)**	1	G-5	Sq.							
	Set up, secure and take down ladder				1.000		40		40	60	72
	Broom roof surface clean				.258		10.35		10.35	15.50	18.70
	Remove / vent damaged membrane				.258		10.35		10.35	15.50	18.70
	Install thermosetting patch				2.597	68	95	13	176	230	270
	Clean up				.390		15.65		15.65	23.50	28.50
	Total				**4.503**	**68**	**171.35**	**13**	**252.35**	**344.50**	**407.90**
0400	**Thermoset flashing repairs - (2 S.F. per sq. repaired)**	1	2 ROFC	S.F.							
	Set up, secure and take down ladder				.010		.40		.40	.60	.72
	Remove flashing				.020		.80		.80	1.20	1.45
	Prime surfaces				.003		.12		.12	.18	.22
	Install 2 ply membrane flashing				.007	.10	.27		.37	.51	.61
	Clean up				.004		.16		.16	.24	.29
	Total				**.044**	**.10**	**1.75**		**1.85**	**2.73**	**3.29**
0500	**Minor thermoset membrane replacement - (25% of roof area)**	10	G-5	Sq.							
	Set up, secure and take down ladder				.078		3.13		3.13	4.69	5.65
	Broom roof surface clean				.258		10.35		10.35	15.50	18.70
	Remove, strip-in & seal joint				.593		24		24	35.50	43
	Install base sheet and 2 plies of glass mopped				3.294	107	124	33.50	264.50	335	395
	Clean up				.390		15.65		15.65	23.50	28.50
	Total				**4.613**	**107**	**177.13**	**33.50**	**317.63**	**414.19**	**490.85**

For customer support on your Facilities Maintenance & Repair Cost Data, call 877.801.5724.

101

B3013 125 Thermosetting

	System Description	Freq. (Years)	Crew	Unit	Labor Hours	2015 Bare Costs				Total In-House	Total w/O&P
						Material	Labor	Equipment	Total		
0600	**Total thermoset roof replacement**	20	G-1	Sq.							
	Set up, secure and take down ladder				.020		.80		.80	1.20	1.45
	Remove existing membrane / insulation				3.501		140		140	210	254
	Remove flashing				.026		1.04		1.04	1.56	1.88
	Install 2″ perlite insulation				.879	100	35		135	163	189
	Install new flashing				.020	.39	.80		1.19	1.62	1.92
	Install thermosetting membrane				2.800	172	105	28	305	375	435
	Clean up				1.000		40		40	60	72.50
	Total				8.246	272.39	322.64	28	623.03	812.38	955.75

B3013 127 EPDM (Ethylene Propylene Diene Monomer)

	System Description	Freq. (Years)	Crew	Unit	Labor Hours	2015 Bare Costs				Total In-House	Total w/O&P
						Material	Labor	Equipment	Total		
0600	**Total EPDM roof replacement**	25	G-5	Sq.							
	Set up, secure and take down ladder				.020		.80		.80	1.20	1.45
	Remove existing membrane / insulation				3.501		140		140	210	254
	Remove flashing				.026		1.04		1.04	1.56	1.88
	Install 2″ perlite insulation				.879	100	35		135	163	189
	Install new flashing				.020	.39	.80		1.19	1.62	1.92
	Fully adhered with adhesive				1.538	111	56	7.45	174.45	214	248
	Clean up				1.000		40		40	60	72.50
	Total				6.985	211.39	273.64	7.45	492.48	651.38	768.75

B3013 130 Metal Panel Roofing

	System Description	Freq. (Years)	Crew	Unit	Labor Hours	2015 Bare Costs				Total In-House	Total w/O&P
						Material	Labor	Equipment	Total		
0100	**Debris removal by hand & visual inspection**	1	2 ROFC	M.S.F.							
	Set up, secure and take down ladder				.100		4.01		4.01	6	7.25
	Visual inspection				.327		13.10		13.10	19.60	23.50
	Total				.427		17.11		17.11	25.60	30.75

For customer support on your Facilities Maintenance & Repair Cost Data, call 877.801.5724.

B3013 130 | **Metal Panel Roofing**

System Description	Freq. (Years)	Crew	Unit	Labor Hours	2015 Bare Costs				Total In-House	Total w/O&P
					Material	Labor	Equipment	Total		
0300 Minor metal roof finish repairs - (2% of roof area)	5	2 ROFC	S.F.							
Set up, secure and take down ladder				.010		.40		.40	.60	.72
Wire brush surface to remove oxide				.020		.80		.80	1.20	1.45
Prime and paint prepared surface				.016	.69	.64		1.33	1.72	2.02
Clean up				.004		.16		.16	.24	.29
Total				.050	.69	2		2.69	3.76	4.48
0400 Metal roof flashing replacement - (2 S.F. per sq. repaired)	1	2 ROFC	S.F.							
Set up, secure and take down ladder				.010		.40		.40	.60	.72
Dismantle seam				.016		.64		.64	.96	1.16
Remove metal roof panels				.060		2.40		2.40	3.60	4.35
Remove damaged metal flashing				.019		.76		.76	1.14	1.38
Install new aluminum flashing				.072	1.35	2.89		4.24	5.80	6.90
Replace metal panels				.129		5.16		5.16	7.75	9.35
Clean up				.004		.16		.16	.24	.29
Total				.310	1.35	12.41		13.76	20.09	24.15
0500 Minor metal roof panel replacement - (2.5% of roof area)	20	2 ROFC	S.F.							
Set up, secure and take down ladder				.010		.40		.40	.60	.72
Dismantle seam				.016		.64		.64	.96	1.16
Remove metal roof panels				.020		.80		.80	1.20	1.45
Install new metal panel				.044	2.90	2.05		4.95	5.85	6.90
Re-assemble seam				.016		.64		.64	.96	1.16
Prime and paint surface				.016	.69	.64		1.33	1.72	2.02
Clean up				.004		.16		.16	.24	.29
Total				.126	3.59	5.33		8.92	11.53	13.70
0600 Total metal roof panel replacement	30	2 ROFC	Sq.							
Set up, secure and take down ladder				.020		.80		.80	1.20	1.45
Remove existing roofing panel				3.600		144		144	216	261
Remove flashing metal				.038		1.52		1.52	2.28	2.76
Install new aluminum flashing				.072	1.35	2.89		4.24	5.80	6.90
Install new metal panel				4.384	290	205		495	585	690
Clean up				.390		15.65		15.65	23.50	28.50
Total				8.504	291.35	369.86		661.21	833.78	990.61

B3013 140 Slate Tile Roofing

System Description	Freq. (Years)	Crew	Unit	Labor Hours	2015 Bare Costs				Total In-House	Total w/O&P
					Material	Labor	Equipment	Total		
0100 Visual inspection	3	2 ROFC	M.S.F.							
Set up, secure and take down ladder				.100		4.01		4.01	6	7.25
Visual inspection				.327		13.10		13.10	19.60	23.50
Total				.427		17.11		17.11	25.60	30.75
0200 Minor slate tile replacement - (2.5% of roof area)	20	2 ROTS	S.F.							
Set up, secure and take down ladder				.010		.40		.40	.60	.72
Remove and replace damaged slates				.800	7	32		39	55.50	67
Clean up				.013		.52		.52	.78	.94
Total				.823	7	32.92		39.92	56.88	68.66
0300 Slate tile flashing repairs - (2 S.F. per sq. repaired)	20	2 ROTS	S.F.							
Set up, secure and take down ladder				.010		.40		.40	.60	.72
Removal of adjacent slate tiles				.100		4.02		4.02	6	7.25
Remove damaged metal flashing				.052		2.08		2.08	3.12	3.76
Install 16 oz. copper flashing				.089	7.65	3.56		11.21	13.75	16
Reinstall tiles in position				.134		5.36		5.36	8.05	9.70
Clean up				.013		.52		.52	.78	.94
Total				.398	7.65	15.94		23.59	32.30	38.37
0400 Total slate tile roof replacement	70	2 ROTS	Sq.							
Set up, secure and take down ladder				.020		.80		.80	1.20	1.45
Remove slate shingles				1.600		61		61	81	100
Install 16 oz. copper flashing				.178	15.30	7.12		22.42	27.50	32
Install 15# felt				.138	9.65	5.55		15.20	18.90	22
Install slate roofing				4.571	475	183		658	795	925
Clean up				1.333		53.50		53.50	80	96.50
Total				7.840	499.95	310.97		810.92	1,003.60	1,176.95

B3013 142 Mineral Fiber Steep Roofing

System Description	Freq. (Years)	Crew	Unit	Labor Hours	2015 Bare Costs				Total In-House	Total w/O&P
					Material	Labor	Equipment	Total		
0300 Visual inspection	3	2 ROFC	M.S.F.							
Set up, secure and take down ladder				.100		4.01		4.01	6	7.25
Visual inspection				.327		13.10		13.10	19.60	23.50
Total				.427		17.11		17.11	25.60	30.75

For customer support on your Facilities Maintenance & Repair Cost Data, call 877.801.5724.

B3013 144 | Clay Tile Roofing

System Description	Freq. (Years)	Crew	Unit	Labor Hours	2015 Bare Costs				Total In-House	Total w/O&P
					Material	Labor	Equipment	Total		
0300 Visual inspection										
Set up, secure and take down ladder	3	2 ROFC	M.S.F.	.100		4.01		4.01	6	7.25
Visual inspection				.327		13.10		13.10	19.60	23.50
Total				.427		17.11		17.11	**25.60**	**30.75**
0400 Minor clay tile repairs - (2% of roof area)										
Set up, secure and take down ladder	20	2 ROFC	S.F.	.010		.40		.40	.60	.72
Remove damaged tiles				.020		.80		.80	1.20	1.45
Install new clay tiles				.073	5.05	2.92		7.97	9.90	11.55
Clean up				.013		.52		.52	.78	.94
Total				.116	5.05	4.64		9.69	**12.48**	**14.66**
0500 Clay tile flashing repairs - (2 S.F. per sq. repaired)										
Set up, secure and take down ladder	1	2 ROFC	S.F.	.010		.40		.40	.60	.72
Remove adjacent tiles				.208		8.32		8.32	12.50	15.05
Remove damaged metal flashing				.052		2.08		2.08	3.12	3.76
Install 16 oz copper flashing				.178	15.30	7.12		22.42	27.50	32
Reinstall clay tiles				.073		2.92		2.92	4.35	5.25
Clean up				.013		.52		.52	.78	.94
Total				.533	15.30	21.36		36.66	**48.85**	**57.72**
0600 Total clay tile roof replacement										
Set up, secure and take down ladder	70	2 ROFC	Sq.	.020		.80		.80	1.20	1.45
Remove clay tile roofing				3.200		128		128	192	232
Remove damaged metal flashing				.040		1.60		1.60	2.40	2.90
Install 16 oz copper flashing				.178	15.30	7.12		22.42	27.50	32
Install felt				.138	9.65	5.55		15.20	18.90	22
Install new clay tiles				7.273	505	292		797	990	1,150
Clean up				1.333		53.50		53.50	80	96.50
Total				12.182	529.95	488.57		1,018.52	**1,312**	**1,536.85**

For customer support on your Facilities Maintenance & Repair Cost Data, call 877.801.5724.

105

B3013 146   Asphalt Shingle Roofing

System Description	Freq. (Years)	Crew	Unit	Labor Hours	2015 Bare Costs				Total In-House	Total w/O&P
					Material	Labor	Equipment	Total		
0100 **Debris removal by hand & visual inspection**	1	2 ROFC	M.S.F.							
Set up, secure and take down ladder				.100		4.01		4.01	6	7.25
Visual inspection				.327		13.10		13.10	19.60	23.50
Total				.427		17.11		17.11	25.60	30.75
0300 **Minor asphalt shingle repair - (2% of roof area)**	1	2 ROFC	S.F.							
Set up, secure and take down ladder				.010		.40		.40	.60	.72
Remove damaged shingles				.024		.89		.89	1.19	1.47
Install asphalt shingles				.036	1.10	1.43		2.53	3.35	3.96
Clean up				.003		.10		.10	.16	.19
Total				.072	1.10	2.82		3.92	5.30	6.34
0500 **Asphalt shingle flashing repairs - (2 S.F. per sq. repaired)**	1	2 ROFC	S.F.							
Set up, secure and take down ladder				.010		.40		.40	.60	.72
Remove adjoining shingles				.046		1.74		1.74	2.30	2.84
Remove damaged metal flashing				.019		.76		.76	1.14	1.38
Install new aluminum flashing				.072	1.35	2.89		4.24	5.80	6.90
Install asphalt shingles				.071	2.20	2.86		5.06	6.70	7.90
Clean up				.003		.10		.10	.16	.19
Total				.221	3.55	8.75		12.30	16.70	19.93
0600 **Install new asphalt shingles over existing**	20	2 ROFC	Sq.							
Set up, secure and take down ladder				.020		.80		.80	1.20	1.45
Remove shingles at eaves and ridge				.460		18.40		18.40	27.50	33.50
Install new aluminum flashing				.144	2.70	5.78		8.48	11.60	13.80
Install asphalt shingles				1.778	110	71.50		181.50	228	267
Clean up				.258		10.35		10.35	15.50	18.70
Total				2.660	112.70	106.83		219.53	283.80	334.45
0700 **Total asphalt shingle roof replacement**	40	2 ROFC	Sq.							
Set up, secure and take down ladder				.020		.80		.80	1.20	1.45
Remove existing shingles				2.353		89		89	119	147
Remove damaged metal flashing				.038		1.52		1.52	2.28	2.76
Install 15# felt				.138	9.65	5.55		15.20	18.90	22
Install new aluminum flashing				.144	2.70	5.78		8.48	11.60	13.80
Install asphalt shingles				1.778	110	71.50		181.50	228	267
Clean up				.258		10.35		10.35	15.50	18.70
Total				4.729	122.35	184.50		306.85	396.48	472.71

System Description	Freq. (Years)	Crew	Unit	Labor Hours	2015 Bare Costs				Total In-House	Total w/O&P
					Material	Labor	Equipment	Total		
0100 Debris removal by hand & visual inspection	1	2 ROFC	M.S.F.							
Set up, secure and take down ladder				.100		4.01		4.01	6	7.25
Pick up trash / debris & clean up				.327		13.10		13.10	19.60	23.50
Visual inspection				.327		13.10		13.10	19.60	23.50
Total				.753		30.21		30.21	45.20	54.25
0300 Minor roofing repairs - (2% of roof area)	1	2 ROFC	S.F.							
Set up, secure and take down ladder				.010		.40		.40	.60	.72
Cut out damaged area				.015		.60		.60	.90	1.09
Install roll roofing with mastic				.027	.80	1.01	.28	2.09	2.67	3.14
Clean up				.003		.10		.10	.16	.19
Total				.055	.80	2.11	.28	3.19	4.33	5.14
0400 Flashing repairs - (2 S.F. per sq. repaired)	1	2 ROFC	S.F.							
Set up, secure and take down ladder				.010		.40		.40	.60	.72
Remove adjacent roofing				.015		.60		.60	.90	1.09
Remove damaged metal flashing				.019		.76		.76	1.14	1.38
Install aluminum flashing				.072	1.35	2.89		4.24	5.80	6.90
Install roll roofing with mastic				.054	1.60	2.02	.55	4.17	5.35	6.25
Clean up				.003		.10		.10	.16	.19
Total				.173	2.95	6.77	.55	10.27	13.95	16.53
0500 Minor replacement - (25% of roof area)	15	2 ROFC	Sq.							
Set up, secure and take down ladder				.078		3.13		3.13	4.69	5.65
Remove strip-in				1.067		43		43	64	77.50
Renew striping with 2 ply felts				2.917	126	109	29.50	264.50	330	390
Clean up				.258		10.35		10.35	15.50	18.70
Total				4.319	126	165.48	29.50	320.98	414.19	491.85
0600 Total roof replacement	20	2 ROFC	Sq.							
Set up, secure and take down ladder				.020		.80		.80	1.20	1.45
Remove existing shingles				1.333		53.50		53.50	80	96.50
Remove flashing				.019		.76		.76	1.14	1.38
Install aluminum flashing				.144	2.70	5.78		8.48	11.60	13.80
Install 4 ply roll roofing				2.917	126	109	29.50	264.50	330	390
Clean up				.258		10.35		10.35	15.50	18.70
Total				4.691	128.70	180.19	29.50	338.39	439.44	521.83

For customer support on your Facilities Maintenance & Repair Cost Data, call 877.801.5724.

107

System Description	Freq. (Years)	Crew	Unit	Labor Hours	Material	Labor	Equipment	Total	Total In-House	Total w/O&P
0100 Debris removal & visual inspection	5	2 ROFC	M.S.F.							
Set up, secure and take down ladder				.100		4.01		4.01	6	7.25
Visual inspection				.327		13.10		13.10	19.60	23.50
Total				.427		17.11		17.11	25.60	30.75
0200 Fiberglass roof flashing repairs - (2 S.F. per sq. repaired)	15	2 ROFC	S.F.							
Set up, secure and take down ladder				.010		.40		.40	.60	.72
Cut flashing				.010		.40		.40	.60	.72
Install new flashing				.036	.35	1.46		1.81	2.57	3.07
Clean up				.004		.16		.16	.24	.29
Total				.060	.35	2.42		2.77	4.01	4.80
0400 Minor fiberglass roof panel replacement - (25% of roof area)	15	2 ROFC	S.F.							
Set up, secure and take down ladder				.010		.40		.40	.60	.72
Remove damaged panel				.030		1.20		1.20	1.80	2.17
Install new fiberglass panel				.064	2.10	2.99		5.09	6.20	7.45
Install strip-in with joint seal				.029	.02	1.33		1.35	1.77	2.18
Clean up				.004		.16		.16	.24	.29
Total				.137	2.12	6.08		8.20	10.61	12.81
0600 Total fiberglass roof replacement	20	2 ROFC	Sq.							
Set up, secure and take down ladder				.020		.80		.80	1.20	1.45
Remove existing fiberglass panels				2.500		100		100	150	181
Install new flashing				.073	.70	2.92		3.62	5.15	6.15
Install new fiberglass panel				6.400	210	299		509	620	745
Apply joint sealer to seams				.029	.02	1.33		1.35	1.77	2.18
Clean up				.390		15.65		15.65	23.50	28.50
Total				9.412	210.72	419.70		630.42	801.62	964.28

For customer support on your Facilities Maintenance & Repair Cost Data, call 877.801.5724.

B3013 170 Roof Edges

System Description	Freq. (Years)	Crew	Unit	Labor Hours	2015 Bare Costs				Total In-House	Total w/O&P
					Material	Labor	Equipment	Total		
1010 Replace Roof edges, aluminum, duranodic, .050" thick, 6" face	25	2 ROFC	L.F.							
Set up, secure, and take down ladder				.010		.40		.40	.60	.72
Remove adjacent roofing				.015		.60		.60	.90	1.09
Remove roof edges				.080		3.01		3.01	4	4.93
2" x 8" Edge blocking				.025	1.05	1.15		2.20	2.70	3.22
Cants, 4" x 4", treated timber, cut diagonally				.025	1.75	.99		2.74	3.41	3.97
Aluminum gravel stop with Duranodic finish				.089	10.88	4.98		15.86	18.30	21.50
Roof covering, splice sheet				.011	3.10	.42	.06	3.58	4.10	4.70
Total				.254	16.78	11.55	.06	28.39	34.01	40.13
1020 Replace Roof edges, copper, 20 oz., 8" face	35	2 ROFC	L.F.							
Set up, secure, and take down ladder				.010		.40		.40	.60	.72
Remove adjacent roofing				.015		.60		.60	.90	1.09
Remove roof edge				.080		3.01		3.01	4	4.93
2" x 8" Edge blocking				.025	1.05	1.15		2.20	2.70	3.22
Cants, 4" x 4", treated timber, cut diagonally				.025	1.75	.99		2.74	3.41	3.97
Copper gravel stop				.139	15.30	5.58		20.88	25	29
Roof covering, splice sheet				.011	3.10	.42	.06	3.58	4.10	4.70
Total				.305	21.20	12.15	.06	33.41	40.71	47.63
1030 Replace Roof edges, galvanized, 20 ga., 6" face	25	2 ROFC	L.F.							
Set up, secure, and take down ladder				.010		.40		.40	.60	.72
Remove adjacent roofing				.015		.60		.60	.90	1.09
Remove roof edge				.080		3.01		3.01	4	4.93
2" x 8" edge blocking				.025	1.05	1.15		2.20	2.70	3.22
Cants, 4" x 4", treated timber, cut diagonally				.025	1.75	.99		2.74	3.41	3.97
Galvanized steel gravel stop				.092	1.88	3.71		5.59	7.65	9.05
Roof covering, splice sheet				.011	3.10	.42	.06	3.58	4.10	4.70
Total				.258	7.78	10.28	.06	18.12	23.36	27.68

For customer support on your Facilities Maintenance & Repair Cost Data, call 877.801.5724.

109

B3013 620 Gutters and Downspouts

System Description	Freq. (Years)	Crew	Unit	Labor Hours	2015 Bare Costs				Total In-House	Total w/O&P
					Material	Labor	Equipment	Total		
1010 Replace aluminum gutter, enameled, 5" K type, .027" thick	40	2 SHEE	L.F.							
Set up, secure and take down ladder				.010		.40		.40	.60	.72
Remove gutter				.033		1.25		1.25	1.66	2.05
Install new gutter				.064	2.66	3.58		6.24	7.45	9
Clean up				.003		.10		.10	.16	.19
Total				.110	2.66	5.33		7.99	9.87	11.96
1020 Replace aluminum gutter, enameled, 5" K type, .032" thick	40	2 SHEE	L.F.							
Set up, secure and take down ladder				.010		.40		.40	.60	.72
Remove gutter				.033		1.25		1.25	1.66	2.05
Install new gutter				.064	3.23	3.58		6.81	8.10	9.70
Clean up				.003		.10		.10	.16	.19
Total				.110	3.23	5.33		8.56	10.52	12.66
1060 Replace aluminum downspout, 2" x 3", .024" thick	25	2 SHEE	L.F.							
Set up, secure and take down ladder				.010		.40		.40	.60	.72
Remove downspout				.023		.86		.86	1.14	1.41
Install new downspout				.044	2.05	2.49		4.54	5.40	6.50
Clean up				.003		.10		.10	.16	.19
Total				.080	2.05	3.85		5.90	7.30	8.82
1070 Replace aluminum downspout, 3" x 4", .024" thick	25	2 SHEE	L.F.							
Set up, secure and take down ladder				.010		.40		.40	.60	.72
Remove downspout				.023		.86		.86	1.14	1.41
Install new downspout				.057	2.33	3.20		5.53	6.60	7.95
Clean up				.003		.10		.10	.16	.19
Total				.093	2.33	4.56		6.89	8.50	10.27
2010 Replace galvanized steel gutter, 5" box type, 28 ga.	40	2 SHEE	L.F.							
Set up, secure and take down ladder				.010		.40		.40	.60	.72
Remove gutter				.033		1.25		1.25	1.66	2.05
Install new gutter				.064	2.10	3.58		5.68	6.85	8.30
Clean up				.003		.10		.10	.16	.19
Total				.110	2.10	5.33		7.43	9.27	11.26

For customer support on your Facilities Maintenance & Repair Cost Data, call 877.801.5724.

B3013 620 Gutters and Downspouts

System Description	Freq. (Years)	Crew	Unit	Labor Hours	2015 Bare Costs				Total In-House	Total w/O&P
					Material	Labor	Equipment	Total		
2020 Replace galvanized steel gutter, 6″ box type, 26 ga.	40	2 SHEE	L.F.							
Set up, secure and take down ladder				.010		.40		.40	.60	.72
Remove gutter				.033		1.25		1.25	1.66	2.05
Install new gutter				.064	2.71	3.58		6.29	7.50	9.05
Clean up				.003		.10		.10	.16	.19
Total				.110	2.71	5.33		8.04	9.92	12.01
2060 Replace round corrugated galvanized downspout, 3″ diameter	25	2 SHEE	L.F.							
Set up, secure and take down ladder				.010		.40		.40	.60	.72
Remove downspout				.023		.86		.86	1.14	1.41
Install new downspout				.042	2.08	2.36		4.44	5.30	6.30
Clean up				.003		.10		.10	.16	.19
Total				.078	2.08	3.72		5.80	7.20	8.62
2070 Replace round corrugated galvanized downspout, 4″ diameter	25	2 SHEE	L.F.							
Set up, secure and take down ladder				.010		.40		.40	.60	.72
Remove downspout				.023		.86		.86	1.14	1.41
Install new downspout				.055	2.52	3.09		5.61	6.70	8
Clean up				.003		.10		.10	.16	.19
Total				.091	2.52	4.45		6.97	8.60	10.32
3020 Replace vinyl box gutter, 5″ wide	20	1 CARP	L.F.							
Set up, secure and take down ladder				.010		.40		.40	.60	.72
Remove gutter				.033		1.25		1.25	1.66	2.05
Install new gutter				.070	1.55	3.27		4.82	6.05	7.30
Clean up				.003		.10		.10	.16	.19
Total				.116	1.55	5.02		6.57	8.47	10.26
3060 Replace vinyl downspout, 2″ x 3″	15	1 CARP	L.F.							
Set up, secure and take down ladder				.010		.40		.40	.60	.72
Remove downspout				.023		.86		.86	1.14	1.41
Install new downspout				.038	2.08	2.13		4.21	4.99	5.95
Clean up				.003		.10		.10	.16	.19
Total				.074	2.08	3.49		5.57	6.89	8.27

For customer support on your Facilities Maintenance & Repair Cost Data, call 877.801.5724.

111

B3013 620   Gutters and Downspouts

System Description	Freq. (Years)	Crew	Unit	Labor Hours	2015 Bare Costs				Total In-House	Total w/O&P
					Material	Labor	Equipment	Total		
4020 **Replace 6" copper box gutters, 16 oz.**	50	1 SHEE	L.F.							
Set up, secure and take down ladder				.010		.40		.40	.60	.72
Remove gutter				.033		1.25		1.25	1.66	2.05
Install new gutter				.064	8.50	3.58		12.08	13.90	16.30
Clean up				.003		.10		.10	.16	.19
Total				.110	8.50	5.33		13.83	**16.32**	**19.26**
4060 **Replace copper downspouts, 2" x 3", 16 oz.**	40	1 SHEE	L.F.							
Set up, secure and take down ladder				.010		.40		.40	.60	.72
Remove gutter				.023		.86		.86	1.14	1.41
Install new gutter				.042	8.55	2.36		10.91	12.40	14.40
Clean up				.003		.10		.10	.16	.19
Total				.078	8.55	3.72		12.27	**14.30**	**16.72**

For customer support on your Facilities Maintenance & Repair Cost Data, call 877.801.5724.

B3023 110 Single Unit Glass Skylight

	System Description	Freq. (Years)	Crew	Unit	Labor Hours	2015 Bare Costs				Total In-House	Total w/O&P
						Material	Labor	Equipment	Total		
1010	**Repair glass skylight glazing**	6	1 CARP	S.F.							
	Set up, secure and take down ladder				.130		6.10		6.10	8.10	10
	Remove glazing				.069		2.61		2.61	3.46	4.27
	Install new glazing				.291	25	13.10		38.10	44.50	53
	Total				**.490**	**25**	**21.81**		**46.81**	**56.06**	**67.27**
1020	**Replace continuous skylight and structure**	40	2 CARP	C.S.F.							
	Set up and secure scaffold				.750		35.25		35.25	47	57.50
	Remove unit skylight				5.202		244		244	325	400
	Inspect / remove flashing				3.152	210	126		336	420	490
	Replace flashing				3.152	210	126		336	420	490
	Install new unit				12.960	2,150	605		2,755	3,150	3,650
	Remove scaffold				.750		35.25		35.25	47	57.50
	Total				**25.966**	**2,570**	**1,171.50**		**3,741.50**	**4,409**	**5,145**
1040	**Replace skylight and structure, double glazed, 10 to 20 S.F.**	40	G-3	S.F.							
	Set up, secure and take down ladder				.010		.40		.40	.60	.72
	Remove skylight, plstc domes,flush/curb mtd				.081		3.79		3.79	4.90	6.05
	10 S.F. to 20 S.F., double				.102	27.50	4.75		32.25	36.50	42
	Total				**.193**	**27.50**	**8.94**		**36.44**	**42**	**48.77**
1080	**Replace skylight and structure, double glazed, 30 to 65 S.F.**	40	G-3	S.F.							
	Set up, secure and take down ladder				.010		.40		.40	.60	.72
	Remove skylight, plstc domes,flush/curb mtd				.081		3.79		3.79	4.90	6.05
	30 S.F. to 65 S.F., double				.069	26.50	3.22		29.72	33.50	38.50
	Total				**.160**	**26.50**	**7.41**		**33.91**	**39**	**45.27**

B3023 120 Hatches

	System Description	Freq. (Years)	Crew	Unit	Labor Hours	2015 Bare Costs				Total In-House	Total w/O&P
						Material	Labor	Equipment	Total		
1020	**Replace aluminum roof hatch & structure**	40	2 CARP	Ea.							
	Set up, secure and take down ladder				.130		6.10		6.10	8.10	10
	Remove roof hatch				1.000		37.50		37.50	50	61.50
	Install 2'-6" x 3'-0" aluminum hatch				3.200	1,025	150		1,175	1,325	1,525
	Total				**4.330**	**1,025**	**193.60**		**1,218.60**	**1,383.10**	**1,596.50**

For customer support on your Facilities Maintenance & Repair Cost Data, call 877.801.5724.

113

B3023 120   Hatches

	System Description	Freq. (Years)	Crew	Unit	Labor Hours	2015 Bare Costs				Total In-House	Total w/O&P
						Material	Labor	Equipment	Total		
1120	**Replace galvanized roof hatch & structure**	40	2 CARP	Ea.							
	Set up, secure and take down ladder				.130		6.10		6.10	8.10	10
	Selective demo, therm & moist protect, roof acc, roof hatch, to 10 S.F.				1.000		37.50		37.50	50	61.50
	Install 2'-6" x 3'-0" galvanized roof hatch				3.200	640	150		790	895	1,050
	Total				4.330	640	193.60		833.60	953.10	1,121.50

B3023 130   Smoke Hatches

	System Description	Freq. (Years)	Crew	Unit	Labor Hours	2015 Bare Costs				Total In-House	Total w/O&P
						Material	Labor	Equipment	Total		
1010	**Replace galvanized smoke hatch single unit 4' x 4'**	40	2 CARP	Ea.							
	Set up and secure scaffold				.750		35.25		35.25	47	57.50
	Inspect / remove flashing				.032	2.10	1.26		3.36	4.20	4.91
	Replace flashing				.032	2.10	1.26		3.36	4.20	4.91
	Galvanized steel cover and frame				2.462	1,800	115		1,915	2,125	2,425
	Remove scaffold				.750		35.25		35.25	47	57.50
	Total				4.025	1,804.20	188.02		1,992.22	2,227.40	2,549.82
1020	**Replace aluminum smoke hatch single unit, 4' x 8'**	40	2 CARP	Ea.							
	Set up and secure scaffold				.750		35.25		35.25	47	57.50
	Smoke vent 4' x 8'				4.000		150		150	200	247
	Inspect / remove flashing				.756	50.40	30.24		80.64	101	118
	Replace flashing				.756	50.40	30.24		80.64	101	118
	4' x 8' aluminum cover and frame				5.246	2,800	245		3,045	3,400	3,900
	Remove scaffold				.750		35.25		35.25	47	57.50
	Total				12.259	2,900.80	525.98		3,426.78	3,896	4,498

For customer support on your Facilities Maintenance & Repair Cost Data, call 877.801.5724.

C10 INTERIOR CONSTRUCTION | C1013 | Partitions

C1013 110 | Concrete Block, Painted

	System Description	Freq. (Years)	Crew	Unit	Labor Hours	2015 Bare Costs				Total In-House	Total w/O&P
						Material	Labor	Equipment	Total		
0010	**Repair 8" concrete block wall - (2% of walls)**	25	1 BRIC	C.S.F.							
	Remove block				6.420		244	37	281	360	440
	Replace block				11.111	286	477		763	945	1,125
	Total				17.531	286	721	37	1,044	**1,305**	**1,565**
0020	**Refinish concrete block wall**	4	1 PORD	C.S.F.							
	Prepare surface				.250		10		10	13	16
	Fill & paint block, brushwork				1.376	20	56		76	94	114
	Place and remove mask and drops				.370		15		15	19	24
	Total				1.996	20	81		101	**126**	**154**
0040	**Replace 8" concrete block wall**	75	2 BRIC	C.S.F.							
	Set up and secure scaffold				.750		35.25		35.25	47	57.50
	Remove block				6.420		244	37	281	360	440
	Replace block				11.111	286	477		763	945	1,125
	Remove scaffold				.750		35.25		35.25	47	57.50
	Total				19.031	286	791.50	37	1,114.50	**1,399**	**1,680**

C1013 115 | Glazed CMU Interior Wall Finish

	System Description	Freq. (Years)	Crew	Unit	Labor Hours	2015 Bare Costs				Total In-House	Total w/O&P
						Material	Labor	Equipment	Total		
0010	**Repair 4" glazed C.M.U. wall - (2% of walls)**	25	1 BRIC	C.S.F.							
	Remove damaged blocks				5.200		198	30	228	293	355
	Install new blocks				15.072	850	645		1,495	1,775	2,125
	Total				20.272	850	843	30	1,723	**2,068**	**2,480**
0020	**Replace 4" glazed C.M.U. wall**	75	2 BRIC	C.S.F.							
	Set up and secure scaffold				.750		35.25		35.25	47	57.50
	Remove blocks				5.200		198	30	228	293	355
	Install new blocks				15.072	850	645		1,495	1,775	2,125
	Remove scaffold				.750		35.25		35.25	47	57.50
	Total				21.772	850	913.50	30	1,793.50	**2,162**	**2,595**

C1013 120 — Glass Block

	System Description	Freq. (Years)	Crew	Unit	Labor Hours	2015 Bare Costs Material	Labor	Equipment	Total	Total In-House	Total w/O&P
0010	**Repair glass block wall - (2% of walls)**	25	1 BRIC	C.S.F.							
	Remove damaged glass block				17.829		680		680	905	1,125
	Repair / replace blocks				45.217	2,650	1,940		4,590	5,450	6,475
	Total				63.046	2,650	2,620		5,270	6,355	**7,600**
0020	**Replace glass block wall**	75	2 BRIC	C.S.F.							
	Set up and secure scaffold				.750		35.25		35.25	47	57.50
	Remove damaged glass block				17.829		680		680	905	1,125
	Repair / replace blocks				45.217	2,650	1,940		4,590	5,450	6,475
	Remove scaffold				.750		35.25		35.25	47	57.50
	Total				64.546	2,650	2,690.50		5,340.50	6,449	**7,715**

C1013 125 — Stone Veneer

	System Description	Freq. (Years)	Crew	Unit	Labor Hours	2015 Bare Costs Material	Labor	Equipment	Total	Total In-House	Total w/O&P
0010	**Repair 4" stone veneer wall - (2% of walls)**	25	1 BRIC	C.S.F.							
	Remove damaged stone				7.429	56	343		399	515	630
	Replace stone				41.600	1,900	1,785		3,685	4,450	5,275
	Total				49.029	1,956	2,128		4,084	4,965	**5,905**
0020	**Replace 4" stone veneer wall**	75	2 BRIC	C.S.F.							
	Set up and secure scaffold				.750		35.25		35.25	47	57.50
	Remove damaged stone				7.429	56	343		399	515	630
	Replace stone				41.600	1,900	1,785		3,685	4,450	5,275
	Remove scaffold				.750		35.25		35.25	47	57.50
	Total				50.529	1,956	2,198.50		4,154.50	5,059	**6,020**

For customer support on your Facilities Maintenance & Repair Cost Data, call 877.801.5724.

C1013 230 Demountable Partitions

	System Description	Freq. (Years)	Crew	Unit	Labor Hours	2015 Bare Costs				Total In-House	Total w/O&P
						Material	Labor	Equipment	Total		
1010	**Remove and reinstall demountable partitions**	5	2 CARP	C.L.F.							
	Remove demountable partitions for reuse				16.000		750		750	1,000	1,225
	Vinly clad gypsum wall system, material only to replace damaged					1,200			1,200	1,325	1,500
	Install new and reused vinyl clad gypsum in new configuration				34.286		1,610		1,610	2,150	2,650
	18 ga. prefinished 3'-0" x 6'-8" to 4-7/8" deep frame				3.000	492	141		633	730	845
	Reinstall doors				3.200		150		150	200	246
	Cleanup				1.000		47		47	62	77
	Total				57.486	1,692	2,698		4,390	5,467	6,543

C1013 544 Channel Frame Wire Mesh Wall

	System Description	Freq. (Years)	Crew	Unit	Labor Hours	2015 Bare Costs				Total In-House	Total w/O&P
						Material	Labor	Equipment	Total		
0010	**Repair channel frame wire mesh wall - (2% of walls)**	20	1 SSWK	C.S.F.							
	Remove damaged 1-1/2" diamond mesh panel				12.000		459		459	610	755
	Replace with new 1-1/2" diamond mesh panel				2.857	475	133.75		608.75	700	815
	Total				14.857	475	592.75		1,067.75	1,310	1,570
0020	**Refinish channel frame wire mesh wall**	5	1 PORD	C.S.F.							
	Prepare surface				.370		15		15	19	24
	Scrape surface				1.250		50		50	65	81
	Refinish surface, brushwork, primer + 1 coat				1.538	44	62		106	129	155
	Total				3.158	44	127		171	213	260

C1013 730 Plate Glass Wall - Interior

	System Description	Freq. (Years)	Crew	Unit	Labor Hours	2015 Bare Costs				Total In-House	Total w/O&P
						Material	Labor	Equipment	Total		
0010	**Repair plate glass interior wall - (2% of total)**	25	1 CARP	C.S.F.							
	Remove plate glass & frame				6.933		261		261	345	425
	Repair damaged glass				23.111	6,900	1,130		8,030	9,200	10,600
	Total				30.044	6,900	1,391		8,291	9,545	11,025

For customer support on your Facilities Maintenance & Repair Cost Data, call 877.801.5724.

117

C1013 730   Plate Glass Wall - Interior

System Description	Freq. (Years)	Crew	Unit	Labor Hours	2015 Bare Costs				Total In-House	Total w/O&P
					Material	Labor	Equipment	Total		
0020 Replace plate glass interior wall	50	2 CARP	C.S.F.							
Set up and secure scaffold				.750		35.25		35.25	47	57.50
Remove plate glass & frame				6.933		261		261	345	425
Repair damaged glass				23.111	6,900	1,130		8,030	9,200	10,600
Remove scaffold				.750		35.25		35.25	47	57.50
Total				31.544	6,900	1,461.50		8,361.50	9,639	11,140

C1023 108 | Fully Glazed Wooden Doors

	System Description	Freq. (Years)	Crew	Unit	Labor Hours	2015 Bare Costs				Total In-House	Total w/O&P
						Material	Labor	Equipment	Total		
1010	**Replace insulating glass (3% of glass)**	1	1 CARP	S.F.							
	Remove damaged glass				.052		1.95		1.95	2.59	3.20
	Install new insulating glass				.277	29.50	12.50		42	49	57.50
	Total				.329	29.50	14.45		43.95	51.59	**60.70**
1020	**Repair fully glazed wood door**	10	1 CARP	Ea.							
	Remove lockset				.520		24.50		24.50	32.50	40
	Replace lockset				1.040	108	49		157	184	215
	Oil hinges				.065		3.03		3.03	4.03	4.96
	Oil door closer				.065		3.03		3.03	4.03	4.96
	Total				1.689	108	79.56		187.56	224.56	**264.92**
1030	**Refinish 3'-0" x 7'-0" fully glazed wood door**	4	1 PORD	Ea.							
	Prepare interior door for painting				.433		17.50		17.50	22.50	28
	Paint door, roller + brush, 1 coat				1.333	2.37	54		56.37	72.50	89.50
	Total				1.767	2.37	71.50		73.87	95	**117.50**
1040	**Replace 3'-0" x 7'-0" fully glazed wood door**	40	1 CARP	Ea.							
	Remove interior door				.520		19.55		19.55	26	32
	Remove interior door frame				1.300		61		61	81	100
	Remove door closer				.167		7.85		7.85	10.40	12.85
	Install new wood door and frame				2.600	712.50	122		834.50	950	1,100
	Reinstall door closer				1.735		81.50		81.50	108	134
	Total				6.322	712.50	291.90		1,004.40	1,175.40	**1,378.85**

C1023 110 | Steel Interior Door, Painted

	System Description	Freq. (Years)	Crew	Unit	Labor Hours	2015 Bare Costs				Total In-House	Total w/O&P
						Material	Labor	Equipment	Total		
1010	**Repair steel painted door**	14	1 CARP	Ea.							
	Remove lockset				.520		24.50		24.50	32.50	40
	Replace lockset				1.040	108	49		157	184	215
	Oil hinges				.065		3.03		3.03	4.03	4.96
	Oil door closer				.065		3.03		3.03	4.03	4.96
	Total				1.689	108	79.56		187.56	224.56	**264.92**

For customer support on your Facilities Maintenance & Repair Cost Data, call 877.801.5724.

119

C1023 110 | Steel Interior Door, Painted

System Description	Freq. (Years)	Crew	Unit	Labor Hours	2015 Bare Costs				Total In-House	Total w/O&P
					Material	Labor	Equipment	Total		
1020 Refinish 3'-0" x 7'-0" steel door, painted	4	1 PORD	Ea.							
Prepare interior door for painting				.433		17.50		17.50	22.50	28
Paint door, roller + brush, 1 coat				.800	4.40	32.50		36.90	46.50	57.50
Total				1.233	4.40	50		54.40	69	85.50
1030 Replace 3'-0" x 7'-0" steel door, painted	60	1 CARP	Ea.							
Remove interior door				.520		19.55		19.55	26	32
Remove interior door frame				1.300		61		61	81	100
Remove door closer				.167		7.85		7.85	10.40	12.85
Install new steel door				1.224	640	57.50		697.50	780	895
Install new steel door frame				1.000	151	47		198	229	266
Reinstall door closer				1.735		81.50		81.50	108	134
Total				5.946	791	274.40		1,065.40	1,234.40	1,439.85
2010 Safety glass replace, (3% of glass)	1	1 CARP	S.F.							
Remove damaged glass				.052		1.95		1.95	2.59	3.20
Install new safety glass				.173	8.85	7.80		16.65	20	24
Total				.225	8.85	9.75		18.60	22.59	27.20
2030 Refinish, 3'-0" x 7'-0" steel door w/ safety glass	4	1 PORD	Ea.							
Prepare interior door for painting				.433		17.50		17.50	22.50	28
Paint door, roller + brush, 1 coat				.800	4.40	32.50		36.90	46.50	57.50
Total				1.233	4.40	50		54.40	69	85.50
2040 Replace 3'-0" x 7'-0" steel door & frame w/ vision lite	60	1 CARP	Ea.							
Remove interior door				.520		19.55		19.55	26	32
Remove interior door frame				1.300		61		61	81	100
Remove door closer				.167		7.85		7.85	10.40	12.85
Install new steel door				1.224	640	57.50		697.50	780	895
Install new steel door frame				1.301	151	61		212	247	289
Reinstall door closer				1.735		81.50		81.50	108	134
Vision lite					94.50			94.50	104	118
Total				6.246	885.50	288.40		1,173.90	1,356.40	1,580.85

For customer support on your Facilities Maintenance & Repair Cost Data, call 877.801.5724.

C1023 110 Steel Interior Door, Painted

	System Description	Freq. (Years)	Crew	Unit	Labor Hours	2015 Bare Costs Material	Labor	Equipment	Total	Total In-House	Total w/O&P
3010	**Repair 3'-0" x 7'-0" steel sliding door**	14	1 CARP	Ea.							
	Oil door closer				.065		3.03		3.03	4.03	4.96
	Oil / inspect track / glide wheels				.065		3.03		3.03	4.03	4.96
	Remove door track / glide wheels (25% of total / 14 years)				.220		10.38		10.38	13.75	16.90
	Install new door track / glide wheels (25% of total / 14 years)				.440	13.13	20.63		33.76	42	50
	Total				.789	13.13	37.07		50.20	63.81	76.82
3020	**Refinish 3'-0" x 7'-0" steel sliding door**	4	1 PORD	Ea.							
	Prepare interior door for painting				.433		17.50		17.50	22.50	28
	Paint door, roller + brush, 1 coat				.800	4.40	32.50		36.90	46.50	57.50
	Total				1.233	4.40	50		54.40	69	85.50
3030	**Replace 3'-0" x 7'-0" steel sliding door & frame**	60	1 CARP	Ea.							
	Remove interior door				.520		19.55		19.55	26	32
	Remove interior door frame				1.300		61		61	81	100
	Remove door closer				.167		7.85		7.85	10.40	12.85
	Install new steel door				1.224	640	57.50		697.50	780	895
	Install new steel sliding door frame				3.200	239	170	14.60	423.60	530	620
	Install new door track / glide wheels				1.760	52.50	82.50		135	168	201
	Install door closer				1.735		81.50		81.50	108	134
	Total				9.905	931.50	479.90	14.60	1,426	1,703.40	1,994.85
4010	**Replace wire glass (3% of glass)**	1	1 CARP	S.F.							
	Remove damaged glass				.052		1.95		1.95	2.59	3.20
	Install new wire glass				.179	27.26	8.06		35.32	40.50	47
	Total				.231	27.26	10.01		37.27	43.09	50.20
4030	**Refinish 3'-0" x 7'-0" steel half glass door**	4	1 PORD	Ea.							
	Prepare interior door for painting				.433		17.50		17.50	22.50	28
	Paint door, roller + brush, 1 coat				.800	4.40	32.50		36.90	46.50	57.50
	Total				1.233	4.40	50		54.40	69	85.50

For customer support on your Facilities Maintenance & Repair Cost Data, call 877.801.5724.

121

C1023 110 — Steel Interior Door, Painted

System Description	Freq. (Years)	Crew	Unit	Labor Hours	2015 Bare Costs				Total In-House	Total w/O&P
					Material	Labor	Equipment	Total		
4040 **Replace 3'-0" x 7'-0" steel half glass door & frame**	60	1 CARP	Ea.							
Remove interior door				.520		19.55		19.55	26	32
Remove door closer				.167		7.85		7.85	10.40	12.85
Remove interior door frame				1.300		61		61	81	100
Install new steel door frame				1.301	151	61		212	247	289
Install new half glass interior door				1.224	625	57.50		682.50	765	875
Install new safety glass				1.040	53.10	46.80		99.90	120	143
Reinstall door closer				1.735		81.50		81.50	108	134
Total				7.287	829.10	335.20		1,164.30	1,357.40	1,585.85

C1023 111 — Steel Interior Door, Unpainted

System Description	Freq. (Years)	Crew	Unit	Labor Hours	2015 Bare Costs				Total In-House	Total w/O&P
					Material	Labor	Equipment	Total		
1010 **Repair steel door, unpainted**	14	1 CARP	Ea.							
Remove lockset				.520		24.50		24.50	32.50	40
Replace lockset				1.040	108	49		157	184	215
Oil hinges				.065		3.03		3.03	4.03	4.96
Oil door closer				.065		3.03		3.03	4.03	4.96
Total				1.689	108	79.56		187.56	224.56	264.92
1020 **Replace 3'-0" x 7'-0" steel door & frame, unpainted**	60	1 CARP	Ea.							
Remove interior door				.520		19.55		19.55	26	32
Remove interior door frame				1.300		61		61	81	100
Remove door closer				.167		7.85		7.85	10.40	12.85
Install new steel door				1.224	640	57.50		697.50	780	895
Install new steel door frame				1.000	151	47		198	229	266
Reinstall door closer				1.735		81.50		81.50	108	134
Total				5.946	791	274.40		1,065.40	1,234.40	1,439.85
2010 **Replace safety glass (3% of glass)**	1	1 CARP	S.F.							
Remove damaged glass				.052		1.95		1.95	2.59	3.20
Install new safety glass				.173	8.85	7.80		16.65	20	24
Total				.225	8.85	9.75		18.60	22.59	27.20

For customer support on your Facilities Maintenance & Repair Cost Data, call 877.801.5724.

System Description	Freq. (Years)	Crew	Unit	Labor Hours	2015 Bare Costs				Total In-House	Total w/O&P
					Material	Labor	Equipment	Total		
2030 Replace 3'-0" x 7'-0" steel half glass door & frame, unpainted	60	1 CARP	Ea.							
Remove interior door				.520		19.55		19.55	26	32
Remove interior door frame				1.300		61		61	81	100
Remove door closer				.167		7.85		7.85	10.40	12.85
Install new half glass interior door				1.224	625	57.50		682.50	765	875
Install new safety glass				1.560	79.65	70.20		149.85	180	214
Install new steel door frame				1.301	151	61		212	247	289
Reinstall door closer				1.735		81.50		81.50	108	134
Total				7.807	855.65	358.60		1,214.25	1,417.40	1,656.85
3010 Repair 3'-0" x 7'-0" steel sliding door, unpainted	14	1 CARP	Ea.							
Oil door closer				.065		3.03		3.03	4.03	4.96
Oil / inspect track / glide wheels				.065		3.03		3.03	4.03	4.96
Remove door track / glide wheels (25% of total / 14 years)				.220		10.38		10.38	13.75	16.90
Install new door track / glide wheels (25% of total / 14 years)				.440	13.13	20.63		33.76	42	50
Total				.789	13.13	37.07		50.20	63.81	76.82
3020 Replace 3'-0" x 7'-0" steel sliding door, unpainted	60	1 CARP	Ea.							
Remove interior door				.520		19.55		19.55	26	32
Remove interior door frame				1.300		61		61	81	100
Remove door closer				.167		7.85		7.85	10.40	12.85
Install new steel door				1.224	640	57.50		697.50	780	895
Install new steel sliding door frame				3.200	239	170	14.60	423.60	530	620
Install new door track / glide wheels				1.760	52.50	82.50		135	168	201
Reinstall door closer				1.735		81.50		81.50	108	134
Total				9.905	931.50	479.90	14.60	1,426	1,703.40	1,994.85
4010 Replace wire glass (3% of glass)	1	1 CARP	S.F.							
Remove damaged glass				.052		1.95		1.95	2.59	3.20
Install new wire glass				.179	27.26	8.06		35.32	40.50	47
Total				.231	27.26	10.01		37.27	43.09	50.20

C1023 112 Bi-Fold Interior Doors

System Description	Freq. (Years)	Crew	Unit	Labor Hours	2015 Bare Costs				Total In-House	Total w/O&P
					Material	Labor	Equipment	Total		
2010 Repair 2'-6" x 6'-8" bi-fold louvered door, wood	15	1 CARP	Ea.							
Remove interior door				.520		19.55		19.55	26	32
Remove damaged stile				.078		3.69		3.69	4.89	6.05
Remove damaged slat				.039		1.82		1.82	2.42	2.98
Install new slat				.077	.70	3.62		4.32	5.60	6.80
Install new stile				.157	2.37	7.38		9.75	12.40	15
Reinstall door				.667		31.50		31.50	41.50	51.50
Oil / lubricate door				.065		3.03		3.03	4.03	4.96
Total				1.602	3.07	70.59		73.66	96.84	119.29
2020 Refinish 2'-6" x 6'-8" louvered door, wood	8	1 PORD	Ea.							
Prepare door for finish				.967		39		39	50.50	62.50
Paint door, roller + brush, 1 coat				1.143	5.30	46		51.30	65.50	80.50
Total				2.110	5.30	85		90.30	116	143
2040 Replace 2'-6" x 6'-8" wood louver bi-fold door & frame	24	1 CARP	Ea.							
Remove doors				.520		19.55		19.55	26	32
Install new door				1.600	278	75		353	405	470
Install new interior door frame				.998	88.56	46.80		135.36	159	188
Total				3.118	366.56	141.35		507.91	590	690

C1023 114 Aluminum Interior Doors

System Description	Freq. (Years)	Crew	Unit	Labor Hours	2015 Bare Costs				Total In-House	Total w/O&P
					Material	Labor	Equipment	Total		
1010 Repair aluminum door	12	1 CARP	Ea.							
Remove lockset				.520		24.50		24.50	32.50	40
Replace lockset				1.040	108	49		157	184	215
Oil hinges				.065		3.03		3.03	4.03	4.96
Oil door closer				.065		3.03		3.03	4.03	4.96
Total				1.689	108	79.56		187.56	224.56	264.92

For customer support on your Facilities Maintenance & Repair Cost Data, call 877.801.5724.

	System Description	Freq. (Years)	Crew	Unit	Labor Hours	2015 Bare Costs				Total In-House	Total w/O&P
						Material	Labor	Equipment	Total		
1020	**Replace 3'-0" x 7'-0" aluminum door & frame**	50	1 CARP	Ea.							
	Remove interior door				.520		19.55		19.55	26	32
	Remove interior door frame				1.300		61		61	81	100
	Remove door closer				.167		7.85		7.85	10.40	12.85
	Install new aluminum door				1.224	1,450	57.50		1,507.50	1,675	1,900
	Install new aluminum door frame				2.971	600	156		756	895	1,025
	Reinstall door closer				1.735		81.50		81.50	108	134
	Total				7.917	2,050	383.40		2,433.40	2,795.40	3,203.85
2010	**Replace safety glass (3% of glass)**	1	1 CARP	S.F.							
	Remove damaged glass				.052		1.95		1.95	2.59	3.20
	Install new safety glass				.173	8.85	7.80		16.65	20	24
	Total				.225	8.85	9.75		18.60	22.59	27.20
2030	**Replace 3'-0" x 7'-0" aluminum door & frame w/ vision lite**	50	1 CARP	Ea.							
	Remove interior door				.520		19.55		19.55	26	32
	Remove interior door frame				1.300		61		61	81	100
	Remove door closer				.167		7.85		7.85	10.40	12.85
	Install new aluminum door				1.224	1,450	57.50		1,507.50	1,675	1,900
	Install new aluminum door frame				2.971	600	156		756	895	1,025
	Reinstall door closer				1.735		81.50		81.50	108	134
	Vision lite					94.50			94.50	104	118
	Total				7.917	2,144.50	383.40		2,527.90	2,899.40	3,321.85
3010	**Replace insulating glass (3% of glass)**	1	1 CARP	S.F.							
	Remove damaged glass				.052		1.95		1.95	2.59	3.20
	Install new insulating glass				.277	29.50	12.50		42	49	57.50
	Total				.329	29.50	14.45		43.95	51.59	60.70
3020	**Repair 3'-0" x 7'-0" aluminum sliding door**	14	1 CARP	Ea.							
	Oil door closer				.065		3.03		3.03	4.03	4.96
	Oil / inspect track / glide wheels				.065		3.03		3.03	4.03	4.96
	Remove door track / glide wheels (25% of total / 14 years)				.220		10.38		10.38	13.75	16.90
	Install new door track / glide wheels (25% of total / 14 years)				.440	13.13	20.63		33.76	42	50
	Total				.789	13.13	37.07		50.20	63.81	76.82

For customer support on your Facilities Maintenance & Repair Cost Data, call 877.801.5724.

125

C10 INTERIOR CONSTRUCTION — C1023 — Interior Doors

C1023 114 | Aluminum Interior Doors

	System Description	Freq. (Years)	Crew	Unit	Labor Hours	2015 Bare Costs — Material	Labor	Equipment	Total	Total In-House	Total w/O&P
3030	**Replace 3'-0" x 7'-0" aluminum sliding door**	50	1 CARP	Ea.							
	Remove interior door				.520		19.55		19.55	26	32
	Remove interior door frame				1.300		61		61	81	100
	Remove door closer				.167		7.85		7.85	10.40	12.85
	Install new aluminum door				1.224	1,450	57.50		1,507.50	1,675	1,900
	Install new aluminum door frame				2.971	600	156		756	895	1,025
	Install new door track / glide wheels				1.760	52.50	82.50		135	168	201
	Install door closer				1.735		81.50		81.50	108	134
	Total				9.677	2,102.50	465.90		2,568.40	2,963.40	3,404.85

C1023 116 | Fully Glazed Aluminum Doors

	System Description	Freq. (Years)	Crew	Unit	Labor Hours	2015 Bare Costs — Material	Labor	Equipment	Total	Total In-House	Total w/O&P
1010	**Replace insulating glass (3% of glass)**	1	1 CARP	S.F.							
	Remove damaged glass				.052		1.95		1.95	2.59	3.20
	Install new insulating glass				.277	29.50	12.50		42	49	57.50
	Total				.329	29.50	14.45		43.95	51.59	60.70
1020	**Repair 3'-0" x 7'-0" fully glazed aluminum door**	12	1 CARP	Ea.							
	Remove lockset				.520		24.50		24.50	32.50	40
	Replace lockset				1.040	108	49		157	184	215
	Oil hinges				.065		3.03		3.03	4.03	4.96
	Oil door closer				.065		3.03		3.03	4.03	4.96
	Total				1.689	108	79.56		187.56	224.56	264.92
1030	**Replace 3'-0" x 7'-0" fully glazed aluminum door**	50	1 CARP	Ea.							
	Remove interior door				.520		19.55		19.55	26	32
	Remove interior door frame				1.300		61		61	81	100
	Remove door closer				.167		7.85		7.85	10.40	12.85
	Install new aluminum door and frame				10.458	850	550		1,400	1,750	2,050
	Install new safety glass				3.640	185.85	163.80		349.65	420	500
	Reinstall door closer				1.735		81.50		81.50	108	134
	Total				17.819	1,035.85	883.70		1,919.55	2,395.40	2,828.85

C1023 120 | Hollow Core Interior Doors

System Description	Freq. (Years)	Crew	Unit	Labor Hours	2015 Bare Costs				Total In-House	Total w/O&P
					Material	Labor	Equipment	Total		
1010 **Repair hollow core wood door**	7	1 CARP	Ea.							
Remove lockset				.520		24.50		24.50	32.50	40
Replace lockset				1.040	108	49		157	184	215
Oil hinges				.065		3.03		3.03	4.03	4.96
Oil door closer				.065		3.03		3.03	4.03	4.96
Total				1.689	108	79.56		187.56	224.56	264.92
1020 **Refinish 3'-0" x 7'-0" hollow core wood door**	4	1 PORD	Ea.							
Prepare interior door for painting				.433		17.50		17.50	22.50	28
Paint door, roller + brush, 1 coat				.533	5.30	21.50		26.80	34	41
Total				.967	5.30	39		44.30	56.50	69
1030 **Replace 3'-0" x 7'-0" hollow core wood door**	30	1 CARP	Ea.							
Remove interior door				.520		19.55		19.55	26	32
Remove interior door frame				1.300		61		61	81	100
Install new interior wood door and frame				.941	155	44		199	229	266
Total				2.761	155	124.55		279.55	336	398

C1023 121 | Solid Core Interior Doors

System Description	Freq. (Years)	Crew	Unit	Labor Hours	2015 Bare Costs				Total In-House	Total w/O&P
					Material	Labor	Equipment	Total		
1010 **Repair solid core wood door**	11	1 CARP	Ea.							
Remove lockset				.520		24.50		24.50	32.50	40
Replace lockset				1.040	108	49		157	184	215
Oil hinges				.065		3.03		3.03	4.03	4.96
Oil door closer				.065		3.03		3.03	4.03	4.96
Total				1.689	108	79.56		187.56	224.56	264.92
1020 **Refinish 3'-0" x 7'-0" solid core wood door**	4	1 PORD	Ea.							
Prepare interior door for painting				.433		17.50		17.50	22.50	28
Paint door, roller + brush, 1 coat				.533	5.30	21.50		26.80	34	41
Total				.967	5.30	39		44.30	56.50	69

For customer support on your Facilities Maintenance & Repair Cost Data, call 877.801.5724.

127

C1023 121 | Solid Core Interior Doors

System Description	Freq. (Years)	Crew	Unit	Labor Hours	2015 Bare Costs				Total In-House	Total w/O&P
					Material	Labor	Equipment	Total		
1030 Replace 3'-0" x 7'-0" solid core wood door	40	1 CARP	Ea.							
Remove interior door				.520		19.55		19.55	26	32
Remove interior door frame				.650		30.55		30.55	40.50	50
Install new interior wood door & frame				.941	335	44		379	425	490
Total				2.111	335	94.10		429.10	**491.50**	572
2010 Repair solid core sliding wood door	14	1 CARP	Ea.							
Oil door closer				.065		3.03		3.03	4.03	4.96
Oil / inspect track / glide wheels				.065		3.03		3.03	4.03	4.96
Remove door track / glide wheels (25% of total / 14 years)				.220		10.38		10.38	13.75	16.90
Install new door track / glide wheels (25% of total / 14 years)				.440	13.13	20.63		33.76	42	50
Total				.789	13.13	37.07		50.20	**63.81**	76.82
2030 Replace 3'-0" x 7'-0" solid core sliding wood door	40	1 CARP	Ea.							
Remove interior door				.520		19.55		19.55	26	32
Remove interior door frame				.650		30.55		30.55	40.50	50
Remove door closer				.167		7.85		7.85	10.40	12.85
Install new interior wood door and frame				.941	335	44		379	425	490
Install new door track / glide wheels				1.760	52.50	82.50		135	168	201
Reinstall door closer				1.735		81.50		81.50	108	134
Total				5.773	387.50	265.95		653.45	**777.90**	919.85
3040 Replace 3'-0" x 7'-0" solid core wood door, w/ safety glass	40	1 CARP	Ea.							
Remove interior door				.520		19.55		19.55	26	32
Remove interior door frame				.650		30.55		30.55	40.50	50
Install new interior wood door and frame				.941	335	44		379	425	490
Install new safety glass lite					79			79	87	99
Total				2.111	414	94.10		508.10	**578.50**	671

For customer support on your Facilities Maintenance & Repair Cost Data, call 877.801.5724.

C1023 129 Interior Gates

System Description	Freq. (Years)	Crew	Unit	Labor Hours	2015 Bare Costs				Total In-House	Total w/O&P
					Material	Labor	Equipment	Total		
1030 Prepare and refinish interior metal gate	5	1 PORD	Ea.							
Lay drop cloth				.090		3.60		3.60	4.70	5.80
Prepare / brush surface				.125		5.04		5.04	6.50	8.10
Paint gate, brushwork, primer + 1 coat				.167	1.44	6.66		8.10	10.25	12.60
Remove drop cloth				.090		3.60		3.60	4.70	5.80
Total				.471	1.44	18.90		20.34	26.15	32.30

C1023 324 Hinges, Brass

System Description	Freq. (Years)	Crew	Unit	Labor Hours	2015 Bare Costs				Total In-House	Total w/O&P
					Material	Labor	Equipment	Total		
0010 Replace 1 1/2 pair brass hinges	60	1 CARP	Ea.							
Remove door				.650		24.50		24.50	32.50	40
Remove hinges				.500		23.55		23.55	31	38.50
Replace hinges				.500		23.55		23.55	31	38.50
Hinges, 1 1/2 pair					183			183	201	229
Install door				1.224		57.50		57.50	76.50	94
Total				2.875	183	129.10		312.10	372	440

C1023 325 Lockset, Brass

System Description	Freq. (Years)	Crew	Unit	Labor Hours	2015 Bare Costs				Total In-House	Total w/O&P
					Material	Labor	Equipment	Total		
0010 Replace brass lockset	30	1 CARP	Ea.							
Remove lockset				.400		18.80		18.80	25	31
Install lockset				1.156	220	54.50		274.50	315	365
Total				1.556	220	73.30		293.30	340	396

C1023 326 Door Closer, Brass

	System Description	Freq. (Years)	Crew	Unit	Labor Hours	2015 Bare Costs				Total In-House	Total w/O&P
						Material	Labor	Equipment	Total		
0010	Replace brass closer	15	1 CARP	Ea.							
	Remove door closer				.167		7.85		7.85	10.40	12.85
	Install new door closer				1.735	192	81.50		273.50	320	375
	Total				1.902	192	89.35		281.35	330.40	387.85

C1023 327 Deadbolt, Brass

	System Description	Freq. (Years)	Crew	Unit	Labor Hours	2015 Bare Costs				Total In-House	Total w/O&P
						Material	Labor	Equipment	Total		
0010	Replace brass deadbolt	20	1 CARP	Ea.							
	Remove deadbolt				.578		27		27	36	44.50
	Replace deadbolt				1.156	175	54.50		229.50	265	310
	Total				1.733	175	81.50		256.50	301	354.50

C1023 328 Weatherstripping, Brass

	System Description	Freq. (Years)	Crew	Unit	Labor Hours	2015 Bare Costs				Total In-House	Total w/O&P
						Material	Labor	Equipment	Total		
0010	Replace brass weatherstripping	20	1 CARP	Ea.							
	Remove weatherstripping				.015		.69		.69	.92	1.13
	Install new weatherstripping				3.478	46.50	163		209.50	268	325
	Total				3.493	46.50	163.69		210.19	268.92	326.13

C1023 329 Panic Bar

	System Description	Freq. (Years)	Crew	Unit	Labor Hours	2015 Bare Costs				Total In-House	Total w/O&P
						Material	Labor	Equipment	Total		
0010	Replace panic bar	25	1 CARP	Ea.							
	Remove panic bar				1.040		49		49	65	80
	Install panic bar				2.080	865	97.50		962.50	1,075	1,250
	Total				3.120	865	146.50		1,011.50	1,140	1,330

C1033 110   Toilet Partitions

	System Description	Freq. (Years)	Crew	Unit	Labor Hours	2015 Bare Costs				Total In-House	Total w/O&P
						Material	Labor	Equipment	Total		
1010	**Replace toilet partitions, laminate clad-overhead braced, per stall**	20	2 CARP	Ea.							
	Remove overhead braced laminate clad partitions				1.333		62.50		62.50	83	103
	Install overhead braced laminate clad partitions				3.478	750	163		913	1,050	1,200
	Cleanup				.500		23.50		23.50	31	38.50
	Total				5.312	750	249		999	1,164	1,341.50
1020	**Replace toilet partitions, painted metal-overhead braced, per stall**	20	2 CARP	Ea.							
	Remove overhead braced painted metal partitions				1.333		62.50		62.50	83	103
	Install overhead braced painted metal partitions				3.478	390	163		553	645	755
	Cleanup				.500		23.50		23.50	31	38.50
	Total				5.312	390	249		639	759	896.50
1030	**Replace toilet partitions, porcelain enamel-overhead braced, per stall**	20	2 CARP	Ea.							
	Remove overhead braced partitions				1.333		62.50		62.50	83	103
	Install overhead braced porcelain enamel partitions				3.478	500	163		663	765	895
	Cleanup				.500		23.50		23.50	31	38.50
	Total				5.312	500	249		749	879	1,036.50
1040	**Replace toilet partitions, phenolic-overhead braced, per stall**	20	2 CARP	Ea.							
	Remove overhead braced partitions				1.333		62.50		62.50	83	103
	Install phenolic overhead braced partition				3.478	750	163		913	1,050	1,200
	Cleanup				.500		23.50		23.50	31	38.50
	Total				5.312	750	249		999	1,164	1,341.50
1050	**Replace toilet partitions, stainless steel-overhead braced, per stall**	30	2 CARP	Ea.							
	Remove overhead braced partitions				1.333		62.50		62.50	83	103
	Install stainless steel overhead braced partition				3.478	1,025	163		1,188	1,350	1,550
	Cleanup				.500		23.50		23.50	31	38.50
	Total				5.312	1,025	249		1,274	1,464	1,691.50
1060	**Replace toilet partitions, stainless steel, ceiling hung, per stall**	30	2 CARP	Ea.							
	Remove ceiling hung partitions				1.333		62.50		62.50	83	103
	Install stainless steel ceiling hung partition				4.000	1,075	188		1,263	1,425	1,650
	Cleanup				.500		23.50		23.50	31	38.50
	Total				5.833	1,075	274		1,349	1,539	1,791.50

C1033 110 Toilet Partitions

	System Description	Freq. (Years)	Crew	Unit	Labor Hours	2015 Bare Costs				Total In-House	Total w/O&P
						Material	Labor	Equipment	Total		
2020	**Replace urinal screen, stainless steel**	30	2 CARP	Ea.							
	Remove stainless steel urinal screen				1.000		37.50		37.50	50	61.50
	Install stainless steel urinal screen				2.000	580	94		674	765	880
	Cleanup				.500		23.50		23.50	31	38.50
	Total				3.500	580	155		735	846	**980**
3010	**Replace shower stall, fiberglass, per stall**	30	2 SHEE	Ea.							
	Remove shower base, partition and door				2.000		75		75	100	123
	Install shower base, partition and door				3.556	795	199		994	1,125	1,300
	Cleanup				.500		23.50		23.50	31	38.50
	Total				6.056	795	297.50		1,092.50	1,256	**1,461.50**

C1033 210 Metal Lockers

	System Description	Freq. (Years)	Crew	Unit	Labor Hours	2015 Bare Costs				Total In-House	Total w/O&P
						Material	Labor	Equipment	Total		
1010	**Replace metal lockers, single tier**	20	1 SHEE	Ea.							
	Remove lockers				.533		20		20	26.50	33
	Install 12" x 18" x 72" single enameled locker				.400	228	22.50		250.50	279	320
	Cleanup				.500		23.50		23.50	31	38.50
	Total				1.433	228	66		294	336.50	**391.50**

C2013 110 Interior Concrete Steps

	System Description	Freq. (Years)	Crew	Unit	Labor Hours	2015 Bare Costs				Total In-House	Total w/O&P
						Material	Labor	Equipment	Total		
0010	**Repair interior concrete steps**	15	1 CEFI	S.F.							
	Repair spalls in steps				.104	12.10	4.68		16.78	19.25	22.50
	Total				.104	12.10	4.68		16.78	**19.25**	**22.50**
0020	**Replace interior concrete steps**	100	2 CEFI	S.F.							
	Remove concrete				.520		19.75	3.03	22.78	29.50	36
	Install new concrete steps				.520	4.63	24	.34	28.97	37	45
	Total				1.040	4.63	43.75	3.37	51.75	**66.50**	**81**

C2013 115 Interior Masonry Steps, Painted

	System Description	Freq. (Years)	Crew	Unit	Labor Hours	2015 Bare Costs				Total In-House	Total w/O&P
						Material	Labor	Equipment	Total		
0010	**Repair interior masonry steps, painted**	20	1 BRIC	S.F.							
	Remove damaged masonry				.083		3.13		3.13	4.16	5.15
	Replace with new masonry				.232	4	9.75		13.75	17.25	21
	Refinish surface				.014	.20	.56		.76	.94	1.14
	Total				.329	4.20	13.44		17.64	**22.35**	**27.29**
0040	**Replace interior masonry steps, painted**	50	2 BRIC	S.F.							
	Remove masonry steps				4.000		150		150	200	247
	Clean work area				.015		.58	.15	.73	.92	1.11
	Install new masonry steps				.891	85	38.50		123.50	144	169
	Total				4.907	85	189.08	.15	274.23	**344.92**	**417.11**

C2013 130 Interior Metal Steps

	System Description	Freq. (Years)	Crew	Unit	Labor Hours	2015 Bare Costs				Total In-House	Total w/O&P
						Material	Labor	Equipment	Total		
0010	**Repair interior metal steps**	15	1 SSWK	S.F.							
	Remove damaged tread				.347		18.25		18.25	27.50	33
	Replace with cast iron tread, 1/2" thick				.120	49.17	6.32		55.49	63.50	73
	Total				.467	49.17	24.57		73.74	**91**	**106**

For customer support on your Facilities Maintenance & Repair Cost Data, call 877.801.5724.

133

C2013 130 Interior Metal Steps

	System Description	Freq. (Years)	Crew	Unit	Labor Hours	2015 Bare Costs				Total In-House	Total w/O&P
						Material	Labor	Equipment	Total		
0030	**Refinish interior metal steps**	9	1 PORD	S.F.							
	Prepare steps for painting				.026		1.05		1.05	1.36	1.69
	Wipe surface				.001		.05		.05	.06	.08
	Refinish surface, roller + brush, 1 coat				.025	.38	1.01		1.39	1.73	2.10
	Total				.052	.38	2.11		2.49	**3.15**	**3.87**
0040	**Replace interior metal steps**	50	2 SSWK	S.F.							
	Remove steps				.264		9.90		9.90	13.20	16.35
	Install pre-erected steel steps				.163	118	8.65	.74	127.39	144	164
	Prepare steps for painting				.026		1.05		1.05	1.36	1.69
	Wipe surface				.001		.05		.05	.06	.08
	Refinish surface				.025	.38	1.01		1.39	1.73	2.10
	Total				.479	118.38	20.66	.74	139.78	**160.35**	**184.22**

C2013 435 Interior Metal Stair Railing

	System Description	Freq. (Years)	Crew	Unit	Labor Hours	2015 Bare Costs				Total In-House	Total w/O&P
						Material	Labor	Equipment	Total		
0030	**Refinish interior metal stair railing**	7	1 PORD	S.F.							
	Prepare surface				.019		.77		.77	1	1.24
	Refinish surface, roller + brush, 1 coat				.015	.11	.62		.73	.93	1.14
	Total				.035	.11	1.39		1.50	**1.93**	**2.38**
0040	**Replace interior metal stair railing**	45	2 SSWK	L.F.							
	Remove railing				.098		5.20	.45	5.65	8.20	9.90
	Install new railing				.195	20	10.40	.89	31.29	38.50	44.50
	Total				.293	20	15.60	1.34	36.94	**46.70**	**54.40**

C2013 437 Interior Iron Stair Railing

	System Description	Freq. (Years)	Crew	Unit	Labor Hours	2015 Bare Costs				Total In-House	Total w/O&P
						Material	Labor	Equipment	Total		
0030	**Refinish interior wrought iron stair railing**	7	1 PORD	L.F.							
	Prepare surface				.019		.77		.77	1	1.24
	Refinish surface, roller + brush, 1 coat				.015	.11	.62		.73	.93	1.14
	Total				.035	.11	1.39		1.50	**1.93**	**2.38**

For customer support on your Facilities Maintenance & Repair Cost Data, call 877.801.5724.

C20 STAIRS C2013 Stair Construction

C2013 437 Interior Iron Stair Railing

	System Description	Freq. (Years)	Crew	Unit	Labor Hours	2015 Bare Costs				Total In-House	Total w/O&P
						Material	Labor	Equipment	Total		
0040	Replace interior wrought iron stair railing	45	2 SSWK	L.F.							
	Remove railing				.087		3.26		3.26	4.33	5.35
	Install new railing				1.300	32	68.50		100.50	138	164
	Total				1.387	32	71.76		103.76	142.33	169.35

C2013 440 Interior Wood Steps

	System Description	Freq. (Years)	Crew	Unit	Labor Hours	2015 Bare Costs				Total In-House	Total w/O&P
						Material	Labor	Equipment	Total		
0030	Refinish interior wood steps	3	1 PORD	S.F.							
	Prepare surface				.026		1.05		1.05	1.36	1.69
	Refinish surface, roller + brush, 1 coat				.026	.07	1.05		1.12	1.44	1.78
	Total				.052	.07	2.10		2.17	2.80	3.47
0040	Replace interior wood steps	40	1 CARP	S.F.							
	Remove steps				.172		6.45		6.45	8.60	10.55
	Replace steps				.325	96	15.25		111.25	126	145
	Total				.497	96	21.70		117.70	134.60	155.55

C2013 445 Interior Wood Stair Railing

	System Description	Freq. (Years)	Crew	Unit	Labor Hours	2015 Bare Costs				Total In-House	Total w/O&P
						Material	Labor	Equipment	Total		
0030	Refinish interior wood stair railing	7	1 PORD	L.F.							
	Prepare railings for refinishing				.026		1.05		1.05	1.36	1.69
	Refinish surface, roller + brush, 1 coat				.026	.07	1.05		1.12	1.44	1.78
	Total				.052	.07	2.10		2.17	2.80	3.47
0040	Replace interior wood stair railing	40	1 CARP	L.F.							
	Remove railing				.087		3.26		3.26	4.33	5.35
	Replace railing				.130	2.56	6.10		8.66	10.95	13.20
	Total				.217	2.56	9.36		11.92	15.28	18.55

C2023 110 Carpeted Steps

	System Description	Freq. (Years)	Crew	Unit	Labor Hours	2015 Bare Costs				Total In-House	Total w/O&P
						Material	Labor	Equipment	Total		
0020	Replace carpeted steps	8	2 TILF	S.F.							
	Remove old carpet				.010		.39		.39	.52	.64
	Install new carpet				.069	14.85	2.94		17.79	20	23
	Additional labor				.229	29.70	9.80		39.50	45	52.50
	Total				.308	44.55	13.13		57.68	65.52	76.14

C2023 120 Rubber Steps

	System Description	Freq. (Years)	Crew	Unit	Labor Hours	2015 Bare Costs				Total In-House	Total w/O&P
						Material	Labor	Equipment	Total		
0020	Replace rubber steps	18	2 TILF	L.F.							
	Remove rubber covering				.015		.56		.56	.74	.92
	Prepare floor for installation				.063		2.72		2.72	3.45	4.29
	Install new rubber tread				.090	14.75	3.87		18.62	21	24.50
	Install new rubber riser				.042	7.70	1.78		9.48	10.75	12.45
	Total				.210	22.45	8.93		31.38	35.94	42.16

C2023 130 Terrazzo Steps

	System Description	Freq. (Years)	Crew	Unit	Labor Hours	2015 Bare Costs				Total In-House	Total w/O&P
						Material	Labor	Equipment	Total		
0020	Replace terrazzo steps	50	2 MSTZ	S.F.							
	Remove old terrazzo tile				.052		1.96		1.96	2.60	3.21
	Clean area				.020		.94		.94	1.24	1.53
	Replace with new tread				.743	3.08	29.25	14	46.33	54.50	65.50
	Total				.815	3.08	32.15	14	49.23	58.34	70.24

For customer support on your Facilities Maintenance & Repair Cost Data, call 877.801.5724.

C3013 106 Vinyl Wall Covering

	System Description	Freq. (Years)	Crew	Unit	Labor Hours	2015 Bare Costs				Total In-House	Total w/O&P
						Material	Labor	Equipment	Total		
0010	**Repair med. wt. vinyl wall covering - (2% of walls)**	1	1 PORD	C.S.F.							
	Remove damaged vinyl wall covering				1.156		47		47	61	76
	Install vinyl wall covering				2.391	90	98		188	226	270
	Total				3.547	90	145		235	**287**	**346**
0020	**Replace medium weight vinyl wall covering**	15	1 PORD	C.S.F.							
	Set up and secure scaffold				.750		35.25		35.25	47	57.50
	Remove vinyl wall covering				1.156		47		47	61	76
	Install vinyl wall covering				2.391	90	98		188	226	270
	Remove scaffold				.750		35.25		35.25	47	57.50
	Total				5.047	90	215.50		305.50	**381**	**461**

C3013 202 Wallpaper

	System Description	Freq. (Years)	Crew	Unit	Labor Hours	2015 Bare Costs				Total In-House	Total w/O&P
						Material	Labor	Equipment	Total		
0010	**Repair wallpaper - (2% of walls)**	8	1 PORD	S.Y.							
	Remove damaged wallpaper				.104		4.23		4.23	5.50	6.85
	Repair / replace wallpaper				.175	9.99	7.20		17.19	20.50	24
	Total				.279	9.99	11.43		21.42	**26**	**30.85**
0020	**Replace wallpaper**	20	1 PORD	S.Y.							
	Set up and secure scaffold				.130		6.10		6.10	8.10	10
	Remove damaged wallpaper				.104		4.23		4.23	5.50	6.85
	Replace wallpaper				.175	9.99	7.20		17.19	20.50	24
	Remove scaffold				.130		6.10		6.10	8.10	10
	Total				.539	9.99	23.63		33.62	**42.20**	**50.85**

C3013 206 — Fabric Interior Wall Finish

	System Description	Freq. (Years)	Crew	Unit	Labor Hours	2015 Bare Costs				Total In-House	Total w/O&P
						Material	Labor	Equipment	Total		
0010	**Repair fabric wall finish**	9	1 PORD	S.Y.							
	Set up, secure and take down ladder				.130		5.30		5.30	6.90	8.55
	Remove fabric wall covering				.104		4.23		4.23	5.50	6.85
	Repair / replace fabric				.234	8.19	9.54		17.73	21.50	25.50
	Total				.468	8.19	19.07		27.26	**33.90**	**40.90**
0020	**Replace fabric wall finish**	50	1 PORD	S.Y.							
	Set up and secure scaffold				1.350		63.45		63.45	84	104
	Remove fabric wall covering				.104		4.23		4.23	5.50	6.85
	Replace fabric				.234	8.19	9.54		17.73	21.50	25.50
	Remove scaffold				1.350		63.45		63.45	84	104
	Total				3.038	8.19	140.67		148.86	**195**	**240.35**

C3013 208 — Cork Tile

	System Description	Freq. (Years)	Crew	Unit	Labor Hours	2015 Bare Costs				Total In-House	Total w/O&P
						Material	Labor	Equipment	Total		
0010	**Repair 12″ x 12″ x 3/16″ cork tile wall - (2% of walls)**	8	1 CARP	C.S.F.							
	Remove old tiles				.620		23		23	31	38
	Replace with new tile				4.426	451	181		632	730	855
	Total				5.046	451	204		655	**761**	**893**
0020	**Replace 12″ x 12″ x 3/16″ cork tile wall**	16	2 CARP	C.S.F.							
	Set up and secure scaffold				.750		35.25		35.25	47	57.50
	Remove old tiles				.620		23		23	31	38
	Replace with new tile				4.426	451	181		632	730	855
	Remove scaffold				.750		35.25		35.25	47	57.50
	Total				6.546	451	274.50		725.50	**855**	**1,008**

For customer support on your Facilities Maintenance & Repair Cost Data, call 877.801.5724.

C3013 210   Acoustical Tile

System Description	Freq. (Years)	Crew	Unit	Labor Hours	2015 Bare Costs				Total In-House	Total w/O&P
					Material	Labor	Equipment	Total		
0010 Repair acoustical tile - (2% of walls)	25	1 CARP	C.S.F.							
Remove damaged tile				1.067		40		40	53	66
Install new tile				2.737	336	129		465	540	630
Total				3.804	336	169		505	**593**	**696**
0030 Refinish acoustical tile	10	1 PORD	C.S.F.							
Wipe surface				.894	7	36		43	55	67
Prepare surface				.894	7	36		43	55	67
Refinish surface				1.529	22	62		84	104	127
Total				3.317	36	134		170	**214**	**261**
0040 Replace acoustical tile	60	2 CARP	C.S.F.							
Set up and secure scaffold				.750		35.25		35.25	47	57.50
Remove old tiles				1.067		40		40	53	66
Install new tile				2.737	336	129		465	540	630
Remove scaffold				.750		35.25		35.25	47	57.50
Total				5.304	336	239.50		575.50	**687**	**811**

C3013 212   Stucco

System Description	Freq. (Years)	Crew	Unit	Labor Hours	2015 Bare Costs				Total In-House	Total w/O&P
					Material	Labor	Equipment	Total		
0010 Repair stucco wall - (2% of walls)	20	1 BRIC	S.Y.							
Prepare surface				.023		.90		.90	1.17	1.44
Repair stucco				.532	3.69	21.78	1.89	27.36	34	41.50
Place and remove mask and drops				.067		2.70		2.70	3.42	4.32
Total				.621	3.69	25.38	1.89	30.96	**38.59**	**47.26**
0030 Refinish stucco wall	4	1 PORD	S.Y.							
Wash surface				.023		.90		.90	1.17	1.44
Prepare surface				.033		1.35		1.35	1.71	2.16
Refinish surface, brushwork, 2 coats				.138	1.98	5.58		7.56	9.35	11.40
Place and remove mask and drops				.033		1.35		1.35	1.71	2.16
Total				.227	1.98	9.18		11.16	**13.94**	**17.16**

For customer support on your Facilities Maintenance & Repair Cost Data, call 877.801.5724.

139

C3013 212 Stucco

System Description	Freq. (Years)	Crew	Unit	Labor Hours	2015 Bare Costs Material	Labor	Equipment	Total	Total In-House	Total w/O&P
0040 Replace stucco wall	75	2 BRIC	S.Y.							
Set up and secure scaffold				.130		6.10		6.10	8.10	10
Remove stucco				.279		13.14		13.14	17.45	21.50
Replace stucco				1.200	6.25	49.50	3.50	59.25	74	90.50
Remove scaffold				.130		6.10		6.10	8.10	10
Total				1.739	6.25	74.84	3.50	84.59	**107.65**	**132**
4120 Spray refinish stucco wall	5	1 PORD	S.Y.							
Wash surface				.023		.90		.90	1.17	1.44
Prepare surface				.033		1.35		1.35	1.71	2.16
Refinish surface, spray, 2 coats				.049	1.98	1.98		3.96	4.77	5.65
Place and remove mask and drops				.033		1.35		1.35	1.71	2.16
Total				.138	1.98	5.58		7.56	**9.36**	**11.41**

C3013 213 Plaster

System Description	Freq. (Years)	Crew	Unit	Labor Hours	2015 Bare Costs Material	Labor	Equipment	Total	Total In-House	Total w/O&P
0010 Repair plaster wall - (2% of walls)	13	1 PLAS	S.Y.							
Remove damage				.468		17.60		17.60	23.50	29
Replace two coat plaster finish, incl. lath				.752	6.65	31	2.20	39.85	49.50	60
Total				1.220	6.65	48.60	2.20	57.45	**73**	**89**
0030 Refinish plaster wall	4	1 PORD	S.Y.							
Prepare surface				.033		1.35		1.35	1.71	2.16
Paint / seal surface, brushwork, 1 coat				.063	.63	2.52		3.15	3.96	4.84
Place and remove mask and drops				.033		1.35		1.35	1.71	2.16
Total				.129	.63	5.22		5.85	**7.38**	**9.16**
0040 Replace plaster wall	75	2 PLAS	S.Y.							
Set up, secure and take down ladder				.130		6.10		6.10	8.10	10
Remove material				.468		17.60		17.60	23.50	29
Replace two coat plaster wall, incl. lath				.752	6.65	31	2.20	39.85	49.50	60
Total				1.350	6.65	54.70	2.20	63.55	**81.10**	**99**

C3013 214 Drywall

System Description	Freq. (Years)	Crew	Unit	Labor Hours	2015 Bare Costs				Total In-House	Total w/O&P
					Material	Labor	Equipment	Total		
0010 **Repair 5/8" drywall - (2% of walls)**	20	1 CARP	S.F.							
Remove damage				.008		.30		.30	.40	.49
Replace 5/8" drywall, taped and finished				.022	.33	1.01		1.34	1.70	2.07
Total				.030	.33	1.31		1.64	2.10	2.56
0030 **Refinish drywall**	4	1 PORD	S.F.							
Place and remove mask and drops				.004		.15		.15	.19	.24
Prepare surface				.004		.15		.15	.19	.24
Paint surface, brushwork, 1 coat				.007	.07	.28		.35	.44	.54
Total				.014	.07	.58		.65	.82	1.02
0040 **Replace 5/8" drywall**	75	2 CARP	S.F.							
Set up, secure and take down ladder				.014		.68		.68	.90	1.11
Remove drywall				.008		.30		.30	.40	.49
Replace 5/8" drywall, taped and finished				.022	.33	1.01		1.34	1.70	2.07
Total				.044	.33	1.99		2.32	3	3.67
0050 **Office painting, 10' x 12', 10' high walls**	5	1 PORD	Ea.							
Spread drop cloths				.004		.15		.15	.19	.24
Prepare drywall partitions				1.628		66		66	83.50	106
Clean drywall partitions				.220	2.02	8.27		10.29	13.20	16.15
Paint drywall partitions, roller + brush, 1 coat				3.062	30.80	123.20		154	194	237
Remove drop cloths				.004		.15		.15	.19	.24
Total				4.918	32.82	197.77		230.59	291.08	359.63
0060 **Office painting, 10' x 15', 10' high walls**	5	1 PORD	Ea.							
Spread drop cloths				.005		.19		.19	.24	.30
Prepare drywall partitions				1.850		75		75	95	120
Clean drywall partitions				.250	2.29	9.40		11.69	15	18.35
Paint drywall partitions, roller + brush, 1 coat				3.480	35	140		175	220	269
Remove drop cloths				.005		.19		.19	.24	.30
Total				5.589	37.29	224.78		262.07	330.48	407.95

For customer support on your Facilities Maintenance & Repair Cost Data, call 877.801.5724.

141

C3013 214   Drywall

System Description	Freq. (Years)	Crew	Unit	Labor Hours	2015 Bare Costs				Total In-House	Total w/O&P
					Material	Labor	Equipment	Total		
2030										
Refinish drywall, 12' to 24' high	5	2 PORD	S.F.							
Set up and secure scaffold				.015		.71		.71	.94	1.15
Place and remove mask and drops				.004		.15		.15	.19	.24
Prepare surface				.004		.15		.15	.19	.24
Paint surface, roller + brushwork, 1 coat				.007	.07	.28		.35	.44	.54
Remove scaffold				.015		.71		.71	.94	1.15
Total				.044	.07	2		2.07	2.70	3.32
3030										
Refinish drywall, over 24' high	5	2 PORD	S.F.							
Set up and secure scaffold				.023		1.06		1.06	1.40	1.73
Place and remove mask and drops				.004		.15		.15	.19	.24
Prepare surface				.004		.15		.15	.19	.24
Paint surface, roller + brushwork, 1 coat				.007	.07	.28		.35	.44	.54
Remove scaffold				.023		1.06		1.06	1.40	1.73
Total				.059	.07	2.70		2.77	3.62	4.48

C3013 215   Fiberglass Panels, Rigid

System Description	Freq. (Years)	Crew	Unit	Labor Hours	2015 Bare Costs				Total In-House	Total w/O&P
					Material	Labor	Equipment	Total		
0010										
Repair glass cloth fiberglass panels - (2% of walls)	9	1 CARP	C.S.F.							
Remove damaged fiberglass panels				2.600		98		98	130	160
Install new fiberglass panel				6.710	835	315		1,150	1,350	1,550
Total				9.310	835	413		1,248	1,480	1,710
0040										
Replace glass cloth fiberglass panels	35	2 CARP	C.S.F.							
Set up and secure scaffold				.750		35.25		35.25	47	57.50
Remove old panels				2.600		98		98	130	160
Remove old furring				.516		19		19	26	32
Install new furring				2.101	42	99		141	177	215
Install new glass cloth fiberglass panels				6.710	835	315		1,150	1,350	1,550
Remove scaffold				.750		35.25		35.25	47	57.50
Total				13.427	877	601.50		1,478.50	1,777	2,072

C30 INTERIOR FINISHES | C3013 | Wall Finishes

C3013 220 | Tile

	System Description	Freq. (Years)	Crew	Unit	Labor Hours	Material	Labor	Equipment	Total	Total In-House	Total w/O&P
							2015 Bare Costs				
0010	**Repair 4" x 4" thin set ceramic tile - (2% of walls)**	10	1 TILF	C.S.F.							
	Remove damaged tiles				2.600		98		98	130	160
	Replace tile / grout				10.947	210	418		628	760	925
	Total				13.547	210	516		726	890	1,085
0020	**Replace 4" x 4" thin set ceramic tile**	75	1 TILF	C.S.F.							
	Set up and secure scaffold				.750		35.25		35.25	47	57.50
	Remove tiles				2.600		98		98	130	160
	Replace tile / grout				10.947	210	418		628	760	925
	Remove scaffold				.750		35.25		35.25	47	57.50
	Total				15.047	210	586.50		796.50	984	1,200

C3013 230 | Plywood Paneling

	System Description	Freq. (Years)	Crew	Unit	Labor Hours	Material	Labor	Equipment	Total	Total In-House	Total w/O&P
							2015 Bare Costs				
0010	**Repair plywood paneling - (2% of walls)**	10	1 CARP	C.S.F.							
	Remove damaged paneling				1.156		43		43	58	71
	Replace paneling				4.952	120	233		353	440	530
	Total				6.108	120	276		396	498	601
0020	**Refinish plywood paneling**	10	1 PORD	C.S.F.							
	Place mask and drops				.500		20		20	26	32
	Sand & paint, brushwork, 1 coat				.990	6	40		46	59	71.50
	Remove mask and drops				.500		20		20	26	32
	Total				1.990	6	80		86	111	135.50
0040	**Replace plywood paneling**	30	2 CARP	C.S.F.							
	Set up and secure scaffold				.750		35.25		35.25	47	57.50
	Remove damaged paneling				1.156		43		43	58	71
	Replace paneling				4.952	120	233		353	440	530
	Remove scaffold				.750		35.25		35.25	47	57.50
	Total				7.608	120	346.50		466.50	592	716

For customer support on your Facilities Maintenance & Repair Cost Data, call 877.801.5724.

C3013 234 Wainscot

System Description	Freq. (Years)	Crew	Unit	Labor Hours	Material	Labor	Equipment	Total	Total In-House	Total w/O&P
0010 Repair wainscot - (2% of walls)	10	1 CARP	C.S.F.							
Remove damaged wainscot				1.733		65		65	87	107
Replace wainscot				16.000	1,575	750		2,325	2,750	3,200
Total				17.733	1,575	815		2,390	2,837	**3,307**
0020 Refinish wainscot	4	1 PORD	C.S.F.							
Place mask and drops				.500		20		20	26	32
Sand & paint, brushwork, 1 coat				.990	6	40		46	59	71.50
Remove mask and drops				.500		20		20	26	32
Total				1.990	6	80		86	111	**135.50**
0040 Replace wainscot	40	2 CARP	C.S.F.							
Remove wainscot				1.733		65		65	87	107
Replace wainscot				16.000	1,575	750		2,325	2,750	3,200
Total				17.733	1,575	815		2,390	2,837	**3,307**

C3013 240 Stainless Steel Interior Finish

System Description	Freq. (Years)	Crew	Unit	Labor Hours	Material	Labor	Equipment	Total	Total In-House	Total w/O&P
0020 Replace stainless steel wall	75	2 SSWK	C.S.F.							
Set up, secure and take down ladder				1.500		70.50		70.50	93.50	115
Remove existing stainless steel sheets				2.342		88		88	117	144
Replace stainless steel sheets				13.419	425	540		965	1,275	1,500
Total				17.261	425	698.50		1,123.50	1,485.50	**1,759**

C3013 410 Fireplace Mantles, Wood

System Description	Freq. (Years)	Crew	Unit	Labor Hours	Material	Labor	Equipment	Total	Total In-House	Total w/O&P
0030 Refinish wood surface - fireplaces	7	1 PORD	S.F.							
Prepare surface				.003		.14		.14	.18	.22
Varnish, 3 coats, brushwork, sanding included				.025	.28	.99		1.27	1.60	1.95
Total				.028	.28	1.13		1.41	1.78	**2.17**

Column header groups: Material, Labor, Equipment, Total fall under "2015 Bare Costs".

C3023 112 Concrete, Finished

	System Description	Freq. (Years)	Crew	Unit	Labor Hours	2015 Bare Costs				Total In-House	Total w/O&P
						Material	Labor	Equipment	Total		
0020	Refinish concrete floor	25	2 CEFI	C.S.F.							
	Add topping to existing floor 1"				6.933	56	281	47	384	475	575
	Total				6.933	56	281	47	384	475	575

C3023 405 Epoxy Flooring

	System Description	Freq. (Years)	Crew	Unit	Labor Hours	2015 Bare Costs				Total In-House	Total w/O&P
						Material	Labor	Equipment	Total		
0020	Replace epoxy flooring	15	2 CEFI	C.S.F.							
	Strip existing flooring				3.200		150		150	200	246
	Repair and seal floor				12.735	405	499	17	921	1,125	1,325
	Total				15.935	405	649	17	1,071	1,325	1,571

C3023 410 Vinyl Tile

	System Description	Freq. (Years)	Crew	Unit	Labor Hours	2015 Bare Costs				Total In-House	Total w/O&P
						Material	Labor	Equipment	Total		
0020	Replace vinyl tile flooring	18	1 TILF	S.Y.							
	Remove damaged floor tile				.234		8.82		8.82	11.70	14.40
	Prepare surface				.645		24.66		24.66	31	39
	Install new tiles				.187	10.62	8.01		18.63	22	26
	Total				1.067	10.62	41.49		52.11	64.70	79.40

C3023 412 Vinyl Sheet

	System Description	Freq. (Years)	Crew	Unit	Labor Hours	2015 Bare Costs				Total In-House	Total w/O&P
						Material	Labor	Equipment	Total		
0020	Replace vinyl sheet flooring	18	1 TILF	S.Y.							
	Remove damaged floor tile				.234		8.82		8.82	11.70	14.40
	Prepare surface				.645		24.66		24.66	31	39
	Install new vinyl sheet				.407	36.90	17.46		54.36	62.50	73.50
	Total				1.286	36.90	50.94		87.84	105.20	126.90

For customer support on your Facilities Maintenance & Repair Cost Data, call 877.801.5724.

145

C3023 414 Rubber Tile

System Description	Freq. (Years)	Crew	Unit	Labor Hours	2015 Bare Costs				Total In-House	Total w/O&P
					Material	Labor	Equipment	Total		
0020 Replace rubber tile floor	18	1 TILF	S.Y.							
Remove damaged floor tile				.234		8.82		8.82	11.70	14.40
Prepare surface				.645		24.66		24.66	31	39
Install new tiles				1.040	112.05	44.55		156.60	180	210
Total				1.920	112.05	78.03		190.08	222.70	**263.40**

C3023 418 Rubber / Vinyl Trim

System Description	Freq. (Years)	Crew	Unit	Labor Hours	2015 Bare Costs				Total In-House	Total w/O&P
					Material	Labor	Equipment	Total		
0010 Replace rubber cove base	9	1 TILF	L.F.							
Remove damaged base				.017		.65		.65	.87	1.07
Clean up debris				.001		.01		.01	.01	.01
Install new base				.025	1.02	1.09		2.11	2.50	3
Total				.044	1.02	1.75		2.77	3.38	**4.08**

C3023 420 Ceramic Tile

System Description	Freq. (Years)	Crew	Unit	Labor Hours	2015 Bare Costs				Total In-House	Total w/O&P
					Material	Labor	Equipment	Total		
0010 Ceramic tile floor repairs - (2% of floors)	15	1 TILF	C.S.F.							
Regrout ceramic tile floors				16.640	15	635		650	825	1,025
Total				16.640	15	635		650	825	**1,025**
0020 Replace 2″ x 2″ thin set ceramic tile floor	50	1 TILF	C.S.F.							
Remove damaged floor tile				2.600		98		98	130	160
Prepare surface				7.172		274		274	345	430
Install new tiles				10.947	590	418		1,008	1,175	1,400
Total				20.719	590	790		1,380	1,650	**1,990**

For customer support on your Facilities Maintenance & Repair Cost Data, call 877.801.5724.

C3023 428 Ceramic Trim

	System Description	Freq. (Years)	Crew	Unit	Labor Hours	2015 Bare Costs				Total In-House	Total w/O&P
						Material	Labor	Equipment	Total		
0020	Replace ceramic trim	50	2 TILF	L.F.							
	Remove ceramic tile trim				.015		.56		.56	.74	.92
	Clean up debris				.010		.46		.46	.61	.75
	Install new ceramic tile trim				.229	3.79	8.75		12.54	15.25	18.55
	Total				.253	3.79	9.77		13.56	16.60	20.22

C3023 430 Terrazzo

	System Description	Freq. (Years)	Crew	Unit	Labor Hours	2015 Bare Costs				Total In-House	Total w/O&P
						Material	Labor	Equipment	Total		
0100	Terrazzo floor repairs - (2% of floors)	15	1 MSTZ	S.F.							
	Chip existing terrazzo				.089		3.35		3.35	4.46	5.50
	Place terrazzo				.166	3.14	6.55	3.16	12.85	14.90	17.70
	Remove debris				.004		.16		.16	.22	.27
	Total				.260	3.14	10.06	3.16	16.36	19.58	23.47
0200	Replace terrazzo floor	75	2 MSTZ	C.S.F.							
	Break-up existing terrazzo				5.943		223		223	297	365
	Remove debris				.433		16.25		16.25	22	27
	Load into truck				.650		25.25	14	39.25	47.50	56.50
	Place terrazzo				16.640	314	655	316	1,285	1,500	1,775
	Total				23.666	314	919.50	330	1,563.50	1,866.50	2,223.50

C3023 438 Terrazzo Trim

	System Description	Freq. (Years)	Crew	Unit	Labor Hours	2015 Bare Costs				Total In-House	Total w/O&P
						Material	Labor	Equipment	Total		
0020	Replace precast terrazzo trim	75	2 MSTZ	L.F.							
	Remove old terrazzo trim				.023		.87		.87	1.16	1.43
	Clean area				.010		.46		.46	.61	.75
	Install new precast terrazzo trim				.416	16.85	17.85		34.70	41	49
	Total				.449	16.85	19.18		36.03	42.77	51.18

C3023 440 Quarry Tile

	System Description	Freq. (Years)	Crew	Unit	Labor Hours	2015 Bare Costs				Total In-House	Total w/O&P
						Material	Labor	Equipment	Total		
0010	**Quarry tile floor repairs - (2% of floors)**	15	1 TILF	S.F.							
	Regrout quarry tile floor				.166	.15	6.35		6.50	8.25	10.25
	Total				.166	.15	6.35		6.50	**8.25**	**10.25**
0020	**Replace quarry tile floor**	50	2 TILF	S.F.							
	Remove damaged floor tile				.026		.98		.98	1.30	1.60
	Prepare surface				.072		2.74		2.74	3.47	4.32
	Install new tiles				.109	5.90	4.18		10.08	11.80	14
	Total				.207	5.90	7.90		13.80	**16.57**	**19.92**

C3023 450 Brick

	System Description	Freq. (Years)	Crew	Unit	Labor Hours	2015 Bare Costs				Total In-House	Total w/O&P
						Material	Labor	Equipment	Total		
0020	**Replace thickset brick floor**	60	2 BRIC	S.F.							
	Remove brick				.035		1.30		1.30	1.73	2.14
	Prepare 2" thick mortar bed				.260	1.06	10.95		12.01	15.60	19.15
	Install new brick				.219	8.35	8.35		16.70	19.80	23.50
	Total				.514	9.41	20.60		30.01	**37.13**	**44.79**

C3023 460 Marble

	System Description	Freq. (Years)	Crew	Unit	Labor Hours	2015 Bare Costs				Total In-House	Total w/O&P
						Material	Labor	Equipment	Total		
0020	**Replace thinset marble floor**	50	2 TILF	S.F.							
	Remove marble				.035		1.30		1.30	1.73	2.14
	Prepare floor				.260	1.06	10.95		12.01	15.60	19.15
	Replace marble flooring				.347	10.15	13.25		23.40	28	33.50
	Total				.641	11.21	25.50		36.71	**45.33**	**54.79**

For customer support on your Facilities Maintenance & Repair Cost Data, call 877.801.5724.

C3023 470 Plywood

	System Description	Freq. (Years)	Crew	Unit	Labor Hours	2015 Bare Costs				Total In-House	Total w/O&P
						Material	Labor	Equipment	Total		
0020	**Replace plywood floor**	40	2 CARP	C.S.F.							
	Remove existing flooring				1.733		81		81	108	133
	Install new plywood flooring 3/4"				1.600	143	75		218	257	300
	Total				3.333	143	156		299	**365**	**433**

C3023 472 Wood Parquet

	System Description	Freq. (Years)	Crew	Unit	Labor Hours	2015 Bare Costs				Total In-House	Total w/O&P
						Material	Labor	Equipment	Total		
0020	**Sand and refinish parquet floor**	10	1 CARP	S.F.							
	Sand floor				.035	.90	1.33		2.23	2.75	3.30
	Refinish existing floor, brushwork, 2 coats				.026	1.34	.98		2.32	2.77	3.28
	Total				.061	2.24	2.31		4.55	**5.52**	**6.58**
0030	**Replace 5/16" oak parquet floor**	40	2 CARP	S.F.							
	Remove flooring				.023		1.09		1.09	1.44	1.78
	Repair underlayment				.014	.97	.65		1.62	1.94	2.28
	Install new parquet flooring				.065	5.25	3.05		8.30	9.85	11.55
	Total				.102	6.22	4.79		11.01	**13.23**	**15.61**

C3023 474 Maple Strip

	System Description	Freq. (Years)	Crew	Unit	Labor Hours	2015 Bare Costs				Total In-House	Total w/O&P
						Material	Labor	Equipment	Total		
0020	**Sand and refinish maple strip floor**	10	1 CARP	S.F.							
	Sand floor				.026	1.34	.98		2.32	2.77	3.28
	Refinish maple strip floor, brushwork, 2 coats				.035	.90	1.33		2.23	2.75	3.30
	Total				.061	2.24	2.31		4.55	**5.52**	**6.58**
0030	**Replace maple floor**	40	2 CARP	S.F.							
	Remove flooring				.023		1.09		1.09	1.44	1.78
	Repair underlayment				.014	.97	.65		1.62	1.94	2.28
	Install new maple flooring				.061	4.95	2.87		7.82	9.25	10.90
	Total				.098	5.92	4.61		10.53	**12.63**	**14.96**

For customer support on your Facilities Maintenance & Repair Cost Data, call 877.801.5724.

149

C3023 478 | Wood Trim

System Description	Freq. (Years)	Crew	Unit	Labor Hours	2015 Bare Costs				Total In-House	Total w/O&P
					Material	Labor	Equipment	Total		
0010 Repair 1" x 3" wood trim	13	1 CARP	L.F.							
Remove wood				.021		.78		.78	1.04	1.28
Install new trim				.173	5.24	8.12		13.36	16.55	19.85
Total				.194	5.24	8.90		14.14	17.59	**21.13**
0030 Refinish 1" x 3" wood trim	7	1 PORD	L.F.							
Prepare surface				.003		.14		.14	.18	.22
Varnish, 3 coats, brushwork, sanding included				.025	.28	.99		1.27	1.60	1.95
Total				.028	.28	1.13		1.41	1.78	**2.17**
0040 Replace 1" x 3" wood trim	75	2 CARP	L.F.							
Remove wood				.021		.78		.78	1.04	1.28
Install new trim				.173	5.24	8.12		13.36	16.55	19.85
Prepare surface				.003		.14		.14	.18	.22
Varnish, 3 coats, brushwork, sanding included				.025	.28	.99		1.27	1.60	1.95
Total				.222	5.52	10.03		15.55	19.37	**23.30**

C3023 510 | Carpet

System Description	Freq. (Years)	Crew	Unit	Labor Hours	2015 Bare Costs				Total In-House	Total w/O&P
					Material	Labor	Equipment	Total		
0020 Replace carpet	8	2 TILF	S.Y.							
Remove damaged carpet				.094		3.51		3.51	4.68	5.75
Install new carpet				.138	42.50	5.90		48.40	54	62.50
Total				.232	42.50	9.41		51.91	58.68	**68.25**

For customer support on your Facilities Maintenance & Repair Cost Data, call 877.801.5724.

C3033 105 Plaster

System Description	Freq. (Years)	Crew	Unit	Labor Hours	2015 Bare Costs				Total In-House	Total w/O&P
					Material	Labor	Equipment	Total		
0010 Repair plaster ceiling - (2% of ceilings)	12	1 PLAS	S.Y.							
Set up, secure and take down ladder				.130		6.10		6.10	8.10	10
Remove damaged ceiling				.267		10.08		10.08	13.30	16.45
Replace plaster, 2 coats on lath				.752	6.25	31	2.20	39.45	49	59.50
Total				1.149	6.25	47.18	2.20	55.63	70.40	85.95
0030 Refinish plaster ceiling	10	1 PORD	S.Y.							
Set up, secure and take down ladder				.130		6.10		6.10	8.10	10
Place mask and drops				.033		1.35		1.35	1.71	2.16
Sand & paint				.089	.54	3.60		4.14	5.30	6.45
Total				.252	.54	11.05		11.59	15.11	18.61
0040 Replace plaster ceiling	75	2 PLAS	S.Y.							
Set up and secure scaffold				.130		6.10		6.10	8.10	10
Remove ceiling				.267		10.08		10.08	13.30	16.45
Replace plaster, 2 coats on lath				.752	6.25	31	2.20	39.45	49	59.50
Remove scaffold				.130		6.10		6.10	8.10	10
Total				1.279	6.25	53.28	2.20	61.73	78.50	95.95

C3033 107 Gypsum Wall Board

System Description	Freq. (Years)	Crew	Unit	Labor Hours	2015 Bare Costs				Total In-House	Total w/O&P
					Material	Labor	Equipment	Total		
0010 Repair gypsum board ceiling - (2% of ceilings)	20	1 CARP	C.S.F.							
Set up, secure and take down ladder				1.300		61		61	81	100
Remove damaged gypsum board				2.737		103		103	137	169
Replace 5/8" gypsum board, taped and finished				3.382	41	159		200	256	310
Total				7.419	41	323		364	474	579
0020 Refinish gypsum board ceiling, up to 12' high	20	1 PORD	C.S.F.							
Set up, secure and take down ladder				1.300		61		61	81	100
Wash surface				.250		10		10	13	16
Sand & paint				.990	6	40		46	59	71.50
Place and remove mask and drops				.370		15		15	19	24
Total				2.910	6	126		132	172	211.50

For customer support on your Facilities Maintenance & Repair Cost Data, call 877.801.5724.

151

C3033 107 Gypsum Wall Board

System Description	Freq. (Years)	Crew	Unit	Labor Hours	2015 Bare Costs				Total In-House	Total w/O&P
					Material	Labor	Equipment	Total		
0040 **Replace gypsum board ceiling, up to 12' high**	40	2 CARP	C.S.F.							
Set up and secure scaffold				.750		35.25		35.25	47	57.50
Remove damaged gypsum board				2.737		103		103	137	169
Replace 5/8" gypsum board ceiling, taped and finished				3.382	41	159		200	256	310
Remove scaffold				.750		35.25		35.25	47	57.50
Total				7.619	41	332.50		373.50	487	594
2020 **Refinish gypsum board ceiling, 12' to 24' high**	5	2 PORD	C.S.F.							
Set up and secure scaffold				1.500		70.50		70.50	93.50	115
Place and remove mask and drops				.370		15		15	19	24
Prepare surface				.370		15		15	19	24
Paint surface, roller + brushwork, 1 coat				.696	7	28		35	44	54
Remove scaffold				1.500		70.50		70.50	93.50	115
Total				4.436	7	199		206	269	332
3020 **Refinish gypsum board ceiling, over 24' high**	5	2 PORD	C.S.F.							
Set up and secure scaffold				2.250		105.75		105.75	140	173
Place and remove mask and drops				.370		15		15	19	24
Prepare surface				.370		15		15	19	24
Paint surface, roller + brushwork, 1 coat				.696	7	28		35	44	54
Remove scaffold				2.250		105.75		105.75	140	173
Total				5.936	7	269.50		276.50	362	448

C3033 108 Acoustic Tile

System Description	Freq. (Years)	Crew	Unit	Labor Hours	2015 Bare Costs				Total In-House	Total w/O&P
					Material	Labor	Equipment	Total		
0010 **Acoustic tile repairs - (2% of ceilings)**	9	1 CARP	C.S.F.							
Set up, secure and take down ladder				1.300		61		61	81	100
Remove damaged tile				1.067		40		40	53	66
Install new tile				2.737	336	129		465	540	630
Total				5.104	336	230		566	674	796

For customer support on your Facilities Maintenance & Repair Cost Data, call 877.801.5724.

C3033 108 Acoustic Tile

System Description	Freq. (Years)	Crew	Unit	Labor Hours	2015 Bare Costs				Total In-House	Total w/O&P
					Material	Labor	Equipment	Total		
0020 Replace acoustic tile ceiling, non fire-rated	20	1 CARP	C.S.F.							
Set up, secure and take down scaffold				.750		35.25		35.25	47	57.50
Remove old ceiling tiles				.500		19		19	25	31
Remove old ceiling grid				.625		24		24	31	39
Install new ceiling grid				1.000	73	47		120	142	168
Install new ceiling tiles				1.280	91	60		151	180	213
Sweep and clean debris				.520		27.50		27.50	35.50	44
Total				4.675	164	212.75		376.75	460.50	552.50
0030 Refinish acoustic tile ceiling/grid (unoccupied area)	5	2 PORD	C.S.F.							
Protect exposed surfaces and mask light fixtures				.080		3.25		3.25	4.20	5.20
Refinish acoustic tiles, roller + brush, 1 coat				.267	.01	10.80		10.81	14	17.30
Total				.347	.01	14.05		14.06	18.20	22.50
0040 Refinish acoustic tile ceiling/grid (occupied area)	5	2 PORD	C.S.F.							
Protect exposed surfaces and mask light fixtures				.320		12.90		12.90	16.80	21
Refinish acoustic tiles, roller + brush, 1 coat				.320	.01	12.90		12.91	16.80	21
Total				.640	.01	25.80		25.81	33.60	42
2030 Refinish acoustic tile ceiling, 12' to 24' high	5	2 PORD	C.S.F.							
Set up and secure scaffold				1.500		70.50		70.50	93.50	115
Place and remove mask and drops				.100		2		2	2.60	3
Paint surface, roller + brushwork, 1 coat				.267	.01	10.80		10.81	14	17.30
Remove scaffold				1.500		70.50		70.50	93.50	115
Total				3.367	.01	153.80		153.81	203.60	250.30
3030 Refinish acoustic tile ceiling, over 24' high	5	2 PORD	C.S.F.							
Set up and secure scaffold				2.250		105.75		105.75	140	173
Place and remove mask and drops				.100		2		2	2.60	3
Paint surface, roller + brushwork, 1 coat				.267	.01	10.80		10.81	14	17.30
Remove scaffold				2.250		105.75		105.75	140	173
Total				4.867	.01	224.30		224.31	296.60	366.30

For customer support on your Facilities Maintenance & Repair Cost Data, call 877.801.5724.

153

C3033 109 Acoustic Tile, Fire-Rated

	System Description	Freq. (Years)	Crew	Unit	Labor Hours	2015 Bare Costs				Total In-House	Total w/O&P
						Material	Labor	Equipment	Total		
0010	**Replace acoustic tile ceiling, fire-rated**	20	1 CARP	C.S.F.							
	Set up, secure and take down scaffold				.750		35.25		35.25	47	57.50
	Remove old ceiling tiles				.500		19		19	25	31
	Remove old grid system				.625		24		24	31	39
	Install new grid system				1.200	87.60	56.40		144	170	202
	Install new ceiling tiles				1.185	108	56		164	193	226
	Sweep & clean debris				.520		27.50		27.50	35.50	44
	Total				4.780	195.60	218.15		413.75	501.50	599.50

C3033 120 Wood

	System Description	Freq. (Years)	Crew	Unit	Labor Hours	2015 Bare Costs				Total In-House	Total w/O&P
						Material	Labor	Equipment	Total		
0010	**Repair wood ceiling - (2% of ceilings)**	10	1 CARP	C.S.F.							
	Set up, secure and take down ladder				1.300		61		61	81	100
	Remove damaged ceiling				1.455		55		55	73	90
	Replace wood				5.200	149	244		393	490	585
	Total				7.955	149	360		509	644	775
0030	**Refinish wood ceiling**	6	1 PORD	C.S.F.							
	Set up and secure scaffold				.375		17.63		17.63	23.50	29
	Wash surface				.286		12		12	15	19
	Sand & paint, brushwork, 1 coat				.990	6	40		46	59	71.50
	Place and remove mask and drops				.370		15		15	19	24
	Remove scaffold				.375		17.63		17.63	23.50	29
	Total				2.396	6	102.26		108.26	140	172.50
0040	**Replace wood ceiling**	50	2 CARP	C.S.F.							
	Set up and secure scaffold				.750		35.25		35.25	47	57.50
	Remove damaged ceiling				1.455		55		55	73	90
	Replace wood				5.200	149	244		393	490	585
	Remove scaffold				.750		35.25		35.25	47	57.50
	Total				8.155	149	369.50		518.50	657	790

System Description	Freq. (Years)	Crew	Unit	Labor Hours	2015 Bare Costs				Total In-House	Total w/O&P
					Material	Labor	Equipment	Total		
0010 Replace flush valve diaphragm	10	1 PLUM	Ea.							
Turn valve off and on				.010		.61		.61	.76	.95
Remove / replace diaphragm				.300	3.41	17.60		21.01	26	32
Check for leaks				.013		.76		.76	.95	1.19
Total				.323	3.41	18.97		22.38	**27.71**	**34.14**
0015 Rebuild flush valve	20	1 PLUM	Ea.							
Turn valve off and on				.010		.61		.61	.76	.95
Remove flush valve				.100		5.85		5.85	7.35	9.15
Rebuild valve				.350	59.50	20.50		80	91	106
Reinstall valve				1.000		58.50		58.50	73.50	91.50
Check operation				.024		1.41		1.41	1.76	2.20
Total				1.484	59.50	86.87		146.37	**174.37**	**209.80**
0020 Unplug clogged line	5	1 PLUM	Ea.							
Turn valve off and on				.010		.61		.61	.76	.95
Remove water closet				2.078		122		122	153	190
Unplug line				.200		11.75		11.75	14.70	18.30
Reinstall water closet				.510		30		30	37.50	46.50
Check operation				.167		9.80		9.80	12.25	15.30
Total				2.965		174.16		174.16	**218.21**	**271.05**
0030 Replace tankless water closet	35	2 PLUM	Ea.							
Turn valve off and on				.010		.61		.61	.76	.95
Remove flush valve				.100		5.85		5.85	7.35	9.15
Remove water closet				.460		27		27	34	42
Install wall-hung water closet				2.759	945	146		1,091	1,225	1,400
Check operation				.150		8.80		8.80	11	13.75
Total				3.479	945	188.26		1,133.26	**1,278.11**	**1,465.85**
0040 Replace tankless flush valve	25	1 PLUM	Ea.							
Turn valve off and on				.010		.61		.61	.76	.95
Remove flush valve				.100		5.85		5.85	7.35	9.15
Install water closet flush valve				1.000	148	58.50		206.50	236	277
Check operation				.024		1.41		1.41	1.76	2.20
Total				1.134	148	66.37		214.37	245.87	**289.30**

For customer support on your Facilities Maintenance & Repair Cost Data, call 877.801.5724.

155

D2013 110 | Tankless Water Closet

System Description	Freq. (Years)	Crew	Unit	Labor Hours	2015 Bare Costs				Total In-House	Total w/O&P
					Material	Labor	Equipment	Total		
0050 Replace wax ring gasket	5	1 PLUM	Ea.							
Turn valve off and on				.010		.61		.61	.76	.95
Remove floor mounted tankless bowl				.667		39		39	49	61
Install new wax gasket				.083	1.38	4.89		6.27	7.65	9.40
Install floor mounted tankless bowl				1.000		58.50		58.50	73.50	91.50
Check operation				.150		8.80		8.80	11	13.75
Total				1.910	1.38	111.80		113.18	141.91	176.60
0080 Replace stainless steel detention water closet flush valve actuator	20	1 SKWK	Ea.							
Turn valve off and on				.010		.61		.61	.76	.95
Remove flush valve actuator				.100		5.85		5.85	7.35	9.15
Remove flush valve				.100		5.85		5.85	7.35	9.15
Detention equipment, replace toilet actuator				1.000	299	48.50		347.50	395	455
Check operation				.024		1.41		1.41	1.76	2.20
Total				1.234	299	62.22		361.22	412.22	476.45
0090 Replace stainless steel detention water closet flush valve	20	1 SKWK	Ea.							
Turn valve off and on				.010		.61		.61	.76	.95
Remove flush valve				.100		5.85		5.85	7.35	9.15
Detention equipment, replace water closet valve				1.000	325	48.50		373.50	420	485
Check operation				.024		1.41		1.41	1.76	2.20
Total				1.134	325	56.37		381.37	429.87	497.30

D2013 130 | Flush-Tank Water Closet

System Description	Freq. (Years)	Crew	Unit	Labor Hours	2015 Bare Costs				Total In-House	Total w/O&P
					Material	Labor	Equipment	Total		
0010 Unplug clogged line	5	1 PLUM	Ea.							
Turn valve off and on				.010		.61		.61	.76	.95
Remove water closet				2.078		122		122	153	190
Unplug line				.200		11.75		11.75	14.70	18.30
Reinstall water closet				.510		30		30	37.50	46.50
Check operation				.167		9.80		9.80	12.25	15.30
Total				2.965		174.16		174.16	218.21	271.05

System Description	Freq. (Years)	Crew	Unit	Labor Hours	2015 Bare Costs				Total In-House	Total w/O&P
					Material	Labor	Equipment	Total		
0020 Replace washer / diaphragm in ball cock	5	1 PLUM	Ea.							
Turn valve on and off				.010		.61		.61	.76	.95
Remove and replace washer / diaphragm				.180	1.76	10.55		12.31	15.15	18.70
Check operation				.024		1.41		1.41	1.76	2.20
Total				.214	1.76	12.57		14.33	**17.67**	**21.85**
0030 Replace valve and ball cock assembly	15	1 PLUM	Ea.							
Turn valve off and on				.010		.61		.61	.76	.95
Remove flush valve and ball cock assembly				.300		17.60		17.60	22	27.50
Install flush valve and ball cock assembly				.610	13.75	36		49.75	60	73
Check operation				.167		9.80		9.80	12.25	15.30
Total				1.087	13.75	64.01		77.76	**95.01**	**116.75**
0040 Install gasket between tank and bowl	20	1 PLUM	Ea.							
Turn valve off and on				.010		.61		.61	.76	.95
Remove tank and gasket				.200		11.75		11.75	14.70	18.30
Install new gasket and reinstall tank				.300	2.38	17.60		19.98	24.50	30.50
Inspect connection				.009		.53		.53	.67	.83
Total				.520	2.38	30.49		32.87	**40.63**	**50.58**
0050 Replace two piece water closet	35	2 PLUM	Ea.							
Turn water off and on				.010		.61		.61	.76	.95
Remove floor-mounted water closet				1.000		58.50		58.50	73.50	91.50
Install new bowl, tank and seat				3.019	237	159		396	460	545
Check for leaks				.013		.76		.76	.95	1.19
Total				4.042	237	218.87		455.87	**535.21**	**638.64**
0060 Replace one piece water closet	35	2 PLUM	Ea.							
Turn water off and on				.010		.61		.61	.76	.95
Remove floor-mounted water closet				1.000		58.50		58.50	73.50	91.50
Install floor-mounted water closet with seat				3.019	1,125	159		1,284	1,425	1,650
Check for leaks				.013		.76		.76	.95	1.19
Total				4.042	1,125	218.87		1,343.87	**1,500.21**	**1,743.64**

For customer support on your Facilities Maintenance & Repair Cost Data, call 877.801.5724.

157

D2013 210 | Urinal

System Description	Freq. (Years)	Crew	Unit	Labor Hours	2015 Bare Costs Material	Labor	Equipment	Total	Total In-House	Total w/O&P
0010 Replace flush valve diaphragm	7	1 PLUM	Ea.							
Turn valve off and on				.010		.61		.61	.76	.95
Remove / replace diaphragm				.300	3.41	17.60		21.01	26	32
Check for leaks				.013		.76		.76	.95	1.19
Total				.323	3.41	18.97		22.38	**27.71**	**34.14**
0015 Rebuild flush valve	20	1 PLUM	Ea.							
Turn valve off and on				.010		.61		.61	.76	.95
Remove flush valve				.100		5.85		5.85	7.35	9.15
Rebuild valve				.350	59.50	20.50		80	91	106
Reinstall valve				1.000		58.50		58.50	73.50	91.50
Check operation				.024		1.41		1.41	1.76	2.20
Total				1.484	59.50	86.87		146.37	**174.37**	**209.80**
0020 Unplug line	5	1 PLUM	Ea.							
Turn valve off and on				.010		.61		.61	.76	.95
Remove and reinstall bowl				1.600		94		94	117	146
Unplug line				.200		11.75		11.75	14.70	18.30
Check operation				.167		9.80		9.80	12.25	15.30
Total				1.977		116.16		116.16	**144.71**	**180.55**
0030 Replace wall-hung urinal	35	2 PLUM	Ea.							
Turn valve off and on				.010		.61		.61	.76	.95
Remove urinal				2.667		157		157	196	244
Fit up urinal to wall				.196		11.50		11.50	14.40	17.95
Install flushometer pipe				.096		5.65		5.65	7.05	8.80
Install urinal				5.333	282	282		564	660	795
Check operation				.334		19.60		19.60	24.50	30.50
Total				8.636	282	476.36		758.36	**902.71**	**1,097.20**

For customer support on your Facilities Maintenance & Repair Cost Data, call 877.801.5724.

D2013 310   Lavatory, Iron, Enamel

System Description	Freq. (Years)	Crew	Unit	Labor Hours	2015 Bare Costs				Total In-House	Total w/O&P
					Material	Labor	Equipment	Total		
0010 Replace washer in spud connection	7	1 PLUM	Ea.							
Turn valve off and on				.022		1.30		1.30	1.62	2.02
Remove gasket or washer				.017		.99		.99	1.24	1.55
Install new gasket / washer set				.026	3.85	1.53		5.38	6.15	7.20
Install spud connection				.026		1.52		1.52	1.91	2.38
Inspect connection				.009		.53		.53	.67	.83
Total				.100	3.85	5.87		9.72	11.59	13.98
0020 Replace washer in faucet	2	1 PLUM	Ea.							
Turn valves on and off				.022		1.30		1.30	1.62	2.02
Remove valve stem and washer				.091		5.35		5.35	6.65	8.30
Install washer				.026	.41	1.53		1.94	2.36	2.90
Install stem				.017		.99		.99	1.24	1.55
Test faucet				.010		.61		.61	.76	.95
Total				.166	.41	9.78		10.19	12.63	15.72
0040 Replace faucets	10	1 PLUM	Ea.							
Turn valves on and off				.022		1.30		1.30	1.62	2.02
Remove faucets and tubing				.381		22.50		22.50	28	35
Install detention sink faucets and tubing				1.040	44.50	61		105.50	125	151
Test faucets				.010		.61		.61	.76	.95
Total				1.453	44.50	85.41		129.91	155.38	188.97
0050 Clean out strainer and P trap	2	1 PLUM	Ea.							
Remove strainer and P trap				.157		9.20		9.20	11.50	14.35
Clean and reinstall strainer and trap				.320		18.80		18.80	23.50	29.50
Total				.477		28		28	35	43.85
0060 Replace lavatory	40	2 PLUM	Ea.							
Shut off water (hot and cold)				.022		1.30		1.30	1.62	2.02
Disconnect and remove wall-hung lavatory				.800		47		47	58.50	73
Install new wall-hung lavatory with trim, (18" x 15")				2.807	405	165		570	650	765
Total				3.629	405	213.30		618.30	710.12	840.02

D2013 330   **Lavatory, Vitreous China**

System Description	Freq. (Years)	Crew	Unit	Labor Hours	2015 Bare Costs				Total In-House	Total w/O&P
					Material	Labor	Equipment	Total		
0010 Replace washer in spud connection	7	1 PLUM	Ea.							
Turn valve off and on				.022		1.30		1.30	1.62	2.02
Loosen locknuts				.011		.65		.65	.81	1.01
Remove spud connection				.017		.99		.99	1.24	1.55
Remove gasket or washer				.017		.99		.99	1.24	1.55
Clean spud seat				.021		1.22		1.22	1.53	1.90
Install new gasket / washer set				.026	3.85	1.53		5.38	6.15	7.20
Install spud connection				.026		1.52		1.52	1.91	2.38
Tighten locknut				.014		.84		.84	1.05	1.31
Inspect connection				.009		.53		.53	.67	.83
Total				.163	3.85	9.57		13.42	16.22	19.75
0020 Replace washer in faucet	2	1 PLUM	Ea.							
Turn valves on and off				.022		1.30		1.30	1.62	2.02
Remove packing nut and install				.023		1.37		1.37	1.72	2.14
Remove valve stem and cap				.017		.99		.99	1.24	1.55
Remove seat washer and install				.023		1.37		1.37	1.72	2.14
Pry out washer				.017		.99		.99	1.24	1.55
Install washer				.026	.41	1.53		1.94	2.36	2.90
Install stem				.017		.99		.99	1.24	1.55
Tighten main gland valves				.012		.69		.69	.86	1.07
Test faucet				.010		.61		.61	.76	.95
Total				.168	.41	9.84		10.25	12.76	15.87
0040 Replace faucets	10	1 PLUM	Ea.							
Turn valves on and off				.022		1.30		1.30	1.62	2.02
Remove faucets and tubing				.381		22.50		22.50	28	35
Install faucets and tubing				1.040	44.50	61		105.50	125	151
Test faucets				.010		.61		.61	.76	.95
Total				1.453	44.50	85.41		129.91	155.38	188.97
0050 Clean out strainer and P trap	2	1 PLUM	Ea.							
Remove strainer and P trap				.157		9.20		9.20	11.50	14.35
Clean and reinstall strainer and trap				.320		18.80		18.80	23.50	29.50
Total				.477		28		28	35	43.85

D2013 330 Lavatory, Vitreous China

	System Description	Freq. (Years)	Crew	Unit	Labor Hours	2015 Bare Costs Material	2015 Bare Costs Labor	2015 Bare Costs Equipment	2015 Bare Costs Total	Total In-House	Total w/O&P
0060	**Replace lavatory**	35	2 PLUM	Ea.							
	Shut off water (hot and cold)				.022		1.30		1.30	1.62	2.02
	Disconnect and remove wall-hung lavatory				1.485		87		87	109	136
	Install new lavatory (to 20" x 17")				2.971	187	157		344	400	480
	Install faucets and tubing				1.040	44.50	61		105.50	125	151
	Total				5.518	231.50	306.30		537.80	635.62	769.02

D2013 350 Lavatory, Enameled Steel

	System Description	Freq. (Years)	Crew	Unit	Labor Hours	2015 Bare Costs Material	2015 Bare Costs Labor	2015 Bare Costs Equipment	2015 Bare Costs Total	Total In-House	Total w/O&P
0010	**Replace washer in spud connection**	7	1 PLUM	Ea.							
	Turn valve off and on				.022		1.30		1.30	1.62	2.02
	Loosen locknut				.014		.84		.84	1.05	1.31
	Remove spud connection				.017		.99		.99	1.24	1.55
	Remove gasket or washer				.017		.99		.99	1.24	1.55
	Clean spud seat				.021		1.22		1.22	1.53	1.90
	Install new gasket / washer set				.026	3.85	1.53		5.38	6.15	7.20
	Install spud connection				.026		1.52		1.52	1.91	2.38
	Tighten locknut				.014		.84		.84	1.05	1.31
	Inspect connection				.009		.53		.53	.67	.83
	Total				.166	3.85	9.76		13.61	16.46	20.05
0020	**Replace washer in faucet**	2	1 PLUM	Ea.							
	Turn valves on and off				.022		1.30		1.30	1.62	2.02
	Remove valve stem and washer				.091		5.35		5.35	6.65	8.30
	Install washer				.026	.41	1.53		1.94	2.36	2.90
	Install stem				.017		.99		.99	1.24	1.55
	Test faucet				.010		.61		.61	.76	.95
	Total				.166	.41	9.78		10.19	12.63	15.72
0040	**Replace faucets**	10	1 PLUM	Ea.							
	Turn valves on and off				.022		1.30		1.30	1.62	2.02
	Remove faucets and tubing				.381		22.50		22.50	28	35
	Install faucets and tubing				1.040	44.50	61		105.50	125	151
	Test faucets				.010		.61		.61	.76	.95
	Total				1.453	44.50	85.41		129.91	155.38	188.97

For customer support on your **Facilities Maintenance & Repair Cost Data**, call 877.801.5724.

161

D2013 350 — Lavatory, Enameled Steel

	System Description	Freq. (Years)	Crew	Unit	Labor Hours	Material	Labor	Equipment	Total	Total In-House	Total w/O&P
0050	**Clean out strainer and P trap**	2	1 PLUM	Ea.							
	Remove strainer and P trap				.157		9.20		9.20	11.50	14.35
	Clean and reinstall strainer and trap				.320		18.80		18.80	23.50	29.50
	Total				.477		28		28	**35**	**43.85**
0060	**Replace lavatory**	35	2 PLUM	Ea.							
	Shut off water (hot and cold)				.022		1.30		1.30	1.62	2.02
	Disconnect and remove wall-hung lavatory				1.793		105		105	132	164
	Install new lavatory (to 20" x 17")				3.586	116	189		305	365	440
	Install faucets and tubing				1.040	44.50	61		105.50	125	151
	Total				6.441	160.50	356.30		516.80	**623.62**	**757.02**

D2013 410 — Sink, Iron Enamel

	System Description	Freq. (Years)	Crew	Unit	Labor Hours	Material	Labor	Equipment	Total	Total In-House	Total w/O&P
0010	**Replace faucet washer**	2	1 PLUM	Ea.							
	Turn valves on and off				.022		1.30		1.30	1.62	2.02
	Remove valve stem and washer				.091		5.35		5.35	6.65	8.30
	Install washer				.026	.41	1.53		1.94	2.36	2.90
	Install stem				.017		.99		.99	1.24	1.55
	Test faucet				.010		.61		.61	.76	.95
	Total				.166	.41	9.78		10.19	**12.63**	**15.72**
0020	**Clean trap**	3	1 PLUM	Ea.							
	Drain trap				.021		1.24		1.24	1.55	1.93
	Remove / install trap				.043		2.54		2.54	3.17	3.96
	Inspect trap				.031		1.83		1.83	2.29	2.86
	Clean trap				.021		1.22		1.22	1.53	1.90
	Total				.116		6.83		6.83	**8.54**	**10.65**
0030	**Replace faucets**	10	1 PLUM	Ea.							
	Turn valves on and off				.022		1.30		1.30	1.62	2.02
	Remove faucets and tubing				.381		22.50		22.50	28	35
	Install faucets and tubing				1.040	44.50	61		105.50	125	151
	Test faucets				.010		.61		.61	.76	.95
	Total				1.453	44.50	85.41		129.91	**155.38**	**188.97**

D2013 410 Sink, Iron Enamel

System Description	Freq. (Years)	Crew	Unit	Labor Hours	2015 Bare Costs				Total In-House	Total w/O&P
					Material	Labor	Equipment	Total		
0040 Unstop sink										
Unstop plugged drain	2	1 PLUM	Ea.	.571		33.50		33.50	42	52.50
Total				.571		33.50		33.50	**42**	**52.50**
0060 Replace kitchen sink, P.E.C.I.										
Remove PE sink	35	2 PLUM	Ea.	1.625		86		86	107	134
Install countertop porcelain enameled kitchen sink, 24" x 21"				2.857	285	151		436	500	590
Total				4.482	285	237		522	**607**	**724**
0070 Replace laundry sink, P.E.C.I.										
Remove PE sink	35	2 PLUM	Ea.	1.625		86		86	107	134
Install free-standing porcelain enameled laundry sink, 24" x 21"				2.667	580	141		721	815	945
Total				4.292	580	227		807	**922**	**1,079**

D2013 420 Sink, Enameled Steel

System Description	Freq. (Years)	Crew	Unit	Labor Hours	2015 Bare Costs				Total In-House	Total w/O&P
					Material	Labor	Equipment	Total		
0010 Replace faucet washer										
Turn valves on and off	2	1 PLUM	Ea.	.022		1.30		1.30	1.62	2.02
Remove valve stem and washer				.091		5.35		5.35	6.65	8.30
Install washer				.026	.41	1.53		1.94	2.36	2.90
Install stem				.017		.99		.99	1.24	1.55
Test faucet				.010		.61		.61	.76	.95
Total				.166	.41	9.78		10.19	**12.63**	**15.72**
0020 Clean trap										
Drain trap	3	1 PLUM	Ea.	.021		1.24		1.24	1.55	1.93
Remove / install trap				.043		2.54		2.54	3.17	3.96
Inspect trap				.031		1.83		1.83	2.29	2.86
Clean trap				.021		1.22		1.22	1.53	1.90
Total				.116		6.83		6.83	**8.54**	**10.65**

D2013 420 Sink, Enameled Steel

System Description	Freq. (Years)	Crew	Unit	Labor Hours	Material	Labor	Equipment	Total	Total In-House	Total w/O&P
					2015 Bare Costs					
0030 Replace faucets	10	1 PLUM	Ea.							
Turn valves on and off				.022		1.30		1.30	1.62	2.02
Remove faucets and tubing				.381		22.50		22.50	28	35
Install faucets and tubing				1.040	44.50	61		105.50	125	151
Test faucets				.010		.61		.61	.76	.95
Total				1.453	44.50	85.41		129.91	155.38	188.97
0040 Unstop sink	2	1 PLUM	Ea.							
Unstop plugged drain				.571		33.50		33.50	42	52.50
Total				.571		33.50		33.50	42	52.50
0060 Replace sink, enameled steel	35	2 PLUM	Ea.							
Remove sink				1.793		94.50		94.50	118	148
Install countertop enameled steel sink, 24" x 21"				3.714	510	196		706	805	945
Total				5.507	510	290.50		800.50	923	1,093

D2013 430 Sink, Stainless Steel

System Description	Freq. (Years)	Crew	Unit	Labor Hours	Material	Labor	Equipment	Total	Total In-House	Total w/O&P
					2015 Bare Costs					
0010 Replace faucet washer	2	1 PLUM	Ea.							
Turn valves on and off				.022		1.30		1.30	1.62	2.02
Remove valve stem and washer				.091		5.35		5.35	6.65	8.30
Install washer				.026	.41	1.53		1.94	2.36	2.90
Install stem				.017		.99		.99	1.24	1.55
Test faucet				.010		.61		.61	.76	.95
Total				.166	.41	9.78		10.19	12.63	15.72
0020 Clean trap	3	1 PLUM	Ea.							
Drain trap				.021		1.24		1.24	1.55	1.93
Remove / install trap				.043		2.54		2.54	3.17	3.96
Inspect trap				.031		1.83		1.83	2.29	2.86
Clean trap				.021		1.22		1.22	1.53	1.90
Total				.116		6.83		6.83	8.54	10.65

D2013 430 Sink, Stainless Steel

System Description	Freq. (Years)	Crew	Unit	Labor Hours	2015 Bare Costs				Total In-House	Total w/O&P
					Material	Labor	Equipment	Total		
0030 **Replace faucets**	10	1 PLUM	Ea.							
Turn valves on and off				.022		1.30		1.30	1.62	2.02
Remove faucets and tubing				.381		22.50		22.50	28	35
Install faucets and tubing				1.040	44.50	61		105.50	125	151
Test faucets				.010		.61		.61	.76	.95
Total				1.453	44.50	85.41		129.91	**155.38**	**188.97**
0040 **Unstop sink**	2	1 PLUM	Ea.							
Unstop plugged drain				.571		33.50		33.50	42	52.50
Total				.571		33.50		33.50	**42**	**52.50**
0060 **Replace sink, stainless steel**	40	2 PLUM	Ea.							
Remove sink				1.625		86		86	107	134
Install countertop stainless steel sink, 25" x 22"				3.714	660	196		856	970	1,125
Total				5.339	660	282		942	**1,077**	**1,259**
0070 **Replace stainless steel detention sink faucets**	15	1 SKWK	Ea.							
Turn valves on and off				.022		1.30		1.30	1.62	2.02
Remove faucets and tubing				.381		22.50		22.50	28	35
Detention equipment, replace faucet valve and tubing				1.000	625	48.50		673.50	750	860
Test faucets				.010		.61		.61	.76	.95
Total				1.413	625	72.91		697.91	**780.38**	**897.97**

D2013 440 Sink, Plastic

System Description	Freq. (Years)	Crew	Unit	Labor Hours	2015 Bare Costs				Total In-House	Total w/O&P
					Material	Labor	Equipment	Total		
0010 **Replace faucet washer**	2	1 PLUM	Ea.							
Turn valves on and off				.022		1.30		1.30	1.62	2.02
Remove valve stem and washer				.091		5.35		5.35	6.65	8.30
Install washer				.026	.41	1.53		1.94	2.36	2.90
Install stem				.017		.99		.99	1.24	1.55
Test faucet				.010		.61		.61	.76	.95
Total				.166	.41	9.78		10.19	**12.63**	**15.72**

For customer support on your Facilities Maintenance & Repair Cost Data, call 877.801.5724.

165

	System Description	Freq. (Years)	Crew	Unit	Labor Hours	2015 Bare Costs				Total In-House	Total w/O&P
						Material	Labor	Equipment	Total		
0020	**Clean trap**	3	1 PLUM	Ea.							
	Drain trap				.021		1.24		1.24	1.55	1.93
	Remove / install trap				.043		2.54		2.54	3.17	3.96
	Inspect trap				.031		1.83		1.83	2.29	2.86
	Clean trap				.021		1.22		1.22	1.53	1.90
	Total				.116		6.83		6.83	8.54	10.65
0030	**Replace faucets**	10	1 PLUM	Ea.							
	Turn valves on and off				.022		1.30		1.30	1.62	2.02
	Remove faucets and tubing				.381		22.50		22.50	28	35
	Install faucets and tubing				1.040	44.50	61		105.50	125	151
	Test faucets				.010		.61		.61	.76	.95
	Total				1.453	44.50	85.41		129.91	155.38	188.97
0040	**Unstop, sink**	2	1 PLUM	Ea.							
	Unstop plugged drain				.571		33.50		33.50	42	52.50
	Total				.571		33.50		33.50	42	52.50
0060	**Replace sink and fittings, polyethylene**	15	2 PLUM	Ea.							
	Remove sink				3.467		183		183	229	286
	Install bench mounted polyethylene sink, 18-1/2" x 18-1/2"				4.000	335	211		546	635	750
	Total				7.467	335	394		729	864	1,036
0080	**Replace lab sink and fittings, polyethylene**	15	2 PLUM	Ea.							
	Remove sink				3.467		183		183	229	286
	Install bench mounted PE lab sink, 23-1/2" x 20-1/2"				5.333	1,875	282		2,157	2,425	2,775
	Total				8.800	1,875	465		2,340	2,654	3,061

For customer support on your Facilities Maintenance & Repair Cost Data, call 877.801.5724.

D2013 444 — Laboratory Sink

	System Description	Freq. (Years)	Crew	Unit	Labor Hours	2015 Bare Costs				Total In-House	Total w/O&P
						Material	Labor	Equipment	Total		
0010	**Replace faucet washer**	2	1 PLUM	Ea.							
	Turn valves on and off				.022		1.30		1.30	1.62	2.02
	Remove valve stem and washer				.091		5.35		5.35	6.65	8.30
	Install washer				.026	.41	1.53		1.94	2.36	2.90
	Install stem				.017		.99		.99	1.24	1.55
	Test faucet				.010		.61		.61	.76	.95
	Total				.166	.41	9.78		10.19	12.63	15.72
0030	**Replace faucets**	10	1 PLUM	Ea.							
	Turn valves on and off				.022		1.30		1.30	1.62	2.02
	Remove faucets and tubing				.381		22.50		22.50	28	35
	Install faucets and tubing				1.040	44.50	61		105.50	125	151
	Test faucets				.010		.61		.61	.76	.95
	Total				1.453	44.50	85.41		129.91	155.38	188.97
0060	**Replace sink and fittings, stainless steel, lab.**	15	2 PLUM	Ea.							
	Remove lab. sink				3.467		183		183	229	286
	Install bench mounted stainless steel lab. sink, 47" x 24"				5.333	1,875	282		2,157	2,425	2,775
	Total				8.800	1,875	465		2,340	2,654	3,061

D2013 450 — Laundry Sink, Plastic

	System Description	Freq. (Years)	Crew	Unit	Labor Hours	2015 Bare Costs				Total In-House	Total w/O&P
						Material	Labor	Equipment	Total		
0020	**Replace washer in faucet**	2	1 PLUM	Ea.							
	Turn valves on and off				.022		1.30		1.30	1.62	2.02
	Remove valve stem and washer				.091		5.35		5.35	6.65	8.30
	Install washer				.026	.41	1.53		1.94	2.36	2.90
	Install stem				.017		.99		.99	1.24	1.55
	Test faucet				.010		.61		.61	.76	.95
	Total				.166	.41	9.78		10.19	12.63	15.72

For customer support on your Facilities Maintenance & Repair Cost Data, call 877.801.5724.

167

D2013 450 Laundry Sink, Plastic

System Description	Freq. (Years)	Crew	Unit	Labor Hours	2015 Bare Costs				Total In-House	Total w/O&P
					Material	Labor	Equipment	Total		
0040 Replace faucets	10	1 PLUM	Ea.							
Turn valves on and off				.022		1.30		1.30	1.62	2.02
Remove faucets and tubing				.381		22.50		22.50	28	35
Install faucets and tubing				1.040	44.50	61		105.50	125	151
Test faucets				.010		.61		.61	.76	.95
Total				1.453	44.50	85.41		129.91	155.38	188.97
0050 Clean out strainer and P trap	2	1 PLUM	Ea.							
Remove strainer and P trap				.157		9.20		9.20	11.50	14.35
Clean and reinstall strainer and trap				.320		18.80		18.80	23.50	29.50
Total				.477		28		28	35	43.85
0060 Replace laundry sink	20	2 PLUM	Ea.							
Shut off water (hot and cold)				.022		1.30		1.30	1.62	2.02
Disconnect and remove laundry sink				1.600		84.50		84.50	106	132
Install sink (23" x 18", single)				3.200	135	169		304	360	435
Install faucets and tubing				1.040	44.50	61		105.50	125	151
Total				5.862	179.50	315.80		495.30	592.62	720.02

D2013 455 Laundry Sink, Stone

System Description	Freq. (Years)	Crew	Unit	Labor Hours	2015 Bare Costs				Total In-House	Total w/O&P
					Material	Labor	Equipment	Total		
0060 Replace laundry sink, molded stone	40	2 PLUM	Ea.							
Remove molded stone sink				1.625		86		86	107	134
Install laundry sink, w/trim, molded stone, 22"x23", sgl compt				2.667	167	141		308	360	430
Total				4.292	167	227		394	467	564

	System Description	Freq. (Years)	Crew	Unit	Labor Hours	2015 Bare Costs				Total In-House	Total w/O&P
						Material	Labor	Equipment	Total		
0010	**Replace faucet washer**	2	1 PLUM	Ea.							
	Turn valves on and off				.022		1.30		1.30	1.62	2.02
	Remove valve stem and washer				.091		5.35		5.35	6.65	8.30
	Install washer				.026	.41	1.53		1.94	2.36	2.90
	Install stem				.017		.99		.99	1.24	1.55
	Test faucet				.010		.61		.61	.76	.95
	Total				.166	.41	9.78		10.19	12.63	15.72
0020	**Clean trap**	3	1 PLUM	Ea.							
	Drain trap				.021		1.24		1.24	1.55	1.93
	Remove / install trap				.043		2.54		2.54	3.17	3.96
	Inspect trap				.031		1.83		1.83	2.29	2.86
	Clean trap				.021		1.22		1.22	1.53	1.90
	Total				.116		6.83		6.83	8.54	10.65
0030	**Replace faucets**	10	1 PLUM	Ea.							
	Turn valves on and off				.022		1.30		1.30	1.62	2.02
	Remove faucets and tubing				.381		22.50		22.50	28	35
	Install faucets and tubing				1.040	44.50	61		105.50	125	151
	Test faucets				.010		.61		.61	.76	.95
	Total				1.453	44.50	85.41		129.91	155.38	188.97
0040	**Unstop sink**	2	1 PLUM	Ea.							
	Unstop plugged drain				.571		33.50		33.50	42	52.50
	Total				.571		33.50		33.50	42	52.50
0060	**Replace service sink, P.E.C.I.**	35	1 PLUM	Ea.							
	Remove sink				1.625		86		86	107	134
	Install service/utility sink, 22" x 18"				5.212	770	275		1,045	1,200	1,400
	Total				6.837	770	361		1,131	1,307	1,534

D2013 470 Group Washfountain

	System Description	Freq. (Years)	Crew	Unit	Labor Hours	2015 Bare Costs				Total In-House	Total w/O&P
						Material	Labor	Equipment	Total		
0060	**Replace group wash fountain, 54" diameter**	20	Q-2	Ea.							
	Shut off water (hot and cold)				.022		1.30		1.30	1.62	2.02
	Disconnect and remove group wash fountain, 54" dia.				4.800		263		263	330	410
	Install group wash fountain, terrazzo, 54" diameter				9.600	9,050	525		9,575	10,600	12,100
	Install faucets and tubing				1.040	44.50	61		105.50	125	151
	Total				15.462	9,094.50	850.30		9,944.80	11,056.62	12,663.02

D2013 510 Bathtub, Cast Iron Enamel

	System Description	Freq. (Years)	Crew	Unit	Labor Hours	2015 Bare Costs				Total In-House	Total w/O&P
						Material	Labor	Equipment	Total		
0010	**Inspect / clean shower head**	3	1 PLUM	Ea.							
	Turn water off and on				.010		.61		.61	.76	.95
	Remove shower head				.217		12.70		12.70	15.90	19.85
	Check shower head				.009		.53		.53	.67	.83
	Install shower head				.433		25.50		25.50	32	39.50
	Total				.669		39.34		39.34	49.33	61.13
0020	**Replace mixing valve barrel**	2	1 PLUM	Ea.							
	Turn valves on and off				.022		1.30		1.30	1.62	2.02
	Remove mixing valve cover / barrel				.650		38		38	47.50	59.50
	Install new mixing valve barrel / replace cover				.650	78.50	38		116.50	134	158
	Test valve				.010		.61		.61	.76	.95
	Total				1.332	78.50	77.91		156.41	183.88	220.47
0030	**Replace mixing valve**	10	1 PLUM	Ea.							
	Turn valves off and on				.022		1.30		1.30	1.62	2.02
	Remove mixing valve				.867		51		51	63.50	79.50
	Install new mixing valve				1.733	153	102		255	296	350
	Test valve				.010		.61		.61	.76	.95
	Total				2.633	153	154.91		307.91	361.88	432.47
0070	**Replace tub**	40	2 PLUM	Ea.							
	Remove cast iron tub				1.891		100		100	125	156
	Install cast iron enameled bathtub				3.782	1,125	200		1,325	1,475	1,725
	Total				5.673	1,125	300		1,425	1,600	1,881

D2013 530 Bathtub, Enameled Steel

	System Description	Freq. (Years)	Crew	Unit	Labor Hours	2015 Bare Costs				Total In-House	Total w/O&P
						Material	Labor	Equipment	Total		
0010	**Inspect / clean shower head**	3	1 PLUM	Ea.							
	Turn water off and on				.010		.61		.61	.76	.95
	Remove shower head				.217		12.70		12.70	15.90	19.85
	Check shower head				.009		.53		.53	.67	.83
	Install shower head				.433		25.50		25.50	32	39.50
	Total				.669		39.34		39.34	**49.33**	**61.13**
0020	**Replace mixing valve barrel**	2	1 PLUM	Ea.							
	Turn valves on and off				.022		1.30		1.30	1.62	2.02
	Remove mixing valve cover / barrel				.650		38		38	47.50	59.50
	Install new mixing valve barrel / replace cover				.650	78.50	38		116.50	134	158
	Test valve				.010		.61		.61	.76	.95
	Total				1.332	78.50	77.91		156.41	**183.88**	**220.47**
0030	**Replace mixing valve**	10	1 PLUM	Ea.							
	Turn valves off and on				.022		1.30		1.30	1.62	2.02
	Remove mixing valve				.867		51		51	63.50	79.50
	Install new mixing valve				1.733	153	102		255	296	350
	Test valve				.010		.61		.61	.76	.95
	Total				2.633	153	154.91		307.91	**361.88**	**432.47**
0070	**Replace tub**	35	2 PLUM	Ea.							
	Remove tub				1.891		100		100	125	156
	Install enameled steel bathtub				3.782	495	200		695	795	930
	Total				5.673	495	300		795	**920**	**1,086**

D2013 550 Bathtub, Fiberglass

	System Description	Freq. (Years)	Crew	Unit	Labor Hours	2015 Bare Costs				Total In-House	Total w/O&P
						Material	Labor	Equipment	Total		
0010	**Inspect / clean shower head**	3	1 PLUM	Ea.							
	Turn water off and on				.010		.61		.61	.76	.95
	Remove shower head				.217		12.70		12.70	15.90	19.85
	Check shower head				.009		.53		.53	.67	.83
	Install shower head				.433		25.50		25.50	32	39.50
	Total				.669		39.34		39.34	**49.33**	**61.13**

For customer support on your Facilities Maintenance & Repair Cost Data, call 877.801.5724.

171

D2013 550 Bathtub, Fiberglass

	System Description	Freq. (Years)	Crew	Unit	Labor Hours	2015 Bare Costs				Total In-House	Total w/O&P
						Material	Labor	Equipment	Total		
0020	**Replace mixing valve barrel**	2	1 PLUM	Ea.							
	Turn valves on and off				.022		1.30		1.30	1.62	2.02
	Remove mixing valve cover / barrel				.650		38		38	47.50	59.50
	Install new mixing valve barrel / replace cover				.650	78.50	38		116.50	134	158
	Test valve				.010		.61		.61	.76	.95
	Total				1.332	78.50	77.91		156.41	183.88	220.47
0030	**Replace mixing valve**	10	1 PLUM	Ea.							
	Turn valves off and on				.022		1.30		1.30	1.62	2.02
	Remove mixing valve				.867		51		51	63.50	79.50
	Install new mixing valve				1.733	153	102		255	296	350
	Test valve				.010		.61		.61	.76	.95
	Total				2.633	153	154.91		307.91	361.88	432.47
0070	**Replace tub**	20	2 PLUM	Ea.							
	Remove bathtub				2.600		137		137	172	214
	Install fiberglass bathtub with shower wall surround				5.200	840	275		1,115	1,275	1,475
	Total				7.800	840	412		1,252	1,447	1,689

D2013 710 Shower, Terrazzo

	System Description	Freq. (Years)	Crew	Unit	Labor Hours	2015 Bare Costs				Total In-House	Total w/O&P
						Material	Labor	Equipment	Total		
0010	**Inspect / clean shower head**	3	1 PLUM	Ea.							
	Turn water off and on				.010		.61		.61	.76	.95
	Remove shower head				.217		12.70		12.70	15.90	19.85
	Check shower head				.009		.53		.53	.67	.83
	Install shower head				.433		25.50		25.50	32	39.50
	Total				.669		39.34		39.34	49.33	61.13
0020	**Replace mixing valve barrel**	2	1 PLUM	Ea.							
	Turn valves on and off				.022		1.30		1.30	1.62	2.02
	Remove mixing valve cover / barrel				.650		38		38	47.50	59.50
	Install new mixing valve barrel / replace cover				.650	78.50	38		116.50	134	158
	Test valve				.010		.61		.61	.76	.95
	Total				1.332	78.50	77.91		156.41	183.88	220.47

For customer support on your Facilities Maintenance & Repair Cost Data, call 877.801.5724.

D2013 710 Shower, Terrazzo

System Description	Freq. (Years)	Crew	Unit	Labor Hours	2015 Bare Costs				Total In-House	Total w/O&P
					Material	Labor	Equipment	Total		
0030 Replace mixing valve	10	1 PLUM	Ea.							
Turn valves off and on				.022		1.30		1.30	1.62	2.02
Remove mixing valve				.867		51		51	63.50	79.50
Install new mixing valve				1.733	153	102		255	296	350
Test valve				.010		.61		.61	.76	.95
Total				2.633	153	154.91		307.91	361.88	432.47
0060 Replace terrazzo shower surface	30	2 PLUM	Ea.							
Remove old terrazzo surface				3.004		113.10		113.10	150	185
Install precast terrazzo tiles				9.014	311.35	386.75		698.10	830	995
Install curb				2.080	114	90		204	239	284
Total				14.098	425.35	589.85		1,015.20	1,219	1,464

D2013 730 Shower, Enameled Steel

System Description	Freq. (Years)	Crew	Unit	Labor Hours	2015 Bare Costs				Total In-House	Total w/O&P
					Material	Labor	Equipment	Total		
0010 Inspect / clean shower head	3	1 PLUM	Ea.							
Turn water off and on				.010		.61		.61	.76	.95
Remove shower head				.217		12.70		12.70	15.90	19.85
Check shower head				.009		.53		.53	.67	.83
Install shower head				.433		25.50		25.50	32	39.50
Total				.669		39.34		39.34	49.33	61.13
0020 Replace mixing valve barrel	2	1 PLUM	Ea.							
Turn valves on and off				.022		1.30		1.30	1.62	2.02
Remove mixing valve cover / barrel				.650		38		38	47.50	59.50
Install new mixing valve barrel / replace cover				.650	78.50	38		116.50	134	158
Test valve				.010		.61		.61	.76	.95
Total				1.332	78.50	77.91		156.41	183.88	220.47
0030 Replace mixing valve	10	1 PLUM	Ea.							
Turn valves off and on				.022		1.30		1.30	1.62	2.02
Remove mixing valve				.867		51		51	63.50	79.50
Install new mixing valve				1.733	153	102		255	296	350
Test valve				.010		.61		.61	.76	.95
Total				2.633	153	154.91		307.91	361.88	432.47

For customer support on your Facilities Maintenance & Repair Cost Data, call 877.801.5724.

173

D2013 730 Shower, Enameled Steel

	System Description	Freq. (Years)	Crew	Unit	Labor Hours	2015 Bare Costs				Total In-House	Total w/O&P
						Material	Labor	Equipment	Total		
0060	**Replace shower**	35	2 PLUM	Ea.							
	Remove shower				5.200		275		275	345	430
	Install shower				4.522	1,150	239		1,389	1,575	1,825
	Total				9.722	1,150	514		1,664	1,920	2,255

D2013 750 Shower, Fiberglass

	System Description	Freq. (Years)	Crew	Unit	Labor Hours	2015 Bare Costs				Total In-House	Total w/O&P
						Material	Labor	Equipment	Total		
0010	**Inspect / clean shower head**	3	1 PLUM	Ea.							
	Turn water off and on				.010		.61		.61	.76	.95
	Remove shower head				.217		12.70		12.70	15.90	19.85
	Check shower head				.009		.53		.53	.67	.83
	Install shower head				.433		25.50		25.50	32	39.50
	Total				.669		39.34		39.34	49.33	61.13
0020	**Replace mixing valve barrel**	2	1 PLUM	Ea.							
	Turn valves on and off				.022		1.30		1.30	1.62	2.02
	Remove mixing valve cover / barrel				.650		38		38	47.50	59.50
	Install new mixing valve barrel / replace cover				.650	78.50	38		116.50	134	158
	Test valve				.010		.61		.61	.76	.95
	Total				1.332	78.50	77.91		156.41	183.88	220.47
0030	**Replace mixing valve**	10	1 PLUM	Ea.							
	Turn valves off and on				.022		1.30		1.30	1.62	2.02
	Remove mixing valve				.867		51		51	63.50	79.50
	Install new mixing valve				1.733	153	102		255	296	350
	Test valve				.010		.61		.61	.76	.95
	Total				2.633	153	154.91		307.91	361.88	432.47
0060	**Replace shower and fittings**	20	2 PLUM	Ea.							
	Remove shower				4.334		229		229	286	355
	Install shower, fiberglass				3.302	505	174		679	775	905
	Total				7.635	505	403		908	1,061	1,260

System Description	Freq. (Years)	Crew	Unit	Labor Hours	2015 Bare Costs				Total In-House	Total w/O&P
					Material	Labor	Equipment	Total		
0010 Inspect / clean shower head	3	1 PLUM	Ea.							
Turn water off and on				.010		.61		.61	.76	.95
Remove shower head				.217		12.70		12.70	15.90	19.85
Check shower head				.009		.53		.53	.67	.83
Install shower head				.433		25.50		25.50	32	39.50
Total				.669		39.34		39.34	49.33	61.13
0020 Replace mixing valve barrel	2	1 PLUM	Ea.							
Turn valves on and off				.022		1.30		1.30	1.62	2.02
Remove mixing valve cover / barrel				.650		38		38	47.50	59.50
Install new mixing valve barrel / replace cover				.650	78.50	38		116.50	134	158
Test valve				.010		.61		.61	.76	.95
Total				1.332	78.50	77.91		156.41	183.88	220.47
0030 Replace mixing valve	10	1 PLUM	Ea.							
Turn valves off and on				.022		1.30		1.30	1.62	2.02
Remove mixing valve				.867		51		51	63.50	79.50
Install new mixing valve				1.733	153	102		255	296	350
Test valve				.010		.61		.61	.76	.95
Total				2.633	153	154.91		307.91	361.88	432.47
0060 Replace shower and fittings, aluminum	25	2 PLUM	Ea.							
Remove shower				4.334		229		229	286	355
Install shower				3.302	505	174		679	775	905
Total				7.635	505	403		908	1,061	1,260
0070 Replace shower head with water conserving head	10	1 PLUM	Ea.							
Remove shower head				.217		12.70		12.70	15.90	19.85
Install new water conserving head				.400	87.50	23.50		111	126	146
Turn water on and off				.010		.61		.61	.76	.95
Total				.627	87.50	36.81		124.31	142.66	166.80
0120 Replace shower, C.M.U.	20	D-8	Ea.							
Remove C.M.U. shower enclosure				3.595		136.64	20.72	157.36	202	247
Install C.M.U. shower enclosure				6.772	152.32	291.20		443.52	550	665
Install built-in head, arm and valve				2.000	80	117		197	235	283
Total				12.367	232.32	544.84	20.72	797.88	987	1,195

For customer support on your Facilities Maintenance & Repair Cost Data, call 877.801.5724.

175

D2013 770 Shower, Misc.

	System Description	Freq. (Years)	Crew	Unit	Labor Hours	2015 Bare Costs				Total In-House	Total w/O&P
						Material	Labor	Equipment	Total		
0200	**Replace shower, glazed C.M.U.**	25	D-8	Ea.							
	Remove C.M.U. shower enclosure				4.697		178	27.50	205.50	264	325
	Install glazed C.M.U. shower enclosure				9.393	540.40	403.20		943.60	1,125	1,325
	Install built-in head, arm and valve				2.000	80	117		197	235	283
	Total				16.090	620.40	698.20	27.50	1,346.10	**1,624**	**1,933**
0280	**Replace shower surface, ceramic tile**	30	D-7	Ea.							
	Remove ceramic tile shower enclosure surface				3.066		115		115	153	189
	Install ceramic tile in shower enclosure				6.130	117.60	234.08		351.68	425	515
	Install built-in head, arm and valve				2.000	80	117		197	235	283
	Total				11.197	197.60	466.08		663.68	**813**	**987**

D2013 810 Drinking Fountain

	System Description	Freq. (Years)	Crew	Unit	Labor Hours	2015 Bare Costs				Total In-House	Total w/O&P
						Material	Labor	Equipment	Total		
0010	**Check / minor repairs**	1	1 PLUM	Ea.							
	Check unit, make minor adjustments				.667		39		39	49	61
	Total				.667		39		39	**49**	**61**
0020	**Repair internal leaks**	4	1 PLUM	Ea.							
	Open unit, repair leaks, close				.615		36		36	45	56.50
	Total				.615		36		36	**45**	**56.50**
0030	**Correct water pressure**	2	1 PLUM	Ea.							
	Adjust water pressure, close				.571		33.50		33.50	42	52.50
	Total				.571		33.50		33.50	**42**	**52.50**
0040	**Replace refrigerant**	2	1 PLUM	Ea.							
	Add freon to unit				.133	32	7.95		39.95	45	52.50
	Total				.133	32	7.95		39.95	**45**	**52.50**
0050	**Repair drain leak**	4	1 PLUM	Ea.							
	Remove old pipe				.052		3.05		3.05	3.82	4.76
	Install new pipe				.186	10.80	10.90		21.70	25.50	30.50
	Total				.238	10.80	13.95		24.75	**29.32**	**35.26**

For customer support on your Facilities Maintenance & Repair Cost Data, call 877.801.5724.

D2013 810 Drinking Fountain

	System Description	Freq. (Years)	Crew	Unit	Labor Hours	2015 Bare Costs				Total In-House	Total w/O&P
						Material	Labor	Equipment	Total		
0070	**Replace fountain**	10	2 PLUM	Ea.							
	Remove old water cooler				2.600		137		137	172	214
	Install new wall mounted water cooler				5.200	680	275		955	1,100	1,275
	Total				7.800	680	412		1,092	1,272	**1,489**

D2013 910 Emergency Shower Station

	System Description	Freq. (Years)	Crew	Unit	Labor Hours	2015 Bare Costs				Total In-House	Total w/O&P
						Material	Labor	Equipment	Total		
0020	**Inspect and clean shower head**	3	1 PLUM	Ea.							
	Turn water off and on				.010		.61		.61	.76	.95
	Remove shower head				.217		12.70		12.70	15.90	19.85
	Check shower head				.009		.53		.53	.67	.83
	Install shower head				.433		25.50		25.50	32	39.50
	Total				.669		39.34		39.34	49.33	**61.13**
0030	**Replace shower**	25	2 PLUM	Ea.							
	Remove emergency shower				2.600		137		137	172	214
	Install emergency shower				5.200	380	275		655	760	905
	Total				7.800	380	412		792	932	**1,119**

D2013 920 Emergency Eye Wash

	System Description	Freq. (Years)	Crew	Unit	Labor Hours	2015 Bare Costs				Total In-House	Total w/O&P
						Material	Labor	Equipment	Total		
0020	**Inspect and clean spray heads**	3	1 PLUM	Ea.							
	Turn water off and on				.010		.61		.61	.76	.95
	Remove spray head				.217		12.70		12.70	15.90	19.85
	Check shower head				.009		.53		.53	.67	.83
	Install spray head				.433		25.50		25.50	32	39.50
	Total				.669		39.34		39.34	49.33	**61.13**
0030	**Replace eye wash station**	25	2 PLUM	Ea.							
	Remove old eye wash station				2.600		137		137	172	214
	Install eye wash station				5.200	455	275		730	845	1,000
	Total				7.800	455	412		867	1,017	**1,214**

For customer support on your Facilities Maintenance & Repair Cost Data, call 877.801.5724.

177

D2023 110 — Pipe & Fittings, Copper

	System Description	Freq. (Years)	Crew	Unit	Labor Hours	2015 Bare Costs				Total In-House	Total w/O&P
						Material	Labor	Equipment	Total		
0010	**Resolder joint**	10	1 PLUM	Ea.							
	Measure, cut & ream both ends				.053		3.13		3.13	3.91	4.88
	Solder fitting (3/4" coupling)				.495	1.67	29		30.67	38	47.50
	Total				.549	1.67	32.13		33.80	41.91	52.38
0020	**Replace 3/4" copper pipe and fittings**	20	2 PLUM	L.F.							
	Remove old pipe				.071		4.17		4.17	5.20	6.50
	Install 3/4" copper tube with couplings and hangers				.141	7.80	8.25		16.05	18.90	22.50
	Total				.212	7.80	12.42		20.22	24.10	29
0030	**Replace 1" copper pipe and fittings**	25	2 PLUM	L.F.							
	Remove old pipe				.073		4.26		4.26	5.35	6.65
	Install 1" copper tube with couplings and hangers				.158	10.35	9.25		19.60	23	27.50
	Total				.230	10.35	13.51		23.86	28.35	34.15
0050	**Replace 1-1/2" copper pipe and fittings**	25	2 PLUM	L.F.							
	Remove old pipe				.104		6.10		6.10	7.65	9.50
	Install 1 1/2" copper tubing, couplings, hangers				.208	15.60	12.20		27.80	32.50	38.50
	Total				.312	15.60	18.30		33.90	40.15	48
0060	**Replace 2" copper pipe and fittings**	25	2 PLUM	L.F.							
	Remove old pipe				.130		7.65		7.65	9.55	11.90
	Install 2" copper tube with couplings and hangers				.260	23.50	15.25		38.75	45	53.50
	Total				.390	23.50	22.90		46.40	54.55	65.40
0070	**Replace 4" copper pipe and fittings**	25	2 PLUM	L.F.							
	Remove old pipe				.274		14.50		14.50	18.10	22.50
	Install 4" copper tube with couplings and hangers				.547	80	29		109	124	145
	Total				.821	80	43.50		123.50	142.10	167.50
0080	**Replace 8" copper pipe and fittings**	25	2 PLUM	L.F.							
	Remove old pipe				.694		38		38	47.50	59.50
	Install 8" copper tube with couplings and hangers				.918	455	50.50		505.50	565	645
	Total				1.612	455	88.50		543.50	612.50	704.50

For customer support on your Facilities Maintenance & Repair Cost Data, call 877.801.5724.

D2023 120　Pipe & Fittings, Steel/Iron, Threaded

	System Description	Freq. (Years)	Crew	Unit	Labor Hours	2015 Bare Costs				Total In-House	Total w/O&P
						Material	Labor	Equipment	Total		
0030	Replace 3/4" threaded steel pipe and fittings	75	2 PLUM	L.F.							
	Remove old pipe				.086		5.05		5.05	6.30	7.85
	Install 3/4" diameter steel pipe, couplings, hangers				.131	3.91	7.70		11.61	13.95	16.90
	Total				.217	3.91	12.75		16.66	20.25	24.75
0040	Replace 1-1/2" threaded steel pipe and fittings	75	2 PLUM	L.F.							
	Remove old pipe				.130		6.85		6.85	8.60	10.70
	Install 1-1/2" diameter steel pipe, couplings, hangers				.260	7.10	13.75		20.85	25	30.50
	Total				.390	7.10	20.60		27.70	33.60	41.20
0050	Replace 2" threaded steel pipe and fittings	75	2 PLUM	L.F.							
	Remove old pipe				.163		8.60		8.60	10.75	13.45
	Install 2" diameter steel pipe, couplings, hangers				.325	9	17.15		26.15	31.50	38.50
	Total				.488	9	25.75		34.75	42.25	51.95
0060	Replace 4" threaded steel pipe and fittings	75	2 PLUM	L.F.							
	Remove old pipe				.289		15.25		15.25	19.10	24
	Install 4" diameter steel pipe, couplings, hangers				.578	27	30.50		57.50	68	81.50
	Total				.867	27	45.75		72.75	87.10	105.50

D2023 130　Solar Piping: Pipe & Fittings, PVC

	System Description	Freq. (Years)	Crew	Unit	Labor Hours	2015 Bare Costs				Total In-House	Total w/O&P
						Material	Labor	Equipment	Total		
0010	Reglue joint, install 1" tee	10	1 PLUM	Ea.							
	Cut exist pipe, install tee, 1"				.867	1.29	51		52.29	65	81
	Inspect joints				.126		7.40		7.40	9.25	11.55
	Total				.993	1.29	58.40		59.69	74.25	92.55
0110	Reglue joint, install 1-1/4" tee	10	1 PLUM	Ea.							
	Cut exist pipe, install tee, 1-1/4"				.945	2.01	55.50		57.51	71.50	89
	Inspect joints				.126		7.40		7.40	9.25	11.55
	Total				1.071	2.01	62.90		64.91	80.75	100.55

D2023 130   Solar Piping: Pipe & Fittings, PVC

	System Description	Freq. (Years)	Crew	Unit	Labor Hours	2015 Bare Costs				Total In-House	Total w/O&P
						Material	Labor	Equipment	Total		
0210	**Reglue joint, install 1-1/2″ tee**	10	1 PLUM	Ea.							
	Cut exist pipe, install tee, 1-1/2″				1.040	2.45	61		63.45	79	98
	Inspect joints				.126		7.40		7.40	9.25	11.55
	Total				1.166	2.45	68.40		70.85	88.25	109.55
0310	**Reglue joint, install 2″ tee**	10	Q-1	Ea.							
	Cut exist pipe, install tee, 2″				1.224	3.56	64.50		68.06	84.50	105
	Inspect joints				.126		7.40		7.40	9.25	11.55
	Total				1.350	3.56	71.90		75.46	93.75	116.55
1020	**Install 10′ section PVC 1″ diameter**	20	1 PLUM	Ea.							
	Remove broken pipe				1.131		66.50		66.50	83	104
	Install 10′ new PVC pipe 1″ diameter				2.939	76.05	172.25		248.30	299	360
	Inspect joints				.084		4.93		4.93	6.15	7.70
	Total				4.154	76.05	243.68		319.73	388.15	471.70
1120	**Install 10′ PVC 1-1/4″ diameter**	20	1 PLUM	Ea.							
	Remove broken pipe				1.239		72.50		72.50	91	114
	Install 10′ new PVC pipe 1-1/4″ diameter				3.219	85.15	189.15		274.30	330	400
	Inspect joints				.084		4.93		4.93	6.15	7.70
	Total				4.542	85.15	266.58		351.73	427.15	521.70
1220	**Install 10′ PVC 1-1/2″ diameter**	20	1 PLUM	Ea.							
	Remove broken pipe				1.445		85		85	106	133
	Install 10′ new PVC pipe 1-1/2″ diameter				3.756	89.05	220.35		309.40	370	455
	Inspect joints				.084		4.93		4.93	6.15	7.70
	Total				5.284	89.05	310.28		399.33	482.15	595.70
1320	**Install 10′ section PVC 2″ diameter**	20	Q-1	Ea.							
	Remove broken pipe				1.763		93		93	117	146
	Install 10′ new PVC pipe 2″ diameter				4.583	104.65	241.80		346.45	415	510
	Inspect joints				.084		4.93		4.93	6.15	7.70
	Total				6.430	104.65	339.73		444.38	538.15	663.70

D2023 130 | Solar Piping: Pipe & Fittings, PVC

	System Description	Freq. (Years)	Crew	Unit	2015 Bare Costs					Total In-House	Total w/O&P
					Labor Hours	Material	Labor	Equipment	Total		
2030	**Replace 1000' PVC pipe 1" diameter**	30	1 PLUM	M.L.F.							
	Remove broken pipe				113.080		6,650		6,650	8,300	10,400
	Install 1000' new PVC pipe 1" diameter				226.080	5,850	13,250		19,100	23,000	27,800
	Inspect joints				8.399		495		495	615	770
	Total				347.559	5,850	20,395		26,245	31,915	38,970
2130	**Replace 1000' PVC pipe 1-1/4" diameter**	30	1 PLUM	M.L.F.							
	Remove broken pipe				123.850		7,250		7,250	9,100	11,400
	Install 1000' new PVC pipe 1-1/4" diameter				247.620	6,550	14,550		21,100	25,500	30,700
	Inspect joints				8.399		495		495	615	770
	Total				379.869	6,550	22,295		28,845	35,215	42,870
2230	**Replace 1000' PVC pipe 1-1/2" diameter**	30	1 PLUM	M.L.F.							
	Remove broken pipe				144.490		8,500		8,500	10,600	13,300
	Install 1000' new PVC pipe 1-1/2" diameter				288.890	6,850	16,950		23,800	28,500	35,100
	Inspect joints				8.399		495		495	615	770
	Total				441.779	6,850	25,945		32,795	39,715	49,170
2330	**Replace 1000' PVC pipe 2" diameter**	30	Q-1	M.L.F.							
	Remove broken pipe				176.320		9,300		9,300	11,700	14,600
	Install 1000' new PVC pipe 2" diameter				352.540	8,050	18,600		26,650	32,000	39,100
	Inspect joints				8.399		495		495	615	770
	Total				537.259	8,050	28,395		36,445	44,315	54,470

D2023 150 | Valve, Non-Drain, Less Than 1-1/2"

	System Description	Freq. (Years)	Crew	Unit	2015 Bare Costs					Total In-House	Total w/O&P
					Labor Hours	Material	Labor	Equipment	Total		
0020	**Replace old valve**	10	1 PLUM	Ea.							
	Remove old valve				.400		23.50		23.50	29.50	36.50
	Install new valve				.800	300	47		347	390	450
	Total				1.200	300	70.50		370.50	419.50	486.50

D2023 152 Valve, Non-Drain, 2"

						2015 Bare Costs				Total In-House	Total w/O&P
	System Description	Freq. (Years)	Crew	Unit	Labor Hours	Material	Labor	Equipment	Total		
0020	**Replace old valve**	10	1 PLUM	Ea.							
	Remove old valve (2")				.800		42.50		42.50	53	66
	Install new valve (to 2")				.727	480	42.50		522.50	580	665
	Total				1.527	480	85		565	633	731

D2023 154 Valve, Non-Drain, 3"

						2015 Bare Costs				Total In-House	Total w/O&P
	System Description	Freq. (Years)	Crew	Unit	Labor Hours	Material	Labor	Equipment	Total		
0020	**Replace old valve**	10	1 PLUM	Ea.							
	Remove old valve				.800		42.50		42.50	53	66
	Install new valve				1.600	1,575	84.50		1,659.50	1,850	2,100
	Total				2.400	1,575	127		1,702	1,903	2,166

D2023 156 Valve, Non-Drain, 4" and Larger

						2015 Bare Costs				Total In-House	Total w/O&P
	System Description	Freq. (Years)	Crew	Unit	Labor Hours	Material	Labor	Equipment	Total		
0020	**Replace old valve, 4"**	10	2 PLUM	Ea.							
	Remove old flanged gate valve, 4"				5.200		285		285	355	445
	Install new flanged gate valve, 4"				7.038	1,175	385		1,560	1,775	2,075
	Total				12.239	1,175	670		1,845	2,130	2,520
0030	**Replace old valve, 6"**	10	2 PLUM	Ea.							
	Remove old flanged gate valve, 6"				4.615		253		253	315	395
	Install new flanged gate valve, 6"				10.399	2,025	570		2,595	2,950	3,425
	Total				15.014	2,025	823		2,848	3,265	3,820
0040	**Replace old valve, 8"**	10	2 PLUM	Ea.							
	Remove old flanged gate valve, 8"				3.871		212		212	265	330
	Install new flanged gate valve, 8"				12.371	3,450	680		4,130	4,650	5,375
	Total				16.242	3,450	892		4,342	4,915	5,705

For customer support on your Facilities Maintenance & Repair Cost Data, call 877.801.5724.

D2023 156   Valve, Non-Drain, 4" and Larger

	System Description	Freq. (Years)	Crew	Unit	Labor Hours	2015 Bare Costs				Total In-House	Total w/O&P
						Material	Labor	Equipment	Total		
0050	**Replace old valve, 10"**	10	3 PLUM	Ea.							
	Remove old flanged gate valve, 10"				3.380		185		185	232	289
	Install new flanged gate valve, 10"				14.201	6,075	780		6,855	7,650	8,825
	Total				17.581	6,075	965		7,040	7,882	9,114

D2023 160   Insulation, Pipe

	System Description	Freq. (Years)	Crew	Unit	Labor Hours	2015 Bare Costs				Total In-House	Total w/O&P
						Material	Labor	Equipment	Total		
0020	**Remove old insulation & replace with new (1/2" pipe size, 1" wall)**	15	1 PLUM	L.F.							
	Remove pipe insulation				.025		1.45		1.45	1.82	2.27
	Install fiberglass pipe insulation (1" wall, 1/2" pipe size)				.087	.83	4.08		4.91	6.20	7.60
	Total				.111	.83	5.53		6.36	8.02	9.87
0030	**Remove old insulation & replace with new (3/4" pipe size, 1" wall)**	15	1 PLUM	L.F.							
	Remove pipe insulation				.025		1.45		1.45	1.82	2.27
	Install fiberglass pipe insulation (1" wall, 3/4" pipe size)				.090	.90	4.26		5.16	6.50	8
	Total				.115	.90	5.71		6.61	8.32	10.27
0040	**Remove old insulation & replace with new (1-1/2" pipe size, 1" wall)**	15	1 PLUM	L.F.							
	Remove pipe insulation				.025		1.45		1.45	1.82	2.27
	Install fiberglass pipe insulation (1" wall, 1-1/2" pipe size)				.099	1.13	4.67		5.80	7.30	8.90
	Total				.124	1.13	6.12		7.25	9.12	11.17
0050	**Remove old insulation & replace with new (1/2" pipe size, 3/4" wall)**	15	1 PLUM	L.F.							
	Remove pipe insulation				.025		1.45		1.45	1.82	2.27
	Install flexible rubber tubing (3/4" wall, 1/2" pipe size)				.117	1.10	6.10		7.20	9.15	11.25
	Total				.142	1.10	7.55		8.65	10.97	13.52
0060	**Remove old insulation & replace with new (3/4" pipe size, 3/4" wall)**	15	1 PLUM	L.F.							
	Remove pipe insulation				.025		1.45		1.45	1.82	2.27
	Install flexible rubber tubing (3/4" wall, 3/4" pipe size)				.117	1.35	6.10		7.45	9.40	11.55
	Total				.142	1.35	7.55		8.90	11.22	13.82

D2023 160 Insulation, Pipe

System Description	Freq. (Years)	Crew	Unit	Labor Hours	2015 Bare Costs				Total In-House	Total w/O&P
					Material	Labor	Equipment	Total		
0070 **Remove old insulation & replace w/new (1-1/2" pipe size, 3/4" wall)**	15	1 PLUM	L.F.							
Remove pipe insulation				.025		1.45		1.45	1.82	2.27
Install flexible rubber tubing (3/4" wall, 1-1/2" pipe size)				.120	2.32	6.25		8.57	10.65	12.95
Total				.144	2.32	7.70		10.02	12.47	15.22

D2023 210 Water Heater, Gas / Oil, 30 Gallon

System Description	Freq. (Years)	Crew	Unit	Labor Hours	2015 Bare Costs				Total In-House	Total w/O&P
					Material	Labor	Equipment	Total		
0010 **Overhaul heater**	5	1 PLUM	Ea.							
Turn valves off and on				.022		1.30		1.30	1.62	2.02
Remove thermostat, check, and insulate				.182		10.65		10.65	13.35	16.65
Remove burner, clean and install				.229		13.40		13.40	16.80	21
Clean pilot and thermocouple				.296		17.40		17.40	22	27
Remove burner door and install				.033		1.91		1.91	2.39	2.98
Check after maintenance				.216		12.70		12.70	15.85	19.80
Drain and fill tank				.615		36		36	45	56.50
Total				1.593		93.36		93.36	117.01	145.95
0020 **Clean and service**	1	1 PLUM	Ea.							
Check / adjust fuel burner				2.667		157		157	196	244
Total				2.667		157		157	196	244
0030 **Replace water heater**	10	2 PLUM	Ea.							
Remove old heater				2.600		153		153	191	238
Install 30 gallon gas water heater				5.202	895	305		1,200	1,375	1,600
Check after installation				.216		12.70		12.70	15.85	19.80
Total				8.018	895	470.70		1,365.70	1,581.85	1,857.80

D2023 212 | Water Heater, Gas / Oil, 70 Gallon

	System Description	Freq. (Years)	Crew	Unit	Labor Hours	2015 Bare Costs				Total In-House	Total w/O&P
						Material	Labor	Equipment	Total		
0010	**Overhaul heater**	5	1 PLUM	Ea.							
	Remove hot water heater and insulate				.667		39		39	49	61
	Remove thermostat, check and install				.182		10.65		10.65	13.35	16.65
	Remove burner, clean and install				.229		13.40		13.40	16.80	21
	Clean pilot and thermocouple				.296		17.40		17.40	22	27
	Remove burner door and install				.033		1.91		1.91	2.39	2.98
	Drain and refill tank				.615		36		36	45	56.50
	Turn valve off and on				.211		12.35		12.35	15.45	19.25
	Total				2.232		130.71		130.71	**163.99**	**204.38**
0020	**Clean & service**	1	1 PLUM	Ea.							
	Check / adjust fuel burner				2.667		157		157	196	244
	Total				2.667		157		157	**196**	**244**
0030	**Replace hot water heater**	12	2 PLUM	Ea.							
	Remove old heater				2.600		153		153	191	238
	Install water heater, oil fired, 70 gallon				6.932	1,975	405		2,380	2,675	3,100
	Total				9.532	1,975	558		2,533	**2,866**	**3,338**

D2023 214 | Water Heater, Gas / Oil, 1150 GPH

	System Description	Freq. (Years)	Crew	Unit	Labor Hours	2015 Bare Costs				Total In-House	Total w/O&P
						Material	Labor	Equipment	Total		
0010	**Minor repairs, adjustments**	2	2 PLUM	Ea.							
	Reset, adjust controls				1.333		78.50		78.50	98	122
	Total				1.333		78.50		78.50	**98**	**122**
0020	**Clean & service**	2	2 PLUM	Ea.							
	Drain tank, clean, and service				8.000		470		470	585	730
	Total				8.000		470		470	**585**	**730**
0030	**Replace hot water heater**	20	2 PLUM	Ea.							
	Remove water heater				26.016		1,375		1,375	1,725	2,150
	Install water heater, gas fired, 1150 GPH				51.948	27,200	2,750		29,950	33,400	38,300
	Total				77.964	27,200	4,125		31,325	**35,125**	**40,450**

For customer support on your Facilities Maintenance & Repair Cost Data, call 877.801.5724.

185

D2023 220 — Water Heater, Electric, 120 Gallon

System Description	Freq. (Years)	Crew	Unit	Labor Hours	2015 Bare Costs Material	Labor	Equipment	Total	Total In-House	Total w/O&P
0010 Drain and flush										
Drain, clean and readjust	7	1 PLUM	Ea.	4.000		235		235	294	365
Total				4.000		235		235	294	365
0020 Check operation										
Check for relief valve	3	1 PLUM	Ea.	.010		.61		.61	.76	.95
Check for proper water temperature				.003		.15		.15	.19	.24
Fill out maintenance report				.022		1.30		1.30	1.62	2.02
Total				.035		2.06		2.06	**2.57**	**3.21**
0030 Replace water heater										
Remove old heater	15	2 PLUM	Ea.	4.334		254		254	320	395
Install water heater				8.667	8,050	510		8,560	9,500	10,900
Total				13.001	8,050	764		8,814	**9,820**	**11,295**

D2023 222 — Water Heater, Electric, 300 Gallon

System Description	Freq. (Years)	Crew	Unit	Labor Hours	2015 Bare Costs Material	Labor	Equipment	Total	Total In-House	Total w/O&P
0010 Drain and flush										
Drain, clean and adjust	7	1 PLUM	Ea.	4.000		235		235	294	365
Total				4.000		235		235	294	365
0020 Check operation										
Check relief valve	3	1 PLUM	Ea.	.010		.61		.61	.76	.95
Check for proper water temperature				.003		.15		.15	.19	.24
Fill out maintenance report				.022		1.30		1.30	1.62	2.02
Total				.035		2.06		2.06	**2.57**	**3.21**
0030 Replace water heater										
Remove old heater	15	2 PLUM	Ea.	8.000		425		425	530	660
Install water heater				16.000	41,700	845		42,545	46,900	53,500
Total				24.000	41,700	1,270		42,970	**47,430**	**54,160**

D20 PLUMBING   D2023   Domestic Water Distribution

D2023 224   Water Heater, Electric, 1000 Gallon

	System Description	Freq. (Years)	Crew	Unit	Labor Hours	Material	Labor	Equipment	Total	Total In-House	Total w/O&P
0010	**Drain and flush**	7	1 PLUM	Ea.							
	Drain, clean and adjust				4.000		235		235	294	365
	Total				4.000		235		235	294	365
0020	**Check operation**	3	1 PLUM	Ea.							
	Check relief valve				.010		.61		.61	.76	.95
	Check for proper water temperature				.003		.15		.15	.19	.24
	Fill out maintenance report				.022		1.30		1.30	1.62	2.02
	Total				.035		2.06		2.06	2.57	3.21
0030	**Replace water heater**	15	2 PLUM	Ea.							
	Remove heater				22.284		1,225		1,225	1,525	1,900
	Install water heater				44.610	51,000	2,450		53,450	59,000	67,500
	Total				66.894	51,000	3,675		54,675	60,525	69,400

D2023 226   Water Heater, Electric, 2000 Gallon

	System Description	Freq. (Years)	Crew	Unit	Labor Hours	Material	Labor	Equipment	Total	Total In-House	Total w/O&P
0010	**Drain and flush**	7	1 PLUM	Ea.							
	Drain, clean and adjust				4.000		235		235	294	365
	Total				4.000		235		235	294	365
0020	**Check operation**	3	1 PLUM	Ea.							
	Check relief valve				.010		.61		.61	.76	.95
	Check for proper water temperature				.003		.15		.15	.19	.24
	Fill out maintenance report				.022		1.30		1.30	1.62	2.02
	Total				.035		2.06		2.06	2.57	3.21
0030	**Replace water heater**	15	2 PLUM	Ea.							
	Remove heater				31.209		1,700		1,700	2,150	2,675
	Install water heater				62.338	95,500	3,425		98,925	109,500	124,500
	Total				93.547	95,500	5,125		100,625	111,650	127,175

D2023 230 Steam Converter, Domestic Hot Water

	System Description	Freq. (Years)	Crew	Unit	Labor Hours	2015 Bare Costs				Total In-House	Total w/O&P
						Material	Labor	Equipment	Total		
0020	Inspect for leaks	1	1 PLUM	Ea.							
	Check relief valve operation				.038		2.22		2.22	2.77	3.46
	Check for proper water temperature				.054		3.19		3.19	3.99	4.98
	Total				.092		5.41		5.41	6.76	8.44
0030	Replace steam converter	20	2 PLUM	Ea.							
	Remove heat exchanger				1.733		102		102	127	159
	Install new heat exchanger				3.467	2,150	186		2,336	2,600	2,975
	Total				5.200	2,150	288		2,438	2,727	3,134

D2023 240 Storage Tank, Domestic Hot Water

	System Description	Freq. (Years)	Crew	Unit	Labor Hours	2015 Bare Costs				Total In-House	Total w/O&P
						Material	Labor	Equipment	Total		
0020	Replace storage tank, glass lined, P.E., 80 gal.	50	2 PLUM	Ea.							
	Remove hot water storage tank				.578		34		34	42.50	53
	Install new glass lined, P.E., 80 Gal. hot water storage tank				1.156	2,800	68		2,868	3,175	3,600
	Total				1.733	2,800	102		2,902	3,217.50	3,653

D2023 245 Solar Storage Tank, 1000 Gallon

	System Description	Freq. (Years)	Crew	Unit	Labor Hours	2015 Bare Costs				Total In-House	Total w/O&P
						Material	Labor	Equipment	Total		
0010	Replace 1000 gallon solar storage tank	20	Q-9	Ea.							
	Remove tank				8.000		405		405	510	635
	Install 1000 gallon solar storage tank				16.000	7,275	805		8,080	9,025	10,400
	Total				24.000	7,275	1,210		8,485	9,535	11,035

For customer support on your Facilities Maintenance & Repair Cost Data, call 877.801.5724.

D2023 250 Expansion Chamber

	System Description	Freq. (Years)	Crew	Unit	Labor Hours	2015 Bare Costs				Total In-House	Total w/O&P
						Material	Labor	Equipment	Total		
0010	**Refill**	5	1 PLUM	Ea.							
	Refill air chamber				.039		2.29		2.29	2.86	3.57
	Total				.039		2.29		2.29	2.86	3.57
0020	**Remove old chamber, install new**	10	1 PLUM	Ea.							
	Remove / install expansion device, 15 gal.				.941	640	50.50		690.50	765	880
	Total				.941	640	50.50		690.50	765	880

D2023 260 Circulation Pump, 1/12 HP

	System Description	Freq. (Years)	Crew	Unit	Labor Hours	2015 Bare Costs				Total In-House	Total w/O&P
						Material	Labor	Equipment	Total		
0020	**Pump maintenance, 1/12 H.P.**	0.50	1 PLUM	Ea.							
	Check for leaks				.010		.61		.61	.76	.95
	Check pump operation				.012		.69		.69	.86	1.07
	Lubricate pump and motor				.047		2.75		2.75	3.43	4.28
	Check suction or discharge pressure				.004		.23		.23	.29	.36
	Check packing glands, tighten				.014		.84		.84	1.05	1.31
	Fill out inspection report				.022		1.30		1.30	1.62	2.02
	Total				.109		6.42		6.42	8.01	9.99
0030	**Replace pump / motor assembly**	10	2 PLUM	Ea.							
	Remove flanged connection pump				1.733		91.50		91.50	114	143
	Install new flanged connection pump (1/12 H.P.)				3.467	565	183		748	850	990
	Total				5.200	565	274.50		839.50	964	1,133

For customer support on your Facilities Maintenance & Repair Cost Data, call 877.801.5724.

189

D2023 261 Circulation Pump, 1/8 HP

	System Description	Freq. (Years)	Crew	Unit	Labor Hours	2015 Bare Costs				Total In-House	Total w/O&P
						Material	Labor	Equipment	Total		
0020	**Pump maintenance, 1/8 H.P.**	1	1 PLUM	Ea.							
	Check for leaks				.010		.61		.61	.76	.95
	Check pump operation				.012		.69		.69	.86	1.07
	Lubricate pump and motor				.047		2.75		2.75	3.43	4.28
	Check suction or discharge pressure				.004		.23		.23	.29	.36
	Check packing glands, tighten				.014		.84		.84	1.05	1.31
	Fill out inspection report				.022		1.30		1.30	1.62	2.02
	Total				.109		6.42		6.42	8.01	9.99
0030	**Replace pump / motor assembly**	20	2 PLUM	Ea.							
	Remove flanged connection pump				1.733		91.50		91.50	114	143
	Install new flanged connection pump (1/8 H.P.)				3.467	970	183		1,153	1,300	1,500
	Total				5.200	970	274.50		1,244.50	1,414	1,643

D2023 262 Circulation Pump, 1/6 HP

	System Description	Freq. (Years)	Crew	Unit	Labor Hours	2015 Bare Costs				Total In-House	Total w/O&P
						Material	Labor	Equipment	Total		
0020	**Pump maintenance, 1/6 H.P.**	1	1 PLUM	Ea.							
	Check for leaks				.010		.61		.61	.76	.95
	Check pump operation				.012		.69		.69	.86	1.07
	Lubricate pump and motor				.047		2.75		2.75	3.43	4.28
	Check suction or discharge pressure				.004		.23		.23	.29	.36
	Check packing glands, tighten				.014		.84		.84	1.05	1.31
	Fill out inspection report				.022		1.30		1.30	1.62	2.02
	Total				.109		6.42		6.42	8.01	9.99
0030	**Replace pump / motor assembly**	20	2 PLUM	Ea.							
	Remove flanged connection pump				2.080		122		122	153	190
	Install new flanged connection pump (1/6 H.P.)				4.160	1,400	220		1,620	1,825	2,100
	Total				6.240	1,400	342		1,742	1,978	2,290

D2023 264 Circulation Pump, 1/2 HP

	System Description	Freq. (Years)	Crew	Unit	Labor Hours	2015 Bare Costs Material	Labor	Equipment	Total	Total In-House	Total w/O&P
0020	**Pump maintenance, 1/2 H.P.**	1	1 PLUM	Ea.							
	Check for leaks				.010		.61		.61	.76	.95
	Check pump operation				.012		.69		.69	.86	1.07
	Check alignment				.053		3.13		3.13	3.91	4.88
	Lubricate pump and motor				.047		2.75		2.75	3.43	4.28
	Check suction or discharge pressure				.004		.23		.23	.29	.36
	Check packing glands, tighten				.014		.84		.84	1.05	1.31
	Fill out inspection report				.022		1.30		1.30	1.62	2.02
	Total				.163		9.55		9.55	11.92	14.87
0030	**Replace pump / motor assembly, partial**	20	2 PLUM	Ea.							
	Remove flanged connection pump				2.600		137		137	172	214
	Install new flanged connection pump (1/2 H.P.)				5.200	2,325	275		2,600	2,900	3,325
	Total				7.800	2,325	412		2,737	3,072	3,539

D2023 266 Circulation Pump, Bronze 1 HP

	System Description	Freq. (Years)	Crew	Unit	Labor Hours	2015 Bare Costs Material	Labor	Equipment	Total	Total In-House	Total w/O&P
0020	**Pump maintenance, 1 H.P.**	0.50	1 PLUM	Ea.							
	Check for leaks				.010		.61		.61	.76	.95
	Check pump operation				.012		.69		.69	.86	1.07
	Lubricate pump and motor				.047		2.75		2.75	3.43	4.28
	Check suction or discharge pressure				.004		.23		.23	.29	.36
	Check packing glands, tighten				.014		.84		.84	1.05	1.31
	Fill out inspection report				.022		1.30		1.30	1.62	2.02
	Total				.109		6.42		6.42	8.01	9.99
0040	**Replace pump / motor assembly**	20	2 PLUM	Ea.							
	Remove pump / motor				2.600		137		137	172	214
	Install new flanged connection pump, bronze (1 H.P.)				5.200	4,100	275		4,375	4,850	5,550
	Total				7.800	4,100	412		4,512	5,022	5,764

For customer support on your Facilities Maintenance & Repair Cost Data, call 877.801.5724.

191

D2023 267 | Circulation Pump, C.I. 1-1/2 HP

System Description	Freq. (Years)	Crew	Unit	Labor Hours	2015 Bare Costs				Total In-House	Total w/O&P
					Material	Labor	Equipment	Total		
0020 Pump maintenance, 1-1/2 H.P.	0.50	1 PLUM	Ea.							
Check for leaks				.010		.61		.61	.76	.95
Check pump operation				.012		.69		.69	.86	1.07
Check alignment				.053		3.13		3.13	3.91	4.88
Lubricate pump and motor				.047		2.75		2.75	3.43	4.28
Check suction / discharge pressure				.004		.23		.23	.29	.36
Check packing glands, tighten				.014		.84		.84	1.05	1.31
Fill out inspection report				.022		1.30		1.30	1.62	2.02
Total				.163		9.55		9.55	11.92	14.87
0040 Replace pump / motor assembly	20	2 PLUM	Ea.							
Remove flanged connection pump				2.600		137		137	172	214
Install new flanged connection pump, C.I. (1-1/2 H.P.)				5.200	1,900	275		2,175	2,425	2,800
Total				7.800	1,900	412		2,312	2,597	3,014

D2023 268 | Circulation Pump, 3 HP

System Description	Freq. (Years)	Crew	Unit	Labor Hours	2015 Bare Costs				Total In-House	Total w/O&P
					Material	Labor	Equipment	Total		
0020 Pump maintenance, 3 H.P.	0.50	Q-1	Ea.							
Check for leaks				.012		.68		.68	.84	1.06
Check pump operation				.012		.69		.69	.86	1.07
Check alignment				.059		3.48		3.48	4.34	5.40
Lubricate pump and motor				.052		3.06		3.06	3.81	4.76
Check suction / discharge pressure				.004		.23		.23	.29	.36
Check packing glands, tighten				.016		.93		.93	1.17	1.46
Fill out inspection report				.022		1.30		1.30	1.62	2.02
Total				.176		10.37		10.37	12.93	16.13
0040 Replace pump / motor assembly	20	Q-1	Ea.							
Remove flanged connection pump				4.400		232.10		232.10	290	365
Install new flanged connection pump, 3 HP				8.889	7,900	470		8,370	9,275	10,600
Total				13.289	7,900	702.10		8,602.10	9,565	10,965

D2023 310 Hose Bibb

System Description	Freq. (Years)	Crew	Unit	Labor Hours	2015 Bare Costs Material	Labor	Equipment	Total	Total In-House	Total w/O&P
0020 Replace old valve with new										
Remove hose bibb		1 PLUM	Ea.	.217		12.70		12.70	15.90	19.85
Install new hose bibb				.433	9.70	25.50		35.20	42.50	51.50
Total	10			.650	9.70	38.20		47.90	**58.40**	**71.35**

D2023 320 Water Meter

System Description	Freq. (Years)	Crew	Unit	Labor Hours	2015 Bare Costs Material	Labor	Equipment	Total	Total In-House	Total w/O&P
0010 Overhaul										
Open meter, replace worn parts	13	1 PLUM	Ea.	.300	3.41	17.60		21.01	26	32
Total				.300	3.41	17.60		21.01	**26**	**32**
0020 Remove old 5/8" meter, install new										
Remove 5/8" water meter	25	1 PLUM	Ea.	.325		19.10		19.10	24	30
Install 5/8" threaded water meter				.650	50	38		88	103	122
Total				.975	50	57.10		107.10	**127**	**152**
0025 Remove old 3/4" meter, install new										
Remove 3/4" water meter	25	1 PLUM	Ea.	.371		22		22	27.50	34
Install 3/4" threaded water meter				.743	91	43.50		134.50	155	182
Total				1.114	91	65.50		156.50	**182.50**	**216**
0030 Remove old 1" meter, install new										
Remove 1" water meter	25	1 PLUM	Ea.	.433		25.50		25.50	32	39.50
Install 1" threaded water meter				.867	138	51		189	215	252
Total				1.300	138	76.50		214.50	**247**	**291.50**
0035 Remove old 1-1/2" meter, install new										
Remove 1-1/2" water meter	25	1 PLUM	Ea.	.650		38		38	47.50	59.50
Install 1-1/2" threaded water meter				1.300	340	76.50		416.50	470	545
Total				1.950	340	114.50		454.50	**517.50**	**604.50**
0040 Remove old 2" meter, install new										
Remove 2" water meter	25	1 PLUM	Ea.	.867		51		51	63.50	79.50
Install 2" threaded water meter				1.733	460	102		562	635	735
Total				2.600	460	153		613	**698.50**	**814.50**

For customer support on your Facilities Maintenance & Repair Cost Data, call 877.801.5724.

193

D2023 320 Water Meter

System Description	Freq. (Years)	Crew	Unit	Labor Hours	2015 Bare Costs				Total In-House	Total w/O&P
					Material	Labor	Equipment	Total		
0045 Remove old 3" meter, install new	25	Q-1	Ea.							
Remove 3" water meter				3.467		183		183	229	286
Install 3" flanged water meter				6.932	3,125	365		3,490	3,900	4,475
Total				10.399	3,125	548		3,673	4,129	**4,761**
0050 Remove old 4" meter, install new	25	Q-1	Ea.							
Remove 4" water meter				6.932		365		365	460	570
Install 4" flanged water meter				13.865	5,000	730		5,730	6,425	7,400
Total				20.797	5,000	1,095		6,095	6,885	**7,970**
0055 Remove old 6" meter, install new	25	Q-1	Ea.							
Remove 6" water meter				10.403		550		550	685	855
Install 6" flanged water meter				20.806	7,975	1,100		9,075	10,100	11,700
Total				31.209	7,975	1,650		9,625	10,785	**12,555**
0060 Remove old 8" meter, install new	25	Q-1	Ea.							
Remove 8" water meter				12.998		685		685	860	1,075
Install 8" flanged water meter				26.016	12,500	1,375		13,875	15,500	17,800
Total				39.014	12,500	2,060		14,560	16,360	**18,875**

D2023 370 Water Softener

System Description	Freq. (Years)	Crew	Unit	Labor Hours	2015 Bare Costs				Total In-House	Total w/O&P
					Material	Labor	Equipment	Total		
0030 Replace softener	15	2 PLUM	Ea.							
Remove old unit				2.600		153		153	191	238
Install new water softener				5.200	660	305		965	1,100	1,300
Total				7.800	660	458		1,118	1,291	**1,538**

For customer support on your Facilities Maintenance & Repair Cost Data, call 877.801.5724.

D2033 110 Pipe & Fittings, Cast Iron

	System Description	Freq. (Years)	Crew	Unit	Labor Hours	2015 Bare Costs				Total In-House	Total w/O&P
						Material	Labor	Equipment	Total		
0020	**Unclog main drain**	10	1 PLUM	Ea.							
	Unclog main drain using auger				.630		37		37	46.50	57.50
	Total				.630		37		37	**46.50**	**57.50**
0030	**Replace cast iron pipe & fittings, 4"**	40	2 PLUM	L.F.							
	Remove old pipe				.189		10		10	12.50	15.55
	Install new 4" cast iron pipe				.378	17.45	20		37.45	44	53
	Total				.567	17.45	30		47.45	**56.50**	**68.55**

D2033 130 Pipe & Fittings, PVC

	System Description	Freq. (Years)	Crew	Unit	Labor Hours	2015 Bare Costs				Total In-House	Total w/O&P
						Material	Labor	Equipment	Total		
0010	**Unclog floor drain**	20	1 PLUM	Ea.							
	Unclog plugged floor drain				.650		38		38	47.50	59.50
	Total				.650		38		38	**47.50**	**59.50**
0020	**Unclog 4" – 12" diameter PVC main drain per L.F.**	10	1 PLUM	L.F.							
	Unclog main drain using auger				.052		3.03		3.03	3.79	4.72
	Total				.052		3.03		3.03	**3.79**	**4.72**
0040	**Repair joint**	10	1 PLUM	Ea.							
	Remove plastic fitting (2 connections)				.650		34.50		34.50	43	53.50
	Connect fitting (2 connections)				1.300	8.65	68.50		77.15	95.50	118
	Inspect joints				.163		9.60		9.60	12	14.95
	Total				2.113	8.65	112.60		121.25	**150.50**	**186.45**
0060	**Replace pipe, 1-1/2"**	30	2 PLUM	L.F.							
	Remove broken pipe				.145		7.95		7.95	9.95	12.40
	Install new pipe, couplings, hangers				.289	6.85	16.95		23.80	28.50	35
	Inspect joints				.327		19.20		19.20	24	30
	Total				.760	6.85	44.10		50.95	**62.45**	**77.40**

D2033 130 — Pipe & Fittings, PVC

	System Description	Freq. (Years)	Crew	Unit	Labor Hours	2015 Bare Costs — Material	Labor	Equipment	Total	Total In-House	Total w/O&P
0080	**Replace pipe, 2"**	30	2 PLUM	L.F.							
	Remove broken pipe				.176		9.30		9.30	11.65	14.50
	Install new pipe, couplings, hangers				.353	8.05	18.60		26.65	32	39
	Inspect joints				.327		19.20		19.20	24	30
	Total				.855	8.05	47.10		55.15	67.65	83.50
0100	**Replace pipe, 4"**	30	2 PLUM	L.F.							
	Remove broken pipe				.217		11.45		11.45	14.30	17.85
	Install new pipe, couplings, hangers				.433	16.10	23		39.10	46.50	55.50
	Inspect joints				.327		19.20		19.20	24	30
	Total				.977	16.10	53.65		69.75	84.80	103.35
0120	**Replace pipe, 6"**	30	2 PLUM	L.F.							
	Remove broken pipe				.267		14.10		14.10	17.60	22
	New pipe, couplings, hangers				.533	26	31.50		57.50	67.50	81.50
	Inspect joints				.327		19.20		19.20	24	30
	Total				1.126	26	64.80		90.80	109.10	133.50
0140	**Replace pipe, 8"**	30	2 PLUM	L.F.							
	Remove broken pipe				.325		17.80		17.80	22.50	28
	Install new pipe, couplings, hangers				.650	36.50	35.50		72	84.50	101
	Inspect joints				.327		19.20		19.20	24	30
	Total				1.302	36.50	72.50		109	131	159

D2033 305 — Pipe & Fittings

	System Description	Freq. (Years)	Crew	Unit	Labor Hours	2015 Bare Costs — Material	Labor	Equipment	Total	Total In-House	Total w/O&P
3010	**Unclog floor drain**	10	1 PLUM	Ea.							
	Unclog plugged floor drain				4.000		239		239	299	375
	Solvent					28.50			28.50	31.50	35.50
	Total				4.000	28.50	239		267.50	330.50	410.50
3020	**Unclog 4" – 12" diameter main drain per L.F.**	10	2 PLUM	L.F.							
	Unclog main drain using auger				.052		3.03		3.03	3.79	4.72
	Total				.052		3.03		3.03	3.79	4.72

For customer support on your Facilities Maintenance & Repair Cost Data, call 877.801.5724.

D2033 310 Floor Drain W/O Bucket

	System Description	Freq. (Years)	Crew	Unit	Labor Hours	2015 Bare Costs				Total In-House	Total w/O&P
						Material	Labor	Equipment	Total		
0010	**Clean drain** Remove debris from drain	4	1 PLUM	Ea.	1.590		93.50		93.50	117	146
	Total				1.590		93.50		93.50	**117**	**146**
0030	**Replace floor drain** Remove floor drain Install floor drain	40	1 PLUM	Ea.	1.040 2.080	970	61 110		61 1,080	76.50 1,200	95 1,375
	Total				3.120	970	171		1,141	**1,276.50**	**1,470**

D2033 330 Floor Drain With Bucket

	System Description	Freq. (Years)	Crew	Unit	Labor Hours	2015 Bare Costs				Total In-House	Total w/O&P
						Material	Labor	Equipment	Total		
0010	**Clean out bucket** Remove debris from bucket	5	1 PLUM	Ea.	4.000		235		235	294	365
	Total				4.000		235		235	**294**	**365**
0030	**Replace floor drain** Remove floor drain Install floor drain	40	1 PLUM	Ea.	1.040 2.080	970	61 110		61 1,080	76.50 1,200	95 1,375
	Total				3.120	970	171		1,141	**1,276.50**	**1,470**

For customer support on your Facilities Maintenance & Repair Cost Data, call 877.801.5724.

197

D20 PLUMBING — D2043 Rain Water Drainage

D2043 110 Distribution: Gutters, Pipe

System Description	Freq. (Years)	Crew	Unit	Labor Hours	2015 Bare Costs Material	2015 Bare Costs Labor	2015 Bare Costs Equipment	2015 Bare Costs Total	Total In-House	Total w/O&P
1010 General maintenance & repair	1	1 PLUM	M.L.F.							
Inspect rainwater pipes				.104		6.10		6.10	7.65	9.50
Clean out pipes				4.000		235		235	294	365
Total				4.104		241.10		241.10	301.65	374.50
1020 Replace pipe or gutter	20	1 PLUM	L.F.							
Remove 4" PVC roof drain system				.217		11.45		11.45	14.30	17.85
Install 4" roof drain system				.333	10	17.60		27.60	33	40
Total				.550	10	29.05		39.05	47.30	57.85

D2043 210 Drain: Roof, Scupper, Area

System Description	Freq. (Years)	Crew	Unit	Labor Hours	2015 Bare Costs Material	2015 Bare Costs Labor	2015 Bare Costs Equipment	2015 Bare Costs Total	Total In-House	Total w/O&P
1010 General maintenance & repair	1	1 PLUM	Ea.							
Inspect drain				.104		6.10		6.10	7.65	9.50
Clean drain				.400		23.50		23.50	29.50	36.50
Total				.504		29.60		29.60	37.15	46
1020 Replace drain	40	1 PLUM	Ea.							
Remove drain				.800		42.50		42.50	53	66
Install roof drain				1.600	395	84.50		479.50	540	625
Total				2.400	395	127		522	593	691

For customer support on your Facilities Maintenance & Repair Cost Data, call 877.801.5724.

	System Description	Freq. (Years)	Crew	Unit	Labor Hours	2015 Bare Costs				Total In-House	Total w/O&P
						Material	Labor	Equipment	Total		
1020	**Pump maintenance, 87 GPM**	1	1 PLUM	Ea.							
	Check for leaks				.010		.61		.61	.76	.95
	Check pump operation				.012		.69		.69	.86	1.07
	Check alignment				.053		3.13		3.13	3.91	4.88
	Lubricate pump and motor				.047		2.75		2.75	3.43	4.28
	Check suction or discharge pressure				.004		.23		.23	.29	.36
	Check packing glands, tighten				.014		.84		.84	1.05	1.31
	Fill out inspection report				.022		1.30		1.30	1.62	2.02
	Total				.163		9.55		9.55	11.92	14.87
1030	**Replace sump pump / motor assembly**	20	2 PLUM	Ea.							
	Remove sump pump assembly				1.300		76.50		76.50	95.50	119
	Install new cast iron sump pump (87 GPM)				2.078	277	122		399	455	535
	Total				3.378	277	198.50		475.50	550.50	654

D2093 910   Pipe & Fittings, Industrial Gas

System Description	Freq. (Years)	Crew	Unit	Labor Hours	2015 Bare Costs				Total In-House	Total w/O&P
					Material	Labor	Equipment	Total		
1010 General maintenance										
General maintenance	2	1 PLUM	M.L.F.	.500		29.50		29.50	36.50	46
Total				.500		29.50		29.50	**36.50**	**46**
1030 Replace pipe and fittings										
Shut off valve	75	2 PLUM	L.F.	1.067		62.50		62.50	78.50	97.50
Remove old pipe				.162		8.55		8.55	10.70	13.35
Install 2" black steel pipe, couplings, and hangers				.325	9	17.15		26.15	31.50	38.50
Check for leaks				.889		52		52	65	81.50
Total				2.443	9	140.20		149.20	**185.70**	**230.85**

D2093 920   Pipe & Fittings, Anesthesia

System Description	Freq. (Years)	Crew	Unit	Labor Hours	2015 Bare Costs				Total In-House	Total w/O&P
					Material	Labor	Equipment	Total		
1010 Resolder joint										
Measure, cut & clean both ends	12	1 PLUM	Ea.	.053		3.13		3.13	3.91	4.88
Attach fitting				.578	5.55	34		39.55	48.50	60
Total				.631	5.55	37.13		42.68	**52.41**	**64.88**
1030 Replace pipe and fittings										
Remove old pipe	25	2 PLUM	L.F.	.069		4.07		4.07	5.10	6.35
Install new 3/4" copper tubing, couplings, hangers				.141	7.80	8.25		16.05	18.90	22.50
Total				.210	7.80	12.32		20.12	**24**	**28.85**

D2093 930   Pipe & Fittings, Oxygen

System Description	Freq. (Years)	Crew	Unit	Labor Hours	2015 Bare Costs				Total In-House	Total w/O&P
					Material	Labor	Equipment	Total		
1010 Resolder joint										
Measure, cut & clean both ends	12	1 PLUM	Ea.	.289		16.95		16.95	21	26.50
Attach fitting				.578	5.55	34		39.55	48.50	60
Total				.867	5.55	50.95		56.50	**69.50**	**86.50**

D2093 930 Pipe & Fittings, Oxygen

	System Description	Freq. (Years)	Crew	Unit	Labor Hours	2015 Bare Costs				Total In-House	Total w/O&P
						Material	Labor	Equipment	Total		
1030	**Replace pipe and fittings**	25	2 PLUM	L.F.							
	Remove old pipe				.069		4.07		4.07	5.10	6.35
	Install new 3/4" copper tubing, couplings, hangers				.141	7.80	8.25		16.05	18.90	22.50
	Total				.210	7.80	12.32		20.12	24	28.85

D2093 940 Pipe & Fittings, Compressed Air

	System Description	Freq. (Years)	Crew	Unit	Labor Hours	2015 Bare Costs				Total In-House	Total w/O&P
						Material	Labor	Equipment	Total		
1010	**General maintenance**	2	1 PLUM	M.L.F.							
	General maintenance				.500		29.50		29.50	36.50	46
	Total				.500		29.50		29.50	36.50	46
1030	**Replace pipe and fittings**	75	2 PLUM	L.F.							
	Shut off valve				1.067		62.50		62.50	78.50	97.50
	Remove old pipe				.052		3.05		3.05	3.82	4.76
	Install 2" black steel pipe, couplings, and hangers				.325	9	17.15		26.15	31.50	38.50
	Check for leaks				.889		52		52	65	81.50
	Total				2.333	9	134.70		143.70	178.82	222.26

For customer support on your Facilities Maintenance & Repair Cost Data, call 877.801.5724.

201

D2093 946   **Compressed Air Systems, Compressors**

System Description	Freq. (Years)	Crew	Unit	Labor Hours	2015 Bare Costs				Total In-House	Total w/O&P
					Material	Labor	Equipment	Total		
1004										
Check and adjust 3/4 H.P. compressor	1	1 SPRI	Ea.							
Check compressor oil level				.022		1.30		1.30	1.62	2.02
Check V-belt tension				.038		2.22		2.22	2.77	3.46
Drain moisture from air tank				.033		1.91		1.91	2.39	2.98
Clean air intake filter on compressor				.178		10.45		10.45	13.05	16.30
Clean reusable oil filter				.182		10.65		10.65	13.35	16.65
Check pressure relief valve				.108		6.35		6.35	7.95	9.90
Clean cooling fins				.105		6.20		6.20	7.75	9.65
Check motor				.060		3.50		3.50	4.38	5.45
Check control circuits				.182		10.65		10.65	13.35	16.65
Operation checks				.222		13.05		13.05	16.30	20.50
Fill out maintenance report				.022		1.30		1.30	1.62	2.02
Total				1.151		67.58		67.58	84.53	105.58
1006										
Replace 3/4 H.P. compressor	25	1 SPRI	Ea.							
Remove air compressor				4.457		249		249	310	390
Install new 3/4 H.P. compressor				6.250	765	350		1,115	1,275	1,500
Total				10.707	765	599		1,364	1,585	1,890
1010										
Check and adjust 2 H.P. compressor	1	1 PLUM	Ea.							
Check compressor oil level				.022		1.30		1.30	1.62	2.02
Check V-belt tension				.038		2.22		2.22	2.77	3.46
Drain moisture from air tank				.033		1.91		1.91	2.39	2.98
Clean air intake filter on compressor				.178		10.45		10.45	13.05	16.30
Clean reusable oil filter				.182		10.65		10.65	13.35	16.65
Check pressure relief valve				.108		6.35		6.35	7.95	9.90
Clean cooling fins				.105		6.20		6.20	7.75	9.65
Check motor				.060		3.50		3.50	4.38	5.45
Check control circuits				.182		10.65		10.65	13.35	16.65
Operation checks				.222		13.05		13.05	16.30	20.50
Fill out maintenance report				.022		1.30		1.30	1.62	2.02
Total				1.151		67.58		67.58	84.53	105.58
1030										
Replace 2 H.P. compressor	25	2 PLUM	Ea.							
Remove air compressor				4.457		249		249	310	390
Install new 2 H.P. compressor				8.915	3,775	495		4,270	4,775	5,500
Total				13.372	3,775	744		4,519	5,085	5,890

D2093 946 Compressed Air Systems, Compressors

System Description	Freq. (Years)	Crew	Unit	Labor Hours	Material	2015 Bare Costs — Labor	Equipment	Total	Total In-House	Total w/O&P
3010 Check and adjust 10 H.P. compressor	1	1 PLUM	Ea.							
Check compressor oil level				.022		1.30		1.30	1.62	2.02
Check tension on V-belt				.038		2.22		2.22	2.77	3.46
Drain moisture from air tank				.033		1.91		1.91	2.39	2.98
Clean air intake filter on compressor				.178		10.45		10.45	13.05	16.30
Clean reusable oil filter				.182		10.65		10.65	13.35	16.65
Check operation of pressure valve				.108		6.35		6.35	7.95	9.90
Clean cooling fins				.105		6.20		6.20	7.75	9.65
Check motor				.060		3.50		3.50	4.38	5.45
Check compressor circuits				.182		10.65		10.65	13.35	16.65
Perform operation				.222		13.05		13.05	16.30	20.50
Fill out maintenance report				.022		1.30		1.30	1.62	2.02
Total				1.151		67.58		67.58	84.53	105.58
3030 Replace 10 H.P. compressor	25	2 PLUM	Ea.							
Remove compressor				17.335		930		930	1,175	1,450
Install 2-stage compressor 10 H.P.				34.632	6,825	1,875		8,700	9,825	11,400
Total				51.967	6,825	2,805		9,630	11,000	12,850
4010 Check and adjust 25 H.P. compressor	1	1 PLUM	Ea.							
Check compressor oil level				.022		1.30		1.30	1.62	2.02
Check tension on V-belt				.038		2.22		2.22	2.77	3.46
Drain moisture from air tank				.033		1.91		1.91	2.39	2.98
Clean air intake filter on compressor				.178		10.45		10.45	13.05	16.30
Clean reusable oil filter				.182		10.65		10.65	13.35	16.65
Check operation of pressure valve				.108		6.35		6.35	7.95	9.90
Clean cooling fins				.105		6.20		6.20	7.75	9.65
Check motor				.060		3.50		3.50	4.38	5.45
Check compressor circuits				.182		10.65		10.65	13.35	16.65
Perform operation				.222		13.05		13.05	16.30	20.50
Fill out maintenance report				.022		1.30		1.30	1.62	2.02
Total				1.151		67.58		67.58	84.53	105.58
4030 Replace 25 H.P. compressor	25	2 PLUM	Ea.							
Remove compressor				26.002		1,450		1,450	1,825	2,250
Install 2-stage compressor 25 H.P.				51.948	14,100	2,900		17,000	19,100	22,200
Total				77.950	14,100	4,350		18,450	20,925	24,450

D2093 Other Plumbing Systems

D2093 946 Compressed Air Systems, Compressors

System Description	Freq. (Years)	Crew	Unit	Labor Hours	2015 Bare Costs				Total In-House	Total w/O&P
					Material	Labor	Equipment	Total		
5030 **Check operation** Check operation of unit	1	1 STPI	Ea.	.320		19.10		19.10	24	30
Total				.320		19.10		19.10	**24**	**30**

For customer support on your Facilities Maintenance & Repair Cost Data, call 877.801.5724.

D30 HVAC | D3013 | Energy Supply

D3013 110 | Fuel Oil Storage Tank, 275 Gallon

	System Description	Freq. (Years)	Crew	Unit	Labor Hours	2015 Bare Costs				Total In-House	Total w/O&P
						Material	Labor	Equipment	Total		
0010	**Replace 275 gallon fuel storage tank**	30	Q-5	Ea.							
	Remove 275 gallon tank				2.081		112		112	140	175
	Install 275 gallon tank				4.156	490	224		714	820	965
	Total				6.236	490	336		826	960	**1,140**

D3013 150 | Fuel Level Meter

	System Description	Freq. (Years)	Crew	Unit	Labor Hours	2015 Bare Costs				Total In-House	Total w/O&P
						Material	Labor	Equipment	Total		
0010	**Preventive maintenance**	5	1 STPI	Ea.							
	Check fuel level meter calibration				.500		30		30	37.50	46.50
	Total				.500		30		30	**37.50**	**46.50**
0020	**Replace remote tank fuel gauge**	20	1 STPI	Ea.							
	Remove remote read gauge				2.081		124		124	155	194
	Replace remote read gauge 5" pointer travel				4.167	3,800	249		4,049	4,500	5,150
	Total				6.247	3,800	373		4,173	4,655	**5,344**

D3013 160 | Oil Filter

	System Description	Freq. (Years)	Crew	Unit	Labor Hours	2015 Bare Costs				Total In-House	Total w/O&P
						Material	Labor	Equipment	Total		
0010	**Preventive maintenance**	1	1 STPI	Ea.							
	Replace filter element				.052	3.45	3.10		6.55	7.70	9.15
	Total				.052	3.45	3.10		6.55	**7.70**	**9.15**
0020	**Replace filter housing**	30	1 STPI	Ea.							
	Replace filter housing				.520	34.50	31		65.50	77	91.50
	Total				.520	34.50	31		65.50	**77**	**91.50**

D3013 170 Fuel Oil Storage: Pipe & Fittings, Copper

	System Description	Freq. (Years)	Crew	Unit	Labor Hours	Material	Labor	Equipment	Total	Total In-House	Total w/O&P
							2015 Bare Costs				
0010	**Remake flare type joint**	10	1 STPI	M.L.F.							
	Remake flare type joint				.286		17.05		17.05	21.50	26.50
	Total				.286		17.05		17.05	**21.50**	**26.50**
0020	**Install 10' sect. 3/8" type L copper per M.L.F.**	20	1 PLUM	Ea.							
	Remove section 3/8" diameter copper				.619		36.30		36.30	45.50	56.50
	Install 10' section new 3/8" diameter type L				1.238	35	72.50		107.50	130	157
	Total				1.857	35	108.80		143.80	**175.50**	**213.50**
0030	**Install 10' sect. 1/2" type L copper per M.L.F.**	20	1 PLUM	Ea.							
	Remove section 1/2" diameter copper				.642		37.70		37.70	47	59
	Install 10' section new 1/2" diameter type L				1.284	37	75.50		112.50	135	164
	Total				1.926	37	113.20		150.20	**182**	**223**
0040	**Install 10' sect. 5/8" type L copper per M.L.F.**	20	1 PLUM	Ea.							
	Remove section 5/8" diameter copper				.658		38.60		38.60	48.50	60.50
	Install 10' section new 5/8" diameter type L				1.316	53.50	77.50		131	156	187
	Total				1.975	53.50	116.10		169.60	**204.50**	**247.50**
0050	**Install 10' sect. 3/4" type L copper per M.L.F.**	20	1 PLUM	Ea.							
	Remove section 3/4" diameter copper				.684		40.20		40.20	50	62.50
	Install 10' section new 3/4" diameter type L				1.369	52	80.50		132.50	158	191
	Total				2.053	52	120.70		172.70	**208**	**253.50**
0060	**Install 10' section 1" type L copper per M.L.F.**	20	1 PLUM	Ea.							
	Remove section 1" diameter copper				.765		44.90		44.90	56	70
	Install 10' section new 1" diameter type L				1.529	78.50	90		168.50	199	238
	Total				2.294	78.50	134.90		213.40	**255**	**308**
0130	**Replace 1000' type L 3/8" copper**	25	1 PLUM	M.L.F.							
	Remove section 3/8" diameter copper				61.920		3,630		3,630	4,550	5,650
	Install 1000' section new 3/8" diameter type L				123.800	3,500	7,250		10,750	13,000	15,700
	Total				185.720	3,500	10,880		14,380	**17,550**	**21,350**
0140	**Replace 1000' type L 1/2" copper**	25	1 PLUM	M.L.F.							
	Remove section 1/2" diameter copper				64.220		3,770		3,770	4,700	5,900
	Install 1000' section new 1/2" diameter type L				128.390	3,700	7,550		11,250	13,500	16,400
	Total				192.610	3,700	11,320		15,020	**18,200**	**22,300**

D3013 170 Fuel Oil Storage: Pipe & Fittings, Copper

	System Description	Freq. (Years)	Crew	Unit	Labor Hours	2015 Bare Costs				Total In-House	Total w/O&P
						Material	Labor	Equipment	Total		
0150	**Replace 1000' type L 5/8" copper**	25	1 PLUM	M.L.F.							
	Remove section 5/8" diameter copper				65.840		3,860		3,860	4,825	6,050
	Install 1000' section new 5/8" diameter type L				131.640	5,350	7,750		13,100	15,600	18,700
	Total				197.480	5,350	11,610		16,960	20,425	24,750
0160	**Replace 1000' type L 3/4" copper**	25	1 PLUM	M.L.F.							
	Remove section 3/4" diameter copper				68.440		4,020		4,020	5,000	6,250
	Install 1000' section new 3/4" diameter type L				136.850	5,200	8,050		13,250	15,800	19,100
	Total				205.290	5,200	12,070		17,270	20,800	25,350
0170	**Replace 1000' type L 1" copper**	25	1 PLUM	M.L.F.							
	Remove section 1" diameter copper				76.480		4,490		4,490	5,600	7,000
	Install 1000' section new 1" diameter type L				152.930	7,850	9,000		16,850	19,900	23,800
	Total				229.410	7,850	13,490		21,340	25,500	30,800

D3013 210 Natural Gas: Pipe & Fittings, Steel/Iron

	System Description	Freq. (Years)	Crew	Unit	Labor Hours	2015 Bare Costs				Total In-House	Total w/O&P
						Material	Labor	Equipment	Total		
0010	**Install new 2" gasket, 1 per M.L.F.**	30	1 STPI	Ea.							
	Disconnect joint, remove gasket & clean faces				.803		48.03		48.03	60	75
	Install gasket and make up bolts				.800	5.95	48		53.95	66.50	82
	Total				1.603	5.95	96.03		101.98	126.50	157
0020	**Install new 3" gasket, 1 per M.L.F.**	30	1 STPI	Ea.							
	Disconnect joint, remove gasket & clean faces				.945		56.50		56.50	70.50	88
	Install gasket and make up bolts				.945	6.35	56.50		62.85	77.50	96
	Total				1.890	6.35	113		119.35	148	184
0030	**Install new 4" gasket, 1 per M.L.F.**	30	1 STPI	Ea.							
	Disconnect joint, remove gasket & clean faces				1.300		77.50		77.50	97	121
	Install gasket and make up bolts				1.300	14.50	76.50		91	111	137
	Total				2.600	14.50	154		168.50	208	258

D3013 210 Natural Gas: Pipe & Fittings, Steel/Iron

System Description	Freq. (Years)	Crew	Unit	Labor Hours	2015 Bare Costs				Total In-House	Total w/O&P
					Material	Labor	Equipment	Total		
0040 Install new 6" gasket, 1 per M.L.F.	30	1 STPI	Ea.							
Disconnect joint, remove gasket & clean faces				1.733		104		104	129	162
Install gasket and make up bolts				1.733	21.50	104		125.50	153	189
Total				3.466	21.50	208		229.50	282	351
0110 Replace 10' of buried 2" diam. st. pipe/M.L.F.	12	Q-4	Ea.							
Check break or leak				.016		.96		.96	1.20	1.49
Turn valve on and off				.010		.60		.60	.75	.93
Excavate by hand				5.771		216.65		216.65	289	355
Remove broken 2" diameter section				.743		41.50	1.30	42.80	53	66
Install 10' of pipe coated & wrapped				1.143	74.50	64	2.10	140.60	164	195
Check for leaks, additional section				.900		54		54	67	84
Backfill by hand				2.101		78.88		78.88	104	130
Total				10.684	74.50	456.59	3.40	534.49	678.95	832.42
0120 Replace 10' of buried 3" diam. st. pipe/M.L.F.	12	Q-4	Ea.							
Check break or leak				.016		.96		.96	1.20	1.49
Turn valve on and off				.010		.60		.60	.75	.93
Excavate by hand				5.771		216.65		216.65	289	355
Remove broken 3" diameter section				.800		44.70	1.40	46.10	57.50	71
Install 10' of pipe coated & wrapped				1.231	123.50	69	2.20	194.70	225	264
Check for leaks, additional section				.900		54		54	67	84
Backfill by hand				2.101		78.88		78.88	104	130
Total				10.829	123.50	464.79	3.60	591.89	744.45	906.42
0130 Replace 10' of buried 4" diam. st. pipe/M.L.F.	12	B-35	Ea.							
Check break or leak				.016		.96		.96	1.20	1.49
Turn valve on and off				.010		.60		.60	.75	.93
Excavate by hand				5.771		216.65		216.65	289	355
Remove broken 4" diameter section				1.224		57.50	18.20	75.70	92.50	112
Install 10' of pipe coated & wrapped				1.882	160.50	88.50	27.90	276.90	320	375
Check for leaks, additional section				.900		54		54	67	84
Backfill by hand				2.101		78.88		78.88	104	130
Total				11.905	160.50	497.09	46.10	703.69	874.45	1,058.42

	System Description	Freq. (Years)	Crew	Unit	Labor Hours	2015 Bare Costs				Total In-House	Total w/O&P
						Material	Labor	Equipment	Total		
0140	**Replace 10' of buried 6" diam. st. pipe/M.L.F.**	12	B-35	Ea.							
	Check break or leak				.016		.96		.96	1.20	1.49
	Turn valve on and off				.010		.60		.60	.75	.93
	Excavate by hand				5.771		216.65		216.65	289	355
	Remove broken 6" diameter section				1.734		81.50	25.70	107.20	131	159
	Install 10' of pipe coated & wrapped				2.667	285	125	39.60	449.60	515	600
	Check for leaks, additional section				.900		54		54	67	84
	Backfill by hand				2.101		78.88		78.88	104	130
	Total				13.199	285	557.59	65.30	907.89	1,107.95	1,330.42
0210	**Replace 1000 L.F. of buried 2" diam. st. pipe**	75	Q-4	M.L.F.							
	Check break or leak				1.600		95.50		95.50	120	149
	Turn valve on and off				1.000		60		60	74.50	93
	Excavate with 1/2 C.Y. backhoe				23.111		1,020	566.67	1,586.67	1,900	2,275
	Remove broken 2" diameter section				74.250		4,150	130	4,280	5,300	6,600
	Install 1000' of pipe coated & wrapped				114.290	7,450	6,400	210	14,060	16,400	19,500
	Check for leaks, additional section				9.000		540		540	670	840
	Backfill with 1 C.Y. bucket				8.667		400	215.56	615.56	735	880
	Total				231.918	7,450	12,665.50	1,122.23	21,237.73	25,199.50	30,337
0220	**Replace 1000 L.F. of buried 3" diam. st. pipe**	75	Q-4	M.L.F.							
	Check break or leak				1.600		95.50		95.50	120	149
	Turn valve on and off				1.000		60		60	74.50	93
	Excavate with 1/2 C.Y. backhoe				23.111		1,020	566.67	1,586.67	1,900	2,275
	Remove broken 3" diameter section				80.000		4,470	140	4,610	5,750	7,100
	Install 1000' of pipe coated & wrapped				123.080	12,350	6,900	220	19,470	22,500	26,400
	Check for leaks, additional section				9.000		540		540	670	840
	Backfill with 1 C.Y. bucket				8.667		400	215.56	615.56	735	880
	Total				246.458	12,350	13,485.50	1,142.23	26,977.73	31,749.50	37,737
0230	**Replace 1000 L.F. of buried 4" diam. st. pipe**	75	B-35	M.L.F.							
	Check break or leak				1.600		95.50		95.50	120	149
	Turn valve on and off				1.000		60		60	74.50	93
	Excavate with 1/2 C.Y. backhoe				23.111		1,020	566.67	1,586.67	1,900	2,275
	Remove broken 4" diameter section				122.390		5,750	1,820	7,570	9,250	11,200
	Install 1000' of pipe coated & wrapped				188.240	16,050	8,850	2,790	27,690	32,000	37,300
	Check for leaks, additional section				9.000		540		540	670	840
	Backfill with 1 C.Y. bucket				8.667		400	215.56	615.56	735	880
	Total				354.008	16,050	16,715.50	5,392.23	38,157.73	44,749.50	52,737

System Description	Freq. (Years)	Crew	Unit	Labor Hours	2015 Bare Costs				Total In-House	Total w/O&P
					Material	Labor	Equipment	Total		
0240 Replace 1000 L.F. of buried 6" diam. st. pipe	75	B-35	M.L.F.							
Check break or leak				1.600		95.50		95.50	120	149
Turn valve on and off				1.000		60		60	74.50	93
Excavate with 1/2 C.Y. backhoe				23.111		1,020	566.67	1,586.67	1,900	2,275
Remove broken 6" diameter section				173.390		8,150	2,570	10,720	13,100	15,900
Install 1000' of pipe coated & wrapped				266.670	28,500	12,500	3,960	44,960	51,500	60,000
Check for leaks, additional section				9.000		540		540	670	840
Backfill with 1 C.Y. bucket				8.667		400	215.56	615.56	735	880
Total				483.438	28,500	22,765.50	7,312.23	58,577.73	68,099.50	80,137
0410 Replace 10' of hung 2" diam. st. pipe/M.L.F.	12	Q-1	Ea.							
Check break or leak				.016		.96		.96	1.20	1.49
Turn valve on and off				1.000		60		60	75	93
Remove broken 2" diameter section				1.626		86		86	108	134
Install 10' of black 2" diameter steel pipe				3.250	90	171.50		261.50	315	385
Check for leaks, additional section				.900		54		54	67	84
Total				6.792	90	372.46		462.46	566.20	697.49
0420 Replace 10' of hung 3" diam. st. pipe/M.L.F.	12	Q-15	Ea.							
Check break or leak				.016		.96		.96	1.20	1.49
Turn valve on and off				.010		.60		.60	.75	.93
Remove broken 3" diameter section				2.420		128	8.70	136.70	169	209
Install 10' of black 3" diameter steel pipe				4.837	147.50	255	17.50	420	500	605
Check for leaks, additional section				.900		54		54	67	84
Total				8.182	147.50	438.56	26.20	612.26	737.95	900.42
0430 Replace 10' of hung 4" diam. st. pipe/M.L.F.	12	Q-15	Ea.							
Check break or leak				.016		.96		.96	1.20	1.49
Turn valve on and off				.010		.60		.60	.75	.93
Remove broken 4" diameter section				2.812		148.50	10.20	158.70	196	241
Install 10' of black 4" diameter steel pipe				5.622	215	295	20.30	530.30	630	755
Check for leaks, additional section				.900		54		54	67	84
Total				9.359	215	499.06	30.50	744.56	894.95	1,082.42

For customer support on your Facilities Maintenance & Repair Cost Data, call 877.801.5724.

D3013 210 | Natural Gas: Pipe & Fittings, Steel/Iron

System Description	Freq. (Years)	Crew	Unit	Labor Hours	2015 Bare Costs				Total In-House	Total w/O&P
					Material	Labor	Equipment	Total		
0440 Replace 10' of hung 6" diam. st. pipe/M.L.F.	12	Q-16	Ea.							
Check break or leak				.016		.96		.96	1.20	1.49
Turn valve on and off				.010		.60		.60	.75	.93
Remove broken 6" diameter section				4.335		235	10.40	245.40	305	380
Install 10' of black 6" diameter steel pipe				8.667	440	475	20.80	935.80	1,100	1,325
Check for leaks, additional section				.900		54		54	67	84
Total				13.928	440	765.56	31.20	1,236.76	1,473.95	1,791.42
0510 Replace 1000 L.F. of hung 2" diam. steel pipe	75	Q-1	M.L.F.							
Check break or leak				1.600		95.50		95.50	120	149
Turn valve on and off				100.000		6,000		6,000	7,450	9,300
Remove broken 2" diameter section				162.550		8,600		8,600	10,800	13,400
Install 1000' of black 2" diameter steel pipe				325.000	9,000	17,150		26,150	31,500	38,300
Check for leaks, additional section				9.000		540		540	670	840
Total				598.150	9,000	32,385.50		41,385.50	50,540	61,989
0520 Replace 1000 L.F. of hung 3" diam. steel pipe	75	Q-15	M.L.F.							
Check break or leak				1.600		95.50		95.50	120	149
Turn valve on and off				1.000		60		60	74.50	93
Remove broken 3" diameter section				241.950		12,800	870	13,670	16,900	20,900
Install 1000' of black 3" diameter steel pipe				483.680	14,750	25,500	1,750	42,000	50,000	60,500
Check for leaks, additional section				9.000		540		540	670	840
Total				737.230	14,750	38,995.50	2,620	56,365.50	67,764.50	82,482
0530 Replace 1000 L.F. of hung 4" diam. steel pipe	75	Q-15	M.L.F.							
Check break or leak				1.600		95.50		95.50	120	149
Turn valve on and off				1.000		60		60	74.50	93
Remove broken 4" diameter section				281.150		14,850	1,020	15,870	19,600	24,100
Install 1000' of black 4" diameter steel pipe				562.190	21,500	29,500	2,030	53,030	63,000	75,500
Check for leaks, additional section				9.000		540		540	670	840
Total				854.940	21,500	45,045.50	3,050	69,595.50	83,464.50	100,682

For customer support on your Facilities Maintenance & Repair Cost Data, call 877.801.5724.

211

D3013 210   Natural Gas: Pipe & Fittings, Steel/Iron

	System Description	Freq. (Years)	Crew	Unit	Labor Hours	2015 Bare Costs				Total In-House	Total w/O&P
						Material	Labor	Equipment	Total		
0540	**Replace 1000 L.F. of hung 6" diam. steel pipe**	75	Q-16	M.L.F.							
	Check break or leak				1.600		95.50		95.50	120	149
	Turn valve on and off				1.000		60		60	74.50	93
	Remove broken 6" diameter section				433.450		23,500	1,040	24,540	30,500	38,100
	Install 1000' of black 6" diameter steel pipe				866.740	44,000	47,500	2,080	93,580	110,000	131,500
	Check for leaks, additional section				9.000		540		540	670	840
	Total				1311.790	44,000	71,695.50	3,120	118,815.50	141,364.50	170,682

D3013 240   Natural Gas: Pressure Reducing Valve

	System Description	Freq. (Years)	Crew	Unit	Labor Hours	2015 Bare Costs				Total In-House	Total w/O&P
						Material	Labor	Equipment	Total		
0010	**Check gas pressure**	5	1 STPI	Ea.							
	Check pressure				.130		7.75		7.75	9.70	12.10
	Total				.130		7.75		7.75	9.70	12.10
0110	**Replace pressure regulator 1/2" diam. pipe**	14	1 STPI	Ea.							
	Turn valve on and off				.010		.60		.60	.75	.93
	Remove 1/2" diameter regulator				.217		12.95		12.95	16.20	20
	Replace 1/2" diameter pipe size regulator				.433	49	26		75	86.50	102
	Total				.660	49	39.55		88.55	103.45	122.93
0120	**Replace pressure regulator 1" diam. pipe**	14	1 STPI	Ea.							
	Turn valve on and off				.010		.60		.60	.75	.93
	Remove 1" diameter regulator				.274		16.35		16.35	20.50	25.50
	Replace 1" diameter pipe size regulator				.547	100	32.50		132.50	151	176
	Total				.831	100	49.45		149.45	172.25	202.43
0130	**Replace pressure regulator 1-1/2" diam. pipe**	14	1 STPI	Ea.							
	Turn valve on and off				.010		.60		.60	.75	.93
	Remove 1-1/2" diameter regulator				.400		24		24	30	37.50
	Replace 1-1/2" diameter pipe size regulator				.800	640	48		688	765	875
	Total				1.210	640	72.60		712.60	795.75	913.43

D30 HVAC — D3013 — Energy Supply

D3013 240 Natural Gas: Pressure Reducing Valve

System Description	Freq. (Years)	Crew	Unit	Labor Hours	Material	Labor	Equipment	Total	Total In-House	Total w/O&P
0140 Replace pressure regulator 2″ diam. pipe	14	1 STPI	Ea.							
Turn valve on and off				.010		.60		.60	.75	.93
Remove 2″ diameter regulator				.473		28.50		28.50	35.50	44
Replace 2″ diameter pipe size regulator				.946	640	56.50		696.50	775	890
Total				1.428	640	85.60		725.60	811.25	934.93

D3013 260 LPG Distribution: Pipe & Fittings, Steel/Iron

System Description	Freq. (Years)	Crew	Unit	Labor Hours	Material	Labor	Equipment	Total	Total In-House	Total w/O&P
0120 Replace 10′ st. pipe 1/2″ diam. per M.L.F.	12	1 PLUM	Ea.							
Check break or leak				.016		.96		.96	1.20	1.49
Turn valve on and off				.010		.60		.60	.75	.93
Remove broken 1/2″ diameter section				.826		48.50		48.50	60.50	75.50
Install section of pipe 1/2″ diameter				1.651	33.20	97		130.20	158	193
Check for leaks, additional section				.900		54		54	67	84
Total				3.402	33.20	201.06		234.26	287.45	354.92
0220 Replace 10′ st. pipe 3/4″ diam. per M.L.F.	12	1 PLUM	Ea.							
Check break or leak				.016		.96		.96	1.20	1.49
Turn valve on and off				.010		.60		.60	.75	.93
Remove broken 3/4″ diameter section				.853		50		50	62.50	78
Install section of pipe 3/4″ diameter				1.312	39.10	77		116.10	140	169
Check for leaks, additional section				.900		54		54	67	84
Total				3.090	39.10	182.56		221.66	271.45	333.42
0320 Replace 10′ st. pipe 1″ diam. per M.L.F.	12	1 PLUM	Ea.							
Check break or leak				.016		.96		.96	1.20	1.49
Turn valve on and off				.010		.60		.60	.75	.93
Remove broken 1″ diameter section				.982		57.50		57.50	72	90
Install section of pipe 1″ diameter				1.962	50.50	115		165.50	200	243
Check for leaks, additional section				.900		54		54	67	84
Total				3.870	50.50	228.06		278.56	340.95	419.42

System Description	Freq. (Years)	Crew	Unit	Labor Hours	2015 Bare Costs				Total In-House	Total w/O&P
					Material	Labor	Equipment	Total		
0420										
Replace 10' st. pipe 1-1/4" diam. M.L.F.	12	Q-1	Ea.							
Check break or leak				.016		.96		.96	1.20	1.49
Turn valve on and off				.010		.60		.60	.75	.93
Remove broken 1-1/4" diameter section				1.169		62		62	77	96.50
Install section of pipe 1-1/4" diameter				2.337	62	123.50		185.50	225	270
Check for leaks, additional section				.900		54		54	67	84
Total				4.432	62	241.06		303.06	370.95	452.92
0520										
Replace 10' st. pipe 1-1/2" diam. M.L.F.	12	Q-1	Ea.							
Check break or leak				.016		.96		.96	1.20	1.49
Turn valve on and off				.010		.60		.60	.75	.93
Remove broken 1-1/2" diameter section				1.300		68.50		68.50	86	107
Install section of pipe 1-1/2" diameter				2.600	71	137.50		208.50	250	305
Check for leaks, additional section				.900		54		54	67	84
Total				4.826	71	261.56		332.56	404.95	498.42
0620										
Replace 10' section st. pipe 2" diam. M.L.F.	12	Q-1	Ea.							
Check break or leak				.016		.96		.96	1.20	1.49
Turn valve on and off				.010		.60		.60	.75	.93
Remove broken 2" diameter section				1.626		86		86	108	134
Install section of pipe 2" diameter				3.250	90	171.50		261.50	315	385
Check for leaks, additional section				.900		54		54	67	84
Total				5.802	90	313.06		403.06	491.95	605.42
1130										
Replace 1000' of 1/2" diameter steel pipe	75	Q-1	M.L.F.							
Check break or leak				1.600		95.50		95.50	120	149
Turn valve on and off				1.000		60		60	74.50	93
Remove broken 1/2" diameter section				82.560		4,850		4,850	6,050	7,550
Install section of pipe 1/2" diameter				165.080	3,320	9,700		13,020	15,800	19,300
Check for leaks, additional section				9.000		540		540	670	840
Total				259.240	3,320	15,245.50		18,565.50	22,714.50	27,932

D3013 260 | **LPG Distribution: Pipe & Fittings, Steel/Iron**

	System Description	Freq. (Years)	Crew	Unit	Labor Hours	2015 Bare Costs Material	2015 Bare Costs Labor	2015 Bare Costs Equipment	2015 Bare Costs Total	Total In-House	Total w/O&P
1230	**Replace 1000' of 3/4" diameter steel pipe**	75	Q-1	M.L.F.							
	Check break or leak				1.600		95.50		95.50	120	149
	Turn valve on and off				1.000		60		60	74.50	93
	Remove broken 3/4" diameter section				85.270		5,000		5,000	6,250	7,800
	Install section of pipe 3/4" diameter				131.150	3,910	7,700		11,610	14,000	16,900
	Check for leaks, additional section				9.000		540		540	670	840
	Total				228.020	3,910	13,395.50		17,305.50	21,114.50	25,782
1330	**Replace 1000' of 1" diameter steel pipe**	75	Q-1	M.L.F.							
	Check break or leak				1.600		95.50		95.50	120	149
	Turn valve on and off				1.000		60		60	74.50	93
	Remove broken 1" diameter section				98.150		5,750		5,750	7,200	9,000
	Install section of pipe 1" diameter				196.220	5,050	11,500		16,550	20,000	24,300
	Check for leaks, additional section				9.000		540		540	670	840
	Total				305.970	5,050	17,945.50		22,995.50	28,064.50	34,382
1430	**Replace 1000' of 1-1/4" diameter steel pipe**	75	Q-1	M.L.F.							
	Check break or leak				1.600		95.50		95.50	120	149
	Turn valve on and off				1.000		60		60	74.50	93
	Remove broken 1-1/4" diameter section				116.890		6,200		6,200	7,700	9,650
	Install section of pipe 1-1/4" diameter				233.710	6,200	12,350		18,550	22,500	27,000
	Check for leaks, additional section				9.000		540		540	670	840
	Total				362.200	6,200	19,245.50		25,445.50	31,064.50	37,732
1530	**Replace 1000' of 1-1/2" diameter steel pipe**	75	Q-1	M.L.F.							
	Check break or leak				1.600		95.50		95.50	120	149
	Turn valve on and off				1.000		60		60	74.50	93
	Remove broken 1-1/2" diameter section				130.040		6,850		6,850	8,600	10,700
	Install section of pipe 1-1/2" diameter				260.000	7,100	13,750		20,850	25,000	30,400
	Check for leaks, additional section				9.000		540		540	670	840
	Total				401.640	7,100	21,295.50		28,395.50	34,464.50	42,182

D3013 260 LPG Distribution: Pipe & Fittings, Steel/Iron

System Description	Freq. (Years)	Crew	Unit	Labor Hours	2015 Bare Costs Material	Labor	Equipment	Total	Total In-House	Total w/O&P
1630 Replace 1000' of 2" diameter steel pipe	75	Q-1	M.L.F.							
Check break or leak				1.600		95.50		95.50	120	149
Turn valve on and off				1.000		60		60	74.50	93
Remove broken 2" diameter section				162.550		8,600		8,600	10,800	13,400
Install section of pipe 2" diameter				325.000	9,000	17,150		26,150	31,500	38,300
Check for leaks, additional section				9.000		540		540	670	840
Total				499.150	9,000	26,445.50		35,445.50	43,164.50	52,782

D3013 601 Solar Panel, 3' x 8'

System Description	Freq. (Years)	Crew	Unit	Labor Hours	2015 Bare Costs Material	Labor	Equipment	Total	Total In-House	Total w/O&P
0010 Replace solar panel 3' x 8'	15	Q-1	Ea.							
Remove panel				1.891		100		100	125	156
Install 3' x 8' solar panel aluminum frame				3.783	1,050	200		1,250	1,400	1,625
Total				5.674	1,050	300		1,350	1,525	1,781

For customer support on your Facilities Maintenance & Repair Cost Data, call 877.801.5724.

	System Description	Freq. (Years)	Crew	Unit	Labor Hours	2015 Bare Costs				Total In-House	Total w/O&P
						Material	Labor	Equipment	Total		
1010	**Repair boiler, gas, 250 MBH**	7	1 STPI	Ea.							
	Remove / replace burner blower				.600		36		36	45	56
	Remove burner blower bearing				1.000		60		60	74.50	93
	Replace burner blower bearing				2.000	57.50	120		177.50	213	258
	Remove burner blower motor				.976		58.50		58.50	73	91
	Replace burner blower motor				1.951	190	117		307	355	420
	Remove burner fireye				.195		11.65		11.65	14.60	18.20
	Replace burner fireye				.300	221	17.95		238.95	266	305
	Remove burner gas regulator				.274		16.35		16.35	20.50	25.50
	Replace burner gas regulator				.547	100	32.50		132.50	151	176
	Remove burner auto gas valve				.274		16.35		16.35	20.50	25.50
	Replace burner auto gas valve				.421	172	25		197	221	254
	Remove burner solenoid valve				.279		16		16	19.90	25
	Replace burner solenoid valve				.557	605	32		637	705	805
	Repair controls				.500		30		30	37.50	46.50
	Total				9.874	1,345.50	589.30		1,934.80	2,216.50	2,598.70
1060	**Replace boiler, gas, 250 MBH**	30	Q-7	Ea.							
	Remove boiler				21.888		1,250		1,250	1,550	1,950
	Replace boiler, 250 MBH				43.776	3,700	2,500		6,200	7,175	8,500
	Total				65.663	3,700	3,750		7,450	8,725	10,450
2010	**Repair boiler, gas, 2000 MBH**	7	Q-5	Ea.							
	Remove / replace burner blower				.600		36		36	45	56
	Remove burner blower bearing				1.000		60		60	74.50	93
	Replace burner blower bearing				2.000	57.50	120		177.50	213	258
	Remove burner blower motor				.976		58.50		58.50	73	91
	Replace burner blower motor				1.951	195	107		302	345	410
	Remove burner fireye				.195		11.65		11.65	14.60	18.20
	Replace burner fireye				.300	221	17.95		238.95	266	305
	Remove burner gas regulator				.727		39		39	49	61
	Replace burner gas regulator				1.455	1,250	78		1,328	1,475	1,675
	Remove burner auto gas valve				.800		42.50		42.50	53	66
	Replace burner auto gas valve				1.231	370	65		435	490	565
	Remove burner solenoid valve				2.600		149		149	186	232
	Replace burner solenoid valve				5.199	1,475	298		1,773	2,000	2,300
	Repair controls				.500		30		30	37.50	46.50
	Total				19.534	3,568.50	1,112.60		4,681.10	5,321.60	6,176.70

For customer support on your Facilities Maintenance & Repair Cost Data, call 877.801.5724.

217

	System Description	Freq. (Years)	Crew	Unit	Labor Hours	2015 Bare Costs Material	Labor	Equipment	Total	Total In-House	Total w/O&P
2070	**Replace boiler, gas, 2000 MBH**	30	Q-7	Ea.							
	Remove boiler				57.762		3,275		3,275	4,100	5,125
	Replace boiler, 2000 MBH				109.000	21,400	6,225		27,625	31,300	36,500
	Total				166.762	21,400	9,500		30,900	35,400	41,625
3010	**Repair boiler, gas, 10,000 MBH**	7	Q-5	Ea.							
	Remove / replace burner blower				1.000		60		60	74.50	93
	Remove burner blower bearing				1.000		60		60	74.50	93
	Replace burner blower bearing				2.000	57.50	120		177.50	213	258
	Remove burner motor				1.000		60		60	74.50	93
	Replace burner motor				1.951	218	107		325	370	440
	Remove burner fireye				.195		11.65		11.65	14.60	18.20
	Replace burner fireye				.300	221	17.95		238.95	266	305
	Remove burner gas regulator				1.300		70		70	87.50	109
	Replace burner gas regulator				2.600	2,225	140		2,365	2,625	3,000
	Remove burner auto gas valve				5.195		285		285	355	445
	Replace burner auto gas valve				8.000	5,875	440		6,315	7,000	8,025
	Remove burner solenoid valve				2.600		149		149	186	232
	Replace burner solenoid valve				10.399	3,725	595		4,320	4,850	5,575
	Repair controls				.500		30		30	37.50	46.50
	Total				38.040	12,321.50	2,145.60		14,467.10	16,228.10	18,732.70
3070	**Replace boiler, gas, 10,000 MBH**	30	Q-7	Ea.							
	Remove boiler				364.000		20,720		20,720	25,900	32,300
	Replace boiler, gas, 10,000 MBH				722.400	165,900	41,160		207,060	234,000	271,500
	Total				1086.400	165,900	61,880		227,780	259,900	303,800

D3023 182 Boiler, Coal

	System Description	Freq. (Years)	Crew	Unit	Labor Hours	2015 Bare Costs				Total In-House	Total w/O&P
						Material	Labor	Equipment	Total		
1010	**Repair boiler, coal, 4600 MBH**	20	4 STPI	Ea.							
	Remove coal feeder				2.703		161.20		161.20	203	252
	Replace coal feeder				4.054	1,601.60	241.80		1,843.40	2,075	2,375
	Remove vibrating grates				33.592		2,007.20		2,007.20	2,500	3,125
	Replace vibrating grates				62.712	20,384	3,754.40		24,138.40	27,100	31,400
	Remove motor				.832		49.92		49.92	62.50	77.50
	Replace motor				.450	234	24.65		258.65	289	330
	Repair controls				5.200		312		312	385	485
	Total				109.543	22,219.60	6,551.17		28,770.77	32,614.50	38,044.50
1050	**Replace boiler, coal, 4600 MBH**	30	5 STPI	Ea.							
	Remove boiler				122.304		7,332		7,332	9,150	11,400
	Replace boiler, coal, 4600 MBH				244.608	103,636	14,612		118,248	132,500	152,500
	Total				366.912	103,636	21,944		125,580	141,650	163,900

D3023 184 Boiler, Oil

	System Description	Freq. (Years)	Crew	Unit	Labor Hours	2015 Bare Costs				Total In-House	Total w/O&P
						Material	Labor	Equipment	Total		
1010	**Repair boiler, oil, 250 MBH**	7	Q-5	Ea.							
	Remove / replace burner blower				.600		36		36	45	56
	Remove burner blower bearing				1.000		60		60	74.50	93
	Replace burner blower bearing				2.000	57.50	120		177.50	213	258
	Remove burner blower motor				.976		58.50		58.50	73	91
	Replace burner blower motor				1.951	190	117		307	355	420
	Remove burner fireye				.195		11.65		11.65	14.60	18.20
	Replace burner fireye				.300	221	17.95		238.95	266	305
	Remove burner ignition transformer				.250		14.95		14.95	18.70	23.50
	Replace burner ignition transformer				.500	80.50	30		110.50	126	147
	Remove burner ignition electrode				.200		11.95		11.95	14.95	18.65
	Replace burner ignition electrode				.400	17.60	24		41.60	49	59.50
	Remove burner oil pump				.350		21		21	26	32.50
	Replace burner oil pump				.700	74.50	42		116.50	134	158
	Remove burner nozzle				.167		9.95		9.95	12.45	15.55
	Replace burner nozzle				.300	8.65	17.95		26.60	32	39
	Repair controls				.600		36		36	45	56
	Total				10.489	649.75	628.90		1,278.65	1,499.20	1,790.90

For customer support on your Facilities Maintenance & Repair Cost Data, call 877.801.5724.

219

	System Description	Freq. (Years)	Crew	Unit	Labor Hours	2015 Bare Costs Material	Labor	Equipment	Total	Total In-House	Total w/O&P
1060	**Replace boiler, oil, 250 MBH**	30	Q-7	Ea.							
	Remove boiler				24.465		1,400		1,400	1,750	2,175
	Replace boiler, oil, 250 MBH				48.930	3,725	2,775		6,500	7,575	9,000
	Total				73.394	3,725	4,175		7,900	9,325	11,175
1066	**Replace boiler, oil, 300 MBH**	30	Q-7	Ea.							
	Remove boiler				29.091		1,650		1,650	2,075	2,575
	Replace boiler, oil, 300 MBH				59.480	4,700	3,375		8,075	9,400	11,200
	Total				88.570	4,700	5,025		9,725	11,475	13,775
1070	**Replace boiler, oil, 420 MBH**	30	Q-7	Ea.							
	Remove boiler				35.955		2,050		2,050	2,550	3,200
	Replace boiler, oil, 420 MBH				71.749	5,300	4,075		9,375	10,900	13,000
	Total				107.704	5,300	6,125		11,425	13,450	16,200
1080	**Replace boiler, oil, 530 MBH**	30	Q-7	Ea.							
	Remove boiler				42.440		2,425		2,425	3,025	3,775
	Replace boiler, oil, 530 MBH				84.881	6,700	4,825		11,525	13,400	15,900
	Total				127.321	6,700	7,250		13,950	16,425	19,675
2010	**Repair boiler, oil, 2000 MBH**	7	1 STPI	Ea.							
	Remove / replace burner blower				.600		36		36	45	56
	Remove burner blower bearing				1.000		60		60	74.50	93
	Replace burner blower bearing				2.000	57.50	120		177.50	213	258
	Remove burner blower motor				.976		53.50		53.50	66	82.50
	Replace burner blower motor				1.951	195	107		302	345	410
	Remove burner fireye				.195		11.65		11.65	14.60	18.20
	Replace burner fireye				.300	221	17.95		238.95	266	305
	Remove burner ignition transformer				.286		17.05		17.05	21.50	26.50
	Replace burner ignition transformer				.571	80.50	34		114.50	131	154
	Remove burner ignition electrode				.200		11.95		11.95	14.95	18.65
	Replace burner ignition electrode				.400	18	24		42	49.50	60
	Remove burner oil pump				.400		24		24	30	37.50
	Replace burner oil pump				.800	77.50	48		125.50	145	171
	Remove burner nozzle				.167		9.95		9.95	12.45	15.55
	Replace burner nozzle				.333	13.50	19.90		33.40	40	48
	Repair controls				.600		36		36	45	56
	Total				10.779	663	630.95		1,293.95	1,513.50	1,809.90

For customer support on your **Facilities Maintenance & Repair Cost Data**, call 877.801.5724.

	System Description	Freq. (Years)	Crew	Unit	Labor Hours	2015 Bare Costs				Total In-House	Total w/O&P
						Material	Labor	Equipment	Total		
2060	**Replace boiler, oil, 2000 MBH**	30	Q-7	Ea.							
	Remove boiler				57.762		3,275		3,275	4,100	5,125
	Replace boiler, oil, 2000 MBH				148.000	18,100	8,475		26,575	30,400	35,800
	Total				205.762	18,100	11,750		29,850	34,500	**40,925**
2070	**Replace boiler, oil, 3000 MBH**	30	Q-7	Ea.							
	Remove boiler				110.000		6,275		6,275	7,825	9,800
	Replace boiler, oil, 3000 MBH				220.000	23,700	12,600		36,300	41,700	49,200
	Total				330.000	23,700	18,875		42,575	49,525	**59,000**
3010	**Repair boiler, oil, 10,000 MBH**	7	1 STPI	Ea.							
	Remove / replace burner blower				.600		36		36	45	56
	Remove burner blower bearing				1.000		60		60	74.50	93
	Replace burner blower bearing				2.000	57.50	120		177.50	213	258
	Remove burner blower motor				.800		48		48	60	74.50
	Replace burner blower motor				1.951	218	107		325	370	440
	Remove burner fireye				.195		11.65		11.65	14.60	18.20
	Replace burner fireye				.300	221	17.95		238.95	266	305
	Remove burner ignition transformer				.333		19.90		19.90	25	31
	Replace burner ignition transformer				.667	80.50	40		120.50	138	163
	Remove burner ignition electrode				.200		11.95		11.95	14.95	18.65
	Replace burner ignition electrode				.444	16	26.50		42.50	51	61.50
	Remove burner oil pump				.500		30		30	37.50	46.50
	Replace burner oil pump				1.000	102	60		162	187	221
	Remove burner nozzle				.167		9.95		9.95	12.45	15.55
	Replace burner nozzle				.400	13.50	24		37.50	44.50	54.50
	Repair controls				.600		36		36	45	56
	Total				11.158	708.50	658.90		1,367.40	1,598.50	**1,912.40**
3060	**Replace boiler, oil, 10,000 MBH**	30	Q-7	Ea.							
	Remove boiler				373.100		21,238		21,238	26,500	33,100
	Replace boiler, oil, 10,000 MBH				520.905	144,935	29,704.50		174,639.50	196,500	227,500
	Total				894.005	144,935	50,942.50		195,877.50	223,000	**260,600**

	System Description	Freq. (Years)	Crew	Unit	Labor Hours	2015 Bare Costs				Total In-House	Total w/O&P
						Material	Labor	Equipment	Total		
1010	**Repair boiler, gas/oil, 2000 MBH**	7	Q-5	Ea.							
	Remove/replace burner blower				.600		36		36	45	56
	Remove burner blower bearing				1.000		60		60	74.50	93
	Replace burner blower bearing				2.000	57.50	120		177.50	213	258
	Remove burner blower motor				.976		58.50		58.50	73	91
	Replace burner blower motor				1.951	195	107		302	345	410
	Remove burner fireye				.195		11.65		11.65	14.60	18.20
	Replace burner fireye				.300	221	17.95		238.95	266	305
	Remove burner ignition transformer				.333		19.90		19.90	25	31
	Replace burner ignition transformer				.571	80.50	34		114.50	131	154
	Remove burner ignition electrode				.200		11.95		11.95	14.95	18.65
	Replace burner ignition electrode				.400	18	24		42	49.50	60
	Remove burner oil pump				.500		30		30	37.50	46.50
	Replace burner oil pump				.800	77.50	48		125.50	145	171
	Remove burner nozzle				.167		9.95		9.95	12.45	15.55
	Replace burner nozzle				.333	13.50	19.90		33.40	40	48
	Remove burner gas regulator				.727		39		39	49	61
	Replace burner gas regulator				1.455	1,250	78		1,328	1,475	1,675
	Remove burner auto gas valve				.800		42.50		42.50	53	66
	Replace burner auto gas valve				1.231	370	65		435	490	565
	Remove burner solenoid valve				2.600		149		149	186	232
	Replace burner solenoid valve				5.199	1,475	298		1,773	2,000	2,300
	Repair controls				.600		36		36	45	56
	Total				22.939	3,758	1,316.30		5,074.30	5,784.50	6,730.90
1050	**Replace boiler, gas/oil, 2000 MBH**	30	Q-7	Ea.							
	Remove boiler				57.762		3,275		3,275	4,100	5,125
	Replace boiler, gas/oil, 2000 MBH				115.000	41,300	6,575		47,875	53,500	62,000
	Total				172.762	41,300	9,850		51,150	57,600	67,125

For customer support on your Facilities Maintenance & Repair Cost Data, call 877.801.5724.

D3023 186 Boiler, Gas/Oil

	System Description	Freq. (Years)	Crew	Unit	Labor Hours	2015 Bare Costs Material	2015 Bare Costs Labor	2015 Bare Costs Equipment	2015 Bare Costs Total	Total In-House	Total w/O&P
2010	**Repair boiler, gas/oil, 20,000 MBH**	7	Q-5	Ea.							
	Remove/replace burner blower				.600		36		36	45	56
	Remove burner blower bearing				1.000		60		60	74.50	93
	Replace burner blower bearing				2.000	57.50	120		177.50	213	258
	Remove burner blower motor				.976		53.50		53.50	66	82.50
	Replace burner blower motor				1.951	248	107		355	405	475
	Remove burner fireye				.195		11.65		11.65	14.60	18.20
	Replace burner fireye				.300	221	17.95		238.95	266	305
	Remove burner ignition transformer				.333		19.90		19.90	25	31
	Replace burner ignition transformer				.667	80.50	40		120.50	138	163
	Remove burner ignition electrode				.267		15.95		15.95	19.90	25
	Replace burner ignition electrode				.444	16	26.50		42.50	51	61.50
	Remove burner oil pump				.500		30		30	37.50	46.50
	Replace burner oil pump				2.602	162	140		302	355	420
	Remove burner nozzle				.167		9.95		9.95	12.45	15.55
	Replace burner nozzle				.571	13.50	34		47.50	57.50	70.50
	Remove burner gas regulator				1.300		70		70	87.50	109
	Replace burner gas regulator				2.600	2,225	140		2,365	2,625	3,000
	Remove burner auto gas valve				5.195		285		285	355	445
	Replace burner auto gas valve				8.000	5,875	440		6,315	7,000	8,025
	Remove burner solenoid valve				2.600		149		149	186	232
	Replace burner solenoid valve				10.399	3,725	595		4,320	4,850	5,575
	Repair controls				.600		36		36	45	56
	Total				43.266	12,623.50	2,437.40		15,060.90	16,928.95	19,562.75
2050	**Replace boiler, gas/oil, 20,000 MBH**	30	Q-7	Ea.							
	Remove boiler				1527.360		86,580		86,580	109,000	135,500
	Replace boiler, gas/oil, 20,000 MBH				3156.840	281,940	179,820		461,760	534,500	633,000
	Total				4684.200	281,940	266,400		548,340	643,500	768,500

D3023 198 Blowoff System

	System Description	Freq. (Years)	Crew	Unit	Labor Hours	2015 Bare Costs Material	2015 Bare Costs Labor	2015 Bare Costs Equipment	2015 Bare Costs Total	Total In-House	Total w/O&P
1010	**Repair boiler blowoff system**	10	1 STPI	Ea.							
	Repair leak				1.000		60		60	74.50	93
	Total				1.000		60		60	74.50	93

For customer support on your Facilities Maintenance & Repair Cost Data, call 877.801.5724.

223

D30 HVAC | D3023 | Heat Generating Systems

D3023 198 Blowoff System

	System Description	Freq. (Years)	Crew	Unit	Labor Hours	2015 Bare Costs				Total In-House	Total w/O&P
						Material	Labor	Equipment	Total		
1020	Replace boiler blowoff system	15	Q-5	Ea.							
	Remove boiler blowoff				2.773		149		149	186	233
	Replace boiler blowoff				5.556	4,850	299		5,149	5,700	6,525
	Total				8.329	4,850	448		5,298	5,886	6,758

D3023 292 Chemical Feed System

	System Description	Freq. (Years)	Crew	Unit	Labor Hours	2015 Bare Costs				Total In-House	Total w/O&P
						Material	Labor	Equipment	Total		
1010	Repair chemical feed	15	1 STPI	Ea.							
	Repair controls				.040		2.39		2.39	2.99	3.73
	Remove / replace agitator motor				1.951	190	117		307	355	420
	Remove / replace pump seals / bearings				2.078	33	122		155	189	231
	Remove / replace pump motor				1.951	195	107		302	345	410
	Total				6.020	418	348.39		766.39	891.99	1,064.73
1030	Replace chemical feed	15	2 STPI	Ea.							
	Remove / replace feeder				2.500	580	147		727	820	955
	Total				2.500	580	147		727	820	955

D3023 294 Feed Water Supply

	System Description	Freq. (Years)	Crew	Unit	Labor Hours	2015 Bare Costs				Total In-House	Total w/O&P
						Material	Labor	Equipment	Total		
1010	Repair feed water supply pump	15	1 STPI	Ea.							
	Repair controls				.600		36		36	45	56
	Remove / replace pump seals / bearings				1.600		95.50		95.50	120	149
	Remove pump motor				2.078		124		124	155	194
	Replace pump motor				4.167	2,575	249		2,824	3,150	3,600
	Remove / replace pump coupling				1.000		60		60	74.50	93
	Total				9.445	2,575	564.50		3,139.50	3,544.50	4,092
1030	Replace feed water pump	15	Q-2	Ea.							
	Remove pump				11.111		610		610	760	950
	Replace pump				22.222	19,700	1,225		20,925	23,200	26,500
	Total				33.333	19,700	1,835		21,535	23,960	27,450

D3023 296 Deaerator

	System Description	Freq. (Years)	Crew	Unit	Labor Hours	2015 Bare Costs				Total In-House	Total w/O&P
						Material	Labor	Equipment	Total		
1010	**Repair deaerator**										
	Repair controls	10	1 STPI	Ea.	1.000		60		60	74.50	93
	Total				1.000		60		60	**74.50**	**93**
1030	**Replace deaerator**										
	Remove unit	20	4 STPI	Ea.	7.500		420		420	525	650
	Replace unit				179.000	30,800	10,700		41,500	47,300	55,500
	Total				186.500	30,800	11,120		41,920	47,825	56,150

D3023 298 Separators for 9000 Ton Chilled Water System

	System Description	Freq. (Years)	Crew	Unit	Labor Hours	2015 Bare Costs				Total In-House	Total w/O&P
						Material	Labor	Equipment	Total		
1010	**Clean separator strainer**										
	Clean separator strainer	10	Q-6	Ea.	10.000		600		600	745	930
	Total				10.000		600		600	**745**	**930**
1030	**Replace separator**										
	Remove unit	20	Q-6	Ea.	22.500		1,260		1,260	1,575	1,950
	Replace separator				42.857	51,900	2,385		54,285	60,000	68,500
	Total				65.357	51,900	3,645		55,545	61,575	70,450

D3023 310 Metal Flue / Chimney

	System Description	Freq. (Years)	Crew	Unit	Labor Hours	2015 Bare Costs				Total In-House	Total w/O&P
						Material	Labor	Equipment	Total		
0010	**Replace metal flue, all fuel SS, 6" diameter**										
	Replace flue, all fuel stainless steel, 6" diameter	15	Q-9	L.F.	.387	78.30	19.50		97.80	111	128
	Total				.387	78.30	19.50		97.80	**111**	**128**
0020	**Replace metal flue, all fuel SS, 10" diameter**										
	Replace flue, all fuel stainless steel, 10" diameter	15	Q-9	L.F.	.483	98.60	24.36		122.96	139	162
	Total				.483	98.60	24.36		122.96	**139**	**162**

For customer support on your Facilities Maintenance & Repair Cost Data, call 877.801.5724.

225

D3023 310 Metal Flue / Chimney

	System Description	Freq. (Years)	Crew	Unit	Labor Hours	2015 Bare Costs				Total In-House	Total w/O&P
						Material	Labor	Equipment	Total		
0030	**Replace metal flue, all fuel SS, 20" diameter**										
	Replace flue, all fuel stainless steel, 20" diameter	15	Q-10	L.F.	.967	184.15	50.75		234.90	267	310
	Total				.967	184.15	50.75		234.90	**267**	**310**
0040	**Replace metal flue, all fuel SS, 32" diameter**										
	Replace flue, all fuel stainless steel, 32" diameter	15	Q-10	L.F.	1.289	305.95	67.43		373.38	420	490
	Total				1.289	305.95	67.43		373.38	**420**	**490**
0050	**Replace metal flue, all fuel SS, 48" diameter**										
	Replace flue, all fuel stainless steel, 48" diameter	15	Q-10	L.F.	1.832	456.75	95.70		552.45	625	720
	Total				1.832	456.75	95.70		552.45	**625**	**720**
2010	**Replace metal flue, gas vent, galvanized, 48" diameter**										
	Replace flue, gas vent, galvanized, 48" diameter	10	Q-10	L.F.	1.895	1,485	99		1,584	1,775	2,000
	Total				1.895	1,485	99		1,584	**1,775**	**2,000**

D3023 388 Pneumatic Coal Spreader

	System Description	Freq. (Years)	Crew	Unit	Labor Hours	2015 Bare Costs				Total In-House	Total w/O&P
						Material	Labor	Equipment	Total		
1010	**Repair spreader, pneumatic coal**	10	2 STPI	Ea.							
	Repair controls				1.000		60		60	74.50	93
	Remove conveyor bearing				1.111		66.50		66.50	83	104
	Replace conveyor bearing				2.222	64.50	133		197.50	237	288
	Remove conveyor motor				2.597		155		155	194	242
	Replace conveyor motor				5.195	2,925	284		3,209	3,575	4,100
	Remove / replace blower				1.200		71.50		71.50	89.50	112
	Remove blower bearings				1.000		60		60	74.50	93
	Replace blower bearings				2.000	57.50	120		177.50	213	258
	Remove blower motor				2.000		120		120	149	186
	Replace blower motor				4.000	1,750	219		1,969	2,200	2,525
	Total				22.326	4,797	1,289		6,086	**6,889.50**	**8,001**
1060	**Replace coal spreader**	12	4 STPI	Ea.							
	Remove spreader				65.041		3,875		3,875	4,850	6,050
	Replace spreader				130.000	7,275	7,775		15,050	17,700	21,200
	Total				195.041	7,275	11,650		18,925	**22,550**	**27,250**

For customer support on your Facilities Maintenance & Repair Cost Data, call 877.801.5724.

	System Description	Freq. (Years)	Crew	Unit	Labor Hours	2015 Bare Costs				Total In-House	Total w/O&P
						Material	Labor	Equipment	Total		
1010	**Repair fuel oil equipment, pump**	10	1 STPI	Ea.							
	Remove pump seals / bearings				.600		36		36	45	56
	Replace pump seals / bearings				1.500	30	89.50		119.50	145	178
	Remove / replace pump coupling				.900		54		54	67	84
	Remove / replace impeller / shaft				.900		54		54	67	84
	Remove / replace motor				.650		39		39	48.50	60.50
	Remove strainers				.289		17.25		17.25	21.50	27
	Replace strainers				.578	12.95	34.50		47.45	57.50	70
	Total				5.417	42.95	324.25		367.20	451.50	559.50
1030	**Replace fuel oil 25 GPH pump / motor set**	15	Q-5	Ea.							
	Remove equipment				1.733		93		93	117	145
	Replace equipment				3.468	1,050	186		1,236	1,400	1,600
	Total				5.201	1,050	279		1,329	1,517	1,745
1040	**Replace fuel oil 45 GPH pump / motor set**	15	Q-5	Ea.							
	Remove equipment				1.733		93		93	117	145
	Replace equipment				3.468	1,050	186		1,236	1,400	1,600
	Total				5.201	1,050	279		1,329	1,517	1,745
1050	**Replace fuel oil 90 GPH pump / motor set**	15	Q-5	Ea.							
	Remove equipment				2.081		112		112	140	175
	Replace equipment				4.161	1,050	224		1,274	1,425	1,675
	Total				6.242	1,050	336		1,386	1,565	1,850
1060	**Replace fuel oil 160 GPH pump / motor set**	15	Q-5	Ea.							
	Remove equipment				2.597		140		140	175	218
	Replace equipment				5.195	1,100	279		1,379	1,550	1,800
	Total				7.792	1,100	419		1,519	1,725	2,018

For customer support on your Facilities Maintenance & Repair Cost Data, call 877.801.5724.

227

System Description	Freq. (Years)	Crew	Unit	Labor Hours	2015 Bare Costs				Total In-House	Total w/O&P
					Material	Labor	Equipment	Total		
1010 Repair cooling tower, 50 ton	10	2 STPI	Ea.							
Repair controls				1.000		60		60	74.50	93
Remove bearings				1.200		71.50		71.50	89.50	112
Replace bearings				2.000	57.50	120		177.50	213	258
Remove fan motor				1.155		69		69	86.50	108
Replace fan motor				2.312	560	126		686	775	895
Remove / replace float valve				.500	19.85	27		46.85	55.50	67
Total				8.167	637.35	473.50		1,110.85	1,294	1,533
1030 Replace cooling tower, 50 ton	15	Q-6	Ea.							
Remove cooling tower				10.397		579		579	725	905
Replace cooling tower, 50 ton				20.779	12,900	1,158		14,058	15,700	17,900
Total				31.177	12,900	1,737		14,637	16,425	18,805
2010 Repair cooling tower, 100 ton	10	2 STPI	Ea.							
Repair controls				1.000		60		60	74.50	93
Remove bearings				4.800		286		286	360	450
Replace bearings				8.000	230	480		710	850	1,025
Remove fan motors				4.621		276		276	345	430
Replace fan motors				9.249	2,240	504		2,744	3,100	3,575
Remove / replace float valve				.575	22.83	31.05		53.88	64	77
Total				28.245	2,492.83	1,637.05		4,129.88	4,793.50	5,650
2030 Replace cooling tower, 100 ton	15	Q-6	Ea.							
Remove cooling tower				14.286		795		795	995	1,250
Replace cooling tower, 100 ton				28.571	18,000	1,595		19,595	21,800	25,000
Total				42.857	18,000	2,390		20,390	22,795	26,250
3010 Repair cooling tower, 300 ton	10	2 STPI	Ea.							
Repair controls				1.000		60		60	74.50	93
Remove bearings				7.200		429		429	535	670
Replace bearings				13.333	387	798		1,185	1,425	1,725
Remove fan motors				6.931		414		414	520	650
Replace fan motors				13.865	3,660	756		4,416	4,950	5,750
Remove / replace float valve				1.250	49.63	67.50		117.13	139	167
Total				43.580	4,096.63	2,524.50		6,621.13	7,643.50	9,055

For customer support on your Facilities Maintenance & Repair Cost Data, call 877.801.5724.

D3033 115 | Cooling Tower

	System Description	Freq. (Years)	Crew	Unit	Labor Hours	2015 Bare Costs				Total In-House	Total w/O&P
						Material	Labor	Equipment	Total		
3030	**Replace cooling tower, 300 ton**	15	Q-6	Ea.							
	Remove cooling tower				36.273		2,025		2,025	2,525	3,150
	Replace cooling tower, 300 ton				72.582	31,500	4,050		35,550	39,600	45,700
	Total				108.855	31,500	6,075		37,575	42,125	**48,850**
4010	**Repair cooling tower, 1000 ton**	10	2 STPI	Ea.							
	Repair controls				1.000		60		60	74.50	93
	Remove bearings				12.000		715		715	895	1,125
	Replace bearings				23.529	1,800	1,410		3,210	3,750	4,450
	Remove fan motors				11.552		690		690	865	1,075
	Replace fan motors				25.974	9,900	1,420		11,320	12,800	14,600
	Remove / replace float valve				2.350	93.30	126.90		220.20	261	315
	Total				76.406	11,793.30	4,421.90		16,215.20	18,645.50	**21,658**
4030	**Replace cooling tower, 1000 ton**	15	Q-6	Ea.							
	Remove cooling tower				104.350		5,800		5,800	7,300	9,100
	Replace cooling tower, 1000 ton				208.700	79,000	11,650		90,650	101,000	117,000
	Total				313.050	79,000	17,450		96,450	108,300	**126,100**

D3033 130 | Chiller, Water Cooled, Reciprocating

	System Description	Freq. (Years)	Crew	Unit	Labor Hours	2015 Bare Costs				Total In-House	Total w/O&P
						Material	Labor	Equipment	Total		
1010	**Repair water cooled chiller, 20 ton**	10	Q-6	Ea.							
	Repair controls				1.000		60		60	74.50	93
	Remove compressor				21.463		1,188		1,188	1,500	1,875
	Replace compressor				42.857	27,200	2,400		29,600	32,900	37,700
	Remove / replace evaporator tube				1.200	176	67		243	277	325
	Remove / replace condenser tube				1.200	176	67		243	277	325
	Replace refrigerant				2.667	640	159		799	900	1,050
	Total				70.387	28,192	3,941		32,133	35,928.50	**41,368**
1030	**Replace chiller, water cooled, 20 ton**	20	Q-7	Ea.							
	Remove chiller				50.633		2,875		2,875	3,600	4,500
	Replace chiller, water cooled, 20 ton				101.000	19,500	5,750		25,250	28,600	33,400
	Total				151.633	19,500	8,625		28,125	32,200	**37,900**

For customer support on your Facilities Maintenance & Repair Cost Data, call 877.801.5724.

229

D3033 130 | **Chiller, Water Cooled, Reciprocating**

System Description	Freq. (Years)	Crew	Unit	Labor Hours	Material	Labor	Equipment	Total	Total In-House	Total w/O&P
						2015 Bare Costs				
2010	**Repair water cooled chiller, 50 ton**	10	Q-6	Ea.						
Repair controls				1.000		60		60	74.50	93
Remove compressor				35.610		1,989.25		1,989.25	2,475	3,100
Replace compressor				71.006	30,500	3,950		34,450	38,500	44,300
Remove / replace evaporator tube				1.200	176	67		243	277	325
Remove / replace condenser tube				1.200	176	67		243	277	325
Replace refrigerant				4.000	960	238.50		1,198.50	1,350	1,575
Total				114.016	31,812	6,371.75		38,183.75	42,953.50	49,718
2030	**Replace chiller, water cooled 50 ton**	20	Q-7	Ea.						
Remove chiller				74.074		4,225		4,225	5,275	6,575
Replace chiller, water cooled, 50 ton				148.000	35,800	8,425		44,225	49,900	58,000
Total				222.074	35,800	12,650		48,450	55,175	64,575
3010	**Repair water cooled chiller, 100 ton**	10	Q-6	Ea.						
Repair controls				1.000		60		60	74.50	93
Remove compressor				77.922		4,350		4,350	5,425	6,775
Replace compressor				153.000	35,500	8,750		44,250	49,900	58,000
Remove / replace evaporator tube				4.798	670	273		943	1,075	1,275
Remove / replace condenser tube				4.798	670	273		943	1,075	1,275
Replace refrigerant				8.000	1,920	477		2,397	2,700	3,150
Total				249.517	38,760	14,183		52,943	60,249.50	70,568
3030	**Replace chiller, water cooled, 100 ton**	20	Q-7	Ea.						
Remove chiller				116.000		6,650		6,650	8,250	10,400
Replace chiller, water cooled, 100 ton				233.000	60,500	13,300		73,800	83,000	96,500
Total				349.000	60,500	19,950		80,450	91,250	106,900
5010	**Repair water cooled chiller, 200 ton**	10	Q-7	Ea.						
Repair controls				1.000		60		60	74.50	93
Remove compressor				152.099		8,662.50		8,662.50	10,800	13,500
Replace compressor				303.380	62,370	17,248		79,618	90,000	105,000
Remove / replace evaporator tube				9.610	1,325	545		1,870	2,150	2,500
Remove / replace condenser tube				9.610	1,325	545		1,870	2,150	2,500
Replace refrigerant				13.333	3,200	795		3,995	4,500	5,250
Total				489.031	68,220	27,855.50		96,075.50	109,674.50	128,843

For customer support on your Facilities Maintenance & Repair Cost Data, call 877.801.5724.

D3033 130 — Chiller, Water Cooled, Reciprocating

	System Description	Freq. (Years)	Crew	Unit	Labor Hours	2015 Bare Costs				Total In-House	Total w/O&P
						Material	Labor	Equipment	Total		
5030	Replace chiller, water cooled, 200 ton	20	Q-7	Ea.							
	Remove chiller				163.000		9,300		9,300	11,600	14,500
	Replace chiller, water cooled, 200 ton				326.000	95,000	18,600		113,600	127,500	148,000
	Total				489.000	95,000	27,900		122,900	139,100	162,500

D3033 135 — Chiller, Air Cooled, Reciprocating

	System Description	Freq. (Years)	Crew	Unit	Labor Hours	2015 Bare Costs				Total In-House	Total w/O&P
						Material	Labor	Equipment	Total		
1010	Repair recip. chiller, air cooled, 20 ton	10	2 STPI	Ea.							
	Repair controls				.800		48		48	60	74.50
	Remove fan bearing				3.600		214.50		214.50	269	335
	Replace fan bearing				6.000	172.50	360		532.50	640	775
	Remove fan motor				2.927		175.50		175.50	219	273
	Replace fan motor				5.854	744	321		1,065	1,225	1,425
	Remove compressor				24.490		1,375		1,375	1,700	2,125
	Replace compressor				48.780	28,300	2,725		31,025	34,500	39,600
	Replace refrigerant				2.667	640	159		799	900	1,050
	Total				95.117	29,856.50	5,378		35,234.50	39,513	45,657.50
1030	Replace chiller, 20 ton	20	Q-7	Ea.							
	Remove chiller				42.857		2,400		2,400	3,000	3,725
	Replace chiller, air cooled, reciprocating, 20 ton				118.000	21,500	6,775		28,275	32,000	37,500
	Total				160.857	21,500	9,175		30,675	35,000	41,225
2010	Repair recip. chiller, air cooled, 50 ton	10	Q-6	Ea.							
	Repair controls				1.000		60		60	74.50	93
	Remove fan bearing				3.600		214.50		214.50	269	335
	Replace fan bearing				6.667	193.50	399		592.50	710	865
	Remove fan motor				3.096		169.50		169.50	210	263
	Replace fan motor				6.936	1,680	378		2,058	2,325	2,700
	Remove compressor				60.976		3,375		3,375	4,250	5,325
	Replace compressor				155.000	32,100	8,700		40,800	46,100	53,500
	Remove / replace compressor motor				7.201		394		394	490	610
	Replace refrigerant				4.000	960	238.50		1,198.50	1,350	1,575
	Total				248.475	34,933.50	13,928.50		48,862	55,778.50	65,266

For customer support on your Facilities Maintenance & Repair Cost Data, call 877.801.5724.

231

D3033 135 Chiller, Air Cooled, Reciprocating

	System Description	Freq. (Years)	Crew	Unit	Labor Hours	2015 Bare Costs				Total In-House	Total w/O&P
						Material	Labor	Equipment	Total		
2030	**Replace chiller, 50 ton**	20	Q-7	Ea.							
	Remove chiller				73.395		4,175		4,175	5,225	6,525
	Replace chiller, air cooled, reciprocating, 50 ton				146.000	38,200	8,350		46,550	52,500	61,000
	Total				219.395	38,200	12,525		50,725	57,725	67,525
3010	**Repair chiller, air cooled, 100 ton**	10	Q-6								
	Repair controls				1.000		60		60	74.50	93
	Remove fan bearing				4.200		250.50		250.50	315	390
	Replace fan bearing				6.667	193.50	399		592.50	710	865
	Remove fan motor				3.900		213		213	264	330
	Replace fan motor				7.792	2,970	426		3,396	3,825	4,375
	Remove compressor				97.403		5,437.50		5,437.50	6,775	8,475
	Replace compressor				191.250	44,375	10,937.50		55,312.50	62,500	72,500
	Remove / replace compressor motor				9.359		512		512	630	790
	Replace refrigerant				6.667	1,600	397.50		1,997.50	2,250	2,625
	Total				328.237	49,138.50	18,633		67,771.50	77,343.50	90,443
3030	**Replace chiller, 100 ton**	20	Q-7	Ea.							
	Remove chiller				84.211		4,800		4,800	6,000	7,475
	Replace chiller, air cooled, reciprocating, 100 ton				168.000	73,500	9,575		83,075	93,000	107,000
	Total				252.211	73,500	14,375		87,875	99,000	114,475

D3033 137 Chiller, Water Cooled, Scroll

	System Description	Freq. (Years)	Crew	Unit	Labor Hours	2015 Bare Costs				Total In-House	Total w/O&P
						Material	Labor	Equipment	Total		
4010	**Repair water cooled chiller, 5 ton**	10	1 STPI	Ea.							
	Repair controls				.500		30		30	37.50	46.50
	Remove fan bearing				1.200		71.50		71.50	89.50	112
	Replace fan bearing				2.000	57.50	120		177.50	213	258
	Remove compressor				2.477		148		148	185	231
	Replace compressor				4.954	505	296		801	925	1,100
	Replace refrigerant				.667	160	39.75		199.75	225	262
	Total				11.797	722.50	705.25		1,427.75	1,675	2,009.50

D3033 137   Chiller, Water Cooled, Scroll

System Description	Freq. (Years)	Crew	Unit	Labor Hours	2015 Bare Costs				Total In-House	Total w/O&P
					Material	Labor	Equipment	Total		
4030										
Replace chiller, water cooled, 5 ton	20	Q-5	Ea.							
Remove chiller				18.182		980		980	1,225	1,525
Replace chiller, water cooled, 5 ton				36.364	3,500	1,950		5,450	6,300	7,425
Total				54.545	3,500	2,930		6,430	7,525	8,950
5010										
Repair water cooled chiller, 10 ton	10	2 STPI	Ea.							
Repair controls				1.000		60		60	74.50	93
Remove fan bearing				3.600		214.50		214.50	269	335
Replace fan bearing				6.000	172.50	360		532.50	640	775
Remove compressor				4.954		296		296	370	460
Replace compressor				9.907	1,010	592		1,602	1,850	2,175
Replace refrigerant				1.333	320	79.50		399.50	450	525
Total				26.794	1,502.50	1,602		3,104.50	3,653.50	4,363
5030										
Replace chiller, water cooled, 10 ton	20	Q-6	Ea.							
Remove chiller				43.478		2,425		2,425	3,025	3,775
Replace chiller, water cooled, 10 ton				86.957	6,625	4,850		11,475	13,400	15,900
Total				130.435	6,625	7,275		13,900	16,425	19,675
6010										
Repair water cooled chiller, 15 ton	10	Q-5	Ea.							
Repair controls				1.000		60		60	74.50	93
Remove fan bearing				24.000		1,170		1,170	1,525	1,900
Replace fan bearing				6.000	172.50	360		532.50	640	775
Remove compressor				6.932		372		372	465	580
Replace compressor				13.865	2,300	750		3,050	3,450	4,025
Replace refrigerant				2.000	480	119.25		599.25	675	785
Total				53.797	2,952.50	2,831.25		5,783.75	6,829.50	8,158
6030										
Replace chiller, water cooled, 15 ton	20	Q-6	Ea.							
Remove chiller				42.857		2,400		2,400	3,000	3,725
Replace chiller, water cooled, 15 ton				85.714	18,900	4,775		23,675	26,800	31,100
Total				128.571	18,900	7,175		26,075	29,800	34,825

System Description	Freq. (Years)	Crew	Unit	Labor Hours	2015 Bare Costs				Total In-House	Total w/O&P
					Material	Labor	Equipment	Total		
1010 Repair centrifugal chiller, 100 ton	10	Q-7	Ea.							
Repair controls				1.000		60		60	74.50	93
Remove compressor				48.780		2,700		2,700	3,400	4,250
Replace compressor				109.600	19,760	6,260		26,020	29,500	34,500
Remove / replace condenser tube				4.800	704	268		972	1,100	1,300
Remove / replace evaporator tube				4.800	704	268		972	1,100	1,300
Remove / replace refrigerant filter				1.600	268	91		359	410	475
Remove refrigerant				.440		26.30		26.30	33	41
Replace refrigerant				8.000	1,920	477		2,397	2,700	3,150
Total				179.020	23,356	10,150.30		33,506.30	38,317.50	45,109
1030 Replace centrifugal chiller, 100 ton	20	Q-7	Ea.							
Remove chiller				111.000		6,375		6,375	7,900	9,925
Replace chiller, centrifugal, 100 ton				223.000	63,000	12,700		75,700	85,000	98,500
Total				334.000	63,000	19,075		82,075	92,900	108,425
2010 Repair centrifugal chiller, 300 ton	10	Q-7	Ea.							
Repair controls				1.000		60		60	74.50	93
Remove compressor				94.390		5,224.50		5,224.50	6,575	8,225
Replace compressor				187.000	25,900	10,600		36,500	41,800	49,000
Remove / replace condenser tube				12.000	1,760	670		2,430	2,775	3,250
Remove / replace evaporator tube				12.000	1,760	670		2,430	2,775	3,250
Remove / replace refrigerant filter				1.200	570	68.50	2.16	640.66	715	820
Remove refrigerant				1.319		79.50		79.50	99	123
Replace refrigerant				21.333	5,120	1,272		6,392	7,200	8,400
Total				330.242	35,110	18,644.50	2.16	53,756.66	62,013.50	73,161
2030 Replace centrifugal chiller, 300 ton	20	Q-7	Ea.							
Remove chiller				150.000		8,550		8,550	10,700	13,300
Replace chiller, 300 ton				301.000	94,500	17,200		111,700	125,500	145,000
Total				451.000	94,500	25,750		120,250	136,200	158,300

For customer support on your Facilities Maintenance & Repair Cost Data, call 877.801.5724.

D3033 140 | **Chiller, Hermetic Centrifugal**

	System Description	Freq. (Years)	Crew	Unit	Labor Hours	2015 Bare Costs				Total In-House	Total w/O&P
						Material	Labor	Equipment	Total		
3010	**Repair centrifugal chiller, 1000 ton**	10	Q-7	Ea.							
	Repair controls				1.000		60		60	74.50	93
	Remove compressor				153.659		8,505		8,505	10,700	13,400
	Replace compressor				296.000	63,500	16,900		80,400	91,000	105,500
	Remove / replace condenser tube				16.800	2,464	1,001		3,465	3,950	4,650
	Remove / replace evaporator tube				12.600	2,464	756		3,220	3,650	4,250
	Remove / replace refrigerant filter				1.600	570	95.50		665.50	745	860
	Remove refrigerant				3.516		212		212	264	330
	Replace refrigerant				29.333	7,040	1,749		8,789	9,900	11,500
	Total				514.508	76,038	29,278.50		105,316.50	120,283.50	140,583
3030	**Replace centrifugal chiller, 1000 ton**	20	Q-7	Ea.							
	Remove chiller				242.000		13,800		13,800	17,200	21,500
	Replace chiller, 1000 ton				484.000	349,500	27,600		377,100	419,000	480,000
	Total				726.000	349,500	41,400		390,900	436,200	501,500
6010	**Repair centrifugal chiller, 9000 ton**	10	Q-7	Ea.							
	Repair controls				1.000		60		60	74.50	93
	Remove compressor				1083.600		60,480		60,480	75,500	94,500
	Replace compressor				2106.000	480,000	120,000		600,000	678,000	787,000
	Remove / replace condenser tube				151.204	22,176	9,009		31,185	35,700	41,800
	Remove / replace evaporator tube				113.399	22,176	6,804		28,980	32,900	38,300
	Remove / replace refrigerant filter				14.400	5,130	859.50		5,989.50	6,700	7,750
	Remove refrigerant				31.648		1,908		1,908	2,375	2,950
	Replace refrigerant				263.993	63,360	15,741		79,101	89,000	104,000
	Total				3765.244	592,842	214,861.50		807,703.50	920,249.50	1,076,393
6030	**Replace centrifugal chiller, 9000 ton**	20	Q-7	Ea.							
	Remove chiller				1452.000		82,800		82,800	103,000	129,000
	Replace chiller, 9000 ton				3306.000	2,814,000	188,400		3,002,400	3,330,000	3,811,500
	Total				4758.000	2,814,000	271,200		3,085,200	3,433,000	3,940,500

D3033 142 — Chiller, Open Centrifugal

	System Description	Freq. (Years)	Crew	Unit	2015 Bare Costs					Total In-House	Total w/O&P
					Labor Hours	Material	Labor	Equipment	Total		
1010	**Repair open centrifugal chiller, 300 ton**	10	Q-7	Ea.							
	Repair controls				1.000		60		60	74.50	93
	Remove compressor				104.878		5,805		5,805	7,300	9,150
	Replace compressor				205.700	28,490	11,660		40,150	46,000	54,000
	Remove / replace condenser tube				12.000	1,760	670		2,430	2,775	3,250
	Remove / replace evaporator tube				12.000	1,760	670		2,430	2,775	3,250
	Remove / replace refrigerant filter				1.200	570	68.50	2.16	640.66	715	820
	Remove refrigerant				20.000		975		975	1,275	1,575
	Replace refrigerant				21.333	5,120	1,272		6,392	7,200	8,400
	Total				378.111	37,700	21,180.50	2.16	58,882.66	68,114.50	80,538
1030	**Replace open centrifugal chiller, 300 ton**	20	Q-7	Ea.							
	Remove chiller				128.000		7,280		7,280	9,100	11,400
	Replace chiller, 300 ton				256.300	103,950	14,630		118,580	132,500	152,500
	Total				384.300	103,950	21,910		125,860	141,600	163,900
2010	**Repair open centrifugal chiller, 1000 ton**	10	Q-7	Ea.							
	Repair controls				1.000		60		60	74.50	93
	Remove compressor				165.854		9,180		9,180	11,600	14,500
	Replace compressor				325.600	69,850	18,590		88,440	100,000	116,000
	Remove / replace condenser tube				16.800	2,464	938		3,402	3,875	4,525
	Remove / replace evaporator tube				16.800	2,464	938		3,402	3,875	4,525
	Remove / replace refrigerant filter				1.600	1,150	91	2.88	1,243.88	1,375	1,575
	Remove refrigerant				35.204		2,104		2,104	2,650	3,275
	Replace refrigerant				34.666	8,320	2,067		10,387	11,700	13,600
	Total				597.523	84,248	33,968	2.88	118,218.88	135,149.50	158,093
2030	**Replace open centrifugal chiller, 1000 ton**	20	Q-7	Ea.							
	Remove chiller				219.100		12,460		12,460	15,600	19,500
	Replace chiller, 1000 ton				409.200	384,450	23,320		407,770	452,000	517,000
	Total				628.300	384,450	35,780		420,230	467,600	536,500

D3033 145 Chiller, Absorption

System Description	Freq. (Years)	Crew	Unit	Labor Hours	2015 Bare Costs Material	Labor	Equipment	Total	Total In-House	Total w/O&P
1010 Repair chiller, absorption, 100 ton	10	2 STPI	Ea.							
Repair controls				1.000		60		60	74.50	93
Remove / replace concentrator tube				.900	176	48.50		224.50	254	296
Remove / replace condenser tube				6.000	880	335		1,215	1,375	1,625
Remove / replace evaporator tube				6.000	880	335		1,215	1,375	1,625
Remove / replace absorber tube				.900	184	48.50		232.50	263	305
Remove / replace gang pump				1.600	7,350	86		7,436	8,200	9,325
Remove gang pump motor				3.704		203		203	251	315
Replace gang pump motor				7.407	5,275	405		5,680	6,300	7,225
Add refrigerant				8.000	263	478		741	885	1,075
Total				35.511	15,008	1,999		17,007	18,977.50	21,884
1030 Replace chiller, absorption, 100 ton	20	Q-7	Ea.							
Remove chiller				155.000		8,825		8,825	11,000	13,800
Replace chiller				310.000	125,000	17,700		142,700	159,500	184,000
Total				465.000	125,000	26,525		151,525	170,500	197,800
2010 Repair chiller, absorption, 350 ton	10	2 STPI	Ea.							
Repair controls				1.000		60		60	74.50	93
Remove / replace concentrator tube				1.800	352	97		449	510	590
Remove / replace condenser tube				7.200	1,056	402		1,458	1,650	1,950
Remove / replace evaporator tube				7.200	1,056	402		1,458	1,650	1,950
Remove / replace absorber tube				1.800	368	97		465	525	610
Remove / replace gang pump				1.600	7,350	86		7,436	8,200	9,325
Remove gang pump motor				3.704		203		203	251	315
Replace gang pump motor				7.407	5,275	405		5,680	6,300	7,225
Add refrigerant				8.000	263	478		741	885	1,075
Total				39.711	15,720	2,230		17,950	20,045.50	23,133
2030 Replace chiller, absorption, 350 ton	20	Q-7	Ea.							
Remove chiller				200.000		11,400		11,400	14,200	17,700
Replace chiller				400.000	335,000	22,800		357,800	397,000	454,500
Total				600.000	335,000	34,200		369,200	411,200	472,200

For customer support on your Facilities Maintenance & Repair Cost Data, call 877.801.5724.

237

D3033 145 Chiller, Absorption

	System Description	Freq. (Years)	Crew	Unit	Labor Hours	2015 Bare Costs				Total In-House	Total w/O&P
						Material	Labor	Equipment	Total		
3010	**Repair chiller, absorption, 950 ton**	10	4 STPI	Ea.							
	Repair controls				1.000		60		60	74.50	93
	Remove / replace concentrator tube				3.600	704	194		898	1,025	1,175
	Remove / replace condenser tube				12.000	1,760	670		2,430	2,775	3,250
	Remove / replace evaporator tube				12.000	1,760	670		2,430	2,775	3,250
	Remove / replace absorber tube				3.600	736	194		930	1,050	1,225
	Remove / replace gang pump				1.600	7,350	86		7,436	8,200	9,325
	Remove gang pump motor				3.704		203		203	251	315
	Replace gang pump motor				8.696	6,425	475		6,900	7,650	8,775
	Add refrigerant				8.000	263	478		741	885	1,075
	Total				54.198	18,998	3,030		22,028	24,685.50	28,483
3030	**Replace chiller, absorption, 950 ton**	20	Q-7	Ea.							
	Remove chiller				266.000		15,200		15,200	18,900	23,700
	Replace chiller				533.000	686,000	30,300		716,300	792,500	905,000
	Total				799.000	686,000	45,500		731,500	811,400	928,700

D3033 210 Air Cooled Condenser

	System Description	Freq. (Years)	Crew	Unit	Labor Hours	2015 Bare Costs				Total In-House	Total w/O&P
						Material	Labor	Equipment	Total		
1010	**Repair condenser, air cooled, 5 ton**	10	1 STPI	Ea.							
	Repair controls				.500		30		30	37.50	46.50
	Remove fan motor				1.800		108		108	134	168
	Replace fan motor				2.311	279	126		405	465	545
	Total				4.611	279	264		543	636.50	759.50
1030	**Replace condenser, air cooled, 5 ton**	15	Q-5	Ea.							
	Remove condenser				5.195		279		279	350	435
	Replace condenser, air cooled, 5 ton				10.390	5,400	560		5,960	6,650	7,625
	Total				15.584	5,400	839		6,239	7,000	8,060
2010	**Repair condenser, air cooled, 20 ton**	10	2 STPI	Ea.							
	Repair controls				1.000		60		60	74.50	93
	Remove fan motor				2.310		138		138	173	216
	Replace fan motor				4.622	670	252		922	1,050	1,225
	Total				7.932	670	450		1,120	1,297.50	1,534

D3033 210 | Air Cooled Condenser

System Description	Freq. (Years)	Crew	Unit	Labor Hours	2015 Bare Costs				Total In-House	Total w/O&P
					Material	Labor	Equipment	Total		
2030 Replace condenser, air cooled, 20 ton	15	Q-5	Ea.							
Remove condenser				10.390		560		560	700	870
Replace condenser, air cooled, 20 ton				20.779	11,100	1,125		12,225	13,600	15,600
Total				31.169	11,100	1,685		12,785	14,300	**16,470**
3010 Repair condenser, air cooled, 50 ton	10	2 STPI	Ea.							
Repair controls				1.000		60		60	74.50	93
Remove fan motor				3.466		207		207	260	325
Replace fan motor				6.936	1,680	378		2,058	2,325	2,700
Total				11.402	1,680	645		2,325	2,659.50	**3,118**
3030 Replace condenser, air cooled, 50 ton	15	Q-6	Ea.							
Remove condenser				38.961		2,175		2,175	2,725	3,400
Replace condenser, air cooled, 50 ton				77.922	25,100	4,350		29,450	33,000	38,200
Total				116.883	25,100	6,525		31,625	35,725	**41,600**
4010 Repair condenser, air cooled, 100 ton	10	2 STPI	Ea.							
Repair controls				1.000		60		60	74.50	93
Remove fan motor				6.931		414		414	520	650
Replace fan motor				13.873	3,360	756		4,116	4,650	5,375
Total				21.804	3,360	1,230		4,590	5,244.50	**6,118**
4030 Replace condenser, air cooled, 100 ton	15	Q-7	Ea.							
Remove condenser				69.264		3,950		3,950	4,925	6,150
Replace condenser, air cooled, 100 ton				138.000	51,000	7,875		58,875	66,000	76,000
Total				207.264	51,000	11,825		62,825	70,925	**82,150**

System Description	Freq. (Years)	Crew	Unit	Labor Hours	2015 Bare Costs				Total In-House	Total w/O&P
					Material	Labor	Equipment	Total		
1010 Repair evaporative condenser, 20 ton	10	2 STPI	Ea.							
Repair controls				1.000		60		60	74.50	93
Remove fan motors				2.310		138		138	173	216
Replace fan motors				4.622	670	252		922	1,050	1,225
Remove pump and motor				20.779		1,100		1,100	1,375	1,700
Replace pump and motor				41.558	3,675	2,200		5,875	6,775	8,025
Remove / replace pump nozzles				5.000	47.50	270		317.50	390	480
Remove / replace float valve				.500	19.85	27		46.85	55.50	67
Total				75.770	4,412.35	4,047		8,459.35	9,893	11,806
1030 Replace evaporative condenser, 20 ton	15	Q-5	Ea.							
Remove condenser				22.130		1,200		1,200	1,500	1,850
Replace evaporative condenser, 20 ton				44.199	8,800	2,375		11,175	12,700	14,700
Total				66.329	8,800	3,575		12,375	14,200	16,550
2010 Repair evaporative condenser, 100 ton	10	2 STPI	Ea.							
Repair controls				1.000		60		60	74.50	93
Remove fan motors				6.931		414		414	520	650
Replace fan motors				13.873	3,360	756		4,116	4,650	5,375
Remove pump and motor				20.779		1,100		1,100	1,375	1,700
Replace pump and motor				41.558	3,675	2,200		5,875	6,775	8,025
Remove / replace nozzles				25.000	237.50	1,350		1,587.50	1,950	2,400
Remove / replace float valve				.500	19.85	27		46.85	55.50	67
Total				109.642	7,292.35	5,907		13,199.35	15,400	18,310
2030 Replace evaporative condenser, 100 ton	15	Q-7	Ea.							
Remove condenser				57.762		3,275		3,275	4,100	5,125
Replace evaporative condenser, 100 ton				115.000	16,600	6,575		23,175	26,400	31,100
Total				172.762	16,600	9,850		26,450	30,500	36,225

	System Description	Freq. (Years)	Crew	Unit	Labor Hours	2015 Bare Costs				Total In-House	Total w/O&P
						Material	Labor	Equipment	Total		
3010	**Repair evaporative condenser, 300 ton**	10	2 STPI	Ea.							
	Repair controls				1.000		60		60	74.50	93
	Remove fan motors				11.552		690		690	865	1,075
	Replace fan motors				23.121	5,600	1,260		6,860	7,750	8,950
	Remove pump and motor				20.779		1,100		1,100	1,375	1,700
	Replace pump and motor				24.000	4,450	1,325		5,775	6,550	7,625
	Remove / replace nozzles				35.000	332.50	1,890		2,222.50	2,725	3,350
	Remove / replace float valve				1.000	39.70	54		93.70	111	134
	Total				116.453	10,422.20	6,379		16,801.20	19,450.50	22,927
3030	**Replace evaporative condenser, 300 ton**	15	Q-7	Ea.							
	Remove condenser				152.471		8,667		8,667	10,900	13,500
	Replace evaporative condenser, 300 ton				304.560	41,958	17,334		59,292	68,000	79,500
	Total				457.031	41,958	26,001		67,959	78,900	93,000

For customer support on your Facilities Maintenance & Repair Cost Data, call 877.801.5724.

241

System Description	Freq. (Years)	Crew	Unit	Labor Hours	2015 Bare Costs				Total In-House	Total w/O&P
					Material	Labor	Equipment	Total		
1010 Repair fan coil unit, 1 ton	10	1 STPI	Ea.							
Remove fan motor				1.200		71.50		71.50	89.50	112
Replace fan motor				1.951	195	107		302	345	410
Total				3.151	195	178.50		373.50	**434.50**	522
1030 Replace fan coil unit, 1 ton	15	Q-5	Ea.							
Remove fan coil unit				1.733		93		93	117	145
Replace fan coil unit, 1 ton				3.467	950	186		1,136	1,275	1,475
Total				5.200	950	279		1,229	**1,392**	**1,620**
2010 Repair fan coil unit, 3 ton	10	1 STPI	Ea.							
Remove fan motor				1.200		71.50		71.50	89.50	112
Replace fan motor				1.951	248	107		355	405	475
Total				3.151	248	178.50		426.50	**494.50**	587
2030 Replace fan coil unit, 3 ton	15	Q-5	Ea.							
Remove fan coil unit				2.597		140		140	175	218
Replace fan coil unit, 3 ton				5.195	2,175	279		2,454	2,750	3,150
Total				7.792	2,175	419		2,594	**2,925**	**3,368**
3010 Repair fan coil unit, 5 ton	10	1 STPI	Ea.							
Remove fan motor				1.200		71.50		71.50	89.50	112
Replace fan motor				2.311	279	126		405	465	545
Total				3.511	279	197.50		476.50	**554.50**	657
3030 Replace fan coil unit, 5 ton	15	Q-5	Ea.							
Remove fan coil unit				2.680		144		144	180	225
Replace fan coil unit, 5 ton				5.371	2,256	286.70		2,542.70	2,850	3,275
Total				8.052	2,256	430.70		2,686.70	**3,030**	**3,500**
4010 Repair fan coil unit, 10 ton	10	1 STPI	Ea.							
Remove fan motor				1.400		83.50		83.50	105	130
Replace fan motor				2.312	560	126		686	775	895
Total				3.712	560	209.50		769.50	880	1,025

D3043 120   Fan Coil

System Description	Freq. (Years)	Crew	Unit	Labor Hours	2015 Bare Costs				Total In-House	Total w/O&P
					Material	Labor	Equipment	Total		
4030 **Replace fan coil unit, 10 ton**	15	Q-6	Ea.							
Remove fan coil unit				11.538		645		645	805	1,000
Replace fan coil unit, 10 ton				23.077	2,850	1,275		4,125	4,750	5,575
Total				34.615	2,850	1,920		4,770	**5,555**	**6,575**
5010 **Repair fan coil unit, 20 ton**	10	1 STPI	Ea.							
Remove fan motor				1.600		95.50		95.50	120	149
Replace fan motor				2.311	610	126		736	825	960
Total				3.911	610	221.50		831.50	**945**	**1,109**
5030 **Replace fan coil unit, 20 ton**	15	Q-6	Ea.							
Remove fan coil unit				19.512		1,100		1,100	1,350	1,700
Replace fan coil unit, 20 ton				39.024	4,600	2,175		6,775	7,775	9,150
Total				58.537	4,600	3,275		7,875	**9,125**	**10,850**
6010 **Repair fan coil unit, 30 ton**	10	1 STPI	Ea.							
Remove fan motor				1.800		108		108	134	168
Replace fan motor				2.597	990	142		1,132	1,275	1,450
Total				4.398	990	250		1,240	**1,409**	**1,618**
6030 **Replace fan coil unit, 30 ton**	15	Q-6	Ea.							
Remove fan coil unit				25.974		1,450		1,450	1,800	2,250
Replace fan coil unit, 30 ton				51.948	7,925	2,900		10,825	12,300	14,400
Total				77.922	7,925	4,350		12,275	**14,100**	**16,650**

D3043 122 Fan Coil, DX Air Conditioner, Cooling Only

System Description	Freq. (Years)	Crew	Unit	Labor Hours	2015 Bare Costs				Total In-House	Total w/O&P
					Material	Labor	Equipment	Total		
1010 Repair fan coil, DX 1-1/2 ton, cooling only	10	1 STPI	Ea.							
Repair controls				.300		17.95		17.95	22.50	28
Remove / replace supply fan				.650		39		39	48.50	60.50
Remove supply fan bearing				1.000		60		60	74.50	93
Replace supply fan bearing				2.000	57.50	120		177.50	213	258
Remove supply fan motor				1.156		63		63	78.50	98
Replace supply fan motor				1.951	190	117		307	355	420
Remove compressor				2.477		148		148	185	231
Replace compressor				3.714	270	222		492	575	685
Replace refrigerant				.267	64	15.90		79.90	90	105
Total				13.515	581.50	802.85		1,384.35	1,642	1,978.50
1040 Replace fan coil, DX 1-1/2 ton, no heat	15	Q-5	Ea.							
Remove fan coil unit				2.080		112		112	140	174
Replace fan coil unit, 1-1/2 ton, no heat				4.160	680	224		904	1,025	1,200
Total				6.240	680	336		1,016	1,165	1,374
2010 Repair fan coil, DX 2 ton, cooling only	10	1 STPI	Ea.							
Repair controls				.300		17.95		17.95	22.50	28
Remove / replace supply fan				.650		39		39	48.50	60.50
Remove supply fan bearing				1.000		60		60	74.50	93
Replace supply fan bearing				2.000	57.50	120		177.50	213	258
Remove supply fan motor				1.156		63		63	78.50	98
Replace supply fan motor				1.951	195	107		302	345	410
Remove compressor				2.477		148		148	185	231
Replace compressor				4.000	278	239		517	605	725
Replace refrigerant				.400	96	23.85		119.85	135	157
Total				13.934	626.50	817.80		1,444.30	1,707	2,060.50
2040 Replace fan coil, DX 2 ton, no heat	15	Q-5	Ea.							
Remove fan coil unit				2.167		117		117	146	182
Replace fan coil unit, DX 2 ton, no heat				4.334	730	233		963	1,100	1,275
Total				6.500	730	350		1,080	1,246	1,457

D3043 122 Fan Coil, DX Air Conditioner, Cooling Only

System Description	Freq. (Years)	Crew	Unit	Labor Hours	2015 Bare Costs				Total In-House	Total w/O&P
					Material	Labor	Equipment	Total		
3010 Repair fan coil, DX 2-1/2 ton	10	1 STPI	Ea.							
Repair controls				.300		17.95		17.95	22.50	28
Remove / replace supply fan				.650		39		39	48.50	60.50
Remove supply fan bearing				1.000		60		60	74.50	93
Replace supply fan bearing				2.000	57.50	120		177.50	213	258
Remove supply fan motor				1.156		63		63	78.50	98
Replace supply fan motor				1.951	218	107		325	370	440
Remove compressor				2.477		148		148	185	231
Replace compressor				4.334	330	259		589	685	820
Replace refrigerant				.533	128	31.80		159.80	180	210
Total				14.401	733.50	845.75		1,579.25	1,857	2,238.50
3040 Replace fan coil, DX 2-1/2 ton, no heat	15	Q-5	Ea.							
Remove fan coil unit				2.363		127		127	159	198
Replace fan coil, DX 2-1/2 ton, no heat				4.727	775	254		1,029	1,175	1,375
Total				7.090	775	381		1,156	1,334	1,573
4010 Repair fan coil, DX 3 ton, cooling only	10	1 STPI	Ea.							
Repair controls				.300		17.95		17.95	22.50	28
Remove / replace supply fan				.650		39		39	48.50	60.50
Remove supply fan bearing				1.000		60		60	74.50	93
Replace supply fan bearing				2.000	57.50	120		177.50	213	258
Remove supply fan motor				1.156		63		63	78.50	98
Replace supply fan motor				1.951	218	107		325	370	440
Remove compressor				2.477		148		148	185	231
Replace compressor				3.844	306	229.50		535.50	625	740
Replace refrigerant				.533	128	31.80		159.80	180	210
Total				13.911	709.50	816.25		1,525.75	1,797	2,158.50
4040 Replace fan coil, DX 3 ton, no heat	15	Q-5	Ea.							
Remove fan coil unit				2.737		147		147	184	230
Replace fan coil, DX 3 ton, no heat				5.474	960	294		1,254	1,425	1,650
Total				8.211	960	441		1,401	1,609	1,880

For customer support on your Facilities Maintenance & Repair Cost Data, call 877.801.5724.

245

D3043 122 Fan Coil, DX Air Conditioner, Cooling Only

	System Description	Freq. (Years)	Crew	Unit	Labor Hours	Material	Labor	Equipment	Total	Total In-House	Total w/O&P
							2015 Bare Costs				
5050	**Repair fan coil, DX 5 ton, cooling only**	10	1 STPI	Ea.							
	Repair controls				.300		17.95		17.95	22.50	28
	Remove / replace supply fan				.650		39		39	48.50	60.50
	Remove supply fan bearing				1.000		60		60	74.50	93
	Replace supply fan bearing				2.000	57.50	120		177.50	213	258
	Remove supply fan motor				1.156		63		63	78.50	98
	Replace supply fan motor				1.951	218	107		325	370	440
	Remove compressor				2.477		148		148	185	231
	Replace compressor				4.954	505	296		801	925	1,100
	Replace refrigerant				.533	128	31.80		159.80	180	210
	Total				15.021	908.50	882.75		1,791.25	2,097	2,518.50
6060	**Replace fan coil, DX 5 ton, no heat**	15	Q-5	Ea.							
	Remove fan coil unit				3.466		186		186	233	291
	Replace fan coil, DX 5 ton, no heat				6.932	1,300	375		1,675	1,900	2,200
	Total				10.399	1,300	561		1,861	2,133	2,491
7070	**Repair fan coil, DX 10 ton, cooling only**	10	Q-6	Ea.							
	Repair controls				.300		17.95		17.95	22.50	28
	Remove / replace supply fan				.650		39		39	48.50	60.50
	Remove supply fan bearing				1.000		60		60	74.50	93
	Replace supply fan bearing				2.000	57.50	120		177.50	213	258
	Remove supply fan motor				1.156		63		63	78.50	98
	Replace supply fan motor				2.311	335	126		461	525	615
	Remove compressor				14.513		804		804	1,000	1,250
	Replace compressor				28.714	18,224	1,608		19,832	22,000	25,300
	Replace refrigerant				.800	192	47.70		239.70	270	315
	Total				51.444	18,808.50	2,885.65		21,694.15	24,232	28,017.50
8080	**Replace fan coil, DX 10 ton, no heat**	15	Q-6	Ea.							
	Remove fan coil unit				6.000		335		335	420	520
	Replace fan coil, DX 10 ton, no heat				12.000	2,800	670		3,470	3,925	4,550
	Total				18.000	2,800	1,005		3,805	4,345	5,070

D3043 122 | Fan Coil, DX Air Conditioner, Cooling Only

System Description	Freq. (Years)	Crew	Unit	Labor Hours	2015 Bare Costs				Total In-House	Total w/O&P
					Material	Labor	Equipment	Total		
9090 Repair fan coil, DX 20 ton, cooling only	10	Q-6	Ea.							
Repair controls				.300		17.95		17.95	22.50	28
Remove / replace supply fan				.650		39		39	48.50	60.50
Remove supply fan bearing				1.000		60		60	74.50	93
Replace supply fan bearing				2.000	57.50	120		177.50	213	258
Remove supply fan motor				1.156		63		63	78.50	98
Replace supply fan motor				2.476	810	135		945	1,050	1,225
Remove compressor				21.929		1,215		1,215	1,525	1,925
Replace compressor				43.902	25,470	2,452.50		27,922.50	31,100	35,700
Replace refrigerant				1.333	320	79.50		399.50	450	525
Total				74.747	26,657.50	4,181.95		30,839.45	34,562	39,912.50
9590 Replace fan coil, DX 20 ton, no heat	15	Q-6	Ea.							
Remove fan coil unit				22.284		1,250		1,250	1,550	1,950
Replace fan coil, DX 20 ton, no heat				44.610	5,400	2,500		7,900	9,050	10,600
Total				66.894	5,400	3,750		9,150	10,600	12,550
9600 Repair fan coil, DX 30 ton, cooling only	10	Q-6	Ea.							
Repair controls				.300		17.95		17.95	22.50	28
Remove / replace supply fan				.650		39		39	48.50	60.50
Remove supply fan bearing				1.000		60		60	74.50	93
Replace supply fan bearing				2.000	57.50	120		177.50	213	258
Remove supply fan motor				1.156		63		63	78.50	98
Replace supply fan motor				2.476	810	135		945	1,050	1,225
Remove compressor				24.490		1,375		1,375	1,700	2,125
Replace compressor				48.780	28,300	2,725		31,025	34,500	39,600
Replace refrigerant				1.733	416	103.35		519.35	585	680
Total				82.586	29,583.50	4,638.30		34,221.80	38,272	44,167.50
9610 Replace fan coil, DX 30 ton, no heat	15	Q-6	Ea.							
Remove fan coil unit				25.974		1,450		1,450	1,800	2,250
Replace fan coil, DX 20 ton, no heat				48.000	7,050	2,675		9,725	11,100	13,000
Total				73.974	7,050	4,125		11,175	12,900	15,250

D3043 124 Fan Coil, DX Air Conditioner W/ Heat

System Description	Freq. (Years)	Crew	Unit	Labor Hours	2015 Bare Costs				Total In-House	Total w/O&P
					Material	Labor	Equipment	Total		
1010 Replace fan coil, DX 1-1/2 ton, with heat	15	Q-5	Ea.							
Remove fan coil unit				2.312		124		124	155	194
Replace fan coil unit, 1-1/2 ton, with heat				6.474	952	348.60		1,300.60	1,475	1,725
Total				8.786	952	472.60		1,424.60	**1,630**	**1,919**
2010 Replace fan coil, DX 2 ton, with heat	15	Q-5	Ea.							
Remove fan coil unit				2.407		129		129	162	202
Replace fan coil unit, DX 2 ton, with heat				6.741	1,022	362.60		1,384.60	1,575	1,850
Total				9.148	1,022	491.60		1,513.60	**1,737**	**2,052**
3010 Replace fan coil, DX 2-1/2 ton, with heat	15	Q-5	Ea.							
Remove fan coil unit				2.626		141		141	177	220
Replace fan coil, DX 2-1/2 ton, with heat				7.354	1,085	396.20		1,481.20	1,675	1,975
Total				9.979	1,085	537.20		1,622.20	**1,852**	**2,195**
4010 Replace fan coil, DX 3 ton, with heat	15	Q-5	Ea.							
Remove fan coil unit				3.042		164		164	205	255
Replace fan coil, DX 3 ton, with heat				8.514	1,344	455		1,799	2,075	2,400
Total				11.556	1,344	619		1,963	**2,280**	**2,655**
5010 Replace fan coil, DX 5 ton, with heat	10	Q-5	Ea.							
Remove fan coil unit				3.852		207		207	259	325
Replace fan coil, DX 5 ton, with heat				10.785	1,820	581		2,401	2,725	3,175
Total				14.637	1,820	788		2,608	**2,984**	**3,500**
6010 Replace fan coil, DX 10 ton, with heat	10	Q-6	Ea.							
Remove fan coil unit				6.667		370		370	465	580
Replace fan coil, DX 10 ton, with heat				18.667	3,920	1,043		4,963	5,600	6,500
Total				25.333	3,920	1,413		5,333	**6,065**	**7,080**
7010 Replace fan coil, DX 20 ton, with heat	10	Q-6	Ea.							
Remove fan coil unit				24.793		1,375		1,375	1,725	2,150
Replace fan coil, DX 20 ton, with heat				69.421	7,560	3,885		11,445	13,200	15,500
Total				94.215	7,560	5,260		12,820	**14,925**	**17,650**

D30 HVAC | D3043 | Distribution Systems

D3043 128 | Unit Ventilator

ID	System Description	Freq. (Years)	Crew	Unit	Labor Hours	Material	Labor	Equipment	Total	Total In-House	Total w/O&P
							2015 Bare Costs				
1010	**Repair unit ventilator, 750 CFM, 2 ton**	10	1 STPI	Ea.							
	Repair controls				.200		11.95		11.95	14.95	18.65
	Remove / replace fan				.650		39		39	48.50	60.50
	Remove fan motor				1.156		63		63	78.50	98
	Replace fan motor				1.951	195	107		302	345	410
	Total				3.957	195	220.95		415.95	486.95	587.15
1030	**Replace unit vent, 750 CFM, heat/cool coils**	15	Q-6	Ea.							
	Remove unit ventilator				7.792		435		435	545	680
	Replace unit, 750 CFM, with heat / cool coils				15.605	4,050	870		4,920	5,550	6,425
	Total				23.397	4,050	1,305		5,355	6,095	7,105
2010	**Repair unit ventilator, 1250 CFM, 3 ton**	10	1 STPI	Ea.							
	Repair controls				.200		11.95		11.95	14.95	18.65
	Remove / replace fan				.650		39		39	48.50	60.50
	Remove fan motor				1.156		63		63	78.50	98
	Replace fan motor				1.951	248	107		355	405	475
	Total				3.957	248	220.95		468.95	546.95	652.15
2030	**Replace unit vent, 1250 CFM, heat/cool coils**	15	Q-6	Ea.							
	Remove unit ventilator				11.142		620		620	775	970
	Replace unit , 1250 CFM, with heat / cool coils				22.284	4,950	1,250		6,200	7,000	8,150
	Total				33.426	4,950	1,870		6,820	7,775	9,120
2040	**Repair unit ventilator, 2000 CFM, 5 ton**	10	1 STPI	Ea.							
	Repair controls				.300		17.95		17.95	22.50	28
	Remove / replace fan				.650		39		39	48.50	60.50
	Remove fan motor				1.156		63		63	78.50	98
	Replace fan motor				2.311	279	126		405	465	545
	Total				4.417	279	245.95		524.95	614.50	731.50
2050	**Replace unit vent., 2000 CFM, heat/cool coils**	15	Q-6	Ea.							
	Remove unit ventilator				31.209		1,750		1,750	2,175	2,725
	Replace unit, 2000 CFM, with heat / cool coils				62.338	6,225	3,475		9,700	11,200	13,200
	Total				93.547	6,225	5,225		11,450	13,375	15,925

D3043 140 Duct Heater

	System Description	Freq. (Years)	Crew	Unit	Labor Hours	2015 Bare Costs				Total In-House	Total w/O&P
						Material	Labor	Equipment	Total		
0020	**Maintenance and inspection**										
	Inspect and clean duct heater	0.50	1 ELEC	Ea.	1.143		62.50		62.50	77.50	97
	Total				1.143		62.50		62.50	**77.50**	**97**
0030	**Replace duct heater**										
	Remove duct heater	15	1 ELEC	Ea.	.667		36.50		36.50	45	56.50
	Duct heater 12 KW				2.000	1,775	109		1,884	2,100	2,400
	Total				2.667	1,775	145.50		1,920.50	2,145	2,456.50

D3043 210 Draft Fan

	System Description	Freq. (Years)	Crew	Unit	Labor Hours	2015 Bare Costs				Total In-House	Total w/O&P
						Material	Labor	Equipment	Total		
1010	**Repair fan, induced draft, 2000 CFM**										
	Remove bearings	10	1 STPI	Ea.	1.000		60		60	74.50	93
	Replace bearings				2.000	57.50	120		177.50	213	258
	Total				3.000	57.50	180		237.50	**287.50**	**351**
1030	**Replace fan, induced draft, 2000 CFM**										
	Remove fan	20	Q-9	Ea.	3.151		159		159	201	250
	Replace fan, induced draft, 2000 CFM				6.304	2,650	315		2,965	3,325	3,825
	Total				9.456	2,650	474		3,124	**3,526**	**4,075**
2010	**Repair fan, induced draft, 6700 CFM**										
	Remove bearings	10	1 STPI	Ea.	1.000		60		60	74.50	93
	Replace bearings				2.000	57.50	120		177.50	213	258
	Total				3.000	57.50	180		237.50	**287.50**	**351**
2030	**Replace fan, induced draft, 6700 CFM**										
	Remove fan	20	Q-9	Ea.	4.522		228		228	289	360
	Replace fan, induced draft, 6700 CFM				9.045	3,250	455		3,705	4,150	4,775
	Total				13.567	3,250	683		3,933	**4,439**	**5,135**
3010	**Repair fan, induced draft, 17,700 CFM**										
	Remove bearings	10	1 STPI	Ea.	1.000		60		60	74.50	93
	Replace bearings				2.222	64.50	133		197.50	237	288
	Total				3.222	64.50	193		257.50	**311.50**	**381**

For customer support on your Facilities Maintenance & Repair Cost Data, call 877.801.5724.

D3043 210 Draft Fan

	System Description	Freq. (Years)	Crew	Unit	Labor Hours	2015 Bare Costs				Total In-House	Total w/O&P
						Material	Labor	Equipment	Total		
3030	**Replace fan, induced draft, 17,700 CFM**	20	Q-9	Ea.							
	Remove fan				12.998		655		655	830	1,025
	Replace fan, induced draft, 17,700 CFM				26.016	8,375	1,300		9,675	10,900	12,500
	Total				39.014	8,375	1,955		10,330	11,730	13,525

D3043 220 Exhaust Fan

	System Description	Freq. (Years)	Crew	Unit	Labor Hours	2015 Bare Costs				Total In-House	Total w/O&P
						Material	Labor	Equipment	Total		
1010	**Replace fan & motor, propeller exh., 375 CFM**	15	Q-20	Ea.							
	Remove fan and motor				1.300		66.50		66.50	84	105
	Replace fan & motor, propeller exhaust, 375 CFM				2.600	192	133		325	380	450
	Total				3.900	192	199.50		391.50	464	555
1030	**Replace fan & motor, propeller exh., 1000 CFM**	15	Q-20	Ea.							
	Remove fan and motor				1.625		83		83	105	131
	Replace fan & motor, propeller exhaust, 1000 CFM				3.250	320	166		486	560	660
	Total				4.875	320	249		569	665	791
1040	**Replace fan & motor, propeller exh., 4700 CFM**	15	Q-20	Ea.							
	Remove fan and motor				2.600		133		133	168	209
	Replace fan & motor, propeller exhaust, 4700 CFM				5.200	1,225	266		1,491	1,675	1,950
	Total				7.800	1,225	399		1,624	1,843	2,159
2030	**Replace roof mounted exhaust fan, 800 CFM**	20	Q-20	Ea.							
	Remove fan & motor				2.600		133		133	168	209
	Replace fan & motor, roof mounted exhaust, 800 CFM				5.200	900	266		1,166	1,325	1,550
	Total				7.800	900	399		1,299	1,493	1,759
2040	**Replace roof mounted exhaust fan, 2000 CFM**	20	Q-20	Ea.							
	Remove fan & motor				3.250		166		166	210	262
	Replace fan & motor, roof mounted exhaust, 2000 CFM				6.500	1,725	335		2,060	2,325	2,675
	Total				9.750	1,725	501		2,226	2,535	2,937

For customer support on your Facilities Maintenance & Repair Cost Data, call 877.801.5724.

251

System Description	Freq. (Years)	Crew	Unit	Labor Hours	Material	Labor	Equipment	Total	Total In-House	Total w/O&P
2050 **Replace roof mounted exhaust fan, 8500 CFM**										
Remove fan & motor	20	Q-20	Ea.	4.333		222		222	280	350
Replace fan & motor, roof mounted exhaust, 8500 CFM				8.666	2,800	445		3,245	3,650	4,200
Total				12.998	2,800	667		3,467	3,930	4,550
2060 **Replace roof mounted exhaust fan, 20,300 CFM**										
Remove fan & motor	20	Q-20	Ea.	13.004		665		665	840	1,050
Replace fan & motor, roof mounted exhaust, 20,300 CFM				26.008	7,875	1,325		9,200	10,300	11,900
Total				39.012	7,875	1,990		9,865	11,140	12,950
3010 **Replace utility set, belt drive, 800 CFM**										
Remove fan and motor	10	Q-20	Ea.	2.167		111		111	140	174
Replace utility set, belt drive, 800 CFM				4.334	980	222		1,202	1,350	1,575
Total				6.500	980	333		1,313	1,490	1,749
3020 **Replace utility set, belt drive, 3600 CFM**										
Remove fan and motor	10	Q-20	Ea.	3.250		166		166	210	262
Replace utility set, belt drive, 3600 CFM				6.500	2,025	335		2,360	2,650	3,050
Total				9.750	2,025	501		2,526	2,860	3,312
3030 **Replace utility set, belt drive, 11,000 CFM**										
Remove fan and motor	10	Q-20	Ea.	6.500		335		335	420	525
Replace utility set, belt drive, 11,000 CFM				13.004	5,325	665		5,990	6,700	7,700
Total				19.504	5,325	1,000		6,325	7,120	8,225
3040 **Replace utility set, belt drive, 20,000 CFM**										
Remove fan and motor	10	Q-20	Ea.	16.247		830		830	1,050	1,300
Replace utility set, belt drive, 20,000 CFM				32.520	7,150	1,675		8,825	9,975	11,600
Total				48.767	7,150	2,505		9,655	11,025	12,900
4010 **Replace axial flow fan, 3900 CFM**										
Remove fan and motor	10	Q-20	Ea.	3.823		196		196	247	310
Replace axial flow fan, 3900 CFM				7.648	1,250	390		1,640	1,875	2,175
Total				11.472	1,250	586		1,836	2,122	2,485

2015 Bare Costs (spanning Material, Labor, Equipment, Total columns)

For customer support on your Facilities Maintenance & Repair Cost Data, call 877.801.5724.

D3043 220 Exhaust Fan

	System Description	Freq. (Years)	Crew	Unit	Labor Hours	2015 Bare Costs				Total In-House	Total w/O&P
						Material	Labor	Equipment	Total		
4020	**Replace axial flow fan, 6400 CFM**	10	Q-20	Ea.							
	Remove fan and motor				4.643		238		238	300	375
	Replace axial flow fan, 6400 CFM				9.285	1,800	475		2,275	2,575	3,000
	Total				13.928	1,800	713		2,513	**2,875**	**3,375**
4030	**Replace axial flow fan, 16,900 CFM**	10	Q-20	Ea.							
	Remove fan and motor				6.502		335		335	420	525
	Replace axial flow fan, 16,900 CFM				13.004	2,475	665		3,140	3,575	4,150
	Total				19.506	2,475	1,000		3,475	**3,995**	**4,675**
4040	**Replace axial flow fan, 29,000 CFM**	10	Q-20	Ea.							
	Remove fan and motor				9.001		460		460	580	725
	Replace axial flow fan, 29,000 CFM				17.986	3,925	920		4,845	5,475	6,350
	Total				26.987	3,925	1,380		5,305	**6,055**	**7,075**

D3043 250 Fireplaces, Clay Flue

	System Description	Freq. (Years)	Crew	Unit	Labor Hours	2015 Bare Costs				Total In-House	Total w/O&P
						Material	Labor	Equipment	Total		
0010	**Replace baked clay flue, architectural**	75	2 BRIC	L.F.							
	Set up and secure scaffold				.022		1.02		1.02	1.34	1.66
	Masonry demolition				.083		3.13		3.13	4.16	5.15
	Remove damaged flue				.002		.10		.10	.13	.17
	Install new flue lining				.224	10.70	9.40		20.10	24	28.50
	Install new chimney				1.518	45.50	64		109.50	134	161
	Remove scaffold				.022		1.02		1.02	1.34	1.66
	Total				1.872	56.20	78.67		134.87	**164.97**	**198.14**

D3043 252 Fireplaces, Metal Flue

	System Description	Freq. (Years)	Crew	Unit	Labor Hours	2015 Bare Costs				Total In-House	Total w/O&P
						Material	Labor	Equipment	Total		
0020	**Replace metal pipe flue, architectural**	50	2 SSWK	L.F.							
	Remove damaged pipe				.008		.44		.44	.66	.80
	Install new vent chimney, prefabricated metal				.306	7.20	15.40		22.60	27.50	33.50
	Total				.314	7.20	15.84		23.04	**28.16**	**34.30**

For customer support on your Facilities Maintenance & Repair Cost Data, call 877.801.5724.

253

D3043 310 Steam Converter, Commercial

System Description	Freq. (Years)	Crew	Unit	Labor Hours	2015 Bare Costs Material	2015 Bare Costs Labor	2015 Bare Costs Equipment	2015 Bare Costs Total	Total In-House	Total w/O&P
0010 Repair steam converter	5	1 STPI	Ea.							
Repair heat exchanger				5.944		355		355	445	555
Total				5.944		355		355	**445**	**555**
0020 Inspect for leaks	2	1 STPI	Ea.							
Check relief valve operation				.038		2.26		2.26	2.82	3.52
Check for proper water temperature				.055		3.26		3.26	4.08	5.10
Total				.092		5.52		5.52	**6.90**	**8.62**
0030 Replace steam converter	30	Q-5	Ea.							
Remove exchanger				2.081		112		112	140	175
Install new exchanger 10 GPM of 40F to 180F				4.160	3,250	224		3,474	3,850	4,425
Total				6.241	3,250	336		3,586	**3,990**	**4,600**

D3043 320 Flash Tank, 24 Gallon

System Description	Freq. (Years)	Crew	Unit	Labor Hours	2015 Bare Costs Material	2015 Bare Costs Labor	2015 Bare Costs Equipment	2015 Bare Costs Total	Total In-House	Total w/O&P
0010 Repair flash tank	5	1 STPI	Ea.							
Repair flash tank				5.944		355		355	445	555
Total				5.944		355		355	**445**	**555**
0030 Replace flash tank 24 gal	15	Q-5	Ea.							
Remove old tank				.743		40		40	50	62.50
Install new tank 24 gallon				1.486	715	80		795	885	1,025
Total				2.229	715	120		835	**935**	**1,087.50**

D3043 330 Steam Regulator Valve

System Description	Freq. (Years)	Crew	Unit	Labor Hours	2015 Bare Costs Material	2015 Bare Costs Labor	2015 Bare Costs Equipment	2015 Bare Costs Total	Total In-House	Total w/O&P
0010 Replace steam regulator valve 1-1/2" diameter	6	1 STPI	Ea.							
Remove steam valve				.400		24		24	30	37.50
Install steam regulator valve 1-1/2" diameter				.800	3,025	48		3,073	3,375	3,850
Total				1.200	3,025	72		3,097	**3,405**	**3,887.50**

D3043 330 | Steam Regulator Valve

System Description	Freq. (Years)	Crew	Unit	Labor Hours	Material	Labor	Equipment	Total	Total In-House	Total w/O&P
0110 Replace steam regulator valve 2″ diameter	6	1 STPI	Ea.							
Remove steam valve				.473		28.50		28.50	35.50	44
Install steam regulator valve 2″ diameter				.945	3,725	56.50		3,781.50	4,175	4,750
Total				1.418	3,725	85		3,810	4,210.50	4,794
0210 Replace steam regulator valve 2-1/2″ diameter	6	Q-5	Ea.							
Remove steam valve				.867		46.50		46.50	58.50	72.50
Install steam regulator valve 2-1/2″ diameter				1.733	4,650	93		4,743	5,225	5,950
Total				2.600	4,650	139.50		4,789.50	5,283.50	6,022.50
0310 Replace steam regulator valve 3″ diameter	6	Q-5	Ea.							
Remove steam valve				.946		51		51	63.50	79.50
Install steam regulator valve 3″ diameter flanged				1.891	5,800	102		5,902	6,500	7,400
Total				2.836	5,800	153		5,953	6,563.50	7,479.50

D3043 340 | Condensate Meter

System Description	Freq. (Years)	Crew	Unit	Labor Hours	Material	Labor	Equipment	Total	Total In-House	Total w/O&P
0010 Repair meter	15	1 STPI	Ea.							
Replace drum				1.538	805	90.50		895.50	1,000	1,150
Replace bearings				2.597	132	152		284	335	405
Total				4.136	937	242.50		1,179.50	1,335	1,555
0030 Replace condensate meter 500 lb./hr.	30	1 STPI	Ea.							
Remove meter				.372		22		22	28	34.50
Replace meter 500#/hr				.743	3,275	44.50		3,319.50	3,650	4,175
Total				1.114	3,275	66.50		3,341.50	3,678	4,209.50
0130 Replace condensate meter 1500 lb./hr.	30	1 STPI	Ea.							
Remove meter				.743		44.50		44.50	55.50	69.50
Replace meter 1500#/hr				1.486	4,100	89		4,189	4,625	5,275
Total				2.229	4,100	133.50		4,233.50	4,680.50	5,344.50

(2015 Bare Costs columns: Material, Labor, Equipment, Total)

For customer support on your Facilities Maintenance & Repair Cost Data, call 877.801.5724.

255

D3043 350   Steam Traps

	System Description	Freq. (Years)	Crew	Unit	Labor Hours	2015 Bare Costs				Total In-House	Total w/O&P
						Material	Labor	Equipment	Total		
1030	**Replace steam trap, 15 PSIG, 3/4" threaded**	7	1 STPI	Ea.							
	Remove old CI body float & thermostatic trap				.325		19.40		19.40	24.50	30.50
	Install new CI body float & thermostatic trap				.650	141	39		180	204	237
	Total				.975	141	58.40		199.40	228.50	267.50
1040	**Replace steam trap, 15 PSIG, 1" threaded**	7	1 STPI	Ea.							
	Remove old CI body float & thermostatic trap				.347		20.50		20.50	26	32.50
	Install new CI body float & thermostatic trap				.693	169	41.50		210.50	238	276
	Total				1.040	169	62		231	264	308.50
1050	**Replace steam trap, 15 PSIG, 1-1/4" threaded**	7	1 STPI	Ea.							
	Remove old CI body float & thermostatic trap				.400		24		24	30	37.50
	Install new CI body float & thermostatic trap				.800	187	48		235	265	310
	Total				1.200	187	72		259	295	347.50
1060	**Replace steam trap, 15 PSIG, 1-1/2" threaded**	7	1 STPI	Ea.							
	Remove old CI body float & thermostatic trap				.578		34.50		34.50	43	54
	Install new CI body float & thermostatic trap				1.156	298	69		367	415	480
	Total				1.733	298	103.50		401.50	458	534
1070	**Replace steam trap, 15 PSIG, 2" threaded**	7	1 STPI	Ea.							
	Remove old CI body float & thermostatic trap				.867		52		52	64.50	81
	Install new CI body float & thermostatic trap				1.733	590	104		694	780	900
	Total				2.600	590	156		746	844.50	981

D3043 410   Radiator Valve

	System Description	Freq. (Years)	Crew	Unit	Labor Hours	2015 Bare Costs				Total In-House	Total w/O&P
						Material	Labor	Equipment	Total		
0010	**Replace radiator valve 1/2" angle union**	50	1 STPI	Ea.							
	Remove radiator valve				.217		12.95		12.95	16.20	20
	Install radiator valve 1/2" angle union				.433	49	26		75	86.50	102
	Total				.650	49	38.95		87.95	102.70	122

D30 HVAC — D3043 Distribution Systems

D3043 410 — Radiator Valve

System Description	Freq. (Years)	Crew	Unit	Labor Hours	2015 Bare Costs Material	2015 Bare Costs Labor	2015 Bare Costs Equipment	2015 Bare Costs Total	Total In-House	Total w/O&P
0020 Replace radiator valve 3/4" angle union	50	1 STPI	Ea.							
Remove radiator valve				.260		15.55		15.55	19.45	24
Install radiator valve 3/4" angle union				.520	55.50	31		86.50	100	118
Total				.780	55.50	46.55		102.05	119.45	**142**
0030 Replace radiator valve 1" angle union	50	1 STPI	Ea.							
Remove radiator valve				.274		16.35		16.35	20.50	25.50
Install radiator valve 1" angle union				.547	64.50	32.50		97	112	132
Total				.821	64.50	48.85		113.35	**132.50**	**157.50**
0040 Replace radiator valve 1-1/4" angle union	50	1 STPI	Ea.							
Remove radiator valve				.347		20.50		20.50	26	32.50
Install radiator valve 1-1/4" angle union				.693	82.50	41.50		124	143	168
Total				1.040	82.50	62		144.50	**169**	**200.50**

D3043 420 — Cast Iron Radiator, 10' Section

System Description	Freq. (Years)	Crew	Unit	Labor Hours	2015 Bare Costs Material	2015 Bare Costs Labor	2015 Bare Costs Equipment	2015 Bare Costs Total	Total In-House	Total w/O&P
0010 Replace C.I. radiator 4 tube 25"H 10' section	50	Q-5	Section							
Remove radiator				1.084		58.50		58.50	73	91
Install C.I. radiator 4 tube 25" high 10' section				2.167	340	116.50		456.50	520	605
Total				3.250	340	175		515	**593**	**696**

D3043 430 — Baseboard Radiation, 10' Section

System Description	Freq. (Years)	Crew	Unit	Labor Hours	2015 Bare Costs Material	2015 Bare Costs Labor	2015 Bare Costs Equipment	2015 Bare Costs Total	Total In-House	Total w/O&P
0010 Replace radiator, baseboard 10'	20	Q-5	Ea.							
Remove radiator				2.262		121.50		121.50	152	190
Install 10' baseboard radiator				4.522	380	245		625	720	855
Total				6.783	380	366.50		746.50	**872**	**1,045**

For customer support on your Facilities Maintenance & Repair Cost Data, call 877.801.5724.

257

D3043 440 Finned Radiator, Wall, 10' Section

System Description	Freq. (Years)	Crew	Unit	Labor Hours	2015 Bare Costs				Total In-House	Total w/O&P
					Material	Labor	Equipment	Total		
0010 Replace finned radiator										
Replace radiator	20	Q-5	Ea.	3.468		186.50		186.50	235	290
Install 10' finned tube, 2 tier, 1-1/4" cop tube				6.933	505	375		880	1,025	1,200
Total				10.401	505	561.50		1,066.50	1,260	**1,490**

D3043 450 Duct Coil, 1-Row, Hot Water

System Description	Freq. (Years)	Crew	Unit	Labor Hours	2015 Bare Costs				Total In-House	Total w/O&P
					Material	Labor	Equipment	Total		
1020 Replace coil, hot water boost, 12" x 24"										
Remove coil	25	Q-5	Ea.	.858		46		46	57.50	72
Replace coil, hot water duct, 12" x 24"				1.111	390	60		450	505	580
Total				1.969	390	106		496	**562.50**	**652**
1030 Replace coil, hot water boost, 24" x 24"										
Remove coil	25	Q-5	Ea.	1.717		92.50		92.50	115	144
Replace coil, hot water duct, 24" x 24"				3.433	480	185		665	760	890
Total				5.150	480	277.50		757.50	**875**	**1,034**
1040 Replace coil, hot water boost, 24" x 36"										
Remove coil	25	Q-5	Ea.	2.581		139		139	174	216
Replace coil, hot water duct, 24" x 36"				5.161	540	278		818	940	1,100
Total				7.742	540	417		957	**1,114**	**1,316**
1050 Replace coil, hot water boost, 36" x 36"										
Remove coil	25	Q-5	Ea.	3.867		208		208	260	325
Replace coil, hot water duct, 36" x 36"				7.729	660	415		1,075	1,250	1,475
Total				11.596	660	623		1,283	**1,510**	**1,800**

258

For customer support on your Facilities Maintenance & Repair Cost Data, call 877.801.5724.

D3043 510 Pipe & Fittings, Steel/Iron, Flanged

	System Description	Freq. (Years)	Crew	Unit	Labor Hours	2015 Bare Costs				Total In-House	Total w/O&P
						Material	Labor	Equipment	Total		
0010	**Install new gasket, 4" pipe size**	25	1 PLUM	Ea.							
	Disconnect joint / remove flange gasket				1.301		76.50		76.50	95.50	119
	Install new gasket				1.300		76.50		91	111	137
	Total				2.601	14.50	153		167.50	**206.50**	**256**
0030	**Replace 2" flanged steel pipe and fittings**	75	2 PLUM	L.F.							
	Remove old pipe				.231		12.20		12.20	15.25	19.05
	Install flanged 2" steel pipe with hangers				.356	13.90	18.80	1.28	33.98	40	48.50
	Total				.587	13.90	31	1.28	46.18	**55.25**	**67.55**
0040	**Replace 4" flanged steel pipe and fittings**	75	2 PLUM	L.F.							
	Remove old pipe				.400		21		21	26.50	33
	Install flanged 4" steel pipe with hangers				.800	31	42.50	2.89	76.39	90	108
	Total				1.200	31	63.50	2.89	97.39	**116.50**	**141**
0050	**Replace 6" flanged steel pipe and fittings**	75	2 PLUM	L.F.							
	Remove old pipe				.624		33		33	41	51.50
	Install flanged 6" steel pipe with hangers				1.248	59	68.50	3	130.50	153	184
	Total				1.872	59	101.50	3	163.50	**194**	**235.50**
0060	**Replace 8" flanged steel pipe and fittings**	75	2 PLUM	L.F.							
	Remove old pipe				.821		45		45	56	70
	Install flanged 8" steel pipe with hangers				1.642	96.50	90	3.94	190.44	223	265
	Total				2.463	96.50	135	3.94	235.44	**279**	**335**

D3043 520 Valves

	System Description	Freq. (Years)	Crew	Unit	Labor Hours	2015 Bare Costs				Total In-House	Total w/O&P
						Material	Labor	Equipment	Total		
1010	**Repack gate valve gland, 3/8" - 1-1/2"**	10	1 STPI	Ea.							
	Remove / replace gate valve packing				.219	10.90	11.75		22.65	26.50	32
	Total				.219	10.90	11.75		22.65	**26.50**	**32**

D3043 520   Valves

	System Description	Freq. (Years)	Crew	Unit	Labor Hours	2015 Bare Costs				Total In-House	Total w/O&P
						Material	Labor	Equipment	Total		
1020	**Replace gate valve, partial, 3/8" - 1-1/2"**	20	1 STPI	Ea.							
	Turn valve off and on				.017		1.02		1.02	1.27	1.58
	Remove old				.530		31.50		31.50	39.50	49.50
	Install new, 3/8" - 1-1/2" size				.800	198	48		246	278	320
	Total				**1.347**	**198**	**80.52**		**278.52**	**318.77**	**371.08**
2010	**Repack gate valve gland, 2" - 3"**	10	1 STPI	Ea.							
	Remove / replace gate valve packing				.291	14.50	15.63		30.13	35.50	42.50
	Total				**.291**	**14.50**	**15.63**		**30.13**	**35.50**	**42.50**
2020	**Replace gate valve, partial, 2" - 3"**	20	Q-1	Ea.							
	Turn valve off and on				.017		1.02		1.02	1.27	1.58
	Remove valve parts				.440		26.50		26.50	33	41
	Replace valve parts, 2" - 3"				1.600	855	84.50		939.50	1,050	1,200
	Total				**2.057**	**855**	**112.02**		**967.02**	**1,084.27**	**1,242.58**
2060	**Repack gate valve gland, 8" - 12"**	5	1 PLUM	Ea.							
	Replace valve packing, 8" to 12" diameter				1.600	43.50	94		137.50	165	200
	Total				**1.600**	**43.50**	**94**		**137.50**	**165**	**200**
3010	**Repack drain valve gland, 3/4"**	3	1 STPI	Ea.							
	Remove / replace drain valve packing				.219	10.90	11.75		22.65	26.50	32
	Total				**.219**	**10.90**	**11.75**		**22.65**	**26.50**	**32**
3020	**Replace drain valve stem assembly, 3/4"**	8	1 STPI	Ea.							
	Turn valve off and on				.017		1.02		1.02	1.27	1.58
	Remove valve stem assembly				.267		15.95		15.95	19.90	25
	Replace valve stem assembly				.267	9.85	15.95		25.80	31	37.50
	Total				**.550**	**9.85**	**32.92**		**42.77**	**52.17**	**64.08**
3030	**Replace drain valve, 3/4"**	20	1 STPI	Ea.							
	Turn valve off and on				.017		1.02		1.02	1.27	1.58
	Remove old valve				.308		18.40		18.40	23	28.50
	Install new valve, 3/4"				.308	9.85	18.40		28.25	34	41
	Total				**.632**	**9.85**	**37.82**		**47.67**	**58.27**	**71.08**

For customer support on your Facilities Maintenance & Repair Cost Data, call 877.801.5724.

System Description	Freq. (Years)	Crew	Unit	Labor Hours	2015 Bare Costs Material	Labor	Equipment	Total	Total In-House	Total w/O&P
1010 **Repair circulator pump, 1/12 - 3/4 H.P.**	5	1 STPI	Ea.							
Loosen coupling from pump shaft				.009		.54		.54	.67	.84
Remove 4 stud bolts in motor				.035		2.09		2.09	2.61	3.26
Remove motor				.052		3.11		3.11	3.88	4.85
Remove impeller nut from shaft				.033		1.97		1.97	2.47	3.08
Remove seal assembly				.019		1.14		1.14	1.42	1.77
Inspect bearing brackets face				.005		.30		.30	.37	.47
Clean bearing				.016		.96		.96	1.20	1.49
Inspect shaft				.040	40.50	2.15		42.65	47	54
Clean shaft				.017		1.02		1.02	1.27	1.58
Install new seal and rubber				.013		.78		.78	.97	1.21
Install seal spring and washers				.006		.36		.36	.45	.56
Install impeller nut on shaft				.040	10.15	2.15		12.30	13.85	16.05
Clean pump body gasket area				.037		2.21		2.21	2.76	3.45
Install new gasket (ready made)				.010		.60		.60	.75	.93
Install impeller				.009		.54		.54	.67	.84
Install 4 stud bolts				.035		2.09		2.09	2.61	3.26
Assemble motor to bearing				.024		1.43		1.43	1.79	2.24
Install 4 stud bolts				.027		1.61		1.61	2.02	2.52
Tighten coupling				.009		.54		.54	.67	.84
Final tightening of bolts				.027		1.61		1.61	2.02	2.52
Test run				.042		2.51		2.51	3.14	3.91
Total				.505	50.65	29.71		80.36	92.59	109.67
1030 **Replace circulator pump, 1/12 - 3/4 H.P.**	15	Q-1	Ea.							
Remove pump and motor				2.597		137		137	172	214
Install new pump and motor, 1/12-3/4 H.P.				5.195	2,550	274		2,824	3,150	3,625
Total				7.792	2,550	411		2,961	3,322	3,839

D3043 530 Circulator Pump

System Description	Freq. (Years)	Crew	Unit	Labor Hours	2015 Bare Costs				Total In-House	Total w/O&P
					Material	Labor	Equipment	Total		
2010										
Repair circulator pump, 1 H.P.	5	1 STPI	Ea.							
Loosen coupling from pump shaft				.009		.54		.54	.67	.84
Remove 4 stud bolts in motor				.035		2.09		2.09	2.61	3.26
Remove motor				.052		3.11		3.11	3.88	4.85
Remove impeller nut from shaft				.033		1.97		1.97	2.47	3.08
Remove seal assembly				.019		1.14		1.14	1.42	1.77
Inspect bearing brackets face				.005		.30		.30	.37	.47
Clean bearing				.016		.96		.96	1.20	1.49
Inspect shaft				.040	40.50	2.15		42.65	47	54
Clean shaft				.017		1.02		1.02	1.27	1.58
Install new seal and rubber				.013		.78		.78	.97	1.21
Install seal spring and washers				.006		.36		.36	.45	.56
Install impeller nut on shaft				.040	10.15	2.15		12.30	13.85	16.05
Clean pump body gasket area				.037		2.21		2.21	2.76	3.45
Install new gasket (ready made)				.013		.78		.78	.97	1.21
Install impeller				.009		.54		.54	.67	.84
Install 4 stud bolts				.035		2.09		2.09	2.61	3.26
Assemble motor to bearing				.024		1.43		1.43	1.79	2.24
Install 4 stud bolts				.027		1.61		1.61	2.02	2.52
Tighten coupling				.009		.54		.54	.67	.84
Final tightening of bolts				.027		1.61		1.61	2.02	2.52
Test run				.042		2.51		2.51	3.14	3.91
Total				.508	50.65	29.89		80.54	92.81	109.95
2030										
Replace circulator. pump, 1 H.P.	15	Q-1	Ea.							
Remove pump				2.597		137		137	172	214
Install new pump, 1 H.P.				5.200	4,100	275		4,375	4,850	5,550
Total				7.797	4,100	412		4,512	5,022	5,764

D3043 540 Expansion Tank

System Description	Freq. (Years)	Crew	Unit	Labor Hours	2015 Bare Costs				Total In-House	Total w/O&P
					Material	Labor	Equipment	Total		
0010										
Refill expansion tank	5	1 STPI	Ea.							
Refill air chamber				.200		11.95		11.95	14.95	18.65
Total				.200		11.95		11.95	14.95	18.65

For customer support on your Facilities Maintenance & Repair Cost Data, call 877.801.5724.

D3043 540 Expansion Tank

System Description	Freq. (Years)	Crew	Unit	Labor Hours	2015 Bare Costs				Total In-House	Total w/O&P
					Material	Labor	Equipment	Total		
0020 Replace expansion tank, 24 gal capacity	50	Q-5	Ea.							
Remove expansion tank				.743		40		40	50	62.50
Install expansion tank, 24 gallon capacity				1.486	1,150	80		1,230	1,375	1,575
Total				2.228	1,150	120		1,270	1,425	1,637.50
0120 Replace expansion tank, 60 gal capacity	50	Q-5	Ea.							
Remove expansion tank				1.300		70		70	87.50	109
Install expansion tank, 60 gallon capacity				2.602	1,825	140		1,965	2,175	2,500
Total				3.901	1,825	210		2,035	2,262.50	2,609
0220 Replace expansion tank, 175 gal capacity	50	Q-5	Ea.							
Remove expansion tank				2.597		140		140	175	218
Install expansion tank, 175 gallon capacity				5.195	4,950	279		5,229	5,800	6,625
Total				7.792	4,950	419		5,369	5,975	6,843
0320 Replace expansion tank, 400 gal capacity	50	Q-5	Ea.							
Remove expansion tank				3.721		200		200	250	310
Install expansion tank, 400 gallon capacity				7.442	10,600	400		11,000	12,200	13,900
Total				11.163	10,600	600		11,200	12,450	14,210

D3043 550 Pipe Insulation

System Description	Freq. (Years)	Crew	Unit	Labor Hours	2015 Bare Costs				Total In-House	Total w/O&P
					Material	Labor	Equipment	Total		
1010 Repair damaged pipe insulation, fiberglass 1/2"	5	Q-14	Ea.							
Remove 2' length old insulation				.086		4.08		4.08	5.30	6.55
Install insulation, fiberglass, 1" wall - 1/2" diam.				.173	1.66	8.16		9.82	12.40	15.20
Total				.260	1.66	12.24		13.90	17.70	21.75
1110 Repair damaged pipe insulation, fiberglass 3/4"	5	Q-14	Ea.							
Remove 2' length old insulation				.090		4.26		4.26	5.50	6.85
Install insulation, fiberglass, 1" wall - 3/4" diam.				.181	1.80	8.52		10.32	13	15.95
Total				.271	1.80	12.78		14.58	18.50	22.80

For customer support on your Facilities Maintenance & Repair Cost Data, call 877.801.5724.

263

D3043 550   Pipe Insulation

System Description	Freq. (Years)	Crew	Unit	Labor Hours	2015 Bare Costs				Total In-House	Total w/O&P
					Material	Labor	Equipment	Total		
1120 Repair damaged pipe insulation, fiberglass 1"	5	Q-14	Ea.							
Remove 2' length old insulation				.095		4.46		4.46	5.80	7.20
Install insulation, fiberglass, 1" wall - 1" diam.				.189	1.94	8.92		10.86	13.70	16.75
Total				.284	1.94	13.38		15.32	19.50	23.95
1130 Repair damaged pipe insulation, fbgs 1-1/4"	5	Q-14	Ea.							
Remove 2' length old insulation				.099		4.66		4.66	6.05	7.50
Install insulation, fiberglass, 1" wall - 1-1/4" diam.				.198	2.10	9.32		11.42	14.40	17.65
Total				.296	2.10	13.98		16.08	20.45	25.15
1140 Repair damaged pipe insulation, fbgs 1-1/2"	5	Q-14	Ea.							
Remove 2' length old insulation				.099		4.66		4.66	6.05	7.50
Install insulation, fiberglass, 1" wall - 1-1/2" diam.				.198	2.26	9.34		11.60	14.60	17.85
Total				.297	2.26	14		16.26	20.65	25.35
1150 Repair damaged pipe insulation, fiberglass 2"	5	Q-14	Ea.							
Remove 2' length old insulation				.104		4.90		4.90	6.35	7.90
Install insulation, fiberglass, 1" wall - 2" diam.				.208	2.44	9.80		12.24	15.40	18.75
Total				.312	2.44	14.70		17.14	21.75	26.65
1160 Repair damaged pipe insulation, fiberglass 3"	5	Q-14	Ea.							
Remove 2' length old insulation				.116		5.46		5.46	7.10	8.80
Install insulation, fiberglass, 1" wall - 3" diam.				.232	2.98	10.90		13.88	17.50	21.50
Total				.348	2.98	16.36		19.34	24.60	30.30
1170 Repair damaged pipe insulation, fiberglass 4"	5	Q-14	Ea.							
Remove 2' length old insulation				.139		6.56		6.56	8.50	10.50
Install insulation, fiberglass, 1" wall - 4" diam.				.278	3.94	13.10		17.04	21.50	26
Total				.417	3.94	19.66		23.60	30	36.50
1180 Repair damaged pipe insulation, fiberglass 6"	5	Q-14	Ea.							
Remove 2' length old insulation				.173		8.16		8.16	10.60	13.10
Install insulation, fiberglass, 1" wall - 6" diam.				.347	4.72	16.30		21.02	26.50	32
Total				.520	4.72	24.46		29.18	37.10	45.10

For customer support on your Facilities Maintenance & Repair Cost Data, call 877.801.5724.

	System Description	Freq. (Years)	Crew	Unit	Labor Hours	2015 Bare Costs				Total In-House	Total w/O&P
						Material	Labor	Equipment	Total		
1220	**Replace pipe insulation, fiberglass 1/2"**	5	Q-14	M.L.F.							
	Remove 1000' length old insulation				33.330		1,570		1,570	2,050	2,525
	Install insulation, fiberglass, 1" wall - 1/2" diam.				66.670	830	3,140		3,970	5,000	6,100
	Total				100.000	830	4,710		5,540	**7,050**	**8,625**
1230	**Replace pipe insulation, fiberglass 3/4"**	5	Q-14	M.L.F.							
	Remove 1000' length old insulation				34.780		1,640		1,640	2,125	2,625
	Install insulation, fiberglass, 1" wall - 3/4" diam.				69.570	900	3,280		4,180	5,250	6,375
	Total				104.350	900	4,920		5,820	**7,375**	**9,000**
1240	**Replace pipe insulation, fiberglass 1"**	5	Q-14	M.L.F.							
	Remove 1000' length old insulation				36.360		1,710		1,710	2,225	2,750
	Install insulation, fiberglass, 1" wall - 1" diam.				72.730	970	3,430		4,400	5,500	6,725
	Total				109.090	970	5,140		6,110	**7,725**	**9,475**
1250	**Replace pipe insulation, fiberglass 1-1/4"**	5	Q-14	M.L.F.							
	Remove 1000' length old insulation				38.100		1,800		1,800	2,325	2,900
	Install insulation, fiberglass, 1" wall - 1-1/4" diam.				76.190	1,050	3,590		4,640	5,800	7,075
	Total				114.290	1,050	5,390		6,440	**8,125**	**9,975**
1260	**Replace pipe insulation, fiberglass 1-1/2"**	5	Q-14	M.L.F.							
	Remove 1000' length old insulation				38.100		1,800		1,800	2,325	2,900
	Install insulation, fiberglass, 1" wall - 1-1/2" diam.				76.190	1,130	3,590		4,720	5,900	7,175
	Total				114.290	1,130	5,390		6,520	**8,225**	**10,075**
1270	**Replace pipe insulation, fiberglass 2"**	5	Q-14	M.L.F.							
	Remove 1000' length old insulation				40.000		1,890		1,890	2,450	3,025
	Install insulation, fiberglass, 1" wall - 2" diam.				80.000	1,220	3,770		4,990	6,250	7,575
	Total				120.000	1,220	5,660		6,880	**8,700**	**10,600**
1280	**Replace pipe insulation, fiberglass 3"**	5	Q-14	M.L.F.							
	Remove 1000' length old insulation				44.440		2,090		2,090	2,725	3,375
	Install insulation, fiberglass, 1" wall - 3" diam.				88.890	1,490	4,190		5,680	7,050	8,625
	Total				133.330	1,490	6,280		7,770	**9,775**	**12,000**

For customer support on your Facilities Maintenance & Repair Cost Data, call 877.801.5724.

265

D3043 550 Pipe Insulation

System Description	Freq. (Years)	Crew	Unit	Labor Hours	2015 Bare Costs				Total In-House	Total w/O&P
					Material	Labor	Equipment	Total		
1290 Replace pipe insulation, fiberglass 4"	5	Q-14	M.L.F.							
Remove 1000' length old insulation				53.330		2,510		2,510	3,250	4,050
Install insulation, fiberglass, 1" wall - 4" diam.				106.670	1,970	5,050		7,020	8,700	10,600
Total				160.000	1,970	7,560		9,530	**11,950**	**14,650**
1300 Replace pipe insulation, fiberglass 6"	5	Q-14	M.L.F.							
Remove 1000' length old insulation				66.670		3,140		3,140	4,075	5,050
Install insulation, fiberglass, 1" wall - 6" diam.				133.330	2,360	6,300		8,660	10,800	13,100
Total				200.000	2,360	9,440		11,800	**14,875**	**18,150**
1410 Repair damaged pipe insulation rubber 1/2"	5	1 ASBE	Ea.							
Remove 2' length old insulation				.117		6.12		6.12	7.90	9.80
Install insulation, foam rubber, 3/4" wall - 1/2" diam.				.234	2.20	12.20		14.40	18.30	22.50
Total				.350	2.20	18.32		20.52	**26.20**	**32.30**
1420 Repair damaged pipe insulation rubber 3/4"	5	1 ASBE	Ea.							
Remove 2' length old insulation				.117		6.12		6.12	7.90	9.80
Install insulation, foam rubber, 3/4" wall - 3/4" diam.				.234	2.70	12.20		14.90	18.80	23
Total				.350	2.70	18.32		21.02	**26.70**	**32.80**
1430 Repair damaged pipe insulation rubber 1"	5	1 ASBE	Ea.							
Remove 2' length old insulation				.118		6.18		6.18	8	9.95
Install insulation, foam rubber, 3/4" wall - 1" diam.				.236	3.08	12.40		15.48	19.40	24
Total				.355	3.08	18.58		21.66	**27.40**	**33.95**
1440 Repair damaged pipe insulation rubber 1-1/4"	5	1 ASBE	Ea.							
Remove 2' length old insulation				.120		6.26		6.26	8.10	10.10
Install insulation, foam rubber, 3/4" wall - 1-1/4" diam.				.239	4.10	12.50		16.60	21	25
Total				.359	4.10	18.76		22.86	**29.10**	**35.10**
1450 Repair damaged pipe insulation rubber 1-1/2"	5	1 ASBE	Ea.							
Remove 2' length old insulation				.120		6.26		6.26	8.10	10.10
Install insulation, foam rubber, 3/4" wall - 1-1/2" diam.				.239	4.64	12.50		17.14	21.50	26
Total				.359	4.64	18.76		23.40	**29.60**	**36.10**

D3043 550 | **Pipe Insulation**

	System Description	Freq. (Years)	Crew	Unit	Labor Hours	2015 Bare Costs				Total In-House	Total w/O&P
						Material	Labor	Equipment	Total		
1460	**Repair damaged pipe insulation rubber 2"**	5	1 ASBE	Ea.							
	Remove 2' length old insulation				.121		6.34		6.34	8.25	10.20
	Install insulation, foam rubber, 3/4" wall - 2" diam.				.242	5.36	12.70		18.06	22.50	27
	Total				.364	5.36	19.04		24.40	30.75	37.20
1470	**Repair damaged pipe insulation rubber 3"**	5	1 ASBE	Ea.							
	Remove 2' length old insulation				.122		6.40		6.40	8.30	10.30
	Install insulation, foam rubber, 3/4" wall - 3" diam.				.245	8.16	12.80		20.96	25.50	31
	Total				.367	8.16	19.20		27.36	33.80	41.30
1480	**Repair damaged pipe insulation rubber 4"**	5	1 ASBE	Ea.							
	Remove 2' length old insulation				.130		6.80		6.80	8.85	10.90
	Install insulation, foam rubber, 3/4" wall - 4" diam.				.260	10.20	13.60		23.80	29	34.50
	Total				.390	10.20	20.40		30.60	37.85	45.40
1490	**Repair damaged pipe insulation rubber 6"**	5	1 ASBE	Ea.							
	Remove 2' length old insulation				.130		6.80		6.80	8.85	10.90
	Install insulation, foam rubber, 3/4" wall - 6" diam.				.260	14.90	13.60		28.50	34	40.50
	Total				.390	14.90	20.40		35.30	42.85	51.40
1510	**Replace pipe insulation foam rubber 1/2"**	5	1 ASBE	L.F.							
	Remove 1000' length old insulation				.090		4.70		4.70	6.10	7.55
	Install insulation, foam rubber, 3/4" wall - 1/2" diam.				.180	2.20	9.42		11.62	14.60	17.85
	Total				.270	2.20	14.12		16.32	20.70	25.40
1520	**Replace pipe insulation foam rubber 3/4"**	5	1 ASBE	L.F.							
	Remove 1000' length old insulation				.090		4.70		4.70	6.10	7.55
	Install insulation, foam rubber, 3/4" wall - 3/4" diam.				.180	2.70	9.42		12.12	15.20	18.50
	Total				.270	2.70	14.12		16.82	21.30	26.05
1530	**Replace pipe insulation foam rubber 1"**	5	1 ASBE	L.F.							
	Remove 1000' length old insulation				.091		4.76		4.76	6.20	7.65
	Install insulation, foam rubber, 3/4" wall - 1" diam.				.182	3.08	9.52		12.60	15.70	19.15
	Total				.273	3.08	14.28		17.36	21.90	26.80

For customer support on your Facilities Maintenance & Repair Cost Data, call 877.801.5724.

267

D3043 550 Pipe Insulation

System Description	Freq. (Years)	Crew	Unit	Labor Hours	2015 Bare Costs				Total In-House	Total w/O&P
					Material	Labor	Equipment	Total		
1540 Replace pipe insulation foam rubber 1-1/4"	5	1 ASBE	L.F.							
Remove 1000' length old insulation				.092		4.82		4.82	6.25	7.75
Install insulation, foam rubber, 3/4" wall - 1-1/4" diam.				.184	4.10	9.62		13.72	17	20.50
Total				.276	4.10	14.44		18.54	23.25	28.25
1550 Replace pipe insulation foam rubber 1-1/2"	5	1 ASBE	L.F.							
Remove 1000' length old insulation				.092		4.82		4.82	6.25	7.75
Install insulation, foam rubber, 3/4" wall - 1-1/2" diam.				.184	4.64	9.62		14.26	17.60	21.50
Total				.276	4.64	14.44		19.08	23.85	29.25
1560 Replace pipe insulation foam rubber 2"	5	1 ASBE	L.F.							
Remove 1000' length old insulation				.093		4.86		4.86	6.30	7.80
Install insulation, foam rubber, 3/4" wall - 2" diam.				.186	5.36	9.74		15.10	18.50	22.50
Total				.279	5.36	14.60		19.96	24.80	30.30
1570 Replace pipe insulation foam rubber 3"	5	1 ASBE	L.F.							
Remove 1000' length old insulation				.094		4.92		4.92	6.40	7.90
Install insulation, foam rubber, 3/4" wall - 3" diam.				.188	8.16	9.86		18.02	22	26
Total				.282	8.16	14.78		22.94	28.40	33.90
1580 Replace pipe insulation foam rubber 4"	5	1 ASBE	L.F.							
Remove 1000' length old insulation				.100		5.24		5.24	6.80	8.40
Install insulation, foam rubber, 3/4" wall - 4" diam.				.200	10.20	10.50		20.70	25	29.50
Total				.300	10.20	15.74		25.94	31.80	37.90
1590 Replace pipe insulation foam rubber 6"	5	1 ASBE	L.F.							
Remove 1000' length old insulation				.100		5.24		5.24	6.80	8.40
Install insulation, foam rubber, 3/4" wall - 6" diam.				.200	14.90	10.50		25.40	30	35.50
Total				.300	14.90	15.74		30.64	36.80	43.90

	System Description	Freq. (Years)	Crew	Unit	Labor Hours	2015 Bare Costs				Total In-House	Total w/O&P
						Material	Labor	Equipment	Total		
1010	**Repair unit heater, 12 MBH, 2 PSI steam**	10	1 STPI	Ea.							
	Repair controls				.200		11.95		11.95	14.95	18.65
	Remove fan motor				.976		58.50		58.50	73	91
	Replace fan motor				1.951	195	107		302	345	410
	Total				3.127	195	177.45		372.45	432.95	519.65
1030	**Replace unit heater, 12 MBH, 2 PSI steam**	15	Q-5	Ea.							
	Remove unit heater				.867		46.50		46.50	58.50	72.50
	Replace unit heater, 12 MBH, 2 PSI steam				1.733	360	93		453	515	595
	Total				2.600	360	139.50		499.50	573.50	667.50
2010	**Repair unit heater, 36 MBH, 2 PSI steam**	10	1 STPI	Ea.							
	Repair controls				.200		11.95		11.95	14.95	18.65
	Remove fan motor				.976		58.50		58.50	73	91
	Replace fan motor				1.951	218	107		325	370	440
	Total				3.127	218	177.45		395.45	457.95	549.65
2030	**Replace unit heater, 36 MBH, 2 PSI steam**	15	Q-5	Ea.							
	Remove unit heater				1.300		70		70	87.50	109
	Replace unit heater, 36 MBH, 2 PSI steam				2.600	545	140		685	775	900
	Total				3.900	545	210		755	862.50	1,009
3010	**Repair unit heater, 85 MBH, 2 PSI steam**	10	1 STPI	Ea.							
	Repair controls				.200		11.95		11.95	14.95	18.65
	Remove fan motor				.976		58.50		58.50	73	91
	Replace fan motor				1.951	248	107		355	405	475
	Total				3.127	248	177.45		425.45	492.95	584.65
3030	**Replace unit heater, 85 MBH, 2 PSI steam**	15	Q-5	Ea.							
	Remove unit heater				1.600		86		86	108	134
	Replace unit heater, 85 MBH, 2 PSI steam				3.200	650	172		822	930	1,075
	Total				4.800	650	258		908	1,038	1,209
4010	**Repair unit heater, 250 MBH, 2 PSI steam**	10	1 STPI	Ea.							
	Repair controls				.200		11.95		11.95	14.95	18.65
	Remove fan motor				1.156		63		63	78.50	98
	Replace fan motor				2.311	279	126		405	465	545
	Total				3.667	279	200.95		479.95	558.45	661.65

For customer support on your Facilities Maintenance & Repair Cost Data, call 877.801.5724.

269

D3053 110 Unit Heater

System Description	Freq. (Years)	Crew	Unit	Labor Hours	2015 Bare Costs Material	Labor	Equipment	Total	Total In-House	Total w/O&P
4030 Replace unit heater, 250 MBH, 2 PSI steam										
Remove unit heater	15	Q-5	Ea.	4.160		224		224	280	350
Replace unit heater, 250 MBH, 2 PSI steam				8.320	1,475	445		1,920	2,175	2,550
Total				12.481	1,475	669		2,144	**2,455**	**2,900**
5010 Repair unit heater, 400 MBH, 2 PSI steam										
Repair controls	10	1 STPI	Ea.	.200		11.95		11.95	14.95	18.65
Remove fan motor				1.156		63		63	78.50	98
Replace fan motor				2.311	335	126		461	525	615
Total				3.667	335	200.95		535.95	**618.45**	**731.65**
5020 Replace unit heater, 400 MBH, 2 PSI steam										
Remove unit heater	15	Q-5	Ea.	6.499		350		350	435	545
Replace unit heater, 400 MBH, 2 PSI steam				13.008	2,150	700		2,850	3,250	3,800
Total				19.507	2,150	1,050		3,200	**3,685**	**4,345**

D3053 112 Infrared Heater Suspended, Commercial

System Description	Freq. (Years)	Crew	Unit	Labor Hours	2015 Bare Costs Material	Labor	Equipment	Total	Total In-House	Total w/O&P
0010 Maintenance and repair										
Repair wiring connections	1	1 ELEC	Ea.	.615		33.50		33.50	41.50	52
Total				.615		33.50		33.50	**41.50**	**52**
0020 Maintenance and inspection										
Inspect and clean infrared heater	0.50	1 ELEC	Ea.	1.143		62.50		62.50	77.50	97
Total				1.143		62.50		62.50	**77.50**	**97**
0030 Replace infrared heater										
Remove infrared heater	15	1 ELEC	Ea.	.667		36.50		36.50	45	56.50
Infrared heater 240 V, 1500 W				2.105	284	115		399	455	535
Total				2.772	284	151.50		435.50	**500**	**591.50**

For customer support on your Facilities Maintenance & Repair Cost Data, call 877.801.5724.

D3053 114 Standard Suspended Heater

					2015 Bare Costs				Total In-House	Total w/O&P
	System Description	Freq. (Years)	Crew	Unit	Labor Hours	Material	Labor	Equipment	Total		
0010	**Maintenance and repair**	2	1 ELEC	Ea.							
	Remove unit heater component				.151		8.25		8.25	10.20	12.80
	Unit heater component				.444	68.50	24.50		93	105	123
	Total				**.595**	**68.50**	**32.75**		**101.25**	**115.20**	**135.80**
0020	**Maintenance and inspection**	0.50	1 ELEC	Ea.							
	Inspect and clean unit heater				1.143		62.50		62.50	77.50	97
	Total				**1.143**		**62.50**		**62.50**	**77.50**	**97**
0030	**Replace heater**	15	1 ELEC	Ea.							
	Remove unit heater				.667		36.50		36.50	45	56.50
	Unit heater, heavy duty, 480 V, 3 KW				1.333	425	73		498	560	645
	Total				**2.000**	**425**	**109.50**		**534.50**	**605**	**701.50**

D3053 116 Explosionproof Industrial Heater

					2015 Bare Costs				Total In-House	Total w/O&P
	System Description	Freq. (Years)	Crew	Unit	Labor Hours	Material	Labor	Equipment	Total		
0010	**Maintenance and repair**	2	1 ELEC	Ea.							
	Remove explosionproof heater part				.200		10.95		10.95	13.55	16.95
	Explosionproof heater part				.615	89	33.50		122.50	140	163
	Total				**.815**	**89**	**44.45**		**133.45**	**153.55**	**179.95**
0020	**Maintenance and inspection**	0.50	1 ELEC	Ea.							
	Inspect / clean explosionproof heater				1.143		62.50		62.50	77.50	97
	Total				**1.143**		**62.50**		**62.50**	**77.50**	**97**
0030	**Replace heater**	15	1 ELEC	Ea.							
	Remove explosionproof heater				.800		44		44	54	68
	Explosionproof heater, 3 KW				3.810	5,325	208		5,533	6,125	6,975
	Total				**4.610**	**5,325**	**252**		**5,577**	**6,179**	**7,043**

For customer support on your Facilities Maintenance & Repair Cost Data, call 877.801.5724.

271

D3053 150 Wall Mounted/Recessed Heater, With Fan

System Description	Freq. (Years)	Crew	Unit	Labor Hours	2015 Bare Costs				Total In-House	Total w/O&P
					Material	Labor	Equipment	Total		
0010 **Maintenance and repair**										
Remove heater fan	5	1 ELEC	Ea.	.250		13.70		13.70	16.95	21
Heater fan				.727	105	40		145	165	193
Total				.977	105	53.70		158.70	181.95	**214**
0020 **Maintenance and inspection**										
Inspect and clean heater with fan	1	1 ELEC	Ea.	1.143		62.50		62.50	77.50	97
Total				1.143		62.50		62.50	**77.50**	**97**
0030 **Replace heater**										
Remove heater with fan	20	1 ELEC	Ea.	.667		36.50		36.50	45	56.50
Heater with fan, commercial 2000 W				2.667	192	146		338	390	465
Total				3.333	192	182.50		374.50	435	**521.50**

D3053 160 Convector Suspended, Commercial

System Description	Freq. (Years)	Crew	Unit	Labor Hours	2015 Bare Costs				Total In-House	Total w/O&P
					Material	Labor	Equipment	Total		
0010 **Maintenance and repair**										
Repair wiring connections	2	1 ELEC	Ea.	.615		33.50		33.50	41.50	52
Total				.615		33.50		33.50	**41.50**	**52**
0020 **Maintenance and inspection**										
Inspect and clean convector heater	0.50	1 ELEC	Ea.	1.143		62.50		62.50	77.50	97
Total				1.143		62.50		62.50	**77.50**	**97**
0030 **Replace heater**										
Remove cabinet convector heater	15	1 ELEC	Ea.	.667		36.50		36.50	45	56.50
Cabinet convector heater, 2' long				2.000	1,900	109		2,009	2,225	2,550
Total				2.667	1,900	145.50		2,045.50	2,270	**2,606.50**

System Description	Freq. (Years)	Crew	Unit	Labor Hours	2015 Bare Costs				Total In-House	Total w/O&P
					Material	Labor	Equipment	Total		
1010 Repair terminal reheat, 12″ x 24″ coil										
Fix leak	10	1 STPI	Ea.	1.200		71.50		71.50	89.50	112
Total				1.200		71.50		71.50	**89.50**	**112**
1040 Replace terminal reheat, 12″ x 24″ coil										
Remove terminal reheat unit	15	Q-5	Ea.	.858		46		46	57.50	72
Replace terminal reheat unit, 12″ x 24″ coil				1.716	835	92.50		927.50	1,025	1,200
Total				2.574	835	138.50		973.50	**1,082.50**	**1,272**
2010 Repair terminal reheat, 18″ x 24″ coil										
Fix leak	10	1 STPI	Ea.	1.400		83.50		83.50	105	130
Total				1.400		83.50		83.50	**105**	**130**
2040 Replace terminal reheat, 18″ x 24″ coil										
Remove terminal reheat unit	15	Q-5	Ea.	1.284		69		69	86.50	108
Replace terminal reheat unit, 18″ x 24″ coil				2.568	1,000	138		1,138	1,275	1,475
Total				3.852	1,000	207		1,207	**1,361.50**	**1,583**
3010 Repair terminal reheat, 36″ x 36″ coil										
Fix leak	10	1 STPI	Ea.	2.200		131		131	164	205
Total				2.200		131		131	**164**	**205**
3040 Replace terminal reheat, 36″ x 36″ coil										
Remove terminal reheat unit	15	Q-5	Ea.	3.852		207		207	259	325
Replace terminal reheat unit, 36″ x 36″ coil				7.703	1,675	415		2,090	2,350	2,750
Total				11.555	1,675	622		2,297	**2,609**	**3,075**
4010 Repair terminal reheat, 48″ x 126″ coil										
Fix leak	10	1 STPI	Ea.	4.000		239		239	299	375
Total				4.000		239		239	**299**	**375**
4040 Replace terminal reheat, 48″ x 126″ coil										
Remove terminal reheat unit	15	Q-5	Ea.	17.937		965		965	1,200	1,500
Replace terminal reheat unit, 48″ x 126″ coil				35.874	4,950	1,925		6,875	7,850	9,200
Total				53.812	4,950	2,890		7,840	**9,050**	**10,700**

System Description	Freq. (Years)	Crew	Unit	Labor Hours	2015 Bare Costs				Total In-House	Total w/O&P
					Material	Labor	Equipment	Total		
1010 Repair heat pump, 1.5 ton, air to air split	10	1 STPI	Ea.							
Repair controls				.600		36		36	45	56
Remove / replace supply fan				.650		39		39	48.50	60.50
Remove supply fan motor				.700		42		42	52.50	65
Replace supply fan motor				1.951	190	117		307	355	420
Remove compressor				1.860		111		111	139	173
Replace compressor				3.714	270	222		492	575	685
Remove / replace condenser fan				1.800		108		108	134	168
Remove condenser fan motor				1.800		108		108	134	168
Replace condenser fan motor				1.951	195	107		302	345	410
Replace refrigerant				.133	32	7.95		39.95	45	52.50
Remove / replace heater				.500		30		30	37.50	46.50
Total				15.661	687	927.95		1,614.95	**1,910.50**	**2,304.50**
1030 Replace heat pump, 1.5 ton, air to air split	20	Q-5	Ea.							
Remove heat pump				4.336		233		233	292	365
Replace heat pump, 1.5 ton, air to air split				8.649	2,175	465		2,640	2,975	3,450
Total				12.985	2,175	698		2,873	**3,267**	**3,815**
2010 Repair heat pump, 5 ton, air to air split	10	1 STPI	Ea.							
Repair controls				.600		36		36	45	56
Remove / replace supply fan				.650		39		39	48.50	60.50
Remove supply fan motor				1.156		63		63	78.50	98
Replace supply fan motor				1.951	195	107		302	345	410
Remove compressor				2.477		148		148	185	231
Replace compressor				4.954	505	296		801	925	1,100
Remove / replace condenser fan				1.800		108		108	134	168
Remove condenser fan motor				.976		58.50		58.50	73	91
Replace condenser fan motor				1.951	218	107		325	370	440
Replace refrigerant				.667	160	39.75		199.75	225	262
Remove / replace heater				.500		30		30	37.50	46.50
Total				17.681	1,078	1,032.25		2,110.25	**2,466.50**	**2,963**
2030 Replace heat pump, 5 ton, air to air split	20	Q-5	Ea.							
Remove heat pump				20.806		1,125		1,125	1,400	1,750
Replace heat pump, 5 ton, air to air split				41.558	3,600	2,225		5,825	6,750	7,975
Total				62.365	3,600	3,350		6,950	**8,150**	**9,725**

For customer support on your Facilities Maintenance & Repair Cost Data, call 877.801.5724.

	System Description	Freq. (Years)	Crew	Unit	Labor Hours	2015 Bare Costs				Total In-House	Total w/O&P
						Material	Labor	Equipment	Total		
3010	**Repair heat pump, 10 ton, air to air split**	10	Q-5	Ea.							
	Repair controls				.600		36		36	45	56
	Remove / replace supply fan				.650		39		39	48.50	60.50
	Remove supply fan motor				.700		42		42	52.50	65
	Replace supply fan motor				2.311	335	126		461	525	615
	Remove compressor				4.000		215		215	269	335
	Replace compressor				8.000	1,350	430		1,780	2,025	2,350
	Remove / replace condenser fan				1.800		108		108	134	168
	Remove condenser fan motor				1.800		108		108	134	168
	Replace condenser fan motor				2.311	279	126		405	465	545
	Replace refrigerant				1.067	256	63.60		319.60	360	420
	Remove / replace heater				.500		30		30	37.50	46.50
	Total				23.739	2,220	1,323.60		3,543.60	4,095.50	4,829
3030	**Replace heat pump, 10 ton, air to air split**	20	Q-6	Ea.							
	Remove heat pump				24.490		1,375		1,375	1,700	2,125
	Replace heat pump, 10 ton, air to air split				48.980	8,500	2,725		11,225	12,800	14,900
	Total				73.469	8,500	4,100		12,600	14,500	17,025
4010	**Repair heat pump, 25 ton, air to air split**	10	Q-5	Ea.							
	Repair controls				.600		36		36	45	56
	Remove / replace supply fan				1.000		60		60	74.50	93
	Remove supply fan motor				1.156		63		63	78.50	98
	Replace supply fan motor				2.311	610	126		736	825	960
	Remove compressor				10.000		540		540	670	840
	Replace compressor				20.000	3,375	1,075		4,450	5,075	5,900
	Remove / replace condenser fan				1.800		108		108	134	168
	Remove condenser fan motor				1.800		108		108	134	168
	Replace condenser fan motor				2.312	560	126		686	775	895
	Replace refrigerant				2.000	480	119.25		599.25	675	785
	Remove / replace heater				.500		30		30	37.50	46.50
	Total				43.479	5,025	2,391.25		7,416.25	8,523.50	10,009.50
4030	**Replace heat pump, 25 ton, air to air split**	20	Q-7	Ea.							
	Remove heat pump				61.538		3,500		3,500	4,375	5,450
	Replace heat pump, 25 ton, air to air split				123.000	20,400	7,000		27,400	31,200	36,400
	Total				184.538	20,400	10,500		30,900	35,575	41,850

For customer support on your Facilities Maintenance & Repair Cost Data, call 877.801.5724.

275

D3053 245 | **Heat Pump**

	System Description	Freq. (Years)	Crew	Unit	Labor Hours	2015 Bare Costs Material	2015 Bare Costs Labor	2015 Bare Costs Equipment	2015 Bare Costs Total	Total In-House	Total w/O&P
5010	**Repair heat pump, 50 ton, air to air split**	10	Q-6	Ea.							
	Repair controls				.600		36		36	45	56
	Remove / replace supply fan				1.000		60		60	74.50	93
	Remove supply fan motor				1.156		63		63	78.50	98
	Replace supply fan motor				4.167	2,050	228		2,278	2,525	2,925
	Remove compressor				77.922		4,350		4,350	5,425	6,775
	Replace compressor				155.000	32,100	8,700		40,800	46,100	53,500
	Remove / replace condenser fan				2.400		143		143	179	224
	Remove condenser fan motor				1.800		108		108	134	168
	Replace condenser fan motor				4.000	1,750	219		1,969	2,200	2,525
	Replace refrigerant				4.667	1,120	278.25		1,398.25	1,575	1,825
	Remove / replace heater				1.200		71.50		71.50	89.50	112
	Total				253.912	37,020	14,256.75		51,276.75	58,425.50	68,301
5030	**Replace heat pump, 50 ton, air to air split**	20	Q-7	Ea.							
	Remove heat pump				96.970		5,525		5,525	6,900	8,600
	Replace heat pump, 50 ton, air to air split				207.000	51,000	11,800		62,800	71,000	82,000
	Total				303.970	51,000	17,325		68,325	77,900	90,600
6010	**Repair heat pump, thru-wall unit, 1.5 ton**	10	1 STPI	Ea.							
	Repair controls				.600		36		36	45	56
	Remove / replace supply fan				.650		39		39	48.50	60.50
	Remove supply fan motor				.700		42		42	52.50	65
	Replace supply fan motor				1.951	190	117		307	355	420
	Remove compressor				1.860		111		111	139	173
	Replace compressor				3.714	270	222		492	575	685
	Remove / replace condenser fan				1.800		108		108	134	168
	Remove condenser fan motor				1.800		108		108	134	168
	Replace condenser fan motor				1.951	195	107		302	345	410
	Replace refrigerant				.133	32	7.95		39.95	45	52.50
	Remove / replace heater				.500		30		30	37.50	46.50
	Total				15.661	687	927.95		1,614.95	1,910.50	2,304.50
6030	**Replace heat pump, thru-wall unit, 1.5 ton**	20	Q-5	Ea.							
	Remove heat pump				6.709		360		360	450	565
	Replace heat pump, thru-wall, 1.5 ton				13.423	3,050	720		3,770	4,250	4,950
	Total				20.131	3,050	1,080		4,130	4,700	5,515

For customer support on your Facilities Maintenance & Repair Cost Data, call 877.801.5724.

D3053 245 Heat Pump

	System Description	Freq. (Years)	Crew	Unit	Labor Hours	2015 Bare Costs				Total In-House	Total w/O&P
						Material	Labor	Equipment	Total		
7010	Repair heat pump, thru-wall unit, 5 ton	10	1 STPI	Ea.							
	Repair controls				.600		36		36	45	56
	Remove / replace supply fan				.650		39		39	48.50	60.50
	Remove supply fan motor				1.156		63		63	78.50	98
	Replace supply fan motor				1.951	195	107		302	345	410
	Remove compressor				2.477		148		148	185	231
	Replace compressor				4.954	505	296		801	925	1,100
	Remove / replace condenser fan				1.800		108		108	134	168
	Remove condenser fan motor				.976		58.50		58.50	73	91
	Replace condenser fan motor				1.951	218	107		325	370	440
	Replace refrigerant				.667	160	39.75		199.75	225	262
	Remove / replace heater				.500		30		30	37.50	46.50
	Total				17.681	1,078	1,032.25		2,110.25	2,466.50	2,963
7030	Replace heat pump, thru-wall unit, 5 ton	20	Q-5	Ea.							
	Remove heat pump				16.000		860		860	1,075	1,350
	Replace heat pump, trhu-wall, 5 ton				32.000	4,800	1,725		6,525	7,425	8,675
	Total				48.000	4,800	2,585		7,385	8,500	10,025

D3053 265 Air Conditioner, Window, 1 Ton

	System Description	Freq. (Years)	Crew	Unit	Labor Hours	2015 Bare Costs				Total In-House	Total w/O&P
						Material	Labor	Equipment	Total		
1010	Repair air conditioner, window, 1 ton	8	1 STPI	Ea.							
	Find / fix leak				1.000		60		60	74.50	93
	Replace refrigerant				.067	16	3.98		19.98	22.50	26
	Total				1.067	16	63.98		79.98	97	119
1030	Replace air conditioner, window, 1 ton	10	L-2	Ea.							
	Remove air conditioner				1.300		53.50		53.50	71.50	88
	Replace air conditioner, 1 ton				2.000	725	82.50		807.50	905	1,050
	Total				3.300	725	136		861	976.50	1,138

For customer support on your Facilities Maintenance & Repair Cost Data, call 877.801.5724.

277

D3053 266 Air Conditioner, Thru-The-Wall

	System Description	Freq. (Years)	Crew	Unit	Labor Hours	2015 Bare Costs				Total In-House	Total w/O&P
						Material	Labor	Equipment	Total		
2010	**Repair air conditioner, thru-the-wall, 2 ton**	8	1 STPI	Ea.							
	Find / fix leak				1.000		60		60	74.50	93
	Replace refrigerant				.133	32	7.95		39.95	45	52.50
	Total				**1.133**	**32**	**67.95**		**99.95**	**119.50**	**145.50**
2030	**Replace air conditioner, thru-the-wall, 2 ton**	10	L-2	Ea.							
	Remove air conditioner				2.600		107		107	143	176
	Replace air conditioner, 2 ton				5.200	1,175	214		1,389	1,575	1,825
	Total				**7.800**	**1,175**	**321**		**1,496**	**1,718**	**2,001**

D3053 272 Air Conditioner, DX Package

	System Description	Freq. (Years)	Crew	Unit	Labor Hours	2015 Bare Costs				Total In-House	Total w/O&P
						Material	Labor	Equipment	Total		
1010	**Repair air conditioner, DX, 5 ton**	10	1 STPI	Ea.							
	Repair controls				.600		36		36	45	56
	Remove / replace supply fan				.650		39		39	48.50	60.50
	Remove supply fan bearing				1.000		60		60	74.50	93
	Replace supply fan bearing				2.000	57.50	120		177.50	213	258
	Remove supply fan motor				1.156		63		63	78.50	98
	Replace supply fan motor				2.311	279	126		405	465	545
	Remove compressor				2.477		148		148	185	231
	Replace compressor				4.954	505	296		801	925	1,100
	Remove / replace condenser fan				1.800		108		108	134	168
	Remove condenser fan bearing				1.200		71.50		71.50	89.50	112
	Replace condenser fan bearing				2.000	57.50	120		177.50	213	258
	Remove condenser fan motor				.976		58.50		58.50	73	91
	Replace condenser fan motor				1.951	248	107		355	405	475
	Replace refrigerant				.667	160	39.75		199.75	225	262
	Remove / replace heating coils				3.900	667.50	217.50		885	1,000	1,175
	Total				**27.641**	**1,974.50**	**1,610.25**		**3,584.75**	**4,174**	**4,982.50**
1030	**Replace air conditioner, DX, 5 ton**	20	Q-6	Ea.							
	Remove air conditioner				13.001		725		725	905	1,125
	Replace air conditioner, 5 ton				26.002	4,450	1,450		5,900	6,700	7,825
	Total				**39.003**	**4,450**	**2,175**		**6,625**	**7,605**	**8,950**

278

For customer support on your Facilities Maintenance & Repair Cost Data, call 877.801.5724.

D3053 272 | Air Conditioner, DX Package

		System Description	Freq. (Years)	Crew	Unit	Labor Hours	2015 Bare Costs				Total In-House	Total w/O&P
							Material	Labor	Equipment	Total		
2010		**Repair air conditioner, DX, 20 ton**	10	Q-6	Ea.							
		Repair controls				.800		48		48	60	74.50
		Remove / replace supply fan				.650		39		39	48.50	60.50
		Remove supply fan bearing				1.000		60		60	74.50	93
		Replace supply fan bearing				2.000	57.50	120		177.50	213	258
		Remove supply fan motor				1.156		63		63	78.50	98
		Replace supply fan motor				2.312	560	126		686	775	895
		Remove compressor				24.390		1,350		1,350	1,700	2,125
		Replace compressor				48.780	28,300	2,725		31,025	34,500	39,600
		Remove / replace condenser fan				1.800		108		108	134	168
		Remove condenser fan bearing				1.200		71.50		71.50	89.50	112
		Replace condenser fan bearing				2.000	57.50	120		177.50	213	258
		Remove condenser fan motor				.976		58.50		58.50	73	91
		Replace condenser fan motor				2.311	279	126		405	465	545
		Replace refrigerant				1.333	320	79.50		399.50	450	525
		Remove / replace heating coils				6.936	1,920	393.75		2,313.75	2,600	3,025
		Total				97.645	31,494	5,488.25		36,982.25	41,474	47,928
2030		**Replace air conditioner, DX, 20 ton**	20	Q-7	Ea.							
		Remove air conditioner				23.105		1,325		1,325	1,650	2,050
		Replace air conditioner, 20 ton				46.243	12,800	2,625		15,425	17,400	20,100
		Total				69.347	12,800	3,950		16,750	19,050	22,150

For customer support on your Facilities Maintenance & Repair Cost Data, call 877.801.5724.

279

D3053 272 Air Conditioner, DX Package

System Description	Freq. (Years)	Crew	Unit	Labor Hours	2015 Bare Costs				Total In-House	Total w/O&P
					Material	Labor	Equipment	Total		
3010 Repair air conditioner, DX, 50 ton	10	Q-6	Ea.							
Repair controls				.800		48		48	60	74.50
Remove / replace supply fan				.650		39		39	48.50	60.50
Remove supply fan bearing				1.200		71.50		71.50	89.50	112
Replace supply fan bearing				2.222	64.50	133		197.50	237	288
Remove supply fan motor				1.156		63		63	78.50	98
Replace supply fan motor				2.597	990	142		1,132	1,275	1,450
Remove compressor				77.922		4,350		4,350	5,425	6,775
Replace compressor				155.000	32,100	8,700		40,800	46,100	53,500
Remove / replace condenser fan				1.800		108		108	134	168
Remove condenser fan bearing				1.200		71.50		71.50	89.50	112
Replace condenser fan bearing				2.222	64.50	133		197.50	237	288
Remove condenser fan motor				.976		58.50		58.50	73	91
Replace condenser fan motor				2.311	279	126		405	465	545
Replace refrigerant				3.333	800	198.75		998.75	1,125	1,300
Remove / replace heating coil				12.468	7,080	708.75		7,788.75	8,700	9,950
Total				265.857	41,378	14,951		56,329	64,137	74,812
3030 Replace air conditioner, DX, 50 ton	20	Q-7	Ea.							
Remove air conditioner				41.558		2,375		2,375	2,950	3,700
Replace air conditioner, 50 ton				83.117	47,200	4,725		51,925	58,000	66,500
Total				124.675	47,200	7,100		54,300	60,950	70,200

	System Description	Freq. (Years)	Crew	Unit	Labor Hours	2015 Bare Costs Material	Labor	Equipment	Total	Total In-House	Total w/O&P
1010	**Repair computer room A/C, air cooled, 5 ton**	10	1 STPI	Ea.							
	Repair controls				.600		36		36	45	56
	Remove / replace supply fan				.650		39		39	48.50	60.50
	Remove supply fan bearing				1.000		60		60	74.50	93
	Replace supply fan bearing				2.000	57.50	120		177.50	213	258
	Remove supply fan motor				1.156		63		63	78.50	98
	Replace supply fan motor				2.311	279	126		405	465	545
	Replace fan belt for supply fan				.400	35	24		59	68.50	81.50
	Remove compressor				2.477		148		148	185	231
	Replace compressor				4.954	505	296		801	925	1,100
	Remove / replace condenser fan				1.800		108		108	134	168
	Remove condenser fan bearing				1.200		71.50		71.50	89.50	112
	Replace condenser fan bearing				2.000	57.50	120		177.50	213	258
	Remove condenser fan motor				.976		58.50		58.50	73	91
	Replace condenser fan motor				1.951	248	107		355	405	475
	Replace fan belt for condenser fan				.400	35	24		59	68.50	81.50
	Replace refrigerant				.667	160	39.75		199.75	225	262
	Remove / replace heating coils				3.900	667.50	217.50		885	1,000	1,175
	Total				28.441	2,044.50	1,658.25		3,702.75	4,311	**5,145.50**
1015	**Replace computer room A/C, air cooled, 5 ton**	20	Q-6	Ea.							
	Remove air conditioner and condenser				13.001		725		725	905	1,125
	Replace air conditioner and remote condenser, 5 ton				35.556	20,100	1,900		22,000	24,500	28,100
	Total				48.557	20,100	2,625		22,725	25,405	**29,225**

D3053 274 Computer Room A/C Units, Air Cooled

System Description	Freq. (Years)	Crew	Unit	Labor Hours	2015 Bare Costs Material	Labor	Equipment	Total	Total In-House	Total w/O&P
1020										
Repair computer room A/C, air cooled, 10 ton	10	1 STPI	Ea.							
Repair controls				.400		24		24	30	37.50
Remove / replace supply fan				.650		39		39	48.50	60.50
Remove supply fan bearing				1.200		71.50		71.50	89.50	112
Replace supply fan bearing				2.000	57.50	120		177.50	213	258
Remove supply fan motor				1.156		63		63	78.50	98
Replace supply fan motor				2.311	335	126		461	525	615
Replace fan belt for supply fan				.400	35	24		59	68.50	81.50
Remove compressor				14.513		804		804	1,000	1,250
Replace compressor				28.714	18,224	1,608		19,832	22,000	25,300
Remove / replace condenser fan				1.800		108		108	134	168
Remove condenser fan bearing				1.200		71.50		71.50	89.50	112
Replace condenser fan bearing				2.000	57.50	120		177.50	213	258
Remove condenser fan motor				.976		58.50		58.50	73	91
Replace condenser fan motor				1.951	248	107		355	405	475
Replace fan belt for condenser fan				.400	35	24		59	68.50	81.50
Replace refrigerant				1.333	320	79.50		399.50	450	525
Remove / replace heating coils				5.200	890	290		1,180	1,350	1,575
Total				66.204	20,202	3,738		23,940	26,836	31,098
1025										
Replace computer room A/C, air cooled, 10 ton	20	Q-6	Ea.							
Remove air conditioner and condenser				33.898		1,900		1,900	2,375	2,950
Replace air conditioner and remote condenser, 10 ton				64.000	39,100	3,450		42,550	47,300	54,500
Total				97.898	39,100	5,350		44,450	49,675	57,450

D3053 274 **Computer Room A/C Units, Air Cooled**

	System Description	Freq. (Years)	Crew	Unit	Labor Hours	2015 Bare Costs				Total In-House	Total w/O&P
						Material	Labor	Equipment	Total		
1030	**Repair computer room A/C, air cooled, 15 ton**	10	1 STPI	Ea.							
	Repair controls				.500		30		30	37.50	46.50
	Remove / replace supply fan				.650		39		39	48.50	60.50
	Remove supply fan bearing				1.200		71.50		71.50	89.50	112
	Replace supply fan bearing				2.000	57.50	120		177.50	213	258
	Remove supply fan motor				1.156		63		63	78.50	98
	Replace supply fan motor				2.312	560	126		686	775	895
	Replace fan belt for supply fan				.400	35	24		59	68.50	81.50
	Remove compressor				21.661		1,200		1,200	1,500	1,875
	Replace compressor				42.857	27,200	2,400		29,600	32,900	37,700
	Remove / replace condenser fan				1.800		108		108	134	168
	Remove condenser fan bearing				1.200		71.50		71.50	89.50	112
	Replace condenser fan bearing				2.000	57.50	120		177.50	213	258
	Remove condenser fan motor				.976		58.50		58.50	73	91
	Replace condenser fan motor				1.951	248	107		355	405	475
	Replace fan belt for condenser fan				.400	35	24		59	68.50	81.50
	Replace refrigerant				2.000	480	119.25		599.25	675	785
	Remove / replace heating coils				6.501	1,112.50	362.50		1,475	1,675	1,950
	Total				89.563	29,785.50	5,044.25		34,829.75	39,043.50	45,047
1035	**Replace computer room A/C, air cooled, 15 ton**	20	Q-6	Ea.							
	Remove air conditioner and condenser				50.314		2,800		2,800	3,500	4,375
	Replace air conditioner and remote condenser, 15 ton				72.727	43,100	3,900		47,000	52,500	60,000
	Total				123.042	43,100	6,700		49,800	56,000	64,375

D3053 274 Computer Room A/C Units, Air Cooled

	System Description	Freq. (Years)	Crew	Unit	Labor Hours	2015 Bare Costs				Total In-House	Total w/O&P
						Material	Labor	Equipment	Total		
1040	Repair computer room A/C, air cooled, 20 ton	10	1 STPI	Ea.							
	Repair controls				.800		48		48	60	74.50
	Remove / replace supply fan				.650		39		39	48.50	60.50
	Remove supply fan bearing				1.000		60		60	74.50	93
	Replace supply fan bearing				2.000	57.50	120		177.50	213	258
	Remove supply fan motor				1.156		63		63	78.50	98
	Replace supply fan motor				2.312	560	126		686	775	895
	Replace fan belt for supply fan				.400	35	24		59	68.50	81.50
	Remove compressor				24.390		1,350		1,350	1,700	2,125
	Replace compressor				48.780	28,300	2,725		31,025	34,500	39,600
	Remove / replace condenser fan				1.800		108		108	134	168
	Remove condenser fan bearing				1.200		71.50		71.50	89.50	112
	Replace condenser fan bearing				2.000	57.50	120		177.50	213	258
	Remove condenser fan motor				.976		58.50		58.50	73	91
	Replace condenser fan motor				2.311	279	126		405	465	545
	Replace fan belt for condenser fan				.400	35	24		59	68.50	81.50
	Replace refrigerant				2.667	640	159		799	900	1,050
	Remove / replace heating coils				6.936	1,920	393.75		2,313.75	2,600	3,025
	Total				99.778	31,884	5,615.75		37,499.75	42,061	48,616
1045	Replace computer room A/C, air cooled, 20 ton	20	Q-6	Ea.							
	Remove air conditioner and condenser				23.105		1,325		1,325	1,650	2,050
	Replace air conditioner and remote condenser, 20 ton				100.000	52,000	5,575		57,575	64,000	73,500
	Total				123.105	52,000	6,900		58,900	65,650	75,550

For customer support on your Facilities Maintenance & Repair Cost Data, call 877.801.5724.

D3053 276 Computer Room A/C Units, Chilled Water

	System Description	Freq. (Years)	Crew	Unit	Labor Hours	2015 Bare Costs				Total In-House	Total w/O&P
						Material	Labor	Equipment	Total		
1010	**Repair computer room A/C, chilled water, 5 ton**	10	1 STPI	Ea.							
	Repair controls				.600		36		36	45	56
	Remove / replace supply fan				.650		39		39	48.50	60.50
	Remove supply fan bearing				1.000		60		60	74.50	93
	Replace supply fan bearing				2.000	57.50	120		177.50	213	258
	Remove supply fan motor				1.156		63		63	78.50	98
	Replace supply fan motor				2.311	279	126		405	465	545
	Replace fan belt for supply fan				.400	35	24		59	68.50	81.50
	Remove / replace heating coils				26.002	4,450	1,450		5,900	6,700	7,825
	Total				34.119	4,821.50	1,918		6,739.50	7,693	9,017
1015	**Replace computer room A/C, chilled water, 5 ton**	20	Q-6	Ea.							
	Remove air conditioner				13.001		725		725	905	1,125
	Replace air conditioner, 5 ton				28.070	14,300	1,500		15,800	17,600	20,200
	Total				41.071	14,300	2,225		16,525	18,505	21,325
1020	**Repair computer room A/C, chilled water, 10 ton**	10	1 STPI	Ea.							
	Repair controls				.400		24		24	30	37.50
	Remove / replace supply fan				.650		39		39	48.50	60.50
	Remove supply fan bearing				1.200		71.50		71.50	89.50	112
	Replace supply fan bearing				2.000	57.50	120		177.50	213	258
	Remove supply fan motor				1.156		63		63	78.50	98
	Replace supply fan motor				2.311	335	126		461	525	615
	Replace fan belt for supply fan				.400	35	24		59	68.50	81.50
	Remove / replace heating coils				26.002	4,450	1,450		5,900	6,700	7,825
	Total				34.119	4,877.50	1,917.50		6,795	7,753	9,087.50
1025	**Replace computer room A/C, chilled water, 10 ton**	20	Q-6	Ea.							
	Remove air conditioner				33.898		1,900		1,900	2,375	2,950
	Replace air conditioner, 10 ton				42.105	14,700	2,275		16,975	19,000	21,900
	Total				76.004	14,700	4,175		18,875	21,375	24,850

For customer support on your Facilities Maintenance & Repair Cost Data, call 877.801.5724.

285

	System Description	Freq. (Years)	Crew	Unit	Labor Hours	2015 Bare Costs				Total In-House	Total w/O&P
						Material	Labor	Equipment	Total		
1030	**Repair computer room A/C, chilled water, 15 ton**	10	1 STPI	Ea.							
	Repair controls				.500		30		30	37.50	46.50
	Remove / replace supply fan				.650		39		39	48.50	60.50
	Remove supply fan bearing				1.200		71.50		71.50	89.50	112
	Replace supply fan bearing				2.000	57.50	120		177.50	213	258
	Remove supply fan motor				1.156		63		63	78.50	98
	Replace supply fan motor				2.312	560	126		686	775	895
	Replace fan belt for supply fan				.400	35	24		59	68.50	81.50
	Remove / replace heating coils				26.002	4,450	1,450		5,900	6,700	7,825
	Total				34.220	5,102.50	1,923.50		7,026	8,010.50	9,376.50
1035	**Replace computer room A/C, chilled water, 15 ton**	20	Q-6	Ea.							
	Remove air conditioner				50.314		2,800		2,800	3,500	4,375
	Replace air conditioner, 15 ton				44.444	15,400	2,400		17,800	19,900	23,000
	Total				94.759	15,400	5,200		20,600	23,400	27,375
1040	**Repair computer room A/C, chilled water, 20 ton**	10	1 STPI	Ea.							
	Repair controls				.500		30		30	37.50	46.50
	Remove / replace supply fan				.650		39		39	48.50	60.50
	Remove supply fan bearing				1.200		71.50		71.50	89.50	112
	Replace supply fan bearing				2.000	57.50	120		177.50	213	258
	Remove supply fan motor				1.156		63		63	78.50	98
	Replace supply fan motor				2.312	560	126		686	775	895
	Replace fan belt for supply fan				.400	35	24		59	68.50	81.50
	Remove / replace heating coils				46.243	12,800	2,625		15,425	17,400	20,100
	Total				54.461	13,452.50	3,098.50		16,551	18,710.50	21,651.50
1045	**Replace computer room A/C, chilled water, 20 ton**	20	Q-6	Ea.							
	Remove air conditioner				23.105		1,325		1,325	1,650	2,050
	Replace air conditioner, 20 ton				47.059	16,300	2,525		18,825	21,100	24,300
	Total				70.164	16,300	3,850		20,150	22,750	26,350

D3053 278   **Multi-Zone Air Conditioner**

System Description	Freq. (Years)	Crew	Unit	Labor Hours	2015 Bare Costs Material	Labor	Equipment	Total	Total In-House	Total w/O&P
1010										
Repair multi-zone rooftop unit, 15 ton	10	Q-6	Ea.							
Repair controls				1.000		60		60	74.50	93
Remove fan bearings				2.400		143		143	179	224
Replace fan bearings				4.000	115	240		355	425	515
Remove fan motor				1.156		63		63	78.50	98
Replace fan motor				2.312	560	126		686	775	895
Remove compressor				17.778		955		955	1,200	1,500
Replace compressor				42.857	27,200	2,400		29,600	32,900	37,700
Replace refrigerant				1.333	320	79.50		399.50	450	525
Remove / replace heater igniter				.500	29.50	30		59.50	70	83.50
Total				73.336	28,224.50	4,096.50		32,321	36,152	41,633.50
1040										
Replace multi-zone rooftop unit, 15 ton	15	Q-7	Ea.							
Remove multi-zone rooftop unit				94.675		5,375		5,375	6,725	8,400
Replace multi-zone rooftop, 15 ton				70.022	65,000	3,975		68,975	76,500	87,500
Total				164.696	65,000	9,350		74,350	83,225	95,900
2010										
Repair multi - zone rooftop unit, 25 ton	10	Q-6	Ea.							
Repair controls				1.000		60		60	74.50	93
Remove fan bearings				2.400		143		143	179	224
Replace fan bearings				4.000	115	240		355	425	515
Remove fan motor				1.156		63		63	78.50	98
Replace fan motor				2.311	610	126		736	825	960
Remove compressor				22.222		1,193.75		1,193.75	1,500	1,875
Replace compressor				60.976	35,375	3,406.25		38,781.25	43,100	49,500
Replace refrigerant				2.000	480	119.25		599.25	675	785
Remove / replace heater igniter				1.000	59	60		119	140	167
Total				97.065	36,639	5,411.25		42,050.25	46,997	54,217
2040										
Replace multi-zone rooftop unit, 25 ton	15	Q-7	Ea.							
Remove multi-zone rooftop unit				115.000		6,575		6,575	8,175	10,300
Replace multi-zone rooftop, 25 ton				95.808	83,500	5,450		88,950	98,500	113,000
Total				210.808	83,500	12,025		95,525	106,675	123,300

D3053 278   Multi-Zone Air Conditioner

	System Description	Freq. (Years)	Crew	Unit	Labor Hours	Material	Labor	Equipment	Total	Total In-House	Total w/O&P
							2015 Bare Costs				
3010	**Repair multi-zone rooftop unit, 40 ton**	10	Q-6	Ea.							
	Repair controls				1.000		60		60	74.50	93
	Remove fan bearings				2.400		143		143	179	224
	Replace fan bearings				4.444	129	266		395	475	575
	Remove fan motor				2.000		109		109	135	169
	Replace fan motor				4.000	1,750	219		1,969	2,200	2,525
	Remove compressor				35.451		1,975		1,975	2,475	3,075
	Replace compressor				71.006	30,500	3,950		34,450	38,500	44,300
	Replace refrigerant				2.667	640	159		799	900	1,050
	Remove / replace heater igniter				1.000	59	60		119	140	167
	Total				123.967	33,078	6,941		40,019	45,078.50	52,178
3040	**Replace multi-zone rooftop unit, 40 ton**	15	Q-7	Ea.							
	Remove multi-zone				172.000		9,850		9,850	12,200	15,300
	Replace multi-zone rooftop, 40 ton				151.000	122,500	8,625		131,125	145,500	166,500
	Total				323.000	122,500	18,475		140,975	157,700	181,800
4010	**Repair multi-zone rooftop unit, 70 ton**	10	Q-7	Ea.							
	Repair controls				1.000		60		60	74.50	93
	Remove fan bearings				2.400		143		143	179	224
	Replace fan bearings				4.444	129	266		395	475	575
	Remove fan motor				2.601		142		142	176	220
	Replace fan motor				5.195	2,925	284		3,209	3,575	4,100
	Remove compressor				76.923		4,375		4,375	5,475	6,825
	Replace compressor				153.000	35,500	8,750		44,250	49,900	58,000
	Replace refrigerant				4.667	1,120	278.25		1,398.25	1,575	1,825
	Remove / replace heater igniter				1.000	59	60		119	140	167
	Total				251.230	39,733	14,358.25		54,091.25	61,569.50	72,029
4040	**Replace multi-zone rooftop unit, 70 ton**	15	Q-7	Ea.							
	Remove multi-zone rooftop unit				231.000		13,200		13,200	16,400	20,600
	Replace multi-zone rooftop, 70 ton				264.000	164,500	15,000		179,500	199,500	229,000
	Total				495.000	164,500	28,200		192,700	215,900	249,600

For customer support on your Facilities Maintenance & Repair Cost Data, call 877.801.5724.

D30 HVAC | D3053 | Terminal and Package Units

D3053 278 Multi-Zone Air Conditioner

	System Description	Freq. (Years)	Crew	Unit	Labor Hours	2015 Bare Costs				Total In-House	Total w/O&P
						Material	Labor	Equipment	Total		
5010	**Repair multi-zone rooftop unit, 105 ton**	10	Q-7	Ea.							
	Repair controls				1.000		60		60	74.50	93
	Remove fan bearings				4.800		286		286	360	450
	Replace fan bearings				9.412	720	564		1,284	1,500	1,775
	Remove fan motor				4.334		237		237	293	365
	Replace fan motor				8.696	4,150	475		4,625	5,150	5,925
	Remove compressor				79.012		4,500		4,500	5,625	7,025
	Replace compressor				157.600	32,400	8,960		41,360	46,800	54,500
	Replace refrigerant				6.667	1,600	397.50		1,997.50	2,250	2,625
	Remove / replace heater igniter				1.500	88.50	90		178.50	210	250
	Total				273.020	38,958.50	15,569.50		54,528	62,262.50	73,008
5040	**Replace multi-zone rooftop unit, 105 ton**	15	Q-7	Ea.							
	Remove multi-zone rooftop unit				347.000		19,800		19,800	24,700	30,900
	Replace multi-zone rooftop, 105 ton				390.000	217,500	22,200		239,700	267,000	306,500
	Total				737.000	217,500	42,000		259,500	291,700	337,400

D3053 280 Single Zone Air Conditioner

	System Description	Freq. (Years)	Crew	Unit	Labor Hours	2015 Bare Costs				Total In-House	Total w/O&P
						Material	Labor	Equipment	Total		
1001	**Repair single zone rooftop unit, 3 ton**	10	2 STPI	Ea.							
	Repair controls				.300		17.95		17.95	22.50	28
	Remove fan bearings				2.400		143		143	179	224
	Replace fan bearings				4.000	115	240		355	425	515
	Remove fan motor				1.156		63		63	78.50	98
	Replace fan motor				1.951	218	107		325	370	440
	Remove compressor				2.477		148		148	185	231
	Replace compressor				4.522	360	270		630	735	870
	Replace refrigerant				.533	128	31.80		159.80	180	210
	Remove/replace heater igniter				.500	29.50	30		59.50	70	83.50
	Total				17.840	850.50	1,050.75		1,901.25	2,245	2,699.50
1002	**Replace single zone rooftop unit, 3 ton**	15	Q-5	Ea.							
	Remove single zone				8.000		430		430	540	670
	Replace single zone rooftop, 3 ton				30.476	3,925	1,650		5,575	6,375	7,450
	Total				38.476	3,925	2,080		6,005	6,915	8,120

For customer support on your Facilities Maintenance & Repair Cost Data, call 877.801.5724.

289

D3053 280 Single Zone Air Conditioner

	System Description	Freq. (Years)	Crew	Unit	Labor Hours	2015 Bare Costs Material	Labor	Equipment	Total	Total In-House	Total w/O&P
1003	**Repair single zone rooftop unit, 5 ton**	10	2 STPI	Ea.							
	Repair controls				.300		17.95		17.95	22.50	28
	Remove fan bearings				2.400		143		143	179	224
	Replace fan bearings				4.000	115	240		355	425	515
	Remove fan motor				1.156		63		63	78.50	98
	Replace fan motor				1.951	218	107		325	370	440
	Remove compressor				2.477		148		148	185	231
	Replace compressor				4.954	505	296		801	925	1,100
	Replace refrigerant				.533	128	31.80		159.80	180	210
	Remove/replace heater igniter				.500	29.50	30		59.50	70	83.50
	Total				18.271	995.50	1,076.75		2,072.25	2,435	2,929.50
1004	**Replace single zone rooftop unit, 5 ton**	15	Q-5	Ea.							
	Remove single zone				18.561		1,000		1,000	1,250	1,550
	Replace single zone rooftop, 5 ton				38.005	5,125	2,050		7,175	8,200	9,600
	Total				56.566	5,125	3,050		8,175	9,450	11,150
1005	**Repair single zone rooftop unit, 7.5 ton**	10	2 STPI	Ea.							
	Repair controls				.400		24		24	30	37.50
	Remove fan bearings				2.400		143		143	179	224
	Replace fan bearings				4.000	115	240		355	425	515
	Remove fan motor				1.156		63		63	78.50	98
	Replace fan motor				2.311	279	126		405	465	545
	Remove compressor				3.467		186		186	233	291
	Replace compressor				6.932	1,150	375		1,525	1,725	2,025
	Replace refrigerant				.667	160	39.75		199.75	225	262
	Remove/replace heater igniter				.500	29.50	30		59.50	70	83.50
	Total				21.833	1,733.50	1,226.75		2,960.25	3,430.50	4,081
1006	**Replace single zone rooftop unit, 7.5 ton**	15	Q-5	Ea.							
	Remove single zone				26.016		1,400		1,400	1,750	2,175
	Replace single zone rooftop, 7.5 ton				43.011	6,825	2,325		9,150	10,400	12,100
	Total				69.027	6,825	3,725		10,550	12,150	14,275

D3053 280 Single Zone Air Conditioner

	System Description	Freq. (Years)	Crew	Unit	Labor Hours	2015 Bare Costs				Total In-House	Total w/O&P
						Material	Labor	Equipment	Total		
1007	**Repair single zone rooftop unit, 10 ton**	10	2 STPI	Ea.							
	Repair controls				.400		24		24	30	37.50
	Remove fan bearings				2.400		143		143	179	224
	Replace fan bearings				4.000	115	240		355	425	515
	Remove fan motor				1.156		63		63	78.50	98
	Replace fan motor				2.311	335	126		461	525	615
	Remove compressor				14.513		804		804	1,000	1,250
	Replace compressor				28.714	18,224	1,608		19,832	22,000	25,300
	Replace refrigerant				.800	192	47.70		239.70	270	315
	Remove/replace heater igniter				.500	29.50	30		59.50	70	83.50
	Total				54.794	18,895.50	3,085.70		21,981.20	24,577.50	28,438
1008	**Replace single zone rooftop unit, 10 ton**	15	Q-6	Ea.							
	Remove single zone				33.898		1,900		1,900	2,375	2,950
	Replace single zone rooftop, 10 ton				48.000	9,425	2,675		12,100	13,700	16,000
	Total				81.898	9,425	4,575		14,000	16,075	18,950
1010	**Repair single zone rooftop unit, 15 ton**	10	2 STPI	Ea.							
	Repair controls				.500		30		30	37.50	46.50
	Remove fan bearings				2.400		143		143	179	224
	Replace fan bearings				4.000	115	240		355	425	515
	Remove fan motor				1.156		63		63	78.50	98
	Replace fan motor				2.312	560	126		686	775	895
	Remove compressor				21.661		1,200		1,200	1,500	1,875
	Replace compressor				42.857	27,200	2,400		29,600	32,900	37,700
	Replace refrigerant				1.333	320	79.50		399.50	450	525
	Remove / replace heater igniter				.500	29.50	30		59.50	70	83.50
	Total				76.719	28,224.50	4,311.50		32,536	36,415	41,962
1040	**Replace single zone rooftop unit, 15 ton**	15	Q-6	Ea.							
	Remove single zone rooftop unit				50.314		2,800		2,800	3,500	4,375
	Replace single zone rooftop, 15 ton				56.075	13,600	3,125		16,725	18,900	21,900
	Total				106.389	13,600	5,925		19,525	22,400	26,275

For customer support on your Facilities Maintenance & Repair Cost Data, call 877.801.5724.

291

D3053 280 | **Single Zone Air Conditioner**

	System Description	Freq. (Years)	Crew	Unit	Labor Hours	2015 Bare Costs				Total In-House	Total w/O&P
						Material	Labor	Equipment	Total		
2010	**Repair single zone rooftop unit, 25 ton**	10	Q-6	Ea.							
	Repair controls				.500		30		30	37.50	46.50
	Remove fan bearings				2.400		143		143	179	224
	Replace fan bearings				4.000	115	240		355	425	515
	Remove fan motor				1.156		63		63	78.50	98
	Replace fan motor				2.311	610	126		736	825	960
	Remove compressor				22.222		1,193.75		1,193.75	1,500	1,875
	Replace compressor				60.976	35,375	3,406.25		38,781.25	43,100	49,500
	Replace refrigerant				2.000	480	119.25		599.25	675	785
	Remove / replace heater igniter				1.000	59	60		119	140	167
	Total				96.565	36,639	5,381.25		42,020.25	**46,960**	**54,170.50**
2040	**Replace single zone rooftop unit, 25 ton**	15	Q-7	Ea.							
	Remove single zone rooftop unit				76.923		4,375		4,375	5,475	6,825
	Replace single zone rooftop, 25 ton				76.739	27,100	4,375		31,475	35,300	40,700
	Total				153.662	27,100	8,750		35,850	**40,775**	**47,525**
3010	**Repair single zone rooftop unit, 60 ton**	10	Q-6	Ea.							
	Repair controls				.500		30		30	37.50	46.50
	Remove fan bearings				2.400		143		143	179	224
	Replace fan bearings				4.444	129	266		395	475	575
	Remove fan motor				2.000		109		109	135	169
	Replace fan motor				4.167	2,050	228		2,278	2,525	2,925
	Remove compressor				93.507		5,220		5,220	6,500	8,125
	Replace compressor				186.000	38,520	10,440		48,960	55,500	64,500
	Replace refrigerant				4.667	1,120	278.25		1,398.25	1,575	1,825
	Remove / replace heater igniter				1.000	59	60		119	140	167
	Total				298.684	41,878	16,774.25		58,652.25	**67,066.50**	**78,556.50**
3040	**Replace single zone rooftop unit, 60 ton**	15	Q-7	Ea.							
	Remove single zone rooftop unit				188.000		10,700		10,700	13,400	16,700
	Replace single zone rooftop, 60 ton				181.000	57,000	10,300		67,300	75,500	87,500
	Total				369.000	57,000	21,000		78,000	**88,900**	**104,200**

D30 HVAC   D3053   Terminal and Package Units

D3053 280   Single Zone Air Conditioner

	System Description	Freq. (Years)	Crew	Unit	Labor Hours	Material	Labor	Equipment	Total	Total In-House	Total w/O&P
4010	**Repair single zone rooftop unit, 100 ton**	10	Q-7	Ea.							
	Repair controls				.500		30		30	37.50	46.50
	Remove fan bearings				4.800		286		286	360	450
	Replace fan bearings				9.412	720	564		1,284	1,500	1,775
	Remove fan motor				4.334		237		237	293	365
	Replace fan motor				7.407	4,075	405		4,480	4,975	5,725
	Remove compressor				98.765		5,625		5,625	7,025	8,775
	Replace compressor				197.000	40,500	11,200		51,700	58,500	68,000
	Replace refrigerant				6.667	1,600	397.50		1,997.50	2,250	2,625
	Remove / replace heater igniter				1.500	88.50	90		178.50	210	250
	Total				330.385	46,983.50	18,834.50		65,818	75,150.50	88,011.50
4040	**Replace single zone rooftop unit, 100 ton**	15	Q-7	Ea.							
	Remove single zone rooftop unit				296.000		16,900		16,900	21,000	26,300
	Replace single zone rooftop, 100 ton				304.000	115,000	17,300		132,300	148,000	171,000
	Total				600.000	115,000	34,200		149,200	169,000	197,300

D3053 282   Multi-Zone Variable Volume

	System Description	Freq. (Years)	Crew	Unit	Labor Hours	Material	Labor	Equipment	Total	Total In-House	Total w/O&P
1010	**Repair multi-zone variable volume, 50 ton**	10	Q-6	Ea.							
	Repair controls				1.300		77.50		77.50	97	121
	Remove fan bearings				2.400		143		143	179	224
	Replace fan bearings				4.444	129	266		395	475	575
	Remove fan motor				2.000		109		109	135	169
	Replace fan motor				4.000	1,750	219		1,969	2,200	2,525
	Remove compressor				77.922		4,350		4,350	5,425	6,775
	Replace compressor				155.000	32,100	8,700		40,800	46,100	53,500
	Replace refrigerant				3.333	800	198.75		998.75	1,125	1,300
	Remove / replace heater igniter				1.000	59	60		119	140	167
	Total				251.400	34,838	14,123.25		48,961.25	55,876	65,356
1040	**Replace multi-zone variable volume, 50 ton**	15	Q-7	Ea.							
	Remove multi-zone variable volume unit				160.000		9,100		9,100	11,400	14,200
	Replace multi-zone variable volume unit, 50 ton				219.000	113,000	12,500		125,500	140,000	160,500
	Total				379.000	113,000	21,600		134,600	151,400	174,700

2015 Bare Costs

System Description	Freq. (Years)	Crew	Unit	Labor Hours	2015 Bare Costs				Total In-House	Total w/O&P
					Material	Labor	Equipment	Total		
2010 Repair multi-zone variable volume, 70 ton	10	Q-7	Ea.							
Repair controls				1.300		77.50		77.50	97	121
Remove fan bearings				2.400		143		143	179	224
Replace fan bearings				4.444	129	266		395	475	575
Remove fan motor				2.601		142		142	176	220
Replace fan motor				5.195	2,925	284		3,209	3,575	4,100
Remove compressor				76.923		4,375		4,375	5,475	6,825
Replace compressor				153.000	35,500	8,750		44,250	49,900	58,000
Replace refrigerant				4.667	1,120	278.25		1,398.25	1,575	1,825
Remove / replace heater igniter				1.000	59	60		119	140	167
Total				251.530	39,733	14,375.75		54,108.75	61,592	72,057
2040 Replace multi-zone variable volume, 70 ton	15	Q-7	Ea.							
Remove multi-zone variable volume unit				231.000		13,200		13,200	16,400	20,600
Replace multi-zone variable volume unit, 70 ton				304.000	158,000	17,300		175,300	195,500	224,500
Total				535.000	158,000	30,500		188,500	211,900	245,100
3010 Repair multi-zone variable volume, 90 ton	10	Q-7	Ea.							
Repair controls				1.300		77.50		77.50	97	121
Remove fan bearings				2.400		143		143	179	224
Replace fan bearings				4.444	129	266		395	475	575
Remove fan motor				2.601		142		142	176	220
Replace fan motor				6.504	3,200	355		3,555	3,950	4,550
Remove compressor				92.308		5,250		5,250	6,575	8,200
Replace compressor				183.600	42,600	10,500		53,100	60,000	69,500
Replace refrigerant				6.667	1,600	397.50		1,997.50	2,250	2,625
Remove / replace heater igniter				1.500	88.50	90		178.50	210	250
Total				301.323	47,617.50	17,221		64,838.50	73,912	86,265
3040 Replace multi-zone variable volume, 90 ton	15	Q-7	Ea.							
Remove multi-zone variable volume unit				296.000		16,900		16,900	21,000	26,300
Replace multi-zone variable volume unit, 90 ton				395.000	181,500	22,500		204,000	227,500	262,000
Total				691.000	181,500	39,400		220,900	248,500	288,300

D3053 282 Multi-Zone Variable Volume

System Description	Freq. (Years)	Crew	Unit	Labor Hours	2015 Bare Costs				Total In-House	Total w/O&P
					Material	Labor	Equipment	Total		
4010 **Repair multi-zone variable volume, 105 ton**	10	Q-7	Ea.							
Repair controls				1.300		77.50		77.50	97	121
Remove fan bearings				4.800		286		286	360	450
Replace fan bearings				9.412	720	564		1,284	1,500	1,775
Remove fan motor				4.334		237		237	293	365
Replace fan motor				8.696	4,150	475		4,625	5,150	5,925
Remove compressor				79.012		4,500		4,500	5,625	7,025
Replace compressor				157.600	32,400	8,960		41,360	46,800	54,500
Replace refrigerant				7.333	1,760	437.25		2,197.25	2,475	2,875
Remove / replace heater igniter				1.500	88.50	90		178.50	210	250
Total				273.987	39,118.50	15,626.75		54,745.25	62,510	73,286
4040 **Replace multi-zone variable volume, 105 ton**	15	Q-7	Ea.							
Remove multi-zone variable volume unit				347.000		19,800		19,800	24,700	30,900
Replace multi-zone variable volume unit, 105 ton				444.000	203,000	25,300		228,300	255,000	293,000
Total				791.000	203,000	45,100		248,100	279,700	323,900
5010 **Repair multi-zone variable volume, 140 ton**	10	Q-7	Ea.							
Repair controls				1.300		77.50		77.50	97	121
Remove fan bearings				7.200		429		429	535	670
Replace fan bearings				14.118	1,080	846		1,926	2,250	2,675
Remove fan motor				4.334		237		237	293	365
Replace fan motor				11.594	5,350	635		5,985	6,675	7,675
Remove compressor				106.370		6,058.13		6,058.13	7,575	9,450
Replace compressor				212.169	43,618.50	12,062.40		55,680.90	63,000	73,500
Replace refrigerant				8.666	2,080	516.75		2,596.75	2,925	3,400
Remove / replace heater igniter				2.000	118	120		238	280	335
Total				367.751	52,246.50	20,981.78		73,228.28	83,630	98,191
5040 **Replace multi-zone variable volume, 140 ton**	15	Q-7	Ea.							
Remove multi-zone variable volume unit				516.000		29,400		29,400	36,700	45,800
Replace multi-zone variable volume unit, 140 ton				592.000	248,500	33,700		282,200	315,500	363,000
Total				1108.000	248,500	63,100		311,600	352,200	408,800

System Description	Freq. (Years)	Crew	Unit	Labor Hours	2015 Bare Costs				Total In-House	Total w/O&P
					Material	Labor	Equipment	Total		
1010 Repair single zone variable volume, 20 ton	10	Q-6	Ea.							
Repair controls				1.000		60		60	74.50	93
Remove fan bearings				2.400		143		143	179	224
Replace fan bearings				4.000	115	240		355	425	515
Remove fan motor				1.156		63		63	78.50	98
Replace fan motor				2.312	560	126		686	775	895
Remove compressor				17.778		955		955	1,200	1,500
Replace compressor				48.780	28,300	2,725		31,025	34,500	39,600
Replace refrigerant				1.333	320	79.50		399.50	450	525
Remove / replace heater igniter				1.000	59	60		119	140	167
Total				79.760	29,354	4,451.50		33,805.50	37,822	43,617
1040 Replace single zone variable volume, 20 ton	15	Q-7	Ea.							
Remove single zone variable volume unit				71.545		4,070		4,070	5,100	6,350
Replace single zone variable volume unit, 20 ton				64.945	28,710	3,685		32,395	36,200	41,700
Total				136.489	28,710	7,755		36,465	41,300	48,050
2010 Repair single zone variable volume, 30 ton	10	Q-6	Ea.							
Repair controls				1.000		60		60	74.50	93
Remove fan bearings				2.400		143		143	179	224
Replace fan bearings				4.000	115	240		355	425	515
Remove fan motor				1.156		63		63	78.50	98
Replace fan motor				2.476	810	135		945	1,050	1,225
Remove compressor				26.667		1,432.50		1,432.50	1,800	2,250
Replace compressor				73.171	42,450	4,087.50		46,537.50	52,000	59,500
Replace refrigerant				2.667	640	159		799	900	1,050
Remove / replace heater igniter				1.000	59	60		119	140	167
Total				114.536	44,074	6,380		50,454	56,647	65,122
2040 Replace single zone variable volume, 30 ton	15	Q-7	Ea.							
Remove single zone variable volume unit				104.142		5,912.50		5,912.50	7,400	9,250
Replace single zone variable volume unit, 30 ton				92.147	36,410	5,252.50		41,662.50	46,600	53,500
Total				196.289	36,410	11,165		47,575	54,000	62,750

For customer support on your Facilities Maintenance & Repair Cost Data, call 877.801.5724.

	System Description	Freq. (Years)	Crew	Unit	Labor Hours	2015 Bare Costs Material	Labor	Equipment	Total	Total In-House	Total w/O&P
3010	**Repair single zone variable volume, 40 ton**	10	Q-6	Ea.							
	Repair controls				1.000		60		60	74.50	93
	Remove fan bearings				2.400		143		143	179	224
	Replace fan bearings				4.000	115	240		355	425	515
	Remove fan motor				1.156		63		63	78.50	98
	Replace fan motor				4.000	1,750	219		1,969	2,200	2,525
	Remove compressor				35.451		1,975		1,975	2,475	3,075
	Replace compressor				71.006	30,500	3,950		34,450	38,500	44,300
	Replace refrigerant				3.333	800	198.75		998.75	1,125	1,300
	Remove / replace heater igniter				1.000	59	60		119	140	167
	Total				123.346	33,224	6,908.75		40,132.75	45,197	52,297
3040	**Replace single zone variable volume, 40 ton**	15	Q-7	Ea.							
	Remove single zone variable volume unit				143.000		8,140		8,140	10,200	12,700
	Replace single zone variable volume unit, 40 ton				122.100	41,910	7,012.50		48,922.50	55,000	63,500
	Total				265.100	41,910	15,152.50		57,062.50	65,200	76,200
4010	**Repair single zone variable volume, 60 ton**	10	Q-6	Ea.							
	Repair controls				.500		30		30	37.50	46.50
	Remove fan bearings				2.400		143		143	179	224
	Replace fan bearings				4.444	129	266		395	475	575
	Remove fan motor				2.000		109		109	135	169
	Replace fan motor				4.167	2,050	228		2,278	2,525	2,925
	Remove compressor				93.507		5,220		5,220	6,500	8,125
	Replace compressor				186.000	38,520	10,440		48,960	55,500	64,500
	Replace refrigerant				4.000	960	238.50		1,198.50	1,350	1,575
	Remove / replace heater igniter				1.000	59	60		119	140	167
	Total				298.018	41,718	16,734.50		58,452.50	66,841.50	78,306.50
4040	**Replace single zone variable volume, 60 ton**	15	Q-7	Ea.							
	Remove single zone variable volume unit				189.200		10,835		10,835	13,400	16,800
	Replace single zone variable volume unit, 60 ton				184.800	58,850	10,532.50		69,382.50	78,000	90,000
	Total				374.000	58,850	21,367.50		80,217.50	91,400	106,800

For customer support on your Facilities Maintenance & Repair Cost Data, call 877.801.5724.

297

	System Description	Freq. (Years)	Crew	Unit	Labor Hours	2015 Bare Costs Material	Labor	Equipment	Total	Total In-House	Total w/O&P
1010	**Repair central station A.H.U., 1300 CFM**	10	1 STPI	Ea.							
	Repair controls				.300		17.95		17.95	22.50	28
	Remove blower motor				1.200		71.50		71.50	89.50	112
	Replace blower motor				1.951	248	107		355	405	475
	Total				3.451	248	196.45		444.45	517	615
1040	**Replace central station A.H.U., 1300 CFM**	15	Q-5	Ea.							
	Remove central station A.H.U.				8.667		465		465	585	725
	Replace central station A.H.U., 1300 CFM				17.335	7,125	930		8,055	9,000	10,400
	Total				26.002	7,125	1,395		8,520	9,585	11,125
2010	**Repair central station A.H.U., 1900 CFM**	10	1 STPI	Ea.							
	Repair controls				.300		17.95		17.95	22.50	28
	Remove blower motor				1.200		71.50		71.50	89.50	112
	Replace blower motor				2.311	279	126		405	465	545
	Total				3.811	279	215.45		494.45	577	685
2040	**Replace central station A.H.U., 1900 CFM**	15	Q-5	Ea.							
	Remove central station A.H.U.				9.456		510		510	635	795
	Replace central station A.H.U., 1900 CFM				18.913	11,000	1,025		12,025	13,400	15,300
	Total				28.369	11,000	1,535		12,535	14,035	16,095
3010	**Repair central station A.H.U., 5400 CFM**	10	1 STPI	Ea.							
	Repair controls				.300		17.95		17.95	22.50	28
	Remove blower motor				1.400		83.50		83.50	105	130
	Replace blower motor				2.312	560	126		686	775	895
	Total				4.012	560	227.45		787.45	902.50	1,053
3040	**Replace central station A.H.U., 5400 CFM**	15	Q-6	Ea.							
	Remove central station A.H.U.				19.496		1,075		1,075	1,350	1,700
	Replace central station A.H.U., 5400 CFM				39.024	17,600	2,175		19,775	22,100	25,400
	Total				58.521	17,600	3,250		20,850	23,450	27,100
4010	**Repair central station A.H.U., 8000 CFM**	10	1 STPI	Ea.							
	Repair controls				.300		17.95		17.95	22.50	28
	Remove blower motor				1.600		95.50		95.50	120	149
	Replace blower motor				2.311	610	126		736	825	960
	Total				4.211	610	239.45		849.45	967.50	1,137

D3053 286 Central Station Air Conditioning Air Handling Unit

System Description	Freq. (Years)	Crew	Unit	Labor Hours	2015 Bare Costs				Total In-House	Total w/O&P
					Material	Labor	Equipment	Total		
4040 Replace central station A.H.U., 8000 CFM	15	Q-6	Ea.							
Remove central station A.H.U.				25.974		1,450		1,450	1,800	2,250
Replace central station A.H.U., 8000 CFM				51.948	28,100	2,900		31,000	34,500	39,700
Total				77.922	28,100	4,350		32,450	36,300	41,950
5010 Repair central station A.H.U., 16,000 CFM	10	1 STPI	Ea.							
Repair controls				.300		17.95		17.95	22.50	28
Remove blower motor				1.800		108		108	134	168
Replace blower motor				2.597	990	142		1,132	1,275	1,450
Total				4.698	990	267.95		1,257.95	1,431.50	1,646
5040 Replace central station A.H.U., 16,000 CFM	15	Q-6	Ea.							
Remove central station A.H.U.				41.026		2,300		2,300	2,850	3,575
Replace central station A.H.U., 16,000 CFM				82.192	53,000	4,575		57,575	64,000	73,500
Total				123.217	53,000	6,875		59,875	66,850	77,075
6010 Repair central station A.H.U., 33,500 CFM	10	1 STPI	Ea.							
Repair controls				.300		17.95		17.95	22.50	28
Remove blower motor				2.601		142		142	176	220
Replace blower motor				5.195	2,925	284		3,209	3,575	4,100
Total				8.096	2,925	443.95		3,368.95	3,773.50	4,348
6040 Replace central station A.H.U., 33,500 CFM	15	Q-6	Ea.							
Remove central station A.H.U.				82.192		4,575		4,575	5,725	7,150
Replace central station A.H.U., 33,500 CFM				164.000	112,500	9,175		121,675	135,000	155,000
Total				246.192	112,500	13,750		126,250	140,725	162,150
7010 Repair central station A.H.U., 63,000 CFM	10	1 STPI	Ea.							
Repair controls				.300		17.95		17.95	22.50	28
Remove blower motor				4.334		237		237	293	365
Replace blower motor				8.696	4,150	475		4,625	5,150	5,925
Total				13.329	4,150	729.95		4,879.95	5,465.50	6,318
7040 Replace central station A.H.U., 63,000 CFM	15	Q-7	Ea.							
Remove central station A.H.U.				160.000		9,100		9,100	11,400	14,200
Replace central station A.H.U., 63,000 CFM				320.000	212,000	18,200		230,200	256,000	293,500
Total				480.000	212,000	27,300		239,300	267,400	307,700

For customer support on your Facilities Maintenance & Repair Cost Data, call 877.801.5724.

299

D3053 310 | Residential Furnace, Gas

System Description	Freq. (Years)	Crew	Unit	Labor Hours	2015 Bare Costs Material	Labor	Equipment	Total	Total In-House	Total w/O&P
1010 Repair furnace, gas, 25 MBH	10	Q-1	Ea.							
Repair controls				.500		30		30	37.50	46.50
Remove fan motor				.976		58.50		58.50	73	91
Replace fan motor				1.951	195	107		302	345	410
Remove auto vent damper				.274		15.30		15.30	19.40	24
Replace auto vent damper				.547	112	30.50		142.50	162	189
Remove burner				1.000		53		53	66	82.50
Replace burner				8.320	720	440		1,160	1,350	1,575
Total				13.568	1,027	734.30		1,761.30	2,052.90	2,418
1030 Replace furnace, gas, 25 MBH	15	Q-9	Ea.							
Remove furnace				2.000		101		101	128	159
Replace furnace				4.000	535	201		736	845	990
Total				6.000	535	302		837	973	1,149
2010 Repair furnace, gas, 100 MBH	10	Q-1	Ea.							
Repair controls				.500		30		30	37.50	46.50
Remove fan motor				.976		58.50		58.50	73	91
Replace fan motor				1.951	218	107		325	370	440
Remove auto vent damper				.289		16.15		16.15	20.50	25.50
Replace auto vent damper				.578	144	32.50		176.50	199	231
Remove burner				1.200		63.50		63.50	79.50	99
Replace burner				10.403	1,200	550		1,750	2,000	2,350
Total				15.897	1,562	857.65		2,419.65	2,779.50	3,283
2030 Replace furnace, gas, 100 MBH	15	Q-9	Ea.							
Remove furnace				3.250		164		164	208	258
Replace furnace				6.499	625	325		950	1,100	1,300
Total				9.749	625	489		1,114	1,308	1,558

D3053 310 Residential Furnace, Gas

System Description	Freq. (Years)	Crew	Unit	Labor Hours	2015 Bare Costs				Total In-House	Total w/O&P
					Material	Labor	Equipment	Total		
3010										
Repair furnace, gas, 200 MBH	10	Q-1	Ea.							
Repair controls				.500		30		30	37.50	46.50
Remove fan motor				.976		58.50		58.50	73	91
Replace fan motor				1.951	248	107		355	405	475
Remove auto vent damper				.325		18.20		18.20	23	28.50
Replace auto vent damper				.650	162	36.50		198.50	224	260
Remove burner				3.077		163		163	203	254
Replace burner				12.298	3,625	650		4,275	4,800	5,550
Total				19.777	4,035	1,063.20		5,098.20	5,765.50	6,705
3030										
Replace furnace, gas, 200 MBH	15	Q-9	Ea.							
Remove furnace				4.000		201		201	255	320
Replace furnace				8.000	2,800	405		3,205	3,600	4,125
Total				12.000	2,800	606		3,406	3,855	4,445

D3053 320 Residential Furnace, Oil

System Description	Freq. (Years)	Crew	Unit	Labor Hours	2015 Bare Costs				Total In-House	Total w/O&P
					Material	Labor	Equipment	Total		
1010										
Repair furnace, oil, 55 MBH	10	Q-1	Ea.							
Repair controls				.500		30		30	37.50	46.50
Remove fan motor				.976		58.50		58.50	73	91
Replace fan motor				1.951	195	107		302	345	410
Remove auto vent damper				.473		24		24	30	37.50
Replace auto vent damper				.945	242	47.50		289.50	325	380
Remove burner				4.334		229		229	286	355
Replace burner				8.667	284	460		744	885	1,075
Total				17.846	721	956		1,677	1,981.50	2,395
1030										
Replace furnace, oil, 55 MBH	15	Q-9	Ea.							
Remove furnace				2.889		145		145	184	230
Replace furnace, oil, 55 mbh				5.778	1,725	291		2,016	2,275	2,625
Total				8.667	1,725	436		2,161	2,459	2,855

For customer support on your Facilities Maintenance & Repair Cost Data, call 877.801.5724.

301

D3053 320 | Residential Furnace, Oil

System Description	Freq. (Years)	Crew	Unit	Labor Hours	2015 Bare Costs				Total In-House	Total w/O&P
					Material	Labor	Equipment	Total		
2010 Repair furnace, oil, 100 MBH	10	Q-1	Ea.							
Repair controls				.500		30		30	37.50	46.50
Remove fan motor				.976		58.50		58.50	73	91
Replace fan motor				1.951	218	107		325	370	440
Remove auto vent damper				.495		25		25	31.50	39.50
Replace auto vent damper				.990	112	50		162	186	219
Remove burner				4.334		229		229	286	355
Replace burner				8.667	284	460		744	885	1,075
Total				17.914	614	959.50		1,573.50	1,869	2,266
2030 Replace furnace, oil, 100 MBH	15	Q-1	Ea.							
Remove furnace				3.059		154		154	195	243
Replace furnace, oil, 100 mbh				6.119	1,875	310		2,185	2,450	2,825
Total				9.177	1,875	464		2,339	2,645	3,068
3010 Repair furnace, oil, 200 MBH	10	Q-1	Ea.							
Repair controls				.500		30		30	37.50	46.50
Remove fan motor				.976		58.50		58.50	73	91
Replace fan motor				1.951	248	107		355	405	475
Remove auto vent damper				.520		26		26	33	41.50
Replace auto vent damper				1.040	144	52.50		196.50	225	263
Remove burner				4.334		229		229	286	355
Replace burner				8.667	284	460		744	885	1,075
Total				17.988	676	963		1,639	1,944.50	2,347
3030 Replace furnace, oil, 200 MBH	15	Q-1	Ea.							
Remove furnace				4.000		201		201	255	320
Replace furnace, oil, 200 mbh				8.000	2,450	405		2,855	3,200	3,700
Total				12.000	2,450	606		3,056	3,455	4,020

For customer support on your Facilities Maintenance & Repair Cost Data, call 877.801.5724.

	System Description	Freq. (Years)	Crew	Unit	Labor Hours	2015 Bare Costs				Total In-House	Total w/O&P
						Material	Labor	Equipment	Total		
1010	**Repair furnace, electric, 25 MBH**	10	Q-20	Ea.							
	Repair controls				.500		30		30	37.50	46.50
	Remove fan motor				.976		58.50		58.50	73	91
	Replace fan motor				1.951	195	107		302	345	410
	Remove heat element				.867		44.50		44.50	56	70
	Replace heat element				1.733	1,275	89		1,364	1,525	1,725
	Total				6.027	1,470	329		1,799	**2,036.50**	**2,342.50**
1030	**Replace furnace, electric, 25 MBH**	15	Q-20	Ea.							
	Remove furnace				2.826		145		145	183	228
	Replace furnace, electric, 25 mbh				5.653	600	290		890	1,025	1,200
	Total				8.479	600	435		1,035	**1,208**	**1,428**
2010	**Repair furnace, electric, 50 MBH**	10	Q-20	Ea.							
	Repair controls				.500		30		30	37.50	46.50
	Remove fan motor				.976		58.50		58.50	73	91
	Replace fan motor				1.951	195	107		302	345	410
	Remove heat element				.929		47.50		47.50	60	75
	Replace heat element				1.857	1,325	95		1,420	1,575	1,800
	Total				6.212	1,520	338		1,858	**2,090.50**	**2,422.50**
2030	**Replace furnace , electric, 50 MBH**	15	Q-20	Ea.							
	Remove furnace				3.095		159		159	200	249
	Replace furnace, electric, 50 mbh				6.190	495	315		810	945	1,125
	Total				9.285	495	474		969	**1,145**	**1,374**
3010	**Repair furnace, electric, 85 MBH**	10	Q-20	Ea.							
	Repair controls				.500		30		30	37.50	46.50
	Remove fan motor				.976		58.50		58.50	73	91
	Replace fan motor				1.951	218	107		325	370	440
	Remove heat element				1.083		55.50		55.50	70	87
	Replace heat element				2.167	2,375	111		2,486	2,750	3,150
	Total				6.677	2,593	362		2,955	**3,300.50**	**3,814.50**
3030	**Replace furnace, electric, 85 MBH**	15	Q-20	Ea.							
	Remove furnace				4.000		201		201	255	320
	Replace furnace, electric, 85 mbh				6.842	520	350		870	1,025	1,200
	Total				10.842	520	551		1,071	**1,280**	**1,520**

For customer support on your Facilities Maintenance & Repair Cost Data, call 877.801.5724.

303

D3053 410 Electric Baseboard Heating Units

System Description	Freq. (Years)	Crew	Unit	Labor Hours	2015 Bare Costs				Total In-House	Total w/O&P
					Material	Labor	Equipment	Total		
0010 Maintenance and repair										
Repair wiring connections	2	1 ELEC	Ea.	.615		33.50		33.50	41.50	52
Total				.615		33.50		33.50	**41.50**	**52**
0020 Maintenance and inspection										
Inspect and clean baseboard heater	0.50	1 ELEC	Ea.	1.143		62.50		62.50	77.50	97
Total				1.143		62.50		62.50	**77.50**	**97**
0030 Replace baseboard heater										
Remove baseboard heater	20	1 ELEC	Ea.	.667		36.50		36.50	45	56.50
Baseboard heater, 5' long				1.860	48	102		150	179	218
Total				2.527	48	138.50		186.50	**224**	**274.50**

D3053 420 Cast Iron Radiator, 10' section, 1 side

System Description	Freq. (Years)	Crew	Unit	Labor Hours	2015 Bare Costs				Total In-House	Total w/O&P
					Material	Labor	Equipment	Total		
0020 Refinish C.I. radiator, 1 side										
Prepare surface	5	1 PORD	Ea.	.296		12		12	15.60	19.20
Prime and paint 1 coat, brushwork				.471	3.60	19		22.60	28.50	35
Total				.767	3.60	31		34.60	**44.10**	**54.20**

D3053 710 Fireplaces, Firebrick

System Description	Freq. (Years)	Crew	Unit	Labor Hours	2015 Bare Costs				Total In-House	Total w/O&P
					Material	Labor	Equipment	Total		
0020 Replace firebrick										
Masonry demolition, fireplace, brick, 30" x 24"	75	2 BRIC	S.F.	1.155		43.51		43.51	57.50	71
Install new firebrick				.208	8.55	8.70		17.25	21	25
Total				1.363	8.55	52.21		60.76	**78.50**	**96**

For customer support on your Facilities Maintenance & Repair Cost Data, call 877.801.5724.

D4013 110 Backflow Preventer

System Description	Freq. (Years)	Crew	Unit	Labor Hours	2015 Bare Costs				Total In-House	Total w/O&P
					Material	Labor	Equipment	Total		
3010 Rebuild 4" diam. reduced pressure backflow preventer	10	1 PLUM	Ea.							
Notify proper authorities prior to closing valves				.250		14.70		14.70	18.35	23
Close valves at 4" diameter BFP and open by-pass				.167		9.80		9.80	12.25	15.25
Rebuild 4" diameter reduced pressure BFP				2.000	355	117		472	535	625
Test backflow preventer				.500		29.50		29.50	36.50	46
Total				2.917	355	171		526	602.10	709.25
3020 Rebuild 6" diam. reduced pressure backflow preventer	10	1 PLUM	Ea.							
Notify proper authorities prior to closing valves				.250		14.70		14.70	18.35	23
Close valves at 6" diameter BFP and open by-pass				.182		10.65		10.65	13.35	16.65
Rebuild 6" diameter reduced pressure BFP				3.008	375	177		552	635	745
Test backflow preventer				.500		29.50		29.50	36.50	46
Total				3.939	375	231.85		606.85	703.20	830.65
3030 Rebuild 8" diam. reduced pressure backflow preventer	10	1 PLUM	Ea.							
Notify proper authorities prior to closing valves				.250		14.70		14.70	18.35	23
Close valves at 8" diameter BFP and open by-pass				.200		11.75		11.75	14.70	18.30
Rebuild 8" diameter reduced pressure BFP				4.000	465	235		700	805	945
Test backflow preventer				.500		29.50		29.50	36.50	46
Total				4.950	465	290.95		755.95	874.55	1,032.30
3040 Rebuild 10" diam. reduced pressure backflow preventer	10	1 PLUM	Ea.							
Notify proper authorities prior to closing valves				.250		14.70		14.70	18.35	23
Close valves at 10" diameter BFP and open by-pass				.222		13.05		13.05	16.30	20.50
Rebuild 10" diameter reduced pressure BFP				5.000	545	294		839	965	1,150
Test backflow preventer				.500		29.50		29.50	36.50	46
Total				5.972	545	351.25		896.25	1,036.15	1,239.50

D4013 310   Sprinkler System, Fire Supression

System Description	Freq. (Years)	Crew	Unit	Labor Hours	2015 Bare Costs				Total In-House	Total w/O&P
					Material	Labor	Equipment	Total		
1020 Inspect sprinkler system	1	1 PLUM	Ea.							
Inspect water pressure gauges				.044		2.60		2.60	3.20	4
Test fire alarm (wet or dry)				.030		1.80		1.80	2.20	2.70
Motor, exposed, inspect				.333		19.60		19.60	24.50	30.50
Inspect, visual				.060		3.50		3.50	4.40	5.50
Total				.467		27.50		27.50	34.30	42.70
1030 Replace sprinkler head	20	1 PLUM	Ea.							
Remove old sprinkler head				.325		18.25		18.25	23	28.50
Install new head				.650	11.90	36.50		48.40	59	72
Total				.975	11.90	54.75		66.65	82	100.50
1040 Rebuild double check 3″ backflow preventer	1	1 PLUM	Ea.							
Rebuild 3″ backflow preventer				3.910	385	230		615	710	840
Test system				1.300		76.50		76.50	95.50	119
Total				5.210	385	306.50		691.50	805.50	959
1050 Rebuild double check 4″ backflow preventer	1	1 PLUM	Ea.							
Rebuild 4″ backflow preventer				4.561	525	268		793	910	1,075
Test system				1.300		76.50		76.50	95.50	119
Total				5.861	525	344.50		869.50	1,005.50	1,194
1060 Rebuild double check 6″ backflow preventer	1	1 PLUM	Ea.							
Rebuild 6″ backflow preventer				5.202	620	305		925	1,075	1,250
Test system				1.300		76.50		76.50	95.50	119
Total				6.502	620	381.50		1,001.50	1,170.50	1,369
1070 Rebuild reduced pressure backflow preventer	1	1 PLUM	Ea.							
Rebuild 3″ backflow preventer				3.910	385	230		615	710	840
Test system				1.300		76.50		76.50	95.50	119
Total				5.210	385	306.50		691.50	805.50	959

D4013 410   Fire Pump

	System Description	Freq. (Years)	Crew	Unit	Labor Hours	2015 Bare Costs				Total In-House	Total w/O&P
						Material	Labor	Equipment	Total		
1030	**Replace fire pump / electric motor assembly 100 H.P.**	25	Q-13	Ea.							
	Remove 100 H.P. electric-drive fire pump				31.373		1,675		1,675	2,100	2,625
	Install new electric-drive fire pump and controls, 100 H.P.				64.000	22,600	3,425		26,025	29,100	33,600
	Total				95.373	22,600	5,100		27,700	31,200	**36,225**

D5013 110 Primary Transformer, Liquid Filled

System Description	Freq. (Years)	Crew	Unit	Labor Hours	2015 Bare Costs				Total In-House	Total w/O&P
					Material	Labor	Equipment	Total		
0010 Repair 500 kVA transformer	10	1 ELEC	Ea.							
Remove transformer component				1.000		54.50		54.50	67.50	84.50
Transformer component				2.963	2,150	162		2,312	2,575	2,950
Total				3.963	2,150	216.50		2,366.50	2,642.50	3,034.50
0020 Maintenance and inspection	0.50	1 ELEC	Ea.							
Test energized transformer				.500		27.50		27.50	34	42.50
Total				.500		27.50		27.50	34	42.50
0030 Replace transformer	30	R-3	Ea.							
Remove transformer				13.793		750	95	845	1,025	1,275
Transformer 500 kVA, 5 KV primary, 277/480 V secondary				50.000	20,400	2,725	345	23,470	26,200	30,100
Total				63.793	20,400	3,475	440	24,315	27,225	31,375

D5013 120 Primary Transformer, Dry

System Description	Freq. (Years)	Crew	Unit	Labor Hours	2015 Bare Costs				Total In-House	Total w/O&P
					Material	Labor	Equipment	Total		
0010 Repair 15 KV primary transformer	15	1 ELEC	Ea.							
Repair transformer terminal				2.000	48.50	109		157.50	189	230
Total				2.000	48.50	109		157.50	189	230
0020 Maintenance and inspection	0.50	1 ELEC	Ea.							
Test energized transformer				.500		27.50		27.50	34	42.50
Total				.500		27.50		27.50	34	42.50
0030 Replace transformer	30	R-3	Ea.							
Remove transformer				13.793		750	95	845	1,025	1,275
Transformer 500 kVA, 15 KV primary, 277/480 V secondary				57.143	50,000	3,100	395	53,495	59,500	68,000
Total				70.936	50,000	3,850	490	54,340	60,525	69,275

For customer support on your Facilities Maintenance & Repair Cost Data, call 877.801.5724.

D5013 210 Switchgear, Mainframe

System Description	Freq. (Years)	Crew	Unit	Labor Hours	2015 Bare Costs				Total In-House	Total w/O&P
					Material	Labor	Equipment	Total		
0010 Repair switchgear 1200 A	5	1 ELEC	Ea.							
Remove relay				1.702		93		93	115	144
Relay				5.000	1,150	274		1,424	1,600	1,875
Total				6.702	1,150	367		1,517	**1,715**	**2,019**
0020 Maintenance and inspection	1	1 ELEC	Ea.							
Check & retighten switchgear connections				.800		44		44	54	68
Total				.800		44		44	**54**	**68**
0030 Replace switchgear 1200 A	20	3 ELEC	Ea.							
Remove switchgear 1200 A				6.000		330		330	405	510
Switchgear 1200 A				21.818	1,800	1,200		3,000	3,450	4,100
Test switchgear				2.000		109		109	135	169
Total				29.818	1,800	1,639		3,439	**3,990**	**4,779**

D5013 216 Fuses

System Description	Freq. (Years)	Crew	Unit	Labor Hours	2015 Bare Costs				Total In-House	Total w/O&P
					Material	Labor	Equipment	Total		
0010 Replace fuse	25	1 ELEC	Ea.							
Remove fuse				.151		8.25		8.25	10.20	12.80
Fuse 600 A				.400	405	22		427	475	540
Total				.551	405	30.25		435.25	**485.20**	**552.80**

D5013 220 Switchgear, Indoor, Less Than 600 V

System Description	Freq. (Years)	Crew	Unit	Labor Hours	2015 Bare Costs				Total In-House	Total w/O&P
					Material	Labor	Equipment	Total		
0010 Repair switchgear, - (5% of total C.B.)	10	1 ELEC	Ea.							
Remove circuit breaker				.400		22		22	27	34
Circuit breaker				1.143	228	62.50		290.50	330	380
Total				1.543	228	84.50		312.50	**357**	**414**
0020 Maintenance and inspection	3	1 ELEC	Ea.							
Inspect and clean panel				.500		27.50		27.50	34	42.50
Total				.500		27.50		27.50	**34**	**42.50**

D5013 220 Switchgear, Indoor, Less Than 600 V

	System Description	Freq. (Years)	Crew	Unit	Labor Hours	2015 Bare Costs				Total In-House	Total w/O&P
						Material	Labor	Equipment	Total		
0030	**Replace switchgear, 225 A**	30	1 ELEC	Ea.							
	Remove panelboard, 4 wire, 120/208 V, 225 A				4.000		219		219	271	340
	Panelboard, 4 wire, 120/208 V, 225 A main circuit breaker				11.940	2,425	655		3,080	3,475	4,025
	Test indoor switchgear				1.509		82.50		82.50	102	128
	Total				17.450	2,425	956.50		3,381.50	3,848	4,493
0200	**Replace switchgear, 400 A**	30	2 ELEC	Ea.							
	Turn power off, later turn on				.200		10.90		10.90	13.50	16.90
	Remove panelboard, 4 wire, 120/208 V, 100 A				3.333		182		182	226	282
	Remove panelboard, 4 wire, 120/208 V, 400 A				8.333		455		455	565	705
	Panelboard, 4 wire, 120/208 V, 100 A main circuit breaker				17.021	1,175	930		2,105	2,450	2,925
	Panelboard, 4 wire, 120/208 V, 400 A main circuit breaker				33.333	3,250	1,825		5,075	5,825	6,900
	Test circuits				.145		7.96		7.96	9.85	12.30
	Total				62.367	4,425	3,410.86		7,835.86	9,089.35	10,841.20
0240	**Replace switchgear, 600 A**	30	2 ELEC	Ea.							
	Turn power off, later turn on				.300		16.35		16.35	20.50	25.50
	Remove panelboard, 4 wire, 120/208 V, 200 A				6.667		365		365	450	565
	Remove switchboard, incoming section, 600 A				4.848		265		265	330	410
	Remove switchboard, distribution section, 600 A				4.000		219		219	271	340
	Panelboard, 4 wire, 120/208 V, 225 A main circuit breaker				28.571	2,425	1,575		4,000	4,600	5,450
	Switchboard, incoming section, 600 A				16.000	4,225	875		5,100	5,725	6,625
	Switchboard, distribution section, 600 A				16.000	1,625	875		2,500	2,875	3,375
	Test circuits				.218		11.94		11.94	14.75	18.45
	Total				76.605	8,275	4,202.29		12,477.29	14,286.25	16,808.95
0280	**Replace switchgear, 800 A**	30	2 ELEC	Ea.							
	Turn power off, later turn on				.400		21.80		21.80	27	34
	Remove panelboard, 4 wire, 120/208 V, 200 A (2 panelboards)				13.333		730		730	900	1,125
	Remove switchboard, incoming section, 800 A				5.517		300		300	375	465
	Remove switchboard, distribution section, 800 A				4.444		243		243	300	375
	Panelboard, 4 wire, 120/208 V, 225 A main lugs (2 panelboards)				47.059	2,750	2,550		5,300	6,200	7,450
	Switchboard, incoming section, 800 A				18.182	4,225	995		5,220	5,875	6,825
	Switchboard, distribution section, 800 A				18.182	2,100	995		3,095	3,550	4,175
	Test circuits				.291		15.92		15.92	19.70	24.50
	Total				107.408	9,075	5,850.72		14,925.72	17,246.70	20,473.50

For customer support on your Facilities Maintenance & Repair Cost Data, call 877.801.5724.

System Description	Freq. (Years)	Crew	Unit	Labor Hours	2015 Bare Costs				Total In-House	Total w/O&P
					Material	Labor	Equipment	Total		
0320										
Replace switchgear, 1200 A	30	3 ELEC	Ea.							
Turn power off, later turn on				.600		32.70		32.70	40.50	50.50
Remove panelboard, 4 wire, 120/208 V, 200 A (4 panelboards)				26.667		1,460		1,460	1,800	2,250
Remove switchboard, incoming section, 1200 A				6.667		365		365	450	565
Remove switchboard, distribution section, 1200 A				5.161		282		282	350	435
Panelboard, 4 wire, 120/208 V, 225 A main lugs (4 panelboards)				94.118	5,500	5,100		10,600	12,400	14,900
Switchboard, incoming section, 1200 A				22.222	5,075	1,225		6,300	7,075	8,225
Switchboard, distribution section, 1200 A				22.222	3,125	1,225		4,350	4,950	5,775
Test circuits				.436		23.88		23.88	29.50	37
Total				178.093	13,700	9,713.58		23,413.58	27,095	32,237.50
0360										
Replace switchgear, 1600 A	30	3 ELEC	Ea.							
Turn power off, later turn on				.800		43.60		43.60	54	67.50
Remove panelboard, 4 wire, 120/208 V, 200 A (6 panelboards)				40.000		2,190		2,190	2,700	3,400
Remove switchboard, incoming section, 1600 A				7.273		400		400	490	615
Remove switchboard, distribution section, 1600 A				5.517		300		300	375	465
Panelboard, 4 wire, 120/208 V, 225 A main lugs (6 panelboards)				141.176	8,250	7,650		15,900	18,600	22,300
Switchboard, incoming section, 1600 A				24.242	5,075	1,325		6,400	7,225	8,400
Switchboard, distribution section, 1600 A				24.242	3,575	1,325		4,900	5,575	6,525
Test circuits				.582		31.84		31.84	39.50	49
Total				243.833	16,900	13,265.44		30,165.44	35,058.50	41,821.50
0400										
Replace switchgear, 2000 A	30	3 ELEC	Ea.							
Turn power off, later turn on				1.000		54.50		54.50	67.50	84.50
Remove panelboard, 4 wire, 120/208 V, 200 A (8 panelboards)				53.333		2,920		2,920	3,600	4,525
Remove switchboard, incoming section, 2000 A				7.619		415		415	515	645
Remove switchboard, distribution section, 2000 A				5.926		325		325	400	500
Panelboard, 4 wire, 120/208 V, 225 A main lugs (8 panelboards)				188.235	11,000	10,200		21,200	24,800	29,800
Switchboard, incoming section, 2000 A				25.806	5,450	1,400		6,850	7,750	9,000
Switchboard, distribution section, 2000 A				25.806	4,200	1,400		5,600	6,375	7,425
Test circuits				.727		39.80		39.80	49	61.50
Total				308.454	20,650	16,754.30		37,404.30	43,556.50	52,041

For customer support on your Facilities Maintenance & Repair Cost Data, call 877.801.5724.

311

D5013 224 Switchgear, Indoor, 600 V

	System Description	Freq. (Years)	Crew	Unit	Labor Hours	2015 Bare Costs				Total In-House	Total w/O&P
						Material	Labor	Equipment	Total		
0010	**Maintenance and repair, - (5% of total fuses)**	10	1 ELEC	Ea.							
	Remove fuse				.151		8.25		8.25	10.20	12.80
	Fuse				.400	405	22		427	475	540
	Total				.551	405	30.25		435.25	485.20	**552.80**
0020	**Maintenance and inspection**	3	1 ELEC	Ea.							
	Inspect and clean panel				.500		27.50		27.50	34	42.50
	Total				.500		27.50		27.50	**34**	**42.50**
0030	**Replace switchgear**	30	2 ELEC	Ea.							
	Remove indoor switchgear				4.000		219		219	271	340
	Switchgear				14.035	1,350	770		2,120	2,425	2,900
	Test indoor switchgear				2.000		109		109	135	169
	Total				20.035	1,350	1,098		2,448	**2,831**	**3,409**

D5013 230 Meters

	System Description	Freq. (Years)	Crew	Unit	Labor Hours	2015 Bare Costs				Total In-House	Total w/O&P
						Material	Labor	Equipment	Total		
0010	**Repair switchboard meter**	10	1 ELEC	Ea.							
	Repair meter				5.000	1,050	274		1,324	1,500	1,750
	Total				5.000	1,050	274		1,324	**1,500**	**1,750**
0020	**Replace switchboard meter**	20	1 ELEC	Ea.							
	Remove meter				.500		27.50		27.50	34	42.50
	Meter				1.600	5,300	87.50		5,387.50	5,950	6,750
	Total				2.100	5,300	115		5,415	**5,984**	**6,792.50**

D5013 240 Inverter

	System Description	Freq. (Years)	Crew	Unit	Labor Hours	2015 Bare Costs				Total In-House	Total w/O&P
						Material	Labor	Equipment	Total		
0010	**Maintenance and repair**	1	2 ELEC	Ea.							
	Repair inverter component				5.000	203	274		477	560	680
	Total				5.000	203	274		477	**560**	**680**

For customer support on your Facilities Maintenance & Repair Cost Data, call 877.801.5724.

D5013 240 Inverter

	System Description	Freq. (Years)	Crew	Unit	Labor Hours	2015 Bare Costs				Total In-House	Total w/O&P
						Material	Labor	Equipment	Total		
0020	**Maintenance and inspection**	0.25	1 ELEC	Ea.	1.600		87.50		87.50	108	136
	Check connections, controls & relays										
	Total				1.600		87.50		87.50	**108**	**136**
0030	**Replace 1 kVA inverter**	20	2 ELEC	Ea.							
	Remove inverter				.500		27.50		27.50	34	42.50
	Inverter				1.600	2,025	87.50		2,112.50	2,325	2,675
	Total				2.100	2,025	115		2,140	**2,359**	**2,717.50**

D5013 250 Rectifier, Up To 600 V

	System Description	Freq. (Years)	Crew	Unit	Labor Hours	2015 Bare Costs				Total In-House	Total w/O&P
						Material	Labor	Equipment	Total		
0010	**Maintenance and repair**	2	1 ELEC	Ea.	4.000	310	219		529	610	730
	Repair rectifier component										
	Total				4.000	310	219		529	**610**	**730**
0020	**Maintenance and inspection**	0.33	1 ELEC	Ea.	1.600		87.50		87.50	108	136
	Check connections, controls & relays										
	Total				1.600		87.50		87.50	**108**	**136**
0030	**Replace rectifier**	20	2 ELEC	Ea.							
	Remove rectifier				.500		27.50		27.50	34	42.50
	Rectifier				1.600	945	87.50		1,032.50	1,150	1,325
	Total				2.100	945	115		1,060	**1,184**	**1,367.50**

D5013 264 Motor Starter, Up To 600 V

	System Description	Freq. (Years)	Crew	Unit	Labor Hours	2015 Bare Costs				Total In-House	Total w/O&P
						Material	Labor	Equipment	Total		
0010	**Maintenance and repair**	5	1 ELEC	Ea.							
	Remove motor starter coil				.500		27.50		27.50	34	42.50
	Motor starter coil				1.600	34.50	87.50		122	146	179
	Total				2.100	34.50	115		149.50	**180**	**221.50**

For customer support on your Facilities Maintenance & Repair Cost Data, call 877.801.5724.

313

D5013 264 Motor Starter, Up To 600 V

System Description	Freq. (Years)	Crew	Unit	Labor Hours	2015 Bare Costs				Total In-House	Total w/O&P
					Material	Labor	Equipment	Total		
0020 Maintenance and inspection	0.50	1 ELEC	Ea.							
Inspect & clean motor starter				.667		36.50		36.50	45	56.50
Total				.667		36.50		36.50	**45**	**56.50**
0030 Replace starter	18	1 ELEC	Ea.							
Remove motor starter				.899		49		49	61	76
Magnetic motor starter 5 H.P., size 0				3.478	272	190		462	535	635
Cut, form, align & connect wires				.151		8.25		8.25	10.20	12.80
Check operation of motor starter				.018		1		1	1.24	1.55
Total				4.546	272	248.25		520.25	**607.44**	**725.35**

D5013 266 Motor Starter, 600 V

System Description	Freq. (Years)	Crew	Unit	Labor Hours	2015 Bare Costs				Total In-House	Total w/O&P
					Material	Labor	Equipment	Total		
0010 Maintenance and repair	3	1 ELEC	Ea.							
Remove motor starter coil				1.702		93		93	115	144
Motor starter coil				5.000	102	274		376	450	555
Total				6.702	102	367		469	**565**	**699**
0020 Maintenance and inspection	0.25	1 ELEC	Ea.							
Inspect & clean motor starter				.667		36.50		36.50	45	56.50
Total				.667		36.50		36.50	**45**	**56.50**
0030 Replace starter	18	2 ELEC	Ea.							
Remove motor starter				3.333		182		182	226	282
Motor starter FVNR 600 V, 3 P, 100 H.P. motor				13.333	2,475	730		3,205	3,625	4,225
Cut, form, align & connect wires				.170		9.30		9.30	11.50	14.40
Check operation of motor starter				.018		1		1	1.24	1.55
Total				16.855	2,475	922.30		3,397.30	**3,863.74**	**4,522.95**

D5013 272 Secondary Transformer, Liquid Filled

System Description	Freq. (Years)	Crew	Unit	Labor Hours	2015 Bare Costs				Total In-House	Total w/O&P
					Material	Labor	Equipment	Total		
0010 Maintenance and repair	25	1 ELEC	Ea.							
Remove transformer terminal				.348		19.05		19.05	23.50	29.50
Transformer terminal				1.000	12.10	54.50		66.60	81	99.50
Total				1.348	12.10	73.55		85.65	104.50	129
0020 Maintenance and inspection	0.50	1 ELEC	Ea.							
Check connections & fluid				.667		36.50		36.50	45	56.50
Total				.667		36.50		36.50	45	56.50

D5013 274 Secondary Transformer, Dry

System Description	Freq. (Years)	Crew	Unit	Labor Hours	2015 Bare Costs				Total In-House	Total w/O&P
					Material	Labor	Equipment	Total		
0010 Maintenance and repair	10	1 ELEC	Ea.							
Remove transformer terminal				.500		27.50		27.50	34	42.50
Transformer terminal				1.000	36.50	54.50		91	108	130
Total				1.500	36.50	82		118.50	142	172.50
0020 Maintenance and inspection	0.50	1 ELEC	Ea.							
Inspect connections & tighten				1.000		54.50		54.50	67.50	84.50
Total				1.000		54.50		54.50	67.50	84.50
0030 Replace 15 kVA transformer	30	2 ELEC	Ea.							
Remove transformer 15 KVA, incl. supp., wire & conduit terminations				4.360		238		238	295	370
Transformer 15 KVA, 240/480 V primary, 120/208 V secondary				14.545	1,300	795		2,095	2,425	2,850
Total				18.905	1,300	1,033		2,333	2,720	3,220
0040 Replace 112.5 kVA transformer	30	R-3	Ea.							
Remove transformer 112.5 KVA, incl. support, wire, conduit termination				6.897		375	47.50	422.50	515	635
Transformer 112.5 KVA, 240/480 V primary, 120/208 V secondary				22.222	3,400	1,200	153	4,753	5,400	6,300
Total				29.119	3,400	1,575	200.50	5,175.50	5,915	6,935
0050 Replace 500 kVA transformer	30	R-3	Ea.							
Remove transformer 500 KVA, incl. support, wire & conduit terminations				14.286		775	98.50	873.50	1,075	1,300
Transformer 500 KVA, 240/480 V primary, 120/208 V secondary				44.444	12,600	2,425	305	15,330	17,200	19,800
Total				58.730	12,600	3,200	403.50	16,203.50	18,275	21,100

For customer support on your **Facilities Maintenance & Repair Cost Data**, call 877.801.5724.

315

D5013 280 Lighting Panel, Indoor

System Description	Freq. (Years)	Crew	Unit	Labor Hours	2015 Bare Costs				Total In-House	Total w/O&P
					Material	Labor	Equipment	Total		
0020 Maintenance and inspection										
Inspect and clean panels	3	1 ELEC	Ea.	.500		27.50		27.50	34	42.50
Total				.500		27.50		27.50	**34**	**42.50**
0030 Replace load center, 100 A										
Turn power off, later turn on	20	2 ELEC	Ea.	.100		5.45		5.45	6.75	8.45
Remove load center, 3 wire, 120/240 V				3.077		168		168	208	261
Load centers, 3 Wire, 120/240V, 100 A				6.667	335	365		700	820	985
Test circuits				.073		3.98		3.98	4.92	6.15
Total				9.916	335	542.43		877.43	1,039.67	1,260.60

D5023 110 Wireway

	System Description	Freq. (Years)	Crew	Unit	Labor Hours	2015 Bare Costs				Total In-House	Total w/O&P
						Material	Labor	Equipment	Total		
0030	**Replace 8" x 8" wireway**	20	2 ELEC	L.F.							
	Remove and reinstall supply box				.050		2.75		2.75	3.40	4.25
	Remove wireway, fittings & support				.100		5.45		5.45	6.75	8.45
	Wireway 8" x 8", fittings & support				.400	33.50	22		55.50	64	76
	Total				.550	33.50	30.20		63.70	74.15	88.70

D5023 112 Conduit EMT

	System Description	Freq. (Years)	Crew	Unit	Labor Hours	2015 Bare Costs				Total In-House	Total w/O&P
						Material	Labor	Equipment	Total		
0010	**Replace 1" EMT conduit**	50	2 ELEC	M.L.F.							
	Remove & reinstall cover plate				1.290		70.50		70.50	87.50	110
	Remove conduit w/ fittings & hangers				20.300		1,110		1,110	1,375	1,725
	Remove junction box				1.000		54.50		54.50	67.50	84.50
	1" EMT with couplings				38.650	1,260	2,110		3,370	4,000	4,850
	Outlet box square				4.000	24.20	220		244.20	295	370
	Box connector				.889	30.50	48.60		79.10	94	114
	Beam clamp				5.000	205	294		499	595	715
	Conduit hanger				5.926	383	324		707	820	980
	Total				77.055	1,902.70	4,231.60		6,134.30	7,334	8,948.50

D5023 120 Cable, Non-Metallic (NM) Sheathed

	System Description	Freq. (Years)	Crew	Unit	Labor Hours	2015 Bare Costs				Total In-House	Total w/O&P
						Material	Labor	Equipment	Total		
0010	**Replace NM cable**	50	2 ELEC	M.L.F.							
	Shut off power, later turn on				.100		5.45		5.45	6.75	8.45
	Remove & reinstall cover plate				1.290		70.50		70.50	87.50	110
	Remove junction box				1.000		54.50		54.50	67.50	84.50
	Remove NM cable #12, 2 wire				12.720		700		700	860	1,075
	NM cable #12, 2 wire				32.000	365	1,750		2,115	2,575	3,175
	Staples, for Romex or BX cable				4.000	4	218		222	274	345
	NM #12-2 wire cable connectors				.602	14.40	32.90		47.30	56.50	69
	Outlet box square 4" for Romex or BX				4.000	76	220		296	355	435
	Total				55.712	459.40	3,051.35		3,510.75	4,282.25	5,301.95

For customer support on your Facilities Maintenance & Repair Cost Data, call 877.801.5724.

317

D5023 122 Cable, Service

	System Description	Freq. (Years)	Crew	Unit	Labor Hours	2015 Bare Costs Material	Labor	Equipment	Total	Total In-House	Total w/O&P
0010	**Replace service cable**	50	2 ELEC	M.L.F.							
	Shut off power, later turn on				.100		5.45		5.45	6.75	8.45
	Remove & reinstall cover plate				1.290		70.50		70.50	87.50	110
	Remove junction box				1.000		54.50		54.50	67.50	84.50
	Remove wire #6 XLPE				2.000		109.50		109.50	136	170
	600 V, copper type XLPE-USE				12.308	725	675		1,400	1,625	1,950
	Staples, for service cable				5.000	8	274		282	345	435
	SER cable connectors				3.333	29	182.50		211.50	260	315
	Outlet box square 4"				4.000	24.20	220		244.20	295	370
	Total				29.031	786.20	1,591.45		2,377.65	2,822.75	3,442.95

D5023 124 Cable, Armored

	System Description	Freq. (Years)	Crew	Unit	Labor Hours	2015 Bare Costs Material	Labor	Equipment	Total	Total In-House	Total w/O&P
0010	**Replace armored cable**	60	2 ELEC	M.L.F.							
	Shut off power, later turn on				.100		5.45		5.45	6.75	8.45
	Remove & reinstall cover plate				1.290		70.50		70.50	87.50	110
	Remove junction box				1.000		54.50		54.50	67.50	84.50
	Remove armored cable				13.220		720		720	890	1,125
	BX #12, 2 wire				34.783	460	1,900		2,360	2,850	3,525
	Staples, for Romex or BX cable				4.000	4	218		222	274	345
	Cable connector				2.000	9.10	109.50		118.60	146	181
	Outlet box square 4" for Romex or BX				4.000	76	220		296	355	435
	Total				60.393	549.10	3,297.95		3,847.05	4,676.75	5,813.95

D5023 126 | Branch Wiring, With Junction Box

	System Description	Freq. (Years)	Crew	Unit	Labor Hours	2015 Bare Costs				Total In-House	Total w/O&P
						Material	Labor	Equipment	Total		
0010	Replace branch wiring	50	2 ELEC	M.L.F.							
	Shut off power, later turn on				.100		5.45		5.45	6.75	8.45
	Remove & reinstall cover plate				1.290		70.50		70.50	87.50	110
	Remove junction box				1.000		54.50		54.50	67.50	84.50
	Remove #12 wire				1.455		79.50		79.50	98.50	123
	Conductor THW-THHN, copper, stranded, #12				7.273	133	400		533	640	780
	Outlet box square 4"				4.000	24.20	220		244.20	295	370
	Wire connectors				1.000	2.10	54.60		56.70	70	87
	Total				16.117	159.30	884.55		1,043.85	1,265.25	1,562.95

D5023 128 | Branch Wiring, 600 V

	System Description	Freq. (Years)	Crew	Unit	Labor Hours	2015 Bare Costs				Total In-House	Total w/O&P
						Material	Labor	Equipment	Total		
0010	Replace branch wiring	50	2 ELEC	M.L.F.							
	Shut off power, later turn on				.100		5.45		5.45	6.75	8.45
	Remove #12 wire				1.455		79.50		79.50	98.50	123
	Conductor THW-THHN, copper, stranded, #12				7.273	133	400		533	640	780
	Cut, form and align 4 wires				.410		22.50		22.50	28	35
	Splice, solder, insulate 2 wires				2.667		146		146	181	226
	Total				11.904	133	653.45		786.45	954.25	1,172.45

D5023 130 | Circuit Breaker

	System Description	Freq. (Years)	Crew	Unit	Labor Hours	2015 Bare Costs				Total In-House	Total w/O&P
						Material	Labor	Equipment	Total		
0010	Maintenance and repair breaker, molded case, 480 V, 1 pole	20	1 ELEC	Ea.							
	Tighten connections				.889		48.50		48.50	60	75.50
	Total				.889		48.50		48.50	60	75.50
0020	Maintenance and inspection C.B., molded case, 480 V, 1 pole	0.50	1 ELEC	Ea.							
	Inspect, tighten C.B. wiring				.400		22		22	27	34
	Total				.400		22		22	27	34

For customer support on your Facilities Maintenance & Repair Cost Data, call 877.801.5724.

319

D5023 130 | Circuit Breaker

	System Description	Freq. (Years)	Crew	Unit	Labor Hours	2015 Bare Costs				Total In-House	Total w/O&P
						Material	Labor	Equipment	Total		
0030	**Replace C.B. molded case, 480 V, 1 pole**	50	1 ELEC	Ea.							
	Remove circuit breaker				.348		19.05		19.05	23.50	29.50
	Circuit breaker molded case, 480 V, 70 to 100 A, 1 P				1.143	228	62.50		290.50	330	380
	Cut, form, align & connect wires				.151		8.25		8.25	10.20	12.80
	Check operation of breaker				.018		1		1	1.24	1.55
	Total				1.660	228	90.80		318.80	364.94	423.85
1010	**Maintenance and repair breaker, molded case, 480 V, 2 pole**	20	1 ELEC	Ea.							
	Tighten connections				.889		48.50		48.50	60	75.50
	Total				.889		48.50		48.50	**60**	**75.50**
1020	**Maintenance and inspection C.B., molded case, 480 V, 2 pole**	0.50	1 ELEC	Ea.							
	Inspect, tighten C.B. wiring				.400		22		22	27	34
	Total				.400		22		22	**27**	**34**
1030	**Replace C.B. molded case, 480 V, 2 pole**	50	1 ELEC	Ea.							
	Remove circuit breaker				.500		27.50		27.50	34	42.50
	Circuit breaker molded case, 480 V, 70 to 100 A, 2 P				1.600	485	87.50		572.50	640	740
	Cut, form, align & connect wires				.151		8.25		8.25	10.20	12.80
	Check operation of breaker				.018		1		1	1.24	1.55
	Total				2.269	485	124.25		609.25	685.44	796.85
1040	**Maintenance and repair breaker, molded case, 480 V, 3 pole**	20	1 ELEC	Ea.							
	Tighten connections				.889		48.50		48.50	60	75.50
	Total				.889		48.50		48.50	**60**	**75.50**
1050	**Maintenance and inspection C.B., molded case, 480 V, 3 pole**	0.50	1 ELEC	Ea.							
	Inspect & tighten C.B. wiring				.400		22		22	27	34
	Total				.400		22		22	**27**	**34**
1060	**Replace C.B. molded case, 480 V, 3 pole**	50	1 ELEC	Ea.							
	Remove circuit breaker				.500		27.50		27.50	34	42.50
	Circuit breaker molded case, 480 V, 70 to 100 A, 3 P				2.000	570	109		679	760	880
	Cut, form, align & connect wires				.151		8.25		8.25	10.20	12.80
	Check operation of breaker				.018		1		1	1.24	1.55
	Total				2.669	570	145.75		715.75	805.44	936.85

D5023 130 **Circuit Breaker**

	System Description	Freq. (Years)	Crew	Unit	Labor Hours	2015 Bare Costs				Total In-House	Total w/O&P
						Material	Labor	Equipment	Total		
2010	**Repair failed breaker, molded case, 600 V, 2 pole**	10	1 ELEC	Ea.							
	Remove trip switch				.500		27.50		27.50	34	42.50
	Trip switch				1.600	140	87.50		227.50	262	310
	Total				2.100	140	115		255	**296**	**352.50**
2020	**Maintenance and inspection C.B., molded case, 600 V, 2 pole**	0.33	1 ELEC	Ea.							
	Inspect & tighten C.B. wiring				.400		22		22	27	34
	Total				.400		22		22	**27**	**34**
2030	**Replace C.B. molded case, 600 V, 2 pole**	50	1 ELEC	Ea.							
	Remove circuit breaker				.500		27.50		27.50	34	42.50
	Circuit breaker molded case, 600 V, 70 to 100 A, 2 P				1.600	550	87.50		637.50	715	825
	Cut, form, align & connect wires				.170		9.30		9.30	11.50	14.40
	Check operation of breakers				.018		1		1	1.24	1.55
	Total				2.289	550	125.30		675.30	**761.74**	**883.45**
3010	**Repair failed breaker, molded case, 600 V, 3 pole**	10	1 ELEC	Ea.							
	Remove trip switch				.500		27.50		27.50	34	42.50
	Trip switch				1.600	140	87.50		227.50	262	310
	Total				2.100	140	115		255	**296**	**352.50**
3020	**Maintenance and inspection C.B., molded case, 600 V, 3 pole**	0.33	1 ELEC	Ea.							
	Inspect & tighten C.B. wiring				.400		22		22	27	34
	Total				.400		22		22	**27**	**34**
3030	**Replace C.B. molded case, 600 V, 3 pole**	50	1 ELEC	Ea.							
	Remove circuit breaker				.889		48.50		48.50	60	75.50
	Circuit breaker molded case, 600 V, 125 to 400 A, 3 P				3.478	2,475	190		2,665	2,950	3,400
	Cut, form, align & connect wires				.170		9.30		9.30	11.50	14.40
	Check operation of breakers				.018		1		1	1.24	1.55
	Total				4.556	2,475	248.80		2,723.80	**3,022.74**	**3,491.45**
3040	**Maintenance and repair breaker, enclosed, 240 V, 1 pole**	25	1 ELEC	Ea.							
	Tighten connections				.889		48.50		48.50	60	75.50
	Total				.889		48.50		48.50	**60**	**75.50**

D5023 130 | Circuit Breaker

System Description	Freq. (Years)	Crew	Unit	Labor Hours	2015 Bare Costs				Total In-House	Total w/O&P
					Material	Labor	Equipment	Total		
3050 Maintenance and inspection C.B., enclosed, 240 V, 1 pole										
Inspect & tighten C.B. wiring	1	1 ELEC	Ea.	.400		22		22	27	34
Total				.400		22		22	**27**	**34**
3060 Replace C.B. enclosed, 240 V, 1 pole	50	1 ELEC	Ea.							
Remove circuit breaker				.500		27.50		27.50	34	42.50
Circuit breaker enclosed, 240 V, 15 to 60 A, 1 P				2.105	243	115		358	410	480
Cut, form, align & connect wires				.151		8.25		8.25	10.20	12.80
Check operation of breaker				.018		1		1	1.24	1.55
Total				2.775	243	151.75		394.75	455.44	536.85
4010 Maintenance and repair breaker, enclosed, 240 V, 2 pole										
Tighten connections	25	1 ELEC	Ea.	.889		48.50		48.50	60	75.50
Total				.889		48.50		48.50	**60**	**75.50**
4020 Maintenance and inspection C.B., enclosed, 240 V, 2 pole										
Inspect & tighten C.B. wiring	1	1 ELEC	Ea.	.400		22		22	27	34
Total				.400		22		22	**27**	**34**
4030 Replace C.B. enclosed, 240 V, 2 pole	50	1 ELEC	Ea.							
Remove circuit breaker				.500		27.50		27.50	34	42.50
Circuit breaker enclosed, 240 V, 15 to 60 A, 2 P				2.286	325	125		450	510	600
Cut, form, align & connect wires				.151		8.25		8.25	10.20	12.80
Check operation of breaker				.018		1		1	1.24	1.55
Total				2.955	325	161.75		486.75	555.44	656.85
4040 Maintenance and repair breaker, enclosed, 240 V, 3 pole										
Tighten connections	25	1 ELEC	Ea.	.889		48.50		48.50	60	75.50
Total				.889		48.50		48.50	**60**	**75.50**
4050 Maintenance and inspection C.B., enclosed, 240 V, 3 pole										
Inspect & tighten C.B. wiring	1	1 ELEC	Ea.	.400		22		22	27	34
Total				.400		22		22	**27**	**34**

For customer support on your Facilities Maintenance & Repair Cost Data, call 877.801.5724.

D5023 130 | Circuit Breaker

	System Description	Freq. (Years)	Crew	Unit	Labor Hours	2015 Bare Costs				Total In-House	Total w/O&P
						Material	Labor	Equipment	Total		
4060	**Replace C.B. enclosed, 240 V, 3 pole**	50	1 ELEC	Ea.							
	Remove circuit breaker				.500		27.50		27.50	34	42.50
	Circuit breaker enclosed, 240 V, 15 to 60 A, 3 P				2.500	430	137		567	640	750
	Cut, form, align & connect wires				.151		8.25		8.25	10.20	12.80
	Check operation of breaker				.018		1		1	1.24	1.55
	Total				3.169	430	173.75		603.75	685.44	806.85
5010	**Repair failed breaker, enclosed, 600 V, 2 pole**	4	1 ELEC	Ea.							
	Repair circuit breaker component				2.105	400	115		515	585	680
	Total				2.105	400	115		515	585	680
5020	**Maintenance and inspection C.B., enclosed, 600 V, 2 pole**	0.33	1 ELEC	Ea.							
	Inspect & tighten C.B. wiring				.400		22		22	27	34
	Total				.400		22		22	27	34
5030	**Replace C.B. enclosed, 600 V, 2 pole**	50	1 ELEC	Ea.							
	Remove circuit breaker				.800		44		44	54	68
	Circuit breaker enclosed, 600 V, 100 A, 2 P				3.200	550	175		725	820	960
	Cut, form, align & connect wires				.170		9.30		9.30	11.50	14.40
	Check operation of breaker				.018		1		1	1.24	1.55
	Total				4.189	550	229.30		779.30	886.74	1,043.95
6010	**Repair failed breaker, enclosed, 600 V, 3 pole**	4	1 ELEC	Ea.							
	Repair circuit breaker component				2.105	425	115		540	610	710
	Total				2.105	425	115		540	610	710
6020	**Maintenance and inspection C.B., enclosed, 600 V, 3 pole**	0.33	1 ELEC	Ea.							
	Inspect & tighten C.B. wiring				.400		22		22	27	34
	Total				.400		22		22	27	34
6030	**Replace C.B. enclosed, 600 V, 3 pole**	50	1 ELEC	Ea.							
	Remove circuit breaker				.889		48.50		48.50	60	75.50
	Circuit breaker enclosed, 600 V, 100 A, 3 P				3.478	705	190		895	1,000	1,175
	Cut, form, align & connect wires				.170		9.30		9.30	11.50	14.40
	Check operation of breaker				.018		1		1	1.24	1.55
	Total				4.556	705	248.80		953.80	1,072.74	1,266.45

D5023 130 Circuit Breaker

System Description	Freq. (Years)	Crew	Unit	Labor Hours	2015 Bare Costs				Total In-House	Total w/O&P
					Material	Labor	Equipment	Total		
6830 Replace C.B. enclosed, 600 V, 3 pole 800 A	50	2 ELEC	Ea.							
Remove circuit breaker				4.444		243		243	300	375
Circuit breaker enclosed, 600 V, 800 A, 3 P				17.021	5,250	930		6,180	6,925	8,025
Cut, form, align & connect wires				.340		18.60		18.60	23	29
Check operation of breaker				.037		2		2	2.48	3.10
Total				21.843	5,250	1,193.60		6,443.60	7,250.48	8,432.10
6930 Replace C.B. enclosed, 600 V, 3 pole 1000 A	50	2 ELEC	Ea.							
Remove circuit breaker				5.161		282		282	350	435
Circuit breaker enclosed, 600 V, 1000 A, 3 P				19.048	6,625	1,050		7,675	8,575	9,900
Cut, form, align & connect wires				.516		28.18		28.18	35	43.50
Check operation of breaker				.055		3.03		3.03	3.76	4.70
Total				24.780	6,625	1,363.21		7,988.21	8,963.76	10,383.20
7130 Replace C.B. enclosed, 600 V, 3 pole 1600 A	50	2 ELEC	Ea.							
Remove circuit breaker				5.714		315		315	385	485
Circuit breaker enclosed, 600 V, 1600 A, 3 P				22.222	15,600	1,225		16,825	18,700	21,400
Cut, form, align & connect wires				.681		37.20		37.20	46	57.50
Check operation of breaker				.073		4		4	4.96	6.20
Total				28.691	15,600	1,581.20		17,181.20	19,135.96	21,948.70
7230 Replace C.B. enclosed, 600 V, 3 pole 2000 A	50	2 ELEC	Ea.							
Remove circuit breaker				6.400		350		350	435	540
Circuit breaker enclosed, 600 V, 2000 A, 3 P				25.000	16,900	1,375		18,275	20,300	23,300
Cut, form, align & connect wires				1.418		77.50		77.50	96	120
Check operation of breaker				.153		8.33		8.33	10.35	12.90
Total				32.971	16,900	1,810.83		18,710.83	20,841.35	23,972.90

D5023 132 Safety Switch, Heavy Duty

System Description	Freq. (Years)	Crew	Unit	Labor Hours	2015 Bare Costs				Total In-House	Total w/O&P
					Material	Labor	Equipment	Total		
0010 Maintenance and repair safety switch, H.D., 3 pole	8	1 ELEC	Ea.							
Resecure lugs				.500		27.50		27.50	34	42.50
Total				.500		27.50		27.50	34	42.50

For customer support on your Facilities Maintenance & Repair Cost Data, call 877.801.5724.

System Description	Freq. (Years)	Crew	Unit	Labor Hours	2015 Bare Costs				Total In-House	Total w/O&P
					Material	Labor	Equipment	Total		
0020 Maintenance and inspection safety switch, 3 pole										
Inspect & clean safety switch	1	1 ELEC	Ea.	.500		27.50		27.50	34	42.50
Total				.500		27.50		27.50	**34**	**42.50**
0030 Replace safety switch, H.D. 30 A	25	2 ELEC	Ea.							
Remove safety switch				.650		35.50		35.50	44	55
Fused safety switch, heavy duty, 600 V, 3 P, 30 A				2.500	201	137		338	390	465
Fuses				.600	56.25	32.85		89.10	102	121
Check operation of switch				.018		1		1	1.24	1.55
Total				3.769	257.25	206.35		463.60	**537.24**	**642.55**
0130 Replace safety switch, H.D. 100 A	25	2 ELEC	Ea.							
Remove safety switch				1.096		60		60	74	93
Fused safety switch, heavy duty, 600 V, 3 P, 100 A				4.211	445	230		675	775	910
Fuses				.667	199.50	36.45		235.95	264	305
Check operation of switch				.018		1		1	1.24	1.55
Total				5.991	644.50	327.45		971.95	**1,114.24**	**1,309.55**
0230 Replace safety switch, H.D. 200 A	25	2 ELEC	Ea.							
Remove safety switch				1.600		87.50		87.50	108	136
Fused safety switch, heavy duty, 600 V, 3 P, 200 A				6.154	640	335		975	1,125	1,325
Fuses				.800	387	43.80		430.80	480	550
Check operation of switch				.018		1		1	1.24	1.55
Total				8.572	1,027	467.30		1,494.30	**1,714.24**	**2,012.55**
0430 Replace safety switch, H.D. 400 A	25	2 ELEC	Ea.							
Remove safety switch				2.353		129		129	159	199
Fused safety switch, heavy duty, 600 V, 3 P, 400 A				8.889	1,675	485		2,160	2,450	2,850
Fuses				1.000	801	54.75		855.75	945	1,075
Check operation of switch				.037		2		2	2.48	3.10
Total				12.278	2,476	670.75		3,146.75	**3,556.48**	**4,127.10**
0630 Replace safety switch, H.D. 600 A	25	2 ELEC	Ea.							
Remove safety switch				3.478		190		190	235	295
Fused safety switch, heavy duty, 600 V, 3 P, 600 A				13.333	2,800	730		3,530	3,975	4,625
Fuses				1.200	1,065	66		1,131	1,250	1,425
Check operation of switch				.037		2		2	2.48	3.10
Total				18.048	3,865	988		4,853	**5,462.48**	**6,348.10**

For customer support on your Facilities Maintenance & Repair Cost Data, call 877.801.5724.

325

	System Description	Freq. (Years)	Crew	Unit	Labor Hours	2015 Bare Costs Material	Labor	Equipment	Total	Total In-House	Total w/O&P
2010	**Maintenance and repair safety switch gen., 2 pole**	8	1 ELEC	Ea.							
	Resecure lugs				.500		27.50		27.50	34	42.50
	Total				.500		27.50		27.50	**34**	**42.50**
2020	**Maintenance and inspection safety switch, 2 pole**	1	1 ELEC	Ea.							
	Inspect & clean safety switch				.500		27.50		27.50	34	42.50
	Total				.500		27.50		27.50	**34**	**42.50**
2030	**Replace safety switch 240 V, 2 pole**	25	2 ELEC	Ea.							
	Remove safety switch				.650		35.50		35.50	44	55
	Fused safety switch, general duty, 240 V, 2 P, 30 A				2.299	88.50	126		214.50	253	305
	Fuse, dual element, time delay, 250 V, 30 A				.320	18.90	17.50		36.40	42	50.50
	Cut, form, align & connect wires				.151		8.25		8.25	10.20	12.80
	Check operation of switch				.018		1		1	1.24	1.55
	Total				3.439	107.40	188.25		295.65	**350.44**	**424.85**
3010	**Maintenance and repair safety switch gen., 3 pole**	8	1 ELEC	Ea.							
	Resecure lugs				.500		27.50		27.50	34	42.50
	Total				.500		27.50		27.50	**34**	**42.50**
3020	**Maintenance and inspection safety switch, 3 pole**	1	1 ELEC	Ea.							
	Inspect & clean safety switch				.500		27.50		27.50	34	42.50
	Total				.500		27.50		27.50	**34**	**42.50**
3030	**Replace safety switch 240 V, 3 pole**	25	2 ELEC	Ea.							
	Remove safety switch				.650		35.50		35.50	44	55
	Fused safety switch, general duty, 240 V, 3 P, 30 A				2.500	118	137		255	299	360
	Fuse, dual element, time delay, 250 V, 30 A				.480	28.35	26.25		54.60	63	76
	Cut, form, align & connect wires				.151		8.25		8.25	10.20	12.80
	Check operation of switch				.018		1		1	1.24	1.55
	Total				3.800	146.35	208		354.35	**417.44**	**505.35**
4010	**Replace low voltage cartridge**	50	1 ELEC	Ea.							
	Remove cartridge fuse cutouts				.100		5.45		5.45	6.75	8.45
	Cartridge fuse cutouts 3 P				.296	17.25	16.20		33.45	39	46.50
	Total				.396	17.25	21.65		38.90	**45.75**	**54.95**

D5023 140 Safety Switch, General Duty

	System Description	Freq. (Years)	Crew	Unit	Labor Hours	2015 Bare Costs				Total In-House	Total w/O&P
						Material	Labor	Equipment	Total		
5010	Replace plug fuse	25	1 ELEC	Ea.							
	Remove plug fuse				.080		4.38		4.38	5.40	6.80
	Fuse, plug, 120 V, 15 to 30 A				.160	4.84	8.75		13.59	16.15	19.60
	Total				.240	4.84	13.13		17.97	21.55	26.40

D5023 150 Receptacles And Plugs

	System Description	Freq. (Years)	Crew	Unit	Labor Hours	2015 Bare Costs				Total In-House	Total w/O&P
						Material	Labor	Equipment	Total		
0010	Maintenance and repair	20	1 ELEC	Ea.							
	Repair receptacle component (cover)				.500	2.56	27.50		30.06	36.50	45.50
	Total				.500	2.56	27.50		30.06	36.50	45.50
0020	Replace receptacle/plug	20	1 ELEC	Ea.							
	Turn power off, later turn on				.100		5.45		5.45	6.75	8.45
	Remove & install cover plates				.170		9.30		9.30	11.50	14.40
	Remove receptacle				.040		2.19		2.19	2.71	3.39
	Wiring devices, receptacle				.381	7.90	21		28.90	34.50	42.50
	Cut, form, align & connect wires				.026		1.42		1.42	1.76	2.20
	Test for operation				.015		.80		.80	.99	1.24
	Total				.732	7.90	40.16		48.06	58.21	72.18

D5023 154 4-Pin Receptacle

	System Description	Freq. (Years)	Crew	Unit	Labor Hours	2015 Bare Costs				Total In-House	Total w/O&P
						Material	Labor	Equipment	Total		
0010	Repair 4-pin receptacle cover	10	1 ELEC	Ea.							
	Repair receptacle(cover)				.615	5.90	33.50		39.40	48	59.50
	Total				.615	5.90	33.50		39.40	48	59.50
0020	Replace 4 - pin receptacle	20	1 ELEC	Ea.							
	Remove 4-pin receptacle				.200		10.95		10.95	13.55	16.95
	Receptacle locking, 4-pin				.615	54.50	33.50		88	102	120
	Total				.815	54.50	44.45		98.95	115.55	136.95

For customer support on your Facilities Maintenance & Repair Cost Data, call 877.801.5724.

327

D5023 210 Contactors And Relays

System Description	Freq. (Years)	Crew	Unit	Labor Hours	2015 Bare Costs				Total In-House	Total w/O&P
					Material	Labor	Equipment	Total		
0010 Maintenance and repair	3	1 ELEC	Ea.							
Remove contactor coil				.602		33		33	40.50	51
Contactor coil				2.000	32.50	109		141.50	171	210
Total				2.602	32.50	142		174.50	211.50	261
0020 Maintenance and inspection	0.50	1 ELEC	Ea.							
Inspect and clean contacts				.250		13.70		13.70	16.95	21
Total				.250		13.70		13.70	16.95	21
0030 Replace contactor	18	1 ELEC	Ea.							
Remove magnetic contactor				.800		44		44	54	68
Magnetic contactor 600 V, 3 P, 25 H.P.				2.667	435	146		581	660	770
Cut, form, align & connect wires				.151		8.25		8.25	10.20	12.80
Check operation of contactor				.018		1		1	1.24	1.55
Total				3.636	435	199.25		634.25	725.44	852.35

D5023 220 Wiring Devices, Switches

System Description	Freq. (Years)	Crew	Unit	Labor Hours	2015 Bare Costs				Total In-House	Total w/O&P
					Material	Labor	Equipment	Total		
0010 Maintenance and repair	10	1 ELEC	Ea.							
Repair switch component (cover)				.500	2.56	27.50		30.06	36.50	45.50
Total				.500	2.56	27.50		30.06	36.50	45.50
0020 Replace switch	15	1 ELEC	Ea.							
Turn power off, later turn on				.100		5.45		5.45	6.75	8.45
Remove & install cover plates				.170		9.30		9.30	11.50	14.40
Remove switch				.040		2.19		2.19	2.71	3.39
Wiring devices, switch				.381	10.80	21		31.80	37.50	46
Cut, form, align & connect wires				.026		1.42		1.42	1.76	2.20
Test for operation				.015		.80		.80	.99	1.24
Total				.732	10.80	40.16		50.96	61.21	75.68

D5023 222 Switch, Pull Cord

System Description	Freq. (Years)	Crew	Unit	Labor Hours	2015 Bare Costs				Total In-House	Total w/O&P
					Material	Labor	Equipment	Total		
0010 Maintenance and repair										
Repair switch component (cover)	5	1 ELEC	Ea.	.500	2.82	27.50		30.32	37	46
Total				.500	2.82	27.50		30.32	**37**	**46**
0020 Replace switch		1 ELEC	Ea.							
Turn power off, later turn on	15			.100		5.45		5.45	6.75	8.45
Remove & install cover plates				.170		9.30		9.30	11.50	14.40
Remove switch, pull cord				.040		2.19		2.19	2.71	3.39
Switch, pull cord				.381	10.25	21		31.25	37	45.50
Cut, form, align & connect wires				.026		1.42		1.42	1.76	2.20
Test for operation				.015		.80		.80	.99	1.24
Total				.732	10.25	40.16		50.41	**60.71**	**75.18**

D5023 230 Light Dimming Panel

System Description	Freq. (Years)	Crew	Unit	Labor Hours	2015 Bare Costs				Total In-House	Total w/O&P
					Material	Labor	Equipment	Total		
0010 Minor repairs to light dimming panel	5	1 ELEC	Ea.							
Check and test components				.400		22		22	27	34
Remove dimming panel component				.222		12.15		12.15	15.05	18.80
Dimming panel component				.667	89.50	36.50		126	144	168
Total				1.289	89.50	70.65		160.15	**186.05**	**220.80**
0020 Maintenance and inspection	1	1 ELEC	Ea.							
Check light dimming panel				.533		29		29	36	45
Total				.533		29		29	**36**	**45**
0030 Replace light dimming panel	15	1 ELEC	Ea.							
Turn power off, later turn on				.100		5.45		5.45	6.75	8.45
Remove dimming panel				1.600		87.50		87.50	108	136
Dimming panel				8.000	635	440		1,075	1,250	1,475
Test circuits				.073		3.98		3.98	4.92	6.15
Total				9.773	635	536.93		1,171.93	**1,369.67**	**1,625.60**

For customer support on your Facilities Maintenance & Repair Cost Data, call 877.801.5724.

329

D5023 240 | **Incandescent Lighting Fixtures**

	System Description	Freq. (Years)	Crew	Unit	Labor Hours	Material	Labor	Equipment	Total	Total In-House	Total w/O&P
							2015 Bare Costs				
0010	**Maintenance and repair**	10	1 ELEC	Ea.							
	Remove glass globe				.100		5.45		5.45	6.75	8.45
	Glass globe				.296	30	16.20		46.20	53	62.50
	Total				.396	30	21.65		51.65	**59.75**	**70.95**
0020	**Replace lamp**	5	1 ELEC	Ea.							
	Remove incand. lighting fixture lamp				.022		1.20		1.20	1.49	1.86
	Incand. lighting fixture lamp				.065	2.56	3.56		6.12	7.20	8.70
	Total				.087	2.56	4.76		7.32	**8.69**	**10.56**
0030	**Replace lighting fixture**	20	1 ELEC	Ea.							
	Turn circuit off and on for fixture				.018		1		1	1.24	1.55
	Remove incand. lighting fixture				.258		14.10		14.10	17.45	22
	Incand. lighting fixture, 75 W				.800	56	44		100	116	138
	Total				1.076	56	59.10		115.10	**134.69**	**161.55**
0120	**Replace lamp, 200 W**	5	1 ELEC	Ea.							
	Remove incand. lighting fixture lamp				.022		1.22		1.22	1.50	1.88
	Incand. lighting fixture lamp, 200 W				.065	3.63	3.56		7.19	8.40	10.05
	Total				.087	3.63	4.78		8.41	**9.90**	**11.93**
0220	**Replace lamp for explosion proof fixture**	5	1 ELEC	Ea.							
	Remove incand. lighting fixture lamp				.044		2.44		2.44	3	3.76
	Incand. lighting fixture lamp for explosion proof fixture				.130	7.26	7.12		14.38	16.80	20
	Total				.175	7.26	9.56		16.82	**19.80**	**23.76**
0230	**Replace incan. lighting fixt., explosion proof, ceiling mtd., 200 W**	20	1 ELEC	Ea.							
	Turn branch circuit off and on				.018		1		1	1.24	1.55
	Remove incandescent lighting fixture, explosion proof				.920		50.50		50.50	62.50	78
	Incan. lighting fixture, explosion proof, ceiling mtd., 200 W				2.000	1,075	109		1,184	1,325	1,525
	Total				2.938	1,075	160.50		1,235.50	**1,388.74**	**1,604.55**
1230	**Replace incan. lighting fixt., high hat can type, ceiling mtd., 200 W**	20	1 ELEC	Ea.							
	Turn branch circuit off and on				.018		1		1	1.24	1.55
	Remove incandescent lighting fixture, can type				.258		14.10		14.10	17.45	22
	Incan. lighting fixture, high hat can type, ceiling mtd., 200 W				1.194	97	65.50		162.50	188	222
	Total				1.470	97	80.60		177.60	**206.69**	**245.55**

D5023 250 Quartz Fixture

	System Description	Freq. (Years)	Crew	Unit	Labor Hours	2015 Bare Costs				Total In-House	Total w/O&P
						Material	Labor	Equipment	Total		
0010	**Maintenance and repair**	10	1 ELEC	Ea.							
	Repair glass lens				.400	25.50	22		47.50	55	66
	Total				.400	25.50	22		47.50	**55**	**66**
0020	**Replace 1500 W quartz lamp**	10	1 ELEC	Ea.							
	Remove quartz lamp				.087		4.76		4.76	5.90	7.35
	Quartz lamp				.258	33.50	14.10		47.60	54.50	64
	Total				.345	33.50	18.86		52.36	**60.40**	**71.35**
0030	**Replace fixture**	20	1 ELEC	Ea.							
	Turn branch circuit off and on				.018		1		1	1.24	1.55
	Remove exterior fixt., quartz, 1500 W				.593		32.50		32.50	40	50
	Exterior fixt. quartz, 1500 W				1.905	103	104		207	242	290
	Total				2.516	103	137.50		240.50	**283.24**	**341.55**

D5023 260 Fluorescent Lighting Fixture

	System Description	Freq. (Years)	Crew	Unit	Labor Hours	2015 Bare Costs				Total In-House	Total w/O&P
						Material	Labor	Equipment	Total		
0010	**Replace fluor. ballast**	10	1 ELEC	Ea.							
	Remove indoor fluor., ballast				.333		18.25		18.25	22.50	28
	Fluorescent, electronic ballast				.667	37	36.50		73.50	86	103
	Test fixture				.018		1		1	1.24	1.55
	Total				1.018	37	55.75		92.75	**109.74**	**132.55**
0020	**Replace lamps (2 lamps), 4', 34 W energy saver**	10	1 ELEC	Ea.							
	Remove fluor. lamps in fixture				.078		4.26		4.26	5.30	6.60
	Fluor. lamp, 4' energy saver				.232	9.48	12.70		22.18	26	31.50
	Total				.310	9.48	16.96		26.44	**31.30**	**38.10**
0030	**Replace fixture, lay-in, recess mtd., 2' x 4', two 40 W**	20	1 ELEC	Ea.							
	Turn branch circuit off and on				.018		1		1	1.24	1.55
	Remove fluor. lighting fixture				.485		26.50		26.50	33	41
	Fluor. 2' x 4', recess mounted, two 40 W				1.509	50	82.50		132.50	157	191
	Total				2.013	50	110		160	**191.24**	**233.55**

For customer support on your Facilities Maintenance & Repair Cost Data, call 877.801.5724.

331

D5023 260 Fluorescent Lighting Fixture

	System Description	Freq. (Years)	Crew	Unit	Labor Hours	2015 Bare Costs Material	Labor	Equipment	Total	Total In-House	Total w/O&P
0040	**Replace fixture, lay-in, recess mtd., 2' x 4', four 32 W**	20	1 ELEC	Ea.							
	Turn branch circuit off and on				.018		1		1	1.24	1.55
	Remove fluor. lighting fixture				.533		29		29	36	45
	Fluor. lay-in, recess mtd., 2' x 4', four 32 W				1.702	71	93		164	193	233
	Total				2.254	71	123		194	230.24	279.55
0120	**Replace lamps (2 lamps), 8', 60 W energy saver**	10	1 ELEC	Ea.							
	Remove fluor. lamps in fixture				.078		4.26		4.26	5.30	6.60
	Fluor. lamps, 8' energy saver				.200	9	10.90		19.90	23.50	28
	Total				.278	9	15.16		24.16	28.80	34.60
0130	**Replace fixture, strip, surface mtd, 8', two 75 W**	20	2 ELEC	Ea.							
	Turn branch circuit off and on				.018		1		1	1.24	1.55
	Remove fluor. lighting fixture				.400		22		22	27	34
	Fluor. strip, surface mounted, 8', two 75 W				1.290	58	70.50		128.50	151	182
	Total				1.709	58	93.50		151.50	179.24	217.55
0140	**Replace fixture, strip, pendent mtd, 8', two 75 W**	20	2 ELEC	Ea.							
	Turn branch circuit off and on				.018		1		1	1.24	1.55
	Remove fluor. lighting fixture				.593		32.50		32.50	40	50
	Fluor. strip, pendent mtd, 8', two 75 W				1.818	98	99.50		197.50	231	277
	Total				2.429	98	133		231	272.24	328.55

D5023 270 Metal Halide Fixture

	System Description	Freq. (Years)	Crew	Unit	Labor Hours	2015 Bare Costs Material	Labor	Equipment	Total	Total In-House	Total w/O&P
2010	**Replace M.H. ballast, 175 W**	10	1 ELEC	Ea.							
	Remove metal halide fixt. ballast				.333		18.25		18.25	22.50	28
	Metal halide fixt. 175 W, ballast				.667	83.50	36.50		120	137	161
	Test fixture				.018		1		1	1.24	1.55
	Total				1.018	83.50	55.75		139.25	160.74	190.55
2020	**Replace lamp, 175 W**	5	1 ELEC	Ea.							
	Remove metal halide fixt. lamp				.116		6.35		6.35	7.85	9.80
	Metal halide lamp, 175 W				.348	28	19.05		47.05	54.50	64.50
	Total				.464	28	25.40		53.40	62.35	74.30

D5023 270 Metal Halide Fixture

	System Description	Freq. (Years)	Crew	Unit	Labor Hours	2015 Bare Costs Material	Labor	Equipment	Total	Total In-House	Total w/O&P
2030	**Replace fixture, 175 W**	20	1 ELEC	Ea.							
	Turn branch circuit off and on				.018		1		1	1.24	1.55
	Remove metal halide fixt.				.800		44		44	54	68
	Metal halide, recessed, square 175 W				2.353	385	129		514	585	680
	Total				3.171	385	174		559	640.24	749.55
2210	**Replace M.H. ballast, 400 W**	10	1 ELEC	Ea.							
	Remove metal halide fixt. ballast				.333		18.25		18.25	22.50	28
	Metal halide fixt. 400 W, ballast				.762	135	41.50		176.50	200	233
	Test fixture				.018		1		1	1.24	1.55
	Total				1.114	135	60.75		195.75	223.74	262.55
2220	**Replace lamp, 400 W**	5	1 ELEC	Ea.							
	Remove metal halide fixt. lamp				.118		6.45		6.45	7.95	9.95
	Metal halide lamp, 400 W				.348	52.50	19.05		71.55	81.50	95
	Total				.465	52.50	25.50		78	89.45	104.95
2230	**Replace fixture, 400 W**	20	2 ELEC	Ea.							
	Turn branch circuit off and on				.018		1		1	1.24	1.55
	Remove metal halide fixt.				1.067		58.50		58.50	72	90.50
	Metal halide, surface mtd, high bay, 400 W				3.478	410	190		600	685	810
	Total				4.563	410	249.50		659.50	758.24	902.05
2310	**Replace M.H. ballast, 1000 W**	10	1 ELEC	Ea.							
	Remove metal halide fixt. ballast				.333		18.25		18.25	22.50	28
	Metal halide fixt. 1000 W, ballast				.889	231	48.50		279.50	315	365
	Test fixture				.018		1		1	1.24	1.55
	Total				1.241	231	67.75		298.75	338.74	394.55
2320	**Replace lamp, 1000 W**	5	1 ELEC	Ea.							
	Remove metal halide fixt. lamp				.118		6.45		6.45	7.95	9.95
	Metal halide lamp, 1000 W				.523	73	28.50		101.50	116	136
	Total				.641	73	34.95		107.95	123.95	145.95

D5023 270 — Metal Halide Fixture

	System Description	Freq. (Years)	Crew	Unit	Labor Hours	2015 Bare Costs				Total In-House	Total w/O&P
						Material	Labor	Equipment	Total		
2330	**Replace fixture, 1000 W**	20	2 ELEC	Ea.							
	Turn branch circuit off and on				.018		1		1	1.24	1.55
	Remove metal halide fixt.				1.333		73		73	90.50	113
	Metal halide, surface mtd, high bay, 1000 W				4.000	590	219		809	920	1,075
	Total				5.352	590	293		883	1,011.74	1,189.55

D5023 280 — H.P. Sodium Fixture, 250 W

	System Description	Freq. (Years)	Crew	Unit	Labor Hours	2015 Bare Costs				Total In-House	Total w/O&P
						Material	Labor	Equipment	Total		
1010	**Replace H.P. sodium ballast**	10	2 ELEC	Ea.							
	Remove H.P. Sodium fixt. ballast				.333		18.25		18.25	22.50	28
	H.P. Sodium fixt., 250 W, ballast				.667	227	36.50		263.50	295	340
	Test fixture				.018		1		1	1.24	1.55
	Total				1.018	227	55.75		282.75	318.74	369.55
1020	**Replace lamp**	10	2 ELEC	Ea.							
	Remove H.P. Sodium fixt. lamp				.116		6.35		6.35	7.85	9.80
	H.P. Sodium fixt. lamp, 250 W				.348	46.50	19.05		65.55	74.50	87.50
	Total				.464	46.50	25.40		71.90	82.35	97.30
1030	**Replace fixture**	20	2 ELEC	Ea.							
	Turn branch circuit off and on				.018		1		1	1.24	1.55
	Remove H.P. Sodium fixt.				.889		48.50		48.50	60	75.50
	H.P. Sodium fixt., recessed, square 250 W				2.667	680	146		826	930	1,075
	Total				3.574	680	195.50		875.50	991.24	1,152.05

For customer support on your Facilities Maintenance & Repair Cost Data, call 877.801.5724.

D5033 310   Telephone Cable

System Description	Freq. (Years)	Crew	Unit	Labor Hours	2015 Bare Costs Material	Labor	Equipment	Total	Total In-House	Total w/O&P
0010 Repair cable, 22 A.W.G., 4 pair	8	1 ELEC	M.L.F.							
Cut telephone cable				.080		4.38		4.38	5.40	6.80
Splice and insulate telephone cable				.471	8.45	25.50		33.95	41	50.50
Total				.551	8.45	29.88		38.33	**46.40**	**57.30**
0020 Replace telephone cable , #22-4 conductor	50	2 ELEC	M.L.F.							
Turn power off, later turn on				.100		5.45		5.45	6.75	8.45
Remove & reinstall cover plates				1.290		70.50		70.50	87.50	110
Remove telephone cable				2.581		141		141	175	219
Telephone twisted, #22-4 conductor				10.000	159	545		704	850	1,050
Total				13.971	159	761.95		920.95	**1,119.25**	**1,387.45**
1020 Replace telephone jack	20	1 ELEC	Ea.							
Remove telephone jack				.059		3.24		3.24	4.01	5
Telephone jack				.250	4.12	13.70		17.82	21.50	26
Total				.309	4.12	16.94		21.06	**25.51**	**31**

D5033 410   Master Clock Control

System Description	Freq. (Years)	Crew	Unit	Labor Hours	2015 Bare Costs Material	Labor	Equipment	Total	Total In-House	Total w/O&P
0010 Maintenance and repair	10	1 ELEC	Ea.							
Remove time control clock motor				.296		16.20		16.20	20	25
Time control clock motor				1.000	72.50	54.50		127	147	175
Total				1.296	72.50	70.70		143.20	**167**	**200**
0020 Check operation	1	1 ELEC	Ea.							
Check time control clock operation				.533		29		29	36	45
Total				.533		29		29	**36**	**45**
0030 Replace time control clock	15	1 ELEC	Ea.							
Remove time control clock				.296		16.20		16.20	20	25
Time control clock				1.333	130	73		203	233	276
Total				1.630	130	89.20		219.20	**253**	**301**

D5033 410 Master Clock Control

	System Description	Freq. (Years)	Crew	Unit	Labor Hours	2015 Bare Costs				Total In-House	Total w/O&P
						Material	Labor	Equipment	Total		
1120	**Maintenance and inspection**	1	2 ELEC	Ea.							
	Check operation of program bell				.320		17.50		17.50	21.50	27
	Total				.320		17.50		17.50	**21.50**	**27**
1130	**Replace program bell**	15	2 ELEC	Ea.							
	Remove program bell				.667		36.50		36.50	45	56.50
	Program bell				1.000	86.50	54.50		141	163	193
	Total				1.667	86.50	91		177.50	**208**	**249.50**

D5033 510 TV Cable Outlet

	System Description	Freq. (Years)	Crew	Unit	Labor Hours	2015 Bare Costs				Total In-House	Total w/O&P
						Material	Labor	Equipment	Total		
0010	**Maintenance and repair**	10	1 ELEC	Ea.							
	Repair TV cable outlet (cover)				.615	2.82	33.50		36.32	45	55.50
	Total				.615	2.82	33.50		36.32	**45**	**55.50**
0020	**Replace TV cable outlet**	20	1 ELEC	Ea.							
	Remove TV cable outlet				.200		10.95		10.95	13.55	16.95
	TV cable outlet				.615	6.30	33.50		39.80	48.50	60
	Total				.815	6.30	44.45		50.75	**62.05**	**76.95**

D5033 610 Door Bell

	System Description	Freq. (Years)	Crew	Unit	Labor Hours	2015 Bare Costs				Total In-House	Total w/O&P
						Material	Labor	Equipment	Total		
0010	**Maintenance and repair**	10	1 ELEC	Ea.							
	Remove door bell transformer				.200		10.95		10.95	13.55	16.95
	Door bell transformer				.667	18.05	36.50		54.55	65	79
	Total				.867	18.05	47.45		65.50	**78.55**	**95.95**
0020	**Maintenance and inspection**	1	1 ELEC	Ea.							
	Check operation of door bell				.320		17.50		17.50	21.50	27
	Total				.320		17.50		17.50	**21.50**	**27**

For customer support on your Facilities Maintenance & Repair Cost Data, call 877.801.5724.

D5033 610 Door Bell

System Description	Freq. (Years)	Crew	Unit	Labor Hours	2015 Bare Costs				Total In-House	Total w/O&P
					Material	Labor	Equipment	Total		
0030										
Replace door bell	15	1 ELEC	Ea.							
Remove door bell system				.667		36.50		36.50	45	56.50
Door bell system				2.667	129	146		275	320	385
Total				3.333	129	182.50		311.50	365	441.50

D5033 620 Sound System Components

System Description	Freq. (Years)	Crew	Unit	Labor Hours	2015 Bare Costs				Total In-House	Total w/O&P
					Material	Labor	Equipment	Total		
0030										
Replace speaker	20	1 ELEC	Ea.							
Turn power off, later turn on				.100		5.45		5.45	6.75	8.45
Remove speaker				.333		18.25		18.25	22.50	28
Speaker				1.000	119	54.50		173.50	199	233
Test for operation				.015		.80		.80	.99	1.24
Total				1.448	119	79		198	229.24	270.69
0120										
Inspect monitor panel	0.50	1 ELEC	Ea.							
Check monitor panel				.533		29		29	36	45
Total				.533		29		29	36	45
0130										
Replace monitor panel	15	1 ELEC	Ea.							
Turn power off, later turn on				.100		5.45		5.45	6.75	8.45
Remove monitor panel				1.667		91.25		91.25	113	140
Monitor panel				2.000	395	109		504	570	665
Test circuits				.073		3.98		3.98	4.92	6.15
Total				3.839	395	209.68		604.68	694.67	819.60
0220										
Inspect volume control	1	1 ELEC	Ea.							
Check operation of volume control				.320		17.50		17.50	21.50	27
Total				.320		17.50		17.50	21.50	27
0230										
Replace volume control	15	2 ELEC	Ea.							
Remove volume control				.333		18.25		18.25	22.50	28
Volume control				1.000	90	54.50		144.50	167	197
Total				1.333	90	72.75		162.75	189.50	225

D5033 620 Sound System Components

System Description	Freq. (Years)	Crew	Unit	Labor Hours	2015 Bare Costs				Total In-House	Total w/O&P
					Material	Labor	Equipment	Total		
0330 Replace amplifier	15	2 ELEC	Ea.							
Turn power off, later turn on				.100		5.45		5.45	6.75	8.45
Remove amplifier				1.667		91.25		91.25	113	140
Amplifier				8.000	1,275	440		1,715	1,950	2,275
Test for operation				.015		.80		.80	.99	1.24
Total				9.781	1,275	537.50		1,812.50	2,070.74	2,424.69

D5033 630 Intercom System Components

System Description	Freq. (Years)	Crew	Unit	Labor Hours	2015 Bare Costs				Total In-House	Total w/O&P
					Material	Labor	Equipment	Total		
0120 Inspect intercom master station	0.50	1 ELEC	Ea.	2.133						
Check intercom master station						116		116	144	180
Total				2.133		116		116	144	180
0130 Replace intercom master station	15	2 ELEC	Ea.							
Turn power off, later turn on				.100		5.45		5.45	6.75	8.45
Remove intercom master station				1.333		73		73	90.50	113
Intercom master station				8.000	2,350	440		2,790	3,125	3,625
Test circuits				.073		3.98		3.98	4.92	6.15
Total				9.506	2,350	522.43		2,872.43	3,227.17	3,752.60
0220 Inspect intercom remote station	1	1 ELEC	Ea.	.320						
Check operation of intercom remote station						17.50		17.50	21.50	27
Total				.320		17.50		17.50	21.50	27
0230 Replace intercom remote station	15	1 ELEC	Ea.							
Remove intercom remote station				.333		18.25		18.25	22.50	28
Intercom remote station				1.000	166	54.50		220.50	250	292
Total				1.333	166	72.75		238.75	272.50	320

D5033 640 Security System Components

System Description	Freq. (Years)	Crew	Unit	Labor Hours	Material	Labor	Equipment	Total	Total In-House	Total w/O&P
						2015 Bare Costs				
0020 Inspect camera and monitor	0.50	1 ELEC	Ea.							
Check camera and monitor				.808		43.94		43.94	54.50	68
Total				.808		43.94		43.94	**54.50**	**68**
0030 Replace camera and monitor	12	2 ELEC	Ea.							
Turn power off, later turn on				.100		5.45		5.45	6.75	8.45
Remove closed circuit TV, one camera & one monitor				2.051		112		112	139	174
Closed circuit TV, one camera & one monitor				6.154	1,350	335		1,685	1,900	2,200
Test for operation				.015		.80		.80	.99	1.24
Total				8.320	1,350	453.25		1,803.25	**2,046.74**	**2,383.69**
0120 Inspect camera	0.50	1 ELEC	Ea.							
Check camera				.533		29		29	36	45
Total				.533		29		29	**36**	**45**
0130 Replace camera	12	1 ELEC	Ea.							
Turn power off, later turn on				.100		5.45		5.45	6.75	8.45
Remove closed circuit TV, one camera				1.000		54.50		54.50	67.50	84.50
Closed circuit TV, one camera				2.963	755	162		917	1,025	1,200
Test for operation				.015		.80		.80	.99	1.24
Total				4.078	755	222.75		977.75	**1,100.24**	**1,294.19**

D5033 710 Smoke Detector

System Description	Freq. (Years)	Crew	Unit	Labor Hours	Material	Labor	Equipment	Total	Total In-House	Total w/O&P
						2015 Bare Costs				
0010 Repair smoke detector	10	1 ELEC	Ea.							
Remove smoke detector lamp				.151		8.25		8.25	10.20	12.80
Smoke detector lamp				.444	4.97	24.50		29.47	35.50	43.50
Total				.595	4.97	32.75		37.72	**45.70**	**56.30**
0020 Check operation	1	1 ELEC	Ea.							
Check smoke detector operation				.200		10.95		10.95	13.55	16.95
Total				.200		10.95		10.95	**13.55**	**16.95**

For customer support on your Facilities Maintenance & Repair Cost Data, call 877.801.5724.

339

D5033 710 Smoke Detector

System Description	Freq. (Years)	Crew	Unit	Labor Hours	2015 Bare Costs				Total In-House	Total w/O&P
					Material	Labor	Equipment	Total		
0030 Replace smoke detector	15	1 ELEC	Ea.							
Turn power off, later turn on				.100		5.45		5.45	6.75	8.45
Remove smoke detector				.348		19.05		19.05	23.50	29.50
Smoke detector, ceiling type				1.290	110	70.50		180.50	208	247
Test for operation				.015		.80		.80	.99	1.24
Total				1.753	110	95.80		205.80	239.24	286.19

D5033 712 Heat Detector

System Description	Freq. (Years)	Crew	Unit	Labor Hours	2015 Bare Costs				Total In-House	Total w/O&P
					Material	Labor	Equipment	Total		
0010 Repair heat detector	10	1 ELEC	Ea.							
Remove heat detector element				.151		8.25		8.25	10.20	12.80
Heat detector element				.444	9.10	24.50		33.60	40	49
Total				.595	9.10	32.75		41.85	50.20	61.80
0020 Check operation	1	1 ELEC	Ea.							
Check heat detector operation				.200		10.95		10.95	13.55	16.95
Total				.200		10.95		10.95	13.55	16.95
0030 Replace heat detector	15	1 ELEC	Ea.							
Turn power off, later turn on				.100		5.45		5.45	6.75	8.45
Remove heat detector				.348		19.05		19.05	23.50	29.50
Heat detector				1.000	51	54.50		105.50	124	148
Test for operation				.015		.80		.80	.99	1.24
Total				1.462	51	79.80		130.80	155.24	187.19

D5033 720 Manual Pull Station

System Description	Freq. (Years)	Crew	Unit	Labor Hours	2015 Bare Costs				Total In-House	Total w/O&P
					Material	Labor	Equipment	Total		
0010 Check and repair manual pull station	10	1 ELEC	Ea.							
Remove pull station component				.200		10.95		10.95	13.55	16.95
Manual pull station component				.667	17.55	36.50		54.05	64.50	78.50
Total				.867	17.55	47.45		65	78.05	95.45

D5033 720 Manual Pull Station

	System Description	Freq. (Years)	Crew	Unit	Labor Hours	Material	Labor	Equipment	Total	Total In-House	Total w/O&P
								2015 Bare Costs			
0020	**Replace manual pull station**	15	1 ELEC	Ea.							
	Turn power off, later turn on				.100		5.45		5.45	6.75	8.45
	Remove manual pull station				.348		19.05		19.05	23.50	29.50
	Manual pull station				1.000	83.50	54.50		138	160	189
	Test for operation				.015		.80		.80	.99	1.24
	Total				1.462	83.50	79.80		163.30	191.24	228.19

D5033 760 Fire Alarm Control Panel

	System Description	Freq. (Years)	Crew	Unit	Labor Hours	Material	Labor	Equipment	Total	Total In-House	Total w/O&P
								2015 Bare Costs			
0010	**Minor repairs to fire alarm**	5	1 ELEC	Ea.							
	Check and test components				.400		22		22	27	34
	Remove relay				.222		12.15		12.15	15.05	18.80
	Relay				.667	41.50	36.50		78	91	108
	Total				1.289	41.50	70.65		112.15	133.05	160.80
0020	**Maintenance and inspection**	0.50	1 ELEC	Ea.							
	Check control panel				.533		29		29	36	45
	Total				.533		29		29	36	45
0030	**Replace fire alarm panel**	15	1 ELEC	Ea.							
	Turn power off, later turn on				.100		5.45		5.45	6.75	8.45
	Remove fire alarm control panel				2.000		109		109	135	169
	Fire alarm control panel				16.000	780	875		1,655	1,950	2,325
	Test circuits				.073		3.98		3.98	4.92	6.15
	Total				18.173	780	993.43		1,773.43	2,096.67	2,508.60

For customer support on your **Facilities Maintenance & Repair Cost Data**, call 877.801.5724.

341

D5033 766   Annunciation Panel

	System Description	Freq. (Years)	Crew	Unit	Labor Hours	2015 Bare Costs				Total In-House	Total w/O&P
						Material	Labor	Equipment	Total		
0010	**Minor repairs to annunciation panel**	5	1 ELEC	Ea.							
	Check and test components				.400		22		22	27	34
	Remove relay				.222		12.15		12.15	15.05	18.80
	Relay				.667	41.50	36.50		78	91	108
	Total				1.289	41.50	70.65		112.15	133.05	**160.80**
0020	**Maintenance and inspection**	0.50	1 ELEC	Ea.							
	Check annunciator panel				.533		29		29	36	45
	Total				.533		29		29	36	**45**
0030	**Replace annunciation panel**	15	1 ELEC	Ea.							
	Turn power off, later turn on				.100		5.45		5.45	6.75	8.45
	Remove annunciation panel				1.600		87.50		87.50	108	136
	Annunciation panel				6.154	335	335		670	785	940
	Test circuits				.073		3.98		3.98	4.92	6.15
	Total				7.927	335	431.93		766.93	904.67	**1,090.60**

D5033 770   Fire Alarm Bell

	System Description	Freq. (Years)	Crew	Unit	Labor Hours	2015 Bare Costs				Total In-House	Total w/O&P
						Material	Labor	Equipment	Total		
0010	**Replace fire alarm bell, 6"**	20	1 ELEC	Ea.							
	Turn power off, later turn on				.100		5.45		5.45	6.75	8.45
	Remove fire alarm bell				.348		19.05		19.05	23.50	29.50
	Fire alarm bell				1.000	78.50	54.50		133	154	183
	Test for operation				.015		.80		.80	.99	1.24
	Total				1.462	78.50	79.80		158.30	185.24	**222.19**

D5093 110 Electrical Service Ground

	System Description	Freq. (Years)	Crew	Unit	Labor Hours	2015 Bare Costs				Total In-House	Total w/O&P
						Material	Labor	Equipment	Total		
0010	**Maintenance and repair**										
	Repair service ground (1 termination)	25	1 ELEC	M.L.F.	1.053	5.25	57.50		62.75	77	95.50
	Total				1.053	5.25	57.50		62.75	**77**	**95.50**
0020	**Replace electrical service ground**	50	1 ELEC	M.L.F.							
	Remove ground wire #6				2.000		109.50		109.50	136	170
	Excavate hole for ground rod				2.222		121.50		121.50	151	188
	Ground rod 5/8", 10'				17.391	240	950		1,190	1,450	1,775
	Ground clamp				2.500	52.50	137		189.50	225	276
	Backfill over top of ground rod				1.053		57.50		57.50	71.50	89
	Insulated copper wire #6				12.308	575	675		1,250	1,475	1,750
	Total				37.474	867.50	2,050.50		2,918	**3,508.50**	**4,248**

D5093 120 Building Structure Ground

	System Description	Freq. (Years)	Crew	Unit	Labor Hours	2015 Bare Costs				Total In-House	Total w/O&P
						Material	Labor	Equipment	Total		
0010	**Maintenance and repair**										
	Repair bldg structure ground (1 termination)	7	1 ELEC	M.L.F.	1.053	5.25	57.50		62.75	77	95.50
	Total				1.053	5.25	57.50		62.75	**77**	**95.50**
0020	**Replace building structure ground**	50	1 ELEC	M.L.F.							
	Remove ground wire #4				3.077		168.50		168.50	210	260
	Excavate hole for ground rod				2.222		121.50		121.50	151	188
	Ground rod 5/8", 10'				17.391	240	950		1,190	1,450	1,775
	Ground clamp				2.500	52.50	137		189.50	225	276
	Backfill over top of ground rod				1.053		57.50		57.50	71.50	89
	Insulated copper wire #4				15.094	875	825		1,700	1,975	2,375
	Total				41.337	1,167.50	2,259.50		3,427	**4,082.50**	**4,963**

D5093 130 Lightning Protection System

	System Description	Freq. (Years)	Crew	Unit	Labor Hours	2015 Bare Costs				Total In-House	Total w/O&P
						Material	Labor	Equipment	Total		
1010	**Maintenance and repair of general wiring**	1	1 ELEC	M.L.F.							
	Repair lightning protection (1 termination)				1.053	11.25	57.50		68.75	83.50	103
	Total				1.053	11.25	57.50		68.75	**83.50**	**103**
1020	**Replace lightning protection general wiring**	25	2 ELEC	M.L.F.							
	Remove lightning protection cable				4.000		219		219	271	340
	Excavate hole for ground rod				2.222		121.50		121.50	151	188
	Ground rod 5/8", 10'				17.391	240	950		1,190	1,450	1,775
	Ground clamp				2.500	52.50	137		189.50	225	276
	Backfill over top of ground rod				1.053		57.50		57.50	71.50	89
	Lightning protection cable				25.000	3,030	1,370		4,400	5,000	5,900
	Air terminals, copper 3/8" x 10"				10.000	220	545		765	920	1,125
	Total				62.166	3,542.50	3,400		6,942.50	**8,088.50**	**9,693**

D5093 140 Lightning Ground Rod

	System Description	Freq. (Years)	Crew	Unit	Labor Hours	2015 Bare Costs				Total In-House	Total w/O&P
						Material	Labor	Equipment	Total		
1010	**Maintenance and repair**	1	1 ELEC	Ea.							
	Repair ground rod connector				1.053	5.25	57.50		62.75	77	95.50
	Total				1.053	5.25	57.50		62.75	**77**	**95.50**
1020	**Replace lightning ground rod**	25	2 ELEC	Ea.							
	Excavate hole for ground rod				.333		18.25		18.25	22.50	28
	Ground rod 5/8", 10'				1.739	24	95		119	144	177
	Ground clamp				.250	5.25	13.70		18.95	22.50	27.50
	Backfill over top of ground rod				.157		8.60		8.60	10.60	13.30
	Total				2.479	29.25	135.55		164.80	**199.60**	**245.80**

D5093 150 Computer Ground System

	System Description	Freq. (Years)	Crew	Unit	Labor Hours	2015 Bare Costs				Total In-House	Total w/O&P
						Material	Labor	Equipment	Total		
0010	**Maintenance and repair**	4	1 ELEC	Ea.							
	Test bonding each connection				.267		14.60		14.60	18.05	22.50
	Total				.267		14.60		14.60	**18.05**	**22.50**

D5093 150 Computer Ground System

	System Description	Freq. (Years)	Crew	Unit	Labor Hours	2015 Bare Costs				Total In-House	Total w/O&P
						Material	Labor	Equipment	Total		
0020	**Replace computer ground system**	50	1 ELEC	M.L.F.							
	Remove ground wire #6				2.000		109.50		109.50	136	170
	Insulated ground wire #6				12.308	575	675		1,250	1,475	1,750
	Grounding connection				2.500	52.50	137		189.50	225	276
	Total				16.808	627.50	921.50		1,549	1,836	2,196

D5093 190 Special Ground System

	System Description	Freq. (Years)	Crew	Unit	Labor Hours	2015 Bare Costs				Total In-House	Total w/O&P
						Material	Labor	Equipment	Total		
0010	**Maintenance and repair**	4	1 ELEC	Ea.							
	Test bonding each connection				.267		14.60		14.60	18.05	22.50
	Total				.267		14.60		14.60	18.05	22.50
0020	**Replace special ground system**	50	2 ELEC	M.L.F.							
	Remove ground wire #6				2.000		109.50		109.50	136	170
	Insulated ground wire #6				12.308	575	675		1,250	1,475	1,750
	Grounding connection				2.500	52.50	137		189.50	225	276
	Total				16.808	627.50	921.50		1,549	1,836	2,196

D5093 210 Generator, Gasoline, 175 KW

	System Description	Freq. (Years)	Crew	Unit	Labor Hours	2015 Bare Costs				Total In-House	Total w/O&P
						Material	Labor	Equipment	Total		
0010	**Maintenance and inspection**	0.08	1 ELEC	Ea.							
	Run test gasoline generator				.800		44		44	54	68
	Total				.800		44		44	54	68
0020	**Replace generator component**	25	2 ELEC	Ea.							
	Remove gasoline generator				36.364		2,000		2,000	2,450	3,075
	Gasoline generator, 175 KW				106.000	71,000	5,825		76,825	85,500	98,000
	Total				142.364	71,000	7,825		78,825	87,950	101,075

For customer support on your Facilities Maintenance & Repair Cost Data, call 877.801.5724.

345

D5093 220 Generator, Diesel, 750 KW

System Description	Freq. (Years)	Crew	Unit	Labor Hours	2015 Bare Costs				Total In-House	Total w/O&P
					Material	Labor	Equipment	Total		
0010 Maintenance and inspection										
Run test diesel generator	0.08	1 ELEC	Ea.	.800		44		44	54	68
Total				.800		44		44	54	68
0020 Replace diesel generator component										
Remove diesel generator	25	4 ELEC	Ea.	57.143		3,125		3,125	3,875	4,850
Diesel generator, 750 KW				160.000	140,000	8,750		148,750	165,000	188,500
Total				217.143	140,000	11,875		151,875	168,875	193,350

D5093 230 Transfer Switch

System Description	Freq. (Years)	Crew	Unit	Labor Hours	2015 Bare Costs				Total In-House	Total w/O&P
					Material	Labor	Equipment	Total		
0010 Maintenance and repair										
Remove transfer switch coil	5	1 ELEC	Ea.	.667		36.50		36.50	45	56.50
Transfer switch coil				2.000	110	109		219	256	305
Total				2.667	110	145.50		255.50	301	361.50
0020 Maintenance and inspection										
Inspect and clean transfer switch	0.50	1 ELEC	Ea.	.500		27.50		27.50	34	42.50
Total				.500		27.50		27.50	34	42.50
0030 Replace transfer switch										
Remove transfer switch	18	1 ELEC	Ea.	.667		36.50		36.50	45	56.50
Transfer switch, 480 V, 3 P, 1200 A				22.857	18,600	1,250		19,850	22,000	25,200
Cut, form, align & connect wires				.151		8.25		8.25	10.20	12.80
Check operation of switch				.018		1		1	1.24	1.55
Total				23.693	18,600	1,295.75		19,895.75	22,056.44	25,270.85

D5093 240 Emergency Lighting Fixture

	System Description	Freq. (Years)	Crew	Unit	Labor Hours	2015 Bare Costs Material	Labor	Equipment	Total	Total In-House	Total w/O&P
0020	**Replace lamp**	2	1 ELEC	Ea.							
	Remove emergency remote lamp				.100		5.45		5.45	6.75	8.45
	Emergency remote lamp				.296	29	16.20		45.20	52	61.50
	Total				.396	29	21.65		50.65	58.75	69.95
0030	**Replace fixture**	20	1 ELEC	Ea.							
	Turn branch circuit off and on				.018		1		1	1.24	1.55
	Remove emergency light units				.667		36.50		36.50	45	56.50
	Emergency light units, battery operated				2.000	560	109		669	750	870
	Total				2.685	560	146.50		706.50	796.24	928.05

D5093 250 Exit Light

	System Description	Freq. (Years)	Crew	Unit	Labor Hours	2015 Bare Costs Material	Labor	Equipment	Total	Total In-House	Total w/O&P
0010	**Maintenance and repair**	20	1 ELEC	Ea.							
	Remove exit light face plate				.100		5.45		5.45	6.75	8.45
	Exit light face plate				.296	5	16.20		21.20	25.50	31.50
	Total				.396	5	21.65		26.65	32.25	39.95
0020	**Replace lamp**	5	1 ELEC	Ea.							
	Remove exit light lamp				.022		1.20		1.20	1.49	1.86
	Exit light lamp				.065	5.40	3.56		8.96	10.35	12.25
	Total				.087	5.40	4.76		10.16	11.84	14.11
0030	**Replace lighting fixture**	20	1 ELEC	Ea.							
	Turn circuit off and on for fixture				.018		1		1	1.24	1.55
	Remove exit light				.333		18.25		18.25	22.50	28
	Exit light				1.000	39	54.50		93.50	111	133
	Total				1.352	39	73.75		112.75	134.74	162.55

For customer support on your Facilities Maintenance & Repair Cost Data, call 877.801.5724.

347

D5093 255 Exit Light L.E.D.

System Description	Freq. (Years)	Crew	Unit	Labor Hours	2015 Bare Costs Material	Labor	Equipment	Total	Total In-House	Total w/O&P
0020 Replace lamp with exit light L.E.D. retrofit kits	15	1 ELEC	Ea.							
Remove exit light lamp				.022		1.20		1.20	1.49	1.86
Exit light L.E.D. retrofit kits				.133	51	7.30		58.30	65	75
Total				.155	51	8.50		59.50	66.49	76.86
0030 Replace lighting fixture with exit light L.E.D. std.	20	1 ELEC	Ea.							
Turn circuit off and on for fixture				.018		1		1	1.24	1.55
Remove exit light				.333		18.25		18.25	22.50	28
Exit light, L.E.D. standard, double face				1.194	77	65.50		142.50	166	197
Total				1.546	77	84.75		161.75	189.74	226.55
0040 Replace lighting fixture with exit light L.E.D. w/battery unit	20	1 ELEC	Ea.							
Turn circuit off and on for fixture				.018		1		1	1.24	1.55
Remove exit light				.333		18.25		18.25	22.50	28
Exit light, L.E.D. with battery unit, double face				2.000	163	109		272	315	375
Total				2.352	163	128.25		291.25	338.74	404.55

D5093 260 Battery, Wet

System Description	Freq. (Years)	Crew	Unit	Labor Hours	2015 Bare Costs Material	Labor	Equipment	Total	Total In-House	Total w/O&P
0010 Maintenance and inspection	0.02	1 ELEC	Ea.							
Clean & maintain wet cell battery				.500		27.50		27.50	34	42.50
Total				.500		27.50		27.50	34	42.50
0020 Replace battery	10	1 ELEC	Ea.							
Remove wet cell battery				.100		5.45		5.45	6.75	8.45
Wet cell battery				.296	620	16.20		636.20	700	800
Total				.396	620	21.65		641.65	706.75	808.45

D5093 265 Battery, Dry

	System Description	Freq. (Years)	Crew	Unit	Labor Hours	2015 Bare Costs				Total In-House	Total w/O&P
						Material	Labor	Equipment	Total		
0010	**Maintenance and inspection** Clean & maintain dry cell battery	0.08	1 ELEC	Ea.	.500		27.50		27.50	34	42.50
	Total				.500		27.50		27.50	**34**	**42.50**
0020	**Replace battery** Remove dry cell battery	5	1 ELEC	Ea.	.100		5.45		5.45	6.75	8.45
	Dry cell battery				.296	175	16.20		191.20	213	244
	Total				.396	175	21.65		196.65	**219.75**	**252.45**

D5093 270 Battery Charger

	System Description	Freq. (Years)	Crew	Unit	Labor Hours	2015 Bare Costs				Total In-House	Total w/O&P
						Material	Labor	Equipment	Total		
0010	**Maintenance and repair** Repair battery charger component	2	1 ELEC	Ea.	3.077	44.50	168		212.50	257	315
	Total				3.077	44.50	168		212.50	**257**	**315**
0020	**Maintenance and inspection** Check battery charger	0.25	1 ELEC	Ea.	1.600		87.50		87.50	108	136
	Total				1.600		87.50		87.50	**108**	**136**
0030	**Replace charger** Remove battery charger	20	1 ELEC	Ea.	.615		33.50		33.50	41.50	52
	Battery charger				1.600	875	87.50		962.50	1,075	1,225
	Total				2.215	875	121		996	**1,116.50**	**1,277**

D5093 280 UPS Battery

	System Description	Freq. (Years)	Crew	Unit	Labor Hours	2015 Bare Costs				Total In-House	Total w/O&P
						Material	Labor	Equipment	Total		
0020	**Maintenance and inspection** UPS battery electrical test	0.17	1 ELEC	Ea.	.800		44		44	54	68
	Total				.800		44		44	**54**	**68**

D5093 280   UPS Battery

	System Description	Freq. (Years)	Crew	Unit	Labor Hours	2015 Bare Costs				Total In-House	Total w/O&P
						Material	Labor	Equipment	Total		
0030	**Replace motor generator UPS battery**	15	1 ELEC	Ea.							
	Remove UPS battery				1.333		73		73	90.50	113
	UPS battery				4.000	495	219		714	815	960
	Total				5.333	495	292		787	**905.50**	**1,073**

D5093 920   Communications Components

	System Description	Freq. (Years)	Crew	Unit	Labor Hours	2015 Bare Costs				Total In-House	Total w/O&P
						Material	Labor	Equipment	Total		
0020	**Maintenance and repair voice/data outlet**	10	1 ELEC	Ea.							
	Repair voice/data outlet				.615	2.82	33.50		36.32	45	55.50
	Total				.615	2.82	33.50		36.32	**45**	**55.50**
0030	**Replace voice/data outlet**	20	1 ELEC	Ea.							
	Remove voice/data outlet				.057		3.13		3.13	3.87	4.84
	Voice/data outlet, single opening, excludes voice/data device				.167	7.85	9.10		16.95	19.90	24
	Total				.224	7.85	12.23		20.08	**23.77**	**28.84**
0120	**Maintenance and inspection patch panel**	0.50	1 ELEC	Ea.							
	Check patch panel				1.067		58		58	72	90
	Total				1.067		58		58	**72**	**90**
0130	**Replace patch panel**	15	1 ELEC	Ea.							
	Remove patch panel				1.010		55.68		55.68	68	85
	Patch panel				4.000	320	219		539	625	740
	Test circuits				.996		54.52		54.52	67.50	84.50
	Total				6.006	320	329.20		649.20	**760.50**	**909.50**

For customer support on your Facilities Maintenance & Repair Cost Data, call 877.801.5724.

E1023 710   Laboratory Equipment

	System Description	Freq. (Years)	Crew	Unit	Labor Hours	2015 Bare Costs Material	Labor	Equipment	Total	Total In-House	Total w/O&P
0020	**Replace glove box gloves**	5	1 SKWK	Ea.							
	Laboratory equipment, remove bacteriological glove box gloves				.500		24.50		24.50	32	39.50
	Lab equipment, replace fiberglass, bacteriological glove box gloves				1.000	191	48.50		239.50	274	320
	Laboratory equipment, test replacement gloves				1.000		48.50		48.50	64	79
	Total				2.500	191	121.50		312.50	370	438.50
0080	**Replace fume hood sash**	20	2 SKWK	Ea.							
	Laboratory equipment, remove fume hood sash				1.000		48.50		48.50	64	79
	Laboratory equipment, replace fume hood sash				2.000	810	97.50		907.50	1,025	1,175
	Total				3.000	810	146		956	1,089	1,254

For customer support on your Facilities Maintenance & Repair Cost Data, call 877.801.5724.

351

E1033 110 Automotive Equipment

	System Description	Freq. (Years)	Crew	Unit	Labor Hours	2015 Bare Costs				Total In-House	Total w/O&P
						Material	Labor	Equipment	Total		
0020	**Remove and replace vehicle lift hydraulic pump**	15	1 SKWK	Ea.							
	Automotive hoist, two posts, adj. frame, remove and replace pump				4.000	3,050	195		3,245	3,600	4,125
	Total				4.000	3,050	195		3,245	**3,600**	**4,125**
0100	**Remove and replace compressor, electric, 5 H.P.**	10	1 ELEC	Ea.							
	Automotive compressors, remove and replace 5 H.P. motor				2.000	895	109		1,004	1,125	1,300
	Total				2.000	895	109		1,004	**1,125**	**1,300**

For customer support on your Facilities Maintenance & Repair Cost Data, call 877.801.5724.

E1093 310   Loading Dock Equipment

System Description	Freq. (Years)	Crew	Unit	Labor Hours	2015 Bare Costs				Total In-House	Total w/O&P
					Material	Labor	Equipment	Total		
0020 Remove and replace hydraulic dock leveler lift cylinder										
Loading dock, dock levelers, 10 ton cap., replace lift cylinder	15	2 SKWK	Ea.	4.000	5,975	195		6,170	6,825	7,775
Total				4.000	5,975	195		6,170	6,825	7,775
0040 Remove and replace hydraulic dock leveler hydraulic pump										
Loading dock, dock levelers, 10 ton cap, replace lift cylinder	20	1 SKWK	Ea.	3.077	1,450	150		1,600	1,800	2,050
Total				3.077	1,450	150		1,600	1,800	2,050

E1093 315   Dishwasher

System Description	Freq. (Years)	Crew	Unit	Labor Hours	2015 Bare Costs				Total In-House	Total w/O&P
					Material	Labor	Equipment	Total		
0020 Replace commercial dishwasher, 10 to 12 racks per hour										
Cleaning and disposal, dishwasher, commercial, rack type	10	2 PLUM	Ea.	12.000		690		690	860	1,075
Install new dishwasher				6.499	3,525	345		3,870	4,300	4,950
Total				18.499	3,525	1,035		4,560	5,160	6,025
0040 Remove and replace dishwasher pump										
Laboratory equip., glassware washer, remove and replace pump	15	1 PLUM	Ea.	1.778	730	104		834	935	1,075
Total				1.778	730	104		834	935	1,075
0080 Replace commercial dishwasher, to 375 racks per hour										
Commercial dishwasher, to 275 racks per hour, selective demo	20	L-4	Ea.	24.000		1,050		1,050	1,400	1,725
Dishwasher, automatic, 235 to 275 racks/hr.				60.000	31,500	3,450		34,950	38,900	44,700
Total				84.000	31,500	4,500		36,000	40,300	46,425

E1093 316   Waste Disposal, Residential

System Description	Freq. (Years)	Crew	Unit	Labor Hours	2015 Bare Costs				Total In-House	Total w/O&P
					Material	Labor	Equipment	Total		
0010 Unstop disposal										
Unstop clogged unit	1	1 PLUM	Ea.	.889		52		52	65	81.50
Total				.889		52		52	65	81.50

E1093 316 Waste Disposal, Residential

System Description	Freq. (Years)	Crew	Unit	Labor Hours	2015 Bare Costs				Total In-House	Total w/O&P
					Material	Labor	Equipment	Total		
0020 Replace waste disposal unit, residential	8	1 PLUM	Ea.							
Remove unit				1.040		59		59	73.50	91.50
Install new disposal				2.080	220	118		338	390	460
Total				3.120	220	177		397	463.50	551.50

E1093 320 Waste Handling Equipment

System Description	Freq. (Years)	Crew	Unit	Labor Hours	2015 Bare Costs				Total In-House	Total w/O&P
					Material	Labor	Equipment	Total		
0020 Remove and replace waste compactor hydraulic cylinder	15	2 SKWK	Ea.							
Replace ram, 5 C.Y. capacity				4.000	5,125	195		5,320	5,900	6,725
Total				4.000	5,125	195		5,320	5,900	6,725
0040 Remove and replace waste compactor hydraulic pump	20	1 SKWK	Ea.							
Replace hydraulic pump, 5 C.Y. capacity				3.200	1,025	156		1,181	1,325	1,525
Total				3.200	1,025	156		1,181	1,325	1,525

E1093 510 Darkroom Dryer

System Description	Freq. (Years)	Crew	Unit	Labor Hours	2015 Bare Costs				Total In-House	Total w/O&P
					Material	Labor	Equipment	Total		
0020 Remove and replace darkroom dryer fan	10	1 SKWK	Ea.							
48" x 25" x 68" high				1.600	178	78		256	298	350
Total				1.600	178	78		256	298	350
0040 Remove and replace darkroom dryer heating element	15	1 SKWK	Ea.							
Darkroom equip., dryers, dehumidifier, replace compressor				1.333	86	65		151	180	213
Total				1.333	86	65		151	180	213

For customer support on your Facilities Maintenance & Repair Cost Data, call 877.801.5724.

E1093 610   Dust Collector

	System Description	Freq. (Years)	Crew	Unit	Labor Hours	2015 Bare Costs				Total In-House	Total w/O&P
						Material	Labor	Equipment	Total		
0100	Remove and replace 20″ dia dust collector bag	5	1 SKWK	Ea.							
	Dust collector bag, 20″ diameter				1.600	440	89.50		529.50	600	690
	Total				1.600	440	89.50		529.50	600	690

E1093 910   Pump Systems

	System Description	Freq. (Years)	Crew	Unit	Labor Hours	2015 Bare Costs				Total In-House	Total w/O&P
						Material	Labor	Equipment	Total		
0500	Remove and replace 50 HP pump motor	25	2 ELEC	Ea.							
	Electrical dml, motors, 230/460 V, 60 Hz, 50 HP				1.667		91		91	113	141
	Install 50 HP motor, 230/460 V, totly encld, 1800 RPM				8.000	3,575	440		4,015	4,475	5,150
	Total				9.667	3,575	531		4,106	4,588	5,291

G1023 210 Underground Storage Tank Removal

System Description	Freq. (Years)	Crew	Unit	Labor Hours	2015 Bare Costs				Total In-House	Total w/O&P
					Material	Labor	Equipment	Total		
1010 Remove 500 gal. underground storage tank(non-leaking)	20	B-34P	Ea.							
Saw cut asphalt				2.700	6	118.50	82.50	207	242	288
Disconnect and remove piping				1.500		88.20		88.20	110	137
Transfer liquids, 10% of volume				.250		14.50		14.50	18.50	23
Insert solid carbon dioxide, 1.5 lbs./100 gal.				.150	8.78	5.63		14.41	17.20	20
Cut accessway to inside of tank				1.501		56.50		56.50	75	92.50
Remove sludge, wash and wipe tank				1.000		58.50		58.50	73.50	91.50
Properly dispose of sludge/water					35			35	38.50	44
Excavate, pull, & load tank, backfill hole				16.000		705	365	1,070	1,275	1,525
Select structural backfill					168			168	184	210
Haul tank to certified dump, 100 miles round trip				8.000		390	245	635	740	885
Total				31.101	217.78	1,436.83	692.50	2,347.11	2,773.70	3,316
1020 Remove 3000 gal. underground storage tank(non-leaking)	20	B-34R	Ea.							
Saw cut asphalt				5.400	12	237	165	414	485	575
Disconnect and remove piping				2.500		147		147	184	229
Transfer liquids, 10% of volume				1.500		87		87	111	138
Insert solid carbon dioxide, 1.5 lbs./100 gal.				.900	52.65	33.75		86.40	103	121
Cut accessway to inside of tank				1.501		56.50		56.50	75	92.50
Remove sludge, wash and wipe tank				1.199		70.50		70.50	88	110
Properly dispose of sludge/water					315			315	345	395
Excavate, pull, & load tank, backfill hole				32.000		1,400	785	2,185	2,625	3,125
Select structural backfill					630			630	690	790
Haul tank to certified dump, 100 miles round trip				8.000		390	305	695	800	950
Total				53.000	1,009.65	2,421.75	1,255	4,686.40	5,506	6,525.50
1030 Remove 5000 gal. underground storage tank(non-leaking)	20	B-34R	Ea.							
Saw cut asphalt				5.940	13.20	260.70	181.50	455.40	530	635
Disconnect and remove piping				2.500		147		147	184	229
Transfer liquids, 10% of volume				2.500		145		145	185	230
Insert solid carbon dioxide, 1.5 lbs./100 gal.				1.500	87.75	56.25		144	172	202
Cut accessway to inside of tank				1.501		56.50		56.50	75	92.50
Remove sludge, wash and wipe tank				1.301		76.50		76.50	95.50	119
Properly dispose of sludge/water					525			525	580	655
Excavate, pull, & load tank, backfill hole				32.000		1,400	785	2,185	2,625	3,125
Select structural backfill					1,050			1,050	1,150	1,325
Haul tank to certified dump, 100 miles round trip				8.000		390	305	695	800	950
Total				55.242	1,675.95	2,531.95	1,271.50	5,479.40	6,396.50	7,562.50

	System Description	Freq. (Years)	Crew	Unit	Labor Hours	2015 Bare Costs				Total In-House	Total w/O&P
						Material	Labor	Equipment	Total		
1040	**Remove 8000 gal. underground storage tank(non-leaking)**	20	B-34R	Ea.							
	Saw cut asphalt				6.480	14.40	284.40	198	496.80	580	690
	Disconnect and remove piping				2.500		147		147	184	229
	Transfer liquids, 10% of volume				4.000		232		232	296	370
	Insert solid carbon dioxide, 1.5 lbs./100 gal.				2.400	140.40	90		230.40	275	325
	Cut accessway to inside of tank				1.501		56.50		56.50	75	92.50
	Remove sludge, wash and wipe tank				1.501		88		88	110	137
	Properly dispose of sludge/water					840			840	925	1,050
	Excavate, pull, & load tank, backfill hole				32.000		1,425	2,350	3,775	4,200	4,875
	Select structural backfill					1,470			1,470	1,600	1,850
	Haul tank to certified dump, 100 miles round trip				8.000		320	950	1,270	1,375	1,575
	Total				58.382	2,464.80	2,642.90	3,498	8,605.70	9,620	11,193.50
1050	**Remove 10000 gal. ugnd. storage tank(non-leaking)**	20	B-34S	Ea.							
	Saw cut asphalt				10.800	24	474	330	828	965	1,150
	Disconnect and remove piping				3.750		220.50		220.50	275	345
	Transfer liquids, 10% of volume				5.000		290		290	370	460
	Insert solid carbon dioxide, 1.5 lbs./100 gal.				3.000	175.50	112.50		288	345	405
	Cut accessway to inside of tank				1.501		56.50		56.50	75	92.50
	Remove sludge, wash and wipe tank				1.751		103		103	128	160
	Properly dispose of sludge/water					1,050			1,050	1,150	1,325
	Excavate, pull, & load tank, backfill hole				32.000		1,425	2,350	3,775	4,200	4,875
	Select structural backfill					1,890			1,890	2,075	2,375
	Haul tank to certified dump, 100 miles round trip				8.000		320	950	1,270	1,375	1,575
	Total				65.801	3,139.50	3,001.50	3,630	9,771	10,958	12,762.50
1060	**Remove 12000 gal. ugnd. storage tank(non-leaking)**	20	B-34S	Ea.							
	Saw cut asphalt				12.150	27	533.25	371.25	931.50	1,075	1,300
	Disconnect and remove piping				5.000		294		294	365	460
	Transfer liquids, 10% of volume				6.000		348		348	445	550
	Insert solid carbon dioxide, 1.5 lbs./100 gal.				3.600	210.60	135		345.60	410	485
	Cut accessway to inside of tank				1.501		56.50		56.50	75	92.50
	Remove sludge, wash and wipe tank				1.900		112		112	139	174
	Properly dispose of sludge/water					1,260			1,260	1,375	1,575
	Excavate, pull, & load tank, backfill hole				32.000		1,425	2,350	3,775	4,200	4,875
	Select structural backfill					2,100			2,100	2,300	2,625
	Haul tank to certified dump, 100 miles round trip				8.000		320	950	1,270	1,375	1,575
	Total				70.151	3,597.60	3,223.75	3,671.25	10,492.60	11,759	13,711.50

For customer support on your Facilities Maintenance & Repair Cost Data, call 877.801.5724.

357

G2023 210 Parking Lot Repairs

System Description	Freq. (Years)	Crew	Unit	Labor Hours	2015 Bare Costs				Total In-House	Total w/O&P
					Material	Labor	Equipment	Total		
1010 **Parking lot repair and sealcoating**	5	B-1	M.S.F.							
Thoroughly clean surface				.312		12.21	3.33	15.54	18.85	22
Patch holes				.240	25.52	9.18	1.22	35.92	41.50	48
Fill cracks				.715	76.50	27.50	10	114	130	151
Install 2 coat petroleum resistant emulsion				3.330	144.30	127.65		271.95	330	390
Restripe lot				.291	17.76	11.12	3.64	32.52	38	44.50
Total				4.887	264.08	187.66	18.19	469.93	558.35	655.50
1020 **Parking lot repair and resurface**	10	B-25B	M.S.F.							
Thoroughly clean surface				.312		12.21	3.33	15.54	18.85	22
Patch holes				.240	25.52	9.18	1.22	35.92	41.50	48
Fill cracks				.715	76.50	27.50	10	114	130	151
Emulsion tack coat, .05 gal. per S.Y.				.710	32.19	32.19	39.96	104.34	117	136
Install 1" thick asphaltic concrete wearing course				1.008	419.58	42.18	31.08	492.84	550	630
Restripe lot				.291	17.76	11.12	3.64	32.52	38	44.50
Total				3.276	571.55	134.38	89.23	795.16	895.35	1,031.50

G2023 310 General

System Description	Freq. (Years)	Crew	Unit	Labor Hours	2015 Bare Costs				Total In-House	Total w/O&P
					Material	Labor	Equipment	Total		
1000 **Remove and replace steel guard rail**	7	B-80	L.F.							
Remove old metal guide/guard rail and posts				.213		8.80	4.80	13.60	16.30	19.55
Install metal guide/guard rail including posts 6'3"				.107	23	4.40	2.40	29.80	33.50	38.50
Total				.320	23	13.20	7.20	43.40	49.80	58.05
1100 **Raise MH or catch basin frame and cover**	10	D-1	Ea.							
Demolish pavement around frame				4.800		191	23.50	214.50	275	335
Raise MH or catch basin frame with brick				1.332	8.33	55.94		64.27	83	102
Total				6.132	8.33	246.94	23.50	278.77	358	437

G2033 130   Asphalt Sidewalk & Curb

	System Description	Freq. (Years)	Crew	Unit	Labor Hours	2015 Bare Costs				Total In-House	Total w/O&P
						Material	Labor	Equipment	Total		
0100	**Remove & replace asphalt sidewalk, 4' wide**	15	B-37	L.F.							
	Remove existing asphalt pavement				.080		3.28	1.94	5.22	6.20	7.45
	Bank run gravel, 3" avg.				.002	.74	.10	.22	1.06	1.16	1.32
	Vibratory plate compaction				.003		.10	.01	.11	.14	.17
	Fine grade gravel base				.026		1.15	1.20	2.35	2.69	3.17
	Place asphalt sidewalk, 2-1/2" thick				.027	2.63	1.06	.09	3.78	4.38	5.10
	Total				.138	3.37	5.69	3.46	12.52	14.57	17.21
1000	**Remove and replace asphalt curb or berm**	10	B-37	L.F.							
	Demolish existing curb or berm				.096		3.82	.47	4.29	5.50	6.70
	Remove broken curb				.032		1.31	.78	2.09	2.49	2.97
	Asphalt curb, 8" wide, 6" high machine formed				.042	.05	1.58	.39	2.02	2.55	3.08
	Total				.170	.05	6.71	1.64	8.40	10.54	12.75

G2033 140   Concrete Sidewalk & Curb

	System Description	Freq. (Years)	Crew	Unit	Labor Hours	2015 Bare Costs				Total In-House	Total w/O&P
						Material	Labor	Equipment	Total		
0100	**Remove and replace concrete sidewalk, 4' wide**	25	B-37	L.F.							
	Remove existing concrete pavement				.107		4.37	2.58	6.95	8.30	9.90
	Bank run gravel, 3" avg.				.002	.74	.10	.22	1.06	1.16	1.32
	Vibratory plate compaction				.003		.10	.01	.11	.14	.17
	Fine grade gravel				.026		1.15	1.20	2.35	2.69	3.17
	Broom finished concrete sidewalk, 4" thick				.209	6.76	9		15.76	19.25	23
	Total				.346	7.50	14.72	4.01	26.23	31.54	37.56
2000	**Remove and replace concrete curb or berm**	25	B-37	L.F.							
	Demolish existing curb or berm				.240		9.55	1.16	10.71	13.75	16.85
	Remove broken curb				.080		3.28	1.94	5.22	6.20	7.45
	Concrete curb, 6" x 18", steel forms				.091	.73	4.14		4.87	6.30	7.70
	Total				.411	.73	16.97	3.10	20.80	26.25	32

For customer support on your Facilities Maintenance & Repair Cost Data, call 877.801.5724.

359

G2033 150 Patios

System Description	Freq. (Years)	Crew	Unit	Labor Hours	2015 Bare Costs Material	Labor	Equipment	Total	Total In-House	Total w/O&P
1030 Refinish Concrete Patio	3	1 PORD	S.F.							
Prepare surface				.003		.13	.03	.16	.20	.24
Paint 1 coat, roller & brush				.008	.09	.33		.42	.53	.64
Total				.012	.09	.46	.03	.58	.73	**.88**
2030 Refinish Masonry Patio	3	1 PORD	S.F.							
Prepare surface				.003		.13	.03	.16	.20	.24
Apply 1 coat of sealer, roller & brush				.005	.25	.19		.44	.53	.62
Total				.008	.25	.32	.03	.60	.73	**.86**
3030 Refinish Stone Patio	3	1 PORD	S.F.							
Prepare surface				.003		.13	.03	.16	.20	.24
Apply 1 coat of sealer, roller & brush				.005	.25	.19		.44	.53	.62
Total				.008	.25	.32	.03	.60	.73	**.86**
4030 Refinish Wood Patio	3	1 PORD	S.F.							
Prepare surface				.015		.60		.60	.78	.96
Primer + 1 top coat, roller & brush				.022	.14	.90		1.04	1.31	1.62
Total				.037	.14	1.50		1.64	2.09	**2.58**

G2033 250 Handicap Ramp

System Description	Freq. (Years)	Crew	Unit	Labor Hours	2015 Bare Costs Material	Labor	Equipment	Total	Total In-House	Total w/O&P
1030 Refinish Metal Handicap Ramp	3	1 PORD	S.F.							
Prepare surface				.015		.60		.60	.78	.96
Prime & paint 1 coat, roller & brush				.015	.27	.59		.86	1.06	1.28
Total				.029	.27	1.19		1.46	1.84	**2.24**
2030 Refinish Wood Handicap Ramp	3	1 PORD	S.F.							
Prepare surface				.015		.60		.60	.78	.96
Prime & paint 1 coat, roller & brush				.022	.14	.90		1.04	1.31	1.62
Total				.037	.14	1.50		1.64	2.09	**2.58**

	System Description	Freq. (Years)	Crew	Unit	Labor Hours	2015 Bare Costs				Total In-House	Total w/O&P
						Material	Labor	Equipment	Total		
1010	**Minor chain link fence repairs (per 10 L.F.)**	1	2 CLAB	Ea.							
	Straighten bent 2" line post				.167		6.25		6.25	8.35	10.30
	Re-tie fence fabric				.167	1.72	6.25		7.97	10.20	12.45
	Refasten loose barbed wire arm				.083		3.13		3.13	4.16	5.15
	Refasten loose barbed wire, 3 strands				.083		3.13		3.13	4.16	5.15
	Total				.500	1.72	18.76		20.48	26.87	33.05
1110	**Replace bent 1-5/8" top rail (per 20 L.F.)**	2	2 CLAB	Ea.							
	Remove fabric ties, top rail & couplings				.333		12.55		12.55	16.65	20.50
	Install new top rail, couplings & fabric ties				.667	37.84	25.08		62.92	75	88.50
	Total				1.000	37.84	37.63		75.47	91.65	109
1120	**Replace broken barbed wire arm**	2	2 CLAB	Ea.							
	Remove fabric ties, top rail & couplings				.333		12.55		12.55	16.65	20.50
	Release barbed wires & remove old arm				.067		2.51		2.51	3.33	4.11
	Install new arm & refasten barbed wires				.167	.52	6.30		6.82	8.90	11
	Install old top rail, couplings & fabric ties				.667		25		25	33.50	41
	Total				1.234	.52	46.36		46.88	62.38	76.61
1130	**Replace barbed wire, 3 strands (per 100 L.F.)**	5	2 CLAB	Ea.							
	Remove old barbed wire				3.002		113		113	150	185
	Install new barbed wire				3.000	103.20	112.80		216	264	315
	Total				6.002	103.20	225.80		329	414	500
1140	**Replace fence fabric, 6' high, 9 ga. (per 10 L.F.)**	10	2 CLAB	Ea.							
	Remove ties from top rail, bottom tension wire & posts				.167		6.25		6.25	8.35	10.30
	Unstitch & remove section of fabric				.667		25		25	33.50	41
	Hang & stretch new fabric, stitch into existing fabric				1.333	68.80	50.20		119	142	168
	Re-tie fence fabric				.167	1.72	6.25		7.97	10.20	12.45
	Total				2.333	70.52	87.70		158.22	194.05	231.75
1150	**Replace double swing gates, 6' high, 20' opng.**	5	2 CLAB	Opng.							
	Remove 2 old gates (6'x10' ea.) & 4 hinges				1.111		42		42	55.50	68.50
	Supply 4 new gate hinges					56.55			56.55	62	70.50
	Install 2 new gates (6'x10' ea.), 4 hinges, adjust alignment				2.412	519.39	90.74		610.13	685	800
	Remove & replace gate latch				.914	13.92	34.40		48.32	61	74
	Remove & replace gate cane bolt				.883	13.92	33.20		47.12	59.50	72
	Total				5.320	603.78	200.34		804.12	923	1,085

For customer support on your Facilities Maintenance & Repair Cost Data, call 877.801.5724.

361

G2043 105 | Chain Link Fence and Gate Repairs

	System Description	Freq. (Years)	Crew	Unit	Labor Hours	2015 Bare Costs				Total In-House	Total w/O&P
						Material	Labor	Equipment	Total		
1160	**Replace 6'x18' cantilever slide gate**	5	2 CLAB	Opng.							
	Remove old gate & 4 rollers				.926		35		35	46.50	57
	Supply 4 new cantilever gate rollers				.003	317.50	.05		317.55	350	395
	Install new 6'x18' cantilever slide gate, 4 rollers, adjust				2.148	1,579.63	80.85		1,660.48	1,850	2,100
	Remove & replace gate latch				.914	13.92	34.40		48.32	61	74
	Total				3.991	1,911.05	150.30		2,061.35	2,307.50	2,626
1210	**Replace 2" line post**	5	B-55	Ea.							
	Remove fabric ties, top rail & couplings				.333		12.55		12.55	16.65	20.50
	Release barbed wires & remove old arm				.067		2.51		2.51	3.33	4.11
	Unstitch & roll up section of fabric				1.000		38	45	83	95.50	112
	Remove 2" post & concrete base from ground				.750		28.50	34	62.50	71.50	83.50
	Install new 2" line post with concrete in old hole				1.500	60.20	57.23	67.73	185.16	210	243
	Install old top rail, couplings & new fabric ties				.667		25		25	33.50	41
	Unroll & stretch fabric, stitch into existing fabric				2.000		76	90	166	191	223
	Re-tie fence fabric				.333	3.44	12.50		15.94	20.50	25
	Total				6.650	63.64	252.29	236.73	552.66	641.98	752.11
1220	**Replace 3" corner post**	10	B-55	Ea.							
	Remove fabric ties, retainer bars & ring fittings				.500		19.05	22.50	41.55	47.50	56
	Roll back fabric				.250		9.55	11.30	20.85	24	28
	Remove top rail & ring fittings				.500		19.05	22.50	41.55	47.50	56
	Remove barbed wires & ring fittings				.750		28.50	34	62.50	71.50	83.50
	Remove diagonal & horizontal braces, ring fittings				.500		19.05	22.50	41.55	47.50	56
	Remove 3" post & concrete base from ground				1.000		38	45	83	95.50	112
	Install new 3" post with concrete in old hole				2.000	85.50	76	90	251.50	285	330
	Install old diagonal & horizontal braces, ring fittings				.500		19.05	22.50	41.55	47.50	56
	Install old barbed wires & ring fittings				.750		28.50	34	62.50	71.50	83.50
	Install old top rail & ring fittings				.500		19.05	22.50	41.55	47.50	56
	Hang & stretch fabric, insert retainer bars & ring fittings				1.500		57	67.50	124.50	143	168
	Re-tie fence fabric				.333	3.44	12.50		15.94	20.50	25
	Total				9.083	88.94	345.30	394.30	828.54	948.50	1,110

For customer support on your Facilities Maintenance & Repair Cost Data, call 877.801.5724.

G20 SITE IMPROVEMENTS

G2043 Site Development

G2043 105 Chain Link Fence and Gate Repairs

	System Description	Freq. (Years)	Crew	Unit	Labor Hours	2015 Bare Costs Material	Labor	Equipment	Total	Total In-House	Total w/O&P
1230	**Replace 3" gate post**	5	B-55	Ea.							
	Remove gate & hinges				.833		32	37.50	69.50	79.50	93.50
	Remove fabric ties, retainer bars & ring fittings				.500		19.05	22.50	41.55	47.50	56
	Roll back fabric				.250		9.55	11.30	20.85	24	28
	Remove top rail & ring fittings				.500		19.05	22.50	41.55	47.50	56
	Remove barbed wires & ring fittings				.750		28.50	34	62.50	71.50	83.50
	Remove diagonal & horizontal braces, ring fittings				.500		19.05	22.50	41.55	47.50	56
	Remove 3" post & concrete base from ground				1.000		38	45	83	95.50	112
	Install new 3" post with concrete in old hole				2.000	85.50	76	90	251.50	285	330
	Install old diagonal & horizontal braces, ring fittings				.500		19.05	22.50	41.55	47.50	56
	Install old barbed wires & ring fittings				.750		28.50	34	62.50	71.50	83.50
	Install old top rail & ring fittings				.500		19.05	22.50	41.55	47.50	56
	Hang & stretch fabric, insert retainer bars & ring fittings				1.500		57	67.50	124.50	143	168
	Re-tie fence fabric				.333	3.44	12.50		15.94	20.50	25
	Install old hinges & gate				1.807		69	81.50	150.50	172	202
	Total				11.724	88.94	446.30	513.30	1,048.54	1,200	1,405.50

G2043 110 Wood Fence

	System Description	Freq. (Years)	Crew	Unit	Labor Hours	2015 Bare Costs Material	Labor	Equipment	Total	Total In-House	Total w/O&P
1020	**Refinish Wood Fence, 6' High**	3	1 PORD	L.F.							
	Prepare surface				.089		3.60		3.60	4.68	5.75
	Prime & 1 top coat, roller & brush				.046	.48	1.86		2.34	2.94	3.60
	Total				.135	.48	5.46		5.94	7.62	9.35

G2043 710 Bleachers, Exterior

	System Description	Freq. (Years)	Crew	Unit	Labor Hours	2015 Bare Costs Material	Labor	Equipment	Total	Total In-House	Total w/O&P
1020	**Refinish Wood Bleachers**	3	1 PORD	S.F.							
	Prepare surface				.015		.60		.60	.78	.96
	Prime & 1 top coat, roller & brush				.022	.14	.90		1.04	1.31	1.62
	Total				.037	.14	1.50		1.64	2.09	2.58

For customer support on your Facilities Maintenance & Repair Cost Data, call 877.801.5724.

363

G2043 750 Tennis Court Resurfacing

System Description	Freq. (Years)	Crew	Unit	2015 Bare Costs					Total In-House	Total w/O&P
				Labor Hours	Material	Labor	Equipment	Total		
1010 Resurface asphalt tennis court	7	2 SKWK	Ea.							
Remove/reinstall tennis net				1.000		37.50		37.50	50	61.50
Thoroughly clean surface				7.760		288		288	385	480
Install 3 coat acrylic emulsion sealcoat				16.000	4,840	600		5,440	6,125	7,025
Total				24.760	4,840	925.50		5,765.50	6,560	7,566.50
1020 Resurface cushioned asphalt tennis court	7	2 SKWK	Ea.							
Remove/reinstall tennis net				1.000		37.50		37.50	50	61.50
Thoroughly clean surface				7.760		288		288	385	480
Two cushion coats				16.000	2,200	600		2,800	3,225	3,725
Install 3 coat acrylic emulsion sealcoat				16.000	4,840	600		5,440	6,125	7,025
Total				40.760	7,040	1,525.50		8,565.50	9,785	11,291.50
2010 Clay tennis court preparation	1	2 SKWK	Ea.							
Remove/reinstall tennis net				1.000		37.50		37.50	50	61.50
Remove and reinstall tapes				6.015		226		226	300	370
Remove weeds and deteriorated surface material				2.000		75		75	100	123
Check court for level				1.000		37.50		37.50	50	61.50
Clay court top dressing, 1.5 tons				7.500	487.50	282		769.50	910	1,075
Clay court, roll after dressing				1.500		69.50	22.50	92	112	136
Total				19.015	487.50	727.50	22.50	1,237.50	1,522	1,827

G2043 810 Flag Pole

System Description	Freq. (Years)	Crew	Unit	2015 Bare Costs					Total In-House	Total w/O&P
				Labor Hours	Material	Labor	Equipment	Total		
1020 Refinish 25' Wood Flag Pole	5	1 PORD	Ea.							
Set up, secure and take down ladder				.295		10.40		10.40	13.50	16.75
Prepare surface				.258		15		15	19.50	24
Prime & 1 top coat, roller & brush				.370	3.50	22.50		26	33	40.50
Total				1.184	3.50	47.90		51.40	66	81.25

	System Description	Freq. (Years)	Crew	Unit	Labor Hours	2015 Bare Costs				Total In-House	Total w/O&P
						Material	Labor	Equipment	Total		
0100	**Prep & paint 100k gal. ground level water stor tank, 30' dia x 19' tall**	10	E-11	Ea.							
	Scaffolding, outer ring				72.000		3,384		3,384	4,500	5,550
	Scaffolding planks				35.666		1,674.55		1,674.55	2,250	2,725
	Sandblast exterior, near white blast (SSPC-SP10), loose scale, fine rust				177.775	2,150	7,525	1,300	10,975	14,400	17,100
	Epoxy primer, spray exterior				13.325	725	550		1,275	1,650	1,925
	Epoxy topcoat, 2 coats, spray exterior				28.550	1,300	1,200		2,500	3,300	3,825
	Clean up				10.000		400		400	600	725
	Total				337.316	4,175	14,733.55	1,300	20,208.55	26,700	31,850
0200	**Prep & paint 250k gal. ground level water stor tank, 40' dia x 27' tall**	10	E-11	Ea.							
	Scaffolding, outer ring				126.000		5,922		5,922	7,850	9,700
	Scaffolding planks				63.333		2,973.50		2,973.50	4,000	4,850
	Sandblast exterior, near white blast (SSPC-SP10), loose scale, fine rust				327.106	3,956	13,846	2,392	20,194	26,500	31,500
	Epoxy primer, spray exterior				24.518	1,334	1,012		2,346	3,025	3,550
	Epoxy topcoat, 2 coats, spray exterior				52.532	2,392	2,208		4,600	6,075	7,050
	Clean up				18.400		736		736	1,100	1,325
	Total				611.889	7,682	26,697.50	2,392	36,771.50	48,550	57,975
0300	**Prep & paint 500k gal. ground level water stor tank, 50' dia x 34' tall**	10	E-11	Ea.							
	Scaffolding, outer ring				192.000		9,024		9,024	12,000	14,800
	Scaffolding planks				96.666		4,538.50		4,538.50	6,100	7,400
	Sandblast exterior, near white blast (SSPC-SP10), loose scale, fine rust				519.103	6,278	21,973	3,796	32,047	42,000	50,000
	Epoxy primer, spray exterior				38.909	2,117	1,606		3,723	4,825	5,650
	Epoxy topcoat, 2 coats, spray exterior				83.366	3,796	3,504		7,300	9,625	11,200
	Clean up				29.200		1,168		1,168	1,750	2,125
	Total				959.244	12,191	41,813.50	3,796	57,800.50	76,300	91,175
0400	**Prep & paint 750k gal. ground level water stor tank, 60' dia x 36' tall**	10	E-11	Ea.							
	Scaffolding, outer ring				234.000		10,998		10,998	14,600	18,000
	Scaffolding planks				117.332		5,508.80		5,508.80	7,400	8,975
	Sandblast exterior, near white blast (SSPC-SP10), loose scale, fine rust				675.545	8,170	28,595	4,940	41,705	54,500	65,000
	Epoxy primer, spray exterior				50.635	2,755	2,090		4,845	6,275	7,350
	Epoxy topcoat, 2 coats, spray exterior				108.490	4,940	4,560		9,500	12,500	14,500
	Clean up				38.000		1,520		1,520	2,275	2,750
	Total				1224.002	15,865	53,271.80	4,940	74,076.80	97,550	116,575

System Description	Freq. (Years)	Crew	Unit	Labor Hours	2015 Bare Costs				Total In-House	Total w/O&P
					Material	Labor	Equipment	Total		
0500 **Prep & paint 1M gal. ground level water stor tank, 70' dia x 35' tall**	10	E-11	Ea.							
Scaffolding, outer ring				261.000		12,267		12,267	16,300	20,100
Scaffolding planks				131.332		6,166.10		6,166.10	8,275	10,000
Sandblast exterior, near white blast (SSPC-SP10), loose scale, fine rust				817.765	9,890	34,615	5,980	50,485	66,000	79,000
Epoxy primer, spray exterior				61.295	3,335	2,530		5,865	7,600	8,875
Epoxy topcoat, 2 coats, spray exterior				131.330	5,980	5,520		11,500	15,200	17,600
Clean up				46.000		1,840		1,840	2,750	3,325
Total				1448.722	19,205	62,938.10	5,980	88,123.10	116,125	138,900
0600 **Prep & paint 2M gal. ground level water stor tank, 100' dia x 34' tall**	10	E-11	Ea.							
Scaffolding, outer ring				354.000		16,638		16,638	22,100	27,300
Scaffolding planks				176.998		8,310.15		8,310.15	11,200	13,500
Sandblast exterior, near white blast (SSPC-SP10), loose scale, fine rust				1315.535	15,910	55,685	9,620	81,215	106,500	127,000
Epoxy primer, spray exterior				98.605	5,365	4,070		9,435	12,200	14,300
Epoxy topcoat, 2 coats, spray exterior				211.270	9,620	8,880		18,500	24,400	28,300
Clean up				74.000		2,960		2,960	4,450	5,375
Total				2230.408	30,895	96,543.15	9,620	137,058.15	180,850	215,775
0700 **Prep & paint 4M gal. ground level water stor tank, 130' dia x 40' tall**	10	E-11	Ea.							
Scaffolding, outer ring				531.000		24,957		24,957	33,100	40,900
Scaffolding planks				266.664		12,520		12,520	16,800	20,400
Sandblast exterior, near white blast (SSPC-SP10), loose scale, fine rust				2133.300	25,800	90,300	15,600	131,700	172,500	205,500
Epoxy primer, spray exterior				159.900	8,700	6,600		15,300	19,800	23,200
Epoxy topcoat, 2 coats, spray exterior				342.600	15,600	14,400		30,000	39,600	45,900
Clean up				120.000		4,800		4,800	7,200	8,700
Total				3553.464	50,100	153,577	15,600	219,277	289,000	344,600

For customer support on your Facilities Maintenance & Repair Cost Data, call 877.801.5724.

G3013 405 Ground Level Water Storage Standpipe

	System Description	Freq. (Years)	Crew	Unit	Labor Hours	2015 Bare Costs				Total In-House	Total w/O&P
						Material	Labor	Equipment	Total		
0100	**Prep & paint 100k gal. grnd. level water standpipe, 24' dia x 30' tall**	10	E-11	Ea.							
	Scaffolding, outer ring				96.000		4,512		4,512	5,975	7,400
	Scaffolding planks				47.666		2,237.95		2,237.95	3,000	3,650
	Sandblast exterior, near white blast (SSPC-SP10), loose scale, fine rust				191.997	2,322	8,127	1,404	11,853	15,500	18,500
	Epoxy primer, spray exterior				14.391	783	594		1,377	1,775	2,075
	Epoxy topcoat, 2 coats, spray exterior				30.834	1,404	1,296		2,700	3,575	4,125
	Clean up				10.800		432		432	650	785
	Total				391.688	4,509	17,198.95	1,404	23,111.95	30,475	36,535
0200	**Prep & paint 250k gal. grnd. level water standpipe, 32' dia x 42' tall**	10	E-11	Ea.							
	Scaffolding, outer ring				165.000		7,755		7,755	10,300	12,700
	Scaffolding planks				82.666		3,881.20		3,881.20	5,200	6,325
	Sandblast exterior, near white blast (SSPC-SP10), loose scale, fine rust				355.550	4,300	15,050	2,600	21,950	28,800	34,300
	Epoxy primer, spray exterior				26.650	1,450	1,100		2,550	3,300	3,875
	Epoxy topcoat, 2 coats, spray exterior				57.100	2,600	2,400		5,000	6,600	7,650
	Clean up				20.000		800		800	1,200	1,450
	Total				706.966	8,350	30,986.20	2,600	41,936.20	55,400	66,300
0300	**Prep & paint 500k gal. grnd. level water standpipe, 40' dia x 53' tall**	10	E-11	Ea.							
	Scaffolding, outer ring				252.000		11,844		11,844	15,700	19,400
	Scaffolding planks				125.665		5,900.05		5,900.05	7,925	9,625
	Sandblast exterior, near white blast (SSPC-SP10), loose scale, fine rust				568.880	6,880	24,080	4,160	35,120	46,000	55,000
	Epoxy primer, spray exterior				42.640	2,320	1,760		4,080	5,275	6,175
	Epoxy topcoat, 2 coats, spray exterior				91.360	4,160	3,840		8,000	10,600	12,200
	Clean up				32.000		1,280		1,280	1,925	2,325
	Total				1112.545	13,360	48,704.05	4,160	66,224.05	87,425	104,725
0400	**Prep & paint 750k gal. grnd. level water standpipe, 46' dia x 60' tall**	10	E-11	Ea.							
	Scaffolding, outer ring				318.000		14,946		14,946	19,800	24,500
	Scaffolding planks				159.665		7,496.35		7,496.35	10,100	12,200
	Sandblast exterior, near white blast (SSPC-SP10), loose scale, fine rust				739.544	8,944	31,304	5,408	45,656	60,000	71,500
	Epoxy primer, spray exterior				55.432	3,016	2,288		5,304	6,875	8,025
	Epoxy topcoat, 2 coats, spray exterior				118.768	5,408	4,992		10,400	13,700	15,900
	Clean up				41.600		1,664		1,664	2,500	3,025
	Total				1433.009	17,368	62,690.35	5,408	85,466.35	112,975	135,150

G3013 405 — Ground Level Water Storage Standpipe

	System Description	Freq. (Years)	Crew	Unit	Labor Hours	2015 Bare Costs Material	Labor	Equipment	Total	Total In-House	Total w/O&P
0500	**Prep & paint 1M gal. grnd. level water standpipe, 50' dia x 68' tall**	10	E-11	Ea.							
	Scaffolding, outer ring				384.000		18,048		18,048	23,900	29,600
	Scaffolding planks				192.998		9,061.35		9,061.35	12,200	14,800
	Sandblast exterior, near white blast (SSPC-SP10), loose scale, fine rust				903.097	10,922	38,227	6,604	55,753	73,000	87,000
	Epoxy primer, spray exterior				67.691	3,683	2,794		6,477	8,375	9,800
	Epoxy topcoat, 2 coats, spray exterior				145.034	6,604	6,096		12,700	16,800	19,400
	Clean up				50.800		2,032		2,032	3,050	3,675
	Total				1743.620	21,209	76,258.35	6,604	104,071.35	137,325	164,275
0600	**Prep & paint 2M gal. grnd. level water standpipe, 64' dia x 83' tall**	10	E-11	Ea.							
	Scaffolding, outer ring				579.000		27,213		27,213	36,100	44,600
	Scaffolding planks				290.664		13,646.80		13,646.80	18,300	22,200
	Sandblast exterior, near white blast (SSPC-SP10), loose scale, fine rust				1422.200	17,200	60,200	10,400	87,800	115,000	137,000
	Epoxy primer, spray exterior				106.600	5,800	4,400		10,200	13,200	15,500
	Epoxy topcoat, 2 coats, spray exterior				228.400	10,400	9,600		20,000	26,400	30,600
	Clean up				80.000		3,200		3,200	4,800	5,800
	Total				2706.864	33,400	118,259.80	10,400	162,059.80	213,800	255,700

G3013 410 — Fire Hydrants

	System Description	Freq. (Years)	Crew	Unit	Labor Hours	2015 Bare Costs Material	Labor	Equipment	Total	Total In-House	Total w/O&P
1010	**Remove and replace fire hydrant**	25	B-21	Ea.							
	Isolate hydrant from system				1.000		37.50		37.50	50	61.50
	Excavate to expose base and feeder pipe				.974		43.05	22.19	65.24	78	93.50
	Disconnect and remove hydrant and associated piping				4.000		150		150	200	247
	Install pipe to new hydrant				3.200	320	146.50		466.50	540	635
	Install hydrant, including lower barrel and shoe				4.000	1,875	173	19.70	2,067.70	2,300	2,650
	Install thrust blocks				1.200	49.80	54	.34	104.14	126	151
	Backfill excavation				5.091		192.50		192.50	256	315
	Fine grade, hand				8.000		300		300	400	490
	Total				27.465	2,244.80	1,096.55	42.23	3,383.58	3,950	4,643

G3013 470 Post Indicator Valve

System Description	Freq. (Years)	Crew	Unit	Labor Hours	2015 Bare Costs				Total In-House	Total w/O&P
					Material	Labor	Equipment	Total		
1010 Remove and replace post indicator valve	35	B-21	Ea.							
Isolate PIV from system				.126		7.40		7.40	9.25	11.55
Excavate to expose base of valve				.139		6.15	3.17	9.32	11.15	13.40
Disconnect and remove post indicator and valve				4.000		150		150	200	247
Install new 4" diameter valve				4.000	1,875	165	60.50	2,100.50	2,350	2,675
Install new post indicator				3.636	815	158	17.90	990.90	1,125	1,300
Backfill excavation				.727		27.50		27.50	36.50	45
Fine grade, hand				8.000		300		300	400	490
Total				20.629	2,690	814.05	81.57	3,585.62	4,131.90	4,781.95

G3013 510 Elevated Water Storage Tank

System Description	Freq. (Years)	Crew	Unit	Labor Hours	2015 Bare Costs				Total In-House	Total w/O&P
					Material	Labor	Equipment	Total		
0100 Prep & paint 50K gal. stg tank, 30' dia x 10' tall, 50' above grnd.	10	E-11	Ea.							
Scaffolding, outer ring				225.000		10,575		10,575	14,000	17,300
Scaffolding, center core under tank				80.000		3,760		3,760	5,000	6,150
Scaffolding planks				58.333		2,738.75		2,738.75	3,675	4,475
Sandblast exterior, near white blast (SSPC-SP10), loose scale, fine rust				167.464	2,025.30	7,088.55	1,224.60	10,338.45	13,500	16,100
Epoxy primer, spray exterior				12.552	682.95	518.10		1,201.05	1,550	1,825
Epoxy topcoat, 2 coats, spray exterior				26.894	1,224.60	1,130.40		2,355	3,100	3,600
Clean up				9.420		376.80		376.80	565	685
Total				579.663	3,932.85	26,187.60	1,224.60	31,345.05	41,390	50,135
0200 Prep & paint 100K gal. stg tank, 30' dia x 20' tall, 50' above grnd.	10	E-11	Ea.							
Scaffolding, outer ring				261.000		12,267		12,267	16,300	20,100
Scaffolding, center core under tank				80.000		3,760		3,760	5,000	6,150
Scaffolding planks				153.332		7,199		7,199	9,650	11,700
Sandblast exterior, near white blast (SSPC-SP10), loose scale, fine rust				227.552	2,752	9,632	1,664	14,048	18,400	21,900
Epoxy primer, spray exterior				17.056	928	704		1,632	2,100	2,475
Epoxy topcoat, 2 coats, spray exterior				36.544	1,664	1,536		3,200	4,225	4,900
Clean up				12.800		512		512	770	930
Total				788.284	5,344	35,610	1,664	42,618	56,445	68,155

G3013 510 Elevated Water Storage Tank

System Description	Freq. (Years)	Crew	Unit	Labor Hours	2015 Bare Costs Material	Labor	Equipment	Total	Total In-House	Total w/O&P
0300 Prep & paint 250K gal. stg tank, 40' dia x 27' tall, 50' above grnd.	10	E-11	Ea.							
Scaffolding, outer ring				360.000		16,920		16,920	22,400	27,700
Scaffolding, center core under tank				134.000		6,298		6,298	8,375	10,300
Scaffolding planks				219.998		10,329		10,329	13,900	16,800
Sandblast exterior, near white blast (SSPC-SP10), loose scale, fine rust				419.549	5,074	17,759	3,068	25,901	33,900	40,400
Epoxy primer, spray exterior				31.447	1,711	1,298		3,009	3,900	4,550
Epoxy topcoat, 2 coats, spray exterior				67.378	3,068	2,832		5,900	7,800	9,025
Clean up				23.600		944		944	1,425	1,700
Total				1255.972	9,853	56,380	3,068	69,301	91,700	110,475
0400 Prep & paint 500K gal. stg tank, 50' dia x 34' tall, 50' above grnd.	10	E-11	Ea.							
Scaffolding, outer ring				480.000		22,560		22,560	29,900	37,000
Scaffolding, center core under tank				202.000		9,494		9,494	12,600	15,600
Scaffolding planks				296.664		13,928.50		13,928.50	18,700	22,700
Sandblast exterior, near white blast (SSPC-SP10), loose scale, fine rust				661.323	7,998	27,993	4,836	40,827	53,500	64,000
Epoxy primer, spray exterior				49.569	2,697	2,046		4,743	6,150	7,175
Epoxy topcoat, 2 coats, spray exterior				106.206	4,836	4,464		9,300	12,300	14,200
Clean up				37.200		1,488		1,488	2,225	2,700
Total				1832.962	15,531	81,973.50	4,836	102,340.50	135,375	163,375
0500 Prep & paint 750K gal. stg tank, 60' dia x 36' tall, 50' above grnd.	10	E-11	Ea.							
Scaffolding, outer ring				570.000		26,790		26,790	35,500	43,900
Scaffolding, center core under tank				283.000		13,301		13,301	17,700	21,800
Scaffolding planks				366.663		17,215		17,215	23,100	28,100
Sandblast exterior, near white blast (SSPC-SP10), loose scale, fine rust				881.764	10,664	37,324	6,448	54,436	71,500	85,000
Epoxy primer, spray exterior				66.092	3,596	2,728		6,324	8,175	9,575
Epoxy topcoat, 2 coats, spray exterior				141.608	6,448	5,952		12,400	16,400	19,000
Clean up				49.600		1,984		1,984	2,975	3,600
Total				2358.727	20,708	105,294	6,448	132,450	175,350	210,975
0600 Prep & paint 1M gal. stg tank, 65' dia x 40' tall, 50' above grnd.	10	E-11	Ea.							
Scaffolding, outer ring				645.000		30,315		30,315	40,200	49,700
Scaffolding, center core under tank				329.000		15,463		15,463	20,600	25,300
Scaffolding planks				413.329		19,406		19,406	26,000	31,600
Sandblast exterior, near white blast (SSPC-SP10), loose scale, fine rust				1059.539	12,814	44,849	7,748	65,411	85,500	102,000
Epoxy primer, spray exterior				79.417	4,321	3,278		7,599	9,825	11,500
Epoxy topcoat, 2 coats, spray exterior				170.158	7,748	7,152		14,900	19,700	22,800
Clean up				59.600		2,384		2,384	3,575	4,325
Total				2756.043	24,883	122,847	7,748	155,478	205,400	247,225

For customer support on your Facilities Maintenance & Repair Cost Data, call 877.801.5724.

G3063 410 | **Ground Level Fuel Storage Tank**

	System Description	Freq. (Years)	Crew	Unit	Labor Hours	2015 Bare Costs				Total In-House	Total w/O&P
						Material	Labor	Equipment	Total		
0100	**Prep & paint 100k gal. ground level fuel storage tank, 24' dia x 30' tall**	25	E-11	Ea.							
	Scaffolding, outer ring				96.000		4,512		4,512	5,975	7,400
	Scaffolding planks				47.333		2,222.30		2,222.30	2,975	3,625
	Sandblast exterior, near white blast (SSPC-SP10), loose scale, fine rust				191.997	2,322	8,127	1,404	11,853	15,500	18,500
	Epoxy primer, spray exterior				14.391	783	594		1,377	1,775	2,075
	Epoxy topcoat, 2 coats, spray exterior				30.834	1,404	1,296		2,700	3,575	4,125
	Clean up				10.800		432		432	650	785
	Total				391.355	4,509	17,183.30	1,404	23,096.30	30,450	36,510

For customer support on your Facilities Maintenance & Repair Cost Data, call 877.801.5724.

371

G4013 210   Overhead Service Cables

	System Description	Freq. (Years)	Crew	Unit	Labor Hours	2015 Bare Costs				Total In-House	Total w/O&P
						Material	Labor	Equipment	Total		
0010	**Repair cable splice**	12	1 ELEC	M.L.F.							
	Repair splice cable, polyethylene jacket				7.619	11.40	415		426.40	530	660
	Total				7.619	11.40	415		426.40	**530**	**660**
0020	**Cable inspection**	5	1 ELEC	M.L.F.							
	Check cables for damage				.667		36.50		36.50	45	56.50
	Total				.667		36.50		36.50	**45**	**56.50**
0030	**Replace service cable**	30	2 ELEC	M.L.F.							
	Remove overhead service cables				1.860		102		102	126	158
	Overhead service 3 ACSR cable				16.000	1,075	875		1,950	2,275	2,700
	Install 3 conductors at terminations				.941		51.50		51.50	63.50	79.50
	Install conductors to insulators and clipping				2.500		137		137	169	212
	Install and remove sag gauge				1.000		54.50		54.50	67.50	84.50
	Make up & install conductor jumper				2.000		109		109	135	169
	Total				24.302	1,075	1,329		2,404	**2,836**	**3,403**

System Description	Freq. (Years)	Crew	Unit	Labor Hours	2015 Bare Costs				Total In-House	Total w/O&P
					Material	Labor	Equipment	Total		
1010 Replace 400W H.P.S. lamp, pole-mounted fixture	10	R-26	Ea.							
Position truck, raise and lower boom bucket				.289		15.80	4.71	20.51	24.50	29.50
Remove & install 400W H.P.S. lamp, pole-mounted fixture				.185	55.25	10.25	3	68.50	76.50	88
Total				.474	55.25	26.05	7.71	89.01	101	117.50
1020 Replace 400W H.P.S. ballast, pole-mounted fixture	10	R-26	Ea.							
Position truck, raise and lower boom bucket				.289		15.80	4.71	20.51	24.50	29.50
Remove & install 400W H.P.S. multi-tap ballast, pole-mounted fixt.				2.974	258	163	48.50	469.50	535	630
Test pole-mounted H.I.D. fixture				.074		4.05	1.21	5.26	6.20	7.60
Total				3.337	258	182.85	54.42	495.27	565.70	667.10
1030 Replace 400W H.P.S. pole-mounted fixt. w/ lamp & blst.	20	R-26	Ea.							
Turn branch cicuit off and on, pole light				.370		20.50	6.05	26.55	31	38
Turn branch cicuit off and on, pole light				.370		20.50	6.05	26.55	31	38
Position truck, raise and lower boom bucket				.289		15.80	4.71	20.51	24.50	29.50
Remove & install pole-mounted 400W H.P.S. fixture w/ lamp & blst.				9.456	380	515	154	1,049	1,200	1,450
Test pole-mounted H.I.D. fixture				.074		4.05	1.21	5.26	6.20	7.60
Total				10.560	380	575.85	172.02	1,127.87	1,292.70	1,563.10
1040 Replace light pole, 2 fixtures (conc. base not incl.)	10	R-3	Ea.							
Turn branch cicuit off and on, pole light				.370		20.50	6.05	26.55	31	38
Remove parking lot light pole with fixtures				2.500		136	17.25	153.25	186	230
Install 2 brackets arms onto aluminum light pole				2.602	486	142		628	710	830
Install two 400W H.P.S. fixtures w/ lamps & ballasts onto light pole				9.467	760	518		1,278	1,475	1,750
Install 30' aluminum light pole (incl. arms & fixtures) onto base				10.000	1,775	545	69	2,389	2,700	3,150
Pull new wires in existing underground conduits (3#10 500')				15.605	279.75	855		1,134.75	1,375	1,675
Test pole-mounted H.I.D. fixture				.074		4.05	1.21	5.26	6.20	7.60
Total				40.618	3,300.75	2,220.55	93.51	5,614.81	6,483.20	7,680.60

G4023 210   Outdoor Pole Lights

					2015 Bare Costs				Total In-House	Total w/O&P
System Description	Freq. (Years)	Crew	Unit	Labor Hours	Material	Labor	Equipment	Total		
1050 Replace concrete base for parking lot light pole	20	B-18	Ea.							
Sawcut asphalt paving				1.296	2.88	56.88	39.60	99.36	116	138
Remove concrete light pole base				10.000		380	58.40	438.40	565	690
Excavate by hand				12.987		487.50		487.50	650	800
Install 1-1/4″ PVC electrical conduit				.612	11.70	33.50		45.20	54.50	66.50
Install 1-1/4″ PVC electrical sweep				.433	44.50	23.50		68	78.50	92
Install new concrete light pole base, 24″ dia. x 8′				12.030	270	535		805	1,000	1,225
Backfill excavation by hand				4.728		177.50		177.50	235	293
Compact backfill				.578		21.75	2.60	24.35	31.50	38.50
Install asphalt patch				.960	68.20	38.20	3.16	109.56	128	151
Total				43.624	397.28	1,753.83	103.76	2,254.87	**2,858.50**	**3,494**

H1043 200 Temporary Cranes

	System Description	Freq. (Years)	Crew	Unit	Labor Hours	2015 Bare Costs				Total In-House	Total w/O&P
						Material	Labor	Equipment	Total		
1010	**Daily use of crane, portal to portal, 12-ton** 12-ton truck-mounted hydraulic crane, daily use for small jobs	1	A-3H	Day	8.000		415	860	1,275	1,400	1,600
	Total				8.000		415	860	1,275	**1,400**	**1,600**
1020	**Daily use of crane, portal to portal, 25-ton** 25-ton truck-mounted hydraulic crane, daily use for small jobs	1	A-3I	Day	8.000		415	990	1,405	1,525	1,750
	Total				8.000		415	990	1,405	**1,525**	**1,750**
1030	**Daily use of crane, portal to portal, 40-ton** 40-ton truck-mounted hydraulic crane, daily use for small jobs	1	A-3J	Day	8.000		415	1,225	1,640	1,750	2,000
	Total				8.000		415	1,225	1,640	**1,750**	**2,000**
1040	**Daily use of crane, portal to portal, 55-ton** 55-ton truck-mounted hydraulic crane, daily use for small jobs	1	A-3K	Day	16.000		775	1,625	2,400	2,625	3,025
	Total				16.000		775	1,625	2,400	**2,625**	**3,025**
1050	**Daily use of crane, portal to portal, 80-ton** 80-ton truck-mounted hydraulic crane, daily use for small jobs	1	A-3L	Day	16.000		775	2,350	3,125	3,325	3,800
	Total				16.000		775	2,350	3,125	**3,325**	**3,800**
1060	**Daily use of crane, portal to portal, 100-ton** 100-ton truck-mounted hydraulic crane, daily use for small jobs	1	A-3M	Day	16.000		775	2,325	3,100	3,325	3,800
	Total				16.000		775	2,325	3,100	**3,325**	**3,800**

For customer support on your Facilities Maintenance & Repair Cost Data, call 877.801.5724.

375

Preventive Maintenance

Table of Contents

Table of Contents (cont.)

How to Use the Preventive Maintenance Cost Tables

The following is a detailed explanation of a sample Preventive Maintenance Cost Table. The Preventive Maintenance Tables are separated into two parts: 1) the components and related labor-hours and frequencies of a typical system and 2) the costs for systems on an annual or annualized basis. Next to each bold number that follows is the described item with the appropriate component of the sample entry following in parentheses.

1 **System/Line Numbers (D3025 130 4950)**
Each Maintenance & Repair Assembly has been assigned a unique identification number based on the UNIFORMAT II classification system.

UNIFORMAT II Major Group
D3025   **130**   **4950**
UNIFORMAT II Level 3
Means Major Classification
Means Individual Line Number

D30 HVAC	D3025 130 **1**	Boiler, Hot Water, Oil/Gas/Comb.					
PM Components		Labor-hrs. **3**	W	M **4**	Q	S	A
PM System D3025 130 4950							
Boiler, hot water; oil, gas or combination fired, over 1000 MBH							
1 Check combustion chamber for air or gas leaks.		.117					√
2 Inspect and clean oil burner gun and ignition assembly where applicable.		.987					√
3 Inspect fuel system for leaks and change fuel filter element, where applicable.		.147					√
4 Check fuel lines and connections for damage.		.035	√	√	√	√	√
5 Check for proper operational response of burner to thermostat controls. **2**		.199		√	√	√	√
6 Check and lubricate burner and blower motors.		.120		√	√	√	√
7 Check main flame failure protection and main flame detection scanner on boiler equipped with spark ignition (oil burner).		.186	√	√	√	√	√
8 Check electrical wiring to burner controls and blower.		.120					√
9 Clean firebox (sweep and vacuum).		.819					√
10 Check operation of mercury control switches (i.e., steam pressure, hot water temperature limit, atomizing or combustion air proving, etc.).		.215		√	√	√	√
11 Check operation and condition of safety pressure relief valve.		.046	√	√	√	√
12 Check operation of boiler low water cut off devices.		.085	√	√	√	√
13 Check hot water pressure gauges.		.109	√	√	√	√
14 Inspect and clean water column sight glass (or replace).		.191	√	√	√	√
15 Clean fire side of water jacket		.433				√
16 Check condition of flue pipe, damper and exhaust stack.		.221		√	√	√
17 Check boiler operation through complete cycle, up to 30 minutes.		.887				√
18 Check fuel level with gauge pole, add as required.		.098	√	√	√	√
19 Clean area around boiler.		.182	√	√	√	√
20 Fill out maintenance checklist and report deficiencies.		.022	√	√	√	√
Total labor-hours/period		**5**		1.169	1.709	1.709	5.219
Total labor-hours/year				9.352	3.418	1.709	5.219

Description **6**	Labor-hrs. **7**	Cost Each 2015 Bare Costs				Total In-House **9**	Total w/O&P **10**
		Material	Labor	Equip. **8**	Total		
4900 Boiler, hot water, O/G/C, over 1000 MBH, annually	5.219	204	310		514	615	740
4950 Annualized	19.698	129	1,175		1,304	1,625	1,975

2 PM Components

The individual preventive maintenance operations required to be performed annually are listed separately to show what has been included in the development of the total system price. The cost table below the PM Components listing contains prices for the system on an annual or annualized basis.

3 Labor-hrs.

The "Labor-hrs." figure represents the number of labor-hours required to perform the individual operations one time.

4 Frequency

The columns marked with "W" for weekly, "M" for monthly, "Q" for quarterly, "S" for semi-annually, and "A" for annually indicate the recommended frequency for performing a task. For example, a check in the "M" column indicates the recommended frequency of the operation as monthly. Note that, as a result, checks also appear in the "Q," "S," and "A" columns, as they would be part of a monthly frequency.

5 Total Labor-hour

"Total Labor-hours" are provided on both a per-period and per-year basis. The per-period totals are derived by summing the hours for each operation recommended to be performed (indicated with a check mark) in a given period. In the example given here:

M	= .035 + .186 + .215 + .046 + .085 + .109 + .191 + .068 + .182 + .022	= 1.169
Similarly		
Q	= .035 + .199 + .120 + . . . + .098 + .182 + .022	= 1.709
S	= .035 + .199 + .120 + . . . + .098 + .182 + .022	= 1.709
A	= .117 + .987 + .147 + .035 + . . . + .887 + .098 + .182 + .022	= 5.219

The per-year totals are derived by multiplying the per-period total by the net frequency for each time period. The net frequency per period is based on the following table using a 50-week work year:

Total Labor-hours		Labor-hours		
Annual	=	Annual	x	1
Semi-Annual	=	Semi-Annual	x	1
Quarterly	=	Quarterly	x	2
Monthly	=	Monthly	x	8
Weekly	=	Weekly	x	38
				50

Based on these values, the total labor-hours/year are calculated for the given example as follows:

M	=	(1.169)	x	8	=	9.352
Q	=	(1.709)	x	2	=	3.418
S	=	(1.709)	x	1	=	1.709
A	=	(5.219)	x	1	=	5.219
						19.698 hrs.

6 System Description (Boiler, Hot Water, O/G/C, etc.)

The "System Description" corresponds to that given at the beginning of the PM Components list. A description of the cost basis, annual or annualized, is given as well. Annual costs are based on performing each operation listed in the PM Components one time in a given year. Annualized costs are based on performing each operation at the recommended periodic frequencies in a given year.

7 Labor-hours: Annual (5.219)/ Annualized (19.698)

Annual labor-hours are based on performing each task once during a given year and are equal to the total labor-hours/period under the "A" column in the PM Components list. Annualized labor-hours are based on the recommended periodic frequencies and are equivalent to the sum of all the values in the total labor-hours/year row at the bottom of the PM Components.

8 Bare Costs:
Annualized Material (129.00)

This figure for the system material cost is the "bare" material cost with no overhead and profit allowances included. *Costs shown reflect national average material prices for the current year and generally include delivery to the facility. Small purchases may require an additional delivery charge. No sales taxes are included.*

Labor (1,175.00)

The labor costs are derived by multiplying bare labor-hour costs by labor-hour units. (The hourly labor costs are listed in the Crew Listings, or for a single trade, the wage rate is found on the inside back cover of the book.)

Equip. (Equipment) (0.00)

The unit equipment cost is derived by multiplying the bare equipment hourly cost by the labor-hour units. Equipment costs for each crew are listed in the description of each crew.

Total (1,304.00)

The total of the bare costs is the arithmetic total of the three previous columns: mat., labor, equip.

Material	+	Labor	+	Equip.	=	Total
$129.00	+	$1,175.00	+	$0.00	=	$1,304.00

9 *Total In-House Costs*
Annualized (1,625)

"Total In-House Costs" include markups to bare costs for the direct overhead requirements of in-house maintenance staff. The figure in this column is the sum of three components: the bare material cost plus 10% (for purchasing and handling small quantities); the bare labor cost plus workers' compensation and average fixed overhead (per the labor rate table on the inside back cover of the book or, if a crew is listed, from the Crew Listings); and the bare equipment cost

10 Total Costs Including O&P Annualized (1,975)

"Total Costs Including O&P" include suggested markups to bare costs for an outside subcontractor's overheard and profit requirements. The figure in this column is the sum of three components: the bare material cost plus 25% for profit; the bare labor cost plus total overhead and profit (per the labor rate table on the inside back cover of this book or, if a crew is listed, from the Crew Listings); and the bare equipment cost plus 10% for profit.

PM Components	Labor-hrs.	W	M	Q	S	A
PM System B2035 110 1950						
Door, sliding, electric						
1 Check with operating or area personnel for deficiencies.	.035				√	√
2 Check for proper operation, binding or misalignment; adjust as necessary.	.062				√	√
3 Check and lubricate door guides, pulleys and hinges.	.142				√	√
4 Inspect and lubricate motor gearbox, drive chain (or belt), and motor; adjust as necessary.	.200				√	√
5 Check operation of limit switch; adjust as necessary.	.224				√	√
6 Check electrical operator, wiring, connections and contacts; adjust as necessary.	.473				√	√
7 Clean area around door.	.066				√	√
8 Fill out maintenance checklist and report deficiencies.	.022				√	√
Total labor-hours/period					1.224	1.224
Total labor-hours/year					1.224	1.224

			Cost Each					
				2015 Bare Costs			Total	Total
Description	Labor-hrs.	Material	Labor	Equip.	Total	In-House	w/O&P
1900	Door, sliding, electric, annually	1.224	15.35	57.50		72.85	93.50	113
1950	Annualized	2.448	21	115		136	176	214

PM Components	Labor-hrs.	W	M	Q	S	A
PM System B2035 110 2950						
Hanger doors, sliding						
1 Check with door operating personnel for deficiencies.	.027			√	√	√
2 Remove debris from door track.	.040			√	√	√
3 Operate door.	.050			√	√	√
4 Check alignment of hanger door, door guides and lubricate.	.650					√
5 Inspect and lubricate wheel drive chain.	.030			√	√	√
6 Adjust brake shoes and check for wear.	.150					√
7 Inspect and lubricate drive wheels, guides, stops and rollers.	.200			√	√	√
8 Fill out maintenance report.	.017			√	√	√
Total labor-hours/period				.364	.364	1.164
Total labor-hours/year				.728	.364	1.164

			Cost Each					
				2015 Bare Costs			Total	Total
Description	Labor-hrs.	Material	Labor	Equip.	Total	In-House	w/O&P
2900	Hanger doors, sliding, annually	1.164	27	61.50		88.50	121	145
2950	Annualized	2.256	47.50	119		166.50	231	274

PM Components	Labor-hrs.	W	M	Q	S	A
PM System B2035 225 1950						
Door, emergency egress, swinging						
1 Remove obstructions that restrict full movement/swing of door.	.013			√	√	√
2 Check swing of door; door must latch on normal closing.	.013			√	√	√
3 Test operation of panic hardware and local alarm battery.	.013			√	√	√
4 Lubricate hardware.	.013			√	√	√
5 Fill out maintenance checklist and report deficiencies.	.013			√	√	√
Total labor-hours/period				.065	.065	.065
Total labor-hours/year				.130	.065	.065

	Description	Labor-hrs.	Cost Each				Total In-House	Total w/O&P
			2015 Bare Costs					
			Material	Labor	Equip.	Total		
1900	Door, emergency egress, swinging, annually	.065	15.35	3.05		18.40	21	24
1950	Annualized	.260	21	12.20		33.20	39.50	46.50

PM Components	Labor-hrs.	W	M	Q	S	A
PM System B2035 300 1950						
Door, revolving, manual						
1 Check with operating or area personnel for deficiencies.	.043				√	√
2 Check for proper operation, binding or misalignment; adjust as necessary.	.043				√	√
3 Check for proper emergency breakaway operation of wings.	.043				√	√
4 Check top, bottom, edge and shaft weatherstripping.	.217				√	√
5 Check and lubricate pivots and bearings.	.217				√	√
6 Clean area around door.	.065				√	√
7 Fill out maintenance checklist and report deficiencies.	.022				√	√
Total labor-hours/period					.650	.650
Total labor-hours/year					.650	.650

				Cost Each				
				2015 Bare Costs			Total	Total
	Description	Labor-hrs.	Material	Labor	Equip.	Total	In-House	w/O&P
1900	Door, revolving, manual, annually	.650	15.35	30.50		45.85	57.50	69
1950	Annualized	1.300	21	61		82	104	126

PM Components	Labor-hrs.	W	M	Q	S	A
PM System B2035 300 2950						
Door, revolving, electric						
1 Check with operating or area personnel for deficiencies.	.043				√	√
2 Check for proper operation, binding or misalignment; adjust as necessary.	.043				√	√
3 Check for proper emergency breakaway operation of wings.	.043				√	√
4 Check top, bottom, edge and shaft weatherstripping.	.217				√	√
5 Check elec. operator, safety devices and activators for proper operation.	.433				√	√
6 Check and lubricate pivots and bearings and operator gearbox.	.325				√	√
7 Clean area around door.	.065				√	√
8 Fill out maintenance checklist and report deficiencies.	.022				√	√
Total labor-hours/period					1.191	1.191
Total labor-hours/year					1.191	1.191

				Cost Each				
				2015 Bare Costs			Total	Total
	Description	Labor-hrs.	Material	Labor	Equip.	Total	In-House	w/O&P
2900	Door, revolving, electric, annually	1.192	49.50	56		105.50	129	153
2950	Annualized	2.382	52.50	112		164.50	206	249

PM Components	Labor-hrs.	W	M	Q	S	A
PM System B2035 400 1950						
Door, overhead, manual, up to 24′ high x 25′ wide						
1 Check with operating or area personnel for deficiencies.	.035				√	√
2 Check for proper operation, binding or misalignment; adjust as necessary.	.135				√	√
3 Check and lubricate door guides, pulleys and hinges.	.880				√	√
4 Clean area around door.	.026				√	√
5 Fill out maintenance checklist and report deficiencies.	.022				√	√
Total labor-hours/period					1.098	1.098
Total labor-hours/year					1.098	1.098

			Cost Each				
			2015 Bare Costs			Total	Total
Description	Labor-hrs.	Material	Labor	Equip.	Total	In-House	w/O&P
1900 Dr., overhead, manual, up to 24′H x 25′W, annually	1.098	15.35	51.50		66.85	85.50	104
1950 Annualized	2.196	21	103		124	160	195

PM Components	Labor-hrs.	W	M	Q	S	A
PM System B2035 410 1950						
Door, overhead, roll-up, electric, up to 24′ high x 25′ wide						
1 Check with operating or area personnel for deficiencies.	.035				√	√
2 Check for proper operation, binding or misalignment; adjust as necessary.	.135				√	√
3 Check and lubricate door guides, pulleys and hinges.	.880				√	√
4 Inspect and lubricate motor gearbox, drive chain (or belt), and motor; adjust as necessary.	.200				√	√
5 Check operation of limit switch; adjust as necessary.	.224				√	√
6 Check electrical operator, wiring, connections and contacts; adjust as necessary.	.473				√	√
7 Clean area around door.	.066				√	√
8 Fill out maintenance checklist and report deficiencies.	.022				√	√
Total labor-hours/period					2.035	2.035
Total labor-hours/year					2.035	2.035

			Cost Each				
			2015 Bare Costs			Total	Total
Description	Labor-hrs.	Material	Labor	Equip.	Total	In-House	w/O&P
1900 Dr., overhead, electric, to 24′H x 25′W, annually	2.035	49.50	95.50		145	182	219
1950 Annualized	4.070	52.50	191		243.50	310	380

PM Components	Labor-hrs.	W	M	Q	S	A
PM System B2035 450 1950						
Shutter, roll-up, electric						
1 Check with operating or area personnel for deficiencies.	.035					√
2 Check for proper operation, binding or misalignment; adjust as necessary.	.135					√
3 Check and lubricate door guides, pulleys and hinges.	.880					√
4 Inspect and lubricate motor gearbox, drive chain (or belt), and motor; adjust as necessary.	.200					√
5 Check operation of limit switch; adjust as necessary.	.224					√
6 Check electrical operator, wiring, connections and contacts; adjust as necessary.	.473					√
7 Clean area around door.	.074					√
8 Fill out maintenance checklist and report deficiencies.	.022					√
Total labor-hours/period						2.043
Total labor-hours/year						2.043

			Cost Each					
			2015 Bare Costs				Total	Total
Description		Labor-hrs.	Material	Labor	Equip.	Total	In-House	w/O&P
1900	Shutter, roll up, electric, annually	2.043	25.50	96		121.50	156	189
1950	Annualized	2.043	32	96		128	163	197

PM Components	Labor-hrs.	W	M	Q	S	A
PM System C1025 110 1950						
Fire Door, Swinging						
1 Remove fusible link hold open devices.	.026			√	√	√
2 Remove obstructions that retard full movement/swing of door.	.013			√	√	√
3 Check swing of door. Door must latch on normal closing.	.013			√	√	√
4 Test operation of panic hardware.	.007			√	√	√
5 Check operation of special devices such as smoke detectors or magnetic door releases.	.013			√	√	√
6 Lubricate hardware.	.013			√	√	√
7 Fill out maintenance checklist and report deficiencies.	.013			√	√	√
Total labor-hours/period				.098	.098	.098
Total labor-hours/year				.196	.098	.098

			Cost Each					
			2015 Bare Costs				Total	Total
Description	Labor-hrs.	Material	Labor	Equip.	Total	In-House	w/O&P
1900 Fire doors, swinging, annually	.098	3.55	4.77		8.32	10.20	12.20
1950 Annualized	.392	7.10	19.05		26.15	33	40

PM Components	Labor-hrs.	W	M	Q	S	A
PM System C1025 110 2950						
Fire Door, Sliding						
1 Clean track.	.013			√	√	√
2 Lubricate pulleys.	.013			√	√	√
3 Inspect cable or chain for wear or damage and proper threading through pulleys.	.078			√	√	√
4 Replace fusible links or other heat activated devices that have been painted. Check operation of other heat activated devices.	.130			√	√	√
5 Permanently remove obstructions that prohibit movement of door.	.013			√	√	√
6 Check operation of door by disconnecting or lifting counterweight or other appropriate means.	.130			√	√	√
7 Check fit of door in binders and fit against stay roll.	.013			√	√	√
8 Check door for breaks in covering and examine for dry rot.	.013			√	√	√
9 Fill out maintenance checklist and report deficiencies.	.013			√	√	√
Total labor-hours/period				.416	.416	.416
Total labor-hours/year				.832	.416	.416

			Cost Each					
			2015 Bare Costs				Total	Total
Description	Labor-hrs.	Material	Labor	Equip.	Total	In-House	w/O&P
2900 Fire doors, sliding, annually	.416	4.35	20		24.35	31.50	38.50
2950 Annualized	1.664	8.70	81		89.70	116	143

PM Components	Labor-hrs.	W	M	Q	S	A
PM System C1025 110 3950						
Fire Door, Rollup						
1 Check with door operating personnel for deficiencies.	.027			√	√	√
2 Drop test heat activated or fusible link roll up fire door.	.070			√	√	√
3 Adjust rate of descent governor to desired closing speed.	.200					√
4 Inspect, clean and lubricate door gear assembly.	.200			√	√	√
5 Check alignment of overhead door guides, clean and lubricate.	.100			√	√	√
6 Check and adjust door counter balance assembly.	.250					√
7 Check electrical wiring and contacts for wear.	.200					√
8 Check and adjust limit switches.	.170					√
9 Inspect general condition of door for need of paint and repairs.	.005			√	√	√
10 Replace missing or tighten loose nuts and bolts.	.040			√	√	√
11 After maintenance clean around door and ensure access.	.050			√	√	√
12 Fill out maintenance report.	.017			√	√	√
Total labor-hours/period				.509	.509	1.329
Total labor-hours/year				1.018	.509	1.329

			Cost Each					
			2015 Bare Costs				Total	Total
Description		Labor-hrs.	Material	Labor	Equip.	Total	In-House	w/O&P
3900	Fire doors, roll-up, annually	1.329	38.50	70		108.50	148	174
3950	Annualized	2.856	50	150		200	280	335

PM Components	Labor-hrs.	W	M	Q	S	A
PM System D1015 100 1950						
Elevator, cable, electric, passenger/freight						
1 Ride car, check for any unusual noise or operation.	.411		✓	✓	✓	✓
2 Inspect machine room equipment.	.286		✓	✓	✓	✓
3 Motor room:						
A) visually inspect controllers and starters.	.012		✓	✓	✓	✓
B) visually inspect selector.	.012		✓	✓	✓	✓
C) lubricate and adjust tension of selector.	.113		✓	✓	✓	✓
D) inspect sleeve bearings of hoist motor.	.012		✓	✓	✓	✓
E) inspect brushes and commutator of hoist motor.	.012		✓	✓	✓	✓
F) inspect sleeve bearings of motor generator.	.012		✓	✓	✓	✓
G) inspect brushes and commutator of motor generator.	.014		✓	✓	✓	✓
H) inspect sleeve bearings of exciter.			✓	✓	✓	✓
I) inspect brushes and commutator of exciter.	.022		✓	✓	✓	✓
J) inspect sleeve bearings of regulator dampening motors and tach generators.	.022		✓	✓	✓	✓
K) inspect brushes and commutator of regulator dampening motors and tach generators.	.044		✓	✓	✓	✓
L) inspect and grease sleeve bearings of geared machines.	.047		✓	✓	✓	✓
M) inspect oil level in worm and gear of geared machines.	.044		✓	✓	✓	✓
N) visually inspect the brakes.	.022		✓	✓	✓	✓
O) inspect and grease, if necessary, sleeve bearings of drive deflectors and secondary sheaves.	.186		✓	✓	✓	✓
P) visually inspect and adjust contacts of controllers and starters.	.338			✓	✓	✓
Q) inspect and adjust main operating contactor and switches of controllers and starters.	.338			✓	✓	✓
R) inspect selector, adjusting contacts, brushes and cams if needed.	.103			✓	✓	✓
S) inspect selsyn and advancer, motor brushes and commutators of selector.	.022			✓	✓	✓
T) clean hoist motor brush rigging and commutator.	.068			✓	✓	✓
U) clean motor generator and tighten brush rigging and clean commutator.	.068			✓	✓	✓
V) clean exciter brush rigging and commutator.	.066			✓	✓	✓
W) clean brush rigging and commutator of regulator dampening motors and tach generators.	.099			✓	✓	✓
X) inspect machine worm and gear thrust and backlash and spider bolts; replace oil.	.047			✓	✓	✓
Y) clean and lubricate brakes, pins and linkage.	.099			✓	✓	✓
Z) inspect governors; lubricate sleeve bearings.	.099			✓	✓	✓
Aa) inspect controllers and starters; adjust setting of overloads.	.169					✓
Ab) tighten connections on controllers and starters; clean fuses and holders.	.169					✓
Ac) set controllers and starter timers.	.169					✓
Ad) clean controllers and starters.	.169					✓
Ae) clean and inspect selector components.	.068					✓
Af) inspect group operation controllers; clean and lubricate contacts, timers, steppers of operation controller.	.099					✓
Ag) change oil and/or grease roller bearings of the hoist motor.	.094					✓
Ah) vacuum and blowout the hoist motor.	.094					✓
Ai) tighten all connections on the hoist motor.	.022					✓

PM Components	Labor-hrs.	W	M	Q	S	A
Aj) change oil and/or grease generator's roller bearings.	.069					√
Ak) vacuum and blowout motor generator.	.094					√
Al) inspect all connections on the motor generator and tighten if necessary.	.022					√
Am) inspect roller bearings of the exciter and change oil and/or grease.	.069					√
An) vacuum and blowout exciter.	.094					√
Ao) change oil and/or grease roller bearings of regulator dampening motors and tach generators.	.094					√
Ap) vacuum/blowout regulator dampening motors and tach generators.	.044					√
Aq) inspect all regulator dampening motors and tach generator connections.	.047					√
Ar) change oil and/or grease roller bearings of the geared machines.	.094					√
As) clean and lubricate brake cores.	.099					√
At) inspect drive deflectors and secondary sheaves; change oil and grease roller bearings if necessary.	.170					√
Au) inspect drive deflectors and secondary sheaves' cable grooves for cracks.	.012					√
Av) inspect governors; change oil and grease roller bearings, if necessary.	.094					√
4 Hatch:						
A) lubricate rails of hoistway.	.733		√	√	√	√
B) inspect car top and grease sleeve bearings if needed.	.151		√	√	√	√
C) inspect counterweight and grease sleeve bearings if needed.	.068		√	√	√	√
D) visually inspect cables, chains, hoist, compensating governor for wear and equalization.	.047		√	√	√	√
E) lubricate cables, chain hoist and compensating governor if necessary.	.186		√	√	√	√
F) inspect door operator; clean and lubricate chain and belt tension.	.177			√	√	√
G) inspect door, clean and adjust safety edge, light ray and cables.	.052			√	√	√
H) clean and adjust proximity devices on door.	.103			√	√	√
I) inspect and lubricate governor tape and compensating sheave in pit.	.068			√	√	√
J) clean and lubricate compensating tie down in pit.	.047			√	√	√
K) inspect pit's run-by on counter weight.	.142			√	√	√
L) inspect governor and compensating sheaves or chains for clearance to pit floor.	.022			√	√	√
M) inspect/lubricate overhead hatch switches and cams.	.096					√
N) clean and lubricate rollers, cables, sight guards, chains, motors, closures, limit and zone switches.	.068					√
O) inspect door gibs and fastening.	.022					√
5 Inspect car top; change oil and grease roller bearings, if necessary.	.044					√
6 Check safety switches, indicators, leveling devices, selector tape, switches and hatches.	.044					√
7 Check and lubricate fan motor.	.072					√
8 Inspect counterweight; change oil and grease roller bearings, if necessary.	.094					√
9 Check safety linkage safety and cwt guides.	.098					√
10 Inspect compensation hatch.	.047					√
11 Inspect chains, hoist and compensating governor.	.094					√
12 Inspect traveling cables for tracking and wear.	.070					√

PM Components	Labor-hrs.	W	M	Q	S	A
13 Inspect door operator commutator and brushes.	.012					√
14 Inspect/clean/lubricate door operator clutch, retiring cam, door gib, rollers, tracks, upthrusts and relating cables.	.103					√
15 Under car:						
A) inspect/clean/lubricate safety devices, linkages, car guides, selector tape, switches and hitches.	.020					√
B) inspect cable hitch and loops; load weighing devices.	.027					√
16 Inspect/lubricate pit hatch switches, cams and pit oil buffer.	.083					√
17 In car:						
A) inspect and clean fixtures and signal in operating panel and car position and direction indicator.	.103		√	√	√	√
B) check operation of emergency lights and bells.	.228		√	√	√	√
C) check handrails, ceiling panels and hang on panels for tightness.	.027		√	√	√	√
D) check for tripping hazards.	.044		√	√	√	√
18 Hallway corridor:						
A) inspect hall buttons, signal lamps, lanterns and hall position indicator.	.046		√	√	√	√
B) inspect starter station, key operation and lamps.	.069		√	√	√	√
19 Clean equipment and surrounding area.	.066		√	√	√	√
20 Fill out maintenance checklist.	.022		√	√	√	√
Total labor-hours/period			3.061	5.019	5.019	8.068
Total labor-hours/year			24.488	10.038	5.019	8.068

	Description	Labor-hrs.	Material	Labor	Equip.	Total	Total In-House	Total w/O&P
1900	Elevator, cable, electric, passenger / freight, annually	8.068	1,600	615		2,215	2,525	2,950
1950	Annualized	47.613	3,175	3,650		6,825	7,975	9,600

PM Components	Labor-hrs.	W	M	Q	S	A
PM System D1015 110 1950						
Elevator, hydraulic, passenger/freight						
1 Ride car, checking for any unusual noise or operation.	.052		√	√	√	√
2 In car:						
A) inspect and clean fixtures and signal in operating panel and car position and direction indicator.	.103		√	√	√	√
B) check operation of emergency lights and bell.	.020		√	√	√	√
C) check handrails, ceiling panels and hang on panels for tightness.	.027		√	√	√	√
D) check for tripping hazards.	.012		√	√	√	√
3 Inspect and lubricate rails of hoistway.	.094		√	√	√	√
4 Hallway corridor:						
A) inspect hall buttons, signal lamps, lanterns and hall position indicator.	.046		√	√	√	√
B) inspect starter station, key operation and lamps.	.069		√	√	√	√
5 Motor room:						
A) inspect machine room equipment.	.077		√	√	√	√
B) lockout and log record.	.022		√	√	√	√
C) inspect tank oil level.	.012		√	√	√	√
D) inspect and adjust controller contacts; main operating contactors and switches.	.196			√	√	√
E) inspect pump and valve unit for leaks.	.051			√	√	√
F) inspect and adjust controller overloads; set timers.	.049					√
G) tighten connections and clean controller fuses and holders.	.020					√
H) inspect and lubricate pump motor bearings.	.099					√
6 Hatch:						
A) check hoistway car rails, brackets and fish plates.	.103					√
B) inspect/lubricate overhead hatch switches and cams.	.070					√
C) inspect/clean/lubricate hatch doors locks, rollers, tracks, upthrusts, relating cables, racks, sight guards and closers, motors, gear boxes, limit and zone switches.	.096					√
D) inspect door gibs and fastening.	.022					√
E) inspect/clean/lubricate cab top guides, steading devices, safety switches, inductors, leveling devices, selector tape, switches, hitches and fan motor.	.068					√
F) check traveling cables for wear.	.070					√
G) inspect/clean/lubricate door operator roller tracks, upthrusts, related cables, clutch, retiring cam and door gib.	.078					√
H) inspect door operator; clean and lubricate chain and belt tension.	.052			√	√	√
I) inspect door, clean and adjust safety edge, light ray and cables.	.103			√	√	√
J) clean and adjust proximity devices on door.	.068			√	√	√
7 Inspect pit gland packing.	.010			√	√	√
8 Inspect/clean/lubricate under car guides, selector tape, traveling cable, switches and platen plate assembly.	.068					√
9 Clean equipment and surrounding area.	.074		√	√	√	√
10 Fill out maintenance checklist.	.022		√	√	√	√
Total labor-hours/period			.630	1.110	1.110	1.853
Total labor-hours/year			5.040	2.220	1.110	1.853

		Cost Each					
		2015 Bare Costs				**Total In-House**	**Total w/O&P**
Description	**Labor-hrs.**	**Material**	**Labor**	**Equip.**	**Total**		
1900 Elevator, hydraulic, passenger / freight, annually	1.853	1,450	142		1,592	1,775	2,025
1950 Annualized	10.223	1,800	780		2,580	2,950	3,450

For customer support on your Facilities Maintenance & Repair Cost Data, call 877.801.5724.

393

PM Components	Labor-hrs.	W	M	Q	S	A
PM System D1015 310 1950						
Wheelchair Lift						
1 Check with operating personnel for any deficiencies.	.033				√	√
2 Check voltage and amperage under load.	.091				√	√
3 Check bolts securing drive cabinet and base, tighten accordingly.	.033				√	√
4 Check belt tension.	.033				√	√
5 Check lift nut assembly.	.026				√	√
6 Check cam rollers.	.026				√	√
7 Check wear pads for excessive wear.	.026				√	√
8 Inspect motor and shaft pulleys.	.039				√	√
9 Check Acme screw alignment.	.033				√	√
10 Inspect and lubricate upper and lower bearings.	.039				√	√
11 Check fastening of cable harness.	.026				√	√
12 Check alignment of platform and doors.	.033				√	√
13 Check door interlock switch for proper operation.	.039				√	√
14 Check operation of final limit switch and emergency stop/alarm.	.033				√	√
15 Check call/send controls at each station and on platform.	.039				√	√
16 Lube hinge of flip up ramp.	.033				√	√
17 Fill out maintenance checklist and report deficiencies.	.022				√	√
Total labor-hours/period					.604	.604
Total labor-hours/year					.604	.604

			Cost Each					
			2015 Bare Costs				Total	Total
	Description	Labor-hrs.	Material	Labor	Equip.	Total	In-House	w/O&P
1900	Wheelchair lift, annually	.604	6.15	33		39.15	47.50	58.50
1950	Annualized	1.208	12.25	66		78.25	95.50	117

PM Components	Labor-hrs.	W	M	Q	S	A
PM System D1025 100 1950						
Escalator						
1 Ride escalator; check operation for smoothness, unusual vibration or noise, condition of handrails, etc.	.104	√	√	√	√	√
2 Deenergize, tag, and lockout the electrical circuit.	.017	√	√	√	√	√
3 Inspect comb plates at both ends of escalator for broken teeth and check for proper clearance between combs and step teeth and check for broken step treads.	.039	√	√	√	√	√
4 Check clearance between steps and skirt panel; look for loose trim, screws or bolts, that could snag or damage clothing or cause injury; check operation of handrail brushes.	.087	√	√	√	√	√
5 Clean escalator machine space.	.113	√	√	√	√	√
6 Lubricate step rollers, step chain, drive gears or chains, handrail drive chains, etc., according to manufacturers' instructions; observe gears and chains for signs of wear, misalignment, etc; adjust as required.	.267	√	√	√	√	√
7 Check motor for signs of overheating.	.004	√	√	√	√	√
8 Inspect controller for loose leads, burned contacts, etc.; repair as required.	.022	√	√	√	√	√
9 Clean handrails as required.	.217	√	√	√	√	√
10 Check escalator lighting; replace bulbs as required.	.099	√	√	√	√	√
11 Operate each emergency stop button and note that the escalator stops; if the escalator has the capabilities of running in both directions, stop buttons should function properly for each direction of travel; observe the stopping distance.	.013	√	√	√	√	√
12 Clean surrounding area.	.066	√	√	√	√	√
13 Fill out maintenance checklist and report deficiencies.	.022	√	√	√	√	√
Total labor-hours/period		1.070	1.070	1.070	1.070	1.070
Total labor-hours/year		40.660	8.560	2.140	1.070	1.070

			Cost Each					
			2015 Bare Costs				Total	Total
Description		Labor-hrs.	Material	Labor	Equip.	Total	In-House	w/O&P
1900	Escalator, annually	1.070	2,325	82		2,407	2,650	3,025
1950	Annualized	53.501	2,925	4,100		7,025	8,275	9,975

PM Components	Labor-hrs.	W	M	Q	S	A
PM System D1095 100 1950						
Dumbwaiter, electric						
1 Check with operating or area personnel for any obvious deficiencies.	.086					✓
2 Operate dumbwaiter through complete cycle and check operation for unusual noises or problems.	.052					✓
3 Inspect and clean operating panel, starter station, key operation and lamps.	.072					✓
4 Check operation of emergency lights and alarms.	.052					✓
5 Inspect, clean and lubricate motor.	.072					✓
6 Check operation of indicator lights and buttons.	.052					✓
7 Inspect and lubricate rails of guides.	.148					✓
8 Clean equipment and surrounding area.	.077					✓
9 Fill out maintenance checklist and report deficiencies.	.022					✓
Total labor-hours/period						.633
Total labor-hours/year						.633

				Cost Each				
				2015 Bare Costs			**Total**	**Total**
	Description	Labor-hrs.	Material	Labor	Equip.	Total	In-House	w/O&P
1900	Dumbwaiter, electric, annually	.633	46.50	48.50		95	111	133
1950	Annualized	.633	93.50	48.50		142	163	191

PM Components	Labor-hrs.	W	M	Q	S	A
PM System D1095 110 1950						
Dumbwaiter, hydraulic						
1 Check with operating or area personnel for any obvious deficiencies.	.086				✓	✓
2 Operate dumbwaiter through complete cycle and check operation for unusual noises or problems.	.052				✓	✓
3 Inspect and clean operating panel, starter station, key operation and lamps.	.072				✓	✓
4 Check operation of emergency lights and alarms.	.052				✓	✓
5 Inspect pump and valve unit for leaks.	.072				✓	✓
6 Inspect tank oil level; add as required.	.039				✓	✓
7 Inspect and lubricate rails of guides.	.148				✓	✓
8 Inspect jack seals.	.051				✓	✓
9 Clean equipment and surrounding area.	.077				✓	✓
10 Fill out maintenance checklist and report deficiencies.	.022				✓	✓
Total labor-hours/period					.671	.671
Total labor-hours/year					.671	.671

				Cost Each				
				2015 Bare Costs			**Total**	**Total**
	Description	Labor-hrs.	Material	Labor	Equip.	Total	In-House	w/O&P
1900	Dumbwaiter, hydraulic, annually	.671	72.50	51.50		124	144	170
1950	Annualized	1.342	145	103		248	287	340

PM Components	Labor-hrs.	W	M	Q	S	A
PM System D1095 200 1950						
Pneumatic Tube System						
1 Check with operating personnel for any deficiencies.	.039			√	√	√
2 Clean air intake screen.	.039			√	√	√
3 Check seal at doors.	.130			√	√	√
4 Inspect check valves	.026			√	√	√
5 Check blower motor and lubricate.	.065			√	√	√
6 Inspect carrier ends, latches and air washers	.026			√	√	√
7 Fill out maintenance checklist and report deficiencies.	.026			√	√	√
Total labor-hours/period				.351	.351	.351
Total labor-hours/year				.702	.351	.351

			Cost Each					
			2015 Bare Costs				**Total**	**Total**
Description		**Labor-hrs.**	**Material**	**Labor**	**Equip.**	**Total**	**In-House**	**w/O&P**
1900	Pneumatic Tube System, annually	.351		19.20		19.20	24	29.50
1950	Annualized	1.404		77		77	95	119

PM Components	Labor-hrs.	W	M	Q	S	A
PM System D2015 100 1950						
Urinals (Time is per fixture)						
1 Urinals - Flush and adjust water flow if required.	.020			√	√	√
2 Inspect for missing or damaged parts/caps and replace.	.020			√	√	√
3 Fill out maintenance checklist and report deficiencies.	.017			√	√	√
Total labor-hours/period				.057	.057	.057
Total labor-hours/year				.114	.057	.057

			Cost Each				
			2015 Bare Costs			Total	Total
Description	Labor-hrs.	Material	Labor	Equip.	Total	In-House	w/O&P
1900 Urinals, annually	.057		3.35		3.35	4.18	5.20
1950 Annualized	.228		13.40		13.40	16.75	21

PM Components	Labor-hrs.	W	M	Q	S	A
PM System D2015 100 2950						
Toilet, vacuum breaker type (Time is per fixture)						
1 Toilet (vacuum breaker type) - Flush and adjust water flow if required.	.020			√	√	√
2 Inspect for missing or damaged parts/caps, seat supports, and replace.	.020			√	√	√
3 Fill out maintenance checklist and report deficiencies.	.017			√	√	√
Total labor-hours/period				.057	.057	.057
Total labor-hours/year				.114	.057	.057

			Cost Each				
			2015 Bare Costs			Total	Total
Description	Labor-hrs.	Material	Labor	Equip.	Total	In-House	w/O&P
2900 Toilet (vacuum breaker type), annually	.057		3.35		3.35	4.18	5.20
2950 Annualized	.228		13.40		13.40	16.75	21

PM Components	Labor-hrs.	W	M	Q	S	A
PM System D2015 100 3950						
Toilet, tank type (Time is per fixture)						
1 Toilet - Clean flapper seat if leaking, adjust fill level if required.	.050			√	√	√
2 Inspect for missing or damaged parts/caps, seat supports, and replace.	.030			√	√	√
3 Fill out maintenance checklist and report deficiencies.	.017			√	√	√
Total labor-hours/period				.097	.097	.097
Total labor-hours/year				.194	.097	.097

			Cost Each				
			2015 Bare Costs			Total	Total
Description	Labor-hrs.	Material	Labor	Equip.	Total	In-House	w/O&P
3900 Toilet (tank type), annually	.097		5.70		5.70	7.10	8.90
3950 Annualized	.388		23		23	28.50	35.50

PM Components	Labor-hrs.	W	M	Q	S	A
PM System D2015 100 4950						
Lavatories (Time is per fixture)						
1 Lavatories - Operate faucets, replace washers/"O" Rings as necessary.	.040			√	√	√
2 Observe drain flow, clean trap if flow is obstructed.	.030			√	√	√
3 Fill out maintenance checklist and report deficiencies.	.017			√	√	√
Total labor-hours/period				.087	.087	.087
Total labor-hours/year				.174	.087	.087

		Cost Each				Total	Total
			2015 Bare Costs			In-House	w/O&P
Description	Labor-hrs.	Material	Labor	Equip.	Total		
4900 Lavatories, annually	.087	7	4.23		11.23	13.25	15.65
4950 Annualized	.348	7	16.95		23.95	30	36.50

PM Components	Labor-hrs.	W	M	Q	S	A
PM System D2015 100 5950						
Showers (Time is per fixture)						
1 Showers - Check for damaged, or missing shower heads/handles and replace if required.	.040			√	√	√
2 Fill out maintenance checklist and report deficiencies.	.017			√	√	√
Total labor-hours/period				.057	.057	.057
Total labor-hours/year				.114	.057	.057

		Cost Each				Total	Total
			2015 Bare Costs			In-House	w/O&P
Description	Labor-hrs.	Material	Labor	Equip.	Total		
5900 Showers, annually	.057	9.15	2.77		11.92	13.70	15.95
5950 Annualized	.228	9.15	11.10		20.25	24.50	29.50

D20 PLUMBING

D2015 800 | Drinking Fountain

PM Components	Labor-hrs.	W	M	Q	S	A
PM System D2015 800 1950						
Drinking fountain						
1 Check unit for proper operation, excessive noise or vibration.	.035					√
2 Clean condenser coils and fan as required.	.222					√
3 Check for water leaks in supply line and drain.	.077					√
4 Check water flow; adjust as necessary.	.039					√
5 Check drinking water temperature to ensure the unit is operating properly.	.039					√
6 Check electrical connections and cord; tighten and repair as necessary.	.120					√
7 Clean area around fountain.	.066					√
8 Fill out maintenance checklist and report deficiencies.	.022					√
Total labor-hours/period						.620
Total labor-hours/year						.620

		Cost Each				Total	Total
			2015 Bare Costs			In-House	w/O&P
Description	Labor-hrs.	Material	Labor	Equip.	Total		
1900 Drinking fountain, annually	.620	17.85	30		47.85	59.50	71.50
1950 Annualized	.620	17.85	30		47.85	59.50	71.50

PM Components	Labor-hrs.	W	M	Q	S	A
PM System D2025 120 1950						
Valve, butterfly, above 4″						
1 Lubricate valves that have grease fittings.	.022					✓
2 Open and close valve using handle, wrench, or hand wheel to check operation.	.049					✓
3 Check for leaks.	.007					✓
4 Clean valve exterior and area around valve.	.066					✓
5 Fill out maintenance report and report deficiencies.	.022					✓
Total labor-hours/period						.166
Total labor-hours/year						.166

			Cost Each					
				2015 Bare Costs			Total	Total
Description	**Labor-hrs.**	**Material**	**Labor**	**Equip.**	**Total**	**In-House**	**w/O&P**
1900	Valve, butterfly, above 4″, annually	.166	8.15	8.10		16.25	19.60	23.50
1950	Annualized	.166	8.15	8.10		16.25	19.60	23.50

PM Components	Labor-hrs.	W	M	Q	S	A
PM System D2025 120 2950						
Valve, butterfly, auto, above 4″						
1 Lubricate valve actuator stem and valve stem, where possible.	.022					✓
2 Check automatic valve for proper operation.	.062					✓
3 Check packing gland for leaks; adjust as required.	.160					✓
4 Check pneumatic operator and tubing for air leaks.	.047					✓
5 Clean valve exterior and area around valve.	.034					✓
6 Fill out maintenance report and report deficiencies.	.022					✓
Total labor-hours/period						.347
Total labor-hours/year						.347

			Cost Each					
				2015 Bare Costs			Total	Total
Description	**Labor-hrs.**	**Material**	**Labor**	**Equip.**	**Total**	**In-House**	**w/O&P**
2900	Valve, butterfly, auto, above 4″, annually	.347	8.15	16.90		25.05	31	37.50
2950	Annualized	.347	8.15	16.90		25.05	31	37.50

PM Components	Labor-hrs.	W	M	Q	S	A
PM System D2025 125 1950						
Valve, check, above 4″						
1 Inspect valve for leaks; repair as necessary.	.009					✓
2 Adjust system pressure/flow to verify that valve is opening and closing properly, if applicable.	.160					✓
3 Clean valve and area around valve.	.066					✓
4 Fill out maintenance report and report deficiencies.	.022					✓
Total labor-hours/period						.257
Total labor-hours/year						.257

			Cost Each					
				2015 Bare Costs			Total	Total
Description	**Labor-hrs.**	**Material**	**Labor**	**Equip.**	**Total**	**In-House**	**w/O&P**
1900	Valve, check, above 4″, annually	.257	8.15	12.50		20.65	25.50	30.50
1950	Annualized	.257	8.15	12.50		20.65	25.50	30.50

PM Components	Labor-hrs.	W	M	Q	S	A
PM System D2025 130 1950						
Valve, ball, above 4″						
1 Lubricate valves that have grease fittings.	.022					√
2 Open and close valve using handle, wrench, or hand wheel to check operation.	.049					√
3 Check for leaks.	.007					√
4 Clean valve and area around valve.	.066					√
5 Fill out maintenance report and report deficiencies.	.022					√
Total labor-hours/period						.166
Total labor-hours/year						.166

			Cost Each					
			2015 Bare Costs				Total	Total
Description	Labor-hrs.	Material	Labor	Equip.	Total	In-House	w/O&P
1900	Valve, ball, above 4″, annually	.166	8.15	8.10		16.25	19.60	23.50
1950	Annualized	.166	8.15	8.10		16.25	19.60	23.50

PM Components	Labor-hrs.	W	M	Q	S	A
PM System D2025 135 1950						
Valve, diaphragm, above 4″						
1 Lubricate valve stem, close and open valve to check operation.	.022					√
2 Check valve for proper operation and leaks; tighten packing and flange bolts as required.	.044					√
3 Clean valve exterior and area around valve.	.034					√
4 Fill out maintenance report and report deficiencies.	.022					√
Total labor-hours/period						.122
Total labor-hours/year						.122

			Cost Each					
			2015 Bare Costs				Total	Total
Description	Labor-hrs.	Material	Labor	Equip.	Total	In-House	w/O&P
1900	Valve, diaphragm, above 4″, annually	.122	8.15	5.95		14.10	16.75	19.85
1950	Annualized	.122	8.15	5.95		14.10	16.75	19.85

PM Components	Labor-hrs.	W	M	Q	S	A
PM System D2025 140 1950						
Valve, gate, above 4″						
1 Lubricate valve stem; close and open valve to check operation.	.049					√
2 Check packing gland for leaks; tighten packing and flange bolts as req.	.022					√
3 Clean valve exterior and area around valve.	.066					√
4 Fill out maintenance report and report deficiencies.	.022					√
Total labor-hours/period						.159
Total labor-hours/year						.159

			Cost Each					
			2015 Bare Costs				Total	Total
Description	Labor-hrs.	Material	Labor	Equip.	Total	In-House	w/O&P
1900	Valve, gate, above 4″, annually	.159	8.15	7.75		15.90	19.15	22.50
1950	Annualized	.159	8.15	7.75		15.90	19.15	22.50

PM Components	Labor-hrs.	W	M	Q	S	A
PM System D2025 145 1950						
Valve, globe, above 4″						
1 Lubricate stem, close and open valve to check operation.	.049					√
2 Check packing gland for leaks; tighten packing and flange bolts as req.	.022					√
3 Clean valve exterior.	.066					√
4 Fill out maintenance report and report deficiencies.	.022					√
Total labor-hours/period						.159
Total labor-hours/year						.159

			Cost Each					
				2015 Bare Costs			Total	Total
Description		Labor-hrs.	Material	Labor	Equip.	Total	In-House	w/O&P
1900	Valve, globe, above 4″, annually	.159	8.15	7.75		15.90	19.15	22.50
1950	Annualized	.159	8.15	7.75		15.90	19.15	22.50

PM Components	Labor-hrs.	W	M	Q	S	A
PM System D2025 145 2950						
Valve, globe, auto, above 4″						
1 Lubricate valve actuator stem and valve stem, where possible.	.022					√
2 Check automatic valve for proper operation.	.062					√
3 Check packing gland for leaks; adjust as required.	.160					√
4 Check penumatic operator and tubing for air leaks.	.025					√
5 Clean valve exterior and area around valve.	.034					√
6 Fill out maintenance report and report deficiencies.	.022					√
Total labor-hours/period						.325
Total labor-hours/year						.325

			Cost Each					
				2015 Bare Costs			Total	Total
Description		Labor-hrs.	Material	Labor	Equip.	Total	In-House	w/O&P
2900	Valve, globe, auto, above 4″, annually	.325	8.15	15.80		23.95	30	35.50
2950	Annualized	.325	8.15	15.80		23.95	30	35.50

D20 PLUMBING | D2025 150 | Valve, Motor Operated

PM Components	Labor-hrs.	W	M	Q	S	A
PM System D2025 150 1950						
Valve, motor operated, above 4″						
1 Lubricate valve actuator stem and valve stem, where possible.	.091					√
2 Check motor and valve for proper operation, including limit switch; adjust as required.	.022				√	√
3 Check packing gland for leaks; adjust as required.	.113				√	√
4 Check electrical wiring, connections and contacts; repair as necessary.	.120				√	√
5 Inspect and lubricate motor gearbox as required.	.099				√	√
6 Clean valve exterior and area around valve.	.034				√	√
7 Fill out maintenance report and report deficiencies.	.022				√	√
Total labor-hours/period						.501	.501
Total labor-hours/year						.501	.501

			Cost Each					
				2015 Bare Costs			Total	Total
Description		Labor-hrs.	Material	Labor	Equip.	Total	In-House	w/O&P
1900	Valve, motor operated, above 4″, annually	.501	8.15	24.50		32.65	41	49.50
1950	Annualized	1.002	16.30	49		65.30	82	99.50

PM Components	Labor-hrs.	W	M	Q	S	A
PM System D2025 155 1950 **Valve, OS&Y, above 4″**						
1 Lubricate stem; open and close valve to check operation.	.049					√
2 Check packing gland for leaks; tighten packing and flange bolts as required.	.022					√
3 Clean valve exterior and area around valve.	.066					√
4 Fill out maintenance report and report deficiencies.	.022					√
Total labor-hours/period						.159
Total labor-hours/year						.159

			Cost Each					
			2015 Bare Costs				Total	Total
Description	Labor-hrs.	Material	Labor	Equip.	Total		In-House	w/O&P
1900 Valve, OS&Y, above 4″, annually	.159	8.15	7.75		15.90		19.15	22.50
1950 Annualized	.159	8.15	7.75		15.90		19.15	22.50

PM Components	Labor-hrs.	W	M	Q	S	A
PM System D2025 190 1950						
Solar, closed loop hot water heating system, up to 6 panels						
1 Check with operating or area personnel for deficiencies.	.035				√	√
2 Inspect interior piping and connections for leaks and damaged insulation; tighten connections and repair damaged insulation as necessary.	.125				√	√
3 Check zone and circulating pump motors for excessive overheating; lubricate motor bearings.	.077				√	√
4 Check pressure and air relief valves for proper operation.	.030				√	√
5 Check control panel and differential thermostat for proper operation.	.094				√	√
6 Clean sight glasses, controls, pumps, and flow indicators on tanks.	.127				√	√
7 Check system pressure on closed loop for loss of fluid.	.046				√	√
8 Check fluid level on drain-back systems; add fluid as necessary.	.029				√	√
9 Test glycol strength in closed systems, as applicable; if required, drain system and replace with new fluid mixture.	.222					√
10 Check heat exchanger for exterior leaks.	.077				√	√
11 Clean strainers and traps.	.181					√
12 Check storage and expansion tanks; for leaks and deteriorated insulation.	.077					√
13 Inspect all collector piping for leaks and damaged insulation; tighten connections and repair as required.	.133				√	√
14 Inspect collector glazing for cracks and seals for tightness; tighten or replace seals as necessary.	.124				√	√
15 Wash/clean glazing on collector panels.	.585					√
16 Inspect ferrule around pipe flashing where solar piping runs through roof; repair as necessary.	.086				√	√
17 Check collector mounting brackets and bolts; tighten as required.	.094				√	√
18 Clean area.	.066				√	√
19 Fill out maintenance checklist and report deficiencies.	.022				√	√
Total labor-hours/period					1.165	2.230
Total labor-hours/year					1.165	2.230

Description		Labor-hrs.	2015 Bare Costs				Total In-House	Total w/O&P
			Material	Labor	Equip.	Total		
1900	Wtr. htng. sys., solar clsd. lp., up to 6 panels, annually	2.230	245	133		378	435	515
1950	Annualized	3.395	246	203		449	525	625

PM Components	Labor-hrs.	W	M	Q	S	A
PM System D2025 260 1950						
Water heater, gas, to 120 gallon						
1 Check with operating or area personnel for deficiencies.	.035				√	√
2 Check for water leaks to tank and piping. check for fuel system leaks.	.077				√	√
3 Check gas burner and pilot for proper flame; adjust if required.	.118				√	√
4 Check operation and condition of pressure relief valve.	.010				√	√
5 Check automatic controls for proper operation (temperature regulators, thermostatic devices, automatic fuel shut off valve, etc.).	.094				√	√
6 Check draft diverter and clear openings, if clogged.	.027				√	√
7 Check electrical wiring for fraying and loose connections on oil burner.	.072				√	√
8 Check for proper water temperature setting; adjust as required.	.029				√	√
9 Check condition of flue pipe, and chimney.	.147				√	√
10 Drain sediment from tank.	.325					√
11 Clean up area around unit.	.066				√	√
12 Fill out maintenance checklist and report deficiencies.	.022				√	√
Total labor-hours/period					.697	1.022
Total labor-hours/year					.697	1.022

			Cost Each				
			2015 Bare Costs			**Total**	**Total**
Description	**Labor-hrs.**	**Material**	**Labor**	**Equip.**	**Total**	**In-House**	**w/O&P**
1900 Water heater, gas, to 120 gal., annually	1.022	65.50	61		126.50	148	177
1950 Annualized	1.719	65.50	103		168.50	200	242

PM Components	Labor-hrs.	W	M	Q	S	A
PM System D2025 260 2950						
Water heater, oil fired, to 100 gallon						
1 Check with operating or area personnel for deficiencies.	.035				√	√
2 Check for water leaks to tank and piping; check for fuel system leaks.	.077				√	√
3 Check burner flame and pilot on oil burner; adjust if required.	.073				√	√
4 Check operation and condition of pressure relief valve.	.010				√	√
5 Check automatic controls for proper operation (temperature regulators, thermostatic devices, automatic fuel shut off valve, etc.).	.094				√	√
6 Check fuel filter element on oil burner.	.068				√	√
7 Check fuel level in tank; check tank, fill pipe and fuel lines and connections for damage.	.022				√	√
8 Inspect, clean, and adjust electrodes and nozzles on oil burners; inspect fire box and flame detection scanner.	.254				√	√
9 Check electrical wiring for fraying and loose connections on oil burner.	.072				√	√
10 Check for proper water temperature setting; adjust as required.	.029				√	√
11 Clean fire box.	.577					√
12 Check for proper draft adjustment; adjust draft meter if necessary.	.005				√	√
13 Check condition of flue pipe, damper and chimney.	.147				√	√
14 Drain sediment from tank.	.325					√
15 Clean up area around unit.	.066				√	√
16 Fill out maintenance checklist and report deficiencies.	.022				√	√
Total labor-hours/period					.974	1.876
Total labor-hours/year					.974	1.876

			Cost Each				
			2015 Bare Costs			**Total**	**Total**
Description	**Labor-hrs.**	**Material**	**Labor**	**Equip.**	**Total**	**In-House**	**w/O&P**
2900 Water heater, oil fired, to 100 gal., annually	1.876	74	112		186	223	268
2950 Annualized	2.850	82	170		252	305	370

PM Components	Labor-hrs.	W	M	Q	S	A
PM System D2025 260 3950						
Water heater, steam coil, to 2500 gallon						
1 Check with operating or area personnel for deficiencies.	.035				√	√
2 Check for water leaks to tank and piping; check steam lines for leaks.	.077				√	√
3 Check operation and condition of pressure relief valve.	.010				√	√
4 Check steam modulating valve and steam condensate trap for proper operation.	.068				√	√
5 Check electrical wiring connections on controls and switches.	.094				√	√
6 Check for proper water temperature setting; adjust as required.	.029				√	√
7 Clean, test and inspect sight gauges, valves and drains.	.040				√	√
8 Check automatic controls for proper operation including temperature regulators and thermostatic devices.	.094				√	√
9 Check insulation on heater; repair as necessary.	.077				√	√
10 Drain sediment from tank.	.325					√
11 Clean up area around unit.	.066				√	√
12 Fill out maintenance checklist and report deficiencies.	.022				√	√
Total labor-hours/period					.612	.937
Total labor-hours/year					.612	.937

Description		Labor-hrs.	Cost Each				Total In-House	Total w/O&P
			2015 Bare Costs					
			Material	Labor	Equip.	Total		
3900	Water heater, steam coil, to 2500 gal., annually	.937	33	56		89	106	129
3950	Annualized	1.549	33	92.50		125.50	153	185

D20 PLUMBING — D2025 262 — Valve, Pressure Relief

PM Components	Labor-hrs.	W	M	Q	S	A
PM System D2025 262 1950						
Valve, pressure relief, above 4″						
1 Inspect valve for leaks, tighten fittings as necessary.	.022					√
2 Manually operate to check operation.	.038					√
3 Clean valve and area around valve.	.066					√
4 Fill out maintenance report and report deficiencies.	.022					√
Total labor-hours/period						.148
Total labor-hours/year						.148

			Cost Each					
				2015 Bare Costs			Total	Total
Description	Labor-hrs.	Material	Labor	Equip.	Total	In-House	w/O&P
1900 Valve, pressure relief, above 4″, annually	.148		7.20		7.20	9.45	11.70
1950 Annualized	.148		7.20		7.20	9.45	11.70

D20 PLUMBING — D2025 265 — Valve, Pressure Regulator

PM Components	Labor-hrs.	W	M	Q	S	A
PM System D2025 265 1950						
Valve, pressure regulator, above 4″						
1 Inspect valve for leaks; repair as necessary.	.160					√
2 Manually operate to check operation; adjust as required.	.049					√
3 Check pressure mechanism for proper opening and closing action.	.062					√
4 Clean valve and area around regulator.	.066					√
5 Fill out maintenance report and report deficiencies.	.022					√
Total labor-hours/period						.359
Total labor-hours/year						.359

			Cost Each					
				2015 Bare Costs			Total	Total
Description	Labor-hrs.	Material	Labor	Equip.	Total	In-House	w/O&P
1900 Valve, pressure regular, above 4″, annually	.359		17.45		17.45	23	28.50
1950 Annualized	.359		17.45		17.45	23	28.50

D20 PLUMBING — D2025 270 — Valve, Sediment Strainer

PM Components	Labor-hrs.	W	M	Q	S	A
PM System D2025 270 1950						
Valve, sediment strainer, above 4″						
1 Inspect valve for leaks, tighten fittings as necessary.	.160					√
2 Open valve drain to remove collected sediment.	.065					√
3 Clean valve exterior and around strainer.	.066					√
4 Fill out maintenance checklist and report deficiencies.	.022					√
Total labor-hours/period						.313
Total labor-hours/year						.313

			Cost Each					
				2015 Bare Costs			Total	Total
Description	Labor-hrs.	Material	Labor	Equip.	Total	In-House	w/O&P
1900 Valve, sediment strainer, above 4″, annually	.313		15.25		15.25	20	24.50
1950 Annualized	.313		15.25		15.25	20	24.50

PM Components	Labor-hrs.	W	M	Q	S	A
PM System D2025 310 1950						
Valve, automatic, above 4″						
1 Lubricate valve actuator stem and valve stem, where possible.	.022					√
2 Check automatic valve for proper operation.	.062					√
3 Check packing gland for leaks; adjust as required.	.022					√
4 Check pneumatic operator and tubing for air leaks.	.025					√
5 Clean valve exterior and area around valve.	.034					√
6 Fill out maintenance report and report deficiencies.	.022					√
Total labor-hours/period						.187
Total labor-hours/year						.187

			Cost Each					
			2015 Bare Costs				Total	Total
	Description	Labor-hrs.	Material	Labor	Equip.	Total	In-House	w/O&P
1900	Valve, automatic, above 4″, annually	.187	8.15	9.10		17.25	21	25
1950	Annualized	.187	8.15	9.10		17.25	21	25

PM Components	Labor-hrs.	W	M	Q	S	A
PM System D2025 310 2950						
Valve, auto diaphragm, above 4″						
1 Lubricate valve actuator mechanism and valve stem.	.022					√
2 Check valve for proper operation and leaks; tighten packing and flange bolts as required.	.044					√
3 Check pneumatic tubing and operator mechanism for proper alignment and damage; adjust as necessary and soap solution test for leaks after maintenance.	.053					√
4 Clean valve exterior and area around valve.	.034					√
5 Fill out maintenance report and report deficiencies.	.022					√
Total labor-hours/period						.175
Total labor-hours/year						.175

			Cost Each					
			2015 Bare Costs				Total	Total
	Description	Labor-hrs.	Material	Labor	Equip.	Total	In-House	w/O&P
2900	Valve, auto diaphragm, above 4″, annually	.175	8.15	8.50		16.65	20	24
2950	Annualized	.175	8.15	8.50		16.65	20	24

PM Components	Labor-hrs.	W	M	Q	S	A
PM System D2095 905 1950						
Duplex sump pump						
1 Check electrical cords, plugs and connections.	.239				✓	✓
2 Activate float switches and check pumps for proper operation.	.078				✓	✓
3 Lubricate pumps as required.	.094				✓	✓
4 Inspect packing and tighten as required.	.062				✓	✓
5 Check pumps for misalignment and bearings for overheating.	.260				✓	✓
6 Clean out trash from sump bottom.	.072				✓	✓
7 Fill out maintenance checklist and report deficiencies.	.022				✓	✓
Total labor-hours/period					.827	.827
Total labor-hours/year					.827	.827

			Cost Each					
			2015 Bare Costs				Total	Total
Description	Labor-hrs.	Material	Labor	Equip.	Total	In-House	w/O&P
1900 Duplex sump pump, annually	.827	27.50	48.50		76	91.50	110
1950 Annualized	1.654	34	97		131	160	194

PM Components	Labor-hrs.	W	M	Q	S	A
PM System D2095 910 1950						
Submersible pump, 1 H.P. and over						
1 Check with operating personnel for any deficiencies.	.035				✓	✓
2 Remove pump from pit.	.585				✓	✓
3 Clean out trash from pump intake.	.338				✓	✓
4 Check electrical plug, cord and connection.	.120				✓	✓
5 Inspect pump body for corrosion; prime and paint as necessary.	.053				✓	✓
6 Check pump and motor operation for excessive vibration, noise and overheating.	.022				✓	✓
7 Lubricate pump and motor, where applicable.	.099				✓	✓
8 Return pump to pit; reset and check float switch for proper operation.	.585				✓	✓
9 Clean area.	.066				✓	✓
10 Fill out maintenance checklist and report deficiencies.	.022				✓	✓
Total labor-hours/period					1.925	1.925
Total labor-hours/year					1.925	1.925

			Cost Each					
			2015 Bare Costs				Total	Total
Description	Labor-hrs.	Material	Labor	Equip.	Total	In-House	w/O&P
1900 Submersible pump, 1 H.P. and over, annually	1.925	17.05	113		130.05	160	197
1950 Annualized	3.850	34	226		260	320	395

PM Components	Labor-hrs.	W	M	Q	S	A
PM System D2095 930 1950						
Oxygen monitor.						
1 Check alarm set points.	.167			√	√	√
2 Check alarm lights for replacement.	.167			√	√	√
3 Check audible annuciator for proper operation.	.167			√	√	√
4 Check battery backup for proper operation.	.167			√	√	√
5 Check and test oxygen sensor for proper operation.	.276			√	√	√
6 Replace oxygen sensor annually.	1.000					√
7 Clean area.	.034			√	√	√
8 Fill out maintenance checklist and report deficiencies.	.022			√	√	√
Total labor-hours/period				1.000	1.000	2.000
Total labor-hours/year				2.000	1.000	2.000

			Cost Each					
			2015 Bare Costs				Total	Total
Description		Labor-hrs.	Material	Labor	Equip.	Total	In-House	w/O&P
1900	Oxygen monitor annually	2.000	209	97.50		306.50	360	420
1950	Annualized	5.000	209	243		452	550	655

PM Components	Labor-hrs.	W	M	Q	S	A
PM System D3025 110 1950						
Boiler, electric, to 1500 gallon						
1 Check with operating or area personnel for deficiencies.	.035			√	√	√
2 Check hot water pressure gauges.	.073			√	√	√
3 Check operation and condition of pressure relief valve.	.030			√	√	√
4 Check for proper operation of primary controls for resistance-type or electrode-type heating elements; check and adjust thermostat.	.133			√	√	√
5 Check electrical wiring to heating elements, blower, motors, overcurrent protective devices, grounding system and other electrical components as required.	.238			√	√	√
6 Check over temperature and over-pressure limit controls for proper operation.	.143			√	√	√
7 Check furnace operation through complete cycle or up to 10 minutes.	.217			√	√	√
8 Clean area around boiler.	.066			√	√	√
9 Fill out maintenance checklist and report deficiencies.	.022			√	√	√
Total labor-hours/period				.957	.957	.957
Total labor-hours/year				1.914	.957	.957

				Cost Each				
				2015 Bare Costs				Total	Total
Description		Labor-hrs.	Material	Labor	Equip.	Total	In-House	w/O&P
1900	Boiler, electric, to 1500 gal., annually	.957	152	57		209	239	279
1950	Annualized	3.828	71.50	229		300.50	365	445

PM Components	Labor-hrs.	W	M	Q	S	A
PM System D3025 130 1950						
Boiler, hot water; oil, gas or combination fired, up to 120 MBH						
1 Check combustion chamber for air or gas leaks.	.077					√
2 Inspect and clean oil burner gun and ignition assembly, where applicable.	.658					√
3 Inspect fuel system for leaks and change fuel filter element, where applicable.	.098					√
4 Check fuel lines and connections for damage.	.023		√	√	√	√
5 Check for proper operational response of burner to thermostat controls.	.133			√	√	√
6 Check and lubricate burner and blower motors.	.079			√	√	√
7 Check main flame failure protection and main flame detection scanner on boiler equipped with spark ignition (oil burner).	.124		√	√	√	√
8 Check electrical wiring to burner controls and blower.	.079					√
9 Clean firebox (sweep and vacuum).	.577					√
10 Check operation of mercury control switches (i.e., steam pressure, hot water temperature limit, atomizing or combustion air proving, etc.).	.143		√	√	√	√
11 Check operation and condition of safety pressure relief valve.	.030		√	√	√	√
12 Check operation of boiler low water cut off devices.	.056		√	√	√	√
13 Check hot water pressure gauges.	.073		√	√	√	√
14 Inspect and clean water column sight glass (or replace).	.127		√	√	√	√
15 Clean fire side of water jacket boiler.	.433					
16 Check condition of flue pipe, damper and exhaust stack.	.147			√	√	√
17 Check boiler operation through complete cycle, up to 30 minutes.	.650					
18 Check fuel level with gauge pole, add as required.	.046		√	√	√	√
19 Clean area around boiler.	.066		√	√	√	√
20 Fill out maintenance checklist and report deficiencies.	.022		√	√	√	√
Total labor-hours/period			.710	1.069	1.069	3.641
Total labor-hours/year			5.680	2.138	1.069	3.641

	Description	Labor-hrs.	2015 Bare Costs				Total In-House	Total w/O&P
			Material	Labor	Equip.	Total		
1900	Boiler, hot water, O/G/C, up to 120 MBH, annually	3.641	161	218		379	450	540
1950	Annualized	12.528	85.50	750		835.50	1,025	1,275

PM Components	Labor-hrs.	W	M	Q	S	A
PM System D3025 130 2950						
Boiler, hot water; oil, gas or combination fired, 120 to 500 MBH						
1 Check combustion chamber for air or gas leaks.	.077					√
2 Inspect and clean oil burner gun and ignition assembly where applicable.	.835					√
3 Inspect fuel system for leaks and change fuel filter element, where applicable.	.125					√
4 Check fuel lines and connections for damage.	.023		√	√	√	√
5 Check for proper operational response of burner to thermostat controls.	.169			√	√	√
6 Check and lubricate burner and blower motors.	.099			√	√	√
7 Check main flame failure protection and main flame detection scanner on boiler equipped with spark ignition (oil burner).	.155		√	√	√	√
8 Check electrical wiring to burner controls and blower.	.100					√
9 Clean firebox (sweep and vacuum).	.793					√
10 Check operation of mercury control switches (i.e., steam pressure, hot water temperature limit, atomizing or combustion air proving, etc.).	.185		√	√	√	√
11 Check operation and condition of safety pressure relief valve.	.038		√	√	√	√
12 Check operation of boiler low water cut off devices.	.070		√	√	√	√
13 Check hot water pressure gauges.	.073		√	√	√	√
14 Inspect and clean water column sight glass (or replace).	.160		√	√	√	√
15 Clean fire side of water jacket	.433					√
16 Check condition of flue pipe, damper and exhaust stack.	.183			√	√	√
17 Check boiler operation through complete cycle, up to 30 minutes.	.806					√
18 Check fuel level with gauge pole, add as required.	.046		√	√	√	√
19 Clean area around boiler.	.137		√	√	√	√
20 Fill out maintenance checklist and report deficiencies.	.022		√	√	√	√
Total labor-hours/period			.909	1.360	1.360	4.529
Total labor-hours/year			7.272	2.720	1.360	4.529

			Cost Each					
			2015 Bare Costs				Total	Total
Description		Labor-hrs.	Material	Labor	Equip.	Total	In-House	w/O&P
2900	Boiler, hot water, O/G/C, 120 to 500 MBH, annually	4.529	182	271		453	540	650
2950	Annualized	15.881	107	950		1,057	1,300	1,600

PM Components	Labor-hrs.	W	M	Q	S	A
PM System D3025 130 3950						
Boiler, hot water; oil, gas or combination fired, 500 to 1000 MBH						
1 Check combustion chamber for air or gas leaks.	.086					√
2 Inspect and clean oil burner gun and ignition assembly where applicable.	.910					√
3 Inspect fuel system for leaks and change fuel filter element, where applicable.	.140					√
4 Check fuel lines and connections for damage.	.026	√	√	√	√	
5 Check for proper operational response of burner to thermostat controls.	.185		√	√	√	√
6 Check and lubricate burner and blower motors.	.109			√	√	√
7 Check main flame failure protection and main flame detection scanner on boiler equipped with spark ignition (oil burner).	.169		√	√	√	√
8 Check electrical wiring to burner controls and blower.	.111					√
9 Clean firebox (sweep and vacuum).	.875					√
10 Check operation of mercury control switches (i.e., steam pressure, hot water temperature limit, atomizing or combustion air proving, etc.).	.203		√	√	√	√
11 Check operation and condition of safety pressure relief valve.	.042		√	√	√	√
12 Check operation of boiler low water cut off devices.	.077		√	√	√	√
13 Check hot water pressure gauges.	.081		√	√	√	√
14 Inspect and clean water column sight glass (or replace).	.176		√	√	√	√
15 Clean fire side of water jacket	.433					√
16 Check condition of flue pipe, damper and exhaust stack.	.202			√	√	√
17 Check boiler operation through complete cycle, up to 30 minutes.	.887					√
18 Check fuel level with gauge pole, add as required.	.049		√	√	√	√
19 Clean area around boiler.	.151		√	√	√	√
20 Fill out maintenance checklist and report deficiencies.	.022		√	√	√	√
Total labor-hours/period			.996	1.492	1.492	4.934
Total labor-hours/year			7.968	2.984	1.492	4.934

			Cost Each					
			2015 Bare Costs				**Total**	**Total**
Description		**Labor-hrs.**	**Material**	**Labor**	**Equip.**	**Total**	**In-House**	**w/O&P**
3900	Boiler, hot water, O/G/C, 500 to 1000 MBH, annually	4.934	189	295		484	575	695
3950	Annualized	17.378	114	1,050		1,164	1,425	1,775

PM Components	Labor-hrs.	W	M	Q	S	A
PM System D3025 130 4950						
Boiler, hot water; oil, gas or combination fired, over 1000 MBH						
1　Check combustion chamber for air or gas leaks.	.117					√
2　Inspect and clean oil burner gun and ignition assembly where applicable.	.987					√
3　Inspect fuel system for leaks and change fuel filter element, where applicable.	.147					√
4　Check fuel lines and connections for damage.	.035		√	√	√	√
5　Check for proper operational response of burner to thermostat controls.	.199			√	√	√
6　Check and lubricate burner and blower motors.	.120			√	√	√
7　Check main flame failure protection and main flame detection scanner on boiler equipped with spark ignition (oil burner).	.186		√	√	√	√
8　Check electrical wiring to burner controls and blower.	.120					√
9　Clean firebox (sweep and vacuum).	.819					√
10　Check operation of mercury control switches (i.e., steam pressure, hot water temperature limit, atomizing or combustion air proving, etc.).	.215		√	√	√	√
11　Check operation and condition of safety pressure relief valve.	.046		√	√	√	√
12　Check operation of boiler low water cut off devices.	.085		√	√	√	√
13　Check hot water pressure gauges.	.109		√	√	√	√
14　Inspect and clean water column sight glass (or replace).	.191		√	√	√	√
15　Clean fire side of water jacket	.433					√
16　Check condition of flue pipe, damper and exhaust stack.	.221			√	√	√
17　Check boiler operation through complete cycle, up to 30 minutes.	.887					√
18　Check fuel level with gauge pole, add as required.	.098		√	√	√	√
19　Clean area around boiler.	.182		√	√	√	√
20　Fill out maintenance checklist and report deficiencies.	.022		√	√	√	√
Total labor-hours/period			1.169	1.709	1.709	5.219
Total labor-hours/year			9.352	3.418	1.709	5.219

		Cost Each						
			2015 Bare Costs				Total	Total
Description	Labor-hrs.	Material	Labor	Equip.	Total	In-House	w/O&P
4900　Boiler, hot water, O/G/C, over 1000 MBH, annually	5.219	204	310		514	615	740
4950　　　Annualized	19.698	129	1,175		1,304	1,625	1,975

PM Components	Labor-hrs.	W	M	Q	S	A
PM System D3025 140 1950						
Boiler, steam; oil, gas or combination fired, up to 120 MBH						
1 Inspect fuel system for leaks or damage.	.098		√	√	√	√
2 Change fuel filter element and clean strainers; repair leaks, where applicable.	.581					√
3 Check main flame failure protection, positive fuel shutoff and main flame detection scanner on boiler equipped with spark ignition (oil burner).	.124		√	√	√	√
4 Check for proper operational response of burner to thermostat controls.	.133			√	√	√
5 Inspect all gas, steam and water lines, valves, connections for leaks or damage; repair as necessary.	.195			√	√	√
6 Check feedwater system and feedwater makeup control and pump.	.056		√	√	√	√
7 Check and lubricate burner and blower motors as required.	.083			√	√	√
8 Check operation and condition of safety pressure relief valve.	.030		√	√	√	√
9 Check combustion controls, combustion blower and damper modulation control.	.133					√
10 Check all indicator lamps and water/steam pressure gauges.	.073		√	√	√	√
11 Check electrical panels and wiring to burner, blowers and other components.	.079			√	√	√
12 Clean blower air-intake dampers, if required.	.055			√		√
13 Check condition of flue pipe, damper and exhaust stack.	.147		√	√	√	√
14 Check boiler operation through complete cycle, up to 30 minutes.	.650			√	√	√
15 Check water column sight glass and water level system; clean or replace sight glass, if required.	.127		√	√	√	√
16 Clean firebox (sweep and vacuum).	.577					√
17 Check fuel level with gauge pole for oil burning boilers.	.046		√	√	√	√
18 Inspect and clean oil burner gun and ignition assembly where applicable.	.650					√
19 Clean area around boiler.	.066		√	√	√	√
20 Fill out maintenance checklist and report deficiencies.	.022		√	√	√	√
Total labor-hours/period			.789	1.984	1.984	3.925
Total labor-hours/year			6.312	3.968	1.984	3.925

			Cost Each					
			2015 Bare Costs				Total	Total
Description		Labor-hrs.	Material	Labor	Equip.	Total	In-House	w/O&P
1900	Boiler, steam, O/G/C, up to 120 MBH, annually	3.925	239	235		474	555	665
1950	Annualized	16.189	164	965		1,129	1,400	1,700

PM Components	Labor-hrs.	W	M	Q	S	A
PM System D3025 140 2950						
Boiler, steam; oil, gas or combination fired, 120 to 500 MBH						
1 Inspect fuel system for leaks or damage.	.098		√	√	√	√
2 Change fuel filter element and clean strainers; repair leaks, where applicable.	.835					√
3 Check main flame failure protection, positive fuel shutoff and main flame detection scanner on boiler equipped with spark ignition (oil burner).	.125		√	√	√	√
4 Check for proper operational response of burner to thermostat controls.	.169			√	√	√
5 Inspect all gas, steam and water lines, valves, connections for leaks or damage; repair as necessary.	.195			√	√	√
6 Check feedwater system and feedwater makeup control and pump.	.069		√	√	√	√
7 Check and lubricate burner, blowers and motors as required.	.099			√	√	√
8 Check operation and condition of safety pressure relief valve.	.039		√	√	√	√
9 Check combustion controls, combustion blower and damper modulation control.	.169					√
10 Check all indicator lamps and water/steam pressure gauges.	.073		√	√	√	√
11 Check electrical panels and wiring to burner, blowers and other components.	.104			√	√	√
12 Clean blower air-intake dampers, if required.	.069			√	√	√
13 Check condition of flue pipe, damper and exhaust stack.	.147		√	√	√	√
14 Check boiler operation through complete cycle, up to 30 minutes.	.806			√	√	√
15 Check water column sight glass and water level system; clean or replace sight glass, if required.	.127		√	√	√	√
16 Clean firebox (sweep and vacuum).	.793					√
17 Check fuel level with gauge pole for oil burning boilers.	.046		√	√	√	√
18 Inspect and clean oil burner gun and ignition assembly where applicable.	.819					√
19 Clean area around boiler.	.137		√	√	√	√
20 Fill out maintenance checklist and report deficiencies.	.022		√	√	√	√
Total labor-hours/period			.883	2.325	2.325	4.941
Total labor-hours/year			7.064	4.650	2.325	4.941

			Cost Each					
			2015 Bare Costs				Total	Total
Description	Labor-hrs.	Material	Labor	Equip.	Total	In-House	w/O&P
2900	Boiler, steam, O/G/C, 120 to 500 MBH, annually	4.941	277	295		572	675	805
2950	Annualized	18.980	207	1,125		1,332	1,650	2,025

PM Components	Labor-hrs.	W	M	Q	S	A
PM System D3025 140 3950						
Boiler, steam; oil, gas or combination fired, 500 to 1000 MBH						
1 Inspect fuel system for leaks or damage.	.098		√	√	√	√
2 Change fuel filter element and clean strainers; repair leaks, where applicable.	.910					√
3 Check main flame failure protection, positive fuel shutoff and main flame detection scanner on boiler equipped with spark ignition (oil burner).	.140		√	√	√	√
4 Check for proper operational response of burner to thermostat controls.	.185			√	√	√
5 Inspect all gas, steam and water lines, valves, connections for leaks or damage; repair as necessary.	.195			√	√	√
6 Check feedwater system and feedwater makeup control and pump.	.075		√	√	√	√
7 Check and lubricate burner, blowers and motors as required.	.109		√	√	√	√
8 Check operation and condition of safety pressure relief valve.	.052		√	√	√	√
9 Check combustion controls, combustion blower and damper modulation control.	.185					√
10 Check all indicator lamps and water/steam pressure gauges.	.091		√	√	√	√
11 Check electrical panels and wiring to burner, blowers and other components.	.117			√	√	√
12 Clean blower air-intake dampers, if required.	.069			√	√	√
13 Check condition of flue pipe, damper and exhaust stack.	.147		√	√	√	√
14 Check boiler operation through complete cycle, up to 30 minutes.	.884		√	√	√	√
15 Check water column sight glass and water level system; clean or replace sight glass, if required.	.127		√	√	√	√
16 Clean firebox (sweep and vacuum).	1.053					√
17 Check fuel level with gauge pole for oil burning boilers.	.046		√	√	√	√
18 Inspect and clean oil burner gun and ignition assembly where applicable.	.923					√
19 Clean area around boiler.	.151		√	√	√	√
20 Fill out maintenance checklist and report deficiencies.	.022		√	√	√	√
Total labor-hours/period			.949	2.508	2.508	5.579
Total labor-hours/year			7.592	5.016	2.508	5.579

			Cost Each					
			2015 Bare Costs				Total	Total
Description		Labor-hrs.	Material	Labor	Equip.	Total	In-House	w/O&P
3900	Boiler, steam, O/G/C, 500 to 1000 MBH, annually	5.579	291	335		626	735	885
3950	Annualized	20.695	221	1,225		1,446	1,800	2,200

PM Components	Labor-hrs.	W	M	Q	S	A
PM System D3025 140 4950						
Boiler, steam; oil, gas or combination fired, over 1000 MBH						
1 Inspect fuel system for leaks or damage.	.098		√	√	√	√
2 Change fuel filter element and clean strainers; repair leaks, where applicable.	1.027					√
3 Check main flame failure protection, positive fuel shutoff and main flame detection scanner on boiler equipped with spark ignition (oil burner).	.147		√	√	√	√
4 Check for proper operational response of burner to thermostat controls.	.199			√	√	√
5 Inspect all gas, steam and water lines, valves, connections for leaks abor damage; repair as necessary.	.195			√	√	√
6 Check feedwater system and feedwater makeup control and pump.	.091		√	√	√	√
7 Check and lubricate burner, blowers and motors as required.	.120		√	√	√	√
8 Check operation and condition of safety pressure relief valve.	.069		√	√	√	√
9 Check combustion controls, combustion blower and damper modulation control.	.199					√
10 Check all indicator lamps and water/steam pressure gauges.	.109		√	√	√	√
11 Check electrical panels and wiring to burner, blowers and other components.	.140			√	√	√
12 Clean blower air-intake dampers, if required.	.069			√	√	√
13 Check condition of flue pipe, damper and exhaust stack.	.147		√	√	√	√
14 Check boiler operation through complete cycle, up to 30 minutes.	.884		√	√	√	√
15 Check water column sight glass and water level system; clean or replace sight glass, if required.	.127		√	√	√	√
16 Clean firebox (sweep and vacuum).	1.144					√
17 Check fuel level with gauge pole for oil burning boilers.	.046		√	√	√	√
18 Inspect and clean oil burner gun and ignition assembly where applicable.	1.196					√
19 Clean area around boiler.	.182		√	√	√	√
20 Fill out maintenance checklist and report deficiencies.	.022		√	√	√	√
Total labor-hours/period			1.038	2.645	2.645	6.211
Total labor-hours/year			8.304	5.290	2.645	6.211

			Cost Each					
			2015 Bare Costs				Total	Total
Description		Labor-hrs.	Material	Labor	Equip.	Total	In-House	w/O&P
4900	Boiler, steam, O/G/C, over 1000 MBH, annually	6.211	315	370		685	810	975
4950	Annualized	22.450	243	1,350		1,593	1,950	2,400

PM Components	Labor-hrs.	W	M	Q	S	A
PM System D3025 210 1950						
Deaerator tank						
1 Check tank and associated piping for leaks.	.013				√	√
2 Bottom - blow deaerator tank.	.022				√	√
3 Perform sulfite test on deaerator water sample.	.013				√	√
4 Check low and high float levels for proper operation and respective water level alarms.	.337				√	√
5 Clean steam and feedwater strainers.	.143				√	√
6 Check steam pressure regulating valve operation.	.007				√	√
7 Check all indicator lights.	.066				√	√
8 Clean unit and surrounding area.	.022				√	√
9 Fill out maintenance checklist and report deficiencies.	.130				√	√
Total labor-hours/period					.753	.753
Total labor-hours/year					.753	.753

			Cost Each					
			2015 Bare Costs				Total	Total
Description		Labor-hrs.	Material	Labor	Equip.	Total	In-House	w/O&P
1900	Deaerator tank, annually	.753	42.50	45		87.50	104	123
1950	Annualized	1.506	42.50	90		132.50	159	193

PM Components	Labor-hrs.	W	M	Q	S	A
PM System D3025 310 1950						
Pump, boiler fuel oil						
1 Check with operating or area personnel for deficiencies.	.035				√	√
2 Check for leaks on discharge piping and seals, etc.	.077				√	√
3 Check pump and motor operation for vibration, noise, overheating, etc.	.022				√	√
4 Check alignment and clearances of shaft and coupler.	.290				√	√
5 Tighten or replace loose, missing, or damaged nuts, bolts, and screws.	.005				√	√
6 Lubricate pump and motor as required.	.099				√	√
7 Clean pump, motor and surrounding area.	.066				√	√
8 Fill out maintenance checklist and report deficiencies.	.022				√	√
Total labor-hours/period					.616	.616
Total labor-hours/year					.616	.616

			Cost Each					
			2015 Bare Costs				Total	Total
Description	Labor-hrs.	Material	Labor	Equip.	Total	In-House	w/O&P
1900	Pump, boiler fuel oil, annually	.616	87	37		124	142	166
1950	Annualized	1.232	55.50	73.50		129	153	184

PM Components	Labor-hrs.	W	M	Q	S	A
PM System D3025 310 2950						
Pump, condensate return, over 1 H.P.						
1 Check for proper operation of pump.	.022				√	√
2 Check for leaks on suction and discharge piping, seals, packing glands, etc.; make minor adjustments as required.	.077				√	√
3 Check pump and motor operation for vibration, noise, overheating, etc.	.022				√	√
4 Check alignment pump and motor; adjust as necessary.	.290				√	√
5 Lubricate pump and motor.	.099				√	√
6 Clean exterior of pump, motor and surrounding area.	.030				√	√
7 Fill out maintenance checklist and report deficiencies.	.022				√	√
Total labor-hours/period					.562	.562
Total labor-hours/year					.562	.562

			Cost Each					
			2015 Bare Costs				Total	Total
Description	Labor-hrs.	Material	Labor	Equip.	Total	In-House	w/O&P
2900	Pump, steam condensate return, over 1 H.P., annually	.562	54	33.50		87.50	101	120
2950	Annualized	1.124	55.50	67		122.50	145	174

PM Components	Labor-hrs.	W	M	Q	S	A
PM System D3025 310 3950						
Pump, steam condensate return, duplex						
1 Check with operating or area personnel for deficiencies.	.035				√	√
2 Check for proper operation.	.022				√	√
3 Check for leaks on suction and discharge piping, seals, packing glands, etc.; make minor adjustments as required.	.153				√	√
4 Check pumps and motors operation for excessive vibration, noise and overheating.	.044				√	√
5 Check pump controller for proper operation.	.130				√	√
6 Lubricate pumps and motors.	.099				√	√
7 Clean condensate return unit and surrounding area.	.066				√	√
8 Fill out maintenance checklist and report deficiencies.	.022				√	√
Total labor-hours/period					.571	.571
Total labor-hours/year					.571	.571

Description	Labor-hrs.	2015 Bare Costs				Total In-House	Total w/O&P
		Material	Labor	Equip.	Total		
3900 Pump, steam condensate return, duplex, annually	.571	108	34		142	161	188
3950 Annualized	1.142	111	68		179	207	245

PM Components	Labor-hrs.	W	M	Q	S	A
PM System D3035 110 1950						
Water cooling tower, forced draft, up to 50 tons						
1 Check with operating or area personnel for deficiencies.	.035				√	√
2 Check operation of unit for water leaks, noise or vibration.	.077				√	√
3 Clean and inspect hot water basin.	.156				√	√
4 Remove access panel.	.048				√	√
5 Check electrical wiring and connections; make appropriate adjustments.	.120				√	√
6 Lubricate all motor and fan bearings.	.047				√	√
7 Check fan blades or blowers for imbalance and tip clearance.	.055				√	√
8 Check belt for wear, tension and alignment; adjust as required.	.029				√	√
9 Drain and flush cold water sump and clean strainer.	.382				√	√
10 Clean inside of water tower using water hose; scrape, brush and wipe as required; heavy deposits of scale should be removed with scale removing compound.	.598				√	√
11 Refill with water, ck. make-up water asm. for leakage, adj. float if nec.	.211				√	√
12 Replace access panel.	.039				√	√
13 Remove, clean and reinstall conductivity and ph electrodes in chemical water treatment system.	.390				√	√
14 Inspect and clean around cooling tower.	.066				√	√
15 Fill out maintenance checklist and report deficiencies.	.022				√	√
Total labor-hours/period					2.275	2.275
Total labor-hours/year					2.275	2.275

			Cost Each					
			2015 Bare Costs				**Total**	**Total**
Description		**Labor-hrs.**	**Material**	**Labor**	**Equip.**	**Total**	**In-House**	**w/O&P**
1900	Cooling tower, forced draft, up to 50 tons, annually	2.275	27.50	136		163.50	201	246
1950	Annualized	4.550	39	272		311	385	475

PM Components	Labor-hrs.	W	M	Q	S	A
PM System D3035 110 2950						
Water cooling tower, forced draft, 51 tons through 500 tons						
1 Check with operating or area personnel for deficiencies.	.074				√	√
2 Check operation of unit for water leaks, noise or vibration.	.163				√	√
3 Clean and inspect hot water basin.	.332				√	√
4 Remove access panel.	.103				√	√
5 Check electrical wiring and connections; make appropriate adjustments.	.255				√	√
6 Lubricate all motor and fan bearings.	.100				√	√
7 Check fan blades or blowers for imbalance and tip clearance.	.247				√	√
8 Check belt for wear, tension and alignment; adjust as required.	.096				√	√
9 Drain and flush cold water sump and clean strainer.	.819				√	√
10 Clean inside of water tower using water hose; scrape, brush and wipe as required; heavy deposits of scale should be removed with scale removing compound.	1.271				√	√
11 Refill with water, chk. make-up water aseb. for leakage, adj. float if nec.	.447				√	√
12 Replace access panel.	.078				√	√
13 Remove, clean and reinstall conductivity and ph electrodes in chemical water treatment system.	.832				√	√
14 Inspect and clean around cooling tower.	.117				√	√
15 Fill out maintenance checklist and report deficiencies.	.022				√	√
Total labor-hours/period					4.956	4.956
Total labor-hours/year					4.956	4.956

			Cost Each					
			2015 Bare Costs				**Total**	**Total**
Description		**Labor-hrs.**	**Material**	**Labor**	**Equip.**	**Total**	**In-House**	**w/O&P**
2900	Cooling tower, forced draft, 50 thru 500 tons, annually	4.956	51	296		347	425	525
2950	Annualized	9.912	70.50	590		660.50	820	1,025

For customer support on your Facilities Maintenance & Repair Cost Data, call 877.801.5724.

423

PM Components	Labor-hrs.	W	M	Q	S	A
PM System D3035 110 3950						
Water cooling tower, forced draft, 500 tons through 1000 tons						
1 Check with operating or area personnel for deficiencies.	.109				√	√
2 Check operation of unit for water leaks, noise or vibration.	.239				√	√
3 Clean and inspect hot water basin.	.488				√	√
4 Remove access panel.	.151				√	√
5 Check electrical wiring and connections; make appropriate adjustments.	.374				√	√
6 Lubricate all motor and fan bearings.	.147				√	√
7 Check fan blades or blowers for imbalance and tip clearance.	.302				√	√
8 Check belt for wear, tension and alignment; adjust as required.	.220				√	√
9 Drain and flush cold water sump and clean strainer.	1.464				√	√
10 Clean inside of water tower using water hose; scrape, brush and wipe as required; heavy deposits of scale should be removed with scale removing compound.	2.259				√	√
11 Refill with water, check make-up water assembly for leakage, adjust float if necessary.	1.578				√	√
12 Replace access panel.	.117				√	√
13 Remove, clean and reinstall conductivity and ph electrodes in chemical water treatment system.	1.222				√	√
14 Inspect and clean around cooling tower.	.172				√	√
15 Fill out maintenance checklist and report deficiencies.	.022				√	√
Total labor-hours/period					8.864	8.864
Total labor-hours/year					8.864	8.864

			Cost Each					
			2015 Bare Costs				Total	Total
	Description	Labor-hrs.	Material	Labor	Equip.	Total	In-House	w/O&P
3900	Cooling tower, forced draft, 500 to 1000 tons, annually	8.864	110	530		640	785	965
3950	Annualized	17.728	157	1,050		1,207	1,500	1,850

PM Components	Labor-hrs.	W	M	Q	S	A
PM System D3035 130 1950						
Chiller, reciprocating, air cooled, up to 25 tons						
1 Check unit for proper operation, excessive noise or vibration.	.033		√	√	√	√
2 Run system diagnostics test.	.325		√	√	√	√
3 Check oil level in sight glass of lead compressor only, add oil as necessary.	.042		√	√	√	√
4 Check superheat and subcooling temperatures.	.325					√
5 Check liquid line sight glass, oil and refrigerant pressures.	.036		√	√	√	√
6 Check contactors, sensors and mechanical safety limits.	.094					√
7 Check electrical wiring and connections; tighten loose connections.	.120					√
8 Clean intake side of condenser coils, fans and intake screens.	1.282					√
9 Inspect fan(s) or blower(s) for bent blades of imbalance.	.086					√
10 Lubricate shaft bearings and motor bearings as required.	.281					√
11 Inspect plumbing and valves for leaks, adjust as necessary.	.077		√	√	√	√
12 Check evaporator and condenser for corrosion.	.026		√	√	√	√
13 Clean chiller and surrounding area.	.066		√	√	√	√
14 Fill out maintenance checklist and report deficiencies.	.022		√	√	√	√
Total labor-hours/period			.627	.627	.627	2.815
Total labor-hours/year			5.016	1.254	.627	2.815

			Cost Each				
			2015 Bare Costs			Total	Total
Description	Labor-hrs.	Material	Labor	Equip.	Total	In-House	w/O&P
1900 Chiller, recip., air cooled, up to 25 tons, annually	2.815	55	168		223	272	330
1950 Annualized	9.712	55	580		635	785	975

PM Components	Labor-hrs.	W	M	Q	S	A
PM System D3035 130 2950						
Chiller, reciprocating, air cooled, over 25 tons						
1 Check unit for proper operation, excessive noise or vibration.	.033		√	√	√	√
2 Run system diagnostics test.	.455		√	√	√	√
3 Check oil level in sight glass of lead compressor only, add oil as necessary.	.042		√	√	√	√
4 Check superheat and subcooling temperatures.	.325					√
5 Check liquid line sight glass, oil and refrigerant pressures.	.036		√	√	√	√
6 Check contactors, sensors and mechanical safety limits.	.094					√
7 Check electrical wiring and connections; tighten loose connections.	.120					√
8 Clean intake side of condenser coils, fans and intake screens.	1.282					√
9 Inspect fan(s) or blower(s) for bent blades of imbalance.	.237					√
10 Lubricate shaft bearings and motor bearings as required.	.341					√
11 Inspect plumbing and valves for leaks, adjust as necessary.	.117		√	√	√	√
12 Check evaporator and condenser for corrosion.	.052		√	√	√	√
13 Clean chiller and surrounding area.	.117		√	√	√	√
14 Fill out maintenance checklist and report deficiencies.	.022		√	√	√	√
Total labor-hours/period			.874	.874	.874	3.273
Total labor-hours/year			6.992	1.748	.874	3.273

			Cost Each				
			2015 Bare Costs			Total	Total
Description	Labor-hrs.	Material	Labor	Equip.	Total	In-House	w/O&P
2900 Chiller, recip., air cooled, over 25 tons, annually	3.273	78.50	196		274.50	330	405
2950 Annualized	12.887	78.50	770		848.50	1,050	1,300

For customer support on your Facilities Maintenance & Repair Cost Data, call 877.801.5724.

425

PM Components	Labor-hrs.	W	M	Q	S	A
PM System D3035 135 1950						
Chiller, reciprocating, water cooled, up to 50 tons						
1 Check unit for proper operation, excessive noise or vibration.	.033		√	√	√	√
2 Run system diagnostics test.	.325		√	√	√	√
3 Check oil level in sight glass of lead compressor only, add oil as necessary.	.042		√	√	√	√
4 Check superheat and subcooling temperatures.	.325					√
5 Check liquid line sight glass, oil and refrigerant pressures.	.036		√	√	√	√
6 Check contactors, sensors and mechanical limits, adjust as necessary.	.094					√
7 Inspect plumbing and valves for leaks, tighten connections as necessary.	.077		√	√	√	√
8 Check condenser and evaporator for corrosion.	.026		√	√	√	√
9 Clean chiller and surrounding area.	.066		√	√	√	√
10 Fill out maintenance checklist and report deficiencies.	.022		√	√	√	√
Total labor-hours/period			.627	.627	.627	1.046
Total labor-hours/year			5.016	1.254	.627	1.046

			Cost Each					
			2015 Bare Costs				Total	Total
Description		Labor-hrs.	Material	Labor	Equip.	Total	In-House	w/O&P
1900	Chiller, recip., water cooled, up to 50 tons, annually	1.046	55	62.50		117.50	139	166
1950	Annualized	7.943	55	475		530	655	810

PM Components	Labor-hrs.	W	M	Q	S	A
PM System D3035 135 2950						
Chiller, reciprocating, water cooled, over 50 tons						
1 Check unit for proper operation, excessive noise or vibration.	.033		√	√	√	√
2 Run system diagnostics test.	.455		√	√	√	√
3 Check oil level in sight glass of lead compressor only, add oil as necessary.	.042		√	√	√	√
4 Check superheat and subcooling temperatures.	.325					√
5 Check liquid line sight glass, oil and refrigerant pressures.	.036		√	√	√	√
6 Check contactors, sensors and mechanical limits, adjust as necessary.	.094					√
7 Inspect plumbing and valves for leaks, tighten connections as necessary.	.117		√	√	√	√
8 Check condenser and evaporator for corrosion.	.052		√	√	√	√
9 Clean chiller and surrounding area.	.117		√	√	√	√
10 Fill out maintenance checklist and report deficiencies.	.022		√	√	√	√
Total labor-hours/period			.874	.874	.874	1.293
Total labor-hours/year			6.992	1.748	.874	1.293

			Cost Each					
			2015 Bare Costs				Total	Total
Description		Labor-hrs.	Material	Labor	Equip.	Total	In-House	w/O&P
2900	Chiller, recip., water cooled, over 50 tons, annually	1.293	62.50	77.50		140	166	199
2950	Annualized	10.907	62.50	650		712.50	885	1,100

426

PM Components	Labor-hrs.	W	M	Q	S	A
PM System D3035 140 1950						
Chiller, centrifugal water cooled, up to 100 tons						
1 Check unit for proper operation.	.035	√	√	√	√	√
2 Check oil level; add oil as necessary.	.022	√	√	√	√	√
3 Check oil temperature.	.038	√	√	√	√	√
4 Check dehydrator or purge system; remove water if observed in sight glass.	.046	√	√	√	√	√
5 Run system control tests.	.325		√	√	√	√
6 Check refrigerant charge/level, add as necessary.	.272		√	√	√	√
7 Check compressor for excessive noise/vibration.	.025		√	√	√	√
8 Check sensor and mechanical safety limits; replace as necessary.	.094				√	√
9 Clean dehydrator float valve.	.195					√
10 Perform spectrochemical analysis of compressor oil; replace oil as necessary.	.039					√
11 Replace oil filters and add oil as necessary.	.081					√
12 Inspect cooler and condenser tubes for leaks; clean screens as necessary.	5.200					√
13 Inspect utility vessel vent piping and safety relief valve; replace as necessary.	.195					√
14 Inspect/clean the economizer (vane) gas line, damper valve and actuator arm.	.650					√
15 Run an insulation test on the centrifugal motor.	1.300					√
16 Clean area around equipment.	.066	√	√	√	√	√
17 Document all maintenance and cleaning procedures.	.022	√	√	√	√	√
Total labor-hours/period		.229	.851	.851	.945	8.605
Total labor-hours/year		8.702	6.808	1.702	.945	8.605

	Description	Labor-hrs.	2015 Bare Costs				Total In-House	Total w/O&P
			Material	Labor	Equip.	Total		
1900	Chiller, centrif., water cooled, up to 100 tons, annually	8.605	255	515		770	925	1,125
1950	Annualized	26.762	259	1,600		1,859	2,275	2,825

PM Components	Labor-hrs.	W	M	Q	S	A
PM System D3035 140 2950						
Chiller, centrifugal water cooled, over 100 tons						
1 Check unit for proper operation.	.035	√	√	√	√	√
2 Check oil level; add oil as necessary.	.022	√	√	√	√	√
3 Check oil temperature.	.038	√	√	√	√	√
4 Check dehydrator or purge system; remove water if observed in sight glass.	.046	√	√	√	√	√
5 Run system control tests.	.455		√	√	√	√
6 Check refrigerant charge/level, add as necessary.	.381		√	√	√	√
7 Check compressor for excessive noise/vibration.	.039		√	√	√	√
8 Check sensor and mechanical safety limits; replace as necessary.	.133				√	√
9 Clean dehydrator float valve.	.351					√
10 Perform spectrochemical analysis of compressor oil; replace oil as necessary.	.039					√
11 Replace oil filters and add oil as necessary.	.161					√
12 Inspect cooler and condenser tubes for leaks; clean screens as necessary.	5.200					√
13 Inspect utility vessel vent piping and safety relief valve; replace as necessary.	.247					√
14 Inspect/clean the economizer (vane), gas line damper valve and actuator arm.	.650					√
15 Run an insulation test on the centrifugal motor.	1.950					√
16 Clean area around equipment.	.117	√	√	√	√	√
17 Document all maintenance and cleaning procedures.	.022	√	√	√	√	√
Total labor-hours/period		.280	1.155	1.155	1.288	9.886
Total labor-hours/year		10.640	9.240	2.310	1.288	9.886

Description		Labor-hrs.	2015 Bare Costs				Total In-House	Total w/O&P
			Material	Labor	Equip.	Total		
2900	Chiller, centrif., water cooled, over 100 tons, annually	9.886	286	590		876	1,050	1,275
2950	Annualized	33.364	290	2,000		2,290	2,800	3,475

D30 HVAC | D3035 150 | Chiller, Absorption Unit

PM Components	Labor-hrs.	W	M	Q	S	A
PM System D3035 150 1950						
Chiller, absorption unit, up to 500 tons						
1 Check with operating personnel for deficiencies; check operating log sheets for indications of increased temperature trends.	.035			√	√	√
2 Check unit for proper operation, excessive noise or vibration.	.026			√	√	√
3 Check and clean strainers in all lines as required.	1.300					√
4 Check pulley alignment and belts for condition, proper tension and misalignment on external purge pump system, if applicable; adjust for proper tension and or alignment.	.043			√	√	√
5 Check purge pump vacuum oil level, as required; add/change oil as necessary.	.650			√	√	√
6 Lubricate pump shaft bearings and motor bearings.	.281			√	√	√
7 Check and service system controls, wirings and connections; tighten loose connections.	.120					√
8 Inspect cooling and system water piping circuits for leakage.	.077			√	√	√
9 Clean area around equipment.	.066			√	√	√
10 Fill out maintenance checklist.	.022			√	√	√
Total labor-hours/period				1.200	1.200	2.620
Total labor-hours/year				2.400	1.200	2.620

			Cost Each					
				2015 Bare Costs			Total	Total
Description	Labor-hrs.	Material	Labor	Equip.	Total	In-House	w/O&P
1900 Chiller, absorption unit, up to 500 tons, annually	2.620	88.50	157		245.50	294	355
1950 Annualized	6.220	94	370		464	570	700

PM Components	Labor-hrs.	W	M	Q	S	A
PM System D3035 150 2950						
Chiller, absorption unit, 500 to 5000 tons						
1 Check with operating personnel for deficiencies; check operating log sheets for indications of increased temperature trends.	.035			√	√	√
2 Check unit for proper operation, excessive noise or vibration.	.026			√	√	√
3 Check and clean strainers in all lines as required.	2.340					√
4 Check pulley alignment and belts for condition, proper tension and misalignment on external purge pump system, if applicable; adjust for proper tension and or alignment.	.086			√	√	√
5 Check purge pump vacuum oil level, as required; add/change oil as necessary.	.975			√	√	√
6 Lubricate pump shaft bearings and motor bearings.	.320			√	√	√
7 Check and service system controls, wirings and connections; tighten loose connections.	.139					√
8 Inspect cooling and system water piping circuits for leakage.	.153			√	√	√
9 Clean area around equipment.	.117			√	√	√
10 Fill out maintenance checklist.	.022			√	√	√
Total labor-hours/period				1.734	1.734	4.213
Total labor-hours/year				3.468	1.734	4.213

			Cost Each					
				2015 Bare Costs			Total	Total
Description	Labor-hrs.	Material	Labor	Equip.	Total	In-House	w/O&P
2900 Chiller, absorption unit, 500 to 5000 tons, annually	4.213	178	252		430	510	620
2950 Annualized	9.415	243	565		808	970	1,175

For customer support on your Facilities Maintenance & Repair Cost Data, call 877.801.5724.

PM Components	Labor-hrs.	W	M	Q	S	A
PM System D3035 160 1950						
Chiller, screw, water cooled, up to 100 tons						
1 Check unit for proper operation, excessive noise or vibration.	.033		√	√	√	√
2 Run system diagnostics test.	.325		√	√	√	√
3 Check oil level and oil temperature; add oil as necessary.	.022		√	√	√	√
4 Check refrigerant pressures; add as necessary.	.272		√	√	√	√
5 Replace oil filters and oil, if applicable.	.081				√	√
6 Check contactors, sensors and mechanical safety limits.	.094				√	√
7 Perform spectrochemical analysis of compressor oil.	.039				√	√
8 Check electrical wiring and connections; tighten loose connections.	.120				√	
9 Inspect cooler and condenser tubes for leaks.	5.200					√
10 Check evaporator and condenser for corrosion.	.026					√
11 Clean chiller and surrounding area.	.066		√	√	√	√
12 Fill out maintenance checklist and report deficiencies.	.022		√	√	√	√
Total labor-hours/period			.740	.740	1.074	6.300
Total labor-hours/year			5.920	1.480	1.074	6.300

				Cost Each				
				2015 Bare Costs			Total	Total
	Description	Labor-hrs.	Material	Labor	Equip.	Total	In-House	w/O&P
1900	Chiller, screw, water cooled, up to 100 tons, annually	6.300	255	375		630	750	905
1950	Annualized	14.774	259	885		1,144	1,400	1,700

PM Components	Labor-hrs.	W	M	Q	S	A
PM System D3035 160 2950						
Chiller, screw, water cooled, over 100 tons						
1 Check unit for proper operation, excessive noise or vibration.	.033		√	√	√	√
2 Run system diagnostics test.	.325		√	√	√	√
3 Check oil level and oil temperature; add oil as necessary.	.022		√	√	√	√
4 Check refrigerant pressures; add as necessary.	.381		√	√	√	√
5 Replace oil filters and oil, if applicable.	.081				√	√
6 Check contactors, sensors and mechanical safety limits.	.094				√	√
7 Perform spectrochemical analysis of compressor oil.	.039				√	√
8 Check electrical wiring and connections; tighten loose connections.	.120				√	√
9 Inspect cooler and condenser tubes for leaks.	5.200					√
10 Check evaporator and condenser for corrosion.	.052					√
11 Clean chiller and surrounding area.	.117		√	√	√	√
12 Fill out maintenance checklist and report deficiencies.	.022		√	√	√	√
Total labor-hours/period			.900	.900	1.234	6.486
Total labor-hours/year			7.200	1.800	1.234	6.486

				Cost Each				
				2015 Bare Costs			Total	Total
	Description	Labor-hrs.	Material	Labor	Equip.	Total	In-House	w/O&P
2900	Chiller, screw, water cooled, over 100 tons, annually	6.486	286	390		676	800	965
2950	Annualized	16.720	290	1,000		1,290	1,575	1,925

D3035 170 | Evaporative Cooler

PM Components	Labor-hrs.	W	M	Q	S	A
PM System D3035 170 1950						
Evaporative cooler						
1 Check with operating or area personnel for deficiencies.	.035					√
2 Check unit for proper operation, noise and vibration.	.016					√
3 Clean evaporation louver panels.	.083					√
4 Lubricate fan, bearings and motor.	.052					√
5 Check belt(s) for excessive wear and deterioration.	.095					√
6 Visually inspect wiring for damage or loose connections; tighten loose connections.	.120					√
7 Clean sump.	.039					√
8 Clean and adjust float.	.257					√
9 Check interior surfaces and check components for loose paint/lime deposits.	.016					√
10 Check/clean, adjust nozzles (6).	.039					√
11 Remove/clean and reinstall evaporative pads.	.352					√
12 Start unit and check for proper operation.	.055					√
13 Remove debris from surrounding area.	.066					√
14 Fill out maintenance checklist.	.022					√
Total labor-hours/period						1.247
Total labor-hours/year						1.247

				Cost Each				
			2015 Bare Costs				**Total**	**Total**
Description		**Labor-hrs.**	**Material**	**Labor**	**Equip.**	**Total**	**In-House**	**w/O&P**
1900	Evaporative cooler, annually	1.247	31.50	74.50		106	128	155
1950	Annualized	1.247	31.50	74.50		106	128	155

D3035 180 | Evaporative Cooler, Rotating Drum

PM Components	Labor-hrs.	W	M	Q	S	A
PM System D3035 180 1950						
Evaporative cooler, rotating drum						
1 Check with operating or area personnel for deficiencies.	.035					√
2 Check unit for proper operation, noise and vibration.	.016					√
3 Clean evaporation louver panels.	.083					√
4 Lubricate rotary and fan, bearings and motor.	.078					√
5 Check belt(s) for excessive wear and deterioration.	.095					√
6 Visually inspect wiring for damage or loose connections; tighten loose connections.	.120					√
7 Clean sump and pump intake.	.039					√
8 Clean and adjust float.	.257					√
9 Check interior surfaces and check components for loose paint/lime deposits.	.055					√
10 Check/clean, adjust nozzles.	.039					√
11 Remove/clean and reinstall evaporative pads.	.352					√
12 Start unit and check for proper operation.	.055					√
13 Remove debris from surrounding area.	.066					√
14 Fill out maintenance checklist.	.022					√
Total labor-hours/period						1.312
Total labor-hours/year						1.312

				Cost Each				
			2015 Bare Costs				**Total**	**Total**
Description		**Labor-hrs.**	**Material**	**Labor**	**Equip.**	**Total**	**In-House**	**w/O&P**
1900	Evaporative cooler, rotating drum, annually	1.312	39	78.50		117.50	141	171
1950	Annualized	1.312	39	78.50		117.50	141	171

PM Components	Labor-hrs.	W	M	Q	S	A
PM System D3035 210 1950						
Condenser, air cooled, 3 tons to 25 tons						
1 Check with operating or area personnel for deficiencies.	.027			√	√	√
2 Check unit for proper operation, excessive noise or vibration.	.025			√	√	√
3 Pressure wash coils and fans with coil cleaning solution.	.470					
4 Check electrical wiring and connections; tighten loose connections.	.092					
5 Lubricate shaft bearings and motor bearings.	.036			√	√	√
6 Inspect fan(s) or blower(s) for bent blades or imbalance; adjust as necessary.	.024			√	√	√
7 Check belt(s) for condition, proper tension, and misalignment; adjust for proper tension and/or alignment, if applicable.	.060			√	√	√
8 Inspect piping and valves for leaks; tighten connections as necessary.	.059			√	√	√
9 Clean area around equipment.	.051			√	√	√
10 Fill out maintenance checklist and report deficiencies.	.017			√	√	√
Total labor-hours/period				.299	.299	.861
Total labor-hours/year				.598	.299	.861

			Cost Each					
				2015 Bare Costs			Total	Total
	Description	Labor-hrs.	Material	Labor	Equip.	Total	In-House	w/O&P
1900	Condenser, air cooled, 3 tons to 25 tons, annually	.861	24.50	51.50		76	91.50	111
1950	Annualized	1.758	41	105		146	176	215

PM Components	Labor-hrs.	W	M	Q	S	A
PM System D3035 210 2950						
Condenser, air cooled, 26 tons through 100 tons						
1 Check with operating or area personnel for deficiencies.	.027			√	√	√
2 Check unit for proper operation, excessive noise or vibration.	.025			√	√	√
3 Pressure wash coils and fans with coil cleaning solution.	.570					√
4 Check electrical wiring and connections; tighten loose connections.	.092					√
5 Lubricate shaft bearings and motor bearings.	.042			√	√	√
6 Inspect fan(s) or blower(s) for bent blades or imbalance; adjust as necessary.	.030			√	√	√
7 Check belt(s) for condition, proper tension, and misalignment; adjust for proper tension and/or alignment, if applicable.	.080			√	√	√
8 Inspect piping and valves for leaks; tighten connections as necessary.	.059			√	√	√
9 Clean area around equipment.	.051			√	√	√
10 Fill out maintenance checklist and report deficiencies.	.017			√	√	√
Total labor-hours/period				.331	.331	.993
Total labor-hours/year				.662	.331	.993

			Cost Each					
				2015 Bare Costs			Total	Total
	Description	Labor-hrs.	Material	Labor	Equip.	Total	In-House	w/O&P
2900	Condenser, air cooled, 26 tons to 100 tons, annually	.993	58.50	59.50		118	139	166
2950	Annualized	1.986	75.50	119		194.50	231	279

PM Components	Labor-hrs.	W	M	Q	S	A
PM System D3035 210 3950						
Condenser, air cooled, over 100 tons						
1 Check with operating or area personnel for deficiencies.	.027			√	√	√
2 Check unit for proper operation, excessive noise or vibration.	.031			√	√	√
3 Pressure wash coils and fans with coil cleaning solution.	.670					√
4 Check electrical wiring and connections; tighten loose connections.	.092					√
5 Lubricate shaft bearings and motor bearings.	.042			√	√	√
6 Inspect fan(s) or blower(s) for bent blades or imbalance; adjust as necessary.	.030			√	√	√
7 Check belt(s) for condition, proper tension, and misalignment; adjust for proper tension and/or alignment, if applicable.	.100			√	√	√
8 Inspect piping and valves for leaks; tighten connections as necessary.	.059			√	√	√
9 Clean area around equipment.	.051			√	√	√
10 Fill out maintenance checklist and report deficiencies.	.017			√	√	√
Total labor-hours/period				.357	.357	1.119
Total labor-hours/year				.714	.357	1.119

	Description	Labor-hrs.	Cost Each				Total In-House	Total w/O&P
			2015 Bare Costs					
			Material	Labor	Equip.	Total		
3900	Condenser, air cooled, over 100 tons, annually	1.119	92	67		159	185	219
3950	Annualized	2.190	109	131		240	283	340

PM Components	Labor-hrs.	W	M	Q	S	A
PM System D3035 220 1950						
Condensing unit, air cooled, 3 tons to 25 tons						
1 Check with operating or area personnel for deficiencies.	.035			√	√	√
2 Check unit for proper operation, excessive noise or vibration.	.033			√	√	√
3 Pressure wash coils and fans with coil cleaning solution.	.611					√
4 Check electrical wiring and connections; tighten loose connections.	.120					√
5 Lubricate shaft bearings and motor bearings.	.047			√	√	√
6 Inspect fan(s) or blower(s) for bent blades or imbalance; adjust as necessary.	.031			√	√	√
7 Check belt(s) for condition, proper tension, and misalignment; adjust for proper tension and/or alignment, if applicable.	.078			√	√	√
8 Inspect piping and valves for leaks; tighten connections as necessary.	.077			√	√	√
9 Check refrigerant pressure; add refrigerant, if necessary.	.272					√
10 Clean area around equipment.	.066			√	√	√
11 Fill out maintenance checklist and report deficiencies.	.022			√	√	√
Total labor-hours/period				.389	.389	1.392
Total labor-hours/year				.778	.389	1.392

			Cost Each					
			2015 Bare Costs				Total	Total
	Description	Labor-hrs.	Material	Labor	Equip.	Total	In-House	w/O&P
1900	Condensing unit, air cooled, 3 to 25 tons, annually	1.392	88.50	83		171.50	202	241
1950	Annualized	2.559	94	153		247	295	355

PM Components	Labor-hrs.	W	M	Q	S	A
PM System D3035 220 2950						
Condensing unit, air cooled, 26 tons through 100 tons						
1 Check with operating or area personnel for deficiencies.	.035			√	√	√
2 Check unit for proper operation, excessive noise or vibration.	.033			√	√	√
3 Pressure wash coils and fans with coil cleaning solution.	.741					√
4 Check electrical wiring and connections; tighten loose connections.	.120					√
5 Lubricate shaft bearings and motor bearings.	.055			√	√	√
6 Inspect fan(s) or blower(s) for bent blades or imbalance; adjust as necessary.	.039			√	√	√
7 Check belt(s) for condition, proper tension, and misalignment; adjust for proper tension and/or alignment, if applicable.	.104			√	√	√
8 Inspect piping and valves for leaks; tighten connections as necessary.	.077			√	√	√
9 Check refrigerant pressure; add refrigerant, if necessary.	.390					√
10 Clean area around equipment.	.066			√	√	√
11 Fill out maintenance checklist and report deficiencies.	.022			√	√	√
Total labor-hours/period				.431	.431	1.682
Total labor-hours/year				.862	.431	1.682

			Cost Each					
			2015 Bare Costs				Total	Total
	Description	Labor-hrs.	Material	Labor	Equip.	Total	In-House	w/O&P
2900	Condensing unit, air cooled, 26 to 100 tons, annually	1.682	120	101		221	258	305
2950	Annualized	2.975	125	178		303	360	435

PM Components	Labor-hrs.	W	M	Q	S	A
PM System D3035 220 3950						
Condensing unit, air cooled, over 100 tons						
1 Check with operating or area personnel for deficiencies.	.035			√	√	√
2 Pressure wash coils and fans with coil cleaning solution.	.871					√
3 Clean intake side of condenser coils, fans and intake screens.	.091			√	√	√
4 Check electrical wiring and connections; tighten loose connections.	.120					√
5 Lubricate shaft bearings and motor bearings.	.055			√	√	√
6 Inspect fan(s) or blower(s) for bent blades or imbalance; adjust as necessary.	.039			√	√	√
7 Check belt(s) for condition, proper tension, and misalignment; adjust for proper tension and/or alignment, if applicable.	.130			√	√	√
8 Inspect piping and valves for leaks; tighten connections as necessary.	.077			√	√	√
9 Check refrigerant pressure; add refrigerant, if necessary.	.455					√
10 Clean area around equipment.	.066			√	√	√
11 Fill out maintenance checklist and report deficiencies.	.022			√	√	√
Total labor-hours/period				.515	.515	1.961
Total labor-hours/year				1.030	.515	1.961

	Description	Labor-hrs.	Cost Each				Total In-House	Total w/O&P
			2015 Bare Costs					
			Material	Labor	Equip.	Total		
3900	Condensing unit, air cooled, over 100 tons, annually	1.961	151	117		268	315	370
3950	Annualized	3.506	157	209		366	435	520

For customer support on your Facilities Maintenance & Repair Cost Data, call 877.801.5724.

435

PM Components	Labor-hrs.	W	M	Q	S	A
PM System D3035 240 1950						
Condensing unit, water cooled, 3 tons to 24 tons						
1 Check with operating or area personnel for deficiencies.	.035			√	√	√
2 Check unit for proper operation, excessive noise or vibration.	.033			√	√	√
3 Check electrical wiring and connections; tighten loose connections.	.120					√
4 Inspect piping and valves for leaks; tighten connections as necessary.	.077			√	√	√
5 Check refrigerant pressure; add refrigerant, if necessary.	.272				√	√
6 Clean area around equipment.	.066			√	√	√
7 Fill out maintenance checklist and report deficiencies.	.022			√	√	√
Total labor-hours/period				.233	.233	.625
Total labor-hours/year				.466	.233	.625

			Cost Each					
			2015 Bare Costs				Total	Total
Description	Labor-hrs.	Material	Labor	Equip.	Total	In-House	w/O&P
1900	Condensing unit, water cooled, 3 to 24 tons, annually	.625	31.50	37.50		69	81.50	98
1950	Annualized	1.324	31.50	79		110.50	134	162

PM Components	Labor-hrs.	W	M	Q	S	A
PM System D3035 240 2950						
Condensing unit, water cooled, 25 tons through 100 tons						
1 Check with operating or area personnel for deficiencies.	.035			√	√	√
2 Check unit for proper operation, excessive noise or vibration.	.033			√	√	√
3 Run system diagnostics test.	.216			√	√	√
4 Check electrical wiring and connections; tighten loose connections.	.120			√	√	√
5 Check oil level in sight glass of lead compressor only; add oil as necessary.	.029			√	√	√
6 Check contactors, sensors and mechanical limits, adjust as necessary.	.156					√
7 Check condenser for corrosion.	.026					√
8 Inspect piping and valves for leaks; tighten connections as necessary.	.077			√	√	√
9 Check refrigerant pressure; add refrigerant, if necessary.	.272					√
10 Clean area around equipment.	.066			√	√	√
11 Fill out maintenance checklist and report deficiencies.	.022			√	√	√
Total labor-hours/period				.598	.598	1.052
Total labor-hours/year				1.196	.598	1.052

			Cost Each					
			2015 Bare Costs				Total	Total
Description	Labor-hrs.	Material	Labor	Equip.	Total	In-House	w/O&P
2900	Condensing unit, WC, 25 to 100 tons, annually	1.052	78.50	63		141.50	165	196
2950	Annualized	2.846	78.50	170		248.50	300	365

PM Components	Labor-hrs.	W	M	Q	S	A
PM System D3035 240 3950						
Condensing unit, water cooled, over 100 tons						
1　Check with operating or area personnel for deficiencies.	.035			√	√	√
2　Check unit for proper operation, excessive noise or vibration.	.033			√	√	√
3　Run system diagnostics test.	.216			√	√	√
4　Check electrical wiring and connections; tighten loose connections.	.120			√	√	√
5　Check oil level in sight glass of lead compressor only; add oil as necessary.	.029			√	√	√
6　Check contactors, sensors and mechanical limits, adjust as necessary.	.156					√
7　Check condenser for corrosion.	.026					√
8　Inspect piping and valves for leaks; tighten connections as necessary.	.077			√	√	√
9　Check refrigerant pressure; add refrigerant, if necessary.	.272					√
10　Clean area around equipment.	.066			√	√	√
11　Fill out maintenance checklist and report deficiencies.	.022			√	√	√
Total labor-hours/period				.598	.598	1.052
Total labor-hours/year				1.196	.598	1.052

			Cost Each					
			2015 Bare Costs				**Total**	**Total**
Description		**Labor-hrs.**	**Material**	**Labor**	**Equip.**	**Total**	**In-House**	**w/O&P**
3900	Condensing unit, WC, over 100 tons, annually	1.052	125	63		188	217	254
3950	Annualized	2.846	125	170		295	350	420

For customer support on your Facilities Maintenance & Repair Cost Data, call 877.801.5724.

437

PM Components	Labor-hrs.	W	M	Q	S	A
PM System D3035 260 1950						
Compressor, DX Refrigeration, to 25 tons						
1 Remove/replace access panel/cover.	.070			√	√	√
2 Check unit for proper operation, excessive noise or vibration.	.030			√	√	√
3 Run systems diagnostics test, if applicable.	.060			√	√	√
4 Check oil level in compressor, if possible, and add oil as required.	.030			√	√	√
5 Check refrigerant pressures; add refrigerant as necessary.	.200			√	√	√
6 Check contactors, sensors and mechanical limits, adjust as necessary.	.100			√	√	√
7 Inspect refrigerant piping and insulation.	.050			√	√	√
8 Clean compressor and surrounding area.	.050			√	√	√
9 Fill out maintenance checklist and report deficiencies.	.020			√	√	√
Total labor-hours/period				.610	.610	.610
Total labor-hours/year				1.220	.610	.610

	Description	Labor-hrs.	Cost Each 2015 Bare Costs Material	Labor	Equip.	Total	Total In-House	Total w/O&P
1900	Compressor, DX refrigeration, to 25 tons, annually	.610	16.40	36.50		52.90	63.50	77.50
1950	Annualized	2.440	16.40	146		162.40	200	248

PM Components	Labor-hrs.	W	M	Q	S	A
PM System D3035 260 2950						
Compressor, DX refrigeration, 25 tons to 100 tons						
1 Check unit for proper operation, excessive noise or vibration.	.030			√	√	√
2 Run system diagnostics test, if applicable.	.098			√	√	√
3 Check oil level in compressor; add oil as necessary.	.033			√	√	√
4 Check refrigerant pressures; add refrigerant as necessary.	.272			√	√	√
5 Check contactors, sensors and mechanical limits, adjust as necessary.	.216			√	√	√
6 Inspect refrigerant piping and insulation.	.077			√	√	√
7 Clean compressor and surrounding area.	.066			√	√	√
8 Fill out maintenance checklist and report deficiencies.	.022			√	√	√
Total labor-hours/period				.814	.814	.814
Total labor-hours/year				1.628	.814	.814

	Description	Labor-hrs.	Cost Each 2015 Bare Costs Material	Labor	Equip.	Total	Total In-House	Total w/O&P
2900	Compressor, DX refrigeration, 25 to 100 tons, annually	.814	15.55	48.50		64.05	78	95.50
2950	Annualized	3.256	15.55	195		210.55	260	325

PM Components	Labor-hrs.	W	M	Q	S	A
PM System D3035 290 1950						
Fluid cooler, 2 fans (no compressor)						
1 Check with operating or area personnel for deficiencies.	.035					√
2 Check unit for proper operation, excessive noise or vibration.	.159					√
3 Clean intake side of condenser coils, fans and intake screens.	.473					√
4 Check electrical wiring and connections; tighten loose connections.	.120					√
5 Inspect fan(s) for bent blades or unbalance; adjust as necessary.	.040					√
6 Check belts for condition, proper tension and misalignment; adjust	.029					√
for proper tension and/or alignment, if required.						
7 Lubricate shaft bearings and motor bearings.	.047					√
8 Inspect piping and valves for leaks; tighten connections as necessary.	.077					√
9 Lubricate and check operation of dampers, if applicable.	.055					√
10 Clean area around fluid cooler.	.066					√
11 Fill out maintenance checklist and report deficiencies.	.022					√
Total labor-hours/period						1.123
Total labor-hours/year						1.123

			2015 Bare Costs				Total	Total
Description		**Labor-hrs.**	**Material**	**Labor**	**Equip.**	**Total**	**In-House**	**w/O&P**
1900	Fluid cooler, 2 fans (no compressor), annually	1.123	15.70	67		82.70	101	125
1950	Annualized	1.123	15.70	67		82.70	101	125

PM Components	Labor-hrs.	W	M	Q	S	A
PM System D3045 110 1950						
Air handling unit, 3 tons through 24 tons						
1 Check with operating or area personnel for deficiencies.	.035			√	√	√
2 Check controls and unit for proper operation.	.033			√	√	√
3 Check for unusual noise or vibration.	.033			√	√	√
4 Check tension, condition and alignment of belts, adjust as necessary.	.029			√	√	√
5 Clean coils, evaporator drain pan, blower, motor and drain piping, as required.	.380					√
6 Lubricate shaft and motor bearings.	.047			√	√	√
7 Replace air filters.	.078			√	√	√
8 Inspect exterior piping and valves for leaks; tighten connections as required.	.077			√	√	√
9 Clean area around equipment.	.066			√	√	√
10 Fill out maintenance checklist and report deficiencies.	.022			√	√	√
Total labor-hours/period				.420	.420	.800
Total labor-hours/year				.840	.420	.800

			Cost Each				
			2015 Bare Costs			Total	Total
Description	Labor-hrs.	Material	Labor	Equip.	Total	In-House	w/O&P
1900 Air handling unit, 3 thru 24 tons, annually	.800	48.50	48		96.50	113	135
1950 Annualized	2.060	104	123		227	268	320

PM Components	Labor-hrs.	W	M	Q	S	A
PM System D3045 110 2950						
Air handling unit, 25 tons through 50 tons						
1 Check with operating or area personnel for deficiencies.	.035			√	√	√
2 Check controls and unit for proper operation.	.033			√	√	√
3 Check for unusual noise or vibration.	.033			√	√	√
4 Clean coils, evaporator drain pan, blower, motor and condensate drain piping, as required.	.380					√
5 Lubricate shaft and motor bearings.	.047			√	√	√
6 Check belts for wear, proper tension, and alignment; adjust as necessary.	.029			√	√	√
7 Inspect exterior piping and valves for leaks; tighten connections as required.	.077			√	√	√
8 Check operation and clean dampers, louvers and shutters; lubricate all pivot points and linkages.	.078					√
9 Replace air filters.	.078			√	√	√
10 Clean area around equipment.	.066			√	√	√
11 Fill out maintenance checklist and report deficiencies.	.022			√	√	√
Total labor-hours/period				.420	.420	.878
Total labor-hours/year				.840	.420	.878

			Cost Each				
			2015 Bare Costs			Total	Total
Description	Labor-hrs.	Material	Labor	Equip.	Total	In-House	w/O&P
2900 Air handling unit, 25 thru 50 tons, annually	.878	151	52.50		203.50	232	271
2950 Annualized	2.138	410	128		538	610	710

PM Components	Labor-hrs.	W	M	Q	S	A
PM System D3045 110 3950						
Air handling unit, over 50 tons						
1 Check with operating or area personnel for deficiencies.	.035			√	√	√
2 Check controls and unit for proper operation.	.033			√	√	√
3 Check for unusual noise or vibration.	.033			√	√	√
4 Clean coils, evaporator drain pan, blower, motor and drain piping, as required.	.510					√
5 Lubricate shaft and motor bearings.	.047			√	√	√
6 Check belts for wear, proper tension, and alignment; adjust as necessary.	.055			√	√	√
7 Inspect exterior piping and valves for leaks; tighten connections as required.	.077			√	√	√
8 Check operation and clean dampers, louvers and shutters; lubricate all pivot points and linkages.	.078					√
9 Clean centrifugal fan.	.065					√
10 Replace air filters.	.286			√	√	√
11 Clean area around equipment.	.066			√	√	√
12 Fill out maintenance checklist and report deficiencies	.022			√	√	√
Total labor-hours/period				.654	.654	1.307
Total labor-hours/year				1.308	.654	1.307

			Cost Each				
			2015 Bare Costs			**Total**	**Total**
Description	**Labor-hrs.**	**Material**	**Labor**	**Equip.**	**Total**	**In-House**	**w/O&P**
3900 Air handling unit, over 50 tons, annually	1.307	216	78		294	335	390
3950 Annualized	3.269	530	195		725	825	970

PM Components	Labor-hrs.	W	M	Q	S	A
PM System D3045 112 1950						
Air handling unit, computer room						
1 Check with operating or area personnel for deficiencies.	.035			√	√	√
2 Run microprocessor check, if available, or check controls and unit for proper operation.	.216			√	√	√
3 Check for unusual noise or vibration.	.033			√	√	√
4 Clean coils, evaporator drain pan, blower, motor and drain piping, as required.	.380					√
5 Lubricate shaft and motor bearings.	.047			√	√	√
6 Check belts for wear, proper tension, and alignment; adjust as necessary.	.029			√	√	√
7 Check humidity lamp, replace if necessary.	.156			√	√	√
8 Inspect exterior piping and valves for leaks; tighten connections as required.	.077			√	√	√
9 Replace air filters.	.078			√	√	√
10 Clean area around equipment.	.066			√	√	√
11 Fill out maintenance checklist and report deficiencies.	.022			√	√	√
Total labor-hours/period				.759	.759	1.139
Total labor-hours/year				1.518	.759	1.139

			Cost Each				
			2015 Bare Costs			**Total**	**Total**
Description	**Labor-hrs.**	**Material**	**Labor**	**Equip.**	**Total**	**In-House**	**w/O&P**
1900 Air handling unit, computer room, annually	1.139	40	68		108	129	156
1950 Annualized	3.416	64	204		268	325	400

For customer support on your Facilities Maintenance & Repair Cost Data, call 877.801.5724.

441

PM Components	Labor-hrs.	W	M	Q	S	A
PM System D3045 120 1950						
Fan coil unit						
1 Check with operating or area personnel for deficiencies.	.035			√	√	√
2 Check coil unit while operating.	.120				√	√
3 Remove access panel and vacuum inside of unit and coils.	.468				√	√
4 Check coils and piping for leaks, damage and corrosion; repair as necessary.	.077			√	√	√
5 Lubricate blower shaft and fan motor bearings.	.047				√	√
6 Clean coil, drip pan, and drain line with solvent.	.473				√	√
7 Replace filters as required.	.009			√	√	√
8 Replace access panel.	.023				√	√
9 Check operation after repairs.	.120				√	√
10 Clean area.	.066			√	√	√
11 Fill out maintenance checklist and report deficiencies.	.022			√	√	√
Total labor-hours/period				.209	1.460	1.460
Total labor-hours/year				.418	1.460	1.460

			Cost Each					
			2015 Bare Costs				Total	Total
Description		Labor-hrs.	Material	Labor	Equip.	Total	In-House	w/O&P
1900	Fan coil unit, annually	1.460	7.05	87		94.05	117	145
1950	Annualized	3.338	28.50	199		227.50	282	345

PM Components	Labor-hrs.	W	M	Q	S	A
PM System D3045 150 1950						
Filters, Electrostatic						
1 Review manufacturers instructions.	.091			√	√	√
2 Check indicators for defective tubes or broken ionizing wires.	.130			√	√	√
3 De-energize power supply,tag and lockout disconnect switch.	.130			√	√	√
4 Ground bus trips, top and bottom.	.065			√	√	√
5 Secure filter and fan unit.	.065			√	√	√
6 Wash each manifold until clean, approximately 4 minutes with hot water, 7 minutes with cold water.	.390			√	√	√
7 Clean or replace dry filters as necessary.	.260			√	√	√
8 Inspect for broken on hum supressors and wipe insulators with soft dry cloth.	.260			√	√	√
9 Disassemble unit as required, check it thoroughly, clean and adjust.	.650			√	√	√
10 Restore to service and check for shorts.	.130			√	√	√
11 Fill out maintenance checklist and report deficiencies.	.022			√	√	√
Total labor-hours/period				2.193	2.193	2.193
Total labor-hours/year				4.386	2.193	2.193

			Cost Each					
			2015 Bare Costs				Total	Total
Description		Labor-hrs.	Material	Labor	Equip.	Total	In-House	w/O&P
1900	Air filter, electrostatic, annually	2.193	12.80	107		119.80	154	189
1950	Annualized	8.772	25.50	425		450.50	590	725

PM Components	Labor-hrs.	W	M	Q	S	A
PM System D3045 160 1950						
VAV Boxes						
1 Open/close VAV control box access.	.060				√	√
2 Check that pneumatic tubing/electrical connections are in place and tight.	.040				√	√
3 Tighten arm on motor output shaft.	.040				√	√
4 Cycle actuator while watching for proper operation.	.200				√	√
Verify that blades fully open and close.	.040				√	√
5 Lubricate actuator linkage and damper blade pivot points.	.070				√	√
6 Fill out maintenance checklist and report deficiencies.	.017				√	√
Total labor-hours/period					.467	.467
Total labor-hours/year					.467	.467

			Cost Each					
				2015 Bare Costs			Total	Total
Description		Labor-hrs.	Material	Labor	Equip.	Total	In-House	w/O&P
1900	VAV Boxes, annually	.467	8.55	28		36.55	44.50	54
1950	Annualized	.934	8.55	56		64.55	79	97.50

PM Components	Labor-hrs.	W	M	Q	S	A
PM System D3045 170 1950						
Fire dampers						
1 Remove/replace access door.	.800					√
2 Clean out debris/dirt blown against damper.	.200					√
3 Remove fusible link and check that blades operate freely.	.070					√
4 Lubricate pivot points.	.050					√
5 Replace fusible link.	.020					√
6 Fill out maintenance checklist and report deficiencies.	.017					√
Total labor-hours/period						1.157
Total labor-hours/year						1.157

			Cost Each					
				2015 Bare Costs			Total	Total
Description		Labor-hrs.	Material	Labor	Equip.	Total	In-House	w/O&P
1900	Fire dampers, annually	1.157	6.65	69		75.65	94	116
1950	Annualized	1.157	6.65	69		75.65	94	116

PM Components	Labor-hrs.	W	M	Q	S	A
PM System D3045 210 1950						
Fan, axial, up to 5,000 CFM						
1 Start and stop fan with local switch.	.012				√	√
2 Check fan for noise and vibration.	.091				√	√
3 Check electrical wiring and connections; tighten loose connections.	.029				√	√
4 Check motor and fan shaft bearings for noise, vibration, overheating; lubricate as required.	.325				√	√
5 Clean area around fan.	.143				√	√
6 Fill out maintenance checklist and report deficiencies.	.022				√	√
Total labor-hours/period					.622	.622
Total labor-hours/year					.622	.622

			Cost Each					
				2015 Bare Costs			Total	Total
Description		Labor-hrs.	Material	Labor	Equip.	Total	In-House	w/O&P
1900	Fan, axial, up to 5,000 CFM, annually	.622	3.93	30.50		34.43	44	54
1950	Annualized	1.244	7.85	60.50		68.35	88	108

PM Components	Labor-hrs.	W	M	Q	S	A
PM System D3045 210 2950						
Fan, axial, 5,000 to 10,000 CFM						
1 Start and stop fan with local switch.	.012				√	√
2 Check motor and fan shaft bearings for noise, vibration, overheating; lubricate bearings.	.325				√	√
3 Check belts for wear, tension, and alignment, if applicable; adjust as required.	.057				√	√
4 Check fan pitch operator, lubricate; if applicable.	.029				√	√
5 Check electrical wiring and connections; tighten loose connections.	.057				√	√
6 Clean fan and surrounding fan.	.143				√	√
7 Fill out maintenance checklist and report deficiencies.	.022				√	√
Total labor-hours/period					.645	.645
Total labor-hours/year					.645	.645

			Cost Each					
				2015 Bare Costs			Total	Total
Description		Labor-hrs.	Material	Labor	Equip.	Total	In-House	w/O&P
2900	Fan, axial, 5,000 to 10,000 CFM, annually	.645	3.93	31.50		35.43	45.50	56
2950	Annualized	1.290	7.85	63		70.85	91	112

PM Components	Labor-hrs.	W	M	Q	S	A
PM System D3045 210 3950						
Fan, axial, 36″ to 48″ dia (over 10,000 CFM)						
1 Start and stop fan with local switch.	.012				√	√
2 Check motor and fan shaft bearings for noise, vibration, overheating; lubricate bearings.	.325				√	√
3 Check belts for wear, tension, and alignment, if applicable; adjust as required.	.086				√	√
4 Check fan pitch operator, lubricate; if applicable.	.029				√	√
5 Check electrical wiring and connections; tighten loose connections.	.078				√	√
6 Clean fan and surrounding area.	.143				√	√
7 Fill out maintenance checklist and report deficiencies.	.022				√	√
Total labor-hours/period					.695	.695
Total labor-hours/year					.695	.695

			Cost Each					
			2015 Bare Costs				Total	Total
Description		Labor-hrs.	Material	Labor	Equip.	Total	In-House	w/O&P
3900	Fan, axial, 36″ to 48″ dia (over 10,000 CFM), annually	.695	3.93	34		37.93	49	60
3950	Annualized	1.390	7.85	67.50		75.35	97.50	120

PM Components	Labor-hrs.	W	M	Q	S	A
PM System D3045 220 1950						
Fan, centrifugal, up to 5,000 CFM						
1 Start and stop fan with local switch.	.012				√	√
2 Check motor and fan shaft bearings for noise, vibration, overheating; lubricate bearings.	.325				√	√
3 Check belts for wear, tension, and alignment, if applicable; adjust as required.	.057				√	√
4 Check blower intake dampers, lubricate; if applicable.	.029				√	√
5 Check electrical wiring and connections; tighten loose connections.	.029				√	√
6 Clean fan and surrounding area.	.066				√	√
7 Fill out maintenance checklist and report deficiencies.	.022				√	√
Total labor-hours/period					.540	.540
Total labor-hours/year					.540	.540

		Labor-hrs.	Cost Each					Total	Total
			2015 Bare Costs					In-House	w/O&P
	Description		Material	Labor	Equip.	Total			
1900	Fan, centrifugal, up to 5,000 CFM, annually	.540	3.93	26.50		30.43		39	47.50
1950	Annualized	1.080	7.85	52.50		60.35		77.50	95.50

PM Components	Labor-hrs.	W	M	Q	S	A
PM System D3045 220 2950						
Fan, centrifugal, 5,000 to 10,000 CFM						
1 Start and stop fan with local switch.	.012				√	√
2 Check motor and fan shaft bearings for noise, vibration, overheating; lubricate bearings.	.325				√	√
3 Check belts for wear, tension, and alignment, if applicable; adjust as required.	.057				√	√
4 Check blower intake dampers, lubricate; if applicable.	.029				√	√
5 Check electrical wiring and connections; tighten loose connections.	.057				√	√
6 Clean fan and surrounding area.	.066				√	√
7 Fill out maintenance checklist and report deficiencies.	.022				√	√
Total labor-hours/period					.568	.568
Total labor-hours/year					.568	.568

		Labor-hrs.	Cost Each					Total	Total
			2015 Bare Costs					In-House	w/O&P
	Description		Material	Labor	Equip.	Total			
2900	Fan, centrifugal, 5,000 to 10,000 CFM, annually	.568	3.93	27.50		31.43		40.50	50
2950	Annualized	1.136	7.85	55.50		63.35		81.50	100

PM Components	Labor-hrs.	W	M	Q	S	A
PM System D3045 220 3950						
Fan, centrifugal, over 10,000 CFM						
1 Start and stop fan with local switch.	.007				√	√
2 Check motor and fan shaft bearings for noise, vibration, overheating; lubricate bearings.	.325				√	√
3 Check belts for wear, tension, and alignment, if applicable; adjust as required.	.086				√	√
4 Check blower intake dampers, lubricate; if applicable.	.029				√	√
5 Check electrical wiring and connections; tighten loose connections.	.057				√	√
6 Clean fan and surrounding area.	.066				√	√
7 Fill out maintenance checklist and report deficiencies.	.022				√	√
Total labor-hours/period					.592	.592
Total labor-hours/year					.592	.592

			Cost Each				
			2015 Bare Costs			Total	Total
Description	Labor-hrs.	Material	Labor	Equip.	Total	In-House	w/O&P
3900 Fan, centrifugal, over 10,000 CFM, annually	.592	3.93	29		32.93	42	52
3950 Annualized	1.184	7.85	57.50		65.35	84.50	103

PM Components	Labor-hrs.	W	M	Q	S	A
PM System D3045 250 1950						
Hood and blower						
1 Check with operating or area personnel for any deficiencies.	.044			√	√	√
2 Check unit for proper operation, including switches, controls and thermostat; calibrate thermostat and repair components as required.	.109			√	√	√
3 Check operation of spray nozzles for spray coverage and drainage, if applicable.	.022					√
4 Inspect soap and spray solution feeder lines.	.007			√	√	√
5 Check components for automatic cleaning system and automatic fire protection system - solenoid valve, line strainer, shut-off valve, detergent tank, etc.	.055			√	√	√
6 Clean spray nozzles.	1.104					√
7 Tighten or replace loose, missing or damaged nuts, bolts or screws.	.005			√	√	√
8 Check operation of exhaust blower on roof; lubricate bearings; adjust tension of fan belts and clean fan blades as required.	.056			√	√	√
9 Fill out maintenance checklist and report deficiencies.	.022			√	√	√
Total labor-hours/period				.298	.298	1.424
Total labor-hours/year				.596	.298	1.424

			Cost Each				
			2015 Bare Costs			Total	Total
Description	Labor-hrs.	Material	Labor	Equip.	Total	In-House	w/O&P
1900 Hood and blower, annually	1.424	62.50	69.50		132	161	191
1950 Annualized	2.318	89.50	113		202.50	248	295

PM Components	Labor-hrs.	W	M	Q	S	A
PM System D3045 410 1950						
Centrifugal pump over 1 H.P.						
1 Check for proper operation of pump.	.022				√	√
2 Check for leaks on suction and discharge piping, seals, packing glands, etc.; make minor adjustments as required.	.077				√	√
3 Check pump and motor operation for excessive vibration, noise and overheating.	.022				√	√
4 Check alignment of pump and motor; adjust as necessary.	.260				√	√
5 Lubricate pump and motor.	.099				√	√
6 Clean exterior of pump, motor and surrounding area.	.096				√	√
7 Fill out maintenance checklist and report deficiencies.	.022				√	√
Total labor-hours/period					.598	.598
Total labor-hours/year					.598	.598

		Cost Each					
		2015 Bare Costs				**Total**	**Total**
Description	**Labor-hrs.**	**Material**	**Labor**	**Equip.**	**Total**	**In-House**	**w/O&P**
1900 Centrifugal pump, over 1 H.P., annually	.598	8.55	35		43.55	53.50	65.50
1950 Annualized	1.196	17.05	70		87.05	107	130

PM Components	Labor-hrs.	W	M	Q	S	A
PM System D3045 410 2950						
Centrifugal pump w/reduction gear, over 1 H.P.						
1 Check with operating or area personnel for deficiencies.	.035				√	√
2 Clean pump exterior and check for corrosion on pump exterior and base plate.	.030				√	√
3 Check for leaks on suction and discharge piping, seals, packing glands, etc.	.077				√	√
4 Check pump, gear and motor operation for vibration, noise, overheating, etc.	.022				√	√
5 Check alignment and clearances of shaft reduction gear and coupler.	.260				√	√
6 Tighten or replace loose, missing, or damaged nuts, bolts and screws.	.005				√	√
7 Lubricate pump and motor as required.	.099				√	√
8 When available, check suction or discharge, pressure gauge readings and flow rate.	.022				√	√
9 Clean area around pump.	.066				√	√
10 Fill out maintenance checklist and report deficiencies.	.022				√	√
Total labor-hours/period					.638	.638
Total labor-hours/year					.638	.638

		Cost Each					
		2015 Bare Costs				**Total**	**Total**
Description	**Labor-hrs.**	**Material**	**Labor**	**Equip.**	**Total**	**In-House**	**w/O&P**
2900 Centrifugal pump, w/ red. gear, over 1H.P., annually	.638	8.55	37.50		46.05	56	69
2950 Annualized	1.276	17.05	75		92.05	112	138

PM Components	Labor-hrs.	W	M	Q	S	A
PM System D3045 420 1950						
Pump w/oil reservoir, over 1 H.P.						
1 Check for proper operation of pump.	.022				√	√
2 Check for leaks on suction and discharge piping, seals, packing glands, etc.; make minor adjustments as required.	.077				√	√
3 Check pump and motor operation for excessive vibration, noise and overheating.	.022				√	√
4 Check alignment of pump and motor; adjust as necessary.	.260				√	√
5 Lubricate motor; check oil level in pump reservoir.	.099				√	√
6 Clean exterior of pump and surrounding area.	.096				√	√
7 Fill out maintenance checklist and report deficiencies.	.022				√	√
Total labor-hours/period					.598	.598
Total labor-hours/year					.598	.598

			Cost Each					
			2015 Bare Costs				**Total**	**Total**
Description		**Labor-hrs.**	**Material**	**Labor**	**Equip.**	**Total**	**In-House**	**w/O&P**
1900	Pump w/ oil reservoir, electric, annually	.598	8.55	35		43.55	53.50	65.50
1950	Annualized	1.196	17.05	70		87.05	107	130

For customer support on your Facilities Maintenance & Repair Cost Data, call 877.801.5724.

449

PM Components	Labor-hrs.	W	M	Q	S	A
PM System D3045 600 1950						
Heat exchanger, steam						
1 Check with operating or area personnel for deficiencies.	.035				√	√
2 Check temperature gauges for proper operating temperatures.	.091				√	√
3 Check steam modulating valve and steam condensate trap for proper operation.	.101				√	√
4 Inspect heat exchanger and adjacent piping for torn or deteriorated insulation.	.147				√	√
5 Clean heat exchanger and surrounding area.	.066				√	√
6 Fill out maintenance checklist and report deficiencies.	.022				√	√
Total labor-hours/period					.462	.462
Total labor-hours/year					.462	.462

			Cost Each					
			2015 Bare Costs				**Total**	**Total**
Description		**Labor-hrs.**	**Material**	**Labor**	**Equip.**	**Total**	**In-House**	**w/O&P**
1900	Heat exchanger, steam, annually	.462	27	27.50		54.50	64	77
1950	Annualized	.924	27	55		82	98.50	120

PM Components	Labor-hrs.	W	M	Q	S	A
PM System D3055 110 1950						
Unit heater, gas radiant						
1 Check with operating or area personnel for deficiencies.	.035					√
2 Inspect, clean and adjust control valves and thermo sensing bulbs on gas burners.	.254					√
3 Inspect fuel system for leaks.	.016					√
4 Check for proper operation of burner controls. check and adjust thermostat.	.133					√
5 Check fan and motor for vibration and noise. lubricate bearings.	.056					√
6 Check electrical wiring to blower motor.	.079					√
7 Check condition of flue pipe, damper and stack.	.147					√
8 Check unit heater operation through complete cycle or up to ten minutes.	.133					√
9 Clean area around unit heater.	.133					√
10 Fill out maintenance checklist and report deficiencies.	.022					√
Total labor-hours/period						1.008
Total labor-hours/year						1.008

				Cost Each			
			2015 Bare Costs			Total	Total
Description	Labor-hrs.	Material	Labor	Equip.	Total	In-House	w/O&P
1900 Unit heater, gas radiant, annually	1.008	3.25	60		63.25	79	98
1950 Annualized	1.008	3.25	60		63.25	79	98

PM Components	Labor-hrs.	W	M	Q	S	A
PM System D3055 110 2950						
Unit heater, gas infrared						
1 Check with operating or area personnel for deficiencies.	.035					√
2 Inspect fuel system for leaks around unit.	.016					√
3 Replace primary air intake filter.	.048					√
4 Blow out burner.	.130					√
5 Clean spark electrode and reset gap, replace if necessary.	.254					√
6 Clean pilot and thermo-sensor bulb, replace if necessary.	.133					√
7 Check alignment of thermo-sensor bulb.	.133					√
8 Check wiring and connections; tighten any loose connections.	.079					√
9 Check and clean pilot sight glass.	.091					√
10 Vacuum pump (fan):						
A) check motor bearings for overheating; adjust or repair as required.	.022					√
B) lubricate motor bearings.	.099					√
C) check motor mounting; adjust as required.	.030					√
D) check fan blade clearance; adjust as required.	.042					√
E) check wiring, connections, switches, etc.; tighten any loose connections.	.079					√
11 Operate unit to ensure that it is in proper working condition.	.225					√
12 Clean equipment and surrounding area.	.066					√
13 Fill out maintenance checklist and report deficiencies.	.022					√
Total labor-hours/period						1.504
Total labor-hours/year						1.504

				Cost Each			
			2015 Bare Costs			Total	Total
Description	Labor-hrs.	Material	Labor	Equip.	Total	In-House	w/O&P
2900 Unit heater, gas infrared, annually	1.504	32.50	90		122.50	148	181
2950 Annualized	1.504	32.50	90		122.50	148	181

PM Components	Labor-hrs.	W	M	Q	S	A
PM System D3055 110 3950						
Unit heater, steam						
1 Check with operating or area personnel for deficiencies.	.035					√
2 Inspect, clean and adjust control valves and thermostat.	.254					√
3 Inspect coils, connections, trap and steam piping for leaks; repair as necessary.	.195					√
4 Check fan and motor for vibration and noise; lubricate bearings.	.056					√
5 Check electrical wiring to motor.	.079					√
6 Check unit heater operation through complete cycle or up to ten minutes.	.133					√
7 Clean equipment and surrounding area.	.066					√
8 Fill out maintenance checklist and report deficiencies.	.022					√
Total labor-hours/period						.840
Total labor-hours/year						.840

			Cost Each					
			2015 Bare Costs				Total	Total
Description		Labor-hrs.	Material	Labor	Equip.	Total	In-House	w/O&P
3900	Unit heater, steam, annually	.840	32.50	50		82.50	99	119
3950	Annualized	.840	32.50	50		82.50	99	119

PM Components	Labor-hrs.	W	M	Q	S	A
PM System D3055 122 1950						
Forced air heater, oil or gas fired, up to 120 MBH						
1 Check with operating or area personnel for deficiencies.	.035			√	√	√
2 Inspect, clean and adjust electrodes and nozzles on oil burners or controls, valves and thermo-sensing bulbs on gas burners; lubricate oil burner motor bearings as applicable.	.254					√
3 Inspect fuel system for leaks.	.016			√	√	√
4 Change fuel filter element on oil burner, where applicable.						
5 Check for proper operation of burner primary controls. check and adjust thermostat.	.133			√	√	√
6 Replace air filters in air handler.	.009			√	√	√
7 Check blower and motor for vibration and noise. lubricate bearings.	.042			√	√	√
8 Check belts for wear and proper tension, tighten if required.	.029			√	√	√
9 Check electrical wiring to burner controls and blower.	.079					√
10 Inspect and clean firebox.	.577					√
11 Clean blower and air plenum.	.294					√
12 Check condition of flue pipe, damper and stack.	.147			√	√	√
13 Check furnace operation through complete cycle or up to 10 minutes.	.650			√	√	√
14 Clean area around furnace.	.066			√	√	√
15 Fill out maintenance checklist and report deficiencies.	.022			√	√	√
Total labor-hours/period				1.149	1.149	2.353
Total labor-hours/year				2.298	1.149	2.353

			Cost Each				
			2015 Bare Costs			Total	Total
Description	Labor-hrs.	Material	Labor	Equip.	Total	In-House	w/O&P
1900 Forced air heater, oil/gas, up to 120 MBH, annually	2.353	103	141		244	289	350
1950 Annualized	5.800	124	345		469	570	695

PM Components	Labor-hrs.	W	M	Q	S	A
PM System D3055 122 2950						
Forced air heater, oil or gas fired, over 120 MBH						
1 Check with operating or area personnel for deficiencies.	.035			√	√	√
2 Inspect, clean and adjust electrodes and nozzles on oil burners or controls valves and thermo-sensing bulbs on gas burners; lubricate oil burner motor bearings as applicable.	.358					√
3 Inspect fuel system for leaks.	.156			√	√	√
4 Change fuel filter element on oil burner, where applicable.	.120					√
5 Check for proper operation of burner primary controls. check and adjust thermostat.	.133			√	√	√
6 Replace air filters in air handler.	.182			√	√	√
7 Check blower and motor for vibration and noise. lubricate bearings.	.047			√	√	√
8 Check belts for wear and proper tension, tighten if required.	.057			√	√	√
9 Check electrical wiring to burner controls and blower.	.079					√
10 Inspect and clean firebox.	.577					√
11 Clean blower and air plenum.	.294					√
12 Check condition of flue pipe, damper and stack.	.147			√	√	√
13 Check furnace operation through complete cycle or up to 10 minutes.	.650			√	√	√
14 Clean area around furnace.	.066			√	√	√
15 Fill out maintenance checklist and report deficiencies.	.022			√	√	√
Total labor-hours/period				1.495	1.495	2.923
Total labor-hours/year				2.990	1.495	2.923

			Cost Each				
			2015 Bare Costs			Total	Total
Description	Labor-hrs.	Material	Labor	Equip.	Total	In-House	w/O&P
2900 Forced air heater, oil/gas, over 120 MBH, annually	2.923	134	175		309	365	440
2950 Annualized	7.408	165	445		610	735	895

PM Components	Labor-hrs.	W	M	Q	S	A
PM System D3055 210 1950						
Package unit, air cooled, 3 tons through 24 tons						
1 Check with operating or area personnel for deficiencies.	.035			√	√	√
2 Check tension, condition, and alignment of belts; adjust as necessary.	.029			√	√	√
3 Lubricate shaft and motor bearings.	.047			√	√	√
4 Replace air filters.	.055			√	√	√
5 Clean electrical wiring and connections; tighten loose connections.	.120					√
6 Clean coils, evaporator drain pan, blowers, fans, motors and drain piping as required.	.385					√
7 Perform operational check of unit; make adjustments on controls and other components as required.	.077			√	√	√
8 During operation of unit, check refrigerant pressure; add refrigerant as necessary.	.135			√	√	√
9 Check compressor oil level; add oil as required.	.033					√
10 Clean area around unit.	.066			√	√	√
11 Fill out maintenance checklist and report deficiencies.	.022			√	√	√
Total labor-hours/period				.466	.466	1.004
Total labor-hours/year				.932	.466	1.004

			Cost Each				
			2015 Bare Costs			Total	Total
Description	Labor-hrs.	Material	Labor	Equip.	Total	In-House	w/O&P
1900 Package unit, air cooled, 3 thru 24 ton, annually	1.004	49	60		109	129	155
1950 Annualized	2.402	78.50	144		222.50	267	320

PM Components	Labor-hrs.	W	M	Q	S	A
PM System D3055 210 2950						
Package unit, air cooled, 25 tons through 50 tons						
1 Check with operating or area personnel for deficiencies.	.035			√	√	√
2 Check tension, condition, and alignment of belts; adjust as necessary.	.029			√	√	√
3 Lubricate shaft and motor bearings.	.047			√	√	√
4 Replace air filters.	.078			√	√	√
5 Clean electrical wiring and connections; tighten loose connections.	.120					√
6 Clean coils, evaporator drain pan, blowers, fans, motors and drain piping as required.	.385					√
7 Perform operational check of unit; make adjustments on controls and other components as required.	.130			√	√	√
8 During operation of unit, check refrigerant pressure; add refrigerant as necessary.	.272			√	√	√
9 Check compressor oil level; add oil as required.	.033					√
10 Clean area around unit.	.066			√	√	√
11 Fill out maintenance checklist and report deficiencies.	.022			√	√	√
Total labor-hours/period				.679	.679	1.217
Total labor-hours/year				1.358	.679	1.217

			Cost Each				
			2015 Bare Costs			Total	Total
Description	Labor-hrs.	Material	Labor	Equip.	Total	In-House	w/O&P
2900 Package unit, air cooled, 25 thru 50 ton, annually	1.217	68.50	72.50		141	166	199
2950 Annualized	3.254	110	194		304	365	445

PM Components	Labor-hrs.	W	M	Q	S	A
PM System D3055 220 1950						
Package unit, water cooled, 3 tons through 24 tons						
1 Check with operating or area personnel for deficiencies.	.035			√	√	√
2 Check tension, condition, and alignment of belts; adjust as necessary.	.029			√	√	√
3 Lubricate shaft and motor bearings.	.047			√	√	√
4 Replace air filters.	.055			√	√	√
5 Clean electrical wiring and connections; tighten loose connections.	.120					√
6 Clean coils, evaporator drain pan, blowers, fans, motors and drain piping as required.	.385					√
7 Perform operational check of unit; make adjustments on controls and other components as required.	.077			√	√	√
8 During operation of unit, check refrigerant pressure; add refrigerant as necessary.	.135			√	√	√
9 Check compressor oil level; add oil as required.	.033					√
10 Clean area around unit.	.066			√	√	√
11 Fill out maintenance checklist and report deficiencies.	.022			√	√	√
Total labor-hours/period				.466	.466	1.004
Total labor-hours/year				.932	.466	1.004

			Cost Each				
			2015 Bare Costs			**Total**	**Total**
Description	Labor-hrs.	Material	Labor	Equip.	Total	In-House	w/O&P
1900 Package unit, water cooled, 3 thru 24 ton, annually	1.004	49	60		109	129	155
1950 Annualized	2.402	78.50	144		222.50	267	320

PM Components	Labor-hrs.	W	M	Q	S	A
PM System D3055 220 2950						
Package unit, water cooled, 25 tons through 50 tons						
1 Check with operating or area personnel for deficiencies.	.035			√	√	√
2 Check tension, condition, and alignment of belts; adjust as necessary.	.029			√	√	√
3 Lubricate shaft and motor bearings.	.047			√	√	√
4 Replace air filters.	.078			√	√	√
5 Clean electrical wiring and connections; tighten loose connections.	.120					√
6 Clean coils, evaporator drain pan, blowers, fans, motors and drain piping as required.	.385					√
7 Perform operational check of unit; make adjustments on controls and other components as required.	.130			√	√	√
8 During operation of unit, check refrigerant pressure; add refrigerant as necessary.	.272			√	√	√
9 Check compressor oil level; add oil as required.	.033					√
10 Clean area around unit.	.066			√	√	√
11 Fill out maintenance checklist and report deficiencies.	.022			√	√	√
Total labor-hours/period				.679	.679	1.217
Total labor-hours/year				1.358	.679	1.217

			Cost Each				
			2015 Bare Costs			**Total**	**Total**
Description	Labor-hrs.	Material	Labor	Equip.	Total	In-House	w/O&P
2900 Package unit, water cooled, 25 thru 50 ton, annually	1.217	68.50	72.50		141	166	199
2950 Annualized	3.254	110	194		304	365	445

PM Components	Labor-hrs.	W	M	Q	S	A
PM System D3055 230 1950						
Package unit, computer room						
1 Check with operating or area personnel for deficiencies.	.035			✓	✓	✓
2 Run microprocessor check, if available, or check controls and unit for proper operation.	.216			✓	✓	✓
3 Check for unusual noise or vibration.	.033			✓	✓	✓
4 Clean coils, evaporator drain pan, humidifier pan, blower, motor and drain piping as required.	.380				✓	✓
5 Replace air filters.	.078			✓	✓	✓
6 Lubricate shaft and motor bearings.	.047			✓	✓	✓
7 Check belts for wear, proper tension, and alignment; adjust as necessary.	.029			✓	✓	✓
8 Check humidity lamp, replace if necessary.	.156			✓	✓	✓
9 During operation of unit, check refrigerant pressures; add refrigerant as necessary.	.270				✓	✓
10 Inspect exterior piping and valves for leaks; tighten connections as required.	.077			✓	✓	✓
11 Clean area around unit.	.066			✓	✓	✓
12 Fill out maintenance checklist and report deficiencies.	.022			✓	✓	✓
Total labor-hours/period				.759	1.409	1.409
Total labor-hours/year				1.518	1.409	1.409

	Description	Labor-hrs.	2015 Bare Costs				Total In-House	Total w/O&P
			Material	Labor	Equip.	Total		
1900	Package unit, computer room, annually	1.409	57	84		141	168	202
1950	Annualized	4.336	62.50	259		321.50	395	485

PM Components	Labor-hrs.	W	M	Q	S	A
PM System D3055 240 1950						
Package unit, with duct gas heater						
1 Check with operating or area personnel for deficiencies.	.035			√	√	√
2 Check tension, condition and alignment of belts; adjust as necessary.	.029			√	√	√
3 Lubricate shaft and motor bearings.	.047			√	√	√
4 Replace air filters.	.078			√	√	√
5 Check electrical wiring and connections; tighten loose connections.	.120					√
6 Clean coils, evaporator drain pan, blowers, fans, motors and drain piping as required.	.385					√
7 Perform operational check of unit; make adjustments on controls and other components as required.	.077			√	√	√
8 During operation of unit, check refrigerant pressures; add refrigerant as necessary.	.272			√	√	√
9 Check compressor oil level; add oil as required.	.033					√
10 Inspect, clean and adjust control valves and thermo-sensing bulbs on gas burners.	.254					√
11 Inspect fuel system for leaks.	.016			√	√	√
12 Check for proper operation of burner primary controls. Check and adjust thermostat.	.133					√
13 Check electrical wiring to burner controls.	.079					√
14 Inspect and clean firebox.	.577					√
15 Check condition of flue pipe, damper and stack.	.147			√	√	√
16 Check heater operation through complete cycle or up to 10 minutes.	.225					√
17 Clean area around entire unit.	.066			√	√	√
18 Fill out maintenance checklist and report deficiencies.	.022			√	√	√
Total labor-hours/period				.789	.789	2.595
Total labor-hours/year				1.578	.789	2.595

Description		Labor-hrs.	Cost Each				Total In-House	Total w/O&P
			2015 Bare Costs					
			Material	Labor	Equip.	Total		
1900	Package unit with duct gas heater, annually	2.595	47	155		202	246	300
1950	Annualized	4.962	70.50	296		366.50	450	550

PM Components	Labor-hrs.	W	M	Q	S	A
PM System D3055 250 1950						
Air conditioning split system, DX, air cooled, up to 10 tons						
1 Check with operating or area personnel for deficiencies.	.035			√	√	√
2 Clean intake side of condenser coils, fans and intake screens.	.055			√	√	√
3 Lubricate shaft and motor bearings.	.047			√	√	√
4 Pressure wash condenser coils with coil clean solution, as required.	.611					√
5 Replace air filters.	.078			√	√	√
6 Clean electrical wiring and connections; tighten loose connections.	.120					√
7 Clean evaporator coils, drain pan, blowers, fans, motors and drain piping as required.	.380					√
8 Perform operational check of unit; make adjustments on controls and other components as required.	.033			√	√	√
9 During operation of unit, check refrigerant pressure; add refrigerant as necessary.	.272			√	√	√
10 Clean area around equipment.	.066			√	√	√
11 Fill out maintenance checklist and report deficiencies.	.022			√	√	√
Total labor-hours/period				.608	.608	1.719
Total labor-hours/year				1.216	.608	1.719

			Cost Each				
			2015 Bare Costs			**Total**	**Total**
Description	**Labor-hrs.**	**Material**	**Labor**	**Equip.**	**Total**	**In-House**	**w/O&P**
1900 A/C, split sys., DX, air cooled, to 10 tons, annually	1.719	50	103		153	183	223
1950 Annualized	3.543	80	212		292	355	430

PM Components	Labor-hrs.	W	M	Q	S	A
PM System D3055 250 2950						
Air conditioning split system, DX, air cooled, over 10 tons						
1 Check with operating or area personnel for deficiencies.	.035			√	√	√
2 Check tension, condition, and alignment of belts; adjust as necessary.	.029			√	√	√
3 Lubricate shaft and motor bearings.	.047			√	√	√
4 Pressure wash condenser coils with coil clean solution, as required.	.611					√
5 Replace air filters.	.078			√	√	√
6 Clean electrical wiring and connections; tighten loose connections.	.120					√
7 Clean evaporator coils, drain pan, blowers, fans, motors and drain piping as required.	.380					√
8 Perform operational check of unit; make adjustments on controls and other components as required.	.033			√	√	√
9 During operation of unit, check refrigerant pressure; add refrigerant as necessary.	.272			√	√	√
10 Check compressor oil level; add oil as required.	.033			√	√	√
11 Clean area around equipment.	.066			√	√	√
12 Fill out maintenance checklist and report deficiencies.	.022			√	√	√
Total labor-hours/period				.615	.615	1.726
Total labor-hours/year				1.230	.615	1.726

			Cost Each				
			2015 Bare Costs			**Total**	**Total**
Description	**Labor-hrs.**	**Material**	**Labor**	**Equip.**	**Total**	**In-House**	**w/O&P**
2900 A/C, split sys., DX, air cooled, over 10 T, annually	1.726	70	103		173	206	249
2950 Annualized	3.571	112	213		325	390	475

PM Components	Labor-hrs.	W	M	Q	S	A
PM System D3055 310 1950						
Heat pump, air cooled, up to 5 tons						
1 Check with operating or area personnel for deficiencies.	.035			√	√	√
2 Check unit for proper operation, excessive noise or vibration.	.033			√	√	√
3 Clean intake side of condenser coils, fans and intake screens.	.055			√	√	√
4 Check electrical wiring and connections; tighten loose connections.	.120					√
5 Inspect fan(s) for bent blades or unbalance; adjust and clean as necessary.	.027					√
6 Check belts for condition, proper tension and misalignment; adjust as required.	.029			√	√	√
7 Lubricate shaft bearings and motor bearings.	.047			√	√	√
8 Inspect piping and valves for leaks; tighten connections as necessary.	.077			√	√	√
9 Replace air filters.	.078			√	√	√
10 Check refrigerant pressure; add refrigerant as necessary.	.135			√	√	√
11 Clean evaporative drain pan, and drain piping as required.	.385					√
12 Cycle the reverse cycle valve to insure proper operation.	.091			√	√	√
13 Clean area around equipment.	.066			√	√	√
14 Fill out maintenance checklist and report deficiencies.	.022			√	√	√
Total labor-hours/period				.668	.668	1.200
Total labor-hours/year				1.336	.668	1.200

			Cost Each				
			2015 Bare Costs			Total	Total
Description	Labor-hrs.	Material	Labor	Equip.	Total	In-House	w/O&P
1900 Heat pump, air cooled, up to 5 ton, annually	1.200	29.50	71.50		101	122	149
1950 Annualized	3.204	47	191		238	291	360

PM Components	Labor-hrs.	W	M	Q	S	A
PM System D3055 310 2950						
Heat pump, air cooled, over 5 tons						
1 Check with operating or area personnel for deficiencies.	.035			√	√	√
2 Check unit for proper operation, excessive noise or vibration.	.033			√	√	√
3 Clean intake side of condenser coils, fans and intake screens.	.055			√	√	√
4 Check electrical wiring and connections; tighten loose connections.	.120					√
5 Inspect fan(s) for bent blades or unbalance; adjust and clean as necessary.	.055					√
6 Check belts for condition, proper tension and misalignment; adjust as required.	.057			√	√	√
7 Lubricate shaft bearings and motor bearings.	.047			√	√	√
8 Inspect piping and valves for leaks; tighten connections as necessary.	.077					√
9 Replace air filters.	.156			√	√	√
10 Check refrigerant pressure; add refrigerant as necessary.	.135			√	√	√
11 Lubricate and check operation of dampers, if applicable.	.029					√
12 Check compressor oil level and add oil, if required.	.033			√	√	√
13 Cycle the reverse cycle valve to insure proper operation.	.091			√	√	√
14 Clean evaporative drain pan, and drain piping as required.	.385					√
15 Clean area around equipment.	.066			√	√	√
16 Fill out maintenance checklist and report deficiencies.	.022			√	√	√
Total labor-hours/period				.730	.730	1.396
Total labor-hours/year				1.460	.730	1.396

			Cost Each				
			2015 Bare Costs			Total	Total
Description	Labor-hrs.	Material	Labor	Equip.	Total	In-House	w/O&P
2900 Heat pump, air cooled, over 5 ton, annually	1.396	55	83.50		138.50	166	199
2950 Annualized	3.586	78.50	214		292.50	355	435

PM Components	Labor-hrs.	W	M	Q	S	A
PM System D3055 320 1950						
Heat pump, water cooled, up to 5 tons						
1 Check with operating or area personnel for deficiencies.	.035			√	√	√
2 Check unit for proper operation, excessive noise or vibration.	.033			√	√	√
3 Clean intake side of evaporator coil, fans and intake screens.	.055			√	√	√
4 Check electrical wiring and connections; tighten loose connections.	.120					√
5 Inspect fan(s) for bent blades or unbalance; adjust and clean as necessary.	.027					√
6 Check belts for condition, proper tension and misalignment; adjust as required.	.029			√	√	√
7 Lubricate shaft bearings and motor bearings.	.047			√	√	√
8 Inspect piping and valves for leaks; tighten connections as necessary.	.077			√	√	√
9 Replace air filters.	.078			√	√	√
10 Check refrigerant pressure; add refrigerant as necessary.	.135			√	√	√
11 Lubricate and check operation of dampers, if applicable.	.029					√
12 Clean evaporator drain pan and drain line with solvent.	.385					√
13 Cycle reverse cycle valve to insure proper operation.	.091			√	√	√
14 Backwash condenser coil to remove sediment.	.325					√
15 Clean area around equipment.	.066			√	√	√
16 Fill out maintenance checklist and report deficiencies.	.022			√	√	√
Total labor-hours/period				.668	.668	1.554
Total labor-hours/year				1.336	.668	1.554

		Cost Each					
		2015 Bare Costs				**Total**	**Total**
Description	**Labor-hrs.**	**Material**	**Labor**	**Equip.**	**Total**	**In-House**	**w/O&P**
1900 Heat pump, water cooled, up to 5 ton, annually	1.554	29.50	93		122.50	150	182
1950 Annualized	3.558	47	213		260	315	390

PM Components	Labor-hrs.	W	M	Q	S	A
PM System D3055 320 2950						
Heat pump, water cooled, over 5 tons						
1 Check with operating or area personnel for deficiencies.	.035			√	√	√
2 Check unit for proper operation, excessive noise or vibration.	.033			√	√	√
3 Clean intake side of condenser coils, fans and intake screens.	.055			√	√	√
4 Check electrical wiring and connections; tighten loose connections.	.120					√
5 Inspect fan(s) for bent blades or unbalance; adjust and clean as necessary.	.027					√
6 Check belts for condition, proper tension and misalignment; adjust as required.	.029			√	√	√
7 Lubricate shaft bearings and motor bearings.	.047			√	√	√
8 Inspect piping and valves for leaks; tighten connections as necessary.	.077			√	√	√
9 Replace air filters.	.078			√	√	√
10 Check refrigerant pressure; add refrigerant as necessary.	.135			√	√	√
11 Lubricate and check operation of dampers, if applicable.	.029					√
12 Clean evaporator drain pan and drain line with solvent.	.385					√
13 Cycle reverse cycle valve to insure proper operation.	.091			√	√	√
14 Clean area around equipment.	.066			√	√	√
15 Fill out maintenance checklist and report deficiencies.	.022			√	√	√
Total labor-hours/period				.668	.668	1.229
Total labor-hours/year				1.336	.668	1.229

		Cost Each					
		2015 Bare Costs				**Total**	**Total**
Description	**Labor-hrs.**	**Material**	**Labor**	**Equip.**	**Total**	**In-House**	**w/O&P**
2900 Heat pump, water cooled, over 5 ton, annually	1.229	55	73.50		128.50	152	184
2950 Annualized	3.233	78.50	193		271.50	330	400

For customer support on your Facilities Maintenance & Repair Cost Data, call 877.801.5724.

PM Components	Labor-hrs.	W	M	Q	S	A
PM System D3065 100 1950						
Controls, central system, electro/pneumatic						
1 With panel disconnected from power source, clean patrol panel compartment with a vacuum.	.468					√
2 Inspect wiring/components for loose connections; tighten, as required.	.120					√
3 Check set point of controls temperature, humidity or pressure.	.033					√
4 Check unit over its range of control.	.033					√
5 Check for correct pressure differential on all two position controllers.	.195					√
6 Check source of the signal and its amplification on electronic controls.	.468					√
7 Check air systems for leaks; repair as necessary.	.156					√
8 Check relays, pilot valves and pressure regulators for proper operation; repair or replace as necessary.	.325					√
9 Replace air filters in sensors, controllers, and thermostats as necessary.	.029					√
10 Clean area around equipment.	.066					√
11 Fill out maintenance checklist and report deficiencies.	.022					√
Total labor-hours/period						1.915
Total labor-hours/year						1.915

		Cost Each					
		2015 Bare Costs				Total	Total
Description	Labor-hrs.	Material	Labor	Equip.	Total	In-House	w/O&P
1900	Controls, central system, electro/pneumatic, annually	1.915	47	114		161	195	237
1950	Annualized	1.915	47	114		161	195	237

For customer support on your Facilities Maintenance & Repair Cost Data, call 877.801.5724.

461

PM Components	Labor-hrs.	W	M	Q	S	A
PM System D3095 110 1950						
Air compressor, gas engine						
1 Check with operating or area personnel for any obvious deficiencies.	.035		✓	✓	✓	✓
2 Check compressor oil level; add oil as required.	.022		✓	✓	✓	✓
3 Replace compressor oil.	.221				✓	✓
4 Clean air intake filter on air compressor(s); replace if necessary.	.177				✓	✓
5 Clean cylinder cooling fins and air cooler on compressor(s).	.156				✓	✓
6 Check tension, condition, and alignment of v-belts; adjust as necessary.	.030				✓	✓
7 Clean oil and water traps.	.178				✓	✓
8 Drain moisture from air storage tank and check discharge for indication of interior corrosion.	.046				✓	✓
9 Perform operation check of compressor; check the operation of low pressure cut-in and high pressure cut-out switches.	.221		✓	✓	✓	✓
10 Check operation of safety pressure relief valve.	.030				✓	✓
11 Check and tighten compressor foundation anchor bolts.	.022				✓	✓
12 Check radiator coolant; add if required.	.012		✓	✓	✓	✓
13 Check battery water; add if required.	.241		✓	✓	✓	✓
14 Check wiring, connections, switches, etc.; tighten loose connections.	.120		✓	✓	✓	✓
15 Check engine oil level; add if required.	.014		✓	✓	✓	✓
16 Change engine oil.	.224				✓	✓
17 Change engine oil filter.	.059				✓	✓
18 Check spark plug and reset cap.	.035				✓	✓
19 Check condition of engine air filter; replace if necessary.	.039				✓	✓
20 Test run engine for proper operation.	.413		✓	✓	✓	✓
21 Wipe dust and dirt from engine and compressor.	.109		✓	✓	✓	✓
22 Check muffler/exhaust system for corrosion.	.020				✓	✓
23 Clean area around equipment.	.066		✓	✓	✓	✓
24 Fill out maintenance checklist and report deficiencies.	.022		✓	✓	✓	✓
Total labor-hours/period			1.155	1.155	2.512	2.512
Total labor-hours/year			9.240	2.310	2.512	2.512

			Cost Each					
			2015 Bare Costs				Total	Total
Description		Labor-hrs.	Material	Labor	Equip.	Total	In-House	w/O&P
1900	Air compressor, gas engine powered, annually	2.512	109	150		259	310	370
1950	Annualized	16.574	265	990		1,255	1,525	1,875

PM Components	Labor-hrs.	W	M	Q	S	A
PM System D3095 114 1950						
Air compressor, centrifugal, to 40 H.P.						
1 Check compressor oil level; add oil as necessary.	.022			√	√	√
2 Perform operational check of compressor system and adjust as required.	.221			√	√	√
3 Check motor for excessive vibration, noise and overheating; lubricate.	.039			√	√	√
4 Check operation of pressure relief valve.	.030			√	√	√
5 Clean cooling fans and air cooler.	.023			√	√	√
6 Check tension, condition, and alignment of V-belts; adjust as necessary.	.030			√	√	√
7 Drain moisture from air storage tank and check low pressure cut-in; while draining, check discharge for indication of interior corrosion.	.046			√	√	√
8 Replace air intake filter, as needed.	.177			√	√	√
9 Clean oil and water trap.	.177			√	√	√
10 Clean compressor and surrounding area.	.066			√	√	√
11 Fill out maintenance checklist and report deficiencies.	.022			√	√	√
Total labor-hours/period				.853	.853	.853
Total labor-hours/year				1.706	.853	.853

				Cost Each				
				2015 Bare Costs			Total	Total
	Description	Labor-hrs.	Material	Labor	Equip.	Total	In-House	w/O&P
1900	Air compressor, centrifugal, to 40 H.P., annually	.853	46.50	51		97.50	115	138
1950	Annualized	3.412	70	204		274	330	410

PM Components	Labor-hrs.	W	M	Q	S	A
PM System D3095 114 2950						
Air compressor, centrifugal, over 40 H.P.						
1 Check with operating or area personnel for any obvious deficiencies.	.035	√	√	√	√	
2 Perform control system check.	.325			√	√	
3 Lubricate main driver coupling if necessary.	.047			√	√	
4 Lubricate prelube pump motor and pump coupling, if necessary.	.047			√	√	
5 Perform operation check of air compressor, adjust as required.	.221	√	√	√	√	
6 Check main driver coupling alignment; realign if necessary.	.325			√	√	
7 Check compressor and motor operation for excessive vibration, noise and overheating; lubricate motor.	.039		√	√	√	√
8 Check compressor scrolls, piping, and impeller housing for all leaks or cracks.	.325	√	√	√	√	
9 Check intercoolers and aftercoolers for high cooling water temperatures or cooling water leakage.	.195	√	√	√	√	
10 Check oil level in compressor oil reservoir; add oil as necessary.	.022	√	√	√	√	
11 Replace compressor oil and oil filters.	.282				√	
12 Record oil pressure and oil temperature.	.013	√	√	√	√	
13 Check compressor for oil leaks.	.026	√	√	√	√	
14 Visually inspect oil mist arrestor, clean housing, lines, and replace element if saturated, if applicable.	.177				√	√
15 Check operation of pressure relief valve.	.030	√	√	√	√	
16 Visually inspect discharge check valve.	.009			√	√	
17 Check oil and water trap.	.022	√	√	√	√	
18 Check indicating lamps or gauges for proper operation if appropriate; replace burned out lamps or repair/replace gauges.	.020					√
19 Visually check all air intake filter elements; replace if necessary.	.022	√	√	√	√	
20 Replace all air intake filter elements.	.177					
21 Lubricate motor.	.047				√	√
22 Clean compressor, motor and surrounding area.	.066	√	√	√	√	
23 Fill out maintenance check and report deficiencies.	.022	√	√	√	√	
Total labor-hours/period			1.038	1.038	2.015	2.494
Total labor-hours/year			8.304	2.076	2.015	2.494

			Cost Each					
			2015 Bare Costs				Total	Total
	Description	Labor-hrs.	Material	Labor	Equip.	Total	In-House	w/O&P
2900	Air compressor, centrifugal, over 40 H.P., annually	2.494	93.50	149		242.50	289	350
2950	Annualized	14.889	218	890		1,108	1,350	1,675

PM Components	Labor-hrs.	W	M	Q	S	A
PM System D3095 118 1950						
Air compressor, reciprocating, less than 5 H.P.						
1 Replace compressor oil.	.341			√	√	√
2 Perform operation check of compressor system and adjust as required.	.221			√	√	√
3 Check motor operation for excessive vibration, noise and overheating.	.042			√	√	√
4 Lubricate motor.	.047			√	√	√
5 Check operation of pressure relief valve.	.030			√	√	√
6 Check tension, condition, and alignment of V-belts; adjust as necessary.	.030			√	√	√
7 Drain moisture from air storage tank and check low pressure cut-in; while draining, check discharge for indication of interior corrosion.	.046			√	√	√
8 Clean air intake filter on compressor.	.177			√	√	√
9 Clean oil and water trap.	.177			√	√	√
10 Clean exterior of compressor, motor and surrounding area.	.066			√	√	√
11 Fill out maintenance checklist and report deficiencies.	.022			√	√	√
Total labor-hours/period				1.199	1.199	1.199
Total labor-hours/year				2.398	1.199	1.199

			Cost Each				
			2015 Bare Costs			**Total**	**Total**
Description	**Labor-hrs.**	**Material**	**Labor**	**Equip.**	**Total**	**In-House**	**w/O&P**
1900 Air compressor, recip., less than 5 H.P., annually	1.199	23.50	71.50		95	115	141
1950 Annualized	4.796	93.50	287		380.50	460	560

PM Components	Labor-hrs.	W	M	Q	S	A
PM System D3095 118 2950						
Air compressor, reciprocating, 5 to 40 H.P.						
1 Replace compressor oil.	.341			√	√	√
2 Perform operation check of compressor system and adjust as required.	.221			√	√	√
3 Check motor operation for excessive vibration, noise and overheating; lubricate motor.	.042			√	√	√
4 Check operation of pressure relief valve.	.043			√	√	√
5 Clean cooling fans and air cooler on compressor.	.023			√	√	√
6 Check tension, condition, and alignment of V-belts; adjust as necessary.	.030			√	√	√
7 Drain moisture from air storage tank and check low pressure cut-in; while draining, check discharge for indication of interior corrosion.	.059			√	√	√
8 Clean air intake filter on compressor.	.177			√	√	√
9 Clean oil and water trap.	.190			√	√	√
10 Clean compressor and surrounding area.	.066			√	√	√
11 Fill out maintenance checklist and report deficiencies.	.022			√	√	√
Total labor-hours/period				1.214	1.214	1.214
Total labor-hours/year				2.428	1.214	1.214

			Cost Each				
			2015 Bare Costs			**Total**	**Total**
Description	**Labor-hrs.**	**Material**	**Labor**	**Equip.**	**Total**	**In-House**	**w/O&P**
2900 Air compressor, reciprocating, 5 to 40 H.P., annually	1.214	23.50	72.50		96	117	142
2950 Annualized	4.856	93.50	290		383.50	465	570

PM Components	Labor-hrs.	W	M	Q	S	A
PM System D3095 118 3950						
Air compressor, reciprocating, over 40 H.P.						
1 Check with operating or area personnel for deficiencies.	.035			√	√	√
2 Perform operation check of compressor system and adjust as required.	.221			√	√	√
3 Replace compressor oil.	.341			√	√	√
4 Check motor(s) operation for excessive vibration, noise and overheating; lubricate motor(s).	.042			√	√	√
5 Clean cylinder cooling fins and air cooler on compressor.	.023			√	√	√
6 Check tension, condition, and alignment of V-belts; adjust as necessary.	.056			√	√	√
7 Check operation of pressure relief valve.	.030			√	√	√
8 Check low pressure cut in and high pressure cut out switches.	.120			√	√	√
9 Drain moisture from air storage tank and check low pressure cut-in; while draining, check discharge for indication of interior corrosion.	.072			√	√	√
10 Clean air intake filter on air compressor(s); replace if necessary.	.177			√	√	√
11 Clean oil and water trap.	.204			√	√	√
12 Check indicating lamps or gauges for proper operation if appropriate; replace burned out lamps or repair/replace gauges.	.020			√	√	√
13 Clean area around equipment.	.066			√	√	√
14 Fill out maintenance checklist and report deficiencies.	.022			√	√	√
Total labor-hours/period				1.429	1.429	1.429
Total labor-hours/year				2.858	1.429	1.429

			Cost Each					
				2015 Bare Costs			Total	Total
	Description	Labor-hrs.	Material	Labor	Equip.	Total	In-House	w/O&P
3900	Air compressor, reciprocating, over 40 H.P., annually	1.429	23.50	85.50		109	133	162
3950	Annualized	5.716	93.50	340		433.50	530	650

PM Components	Labor-hrs.	W	M	Q	S	A
PM System D3095 210 1950						
Steam Humidification System						
1 Operate humidistat through its throttling range to verify activation and deactivation.	.117				√	√
2 Inspect steam trap for proper operation.	.117				√	√
3 Turn off steam supply.	.065				√	√
4 Secure electrical service before servicing humidification unit.	.065				√	√
5 Clean strainer.	.195				√	√
6 Clean and/or replace water/steam nozzles as necessary.	.455				√	√
7 Inspect pneumatic controller for air leaks.	.039				√	√
8 Inspect steam lines for leaks and corrosion and repair leaks.	.195				√	√
9 Fill out maintenance checklist and report deficiencies	.022				√	√
Total labor-hours/period					1.270	1.270
Total labor-hours/year					1.270	1.270

			Cost Each					
			2015 Bare Costs				Total	Total
	Description	Labor-hrs.	Material	Labor	Equip.	Total	In-House	w/O&P
1900	Steam humidification system, annually	1.270	23.50	76		99.50	121	147
1950	Annualized	2.540	47	152		199	241	296

PM Components	Labor-hrs.	W	M	Q	S	A
PM System D3095 210 2950						
Evaporative Pan with Heating Coil Humidification System						
1 Operate humidistat through its throttling range to verify activation and deactivation.	.117				√	√
2 Inspect steam trap for proper operation.	.117				√	√
3 Turn off water and steam supply.	.065				√	√
4 Secure electrical service before servicing humidification unit.	.065				√	√
5 Drain and flush water pans, clean drains, etc.	.260				√	√
6 Check condition of heating element/steam coils and clean.	.065				√	√
7 Inspect pneumatic controller for air leaks.	.039				√	√
8 Inspect steam lines for leaks and corrosion and repair leaks.	.195				√	√
9 Fill out maintenance checklist and report deficiencies.	.022				√	√
Total labor-hours/period					.945	.945
Total labor-hours/year					.945	.945

			Cost Each					
			2015 Bare Costs				Total	Total
	Description	Labor-hrs.	Material	Labor	Equip.	Total	In-House	w/O&P
2900	Evap. pan with heating coil humidif. system, annually	.945	15.70	56.50		72.20	88	108
2950	Annualized	1.890	31.50	113		144.50	176	215

PM Components	Labor-hrs.	W	M	Q	S	A
PM System D3095 220 1950						
Dehumidifier, desiccant wheel						
1 Check with operating or area personnel for deficiencies.	.035		√	√	√	√
2 Check filters for any blockage or fouling, remove and clean as required.	.036		√	√	√	√
3 Check wheel seals for tears or punctures.	.055		√	√	√	√
4 Check the control valve, thermo sensor bulb and burner, if it has heaters; or check for scaling or leaking on steam heating coils, as applicable.	.094				√	√
5 Check that the outlet air temperature is within the proper heat range.	.105		√	√	√	√
6 Clean the desiccant wheel and check for softening of wheel faces.	.031				√	√
7 Check desiccant wheel and motor for vibration and noise, adjust as required.	.109				√	√
8 Check gear reducer oil level; add as required.	.035				√	√
9 Check wheel belt(s) for wear, proper tension and alignment; adjust as required.	.029				√	√
10 Check blower and motor for excessive vibration and noise; adjust as required.	.109				√	√
11 Check blower belt(s) for wear, proper tension and alignment; adjust as required.	.029				√	√
12 Lubricate wheel, blower and motor bearings.	.047				√	√
13 Check electrical wiring and connections; make appropriate adjustments.	.120				√	√
14 Check reactivation ductwork for condensation and air leaks.	.013				√	√
15 Clean the equipment and the surrounding area.	.066		√	√	√	√
16 Fill out maintenance checklist and report deficiencies.	.022		√	√	√	√
Total labor-hours/period			.319	.319	.935	.935
Total labor-hours/year			2.552	.638	.935	.935

	Description	Labor-hrs.	Cost Each 2015 Bare Costs — Material	Labor	Equip.	Total	Total In-House	Total w/O&P
1900	Dehumidifier, desiccant wheel, annually	.935	32.50	56		88.50	107	128
1950	Annualized	5.060	102	300		402	490	600

468

For customer support on your Facilities Maintenance & Repair Cost Data, call 877.801.5724.

PM Components	Labor-hrs.	W	M	Q	S	A
PM System D4015 100 1950						
Backflow prevention device, up to 4″						
NOTE: Test frequency may vary depending on local regulations and application.						
1 Test and calibrate check valve operation of backflow prevention device with test set.	.191					√
2 Bleed air from backflow preventer.	.047					√
3 Inspect for leaks under pressure.	.007					√
4 Clean backflow preventer and surrounding area.	.066					√
5 Fill out maintenance checklist and report deficiencies.	.022					√
Total labor-hours/period						.333
Total labor-hours/year						.333

			Cost Each					
			2015 Bare Costs				Total	Total
Description	Labor-hrs.	Material	Labor	Equip.	Total	In-House	w/O&P
1900	Backflow prevention device, up to 4″, annually	.333		19.55		19.55	24.50	30.50
1950	Annualized	.333		19.55		19.55	24.50	30.50

PM Components	Labor-hrs.	W	M	Q	S	A
PM System D4015 100 2950						
Backflow prevention device, over 4″						
NOTE: Test frequency may vary depending on local regulations and application.						
1 Test and calibrate check valve operation of backflow prevention device with test set.	.334					√
2 Bleed air from backflow preventer.	.065					√
3 Inspect for leaks under pressure.	.007					√
4 Clean backflow preventer and surrounding area.	.066					√
5 Fill out maintenance checklist and report deficiencies.	.022					√
Total labor-hours/period						.494
Total labor-hours/year						.494

			Cost Each					
			2015 Bare Costs				Total	Total
Description	Labor-hrs.	Material	Labor	Equip.	Total	In-House	w/O&P
2900	Backflow prevention device, over 4″, annually	.494		29		29	36.50	45
2950	Annualized	.494		29		29	36.50	45

PM Components	Labor-hrs.	W	M	Q	S	A
PM System D4015 150 1950						
Extinguishing system, wet pipe						
1　Notify proper authorities prior to testing any alarm systems.	.130		√	√	√	√
2　Open and close post indicator valve (PIV) to check operation; make minor adjustments such as lubricating valve stem, cleaning and/or replacing target windows as required.	.176					√
3　Open and close OS&Y (outside stem and yoke) cut-off valve to check operation; make minor repairs such as lubricating stems and tightening packing glands as required.	.176					√
4　Perform operational test of water flow detectors; make minor adjustments and restore system to proper operating condition.	.148					√
5　Check to ensure that alarm drain is open; clean drain line if necessary.	.081		√	√	√	√
6　Open water motor alarm test valve and ensure that outside alarm operates; lubricate alarm, make adjustments as required.	.176		√	√	√	√
7　Conduct main drain test by opening 2″ test valve; maintain a continuous record of drain tests; make minor adjustments if applicable; restore system to proper operating condition.	.333			√	√	√
8　Check general condition of sprinklers and sprinkler system; make minor adjustments as required.	.228					√
9　Check equipment gaskets, piping, packing glands, and valves for leaks; tighten flange bolts and loose connections to stop all leaks.	.013					√
10　Check condition of fire department connections; replace missing or broken covers as required.	.103					√
11　Trip test wet pipe system using test valve furthest from wet pipe valve (control valve); make minor adjustments as necessary and restore system to proper operating condition.	1.733					√
12　Inspect OS&Y and PIV cut-off valves for open position.	.066		√	√	√	√
13　Clean area around system components.	.433			√	√	√
14　Fill out maintenance checklist and report deficiencies.	.022		√	√	√	√
Total labor-hours/period			.475	1.241	1.241	3.818
Total labor-hours/year			3.800	2.482	1.241	3.818

			Cost Each					
			2015 Bare Costs				**Total**	**Total**
Description		**Labor-hrs.**	**Material**	**Labor**	**Equip.**	**Total**	**In-House**	**w/O&P**
1900	Extinguishing system, wet pipe, annually	3.818	88	214		302	365	445
1950	Annualized	11.341	176	635		811	990	1,225

For customer support on your Facilities Maintenance & Repair Cost Data, call 877.801.5724.

PM Components	Labor-hrs.	W	M	Q	S	A
PM System D4015 180 1950						
Extinguishing system, deluge/preaction						
1 Notify proper authorites prior to testing any alarm systems.	.130		√	√	√	√
2 Open and close post indicator valve (PIV) to check operation; make minor repairs such as lubricating valve stem, cleaning and/or replacing target windows as required.	.176					√
3 Open and close outside stem and yoke (OS&Y) cut-off valve to check operation; make minor repairs such as lubricating stems, tightening packing glands as required.	.176					√
4 Perform operational test of supervisory initiating devices and water flow detectors; make minor adjustments and restore system to proper operating condition.	.148					√
5 Check to ensure that alarm drain is open; clean drain if necessary.	.081		√	√	√	√
6 Open water motor alarm test valve and ensure that outside alarm operates; lubricate alarm and adjust as required.	.176		√	√	√	√
7 Visually check water pressure to ensure adequate operating pressure is available; make adjustments as required.	.004		√	√	√	√
8 Conduct main drain test by opening 2″ test valve and observing drop in water pressure on gauge; pressure drop should not exceed 20 PSI; maintain a continuous record of drain tests; make minor adjustments; restore system to proper operating condition.	.333			√	√	√
9 Check general condition of sprinklers and sprinkler system; make minor adjustments as required.	.228					√
10 Check equipment gaskets, piping, packing glands, and valves for leaks; tighten flange bolts and loose connections to stop all leaks.	.013					√
11 Check condition of fire department connections; replace missing or broken covers as required.	.103					√
12 Check and inspect pneumatic system for physical damage and proper operation; make minor adjustments as required.	.320					√
13 Trip test deluge system (control valve closed); make minor adjustments as necessary and restore system to fully operational condition.	1.956					√
14 Inspect OS&Y and PIV cut-off valves for open position.	.066		√	√	√	√
15 Clean area around system components.	.066					√
16 Ensure that system is restored to proper operating condition.	.109		√	√	√	√
17 Fill out maintenance checklist and report deficiencies.	.022		√	√	√	√
Total labor-hours/period			.588	.921	.921	4.107
Total labor-hours/year			4.704	1.842	.921	4.107

		Cost Each					
		2015 Bare Costs				Total	Total
Description	Labor-hrs.	Material	Labor	Equip.	Total	In-House	w/O&P
1900 Extinguishing system, deluge / preaction, annually	4.107	88	231		319	385	470
1950 Annualized	11.574	176	650		826	1,000	1,250

PM Components	Labor-hrs.	W	M	Q	S	A
PM System D4015 210 1950						
Fire pump, electric motor driven						
1 Check control panel and wiring for loose connections; tighten connections as required.	.109			√	√	√
2 Ensure all valves relating to water system are in correct position.	.008	√	√	√	√	√
3 Open and close OS&Y(outside steam and yoke) cut-off valve to check operation; make minor repairs such as lubricating stems and tightening packing glands as required.	.176			√	√	√
4 Centrifugal pump:						
A) perform 10 minute pump test run; check for proper operation and adjust if required.	.217	√	√	√	√	√
B) check for leaks on suction and discharge piping, seals, packing glands, etc.	.077	√	√	√	√	√
C) check for excessive vibration, noise, overheating, etc.	.022	√	√	√	√	√
D) check alignment, clearances, and rotation of shaft and coupler (includes removing and reinstalling safety cover).	.160	√	√	√	√	√
E) tighten or replace loose, missing or damaged nuts, bolts, or screws.	.005	√	√	√	√	√
F) lubricate pump and motor as required.	.099			√	√	√
G) check suction or discharge pressure gauge readings and flow rate.	.078	√	√	√	√	√
H) check packing glands and tighten or repack as required; note that slight dripping is required for proper lubrication of shaft.	.113			√	√	√
5 Inspect and clean strainers after each use and flow test.	.260	√	√	√	√	√
6 Clean equipment and surrounding area.	.066	√	√	√	√	√
7 Fill out maintenance checklist and report deficiencies.	.022	√	√	√	√	√
Total labor-hours/period		.915	.915	1.412	1.412	1.412
Total labor-hours/year		34.770	7.320	2.824	1.412	1.412

			Cost Each					
			2015 Bare Costs				**Total**	**Total**
Description	**Labor-hrs.**	**Material**	**Labor**	**Equip.**	**Total**		**In-House**	**w/O&P**
1900 Fire pump, electric motor driven, annually	1.412	90	79.50		169.50		198	237
1950 Annualized	47.738	139	2,675		2,814		3,500	4,375

PM Components	Labor-hrs.	W	M	Q	S	A
PM System D4015 250 1950						
Fire pump, engine driven						
1 Ensure all valves relating to water system are in correct position.	.008	√	√	√	√	√
2 Open and close OS&Y (outside steam and yoke) cut-off valve to check operation; make minor repairs such as lubricating stems and tightening packing glands as required.	.176			√	√	√
3 Controller panel:						
A) check controller panel for proper operation in accordance with standard procedures for the type of panel installed, replace burned out bulbs and fuses; make other repairs as necessary.	.039			√	√	√
B) check wiring for loose connections and fraying; tighten or replace as required.	.109			√	√	√
4 Diesel engine:						
A) check oil level in crankcase with dipstick; add oil as necessary.	.014	√	√	√	√	√
B) drain and replace engine oil.	.511					√
C) replace engine oil filter.	.059					√
D) check radiator coolant level; add water if low.	.012	√	√	√	√	√
E) drain and replace radiator coolant.	.511					√
F) check battery water level; add water if low.	.012			√	√	√
G) check battery terminals for corrosion; clean if necessary.	.124			√	√	√
H) check belts for tension and wear; adjust or replace if required.	.012			√	√	√
I) check fuel level in tank; refill as required.	.046			√	√	√
J) check electrical wiring, connections, switches, etc.; adjust or tighten as necessary.	.120			√	√	√
K) perform 10 minute pump check for proper operation and adjust as required.	.217	√	√	√	√	√
5 Centrifugal pump:						
A) check for leaks on suction and discharge piping, seals, packing glands, etc.	.077	√	√	√	√	√
B) check pump operation; vibration, noise, overheating, etc.	.022	√	√	√	√	√
C) check alignment, clearances, and rotation of shaft and coupler (includes removing and reinstalling safety cover).	.160	√	√	√	√	√
D) tighten or replace loose, missing or damaged nuts, bolts, or screws.	.005	√	√	√	√	√
E) lubricate pump as required.	.099			√	√	√
F) check suction or discharge pressure gauge readings and flow rate.	.078	√	√	√	√	√
G) check packing glands and tighten or repack as required; note that slight dripping is required for proper lubrication of shaft.	.113			√	√	√
6 Inspect and clean strainers after each use and flow test.	.260	√	√	√	√	√
7 Clean equipment and surrounding area.	.433	√	√	√	√	√
8 Fill out maintenance checklist and report deficiencies.	.022	√	√	√	√	√
Total labor-hours/period		1.308	1.308	2.158	2.158	3.239
Total labor-hours/year		49.704	10.464	4.316	2.158	3.239

			Cost Each					
			2015 Bare Costs				Total	Total
Description	Labor-hrs.	Material	Labor	Equip.	Total	In-House	w/O&P
1900	Fire pump, motor/engine driven, annually	3.239	196	182		378	445	530
1950	Annualized	69.881	229	3,925		4,154	5,175	6,400

For customer support on your Facilities Maintenance & Repair Cost Data, call 877.801.5724.

473

PM Components	Labor-hrs.	W	M	Q	S	A
PM System D4015 310 1950						
Extinguishing system, dry pipe						
1 Notify proper authorities prior to testing any alarm systems.	.130		√	√	√	√
2 Open and close post indicator valve to check operation; make minor repairs such as lubricating valve stem, cleaning and/or replacing target windows as required.	.176					√
3 Open and close OS&Y (outside stem and yoke) cut-off valve to check operation; make minor repairs such as lubricating stems and tightening packing glands as required.	.176					√
4 Perform operational test of water flow detectors; make minor adjustments and restore system to proper operating condition.	.148					√
5 Check to ensure that alarm drain is open; clean drain line if necessary.	.081		√	√	√	√
6 Open water motor alarm test valve and ensure that outside alarm operates; lubricate alarm, make adjustments as required.	.176		√	√	√	√
7 Visually check water pressure to ensure that adequate operating pressure is available; make adjustments as required.	.003		√	√	√	√
8 Conduct main drain test by opening 2″ test valve; maintain a continuous record of drain tests; make minor adjustments if applicable; restore system to proper operating condition.	.333			√	√	√
9 Check general condition of sprinklers and sprinkler system; make minor adjustments as required.	.228					√
10 Check equipment gaskets, piping, packing glands, and valves for leaks; tighten flange bolts and loose connections to stop all leaks.	.013					√
11 Check condition of fire department connections; replace missing or broken covers as required.	.103					√
12 Visually check system air pressure; pump up system and make minor adjustments as required.	.135		√	√	√	√
13 Trip test dry pipe system using test valve furthest from dry pipe valve (control valve partially open); make minor adjustments as necessary and restore system to fully operational condition.	1.733					√
14 Check operation of quick opening device.	.022					√
15 Inspect OS&Y and PIV cut-off valves for open position.	.066		√	√	√	√
16 Clean area around system components.	.433		√	√	√	√
17 Fill out maintenance checklist and report deficiencies.	.022		√	√	√	√
Total labor-hours/period			.613	1.379	1.379	3.978
Total labor-hours/year			4.904	2.758	1.379	3.978

			Cost Each					
			2015 Bare Costs				Total	Total
	Description	Labor-hrs.	Material	Labor	Equip.	Total	In-House	w/O&P
1900	Extinguishing system, dry pipe, annually	3.978	136	223		359	430	520
1950	Annualized	13.019	224	730		954	1,150	1,425

D40 FIRE PROTECTION — D4095 100 — Extinguishing System, CO_2

PM Components	Labor-hrs.	W	M	Q	S	A
PM System D4095 100 1950						
Extinguishing system, CO_2						
1 Check that nozzles, heads and hand hose lines are clear from obstructions, have not been damaged and in proper position.	.098		√	√	√	√
2 Check to ensure that all operating controls are properly set.	.066		√	√	√	√
3 Clean nozzles as required.	.098		√	√	√	√
4 Visually inspect the control panel for obstructions or physical damage; clean dirt and dust from interior and exterior; make sure that all cards are plugged in tightly, tighten loose connections and make other minor adjustments as necessary.	.222		√	√	√	√
5 Check battery voltages where installed, recharge or replace as required.	.012		√	√	√	√
6 Weigh CO_2 cylinders; replace any that show a weight loss greater than 10%.	.650				√	√
7 Blow out entire system to make sure that the system is not plugged or restricted.	.325					√
8 Run out hose reels and check hoses.	.217					√
9 Conduct operational test of actuating devices, both automatic, manual and alarm system. Prior to conducting tests, close valves, disable CO_2 dumping control and manually override computer shutdown feature if applicable, all in accordance with the manufacturer's specifications.	.433					√
10 Check the operation and timing of the time delay control and check operation of the abort station.	.108					√
11 Restore system to proper operating condition and notify personnel upon completion of tests.	.079					√
12 Clean up around system.	.066		√	√	√	√
13 Fill out maintenance checklist and report deficiencies.	.022		√	√	√	√
Total labor-hours/period			.584	.584	1.234	2.396
Total labor-hours/year			4.672	1.168	1.234	2.396

		Cost Each — 2015 Bare Costs				Total	Total
Description	Labor-hrs.	Material	Labor	Equip.	Total	In-House	w/O&P
1900 Extinguishing system, CO_2, annually	2.396	72	135		207	248	300
1950 Annualized	9.470	224	530		754	915	1,100

For customer support on your Facilities Maintenance & Repair Cost Data, call 877.801.5724.

475

PM Components	Labor-hrs.	W	M	Q	S	A
PM System D4095 200 1950						
Extinguishing system, foam bottle						
1 Check that all nozzles, heads and hand held lines are clear from obstructions, have not been damaged and are in proper position.	.195		√	√	√	√
2 Check to ensure that all operating controls are properly set.	.022		√	√	√	√
3 Observe pressure on system to ensure that proper pressure is being maintained.	.004		√	√	√	√
4 Clean area around system.	.066		√	√	√	√
5 Fill out maintenance checklist and report deficiencies.	.022		√	√	√	√
Total labor-hours/period			.309	.309	.309	.309
Total labor-hours/year			2.472	.618	.309	.309

			Cost Each					
			2015 Bare Costs				Total	Total
Description		Labor-hrs.	Material	Labor	Equip.	Total	In-House	w/O&P
1900	Extinguishing system, foam bottle, annually	.309		17.35		17.35	21.50	27
1950	Annualized	3.708		208		208	261	325

PM Components	Labor-hrs.	W	M	Q	S	A
PM System D4095 210 1950						
Extinguishing system, foam, electric pump, deluge system						
1 Check foam concentrate level in tank; add concentrate as required to maintain proper level.	.043		√	√	√	√
2 Ensure all valves relating to foam/water system are in correct position.	.008		√	√	√	√
3 Visually check proportioning devices, pumps, and foam nozzles; correct any observed deficiencies.	.386		√	√	√	√
4 Check water supply pressure; adjust as necessary.	.029		√	√	√	√
5 Open and close OS&Y (outside stem and yoke) cut-off valve to check operation; make minor repairs such as lubricating stems and tightening packing glands as required.	.059			√	√	√
6 Centrifugal pump:						
A) perform 10 minute pump test run, check for proper operation and adjust as required.	.217			√	√	√
B) test start pump without foam discharger.	.022		√	√	√	√
C) check for leaks on suction and discharge piping, seals, packing glands, etc.	.077			√	√	√
D) check for excessive vibration, noise, overheating, etc.	.022			√	√	√
E) check alignment, clearances, and rotation of shaft and coupler (includes removing and reinstalling safety cover).	.160			√	√	√
F) tighten or replace loose, missing or damaged nuts, bolts, or screws.	.005			√	√	√
G) lubricate pump and motor as required.	.099			√	√	√
H) check suction or discharge pressure gauge readings and flow rate.	.078			√	√	√
I) check packing glands and tighten or repack as required; note that slight dripping is required for proper lubrication of shaft.	.113			√	√	√
7 Inspect and clean strainers after each use and flow test.	.260		√	√	√	√
8 Clean equipment and surrounding area.	.433		√	√	√	√
9 Fill out maintenance checklist.	.022		√	√	√	√
Total labor-hours/period			1.203	2.033	2.033	2.033
Total labor-hours/year			9.624	4.066	2.033	2.033

			Cost Each					
			2015 Bare Costs				Total	Total
Description	Labor-hrs.	Material	Labor	Equip.	Total	In-House	w/O&P
1900 Extinguishing system, foam, electric pump, annually	2.033	131	114		245	287	340
1950 Annualized	17.756	180	995		1,175	1,450	1,775

For customer support on your Facilities Maintenance & Repair Cost Data, call 877.801.5724.

477

PM Components	Labor-hrs.	W	M	Q	S	A
PM System D4095 220 1950						
Extinguishing system, foam, diesel pump, deluge system						
1 Check foam concentrate level in tank; add concentrate as required to maintain proper level.	.043	√	√	√	√
2 Ensure all valves relating to foam/water system are in correct position.	.008	√	√	√	√
3 Visually check proportioning devices, pumps, and foam nozzles; correct any observed deficiencies.	.386	√	√	√	√
4 Check water supply pressure; adjust as necessary.	.003	√	√	√	√
5 Open and close OS&Y (outside stem and yoke) cut-off valve to check operation; make minor repairs such as lubricating stems and tightening packing glands as required.	.059			√	√	√
6 Controller panel:						
A) check controller panel for proper operation in accordance with standard procedures for the type of panel installed, replace burned out bulbs and fuses; make other repairs as necessary.	.039			√	√	√
B) check wiring for loose connections and fraying; tighten or replace as required.	.109			√	√	√
7 Diesel engine:						
A) check oil level in crankcase with dipstick; add oil as necessary.	.014	√	√	√	√
B) drain and replace engine oil.	.511					√
C) replace engine oil filter.	.059					√
D) check radiator coolant level; add water if low.	.012	√	√	√	√
E) drain and replace radiator coolant.	.511					√
F) check battery water level; add water if low.	.012		√	√	√
G) check battery terminals for corrosion; clean if necessary.	.124		√	√	√
H) check belts for tension and wear; adjust or replace if required.	.012		√	√	√
I) check fuel level in tank; refill as required.	.046		√	√	√
J) check electrical wiring, connections, switches, etc.; adjust or tighten as necessary.	.120		√	√	√
K) perform 10 minute pump check for proper operation and adjust as required, start pump without foam discharger.	.217	√	√	√	√
8 Centrifugal pump:						
A) check for leaks on suction and discharge piping, seals, packing glands, etc.	.077		√	√	√
B) check pump operation; vibration, noise, overheating, etc.	.022		√	√	√
C) check alignment, clearances, and rotation of shaft and coupler (includes removing and reinstalling safety cover).	.160		√	√	√
D) tighten or replace loose, missing or damaged nuts, bolts, or screws.	.005		√	√	√
E) lubricate pump as required.	.099		√	√	√
F) check suction or discharge pressure gauge readings and flow rate.	.078		√	√	√
G) check packing glands and tighten or repack as as required; note that slight dripping is required for proper lubrication of shaft.	.113		√	√	√
9 Inspect and clean strainers after each use and flow test.	.260	√	√	√	√
10 Clean equipment and surrounding area.	.433	√	√	√	√
11 Fill out maintenance checklist.	.022	√	√	√	√
Total labor-hours/period			1.398	2.473	2.473	3.554
Total labor-hours/year			11.184	4.946	2.473	3.554

			Cost Each					
			2015 Bare Costs				Total	Total
	Description	Labor-hrs.	Material	Labor	Equip.	Total	In-House	w/O&P
1900	Extinguishing system, foam, diesel pump, annually	3.554	237	200		437	510	605
1950	Annualized	22.157	270	1,250		1,520	1,850	2,300

478

PM Components	Labor-hrs.	W	M	Q	S	A
PM System D4095 400 1950						
Extinguishing system, dry chemical						
1 Check nozzles, heads, and fusible links or heat detectors to ensure that they are clean, have adequate clearance from obstructions and have not been damaged.	.098		√	√	√	√
2 Check hand hose lines, if applicable, to ensure that they are clear, in proper position and in good condition.	.098		√	√	√	√
3 Check to ensure that all operating controls are properly set.	.066		√	√	√	√
4 Observe pressure on system to ensure that proper pressure is maintained.	.004		√	√	√	√
5 Check expellant gas cylinders of gas cartridge systems for proper pressure or weight to ensure that sufficient gas is available.	.260				√	√
6 Test all actuating and operating devices; turn valve on dry chemical bottle or cylinder to off position in accordance with manufacturers specifications, and test operation of control head by removing fusible link, activating heat detectors or pulling test control. Restore system to normal operation and make minor adjustments as applicable.	.433					√
7 Open and check dry chemical in cylinder and stored pressure systems to ensure it is free flowing and without lumps.	.325					√
8 When an alarm system is reactivated, ensure that it functions in accordance with manufacturers specifications.	.079					√
9 Clean area around system.	.066		√	√	√	√
10 Fill out maintenance checklist and report deficiencies.	.022		√	√	√	√
Total labor-hours/period			.354	.354	.614	1.451
Total labor-hours/year			2.832	.708	.614	1.451

	Description	Labor-hrs.	Cost Each					Total In-House	Total w/O&P
			2015 Bare Costs						
			Material	Labor	Equip.	Total		In-House	w/O&P
1900	Extinguishing system, dry chemical, annually	1.451	72	81.50		153.50		181	217
1950	Annualized	5.605	224	315		539		640	770

For customer support on your Facilities Maintenance & Repair Cost Data, call 877.801.5724.

479

PM Components	Labor-hrs.	W	M	Q	S	A
PM System D4095 450 1950						
Extinguishing system, FM200						
1 Check that nozzles, heads and hand hose lines are clear from obstructions, have not been damaged and in proper position.	.098		√	√	√	√
2 Check to ensure that all operating controls are properly set.	.066		√	√	√	√
3 Clean nozzles as required.	.098		√	√	√	√
4 Visually inspect the control panel for obstructions or physical damage; clean dirt and dust from interior and exterior; make sure that all cards are plugged in tightly, tighten loose connections and make other minor adjustments as necessary.	.222		√	√	√	√
5 Check battery voltages where installed, recharge or replace as required.	.012	√	√	√	√	
6 Weigh cylinders; replace any that show a weight loss greater than 5% or a pressure loss greater than 10%.	.650				√	√
7 Blow out entire system to make sure that the system is not plugged or restricted.	.333					√
8 Conduct operational test of actuating devices, both automatic, manual and alarm system; prior to conducting tests, close valves disable CO_2 dumping control and manually override computer shutdown feature if applicable, all in accordance with manufacturers specifications.	.436					√
9 Check the operation and timing of the time delay control and check operation of abort station.	.108					√
10 Restore system to proper operating condition and notify cognizant personnel upon completion of tests.	.079					√
11 Clean out around system.	.066		√	√	√	√
12 Fill out maintenance checklist and report deficiencies.	.022		√	√	√	√
Total labor-hours/period			.584	.584	1.234	2.190
Total labor-hours/year			4.672	1.168	1.234	2.190

			Cost Each					
			2015 Bare Costs				Total	Total
	Description	Labor-hrs.	Material	Labor	Equip.	Total	In-House	w/O&P
1900	Extinguishing system, FM200, annually	2.190	48	123		171	207	252
1950	Annualized	9.264	176	520		696	845	1,025

PM Components	Labor-hrs.	W	M	Q	S	A
PM System D5015 210 1950						
Switchboard, electrical						
1 Check with operating or area personnel for deficiencies.	.044					√
2 Check indicating lamps for proper operation, if appropriate; replace burned out lamps.	.018					√
3 Remove and reinstall cover.	.196					√
4 Check for discolorations, hot spots, odors and charred insulation.	.359					√
5 Clean switchboard exterior and surrounding area.	.066					√
6 Fill out maintenance checklist and report deficiencies.	.022					√
Total labor-hours/period						.705
Total labor-hours/year						.705

			Cost Each					
			2015 Bare Costs				Total	Total
Description	Labor-hrs.	Material	Labor	Equip.	Total		In-House	w/O&P
1900 Switchboard, electrical, annually	.705	8.30	38.50		46.80		57	70
1950 Annualized	.705	8.30	38.50		46.80		57	70

PM Components	Labor-hrs.	W	M	Q	S	A
PM System D5015 214 1950						
Switchboard w/air circuit breaker						
1 Check with operating or area personnel for deficiencies.	.035					√
2 Check indicating lamps for proper operation, if appropriate; replace burned out lamps.	.020					√
3 Remove switchboard from service, coordinate with appropriate authority.	.038					√
4 Remove and reinstall cover.	.051					√
5 Check for discolorations, hot spots, odors and charred insulation.	.016					√
6 Check the records on the circuit breaker for any full current rated trips which may indicate a potential problem.	.020					√
7 Examine the exterior of the switchboard for any damage or moisture.	.020					√
8 Check operation of circuit breaker tripping mechanism and relays.	.096					√
9 Check the operation of the circuit breaker charging motor.	.020					√
10 Check hardware for tightness.	.120					√
11 Check interlock for proper operation.	.013					√
12 Check that the charger is operating properly.	.022		√	√	√	√
13 Check for any signs of corrosion on the battery terminals or wires.	.020		√	√	√	√
14 Check the electrolyte level in the batteries; add if required.	.624		√	√	√	√
15 Check the specific gravity of the electrolyte in a 10% sample of the batteries.	.069		√	√	√	√
16 Check 25% of terminal-to-cell connection resistance; rehabilitate connections as required; add anti-corrosion grease to battery terminals and connections.	.113					√
17 Measure and record individual cell and string float voltages.	2.808					√
18 Place switchboard back in service, notify appropriate authority.	.057					√
19 Clean switchboard, charger and the surrounding area.	.066		√	√	√	√
20 Fill out maintenance checklist and report deficiencies.	.022		√	√	√	√
Total labor-hours/period			.823	.823	.823	4.250
Total labor-hours/year			6.584	1.646	.823	4.250

			Cost Each					
			2015 Bare Costs				Total	Total
Description	Labor-hrs.	Material	Labor	Equip.	Total		In-House	w/O&P
1900 Switchboard, with air circuit breaker, annually	4.250	23	232		255		315	390
1950 Annualized	13.303	30	730		760		935	1,175

For customer support on your Facilities Maintenance & Repair Cost Data, call 877.801.5724.

481

PM Components	Labor-hrs.	W	M	Q	S	A
PM System D5015 217 1950						
Switchboard w/air circuit breaker and tie switch						
1 Check with operating or area personnel for deficiencies.	.035					√
2 Check indicating lamps for proper operation, if appropriate; replace burned out lamps.	.020					√
3 Remove switchboard from service, coordinate with appropriate authority.	.038					√
4 Remove and reinstall cover.	.051					√
5 Check for discolorations, hot spots, odors and charred insulation.	.016					√
6 Check the records on the circuit breaker for any full current rated trips which may indicate a potential problem.	.020					√
7 Examine the exterior of the switchboard for any damage or moisture.	.020					√
8 Check operation of circuit breaker tripping mechanism and relays.	.096					√
9 Check the operation of the circuit breaker charging motor.	.020					√
10 Check hardware for tightness.	.120					√
11 Check interlock for proper operation.	.013					√
12 Operate the switching handle and check the locking mechanism.	.052					√
13 Check that the charger is operating properly.	.022		√	√	√	√
14 Check for any signs of corrosion on the battery terminals or wires.	.020		√	√	√	√
15 Check the electrolyte level in the batteries; add if required.	.624		√	√	√	√
16 Check the specific gravity of the electrolyte in a 10% sample of the batteries.	.069		√	√	√	√
17 Check 25% of terminal-to-cell connection resistance; rehabilitate connections as required; add anti-corrosion grease to battery terminals and connections.	.113					√
18 Measure and record individual cell and string float voltages.	2.808					√
19 Place switchboard back in service, notify appropriate authority.	.057					√
20 Clean switchboard, charger and the surrounding area.	.066		√	√	√	√
21 Fill out maintenance checklist and report deficiencies.	.022		√	√	√	√
Total labor-hours/period			.823	.823	.823	4.302
Total labor-hours/year			6.584	1.646	.823	4.302

		Cost Each						
			2015 Bare Costs				Total	Total
Description		Labor-hrs.	Material	Labor	Equip.	Total	In-House	w/O&P
1900	Switchboard, with air C.B. and tie switch, annually	4.302	34	235		269	330	410
1950	Annualized	13.355	48.50	730		778.50	955	1,175

PM Components	Labor-hrs.	W	M	Q	S	A
PM System D5015 220 1950						
Circuit breaker, HV air						
1 Check the records on the circuit breaker for any full current rated trips which may indicate a potential problem.	.066					√
2 Examine the exterior of the circuit breaker for any damage or moisture.	.060					√
3 Remove the circuit breaker from service, coordinate with appropriate authority.	.038					√
4 Check operation of tripping mechanisms and relays.	.096					√
5 Check the operation of the charging motor, if applicable.	.038					√
6 Check hardware for tightness.	.005					√
7 Check interlock for proper operation.	.039					√
8 Clean area around switch.	.066					√
9 Place the circuit breaker back in service, coordinate with appropriate authority.	.040					√
10 Fill out maintenance checklist.	.022					√
Total labor-hours/period						.470
Total labor-hours/year						.470

			Cost Each					
			2015 Bare Costs				Total	Total
Description	Labor-hrs.	Material	Labor	Equip.	Total	In-House	w/O&P
1900	Circuit breaker, high voltage air, annually	.470		25.50		25.50	32	40
1950	Annualized	.470		25.50		25.50	32	40

PM Components	Labor-hrs.	W	M	Q	S	A
PM System D5015 222 1950						
Circuit breaker, HV oil						
1 Check the records on the circuit breaker for any full current rated trips which may indicate a potential problem.	.066					√
2 Examine the exterior of the circuit breaker for any damage, cracks, rust or leaks; check around the gaskets and the drain valves.	.074					√
3 Check the entrance bushings for cracks in the porcelain.	.321					√
4 Check the condition of the ground system.	.170					√
5 Draw oil sample from bottom of circuit breaker; have sample tested; replace removed oil to the proper level.	.042					√
6 Clean the exterior of the circuit breaker and operating mechanism.	.094					√
7 Clean the area around the circuit breaker.	.066					√
8 Fill out maintenance checklist.	.022					√
Total labor-hours/period						.855
Total labor-hours/year						.855

			Cost Each					
			2015 Bare Costs				Total	Total
Description	Labor-hrs.	Material	Labor	Equip.	Total	In-House	w/O&P
1900	Circuit breaker, high voltage oil, annually	.855	18.40	47		65.40	78	95.50
1950	Annualized	.855	18.40	47		65.40	78	95.50

PM Components	Labor-hrs.	W	M	Q	S	A
PM System D5015 230 1950						
Switch, selector, HV, air						
1 Examine the exterior of the selector switch for any damage.	.060					✓
2 Check that the switching handle is locked.	.044					✓
3 Check the condition of the ground system.	.170					✓
4 Clean the exterior of the selector switch and operating mechanism.	.046					✓
5 Clean the area around the selector switch.	.066					✓
6 Fill out maintenance checklist and report deficiencies.	.022					✓
Total labor-hours/period						.408
Total labor-hours/year						.408

			Cost Each					
				2015 Bare Costs			**Total**	**Total**
Description		**Labor-hrs.**	**Material**	**Labor**	**Equip.**	**Total**	**In-House**	**w/O&P**
1900	Switch, selector, high voltage, air, annually	.408	12.45	22.50		34.95	41.50	50
1950	Annualized	.408	12.45	22.50		34.95	41.50	50

PM Components	Labor-hrs.	W	M	Q	S	A
PM System D5015 232 1950						
Switch, selector, HV, oil						
1 Examine the exterior of the interrupt switch for any damage, cracks, rust or leaks; check around the gaskets and the drain valves.	.060					✓
2 Check that the switching handle is locked.	.044					✓
3 Check the condition of the ground system.	.170					✓
4 Draw oil sample from bottom of selector switch; have sample tested; replace removed oil to the proper level.	.042					✓
5 Clean the exterior of the selector switch and operating mechanism.	.046					✓
6 Clean the area around the selector switch.	.066					✓
7 Fill out maintenance checklist and report deficiencies.	.022					✓
Total labor-hours/period						.450
Total labor-hours/year						.450

			Cost Each					
				2015 Bare Costs			**Total**	**Total**
Description		**Labor-hrs.**	**Material**	**Labor**	**Equip.**	**Total**	**In-House**	**w/O&P**
1900	Switch, selector, high voltage, oil, annually	.450	24.50	24.50		49	57.50	68.50
1950	Annualized	.450	24.50	24.50		49	57.50	68.50

PM Components	Labor-hrs.	W	M	Q	S	A
PM System D5015 234 1950						
Switch, automatic transfer						
1 Check with operating or area personnel for deficiencies.	.044		√	√	√	√
2 Inspect wiring, wiring connections, and fuse blocks for looseness,charring, evidence of short circuiting, overheating and tighten all connections.	.263		√	√	√	√
3 Inspect gen. cond. of transf. switch and clean exter. and surrounding area.	.114		√	√	√	√
4 Fill out maintenance checklist and report deficiencies.	.022		√	√	√	√
Total labor-hours/period			.443	.443	.443	.443
Total labor-hours/year			3.544	.886	.443	.443

			Cost Each				
			2015 Bare Costs			Total	Total
Description	Labor-hrs.	Material	Labor	Equip.	Total	In-House	w/O&P
1900 Switch, automatic transfer, annually	.443	12.45	24		36.45	43.50	53
1950 Annualized	5.316	12.45	291		303.45	375	465

PM Components	Labor-hrs.	W	M	Q	S	A
PM System D5015 236 1950						
Switch, interrupt, HV, fused air						
1 Examine the exterior of the interrupt switch for any damage.	.060					√
2 Check the condition of the ground system.	.170					√
3 Clean the exterior of the interrupt switch and operating mechanism.	.046					√
4 Clean the area around the interrupt switch.	.066					√
5 Fill out maintenance checklist and report deficiencies.	.022					√
Total labor-hours/period						.364
Total labor-hours/year						.364

			Cost Each				
			2015 Bare Costs			Total	Total
Description	Labor-hrs.	Material	Labor	Equip.	Total	In-House	w/O&P
1900 Switch, interrupt, high voltage, fused air, annually	.364	12.45	19.90		32.35	38.50	46.50
1950 Annualized	.364	12.45	19.90		32.35	38.50	46.50

PM Components	Labor-hrs.	W	M	Q	S	A
PM System D5015 238 1950						
Switch, interrupt, HV, w/auxiliary fuses, air						
1 Examine the exterior of the interrupt switch for any damage.	.060					√
2 Check the condition of the ground system.	.170					√
3 Clean the exterior of the interrupt switch and operating mechanism.	.046					√
4 Clean the area around the interrupt switch.	.066					√
5 Fill out maintenance checklist and report deficiencies.	.022					√
Total labor-hours/period						.364
Total labor-hours/year						.364

			Cost Each				
			2015 Bare Costs			Total	Total
Description	Labor-hrs.	Material	Labor	Equip.	Total	In-House	w/O&P
1900 Switch, interrupt, HV, w/ aux fuses, air, annually	.364	12.45	19.90		32.35	38.50	46.50
1950 Annualized	.364	12.45	19.90		32.35	38.50	46.50

For customer support on your Facilities Maintenance & Repair Cost Data, call 877.801.5724.

485

PM Components	Labor-hrs.	W	M	Q	S	A
PM System D5015 240 1950						
Transformer, dry type 500 KVA and over						
1 Examine the exterior of the transformer for any damage.	.060					√
2 Remove the covers or open the doors.	.196					√
3 Check for signs of moisture or overheating.	.059					√
4 Check for voltage creeping over insulated surfaces, such as evidenced by tracking or carbonization.	.018					√
5 Check fans, motors and other auxiliary devices for proper operation; where applicable.	.049					√
6 Check the condition of the ground system.	.170					√
7 Replace the covers or close the doors.	.196					√
8 Fill out maintenance checklist and report deficiencies.	.022					√
Total labor-hours/period						.770
Total labor-hours/year						.770

		Cost Each						
			2015 Bare Costs				Total	Total
Description	Labor-hrs.	Material	Labor	Equip.	Total	In-House	w/O&P
1900	Transformer, dry type 500 KVA and over, annually	.770	12.45	42		54.45	66	80.50
1950	Annualized	.770	12.45	42		54.45	66	80.50

PM Components	Labor-hrs.	W	M	Q	S	A
PM System D5015 240 2950						
Transformer, oil, pad mounted						
1 Examine the exterior of the transformer for any damage, cracks, rust or leaks; check around bushings, gaskets and pressure relief device.	.014					√
2 Check the condition of the ground system.	.170					√
3 Check and record the oil level, pressure and temperature readings.	.263					√
4 Check fans, motors and other auxiliary devices for proper operation; where applicable.	.022					√
5 Draw oil sample from top of transformer and have sample tested for dielectric strength; replace removed oil.	.525					√
6 Clean transformer exterior and the surrounding area.	.066					√
7 Fill out maintenance checklist and report deficiencies.	.022					√
Total labor-hours/period						1.082
Total labor-hours/year						1.082

		Cost Each						
			2015 Bare Costs				Total	Total
Description	Labor-hrs.	Material	Labor	Equip.	Total	In-House	w/O&P
2900	Transformer, oil, pad mounted, annually	1.082	24.50	59		83.50	100	122
2950	Annualized	1.082	24.50	59		83.50	100	122

PM Components	Labor-hrs.	W	M	Q	S	A
PM System D5015 240 3950						
Transformer, oil, pad mounted, PCB						
1 Check with operating or area personnel for deficiencies.	.040			√	√	√
2 Examine the exterior of the transformer for any damage, cracks, rust or leaks; check around the gaskets, the pressure relief device and the cooling fins.	.014			√	√	√
3 Check and record the oil levels, pressure and temperature readings: level ____; pressure ____; temperature ____.	.263					√
4 Check fans, motors and other auxiliary devices for proper operation; where applicable.	.051					√
5 Check the exterior bushings for leaking oil.	.321			√	√	√
6 Check the condition of the ground system.	.170					√
7 Draw oil sample from top of transformer and have sample tested; replace removed oil.	.525					√
8 Clean transformer exterior and the surrounding area.	.066					√
9 Verify the PCB warning markings are intact and clearly visible.	.022					√
10 Fill out maintenance checklist and report deficiencies. Note: the EPA PCB regulations under 40 CFR 716.30 shall apply to all inspections, samples, disposal, cleanups, decontamination, reporting and record-keeping.	.022			√	√	√
Total labor-hours/period				.397	.397	1.494
Total labor-hours/year				.794	.397	1.494

			Cost Each					
			2015 Bare Costs				**Total**	**Total**
Description		**Labor-hrs.**	**Material**	**Labor**	**Equip.**	**Total**	**In-House**	**w/O&P**
3900	Transformer, oil, pad mounted, PCB, annually	1.494	49	81.50		130.50	155	188
3950	Annualized	2.655	49	145		194	234	286

PM Components	Labor-hrs.	W	M	Q	S	A
PM System D5015 260 1950						
Panelboard, 225 amps and above						
1 Check with operating or area personnel for deficiencies.	.066					√
2 Check for excessive heat, odors, noise, and vibration.	.239					√
3 Clean and check general condition of panel.	.114					√
4 Fill out maintenance checklist and report deficiencies.	.022					√
Total labor-hours/period						.441
Total labor-hours/year						.441

			Cost Each					
			2015 Bare Costs				Total	Total
Description		Labor-hrs.	Material	Labor	Equip.	Total	In-House	w/O&P
1900	Panelboard, 225 A and above, annually	.441	29	24		53	62	74
1950	Annualized	.441	29	24		53	62	74

PM Components	Labor-hrs.	W	M	Q	S	A
PM System D5015 280 1950						
Motor control center, over 400 amps						
1 Check with operating or area personnel for deficiencies.	.044					√
2 Check starter lights, replace if required.	.018					√
3 Check for excessive heat, odors, noise, and vibration.	.239					√
4 Clean motor control center exterior and surrounding area.	.066					√
5 Fill out maintenance checklist and report deficiencies.	.022					√
Total labor-hours/period						.389
Total labor-hours/year						.389

			Cost Each					
			2015 Bare Costs				Total	Total
Description		Labor-hrs.	Material	Labor	Equip.	Total	In-House	w/O&P
1900	Motor control center, over 400 A, annually	.389	29	21.50		50.50	58	69.50
1950	Annualized	.389	29	21.50		50.50	58	69.50

PM Components	Labor-hrs.	W	M	Q	S	A
PM System D5035 610 1950						
Clocks, Central System						
1 Clean dirt and dust from interior and exterior of cabinet.	.260				√	√
2 Adjust relays and check transmission of signal.	.117				√	√
3 Tighten contacts and terminal screws.	.065				√	√
4 Burnish contacts.	.195				√	√
5 Fill out maintenance checklist and report deficiencies.	.022				√	√
Total labor-hours/period					.659	.659
Total labor-hours/year					.659	.659

			Cost Each					
			2015 Bare Costs				Total	Total
Description		Labor-hrs.	Material	Labor	Equip.	Total	In-House	w/O&P
1900	Central clock systems, annually	.659	2.08	36		38.08	47	58.50
1950	Annualized	1.318	2.08	72		74.08	91.50	115

PM Components	Labor-hrs.	W	M	Q	S	A
PM System D5035 710 1950						
Fire alarm annunciator system						
1 Visually inspect all alarm equipment for obstructions or physical damage, clean dirt and dust from interior and exterior of panel/pull boxes, tighten loose connections.	.222		√	√	√	√
2 Notify cogniant personnel prior to testing.	.112		√	√	√	√
3 Conduct operational test of initiating and signal transmitting devices in populated buildings by building zone/area; for those circuits which do not operate properly, check detectors, control units, and annunciators for dust on defective components; make minor adjustments as required.	.112		√	√	√	√
4 Check battery voltages where installed; replace as required.	.012		√	√	√	√
5 Conduct operational test of 10% of total number of spot type heat detectors and all smoke detectors; for those circuits which do not operate properly, check to determine if problem relates to circuit, device, or control unit, make minor adjustments as required; if detector is defective but no replacement is immediately available, remove detector, re-establish Initiating circuit and tag location until a replacement detector is installed.	2.171				√	√
6 Restore system to proper operating condition and notify personnel upon completion of tests.	.079		√	√	√	√
7 Fill out maintenance checklist and report deficiencies.	.022		√	√	√	√
Total labor-hours/period			.559	.559	2.730	2.730
Total labor-hours/year			4.472	1.118	2.730	2.730

			Cost Each					
			2015 Bare Costs				Total	Total
Description	Labor-hrs.	Material	Labor	Equip.	Total		In-House	w/O&P
1900 Fire alarm annunciator system, annually	2.730	131	153		284		335	405
1950 Annualized	11.050	164	620		784		960	1,175

PM Components	Labor-hrs.	W	M	Q	S	A
PM System D5035 810 1950						
Security / intrusion alarm system						
1 Check in and out with area security officer, notify operating/facility personnel, obtain necessary alarm keys, alarm codes and escort when required.	.520			√	√	√
2 Inspect alarm control panel and conduct operational test of initiating and signal transmitting devices; make minor adjustments as required.	.241			√	√	√
3 Check indicating lamps for proper operation, replace if necessary.	.018			√	√	√
4 Check battery voltages where installed, replace as required.	.012			√	√	√
5 Restore system to proper operating condition and notify personnel upon completion of tests.	.079			√	√	√
6 Clean exterior of cabinet and surrounding area.	.066			√	√	√
7 Fill out maintenance checklist and report deficiencies.	.022			√	√	√
Total labor-hours/period				.958	.958	.958
Total labor-hours/year				1.916	.958	.958

			Cost Each					
			2015 Bare Costs				Total	Total
Description	Labor-hrs.	Material	Labor	Equip.	Total		In-House	w/O&P
1900 Security, intrusion alarm system, annually	.958	32	52.50		84.50		100	121
1950 Annualized	3.832	42.50	210		252.50		305	380

PM Components	Labor-hrs.	W	M	Q	S	A
PM System D5095 210 1950						
Emergency generator up to 15 KVA						
1 Check with the operating or area personnel for any obvious deficiencies.	.044		√	√	√	√
2 Check engine oil level; add as required.	.010		√	√	√	√
3 Change engine oil and oil filter.	.170					√
4 Check battery charge and electrolyte specific gravity, add water as required; check terminals for corrosion, clean as required.	.120		√	√	√	√
5 Check belt tension and wear; adjust as required, if applicable.	.012		√	√	√	√
6 Check engine air filter; change as required.	.042					√
7 Check spark plug or injector nozzle condition; service or replace as required.	.211					√
8 Check wiring, connections, switches, etc., adjust as required.	.036		√	√	√	√
9 Perform 30 minute generator test run; check for proper operation.	.650		√	√	√	√
10 Check fuel level; add as required.	.046		√	√	√	√
11 Wipe dust and dirt from engine and generator.	.056		√	√	√	√
12 Clean area around generator.	.066		√	√	√	√
13 Fill out maintenance checklist and report deficiencies.	.022		√	√	√	√
Total labor-hours/period			1.062	1.062	1.062	1.485
Total labor-hours/year			8.496	2.124	1.062	1.485

			Cost Each				
			2015 Bare Costs			**Total**	**Total**
Description	Labor-hrs.	Material	Labor	Equip.	Total	In-House	w/O&P
1900 Emergency diesel generator, up to 15 KVA, annually	1.485	100	81		181	211	251
1950 Annualized	13.167	129	720		849	1,025	1,275

PM Components	Labor-hrs.	W	M	Q	S	A
PM System D5095 210 2950						
Emergency diesel generator, over 15 KVA						
1 Check with the operating or area personnel for any obvious deficiencies.	.044		√	√	√	√
2 Check engine oil level; add as required.	.010		√	√	√	√
3 Change engine oil and oil filter.	.511					√
4 Check battery charge and electrolyte specific gravity, add water as required; check terminals for corrosion, clean as required.	.241		√	√	√	√
5 Check belts for wear and proper tension; adjust as necessary.	.012		√	√	√	√
6 Check that crank case heater is operating.	.038		√	√	√	√
7 Check engine air filter; change as required.	.042					√
8 Check wiring, connections, switches, etc.; adjust as required.	.036		√	√	√	√
9 Check spark plug or injector nozzle condition; service or replace as required.	.281					√
10 Perform 30 minute generator test run; check for proper operation.	.650		√	√	√	√
11 Check fuel level with gauge pole, add as required.	.046		√	√	√	√
12 Wipe dust and dirt from engine and generator.	.109		√	√	√	√
13 Clean area around generator.	.066		√	√	√	√
14 Fill out maintenance checklist and report deficiencies.	.025		√	√	√	√
Total labor-hours/period			1.277	1.277	1.277	2.111
Total labor-hours/year			10.216	2.554	1.277	2.111

			Cost Each				
			2015 Bare Costs			**Total**	**Total**
Description	Labor-hrs.	Material	Labor	Equip.	Total	In-House	w/O&P
2900 Emergency diesel generator, over 15 KVA, annually	2.111	112	115		227	266	320
2950 Annualized	16.158	141	885		1,026	1,250	1,550

For customer support on your Facilities Maintenance & Repair Cost Data, call 877.801.5724.

491

PM Components	Labor-hrs.	W	M	Q	S	A
PM System D5095 210 3950						
Emergency diesel generator, turbine						
1 Check with the operating or area personnel for any obvious deficiencies.	.044		√	√	√	√
2 Check turbine oil level; add oil as required.	.036		√	√	√	√
3 Change turbine oil and oil filter; check transmission oil level.	.569					√
4 Check that the crankcase heater is operating properly.	.022		√	√	√	√
5 Replace turbine air filter.	.224					√
6 Check wiring, connections, switches, etc.; adjust as required.	.060					√
7 Check starter for proper operation; lubricate as necessary.	.047		√	√	√	√
8 Check fuel nozzles, fuel regulator and ignition device condition; service or replace as required.	.813					√
9 Perform 30 minute generator test run; check for proper operation.	.650		√	√	√	√
10 Check and record transmission oil pressure and temperature, and natural gas pressure: oil press:_____ oil temp:_____ gas press:_____.	.021					√
11 Record running time: beginning _____ hours; ending _____ hours.	.010					√
12 Check that the charger is operating properly.	.127		√	√	√	√
13 Check for any signs of corrosion on battery terminals or wires.	.117		√	√	√	√
14 Check the electrolyte level in the batteries; add if required.	.049		√	√	√	√
15 Check the specific gravity of the electrolyte in a 10% sample of the batteries.	.069					√
16 Check 25% of terminal-to-cell connection resistance; rehabilitate connections as required; add anti-corrosion grease to battery terminals and connections.	.038					√
17 Measure and record individual cell and string float voltages.	.140					√
18 Clean area around generator.	.066		√	√	√	√
19 Fill out maintenance checklist and report deficiencies.	.022		√	√	√	√
Total labor-hours/period			1.180	1.180	1.180	3.124
Total labor-hours/year			9.440	2.360	1.180	3.124

	Description	Labor-hrs.	Cost Each				Total In-House	Total w/O&P
			2015 Bare Costs					
			Material	Labor	Equip.	Total		
3900	Emergency diesel generator, turbine, annually	3.124	170	171		341	400	480
3950	Annualized	16.104	222	880		1,102	1,325	1,650

For customer support on your Facilities Maintenance & Repair Cost Data, call 877.801.5724.

PM Components	Labor-hrs.	W	M	Q	S	A
PM System D5095 220 1950						
Power stabilizer						
1 Remove the covers or open the doors.	.218					✓
2 Check for signs of moisture or overheating.	.022					✓
3 Check fans and other auxiliary devices for proper operation.	.027					✓
4 Check for voltage creeping over transformer insulated surfaces, such as evidenced by tracking or carbonization.	.018					✓
5 Check condition of the ground system.	.170					✓
6 Check wiring, connections, relays, etc; tighten and adjust as necessary.	.060					✓
7 Replace the covers or close the doors.	.022					✓
8 Clean the surrounding area.	.066					✓
9 Fill out maintenance checklist and report deficiencies.	.022					✓
Total labor-hours/period						.625
Total labor-hours/year						.625

	Description	Labor-hrs.	Material	Labor	Equip.	Total	Total In-House	Total w/O&P
1900	Power stablizer, annually	.625		34		34	42.50	53
1950	Annualized	.625		34		34	42.50	53

For customer support on your Facilities Maintenance & Repair Cost Data, call 877.801.5724.

493

PM Components	Labor-hrs.	W	M	Q	S	A
PM System D5095 230 1950						
Uninterruptible power system, up to 200 KVA						
1 Check with operating or area personnel for any obvious deficiencies.	.044		√	√	√	√
2 Check electrolyte level of batteries; add water as required; check terminals for corrosion, clean as required.	.878		√	√	√	√
3 Check 25% of the batteries for charge and electrolyte specific gravity.	.777			√	√	√
4 Check batteries for cracks or leaks.	.059		√	√	√	√
5 Check 25% of the terminal-to-cell connection resistances; rehabilitate connections as required; add anti-corrosion grease to battery terminals and connections.	.143			√	√	√
6 Measure and record individual cell and string float voltages.	.618			√	√	√
7 Check integrity of battery rack.	.176		√	√	√	√
8 Check battery room temperature and ventilation systems.	.010		√	√	√	√
9 Replace air filters on UPS modules.	.036		√	√	√	√
10 Check output voltage and amperages, from control panel.	.007		√	√	√	√
11 Check UPS room temperature and ventilation system.	.062		√	√	√	√
12 Notify personnel and test UPS fault alarm system.	.112		√	√	√	√
13 Clean around batteries and UPS modules.	.066		√	√	√	√
14 Fill out maintenance checklist and report deficiencies.	.022		√	√	√	√
Total labor-hours/period			1.360	3.010	3.010	3.010
Total labor-hours/year			10.880	6.020	3.010	3.010

			Cost Each				
			2015 Bare Costs			**Total**	**Total**
Description	**Labor-hrs.**	**Material**	**Labor**	**Equip.**	**Total**	**In-House**	**w/O&P**
1900 Uninter. power system, up to 200 KVA, annually	3.010	208	165		373	435	515
1950 Annualized	22.920	240	1,250		1,490	1,825	2,250

PM Components	Labor-hrs.	W	M	Q	S	A
PM System D5095 230 2950						
Uninterrupted power system, 200 KVA to 800 KVA						
1 Check with operating or area personnel for any obvious deficiencies.	.044		√	√	√	√
2 Check electrolyte level of batteries; add water as required; check terminals for corrosion, clean as required.	3.510		√	√	√	√
3 Check 25% of the batteries for charge and electrolyte specific gravity.	3.107			√	√	√
4 Check batteries for cracks or leaks.	.585		√	√	√	√
5 Check 25% of the terminal-to-cell connection resistances; rehabilitate connections as required; add anti-corrosion grease to battery terminals and connections.	.234			√	√	√
6 Measure and record individual cell and string float voltages.	.936			√	√	√
7 Check integrity of battery rack.	.702		√	√	√	√
8 Check battery room temperature and ventilation systems.	.010		√	√	√	√
9 Replace air filters on UPS modules.	.036		√	√	√	√
10 Check output voltage and amperages, from control panel.	.007		√	√	√	√
11 Check UPS room temperature and ventilation system.	.062		√	√	√	√
12 Notify personnel and test UPS fault alarm system.	.112		√	√	√	√
13 Clean around batteries and UPS modules.	.066		√	√	√	√
14 Fill out maintenance checklist and report deficiencies.	.022		√	√	√	√
Total labor-hours/period			5.044	9.433	9.433	9.433
Total labor-hours/year			40.352	18.866	9.433	9.433

			Cost Each				
			2015 Bare Costs			**Total**	**Total**
Description	**Labor-hrs.**	**Material**	**Labor**	**Equip.**	**Total**	**In-House**	**w/O&P**
2900 UPS, 200 KVA to 800 KVA, annually	9.433	254	515		769	920	1,125
2950 Annualized	78.087	286	4,275		4,561	5,600	6,975

PM Components	Labor-hrs.	W	M	Q	S	A
PM System D5095 240 1950						
Battery system and charger						
1 Check that the charger is operating properly.	.051		√	√	√	√
2 Check for any signs of corrosion on the battery terminals or wires.	.117		√	√	√	√
3 Check the electrolyte level in the batteries; add if required.	.062		√	√	√	√
4 Check the specific gravity of the electrolyte in a 10% sample of the batteries.	.069		√	√	√	√
5 Check 25% of terminal-to-cell connection resistance; rehabilitate connections as required; add anti-corrosion grease to battery terminals and connections.	.741			√	√	√
6 Measure and record individual cell and string float voltages.	.281			√	√	√
7 Clean area around batteries.	.066		√	√	√	√
8 Fill out maintenance checklist and report deficiencies.	.022		√	√	√	√
Total labor-hours/period			.387	1.409	1.409	1.409
Total labor-hours/year			3.096	2.818	1.409	1.409

	Description	Labor-hrs.	Cost Each				Total In-House	Total w/O&P
			2015 Bare Costs					
			Material	Labor	Equip.	Total		
1900	Battery system and charger, annually	1.409	28.50	77		105.50	127	155
1950	Annualized	8.732	35	480		515	630	785

For customer support on your Facilities Maintenance & Repair Cost Data, call 877.801.5724.

495

PM Components	Labor-hrs.	W	M	Q	S	A
PM System D5095 250 1950						
Light, emergency, hardwired system						
1 Test for proper operation, replace lamps as required.	.013				✓	✓
2 Clean exterior of cabinet and light heads.	.025				✓	✓
3 Check wiring for obvious defects.	.026				✓	✓
4 Adjust lamp heads for maximum illumination of area.	.039				✓	✓
5 Fill out maintenance checklist and report deficiencies.	.022				✓	✓
Total labor-hours/period					.125	.125
Total labor-hours/year					.125	.125

			Cost Each				
			2015 Bare Costs			Total	Total
Description	Labor-hrs.	Material	Labor	Equip.	Total	In-House	w/O&P
1900 Light, emergency, hardwired system, annually	.125		6.85		6.85	8.45	10.60
1950 Annualized	.250		13.70		13.70	16.95	21

PM Components	Labor-hrs.	W	M	Q	S	A
PM System D5095 250 2950						
Light, emergency, dry cell						
1 Test for proper operation, replace lamps as required.	.013				✓	✓
2 Clean exterior of emergency light cabinet and lamp heads.	.025				✓	✓
3 Check condition of batteries and change as required.	.012				✓	✓
4 Clean interior of cabinet, top of battery, and battery terminal.	.009				✓	✓
5 Inspect cabinet, relay, relay contacts, pilot light, wiring, and gen. cond.	.065				✓	✓
6 Adjust lamp heads for maximum illumination of area.	.039				✓	✓
7 Fill out maintenance checklist and report deficiencies.	.022				✓	✓
Total labor-hours/period					.185	.185
Total labor-hours/year					.185	.185

			Cost Each				
			2015 Bare Costs			Total	Total
Description	Labor-hrs.	Material	Labor	Equip.	Total	In-House	w/O&P
2900 Light, emergency, dry cell, annually	.185	35.50	10.10		45.60	51.50	60
2950 Annualized	.370	42.50	20		62.50	72.50	84.50

PM Components	Labor-hrs.	W	M	Q	S	A
PM System D5095 250 3950						
Light, emergency, wet cell						
1 Test for proper operation; correct deficiencies as required.	.013				✓	✓
2 Clean exterior of emergency light cabinet and lamp heads.	.025				✓	✓
3 Clean interior of cabinet, top of battery, and battery terminal.	.009				✓	✓
4 Check battery charge and electrolyte specific gravity; add water if nec.	.012				✓	✓
5 Inspect cabinet, relay, relay contacts, pilot light, wiring, and gen. cond.	.065				✓	✓
6 Apply anti-corrosion coating to battery terminals.	.010				✓	✓
7 Adjust lamp heads for maximum illumination of area.	.039				✓	✓
8 Fill out maintenance checklist and report deficiencies.	.022				✓	✓
Total labor-hours/period					.195	.195
Total labor-hours/year					.195	.195

			Cost Each				
			2015 Bare Costs			Total	Total
Description	Labor-hrs.	Material	Labor	Equip.	Total	In-House	w/O&P
3900 Light, emergency, wet cell, annually	.195	50	10.65		60.65	68	79
3950 Annualized	.390	64.50	21.50		86	97.50	114

For customer support on your Facilities Maintenance & Repair Cost Data, call 877.801.5724.

PM Components	Labor-hrs.	W	M	Q	S	A
PM System E1015 810 1950						
Personal Computers						
1 Check with operating personnel for any deficiencies.	.065				√	√
2 Check all cables for fraying, cracking, or exposed wires; ensure connections are secure.	.065				√	√
3 Remove metal casing and vacuum interior, power supply vents, keyboard, monitor, and surrounding work area.	.130				√	√
4 Check for secure connections of all interior wires and cables. Secure all peripheral cards.	.065				√	√
5 Replace PC cover and clean case, keyboard, monitor screen and housing with an anti-static spray cleaner. Apply cleaner to towel and not the components.	.130				√	√
6 Disassemble the mouse; clean the tracking ball, interior rollers and housing. Blow out the interior of the mouse with clean compressed air.	.065				√	√
7 Turn on PC and check for unusual noises.	.026				√	√
8 Clean floppy drive heads with cleaning diskette and solution.	.039				√	√
9 Defragment the hard drive using a defragmentation utility.	.234				√	√
10 Run diagnostics software to test all the PC's components.	.234				√	√
11 Fill out maintenance checklist and report deficiencies.	.029				√	√
Total labor-hours/period					1.082	1.082
Total labor-hours/year					1.082	1.082

			Cost Each					
			2015 Bare Costs				Total	Total
Description		Labor-hrs.	Material	Labor	Equip.	Total	In-House	w/O&P
1900	Personnel Computers, annually	1.082		59		59	73.50	91.50
1950	Annualized	2.164		118		118	147	183

For customer support on your Facilities Maintenance & Repair Cost Data, call 877.801.5724.

497

PM Components	Labor-hrs.	W	M	Q	S	A
PM System E1035 100 1950						
Hydraulic lift						
1 Check for proper operation of pump.	.035				√	√
2 Check for leaks on suction and discharge piping, seals, packing glands, etc.; make minor adjustments as required.	.077				√	√
3 Check pump and motor operation for excessive vibration, noise and overheating.	.022				√	√
4 Check alignment of pump and motor; adjust as necessary.	.260				√	√
5 Lubricate pump and motor.	.099				√	√
6 Inspect hydraulic lift post(s) for wear or leaks.	.065				√	√
7 Inspect, clean and tighten valves.	.034				√	√
8 Inspect and clean motor contactors.	.026				√	√
9 Inspect and test control relays and check wiring terminals.	.034				√	√
10 Clean pump unit and surrounding area.	.099				√	√
11 Fill out maintenance checklist and report deficiencies.	.022				√	√
Total labor-hours/period					.773	.773
Total labor-hours/year					.773	.773

	Description	Labor-hrs.	Cost Each					Total In-House	Total w/O&P
			2015 Bare Costs						
			Material	Labor	Equip.	Total			
1900	Hydraulic lift, annually	.773	51	45.50		96.50	113	135
1950	Annualized	1.546	68.50	91		159.50	190	228

PM Components	Labor-hrs.	W	M	Q	S	A
PM System E1035 310 1950						
Hydraulic lift, loading dock						
1 Check for proper operation of pump.	.022				√	√
2 Check for leaks on suction and discharge piping, seals, packing glands, etc.; make minor adjustments as required.	.077				√	√
3 Check pump and motor operation for excessive vibration, noise and overheating.	.022				√	√
4 Check alignment of pump and motor; adjust as necessary.	.100				√	√
5 Lubricate pump and motor.	.099				√	√
6 Inspect hydraulic lift post(s) for wear or leaks.	.051				√	√
7 Inspect, clean and tighten valves.	.072				√	√
8 Inspect and clean motor contactors.	.148				√	√
9 Inspect and test control relays and check wiring terminals.	.137				√	√
10 Clean pump unit and surrounding area.	.066				√	√
11 Fill out maintenance checklist and report deficiencies.	.022				√	√
Total labor-hours/period					.816	.816
Total labor-hours/year					.816	.816

	Description	Labor-hrs.	Cost Each					Total In-House	Total w/O&P
			2015 Bare Costs						
			Material	Labor	Equip.	Total			
1900	Hydraulic lift, loading dock, annually	.816	8.80	40		48.80	60.50	74
1950	Annualized	1.632	17.55	80		97.55	121	148

For customer support on your Facilities Maintenance & Repair Cost Data, call 877.801.5724.

PM Components	Labor-hrs.	W	M	Q	S	A
PM System E1095 122 1950						
Crane, electric bridge, up to 5 ton						
1 Check with operating or area personnel for deficiencies.	.035					√
2 Inspect crane for damaged structural members, alignment, worn pins, column fasteners, stops and capacity markings.	.235					√
3 Inspect and lubricate power cable tension reel and inspect condition of cable.	.491					√
4 Check hoist operation - travel function, up, down, speed stepping, upper and lower travel limit switches, etc.	.072					√
5 Inspect, clean and lubricate trolley assembly and reduction gear bearings.	.112					√
6 Inspect and clean hook assembly and chain or wire rope sheaves.	.452					√
7 Inspect, clean and lubricate hoist motor.	.094					√
8 Inspect hoist brake mechanism.	.445					√
9 Lubricate rails.	.114					√
10 Weight capacity test hoisting equipment at 125% of rated capacity in accordance with 29 CFR 1910.179.	.780					√
11 Clean work area.	.066					√
12 Fill out maintenance report and report deficiencies.	.022					√
Total labor-hours/period						2.918
Total labor-hours/year						2.918

			Cost Each				
			2015 Bare Costs			Total	Total
Description	Labor-hrs.	Material	Labor	Equip.	Total	In-House	w/O&P
1900 Crane, electric bridge, up to 5 ton, annually	2.918	46.50	143		189.50	232	283
1950 Annualized	2.918	46.50	143		189.50	232	283

PM Components	Labor-hrs.	W	M	Q	S	A
PM System E1095 122 2950						
Crane, electric bridge 5 to 15 ton						
1 Check with operating or area personnel for deficiencies.	.035			√	√	√
2 Inspect crane for damaged structural members, alignment, worn pins, column fasteners, stops and capacity markings.	.235			√	√	√
3 Inspect and lubricate power cable tension reel and inspect condition of cable.	.491			√	√	√
4 Check operation of upper and lower travel limit switches.	.072			√	√	√
5 Inspect brake mechanism.	.250					√
6 Inspect, clean and lubricate trolley assembly and reduction gear bearings.	.099			√	√	√
7 Drain and replace reduction gear housing oil.	.170					√
8 Inspect and clean hook assembly and chain or wire rope sheaves.	.099			√	√	√
9 Inspect, clean and lubricate hoist motor.	.094			√	√	√
10 Check pendent control for hoist operation - travel function, up, down, speed stepping, etc.	.088			√	√	√
11 Lubricate rails.	.325			√	√	√
12 Weight capacity test hoisting equipment at 125% of rated capacity in accordance with 29 CFR 1910.179.	1.040					√
13 Clean work area.	.066			√	√	√
14 Fill out maintenance report and report deficiencies.	.022			√	√	√
Total labor-hours/period				1.626	1.626	3.086
Total labor-hours/year				3.252	1.626	3.086

			Cost Each				
			2015 Bare Costs			Total	Total
Description	Labor-hrs.	Material	Labor	Equip.	Total	In-House	w/O&P
2900 Crane, electric bridge, 5 to 15 ton, annually	3.086	23.50	152		175.50	217	267
2950 Annualized	7.964	70	390		460	570	705

PM Components	Labor-hrs.	W	M	Q	S	A
PM System E1095 122 3950						
Crane, electric bridge, over 15 tons						
1 Check with operating or area personnel for deficiencies.	.035			√	√	√
2 Inspect crane for damaged structural members, alignment, worn pins, column fasteners, stops and capacity markings.	.235			√	√	√
3 Inspect and lubricate power cable tension reel and inspect condition of cable.	.491			√	√	√
4 Check operation of upper and lower travel limit switches.	.072			√	√	√
5 Inspect brake mechanism.	.250					√
6 Inspect, clean and lubricate trolley assembly and reduction gear bearings.	.099			√	√	√
7 Drain and replace reduction gear housing oil.	.170					√
8 Inspect and clean hook assembly and chain or wire rope sheaves.	.099			√	√	√
9 Inspect, clean and lubricate hoist motor.	.094			√	√	√
10 Check pendent control for hoist operation - travel function, up, down, speed stepping, etc.	.088			√	√	√
11 Lubricate rails.	.325			√	√	√
12 Inspect control cabinet.	.238					√
13 Weight capacity test hoisting equipment, at 125% of rated capacity in accordance with 29 CFR 1910.179.	1.170					√
14 Clean work area.	.066			√	√	√
15 Fill out maintenance report and report deficiencies.	.022			√	√	√
Total labor-hours/period				1.626	1.626	3.454
Total labor-hours/year				3.252	1.626	3.454

			Cost Each					
			2015 Bare Costs				Total	Total
Description		Labor-hrs.	Material	Labor	Equip.	Total	In-House	w/O&P
3900	Crane, electric bridge, over 15 tons, annually	3.454	35	170		205	254	310
3950	Annualized	8.332	93.50	410		503.50	620	760

PM Components	Labor-hrs.	W	M	Q	S	A
PM System E1095 123 1950						
Crane, manual bridge, up to 5 tons						
1 Check with operating or area personnel for deficiencies.	.035					√
2 Operate crane and hoist, inspect underhung hoist monorail for damaged structural members, braces, stops and capacity markings.	.075					√
3 Inspect and lubricate trolley wheels and bearings.	.096					√
4 Inspect, clean and lubricate hooks and chain sheaves.	.099					√
5 Check operation of crane and hoist brakes.	.016					√
6 Make minor adjustments as required for operational integrity; tighten bolts, etc.	.135					√
7 Weight capacity test hoisting equipment at 125% of rated capacity in accordance with 29 CFR 1910.179.	.390					√
8 Clean work area.	.066					√
9 Fill out maintenance report and report deficiencies.	.022					√
Total labor-hours/period						.934
Total labor-hours/year						.934

				Cost Each				
				2015 Bare Costs			Total	Total
Description		Labor-hrs.	Material	Labor	Equip.	Total	In-House	w/O&P
1900	Crane, manual bridge, up to 5 ton, annually	.934	23.50	46		69.50	84	101
1950	Annualized	.934	23.50	46		69.50	84	101

PM Components	Labor-hrs.	W	M	Q	S	A
PM System E1095 123 2950						
Crane, manual bridge, 5 to 15 tons						
1 Check with operating or area personnel for deficiencies.	.035					√
2 Operate crane and hoist, inspect underhung hoist monorail for damaged structural members, braces, stops and capacity markings.	.130					√
3 Inspect and lubricate trolley wheels and bearings.	.096					√
4 Inspect, clean and lubricate hooks and chain sheaves.	.099					√
5 Check operation of crane and hoist brakes.	.016					√
6 Make minor adjustments as required for operational integrity; tighten bolts, etc.	.135					√
7 Weight capacity test hoisting equipment at 125% of rated capacity in accordance with 29 CFR 1910.179.	.772					√
8 Clean work area.	.066					√
9 Fill out maintenance report and report deficiencies.	.022					√
Total labor-hours/period						1.371
Total labor-hours/year						1.371

				Cost Each				
				2015 Bare Costs			Total	Total
Description		Labor-hrs.	Material	Labor	Equip.	Total	In-House	w/O&P
2900	Crane, manual bridge, 5 to 15 ton, annually	1.371	46.50	67.50		114	136	164
2950	Annualized	1.371	46.50	67.50		114	136	164

PM Components	Labor-hrs.	W	M	Q	S	A
PM System E1095 123 3950						
Crane, manual bridge, over 15 tons						
1 Check with operating or area personnel for deficiencies.	.035					√
2 Operate crane and hoist, inspect for damaged structural members, alignment, worn pins, fasteners, stops and capacity markings.	.235					√
3 Inspect and lubricate trolley wheels and bearings.	.096					√
4 Inspect, clean and lubricate hooks and chain sheaves.	.099					√
5 Check operation of crane and hoist brakes.	.016					√
6 Make minor adjustments as required for operational integrity; tighten bolts, etc.	.135					√
7 Weight capacity test hoisting equipment at 125% of rated capacity in accordance with 29 CFR 1910.179.	1.162					√
8 Clean work area.	.066					√
9 Fill out maintenance report and report deficiencies.	.022					√
Total labor-hours/period						1.866
Total labor-hours/year						1.866

			Cost Each					
			2015 Bare Costs				Total	Total
	Description	Labor-hrs.	Material	Labor	Equip.	Total	In-House	w/O&P
3900	Crane, manual bridge, over 15 tons, annually	1.866	70	91.50		161.50	193	232
3950	Annualized	1.866	70	91.50		161.50	193	232

PM Components	Labor-hrs.	W	M	Q	S	A
PM System E1095 125 1950						
Hoist, pneumatic						
1 Check with operating or area personnel for deficiencies.	.035					√
2 Inspect for damaged structural members, alignment, worn pins, column fasteners, stops and capacity markings.	.235					√
3 Inspect and lubricate cable tension reel and inspect condition of cable.	.452					√
4 Check pendent control for hoist operation - travel function, up, down, speed stepping, upper and lower travel limit switches, etc.	.088					√
5 Inspect, clean and lubricate trolley assembly.	.099					√
6 Drain and replace reduction gear housing oil.	.170					√
7 Inspect brake mechanism.	.016					√
8 Inspect and clean hook assembly and chain or wire rope sheaves.	.099					√
9 Inspect, clean and lubricate hoist motor.	.029					√
10 Clean work area.	.066					√
11 Fill out maintenance report and report deficiencies.	.022					√
Total labor-hours/period						1.311
Total labor-hours/year						1.311

			Cost Each					
			2015 Bare Costs				Total	Total
	Description	Labor-hrs.	Material	Labor	Equip.	Total	In-House	w/O&P
1900	Hoist, pneumatic, annually	1.311	93.50	64.50		158	184	218
1950	Annualized	1.311	93.50	64.50		158	184	218

PM Components	Labor-hrs.	W	M	Q	S	A
PM System E1095 126 1950						
Hoist/winch, chain fall/cable, electric						
1 Check with operating or area personnel for deficiencies.	.035					√
2 Inspect for damaged structural members, alignment, worn pins, column fasteners, stops and capacity markings.	.235					√
3 Inspect and lubricate power cable tension reel; inspect condition of cable.	.491					√
4 Check hoist operation - travel function, up, down, speed stepping, upper and lower travel limit switches, etc.	.088					√
5 Inspect, clean and lubricate trolley assembly and reduction gear bearings.	.099					√
6 Drain and replace reduction gear housing oil.	.170					√
7 Inspect brake mechanism.	.250					√
8 Inspect and clean hook assembly and chain or wire rope sheaves.	.099					√
9 Inspect, clean and lubricate hoist motor.	.094					√
10 Clean work area.	.066					√
11 Fill out maintenance report and report deficiencies.	.022					√
Total labor-hours/period						1.649
Total labor-hours/year						1.649

		Cost Each					
		2015 Bare Costs				Total	Total
Description	Labor-hrs.	Material	Labor	Equip.	Total	In-House	w/O&P
1900 Hoist / winch, chain / cable, electric, annually	1.649	23.50	81		104.50	128	156
1950 Annualized	1.649	23.50	81		104.50	128	156

PM Components	Labor-hrs.	W	M	Q	S	A
PM System E1095 302 1950						
Beverage dispensing unit						
1 Check with operating or area personnel for deficiencies.	.044			√	√	√
2 Remove access panels and clean coils, fans, fins, and other areas; vacuum, blow down with compressed air or CO_2, and/or wipe off.	.517			√	√	√
3 Lubricate motor bearings.	.008			√	√	√
4 Check electrical wiring and connections; tighten loose connections.	.120			√	√	√
5 Visually check for refrigerant, water, and other liquid leaks.	.153			√	√	√
6 Check door gaskets for proper sealing; adjust door catch as required.	.033			√	√	√
7 Inspect dispensing valves; adjust as necessary.	.108			√	√	√
8 Perform operational check of unit; make adjustments as necessary.	.040			√	√	√
9 Clean area around equipment.	.066			√	√	√
10 Fill out maintenance checklist and report deficiencies.	.022			√	√	√
Total labor-hours/period				1.111	1.111	1.111
Total labor-hours/year				2.222	1.111	1.111

		Cost Each					
		2015 Bare Costs				Total	Total
Description	Labor-hrs.	Material	Labor	Equip.	Total	In-House	w/O&P
1900 Beverage dispensing unit, annually	1.111	8.95	54		62.95	81	99
1950 Annualized	4.444	17.85	216		233.85	305	370

For customer support on your Facilities Maintenance & Repair Cost Data, call 877.801.5724.

503

PM Components	Labor-hrs.	W	M	Q	S	A
PM System E1095 304 1950						
Braising pan, tilting, gas/electric						
1 Check with operating or area personnel for deficiencies.	.044			√	√	√
2 Check electric transformer, heating elements, insulators, connections and wiring; tighten connections and repair as required.	.239			√	√	√
3 Check surface temperature with meter for hot spots.	.073			√	√	√
4 Check piping and valves on gas operated units for leaks.	.077			√	√	√
5 Check pilot/igniter, burner and combustion chamber for proper operation; adjust as required.	.079			√	√	√
6 Check temperature control system for proper operation; adjust as required.	.458			√	√	√
7 Lubricate tilting gear mechanism and trunnion bearings.	.047			√	√	√
8 Check nuts, bolts and screws for tightness; tighten as required.	.007			√	√	√
9 Check complete operation of unit.	.055			√	√	√
10 Fill out maintenance checklist and report deficiencies.	.022			√	√	√
Total labor-hours/period				1.101	1.101	1.101
Total labor-hours/year				2.202	1.101	1.101

	Description	Labor-hrs.	2015 Bare Costs Material	Labor	Equip.	Total	Total In-House	Total w/O&P
1900	Braising pan, tilting, gas / electric, annually	1.101	167	53.50		220.50	255	296
1950	Annualized	4.404	196	214		410	495	595

PM Components	Labor-hrs.	W	M	Q	S	A
PM System E1095 306 1950						
Bread slicer, electric						
1 Check with operating or area personnel for deficiencies.	.044			√	√	√
2 Check electric motor, controls, wiring connections, insulation and contacts for damage and/or proper operation; adjust as required.	.338			√	√	√
3 Check V-belt for wear, tension and alignment; change/adjust belt as necessary.	.029			√	√	√
4 Lubricate bushings and check oil level in gear drive mechanism; add oil if required.	.065			√	√	√
5 Clean machine of dust, grease and food particles.	.074			√	√	√
6 Check motor and motor bearings for overheating.	.039			√	√	√
7 Check for loose, missing or damaged nuts, bolts, or screws; tighten or replace as necessary.	.005			√	√	√
8 Fill out maintenance checklist and report deficiencies.	.022			√	√	√
Total labor-hours/period				.616	.616	.616
Total labor-hours/year				1.232	.616	.616

	Description	Labor-hrs.	2015 Bare Costs Material	Labor	Equip.	Total	Total In-House	Total w/O&P
1900	Bread slicer, electric, annually	.616	68.50	30		98.50	115	134
1950	Annualized	2.464	98	120		218	265	320

PM Components	Labor-hrs.	W	M	Q	S	A
PM System E1095 310 1950						
Broiler, conveyor type, gas						
1 Check with operating or area personnel for deficiencies.	.044			√	√	√
2 Check broiler pilot/igniters and jets for uniform flame; adjust as required.	.166			√	√	√
3 Check air filter; change as required.	.059			√	√	√
4 Check operation of chains and counterweights.	.012			√	√	√
5 Check piping and valves for leaks.	.077			√	√	√
6 Check electrical connections and wiring for defects; tighten loose connections.	.120			√	√	√
7 Check thermocouple and thermostat; calibrate thermostat as required.	.646			√	√	√
8 Tighten and replace any loose nuts, bolts, or screws.	.005			√	√	√
9 Check operation of unit.	.055			√	√	√
10 Fill out maintenance checklist and report deficiencies.	.022			√	√	√
Total labor-hours/period				1.206	1.206	1.206
Total labor-hours/year				2.412	1.206	1.206

			Cost Each				
			2015 Bare Costs			Total	Total
Description	Labor-hrs.	Material	Labor	Equip.	Total	In-House	w/O&P
1900 Broiler, conveyor type, gas, annually	1.206	177	58.50		235.50	273	315
1950 Annualized	4.824	177	235		412	505	600

PM Components	Labor-hrs.	W	M	Q	S	A
PM System E1095 310 2950						
Broiler, hot dog, electric						
1 Check with operating or area personnel for deficiencies.	.044			√	√	√
2 Check electric motor, controls, contacts, wiring connections for damage and proper operation; adjust as required.	.152			√	√	√
3 Lubricate bushings.	.007			√	√	√
4 Inspect lines and unit for mechanical defects; adjust or repair as required.	.021			√	√	√
5 Clean machine of dust and grease.	.109			√	√	√
6 Check unit for proper operation.	.082			√	√	√
7 Fill out maintenance checklist and report deficiencies.	.022			√	√	√
Total labor-hours/period				.437	.437	.437
Total labor-hours/year				.874	.437	.437

			Cost Each				
			2015 Bare Costs			Total	Total
Description	Labor-hrs.	Material	Labor	Equip.	Total	In-House	w/O&P
2900 Broiler, hot dog type, electric, annually	.437	49	21.50		70.50	82	96
2950 Annualized	1.748	78.50	85		163.50	199	236

PM Components	Labor-hrs.	W	M	Q	S	A
PM System E1095 310 3950						
Broiler/oven (salamander), gas/electric						
1 Check with operating or area personnel for deficiencies.	.044			√	√	√
2 Check broiler pilot and oven for uniform flame; adjust as required.	.166			√	√	√
3 Check piping and valves for leaks.	.077			√	√	√
4 Check doors, shelves for warping, alignment, and seals; adjust as required.	.148			√	√	√
5 Check electrical connections and wiring for defects; tighten loose connections.	.241			√	√	√
6 Tighten and replace any loose nuts, bolts and/or screws.	.012			√	√	√
7 Check operation of broiler arm, adjust as required.	.007			√	√	√
8 Check thermocouple and thermostat; calibrate thermostat as required.	.569			√	√	√
9 Fill out maintenance checklist and report deficiencies.	.022			√	√	√
Total labor-hours/period				1.286	1.286	1.286
Total labor-hours/year				2.572	1.286	1.286

			Cost Each					
			2015 Bare Costs				Total	Total
Description	Labor-hrs.	Material	Labor	Equip.	Total	In-House	w/O&P
3900 Broiler / oven, (salamander), gas / electric, annually	1.286	157	62.50		219.50	256	298
3950 Annualized	5.144	157	250		407	500	600

PM Components	Labor-hrs.	W	M	Q	S	A
PM System E1095 312 1950						
Chopper, electric						
1 Check with operating or area personnel for deficiencies.	.044			√	√	√
2 Check cutting surfaces, clutch assembly and guards for damage and proper alignment; adjust as required.	.065			√	√	√
3 Check nuts, bolts, and screws for tightness; tighten or replace as required.	.005			√	√	√
4 Check switch and controls for proper operation; adjust as required.	.010			√	√	√
5 Clean machine of dust, grease and food particles.	.014			√	√	√
6 Check operation of motor and check motor bearings for overheating; lubricate as required.	.069			√	√	√
7 Check belt tension; replace belt and/or adjust if required.	.315			√	√	√
8 Fill out maintenance checklist and report deficiencies.	.022			√	√	√
Total labor-hours/period				.544	.544	.544
Total labor-hours/year				1.088	.544	.544

			Cost Each					
			2015 Bare Costs				Total	Total
Description	Labor-hrs.	Material	Labor	Equip.	Total	In-House	w/O&P
1900 Chopper, electric, annually	.544	84.50	26.50		111	128	149
1950 Annualized	2.176	104	106		210	254	300

PM Components	Labor-hrs.	W	M	Q	S	A
PM System E1095 314 1950						
Coffee maker/urn, gas/electric/steam						
1 Check thermostat, switch, and temperature gauge; calibrate if required.	.465			√	√	√
2 Check pilots and flame on gas operated units; adjust as required.	.038			√	√	√
3 Check working pressure on steam operated unit.	.005			√	√	√
4 Check electrical connections and wiring for defects; tighten connections.	.120			√	√	√
5 Check for clogged or defective steam trap.	.100			√	√	√
6 Inspect and clean steam strainer.	.207			√	√	√
7 Examine equipment, valves, and piping for leaks.	.077			√	√	√
8 Inspect for leaks at water gauge glasses and at valves; repack valves if necessary.	.046			√	√	√
9 Tighten or replace loose, missing, or damaged nuts, bolts, or screws.	.005			√	√	√
10 Lubricate water filter valve, check rings, tighten as required.	.033			√	√	√
11 Check operation of lights.	.027			√	√	√
12 Check timer mechanism and operation of unit.	.018			√	√	√
13 Fill out maintenance checklist and report deficiencies.	.022			√	√	√
Total labor-hours/period				1.163	1.163	1.163
Total labor-hours/year				2.326	1.163	1.163

	Description	Labor-hrs.	Cost Each				Total In-House	Total w/O&P
			2015 Bare Costs					
			Material	Labor	Equip.	Total		
1900	Coffee maker / urn, gas / electric / steam, annually	1.163	53.50	56.50		110	133	159
1950	Annualized	4.652	53.50	226		279.50	355	435

PM Components	Labor-hrs.	W	M	Q	S	A
PM System E1095 316 1950						
Cooker, gas						
1 Check with operating or area personnel for any deficiencies.	.044			√	√	√
2 Examine doors and door gaskets; make necessary adjustments.	.021			√	√	√
3 Visually inspect hinges and latches; lubricate hinges.	.091			√	√	√
4 Tighten or replace any loose nuts, bolts, screws; replace broken knobs.	.007			√	√	√
5 Check gas piping and valves for leaks.	.077			√	√	√
6 Check pilot and thermosensor bulb.	.079			√	√	√
7 Check gas burner for uniform flame.	.007			√	√	√
8 Check calibration of thermostats.	.458			√	√	√
9 Operate unit to ensure that it is in proper working condition.	.130			√	√	√
10 Fill out maintenance checklist and report deficiencies.	.022			√	√	√
Total labor-hours/period				.936	.936	.936
Total labor-hours/year				1.872	.936	.936

			Cost Each				
			2015 Bare Costs			Total	Total
Description	Labor-hrs.	Material	Labor	Equip.	Total	In-House	w/O&P
1900 Cooker, gas, annually	.936	125	45.50		170.50	198	230
1950 Annualized	3.744	125	182		307	375	450

PM Components	Labor-hrs.	W	M	Q	S	A
PM System E1095 316 2950						
Cooker, vegetable steamer						
1 Check with operating or area personnel for any deficiencies.	.044			√	√	√
2 Examine doors and door gaskets; make necessary adjustments.	.021			√	√	√
3 Visually inspect hinges and latches; lubricate hinges.	.091			√	√	√
4 Tighten or replace any loose nuts, bolts, screws; replace broken knobs.	.005			√	√	√
5 Check working pressure on steam gauge.	.005			√	√	√
6 Inspect piping and valves for leaks; tighten as necessary.	.077			√	√	√
7 Check operation of low water cut-off; adjust as required.	.030			√	√	√
8 Check water level sight glass; adjust as required.	.007			√	√	√
9 Fill out maintenance checklist and report deficiencies.	.022			√	√	√
Total labor-hours/period				.302	.302	.302
Total labor-hours/year				.604	.302	.302

			Cost Each				
			2015 Bare Costs			Total	Total
Description	Labor-hrs.	Material	Labor	Equip.	Total	In-House	w/O&P
2900 Cooker, vegetable steamer, annually	.302	43	14.70		57.70	66.50	78
2950 Annualized	1.208	43	59		102	125	149

508

For customer support on your Facilities Maintenance & Repair Cost Data, call 877.801.5724.

PM Components	Labor-hrs.	W	M	Q	S	A
PM System E1095 316 3950						
Cooker, steam						
1 Check with operating or area personnel for any deficiencies.	.044			√	√	√
2 Examine doors and door gaskets; make necessary adjustments.	.021			√	√	√
3 Visually inspect hinges and latches; lubricate hinges.	.091			√	√	√
4 Tighten or replace any loose nuts, bolts, screws; replace broken knobs.	.005			√	√	√
5 Check for clogged or defective steam trap; clean trap.	.098			√	√	√
6 Inspect and clean steam strainer.	.207			√	√	√
7 Check working pressure on steam gauge.	.005			√	√	√
8 Turn unit on, allow pressure to build, inspect piping, valve and doors for leaks; adjust as required.	.077			√	√	
9 Check operation of pressure relief valve; adjust as required.	.010			√	√	√
10 Check operation of low water cut-off; adjust as required.	.030			√	√	√
11 Check water level sight glass; adjust as required.	.007			√	√	√
12 Check timer mechanism and operation; adjust as required.	.018			√	√	√
13 Fill out maintenance checklist and report deficiencies.	.022			√	√	√
Total labor-hours/period				.635	.635	.635
Total labor-hours/year				1.270	.635	.635

				Cost Each				
				2015 Bare Costs			Total	Total
Description		Labor-hrs.	Material	Labor	Equip.	Total	In-House	w/O&P
3900	Cooker, steam, annually	.635	85.50	31		116.50	135	157
3950	Annualized	2.540	85.50	124		209.50	256	310

PM Components	Labor-hrs.	W	M	Q	S	A
PM System E1095 318 1950						
Cookie maker, electric						
1 Check with operating or area personnel for any deficiencies.	.044			√	√	√
2 Check electrical panel switches, connections, timing device for loose, frayed wiring; tighten connections as required.	.120			√	√	√
3 Check operation of electric motor and bearings for overheating; lubricate motor bearings.	.086		√	√	√	√
4 Check belt; adjust tension or replace belt as required.	.369			√	√	√
5 Check conveyor and rollers by observing movement; lubricate rollers and adjust as required.	.127			√	√	√
6 Tighten or replace loose, missing or damaged nuts, bolts or screws.	.012			√	√	√
7 Wipe off and clean unit.	.096			√	√	√
8 Fill out maintenance checklist and report deficiencies.	.022			√	√	√
Total labor-hours/period				.876	.876	.876
Total labor-hours/year				1.752	.876	.876

				Cost Each				
				2015 Bare Costs			Total	Total
Description		Labor-hrs.	Material	Labor	Equip.	Total	In-House	w/O&P
1900	Cookie maker, electric, annually	.876	62.50	42.50		105	126	148
1950	Annualized	3.504	89.50	170		259.50	325	390

PM Components	Labor-hrs.	W	M	Q	S	A
PM System E1095 320 1950						
Deep fat fryer, pressurized broaster, gas/electric						
1 Check with operating or area personnel for any deficiencies.	.044			√	√	√
2 Check piping and valves for leaks; tighten fittings as required.	.125			√	√	√
3 Check pilot and flame on gas operated unit; adjust as required.	.009			√	√	√
4 Check elements, switches, controls, contacts, and wiring on electrically heated units; repair or adjust as required.	.025			√	√	√
5 Check thermostat; calibrate if necessary.	.439			√	√	√
6 Check basket/racks for bends, breaks or defects; straighten bends or repair as necessary.	.013			√	√	√
7 Check cooking oil filter on circulating system; change as required.	.068			√	√	√
8 Check operation of unit.	.164			√	√	√
9 Check operation of motor for excessive noise and overheating.	.022			√	√	√
10 Check nuts, bolts, and screws for tightness; tighten or replace as necessary.	.005			√	√	√
11 Fill out maintenance checklist and report deficiencies.	.022			√	√	√
Total labor-hours/period				.936	.936	.936
Total labor-hours/year				1.872	.936	.936

			Cost Each				
			2015 Bare Costs			Total	Total
Description	Labor-hrs.	Material	Labor	Equip.	Total	In-House	w/O&P
1900 Fryer, pressurized broaster, gas / electric, annually	.936	125	45.50		170.50	198	230
1950 Annualized	3.744	179	182		361	435	520

PM Components	Labor-hrs.	W	M	Q	S	A
PM System E1095 320 2950						
Deep fat fryer, automatic conveyor belt type, gas/electric						
1 Check with operating or area personnel for any deficiencies.	.044			√	√	√
2 Check compartments, valves and piping for leaks; tighten as required.	.077			√	√	√
3 Check pilot and flame on gas operated fryers; adjust as required.	.168			√	√	√
4 Check elements, switches, controls, contacts and wiring on electrically heated units for defects; repair or adjust as necessary.	.051			√	√	√
5 Check thermostats; calibrate as necessary.	.439			√	√	√
6 Check cooking oil filter on circulating system; change as required.	.068			√	√	√
7 Check operation of motor for excessive noise and overheating; lubricate bearings.	.022			√	√	√
8 Check operation of belt hoist; lubricate and adjust as required.	.052			√	√	√
9 Check nuts, bolts and screws for tightness; replace or tighten as necessary.	.017			√	√	√
10 Clean machine of dust, grease and food particles.	.044			√	√	√
11 Clean area around unit.	.066			√	√	√
12 Fill out maintenance checklist and report deficiencies.	.022			√	√	√
Total labor-hours/period				1.070	1.070	1.070
Total labor-hours/year				2.140	1.070	1.070

			Cost Each				
			2015 Bare Costs			Total	Total
Description	Labor-hrs.	Material	Labor	Equip.	Total	In-House	w/O&P
2900 Fryer, conveyor belt type, gas / electric, annually	1.070	107	52		159	187	218
2950 Annualized	4.280	107	208		315	390	475

PM Components	Labor-hrs.	W	M	Q	S	A
PM System E1095 320 3950						
Deep fat fryer, conventional type, gas/electric						
1 Check with operating or area personnel for any deficiencies.	.044			√	√	√
2 Clean machine thoroughly of dust, grease, and food particles.	.014			√	√	√
3 Check compartments, valves, and piping for leaks; tighten as required.	.077			√	√	√
4 Check pilot and flame on gas operated deep fat fryer; adjust as required.	.005			√	√	√
5 Check thermostat; calibrate if necessary.	.439			√	√	√
6 On electrically heated units check elements, switches, controls, contacts, and wiring for defects; repair or adjust as necessary.	.025			√	√	√
7 Check basket or rack for bends, breaks or defects, straighten bends or repair as necessary.	.013			√	√	√
8 Check nuts, bolts and screws for tightness; tighten or replace as necessary.	.005			√	√	√
9 Fill out maintenance checklist and report deficiencies.	.022			√	√	√
Total labor-hours/period				.644	.644	.644
Total labor-hours/year				1.288	.644	.644

			Cost Each					
			2015 Bare Costs				Total	Total
Description	Labor-hrs.	Material	Labor	Equip.	Total	In-House	w/O&P
3900 Fryer, conventional type, gas / electric, annually	.644	89.50	31.50		121	140	163
3950 Annualized	2.576	89.50	125		214.50	264	315

511

For customer support on your Facilities Maintenance & Repair Cost Data, call 877.801.5724.

PM Components	Labor-hrs.	W	M	Q	S	A
PM System E1095 322 1950						
Dishwasher, electric						
1 Check with operating or area personnel for any deficiencies.	.044			√	√	√
2 Check electric insulators, connections, and wiring (includes removing access panels).	.120			√	√	√
3 Check motor and bearings for excessive noise, vibration and overheating.	.039			√	√	√
4 Check dish conveyor by turning on switch and observing conveyor movement; adjust chain if necessary.	.070			√	√	√
5 Check operation of wash and rinse spray mechanism for spray coverage and drainage.	.022			√	√	√
6 Inspect soap and spray solution feeder lines; clean as necessary.	.013			√	√	√
7 Inspect water lines and fittings for leaks; tighten fittings as necessary.	.077			√	√	√
8 Lubricate conveyor drive bushings and chain drive.	.055			√	√	√
9 Check belt tension; adjust if required.	.100			√	√	√
10 Check doors for: operation of chains and counterweights, warping, alignment and water tightness; adjust if necessary.	.022			√	√	√
11 Check packing glands on wash, rinse, and drain valves; add or replace packing as required.	.147			√	√	√
12 Check lubricant in gear case; add oil if required.	.049			√	√	√
13 Inspect splash curtain for tears, clearance and water tightness; adjust if required.	.016			√	√	√
14 Check proper operation of solenoid valve and float in fill tank; adjust as required.	.096			√	√	√
15 Check pumps for leakage and obstructions; adjust as required.	.120			√	√	√
16 Check proper operation of micro-switch.	.010			√	√	√
17 Check water for proper temperature.	.004			√	√	√
18 Check operation of safety stop and clutch on conveyor belt motor; adjust cut-off switch mechanism as required.	.049			√	√	√
19 Check temperature regulator and adjust if necessary.	.017			√	√	√
20 Clean lime off thermostatic probe and heating elements.	.020			√	√	√
21 Clean area around equipment.	.066			√	√	√
22 Fill out maintenance checklist and report deficiencies.	.022			√	√	√
Total labor-hours/period				1.178	1.178	1.178
Total labor-hours/year				2.356	1.178	1.178

	Description	Labor-hrs.	Cost Each						
			2015 Bare Costs				Total	Total	
			Material	Labor	Equip.	Total	In-House	w/O&P	
1900	Dishwasher, electric, annually	1.178	116	57.50		173.50	204	238	
1950	Annualized	4.712	250	229		479	575	685	

PM Components	Labor-hrs.	W	M	Q	S	A
PM System E1095 322 2950						
Dishwasher, steam						
1 Check with operating or area personnel for any deficiencies.	.044			√	√	√
2 Check electric insulators, connections, and wiring (includes removing access panels).	.120			√	√	√
3 Check motor and bearings for excessive noise, vibration and overheating.	.039			√	√	√
4 Check dish conveyor by turning on switch and observing conveyor movement; adjust chain if necessary.	.070			√	√	√
5 Check operation of wash and rinse spray mechanism for spray coverage and drainage.	.022			√	√	√
6 Inspect soap and spray solution feeder lines; clean as necessary.	.013			√	√	√
7 Inspect steam and water lines and fittings for leaks; tighten fittings as necessary.	.077			√	√	√
8 Lubricate conveyor drive bushings and chain drive.	.055			√	√	√
9 Check belt tension; adjust if required.	.100			√	√	√
10 Check doors for: operation of chains and counterweights, warping, alignment and water tightness; adjust if necessary.	.022			√	√	√
11 Tighten loose nuts, bolts and screws.	.005			√	√	√
12 Check packing glands on wash, rinse, and drain valves; add or replace packing as required.	.147			√	√	√
13 Check lubricant in gear case; add oil if required.	.049			√	√	√
14 Inspect splash curtain for tears, clearance and water tightness; adjust if required.	.016			√	√	√
15 Check proper operation of solenoid valve and float in fill tank; adjust as required.	.096			√	√	√
16 Check pumps for leakage and obstructions; adjust as required.	.120			√	√	√
17 Check proper operation of micro-switch.	.010			√	√	√
18 Check water for proper temperature readings.	.004			√	√	√
19 Check operation of safety stop and clutch on conveyor belt motor; adjust cut-off switch mechanism as required.	.049			√	√	√
20 Check temperature regulator and adjust if necessary.	.017			√	√	√
21 Clean lime off thermostatic probe.	.020			√	√	√
22 Clean area around equipment.	.066			√	√	√
23 Fill out maintenance checklist and report deficienceis.	.022			√	√	√
Total labor-hours/period				1.183	1.183	1.183
Total labor-hours/year				2.366	1.183	1.183

	Description	Labor-hrs.	Cost Each				Total In-House	Total w/O&P
			2015 Bare Costs					
			Material	Labor	Equip.	Total		
2900	Dishwasher, steam, annually	1.183	116	57.50		173.50	204	239
2950	Annualized	4.732	250	230		480	580	690

PM Components	Labor-hrs.	W	M	Q	S	A
PM System E1095 328 1950						
Dispenser, carbonated beverage						
1 Check with operating or area personnel for any deficiencies.	.044			√	√	√
2 Check all lines, connections and valves for leaks.	.077			√	√	√
3 Check pressure gauges; adjust as required.	.018			√	√	√
4 Check water strainer screen on carbonator; clean as required.	.159			√	√	√
5 Check wiring, connectors and switches; repair as required.	.120			√	√	√
6 Check nuts, bolts and screws for tightness; replace or tighten as required.	.005			√	√	√
7 Wipe and clean unit.	.056			√	√	√
8 Fill out maintenance checklist and report deficiencies.	.022			√	√	√
Total labor-hours/period				.501	.501	.501
Total labor-hours/year				1.002	.501	.501

			Cost Each					
			2015 Bare Costs				**Total**	**Total**
Description		**Labor-hrs.**	**Material**	**Labor**	**Equip.**	**Total**	**In-House**	**w/O&P**
1900	Dispenser, carbonated beverage, annually	.501	71.50	24.50		96	111	129
1950	Annualized	2.004	71.50	97.50		169	207	247

PM Components	Labor-hrs.	W	M	Q	S	A
PM System E1095 332 1950						
Disposal, garbage, electric						
1 Check with operating or area personnel for any deficiencies.	.044			√	√	√
2 Check motor and drive shaft for excessive noise, vibration, overheating, etc.	.044			√	√	√
3 Inspect electrical wiring.	.014			√	√	√
4 Visually examine grinder; check for obstructions; adjust cutters if required.	.074			√	√	√
5 Check drive belt(s) for wear and tension; adjust if required.	.014			√	√	√
6 Check for leaks to supply and drain connections; tighten if required.	.103			√	√	√
7 Check electrical switch for proper operation.	.014			√	√	√
8 Lubricate grinder, drive, and motor.	.022			√	√	√
9 Check water sprayer operation.	.030			√	√	√
10 Fill out maintenance checklist and report deficiencies.	.022			√	√	√
Total labor-hours/period				.381	.381	.381
Total labor-hours/year				.762	.381	.381

			Cost Each					
			2015 Bare Costs				**Total**	**Total**
Description		**Labor-hrs.**	**Material**	**Labor**	**Equip.**	**Total**	**In-House**	**w/O&P**
1900	Disposal, garbage, electric, annually	.381	43	18.55		61.55	71.50	84
1950	Annualized	1.524	43	74		117	145	174

PM Components	Labor-hrs.	W	M	Q	S	A
PM System E1095 334 1950						
Dough divider						
1 Check with operating or area personnel for any deficiencies.	.044			√	√	√
2 Check motor and drive shaft for excessive noise, vibration, overheating, etc.	.039			√	√	√
3 Inspect electrical wiring.	.120			√	√	√
4 Visually examine plates; check for obstructions; adjust cutters if required.	.074			√	√	√
5 Check electrical switch for proper operation.	.014			√	√	√
6 Lubricate press, drive, and motor.	.218			√	√	√
7 Fill out maintenance checklist and report deficiencies.	.022			√	√	√
Total labor-hours/period				.531	.531	.531
Total labor-hours/year				1.062	.531	.531

			Cost Each					
			2015 Bare Costs				Total	Total
Description	Labor-hrs.	Material	Labor	Equip.	Total	In-House	w/O&P
1900 Dough divider, annually	.531	89.50	26		115.50	132	154
1950 Annualized	2.124	89.50	103		192.50	235	280

PM Components	Labor-hrs.	W	M	Q	S	A
PM System E1095 334 2950						
Dough roller						
1 Check with operating or area personnel for any deficiencies.	.044			√	√	√
2 Check electric motor, switches, controls, wiring, connections and insulation for defective materials; repair as necessary.	.241			√	√	√
3 Check belt tension; adjust or replace as required.	.446			√	√	√
4 Check operation of conveyor-roller clearance; adjust as necessary.	.109			√	√	√
5 Lubricate all moving parts, motors, pivot points and chain drives.	.017			√	√	√
6 Clean unit and treat for rust and/or corrosion.	.083			√	√	√
7 Tighten or replace loose, missing or damaged nuts, bolts and screws.	.005			√	√	√
8 Fill out maintenance checklist and report deficiencies.	.022			√	√	√
Total labor-hours/period				.967	.967	.967
Total labor-hours/year				1.934	.967	.967

			Cost Each					
			2015 Bare Costs				Total	Total
Description	Labor-hrs.	Material	Labor	Equip.	Total	In-House	w/O&P
2900 Dough roller, annually	.967	134	47		181	209	244
2950 Annualized	3.868	161	188		349	425	505

PM Components	Labor-hrs.	W	M	Q	S	A
PM System E1095 336 1950						
Drink cooler, with external condenser						
1　Check with operating or area personnel for deficiencies.	.035				√	√
2　Clean condenser coils, fans, and intake screens; lubricate motor.	.504				√	√
3　Inspect door gaskets for damage and proper fit; adjust as necessary.	.033				√	√
4　Check starter panels and controls for proper operation, burned or loose contacts, and loose connections.	.094				√	√
5　Clean coils, evaporator drain pan, blowers, fans, motors and drain piping as required; lubricate motor(s).	.473				√	√
6　During operation of unit, check refrigerant pressures and compressor oil level; add refrigerant and/or oil as necessary.	.066				√	√
7　Check operation of low pressure cut-out; adjust or replace as necessary.	.057				√	√
8　Inspect defrost systems for proper operation; adjust as required.	.027				√	√
9　Clean area around equipment.	.066				√	√
10　Fill out maintenance checklist and report deficiencies.	.022				√	√
Total labor-hours/period					1.377	1.377
Total labor-hours/year					1.377	1.377

			Cost Each					
			2015 Bare Costs				Total	Total
	Description	Labor-hrs.	Material	Labor	Equip.	Total	In-House	w/O&P
1900	Drink cooler with external condenser, annually	1.377	17.85	67		84.85	108	131
1950	Annualized	2.754	35.50	134		169.50	215	262

PM Components	Labor-hrs.	W	M	Q	S	A
PM System E1095 338 1950						
Fluid cooler, air cooled condenser						
1　Check with operating or area personnel for deficiencies.	.035					√
2　Check unit for proper operation, excessive noise or vibration.	.159					√
3　Clean intake side of condenser coils, fans and intake screens.	.473					√
4　Check electrical wiring and connections; tighten loose connections.	.120					√
5　Inspect fan(s) for bent blades or unbalance; adjust as necessary.	.014					√
6　Check belts for condition, proper tension and misalignment, if required.	.029					√
7　Lubricate shaft bearings and motor bearings.	.047					√
8　Inspect piping and valves for leaks; tighten connections as necessary.	.077					√
9　Lubricate and check operation of dampers, if applicable.	.055					√
10　Clean area around fluid cooler.	.066					√
11　Fill out maintenance checklist and report deficiencies.	.022					√
Total labor-hours/period						1.097
Total labor-hours/year						1.097

			Cost Each					
			2015 Bare Costs				Total	Total
	Description	Labor-hrs.	Material	Labor	Equip.	Total	In-House	w/O&P
1900	Fluid cooler, annually	1.097	17.85	53.50		71.35	90	109
1950	Annualized	1.097	17.85	53.50		71.35	90	109

For customer support on your Facilities Maintenance & Repair Cost Data, call 877.801.5724.

PM Components	Labor-hrs.	W	M	Q	S	A
PM System E1095 340 1950						
Food saw, electric						
1 Check with operating or area personnel for any deficiencies.	.044			√	√	√
2 Check operation of saw for excessive vibration, blade tension and proper adjustment of safety guards; repair or adjust as required.	.095			√	√	√
3 Check motor and bearings for overheating; lubricate motor bearings.	.039			√	√	√
4 Check switches, motor, controls and wiring for damage and proper operation; repair as required.	.120			√	√	√
5 Clean machine of dust and grease.	.044			√	√	√
6 Check nuts, bolts and screws for tightness; tighten as required.	.005			√	√	√
7 Check belt tension and alignment; adjust as required.	.029			√	√	√
8 Fill out maintenance checklist and report deficiencies.	.022			√	√	√
Total labor-hours/period				.398	.398	.398
Total labor-hours/year				.796	.398	.398

			Cost Each					
			2015 Bare Costs				Total	Total
Description		Labor-hrs.	Material	Labor	Equip.	Total	In-House	w/O&P
1900	Food saw, electric, annually	.398	62.50	19.35		81.85	94.50	110
1950	Annualized	1.592	89.50	77.50		167	201	238

PM Components	Labor-hrs.	W	M	Q	S	A
PM System E1095 340 2950						
Food slicer, electric						
1 Check with operating or area personnel for any deficiencies.	.044			√	√	√
2 Check nuts, bolts and screws for tightness; tighten or replace as required.	.005			√	√	√
3 Lubricate food slicer and add lubricant to gear case if required.	.085			√	√	√
4 Check condition of blade, blade guard, guides, and controls; adjust as required.	.047			√	√	√
5 Check operation of motor and check motor bearings for overheating; lubricate motor bearings.	.039			√	√	√
6 Check sharpening stones; install new stones as required.	.013			√	√	√
7 Clean machine of dust, grease, and food particles.	.014			√	√	√
8 Fill out maintenance checklist and report deficiencies.	.022			√	√	√
Total labor-hours/period				.269	.269	.269
Total labor-hours/year				.538	.269	.269

			Cost Each					
			2015 Bare Costs				Total	Total
Description		Labor-hrs.	Material	Labor	Equip.	Total	In-House	w/O&P
2900	Food slicer, electric, annually	.269	80.50	13.10		93.60	106	122
2950	Annualized	1.076	107	52.50		159.50	188	219

PM Components	Labor-hrs.	W	M	Q	S	A
PM System E1095 346 1950						
Grill, gas/electric						
1 Check with operating or area personnel for any deficiencies.	.044			√	√	√
2 Check nuts, bolts and screws for tightness; tighten or replace as required.	.005			√	√	√
3 On gas operated units, check piping and valves for leaks.	.077			√	√	√
4 On gas operated units, check pilot and gas burners for uniform flame; adjust as required.	.079			√	√	√
5 On electrically operated units, check switches, connections, and wiring for proper operation; adjust as required.	.135			√	√	√
6 Check calibration of thermostats; adjust as required.	.458			√	√	√
7 Fill out maintenance checklist and report deficiencies.	.022			√	√	√
Total labor-hours/period				.820	.820	.820
Total labor-hours/year				1.640	.820	.820

			Cost Each					
			2015 Bare Costs				Total	Total
Description	Labor-hrs.	Material	Labor	Equip.	Total	In-House	w/O&P
1900 Grill, gas / electric, annually	.820	89.50	40		129.50	151	177
1950 Annualized	3.280	89.50	160		249.50	310	370

PM Components	Labor-hrs.	W	M	Q	S	A
PM System E1095 348 1950						
Ice machine, flake or cube						
1 Check with operating or area personnel for any deficiencies.	.044			√	√	√
2 Remove and install access panel.	.044			√	√	√
3 Lubricate all moving parts, pivot points and fan motor(s).	.023			√	√	√
4 Visually check for refrigerant, oil or water leaks.	.022			√	√	√
5 Open and close water valve.	.007			√	√	√
6 Replace in-line water filter.	.009			√	√	√
7 Check and clear ice machine draining system (drain vent and trap).	.198			√	√	√
8 Clean motor, compressor and condenser coil.	.403			√	√	√
9 Check and tighten any loose screw-type electrical connections.	.005			√	√	√
10 Inspect door(s) hinges, gaskets, handles; lubricate as required.	.053			√	√	√
11 Clean area around equipment.	.066			√	√	√
12 Fill out maintenance checklist and report deficiencies.	.022			√	√	√
Total labor-hours/period				.896	.896	.896
Total labor-hours/year				1.792	.896	.896

			Cost Each					
			2015 Bare Costs				Total	Total
Description	Labor-hrs.	Material	Labor	Equip.	Total	In-House	w/O&P
1900 Ice machine, flake or cube, annually	.896	27	43.50		70.50	87	105
1950 Annualized	3.584	107	174		281	345	415

For customer support on your Facilities Maintenance & Repair Cost Data, call 877.801.5724.

PM Components	Labor-hrs.	W	M	Q	S	A
PM System E1095 350 1950						
Kettle, steam, fixed or tilt						
1 Check with operating or area personnel for any deficiencies.	.044			√	√	√
2 Check piping and fittings for leaks; tighten as required.	.085			√	√	√
3 Check operation of electric water valve.	.030			√	√	√
4 Inspect cover, hinges, and seals on units to equipped; lubricate hinge.	.053			√	√	√
5 Lubricate tilting gear mechanism and trunnion bearings, if applicable.	.047			√	√	√
6 Fill out maintenance checklist and report deficiencies.	.022			√	√	√
Total labor-hours/period				.281	.281	.281
Total labor-hours/year				.562	.281	.281

			Cost Each					
			2015 Bare Costs				Total	Total
Description	Labor-hrs.	Material	Labor	Equip.	Total	In-House	w/O&P
1900	Kettle, steam, fixed or tilt, annually	.281	44.50	13.65		58.15	67	77.50
1950	Annualized	1.124	44.50	54.50		99	121	145

For customer support on your Facilities Maintenance & Repair Cost Data, call 877.801.5724.

519

PM Components	Labor-hrs.	W	M	Q	S	A
PM System E1095 354 1950						
Mixer, counter, electric						
1 Check with operating or area personnel for any deficiencies.	.044			√	√	√
2 Check operation of mixer at varying speeds for excessive noise and vibrations; align or adjust as required.	.033			√	√	√
3 Check belt for proper tension and alignment; adjust as required.	.068			√	√	√
4 Lubricate mixer gears.	.042			√	√	√
5 Clean machine thoroughly of dust, grease, and food particles.	.014			√	√	√
6 Check motor, switches, controls and motor bearings for overheating; lubricate motor bearings.	.039			√	√	√
7 Check anchor bolts for tightness; tighten as required.	.010			√	√	√
8 Fill out maintenance checklist and report deficiencies.	.022			√	√	√
Total labor-hours/period				.272	.272	.272
Total labor-hours/year				.544	.272	.272

		Cost Each						
			2015 Bare Costs				Total	Total
Description	Labor-hrs.	Material	Labor	Equip.	Total		In-House	w/O&P
1900 Mixer, counter, electric, annually	.272	44.50	13.25		57.75		66.50	77
1950 Annualized	1.088	71.50	53		124.50		148	175

PM Components	Labor-hrs.	W	M	Q	S	A
PM System E1095 354 2950						
Mixer, floor, electric						
1 Check with operating or area personnel for any deficiencies.	.044			√	√	√
2 Check operation of mixer at varying speeds for excessive noise and vibrations; align or adjust as required.	.120			√	√	√
3 Tighten loose bolts, nuts, and screws.	.005			√	√	√
4 Check oil level in transmission; add or replace oil if required.	.159			√	√	√
5 Lubricate mixer gears and bowl lift mechanism.	.042			√	√	√
6 Check switches and control for damage and proper operation; adjust as required.	.068			√	√	√
7 Check belt for proper tension and alignment; adjust as required.	.029			√	√	√
8 Clean machine of dust and grease.	.030			√	√	√
9 Check anchor bolts for tightness; tighten as required.	.010			√	√	√
10 Clean area around equipment.	.066			√	√	√
11 Fill out maintenance checklist and report deficiencies.	.022			√	√	√
Total labor-hours/period				.595	.595	.595
Total labor-hours/year				1.190	.595	.595

		Cost Each						
			2015 Bare Costs				Total	Total
Description	Labor-hrs.	Material	Labor	Equip.	Total		In-House	w/O&P
2900 Mixer, floor, electric, annually	.595	80.50	29		109.50		127	148
2950 Annualized	2.380	107	116		223		270	320

PM Components	Labor-hrs.	W	M	Q	S	A
PM System E1095 356 1950						
Oven, convection, gas/electric						
1 Check with operating or area personnel for any deficiencies.	.044		√	√	√	√
2 Check doors and seals for warping and misalignment; lubricate hinges and repair as necessary.	.027		√	√	√	√
3 Check piping and valves for leaks.	.077		√	√	√	√
4 Check nuts, bolts, and screws for tightness; replace or tighten as required.	.005		√	√	√	√
5 Check pilot and gas burner for uniform flame; adjust as required.	.083		√	√	√	√
6 Check element, switches, controls and wiring on electrically heated units for defects; repair as required.	.120		√	√	√	√
7 Check fan blades and fan motor for proper operation.	.120		√	√	√	√
8 Check operation of thermostat; calibrate as required.	.439		√	√	√	√
9 Fill out maintenance checklist and report deficiencies.	.022		√	√	√	√
Total labor-hours/period			.937	.937	.937	.937
Total labor-hours/year			7.496	1.874	.937	.937

			Cost Each				
			2015 Bare Costs			Total	Total
Description	Labor-hrs.	Material	Labor	Equip.	Total	In-House	w/O&P
1900 Oven, convection, gas / electric, annually	.937	89.50	45.50		135	158	186
1950 Annualized	11.244	89.50	545		634.50	815	1,000

PM Components	Labor-hrs.	W	M	Q	S	A
PM System E1095 356 2950						
Oven, rotary, electric						
1 Check with operating or area personnel for any deficiencies.	.044		√	√	√	√
2 Remove and reinstall access panel; lubricate main bearing bushings.	.064		√	√	√	√
3 Turn motor on and check operation of chain drive mechanism; lubricate chain.	.064		√	√	√	√
4 Inspect motor and wiring; sensory inspect motor and bearings for overheating; lubricate motor bearings.	.055		√	√	√	√
5 Check V-belt; adjust tension and/or pulley as required.	.029		√	√	√	√
6 Inspect shelves for defects and level if required.	.142		√	√	√	√
7 Tighten or replace any loose nuts, bolts, screws.	.005		√	√	√	√
8 Check thermostat with thermometer; adjust as required.	.439		√	√	√	√
9 Check operation of stop and reverse switch.	.008		√	√	√	√
10 Check timer mechanism and operation.	.018		√	√	√	√
11 Check lubricant in gear case; add as required.	.055		√	√	√	√
12 Check ventilator for proper operation.	.160		√	√	√	√
13 Clean area around equipment.	.066		√	√	√	√
14 Fill out maintenance checklist and report deficiencies.	.022		√	√	√	√
Total labor-hours/period			1.171	1.171	1.171	1.171
Total labor-hours/year			9.368	2.342	1.171	1.171

			Cost Each				
			2015 Bare Costs			Total	Total
Description	Labor-hrs.	Material	Labor	Equip.	Total	In-House	w/O&P
2900 Oven, rotary, electric, annually	1.171	295	57		352	400	460
2950 Annualized	14.052	395	685		1,080	1,325	1,600

PM Components	Labor-hrs.	W	M	Q	S	A
PM System E1095 356 3950						
Oven, rotary, gas						
1 Check with operating or area personnel for any deficiencies.	.044		√	√	√	√
2 Examine doors and door gaskets; make necessary adjustments.	.021		√	√	√	√
3 Visually inspect hinges and latches; lubricate hinges.	.091		√	√	√	√
4 Remove and reinstall access panel; lubricate main bearing bushings.	.025		√	√	√	√
5 Turn motor on and chk. operation of chain drive mechanism; lubricate chain.	.064		√	√	√	√
6 Inspect motor and wiring; sensory inspect motor and bearings for overheating; lubricate motor bearings.	.055		√	√	√	√
7 Check V-belt; adjust tension and/or pulley as required.	.029		√	√	√	√
8 Check pilot and thermosensor bulb.	.042			√	√	√
9 Check gas burner for uniform flame.	.042			√	√	√
10 Check piping and valves for leaks.	.077			√	√	√
11 Inspect shelves for defects and level if required.	.142			√	√	√
12 Tighten or replace any loose nuts, bolts, screws.	.005			√	√	√
13 Check thermostat with thermometer; calibrate if required.	.439			√	√	√
14 Check operation of stop and reverse switch.	.008			√	√	√
15 Check timer mechanism and operation.	.018			√	√	√
16 Check lubricant in gear case, add as required.	.055			√	√	√
17 Check ventilator for proper operation.	.160			√	√	√
18 Clean area around equipment.	.066		√	√	√	√
19 Fill out maintenance checklist and report deficiencies.	.022		√	√	√	√
Total labor-hours/period			1.405	1.405	1.405	1.405
Total labor-hours/year			11.240	2.810	1.405	1.405

			Cost Each				
			2015 Bare Costs			**Total**	**Total**
Description	**Labor-hrs.**	**Material**	**Labor**	**Equip.**	**Total**	**In-House**	**w/O&P**
3900 Oven, rotary, gas, annually	1.405	295	68.50		363.50	415	480
3950 Annualized	16.860	395	820		1,215	1,525	1,825

PM Components	Labor-hrs.	W	M	Q	S	A
PM System E1095 366 1950						
Peeler, vegetable, electric						
1 Check with operating or area personnel for any deficiencies.	.044			√	√	√
2 Check vegetable peeler for proper operation, including switches and controls; lubricate motor bearings.	.156			√	√	√
3 Lubricate vegetable peeler bushings, as required.	.062			√	√	√
4 Check door for loose hinges and latch fittings; adjust if required.	.007			√	√	√
5 Clean peeler of dust, grease, and food particles.	.014			√	√	√
6 Check piping and valves for leaks.	.077			√	√	√
7 Inspect abrasive disk.	.018			√	√	√
8 Tighten or replace loose, missing or damaged nuts, bolts, screws.	.005			√	√	√
9 Check belt for proper tension and alignment; adjust or replace as required.	.029			√	√	√
10 Fill out maintenance checklist and report deficiencies.	.022			√	√	√
Total labor-hours/period				.434	.434	.434
Total labor-hours/year				.868	.434	.434

			Cost Each				
			2015 Bare Costs			**Total**	**Total**
Description	**Labor-hrs.**	**Material**	**Labor**	**Equip.**	**Total**	**In-House**	**w/O&P**
1900 Peeler, vegetable, electric, annually	.434	44.50	21		65.50	76.50	90
1950 Annualized	1.736	71.50	84.50		156	190	226

E1095 368 | Pie Maker, Electric

PM Components	Labor-hrs.	W	M	Q	S	A
PM System E1095 368 1950						
Pie maker, electric						
1 Check with operating or area personnel for any deficiencies.	.044			√	√	√
2 Check electrical system for loose connections and frayed wiring; tighten connections as required.	.120			√	√	√
3 Check operation of motors and bearings for overheating; lubricate motor bearings.	.169			√	√	√
4 Check V-belt for alignment and proper tension; adjust or replace belt as required.	.345			√	√	√
5 Check drain drive and sprocket for excessive wear; lubricate if applicable.	.087			√	√	√
6 Tighten or replace loose, missing or damaged nuts, bolts or screws.	.005			√	√	√
7 Wipe clean the machine.	.096			√	√	√
8 Fill out maintenance checklist and report deficiencies.	.022			√	√	√
Total labor-hours/period				.888	.888	.888
Total labor-hours/year				1.776	.888	.888

			Cost Each				
		2015 Bare Costs				**Total**	**Total**
Description	**Labor-hrs.**	**Material**	**Labor**	**Equip.**	**Total**	**In-House**	**w/O&P**
1900 Pie maker, electric, annually	.888	98	43		141	165	193
1950 Annualized	3.552	125	173		298	365	435

E1095 370 | Proofer, Automatic

PM Components	Labor-hrs.	W	M	Q	S	A
PM System E1095 370 1950						
Proofer, automatic, electric						
1 Check with operating or area personnel for any deficiencies.	.044			√	√	√
2 Check operation of electric motor, switches, controls and bearings; adjust, repair and lubricate as required.	.280			√	√	√
3 Check belt tension; adjust as required.	.029			√	√	√
4 Check chain drive operation for proper alignment and tension; repair as required.	.129			√	√	√
5 Check action of flour sifter mechanism, connecting linkage and conveyor; lubricate bushings as required.	.290			√	√	√
6 Check roller conveyor and connecting drive mechanism.	.113			√	√	√
7 Check door latch; adjust if needed.	.014			√	√	√
8 Check nuts, bolts and screws for tightness; replace or tighten as required.	.029			√	√	√
9 Fill out maintenance checklist and report deficiencies.	.022			√	√	√
Total labor-hours/period				.950	.950	.950
Total labor-hours/year				1.900	.950	.950

			Cost Each				
		2015 Bare Costs				**Total**	**Total**
Description	**Labor-hrs.**	**Material**	**Labor**	**Equip.**	**Total**	**In-House**	**w/O&P**
1900 Proofer, automatic, electric, annually	.950	143	46		189	218	254
1950 Annualized	3.800	170	185		355	430	515

For customer support on your Facilities Maintenance & Repair Cost Data, call 877.801.5724.

523

PM Components	Labor-hrs.	W	M	Q	S	A
PM System E1095 380 1950						
Refrigerator, domestic						
1 Check with operating or area personnel for deficiencies.	.035					✓
2 Clean coils, fans, fan motors, drip pan and other areas with vacuum, brush or wiping as necessary.	.066					✓
3 Inspect door gaskets for damage and proper fit; adjust gaskets as required and lubricate hinges.	.022					✓
4 Check door latch and adjust as necessary.	.023					✓
5 Clean area around equipment.	.066					✓
6 Fill out maintenance checklist and report deficiencies.	.022					✓
Total labor-hours/period						.234
Total labor-hours/year						.234

			Cost Each					
			2015 Bare Costs				**Total**	**Total**
Description		**Labor-hrs.**	**Material**	**Labor**	**Equip.**	**Total**	**In-House**	**w/O&P**
1900	Refrigerator, domestic, annually	.234	17.85	11.40		29.25	34.50	41
1950	Annualized	.234	17.85	11.40		29.25	34.50	41

PM Components	Labor-hrs.	W	M	Q	S	A
PM System E1095 380 2950						
Refrigerator, display case						
1 Check with operating or area personnel for deficiencies.	.035					✓
2 Clean coils, fans, fan motors, drip pan and other areas with vacuum, brush or wiping as necessary.	.066					✓
3 Inspect door gaskets for damage and proper fit; adjust gaskets as required and lubricate hinges.	.044					✓
4 Check door latch and adjust as necessary.	.047					✓
5 Clean area around equipment.	.066					✓
6 Fill out maintenance checklist and report deficiencies.	.022					✓
Total labor-hours/period						.280
Total labor-hours/year						.280

			Cost Each					
			2015 Bare Costs				**Total**	**Total**
Description		**Labor-hrs.**	**Material**	**Labor**	**Equip.**	**Total**	**In-House**	**w/O&P**
2900	Refrigerator, display case, annually	.280	17.85	13.60		31.45	37.50	44.50
2950	Annualized	.280	17.85	13.60		31.45	37.50	44.50

PM Components	Labor-hrs.	W	M	Q	S	A
PM System E1095 380 3950						
Refrigerator/display, walk-in w/external condenser						
1 Check with operating or area personnel for deficiencies.	.035				√	√
2 Clean condenser coils, fans, and intake screens; lubricate motor.	.066				√	√
3 Inspect door gaskets for damage and proper fit; adjust gaskets as required and lubricate hinges.	.137				√	√
4 Check starter panels and controls for proper operation, burned or loose contacts, and loose connections.	.182				√	√
5 Clean coils, evaporator drain pan, blowers, fans, motors and drain piping as required; lubricate motor(s).	.473				√	√
6 During operation of unit, check refrigerant pressures and compressor oil level; add refrigerant and/or oil as necessary.	.066				√	√
7 Check operation of low pressure cut-out; adjust or replace as required.	.047				√	√
8 Inspect defrost systems for proper operation; adjust as required.	.094				√	√
9 Clean area around equipment.	.066				√	√
10 Fill out maintenance checklist and report deficiencies.	.022				√	√
Total labor-hours/period					1.188	1.188
Total labor-hours/year					1.188	1.188

	Description	Labor-hrs.	Cost Each				Total In-House	Total w/O&P
			2015 Bare Costs					
			Material	Labor	Equip.	Total		
3900	Refrig. display, walk-in w/ ext. condenser, annually	1.188	44.50	58		102.50	125	150
3950	Annualized	2.376	53.50	116		169.50	211	255

E10 EQUIPMENT | E1095 382 | Refrigerator/Freezer

PM Components	Labor-hrs.	W	M	Q	S	A
PM System E1095 382 1950						
Refrigerator/freezer, walk-in box w/external condenser						
1 Check with operating or area personnel for deficiencies.	.035				√	√
2 Clean condenser coils, fans, and intake screens; lubricate motor.	.036				√	√
3 Inspect door gaskets for damage and proper fit; adjust gaskets as required and lubricate hinges.	.022				√	√
4 Check starter panels and controls for proper operation, burned or loose contacts, and loose connections.	.182				√	√
5 Clean coils, evaporator drain pan, blowers, fans, motors and drain piping as required; lubricate motor(s).	.114				√	√
6 During operation of unit, check refrigerant pressures and compressor oil level; add refrigerant and/or oil as necessary.	.083				√	√
7 Check operation of low pressure cut-out; adjust or replace as required.	.079				√	√
8 Inspect defrost systems for proper operation; adjust as required.	.094				√	√
9 Clean area around equipment.	.066				√	√
10 Fill out maintenance checklist and report deficiencies.	.022				√	√
Total labor-hours/period					.733	.733
Total labor-hours/year					.733	.733

		Cost Each						
			2015 Bare Costs				Total	Total
Description	Labor-hrs.	Material	Labor	Equip.	Total	In-House	w/O&P
1900	Refrig. freezer, walk-in box w/ext. condenser, annually	.733	44.50	35.50		80	96	114
1950	Annualized	1.466	53.50	71.50		125	153	183

PM Components	Labor-hrs.	W	M	Q	S	A
PM System E1095 382 2950						
Refrigerator unit/display case/freezer w/external condenser						
1 Check with operating or area personnel for deficiencies.	.035				√	√
2 Clean condenser coils, fans, and intake screens; lubricate motor.	.504				√	√
3 Inspect door gaskets for damage and proper fit; adjust gaskets as required and lubricate hinges.	.033				√	√
4 Check starter panels and controls for proper operation, burned or loose contacts, and loose connections.	.213				√	√
5 Clean coils, evaporator drain pan, blowers, fans, motors and drain piping as required; lubricate motor(s).	.473				√	√
6 During operation of unit, check refrigerant pressures and compressor oil level; add refrigerant and/or oil as necessary.	.066				√	√
7 Check operation of low pressure cut-out; adjust or replace as required.	.057				√	√
8 Inspect defrost systems for proper operation; adjust as required.	.027				√	√
9 Clean area around equipment.	.066				√	√
10 Fill out maintenance checklist and report deficiencies.	.022				√	√
Total labor-hours/period					1.496	1.496
Total labor-hours/year					1.496	1.496

		Cost Each						
			2015 Bare Costs				Total	Total
Description	Labor-hrs.	Material	Labor	Equip.	Total	In-House	w/O&P
2900	Refrig., display case, freezer w/ ext. cond., annually	1.496	17.85	73		90.85	115	140
2950	Annualized	2.992	35.50	146		181.50	230	281

For customer support on your Facilities Maintenance & Repair Cost Data, call 877.801.5724.

PM Components	Labor-hrs.	W	M	Q	S	A
PM System E1095 386 1950						
Steam table						
1 Check with operating or area personnel for any deficiencies.	.044			√	√	√
2 Inspect water compartment, steam coil, valves, and piping for leaks.	.117			√	√	√
3 Check steam trap and strainer; clean as required.	.306			√	√	√
4 Check operation of pressure regulating valve and gauge.	.035			√	√	√
5 Check pilots and flame on gas burner units; adjust as required.	.085			√	√	√
6 Check insulators, connections, and wiring, if applicable; tighten connections.	.120			√	√	√
7 Check condition of covers and receptacles; adjust as required.	.025			√	√	√
8 Check thermostat and temperature gauge; calibrate thermostat if necessary.	.152			√	√	√
9 Fill out maintenance checklist and report deficiencies.	.022			√	√	√
Total labor-hours/period				.906	.906	.906
Total labor-hours/year				1.812	.906	.906

				Cost Each				
				2015 Bare Costs			Total	Total
Description		Labor-hrs.	Material	Labor	Equip.	Total	In-House	w/O&P
1900	Steam table, annually	.906	89.50	44		133.50	156	183
1950	Annualized	3.624	89.50	176		265.50	330	400

PM Components	Labor-hrs.	W	M	Q	S	A
PM System E1095 388 1950						
Steamer, vegetable, direct connected units						
1 Check with operating or area personnel for any deficiencies.	.044			√	√	√
2 Inspect doors, door hardware and gaskets; lubricate hinges.	.112			√	√	√
3 Tighten or replace loose, missing, or damaged nuts, bolts, screws.	.005			√	√	√
4 Check for clogged or defective steam and strainers; clean as required.	.304			√	√	√
5 Check working pressure on steam gauge and inspect piping, valves, and doors for leaks; make necessary repairs as required.	.082			√	√	√
6 Check operation of pressure relief valve, low water cut-off, timer operation and water level in sight glass.	.065			√	√	√
7 Fill out maintenance checklist and report deficiencies.	.022			√	√	√
Total labor-hours/period				.634	.634	.634
Total labor-hours/year				1.268	.634	.634

				Cost Each				
				2015 Bare Costs			Total	Total
Description		Labor-hrs.	Material	Labor	Equip.	Total	In-House	w/O&P
1900	Steamer, vegetable, annually	.634	89.50	31		120.50	139	162
1950	Annualized	2.536	89.50	123		212.50	262	310

PM Components	Labor-hrs.	W	M	Q	S	A
PM System E1095 390 1950						
Toaster, rotary, gas/electric						
1 Check with operating or area personnel for any deficiencies.	.044			√	√	√
2 Check operation of toaster by toasting sample slice of bread.	.034			√	√	√
3 Clean toaster.	.014			√	√	√
4 Check valves and piping for leaks on gas units.	.077			√	√	√
5 Check pilot light and burner flame on gas units; adjust when required.	.017			√	√	√
6 Check insulators, heating elements, connections, and wiring on electric units; tighten connections and repair as necessary.	.134			√	√	√
7 Inspect alignment of baskets and conveyor chains; lubricate chains and align as necessary.	.044			√	√	√
8 Check motor and motor bearings for overheating; lubricate motor bearings.	.086			√	√	√
9 Tighten or replace loose, missing or damaged nuts, bolts, screws.	.005			√	√	√
10 Fill out maintenance checklist and report deficiencies.	.022			√	√	√
Total labor-hours/period				.477	.477	.477
Total labor-hours/year				.954	.477	.477

			Cost Each					
			2015 Bare Costs				**Total**	**Total**
Description	**Labor-hrs.**	**Material**	**Labor**	**Equip.**	**Total**	**In-House**	**w/O&P**
1900	Toaster, rotary, gas / electric, annually	.477	44.50	23		67.50	79.50	93
1950	Annualized	1.908	71.50	93		164.50	201	240

PM Components	Labor-hrs.	W	M	Q	S	A
PM System E1095 930 1950						
Vacuum						
1 Check with operating or area personnel for deficiencies.	.035				√	√
2 Check for leaks on suction and discharge piping, seals, etc.	.077				√	√
3 Check pump and motor operation for excessive vibration, noise and overheating.	.022				√	√
4 Check alignment and clearances of shaft and coupler.	.260				√	√
5 Tighten or replace loose, missing, or damaged nuts, bolts and screws.	.005				√	√
6 Lubricate pump and motor as required.	.099				√	√
7 Clean pump, motor and surrounding area.	.096				√	√
8 Fill out maintenance checklist and report deficiencies.	.022				√	√
Total labor-hours/period					.616	.616
Total labor-hours/year					.616	.616

			Cost Each					
			2015 Bare Costs				**Total**	**Total**
Description	**Labor-hrs.**	**Material**	**Labor**	**Equip.**	**Total**	**In-House**	**w/O&P**
1900	Vacuum, annually	.616	8.55	36		44.55	54.50	67
1950	Annualized	1.232	17.05	72.50		89.55	109	134

528

PM Components	Labor-hrs.	W	M	Q	S	A
PM System F1045 110 1950						
Swimming pool						
Quantities of chemicals used vary significantly by climate and pool usage, therefore the price of these materials are not provided.						
1 Clean strainer basket.	.050	√	√	√	√	√
2 Backwash pool water filter.	.120	√	√	√	√	√
3 Add soda ash to feeder.	.080	√	√	√	√	√
4 Clean/flush soda ash pump head.	.400		√	√	√	√
5 Check chlorine, change bottles and check for leaks.	.120	√	√	√	√	√
6 Clean chlorine pump, flush with acid, and check for signs of deterioration.	.200		√	√	√	√
7 Check acid source, add acid, check for leaks.	.100	√	√	√	√	√
8 Clean acid pump head.	.400		√	√	√	√
9 Check circulating pump for leaks and unusual noises. Lubricate as required.	.050	√	√	√	√	√
10 Check diving board, tighten or replace missing hardware. Inspect for structural defects or deterioration.	.080	√	√	√	√	√
11 Wash exposed stainless steel ladders and scum gutters with tap water to remove oil and dirt.	.300	√	√	√	√	√
12 Fill out maintenance report.	.017	√	√	√	√	√
Total labor-hours/period		.917	1.917	1.917	1.917	1.917
Total labor-hours/year		34.846	15.336	3.834	1.917	1.917

			Cost Each					
			2015 Bare Costs				Total	Total
Description		Labor-hrs.	Material	Labor	Equip.	Total	In-House	w/O&P
1900	Swimming pool outdoor, annually	1.917		93.50		93.50	123	152
1950	Annualized	57.849		2,825		2,825	3,700	4,575

PM Components	Labor-hrs.	W	M	Q	S	A
PM System G2015 610 1950						
Traffic signal light						
1 Inspect and clean control cabinet.	.030					√
2 Check wiring for obvious defects and tighten control electrical connections.	.090					√
3 Check operation of cabinet exhaust fan and inspect vents.	.020					√
4 Inspect and tighten power line connections to control cabinet.	.034					√
5 Check operation of cabinet heater.	.037					√
6 Perform the following operational checks of traffic controller.	.082			√	√	√
A) Check controller phase timing with stop watch.						
B) Check controller pedestrian phase timing with stop watch.						
C) Check controller clocks for correct time and log settings.						
7 Walk a circuit of the intersection to ensure all lights are functioning, replace bulbs as required.	.070			√	√	√
8 Check operation of pedestrian push buttons, average two per intersection.	.084			√	√	√
9 Check operation of all informational signs.	.042			√	√	√
10 Check operation of phase detectors (loop amplifiers).	.144			√	√	√
11 Visually inspect road sensors.	.074			√	√	√
12 Check operation of flashers.	.018					√
13 Fill out maintenance checklist and report deficiencies.	.017			√	√	√
Total labor-hours/period				.513	.513	.742
Total labor-hours/year				1.026	.513	.742

	Description	Labor-hrs.	Cost Each				Total In-House	Total w/O&P
			2015 Bare Costs					
			Material	Labor	Equip.	Total		
1900	Traffic signal light, annually	.742		40.50		40.50	50	63
1950	Annualized	2.281		125		125	154	193

PM Components	Labor-hrs.	W	M	Q	S	A
PM System G2045 150 1950						
Manual swing gate						
1 Check gate and hinge alignment.	.325				√	√
2 Tighten diagonal brace turnbuckle.	.108				√	√
3 Check cane bolt alignment.	.108				√	√
4 Grease hinges.	.217				√	√
5 Oil latch.	.108				√	√
Total labor-hours/period					.866	.866
Total labor-hours/year					.866	.866

			Cost Each				
			2015 Bare Costs			Total	Total
Description	Labor-hrs.	Material	Labor	Equip.	Total	In-House	w/O&P
1900 Manual swing gate, annually	.866	8.55	32.50		41.05	52.50	64
1950 Annualized	1.732	8.55	65		73.55	96	118

PM Components	Labor-hrs.	W	M	Q	S	A
PM System G2045 150 2950						
Electric swing gate						
1 Check gate and hinge alignment.	.325				√	√
2 Tighten diagonal brace turnbuckle.	.108				√	√
3 Check cane bolt alignment.	.108				√	√
4 Grease hinges.	.217				√	√
5 Oil latch.	.108				√	√
6 Check electric motor and connections.	.325				√	√
7 Grease gearbox.	.217				√	√
Total labor-hours/period					1.408	1.408
Total labor-hours/year					1.408	1.408

			Cost Each				
			2015 Bare Costs			Total	Total
Description	Labor-hrs.	Material	Labor	Equip.	Total	In-House	w/O&P
2900 Electric swing gate, annually	1.408	8.55	53		61.55	80	97.50
2950 Annualized	2.816	8.55	106		114.55	150	185

PM Components	Labor-hrs.	W	M	Q	S	A
PM System G2045 150 3950						
Manual slide gate						
1 Check alignment of rollers and gate.	.325				√	√
2 Grease rollers.	.325				√	√
3 Oil latch.	.108				√	√
Total labor-hours/period					.758	.758
Total labor-hours/year					.758	.758

			Cost Each				
			2015 Bare Costs			Total	Total
Description	Labor-hrs.	Material	Labor	Equip.	Total	In-House	w/O&P
3900 Manual slide gate, annually	.758	8.55	28.50		37.05	47.50	57
3950 Annualized	1.516	8.55	57		65.55	85	104

PM Components	Labor-hrs.	W	M	Q	S	A
PM System G2045 150 4950						
Electric slide gate						
1 Check alignment of rollers and gate.	.325				√	√
2 Grease rollers.	.325				√	√
3 Oil latch.	.108				√	√
4 Check electric motor and connections.	.325				√	√
5 Clean and grease chain.	.650			√	√	√
Total labor-hours/period				.650	1.733	1.733
Total labor-hours/year				1.300	1.733	1.733

	Description	Labor-hrs.	Cost Each				Total	Total
			2015 Bare Costs				In-House	w/O&P
			Material	Labor	Equip.	Total		
4900	Electric slide gate, annually	1.733	8.55	65		73.55	96	118
4950	Annualized	4.766	8.55	179		187.55	247	305

PM Components	Labor-hrs.	W	M	Q	S	A
PM System G3015 108 1950						
Storage tank, tower						
1 Inspect exterior of tank for leaks or damage, including tank base and base plates.	.151					√
2 Inspect condition of ladders, sway bracing hardware and structural forms.	.624					√
3 Tighten and lubricate altitude valve and inspect valve vault.	.039					√
4 Check operation of storage tank lighting.	.121					√
5 Check cathodic protection system.	.122					√
6 Operate elevated tank heat system, if installed, and inspect insulation and coils.	.107					√
7 Clean up around tank area.	.134					√
8 Fill out maintenance checklist and report deficiencies.	.022					√
Total labor-hours/period						1.320
Total labor-hours/year						1.320

			Cost Each				
			2015 Bare Costs			Total	Total
Description	Labor-hrs.	Material	Labor	Equip.	Total	In-House	w/O&P
1900 Storage tank, tower, annually	1.320	17.05	64		81.05	103	125
1950 Annualized	1.320	17.05	64		81.05	103	125

PM Components	Labor-hrs.	W	M	Q	S	A
PM System G3015 108 2950						
Storage tank, ground level						
1 Inspect exterior of tank for leaks or damage.	.053					√
2 Check condition of roof and roof hatches including locks; remove trash from roof.	.147					√
3 Inspect condition of ladders and tighten or replace missing hardware.	.088					√
4 Inspect valves, fittings, drains and controls for leakage or damage.	.039					√
5 Check for deterioration of tank foundation.	.077					√
6 Inspect ground drainage slope, and vent screens; where applicable.	.043					√
7 Inspect tank lighting system; replace bulbs as needed.	.121					√
8 Clean up around tank.	.134					√
9 Fill out maintenance checklist and report deficiencies.	.022					√
Total labor-hours/period						.724
Total labor-hours/year						.724

			Cost Each				
			2015 Bare Costs			Total	Total
Description	Labor-hrs.	Material	Labor	Equip.	Total	In-House	w/O&P
2900 Storage tank, ground level, annually	.724	17.05	35		52.05	65	78.50
2950 Annualized	.724	17.05	35		52.05	65	78.50

For customer support on your Facilities Maintenance & Repair Cost Data, call 877.801.5724.

533

PM Components	Labor-hrs.	W	M	Q	S	A
PM System G3015 116 1950						
Water flow meter, turbine						
1 Inspect meter for leaks, corrosion or broken sight glass; replace broken glass as required.	.035				√	√
2 Check meter operation for unusual noise.	.034				√	√
3 Clean meter to remove mineral buildup.	.039				√	√
4 Clean out trash or sludge from bottom of pit.	.165				√	√
5 Fill out maintenance report and report deficiencies.	.022				√	√
Total labor-hours/period					.295	.295
Total labor-hours/year					.295	.295

			Cost Each					
			2015 Bare Costs				**Total**	**Total**
Description		**Labor-hrs.**	**Material**	**Labor**	**Equip.**	**Total**	**In-House**	**w/O&P**
1900	Water flow meter, turbine, annually	.295	16.30	14.35		30.65	37	44
1950	Annualized	.590	16.30	28.50		44.80	55.50	67

PM Components	Labor-hrs.	W	M	Q	S	A
PM System G3015 118 1950						
Fresh water distribution reservoir controls						
1 Check with reservoir personnel for any known deficiencies.	.086			√	√	√
2 Visually inspect and clean interior and exterior of controls cabinet.	.061			√	√	√
3 Inspect electrical system for frayed wires or loose connections; repair as required.	.120			√	√	√
4 Inspect water level receiver at reservoir.	.192			√	√	√
5 Inspect and adjust chronoflo transmitter.	.467			√	√	√
6 Perform operational check of relays, safety switches and electrical contactors.	.542			√	√	√
7 Perform operational check of controls system.	.659			√	√	√
8 Clean up controls area and dispose of debris.	.066			√	√	√
9 Fill out maintenance checklist and report deficiencies.	.022			√	√	√
Total labor-hours/period				2.215	2.215	2.215
Total labor-hours/year				4.430	2.215	2.215

			Cost Each					
			2015 Bare Costs				**Total**	**Total**
Description		**Labor-hrs.**	**Material**	**Labor**	**Equip.**	**Total**	**In-House**	**w/O&P**
1900	Reservoir controls, fresh water distribution, annually	2.215	34	108		142	180	218
1950	Annualized	8.860	34	430		464	610	745

534

For customer support on your Facilities Maintenance & Repair Cost Data, call 877.801.5724.

PM Components	Labor-hrs.	W	M	Q	S	A
PM System G3015 126 0950						
Pump, air lift, well						
1 Check with pump operating personnel for any known deficiencies.	.035				√	√
2 Check operation of compressor and pump.	.033				√	√
3 Check water sample for air or oil contamination.	.009				√	√
4 Check compressor oil level; add oil as necessary and lubricate pump and compressor shaft bearings as applicable.	.121				√	
5 Inspect electrical system for frayed wires or loose connections; repair as required.	.120				√	√
6 Check and cycle high and low shut off valves.	.130				√	√
7 Inspect and clean air intake filter.	.094				√	√
8 Clean pump body and tighten or replace loose hardware.	.086				√	√
9 Calibrate and adjust pressure gauge.	.191				√	√
10 Clean surrounding area.	.066				√	√
11 Fill out maintenance checklist and report deficiencies.	.022				√	√
Total labor-hours/period					.907	.907
Total labor-hours/year					.907	.907

			Cost Each				
			2015 Bare Costs			**Total**	**Total**
Description	**Labor-hrs.**	**Material**	**Labor**	**Equip.**	**Total**	**In-House**	**w/O&P**
0900 Pump, air lift, well, annually	.907	17.05	44		61.05	77	93
0950 Annualized	1.814	34	88.50		122.50	154	186

PM Components	Labor-hrs.	W	M	Q	S	A
PM System G3015 126 1950						
Pump, centrifugal ejector						
1 Check for proper operation of pump.	.022				√	√
2 Check for leaks on suction and discharge piping, seals, packing glands, etc.; make minor adjustments as required.	.077				√	√
3 Check pump and motor operation for excessive vibration, noise and overheating.	.022				√	√
4 Check alignment of pump and motor; adjust as necessary.	.260				√	√
5 Lubricate pump and motor.	.099				√	√
6 Clean exterior of pump, motor and surrounding area.	.096				√	√
7 Fill out maintenance checklist and report deficiencies.	.022				√	√
Total labor-hours/period					.598	.598
Total labor-hours/year					.598	.598

			Cost Each				
			2015 Bare Costs			**Total**	**Total**
Description	**Labor-hrs.**	**Material**	**Labor**	**Equip.**	**Total**	**In-House**	**w/O&P**
1900 Pump, centrifugal ejector, annually	.598	8.55	29		37.55	47.50	58
1950 Annualized	1.196	17.05	58		75.05	95	116

For customer support on your Facilities Maintenance & Repair Cost Data, call 877.801.5724.

535

PM Components	Labor-hrs.	W	M	Q	S	A
PM System G3015 126 3950						
Pump, metering (slurry)						
1 Check with pump operating personnel for any obvious deficiencies.	.035				√	√
2 Obtain and put on safety equipment for working with fluoride: goggles, apron, gloves and respirator.	.014				√	√
3 Check oil level in pump with dipstick; add oil as necessary.	.009				√	√
4 Inspect belt condition and tension on pump; adjust as required.	.029				√	√
5 Flush slurry pump with fresh water.	.139				√	√
6 Pressure test pump.	.012				√	√
7 Inspect pump motor.	.039				√	√
8 Operate pump and check for leaks, excessive noise and vibration.	.099				√	√
9 Clean up area.	.066				√	√
10 Fill out maintenance checklist and report deficiencies.	.022				√	√
Total labor-hours/period					.464	.464
Total labor-hours/year					.464	.464

		Cost Each					
			2015 Bare Costs			Total	Total
Description	**Labor-hrs.**	**Material**	**Labor**	**Equip.**	**Total**	**In-House**	**w/O&P**
3900	Pump, metering (slurry), annually	.464	17.05	22.50		39.55	48.50	58
3950	Annualized	.928	34	45		79	96.50	116

PM Components	Labor-hrs.	W	M	Q	S	A
PM System G3015 126 4950						
Pump, mixed or axial flow velocity						
1 Check with pump operating personnel for any obvious deficiencies.	.035				√	√
2 Clean pump exterior.	.014				√	√
3 Check impeller and shaft for correct alignment and clearances.	.031				√	√
4 Check oil level; add oil as needed.	.009				√	√
5 Inspect packing for leaks and tighten as needed.	.031				√	√
6 Inspect bearings for wear, damage or overheating; lubricate as required.	.053				√	√
7 Inspect float operation and adjust as needed.	.098				√	√
8 Inspect V-belt tension and condition; adjust as required.	.029				√	√
9 Operate pump and check for leaks, excessive noise and vibration.	.099				√	√
10 Tighten loose nuts and bolts and replace missing hardware.	.005				√	√
11 Check electrical system for frayed wires and loose connections.	.120				√	√
12 Clean up area.	.066				√	√
13 Fill out maintenance checklist and report deficiencies.	.022				√	√
Total labor-hours/period					.612	.612
Total labor-hours/year					.612	.612

		Cost Each					
			2015 Bare Costs			Total	Total
Description	**Labor-hrs.**	**Material**	**Labor**	**Equip.**	**Total**	**In-House**	**w/O&P**
4900	Pump, mixed or axial flow, annually	.612	17.05	30		47.05	58	70
4950	Annualized	1.224	34	59.50		93.50	116	140

PM Components	Labor-hrs.	W	M	Q	S	A
PM System G3015 126 5950						
Pump, reciprocating positive displacement						
1 Check for proper operation of pump.	.022				√	√
2 Check for leaks on suction and discharge piping, seals, packing glands, etc.; make minor adjustments as required.	.077				√	√
3 Check pump and motor operation for excessive vibration, noise and overheating.	.022				√	√
4 Check operation of pressure controls.	.133				√	√
5 Check alignment of pump and motor; adjust as necessary.	.260				√	√
6 Lubricate pump and motor.	.099				√	√
7 Clean exterior of pump, motor and surrounding area.	.094				√	√
8 Fill out maintenance checklist and report deficiencies.	.022				√	√
Total labor-hours/period					.729	.729
Total labor-hours/year					.729	.729

			Cost Each					
			2015 Bare Costs				Total	Total
	Description	Labor-hrs.	Material	Labor	Equip.	Total	In-House	w/O&P
5900	Pump, reciprocating displacement, annually	.729	8.55	35.50		44.05	56	68
5950	Annualized	1.458	17.05	71		88.05	112	136

PM Components	Labor-hrs.	W	M	Q	S	A
PM System G3015 126 6950						
Pump, rotary positive displacement						
1 Check with operating personnel for obvious defects.	.035				√	√
2 Clean surface of pump with solvent.	.014				√	√
3 Operate pump and check for leaks at pipes and connections and for excessive noise and vibration.	.099				√	√
4 Lubricate pump.	.047				√	√
5 Check shaft alignment and clearances and pump rotation.	.031				√	√
6 Inspect packing for leaks and tighten.	.031				√	√
7 Tighten loose bolts and nuts; replace loose and missing hardware as needed.	.009				√	√
8 Inspect electrical system for loose connections and frayed wires; repair as necessary.	.120				√	√
9 Operate pump upon completion of maintenance, read and record pressure gauge readings and check discharge pressure.	.221				√	√
10 Clean up area around pump after repairs.	.066				√	√
11 Fill out maintenance checklist and report deficiencies.	.022				√	√
Total labor-hours/period					.695	.695
Total labor-hours/year					.695	.695

			Cost Each					
			2015 Bare Costs				Total	Total
	Description	Labor-hrs.	Material	Labor	Equip.	Total	In-House	w/O&P
6900	Pump, rotary displacement, annually	.695	8.55	34		42.55	54	65.50
6950	Annualized	1.390	17.05	67.50		84.55	108	131

PM Components	Labor-hrs.	W	M	Q	S	A
PM System G3015 126 7950						
Pump, sump, up to 1 H.P.						
1 Check electrical plug, cord and connections.	.120				✓	✓
2 Activate float switch and check pump for proper operation.	.039				✓	✓
3 Lubricate pump as required.	.047				✓	✓
4 Inspect packing and tighten as required.	.031				✓	✓
5 Check pump for misalignment and bearings for overheating.	.199				✓	✓
6 Clean out trash from sump.	.065				✓	✓
7 Fill out maintenance checklist and report deficiencies.	.022				✓	✓
Total labor-hours/period					.523	.523
Total labor-hours/year					.523	.523

			Cost Each				
			2015 Bare Costs			**Total**	**Total**
Description	**Labor-hrs.**	**Material**	**Labor**	**Equip.**	**Total**	**In-House**	**w/O&P**
7900 Pump, sump, up to 1 H.P., annually	.523	8.55	25.50		34.05	43	52
7950 Annualized	1.046	17.05	51		68.05	85.50	104

PM Components	Labor-hrs.	W	M	Q	S	A
PM System G3015 126 8950						
Pump, turbine, well						
1 Check with pump operating personnel for any known deficiencies.	.035				✓	✓
2 Remove turbine and pump from deep well.	.585				✓	✓
3 Inspect pump body, bowls, water passages and impeller for corrosion, wear or pitting; repair as necessary.	.057				✓	✓
4 Check packing for leaks; tighten or replace as necessary.	.325				✓	✓
5 Adjust impeller for optimum operation.	.228				✓	✓
6 Inspect wear ring clearances, thrust and pump bearings for wear; repair as necessary.	.142				✓	✓
7 Lubricate all pump parts as required.	.047				✓	✓
8 Clean suction strainer or replace if damaged.	.338				✓	✓
9 Inspect operation of discharge valve.	.026				✓	✓
10 Inspect electrical system for frayed wires or loose connections; repair as necessary.	.120				✓	✓
11 Check operation of pressure and thermal controls.	.052				✓	✓
12 Paint pump with underwater paint.	.259				✓	✓
13 Run pump and check for noise and vibration.	.022				✓	✓
14 Reinstall pump in deep well and test operation.	.585				✓	✓
15 Fill out maintenance checklist and report deficiencies.	.022				✓	✓
Total labor-hours/period					2.843	2.843
Total labor-hours/year					2.843	2.843

			Cost Each				
			2015 Bare Costs			**Total**	**Total**
Description	**Labor-hrs.**	**Material**	**Labor**	**Equip.**	**Total**	**In-House**	**w/O&P**
8900 Pump, turbine, well, annually	2.843	154	138		292	350	420
8950 Annualized	5.686	171	277		448	550	665

PM Components	Labor-hrs.	W	M	Q	S	A
PM System G3015 126 9950						
Pump, vacuum						
1 Check with pump operating personnel for any known deficiencies.	.035				√	√
2 Check for leaks on suction and discharge piping, seals, etc.	.077				√	√
3 Check pump and motor operation for vibration, noise, overheating, etc.	.022				√	√
4 Check alignment and clearances of shaft and coupler.	.260				√	√
5 Tighten or replace loose, missing, or damaged nuts, bolts, and screws.	.005				√	√
6 Lubricate pump and motor as required.	.099				√	√
7 Clean pump, motor and surrounding area.	.099				√	√
8 Fill out maintenance checklist and report deficiencies.	.022				√	√
Total labor-hours/period					.619	.619
Total labor-hours/year					.619	.619

			Cost Each					
				2015 Bare Costs			Total	Total
	Description	Labor-hrs.	Material	Labor	Equip.	Total	In-House	w/O&P
9900	Pump, vacuum, annually	.619	8.55	30		38.55	49	59.50
9950	Annualized	1.238	17.05	60		77.05	98	119

PM Components	Labor-hrs.	W	M	Q	S	A
PM System G3015 310 1950						
Turbine well pump						
1　Check with operating personnel for any deficiencies.	.035				√	√
2　Remove turbine pump from deep well.	.585				√	√
3　Inspect pump body, bowls, water passages and impeller for corrosion, wear or pitting; repair as necessary.	.057				√	√
4　Check packing for leaks; tighten or repack as necessary.	.330				√	√
5　Adjust impeller for optimun operation.	.228				√	√
6　Inspect wear ring clearances, thrust and pump bearings for wear; repair as necessary.	.142				√	√
7　Lubricate all pump parts as required.	.047				√	√
8　Clean suction strainer or replace if damaged.	.338				√	√
9　Inspect operation of discharge valve.	.030				√	√
10　Inspect electrical system for frayed wires or loose connections; repair as necessary.	.120				√	√
11　Check operation of pressure and thermal controls.	.124				√	√
12　Paint pump with underwater paint.	.259				√	√
13　Run pump and check for noise and vibration.	.022				√	√
14　Reinstall pump in deep well and test operation.	.585				√	√
15　Fill out maintenance checklist and report deficiencies.	.022				√	√
Total labor-hours/period					2.924	2.924
Total labor-hours/year					2.924	2.924

			Cost Each					
				2015 Bare Costs			Total	Total
	Description	Labor-hrs.	Material	Labor	Equip.	Total	In-House	w/O&P
1900	Turbine well pump, annually	2.924	85.50	172		257.50	310	375
1950	Annualized	5.848	171	345		516	615	750

PM Components	Labor-hrs.	W	M	Q	S	A
PM System G3015 320 1950						
Vertical lift pump, over 1 H.P.						
1　Check for proper operation of pump.	.130				√	√
2　Check for leaks on suction and discharge piping, seals, packing glands, etc.; make minor adjustments as required.	.077				√	√
3　Check pump and motor operation for excessive vibration, noise and overheating.	.022				√	√
4　Check alignment of pump and motor; adjust as necessary.	.260				√	√
5　Lubricate pump and motor.	.099				√	√
6　When available, check and record suction or discharge gauge pressure and flow rate.	.022				√	√
7　Clean exterior of pump and surrounding area.	.096				√	√
8　Fill out maintenance checklist and report deficiencies.	.022				√	√
Total labor-hours/period					.728	.728
Total labor-hours/year					.728	.728

			Cost Each					
				2015 Bare Costs			Total	Total
	Description	Labor-hrs.	Material	Labor	Equip.	Total	In-House	w/O&P
1900	Vertical lift pump, over 1 H.P., annually	.728	34	42.50		76.50	91	109
1950	Annualized	1.456	68.50	85.50		154	183	219

PM Components	Labor-hrs.	W	M	Q	S	A
PM System G3015 410 1950						
Fire hydrant						
1 Remove hydrant caps and check condition of gaskets; replace as required.	.241					√
2 Flush hydrant, lubricate cap threads and reinstall cap.	.374					√
3 Fill out maintenance checklist and report deficiencies.	.022					√
Total labor-hours/period						.637
Total labor-hours/year						.637

			Cost Each					
			2015 Bare Costs				Total	Total
Description	Labor-hrs.	Material	Labor	Equip.	Total		In-House	w/O&P
1900 Fire hydrant, annually	.637	8.20	36		44.20		54	66.50
1950 Annualized	.637	8.20	36		44.20		54	66.50

PM Components	Labor-hrs.	W	M	Q	S	A
PM System G3015 420 1950						
Valve, post indicator						
1 Remove set screw, cap and wire seal.	.111				√	√
2 Apply lubricant to threads, open and close valve.	.293				√	√
3 Check operation of and sign and clean glass indicator windows.	.130				√	√
4 Install cap and tighten screw, install handle and wire seal with valve open.	.088				√	√
5 Clean valve exterior and area around valve.	.066				√	√
6 Fill out maintenance checklist and report deficiencies.	.022				√	√
Total labor-hours/period					.710	.710
Total labor-hours/year					.710	.710

			Cost Each					
			2015 Bare Costs				Total	Total
Description	Labor-hrs.	Material	Labor	Equip.	Total		In-House	w/O&P
1900 Valve, post indicator, annually	.710	8.20	40		48.20		59	73
1950 Annualized	1.420	8.20	79.50		87.70		109	135

PM Components	Labor-hrs.	W	M	Q	S	A
PM System G3015 612 1950						
Filter plant						
1 Check daily operating records for unusual occurances.	.035			√	√	√
2 Inspect plant sand or gravel filters.	.694			√	√	√
3 Inspect clear well for cleanliness.	.018			√	√	√
4 Inspect and lubricate flocculator chain and baffle mechanism and	.163			√	√	√
waste-line valves.						
5 Inspect and adjust rate-of-flow controller.	.036			√	√	√
6 Calibrate and adjust loss-of-head gauge.	.191			√	√	√
7 Inspect clear well vent screen and manhole.	.143			√	√	√
8 Inspect operation of pumps, water mixers, chemical feeders and flow	1.145			√	√	√
measuring devices.						
9 Check aerator nozzles for obstructions; clean as required.	.065			√	√	√
10 Drain, clean and inspect mixing chambers.	1.455					√
11 Drain sedimentation basins and clean with fire hose.	2.220					√
12 Drain, inspect and clean flocculator.	2.401					√
13 Clean up filter plant area.	.334			√	√	√
14 Fill out maintenance checklist and report deficiencies.	.022			√	√	√
Total labor-hours/period				2.846	2.846	8.922
Total labor-hours/year				5.692	2.846	8.922

			Cost Each					
				2015 Bare Costs			Total	Total
Description		Labor-hrs.	Material	Labor	Equip.	Total	In-House	w/O&P
1900	Filter plant, annually	8.922	8.55	525		533.55	665	825
1950	Annualized	17.460	34	1,025		1,059	1,350	1,650

PM Components	Labor-hrs.	W	M	Q	S	A
PM System G3015 630 1950						
Water de-ionization station						
1 Check with operating or area personnel for deficiencies.	.035				√	√
2 Check bulk storage tanks, pumps, valves and piping for damage,	.130				√	√
corrosion or leaks.						
3 Lubricate pumps.	.047				√	√
4 Check venting of bulk storage tanks.	.043				√	√
5 Check operation of sodium hydroxide tank and piping heating system.	.124				√	√
6 Check operation of resin vent, blower and motor; lubricate as	.033				√	√
required.						
7 Check calibration of meters.	.577				√	√
8 Clean and check probes of metering system.	.040				√	√
9 Check condition of carbon filter bed; replace	.694				√	√
annually.						
10 Check water softener system for proper operation.	.195				√	√
11 Clean equpment and surrounding area.	.066				√	√
12 Fill out maintenance checklist and report deficiencies.	.022				√	√
Total labor-hours/period					2.006	2.006
Total labor-hours/year					2.006	2.006

			Cost Each					
				2015 Bare Costs			Total	Total
Description		Labor-hrs.	Material	Labor	Equip.	Total	In-House	w/O&P
1900	Water de-ionization system, annually	2.006	335	118		453	515	605
1950	Annualized	4.012	340	236		576	670	790

542

PM Components	Labor-hrs.	W	M	Q	S	A
PM System G3015 632 1950						
Reverse osmosis system, 750 gallons/month						
1 Check with operating personnel for report of treatment effectiveness.	.046			√	√	√
2 Inspect unit, piping and connections for leaks.	.077			√	√	√
3 Check operation of pumps and motors; check shafts for alignment and lubricate as required.	.330			√	√	√
4 Inspect electrical system for frayed wires or loose connections; repair as required.	.120			√	√	√
5 Check filters; replace as required.	.036			√	√	√
6 Check water level controls; adjust as required.	.030			√	√	√
7 Check holding tank temperature control valve operation.	.030			√	√	√
8 Replace charcoal filters as indicated by water test.	.143			√	√	√
9 Check to ensure that chemical supply level is properly maintained in the chemical feed system.	.120			√	√	√
Total labor-hours/period				.932	.932	.932
Total labor-hours/year				1.864	.932	.932

		Cost Each					
		2015 Bare Costs				Total	Total
Description	Labor-hrs.	Material	Labor	Equip.	Total	In-House	w/O&P
1900 Reverse osmosis system, annually	.932	59.50	45.50		105	125	148
1950 Annualized	3.728	188	181		369	445	530

PM Components	Labor-hrs.	W	M	Q	S	A
PM System G3015 634 1950						
Water ion exchange system						
1 Check with operating personnel for deficiencies.	.035			√	√	√
2 Clean unit and piping and check for leaks.	.077			√	√	√
3 Clean brine pump and check for proper operation.	.120			√	√	√
4 Clean and check strainers pilot valve, vents and orifices.	.595					√
5 Inspect electrical system including pump contactor and contacts; repair as required.	.120			√	√	√
6 Check regeneration cycle; make adjustments as required.	.325			√	√	√
7 Clean surrounding area.	.066			√	√	√
8 Fill out maintenance report and report deficiencies.	.022			√	√	√
Total labor-hours/period				.765	.765	1.360
Total labor-hours/year				1.530	.765	1.360

		Cost Each					
		2015 Bare Costs				Total	Total
Description	Labor-hrs.	Material	Labor	Equip.	Total	In-House	w/O&P
1900 Water ion exchange system, annually	1.360		66		66	87	108
1950 Annualized	3.655		178		178	234	289

PM Components	Labor-hrs.	W	M	Q	S	A
PM System G3015 650 1950						
Water softener						
1 Check with building personnel for report of water softener effectiveness.	.086			√	√	√
2 Check pressure gauges for proper operation.	.191			√	√	√
3 Check density of brine solution in salt tank.	.124			√	√	√
4 Check operation of float control in brine.	.040			√	√	√
5 Inspect water softener piping, fittings and valves for leaks.	.077			√	√	√
6 Lubricate valves and motors.	.330			√	√	√
7 Make minor adjustments to water softener controls if required.	.014			√	√	√
8 Inspect softener base and brine tank for corrosion and repair as needed.	.259			√	√	√
9 Check operation of automatic fill valve in brine tank.	.030			√	√	√
10 Check softener electrical wiring and phasing.	.120			√	√	√
11 Clean up area around softener.	.066			√	√	√
12 Fill out maintenance report and report deficiencies.	.022			√	√	√
Total labor-hours/period				1.359	1.359	1.359
Total labor-hours/year				2.718	1.359	1.359

			Cost Each					
			2015 Bare Costs				Total	Total
	Description	Labor-hrs.	Material	Labor	Equip.	Total	In-House	w/O&P
1900	Water softener, annually	1.359	16.30	66		82.30	105	127
1950	Annualized	5.436	16.30	264		280.30	365	450

PM Components	Labor-hrs.	W	M	Q	S	A
PM System G3015 660 1950						
Fluoride saturator						
1 Check with operating personnel for any obvious discrepancies.	.035		√	√	√	√
2 Obtain and put on safety equipment such as goggles, apron, gloves and respirator prior to working on the fluoride saturator.	.014		√	√	√	√
3 Check oil level in gearbox, agitator and injector pump; add oil as necessary.	.018		√	√	√	√
4 Check water level in fluoride mixing tank.	.007		√	√	√	√
5 Check operation of fluoride flow regulator, float switch and solenoid.	.127		√	√	√	√
6 Check saturator for leaks.	.077		√	√	√	√
7 Check operation of unit.	.098		√	√	√	√
8 Clean fluoride tank with brush and detergent.	.085		√	√	√	√
9 Clean surrounding area.	.066		√	√	√	√
10 Fill out maintenance checklist and report deficiencies.	.022		√	√	√	√
Total labor-hours/period			.549	.549	.549	.549
Total labor-hours/year			4.392	1.098	.549	.549

			Cost Each					
			2015 Bare Costs				Total	Total
	Description	Labor-hrs.	Material	Labor	Equip.	Total	In-House	w/O&P
1900	Fluoride saturator, annually	.549	8.55	32		40.55	49.50	61
1950	Annualized	6.588	34	385		419	525	650

PM Components	Labor-hrs.	W	M	Q	S	A
PM System G3015 670 1950						
Chlorinator						
1 Check with operating or area personnel for deficiencies.	.035			√	√	√
2 Put on safety equipment prior to checking equipment.	.014			√	√	√
3 Clean chlorinator and check for leaks.	.057			√	√	√
4 Check water trap for proper level and bleed air.	.017			√	√	√
5 Check and clean water strainers.	.675			√	√	√
6 Grease fittings.	.023			√	√	√
7 Make minor adjustments to chlorinator as needed.	.094			√	√	√
8 Fill out maintenance checklist and report deficiencies.	.022			√	√	√
Total labor-hours/period				.937	.937	.937
Total labor-hours/year				1.874	.937	.937

			Cost Each				
			2015 Bare Costs			**Total**	**Total**
Description	**Labor-hrs.**	**Material**	**Labor**	**Equip.**	**Total**	**In-House**	**w/O&P**
1900 Chlorinator, annually	.937	8.55	55		63.55	78	96.50
1950 Annualized	3.748	17.05	220		237.05	294	365

PM Components	Labor-hrs.	W	M	Q	S	A
PM System G3015 670 2950						
Chlorine detector						
1 Put on safety equipment prior to checking equipment.	.014			√	√	√
2 Test alarm and inspect fan and sensor wiring.	.252			√	√	√
3 Test detector with chlorine test kit.	.044			√	√	√
4 Flush sensor unit with fresh water and drain and fill reservoir.	.068			√	√	√
5 Check drip rate and reservoir level in sight glass.	.013			√	√	√
6 Check "O" rings; replace as required.	.029			√	√	√
7 Inspect sensor mounting grommet, fan operation, detector mounting bolts and drain tube connection.	.147			√	√	√
8 Check charcoal filter cartridge; replace as needed.	.009			√	√	√
9 Check basket for cracked or worn areas.	.059			√	√	√
10 Check detector for leaks under pressure.	.077			√	√	√
11 Clean unit and surrounding area.	.066			√	√	√
12 Fill out maintenance checklist and report deficiencies.	.022			√	√	√
Total labor-hours/period				.800	.800	.800
Total labor-hours/year				1.600	.800	.800

			Cost Each				
			2015 Bare Costs			**Total**	**Total**
Description	**Labor-hrs.**	**Material**	**Labor**	**Equip.**	**Total**	**In-House**	**w/O&P**
2900 Chlorine detector, annually	.800	34	47		81	96	116
2950 Annualized	3.200	137	188		325	385	465

PM Components	Labor-hrs.	W	M	Q	S	A
PM System G3025 410 1950						
Ejector, pump, sewage						
1 Check for proper operation of pump.	.130				√	√
2 Check for leaks on suction and discharge piping, seals, packing glands, etc.; make minor adjustments as required.	.069				√	√
3 Check ejector and motor operation for excessive vibration, noise and overheating.	.022				√	√
4 Check float or pressure controls for proper operation.	.039				√	√
5 Lubricate ejector and motor.	.099				√	√
6 Clean pump unit and surrounding area.	.096				√	√
7 Fill out maintenance checklist and report deficiencies.	.022				√	√
Total labor-hours/period					.477	.477
Total labor-hours/year					.477	.477

			Cost Each					
			2015 Bare Costs				Total	Total
Description		Labor-hrs.	Material	Labor	Equip.	Total	In-House	w/O&P
1900	Ejector pump, sewage, annually	.477	8.55	28		36.55	44.50	54
1950	Annualized	.954	17.05	56		73.05	89	109

PM Components	Labor-hrs.	W	M	Q	S	A
PM System G3025 412 1950						
Ejector, pump						
1 Check for proper operation of pump.	.130				√	√
2 Check for leaks on suction and discharge piping, seals, packing glands, etc.; make minor adjustments as required.	.077				√	√
3 Check ejector and motor operation for excessive vibration, noise and overheating.	.022				√	√
4 Check float or pressure controls for proper operation.	.039				√	√
5 Lubricate ejector and motor.	.099				√	√
6 Clean pump unit and surrounding area.	.096				√	√
7 Fill out maintenance checklist and report deficiencies.	.022				√	√
Total labor-hours/period					.485	.485
Total labor-hours/year					.485	.485

			Cost Each					
			2015 Bare Costs				**Total**	**Total**
Description	**Labor-hrs.**	**Material**	**Labor**	**Equip.**	**Total**		**In-House**	**w/O&P**
1900 Ejector pump, annually	.485	9.65	23.50		33.15		41.50	50.50
1950 Annualized	.970	19.30	47		66.30		83.50	101

PM Components	Labor-hrs.	W	M	Q	S	A
PM System G3025 412 2950						
Ejector pump, sump type						
1 Operate float switch, if applicable, and check for proper operation of pump.	.130				√	√
2 Check for leaks on suction and discharge piping, seals, packing glands, etc.; make minor adjustments as required.	.077				√	√
3 Check pump and motor operation for excessive vibration, noise and overheating.	.022				√	√
4 Check alignment of pump and motor; adjust as necessary.	.260				√	√
5 Lubricate motor.	.099				√	√
6 Check electrical wiring and connections and repair as applicable.	.120				√	√
7 Clean exterior of pump and surrounding area.	.096				√	√
8 Fill out maintenance checklist and report deficiencies.	.022				√	√
Total labor-hours/period					.826	.826
Total labor-hours/year					.826	.826

			Cost Each					
			2015 Bare Costs				**Total**	**Total**
Description	**Labor-hrs.**	**Material**	**Labor**	**Equip.**	**Total**		**In-House**	**w/O&P**
2900 Ejector pump, sump type, annually	.826	9.65	40		49.65		63.50	77.50
2950 Annualized	1.652	19.30	80.50		99.80		127	155

PM Components	Labor-hrs.	W	M	Q	S	A
PM System G3025 420 1950						
Sewage lift pump, over 1 H.P.						
1 Check for proper operation of pump.	.130				√	√
2 Check for leaks on suction and discharge piping, seals, packing glands, etc.; make minor adjustments as required.	.077				√	√
3 Check pump and motor operation for excessive vibration, noise and overheating.	.022				√	√
4 Check alignment of pump and motor; adjust as necessary.	.260				√	√
5 Lubricate pump and motor.	.099				√	√
6 When available, check and record suction or discharge gauge pressure and flow rate.	.022				√	√
7 Clean exterior of pump, motor and surrounding area.	.096				√	√
8 Fill out maintenance checklist and report deficiencies.	.022				√	√
Total labor-hours/period					.728	.728
Total labor-hours/year					.728	.728

			Cost Each					
			2015 Bare Costs				Total	Total
Description		Labor-hrs.	Material	Labor	Equip.	Total	In-House	w/O&P
1900	Sewage lift pump, over 1 H.P., annually	.728	8.55	42.50		51.05	63	77
1950	Annualized	1.456	17.05	85.50		102.55	126	154

PM Components	Labor-hrs.	W	M	Q	S	A
PM System G3025 530 1950						
Aerator, floating						
1 Check with operating or area personnel for deficiencies.	.035			√	√	√
2 Remove aerator from pond.	.390			√	√	√
3 Clean exterior of pump and motor.	.066			√	√	√
4 Check pump and motor operation for excessive vibration, noise and overheating.	.022			√	√	√
5 Check alignment of pump and motor; adjust as necessary.	.260			√	√	√
6 Lubricate pump and motor.	.099			√	√	√
7 Position aerator back in pond.	.390			√	√	√
8 Fill out maintenance checklist and report deficiencies.	.022			√	√	√
Total labor-hours/period				1.284	1.284	1.284
Total labor-hours/year				2.568	1.284	1.284

			Cost Each				
			2015 Bare Costs			**Total**	**Total**
Description	**Labor-hrs.**	**Material**	**Labor**	**Equip.**	**Total**	**In-House**	**w/O&P**
1900 Aerator, floating, annually	1.284	9.65	62.50		72.15	92.50	113
1950 Annualized	5.136	19.30	250		269.30	350	430

PM Components	Labor-hrs.	W	M	Q	S	A
PM System G3025 532 1950						
Barminutor						
1 Check with operating or area personnel for deficiencies.	.044		√	√	√	√
2 Check condition and clearance of cutting knives and inspect base seal.	.217		√	√	√	√
3 Check oil level in gearbox; add oil as necessary.	.035		√	√	√	√
4 Change oil in gearbox.	.681					√
5 Wire brush and lubricate directional flow valve stem.	.060		√	√	√	√
6 Check for rust and corrosion; scrape, wire brush and spot paint as necessary.	.105		√	√	√	√
7 Fill out maintenance checklist and report deficiencies.	.022		√	√	√	√
Total labor-hours/period			.483	.483	.483	1.164
Total labor-hours/year			3.864	.966	.483	1.164

			Cost Each				
			2015 Bare Costs			**Total**	**Total**
Description	**Labor-hrs.**	**Material**	**Labor**	**Equip.**	**Total**	**In-House**	**w/O&P**
1900 Barminutor, annually	1.164	34	56.50		90.50	112	135
1950 Annualized	6.477	77	315		392	500	605

PM Components	Labor-hrs.	W	M	Q	S	A
PM System G3025 534 1950						
Blower, aerator						
1 Check with operating or area personnel for deficiencies.	.035		√	√	√	√
2 Check blower oil level; add oil as necessary.	.046		√	√	√	√
3 Change blower oil.	.390				√	√
4 Check bolts and tighten as necessary: foundation, cylinder head, belt guard, etc.	.155		√	√	√	√
5 Check tension, condition, and alignment of V-belts on blower; adjust or replace as necessary.	.030		√	√	√	√
6 Check pump for excessive heat or vibration.	.042		√	√	√	√
7 Clean air intake filter on blower; replace as necessary.	.177		√	√	√	√
8 Perform operation check of air blower and adjust as required.	.221		√	√	√	√
9 Clean area around equipment.	.066		√	√	√	√
10 Fill out maintenance checklist and report deficiencies.	.022		√	√	√	√
Total labor-hours/period			.794	.794	1.184	1.184
Total labor-hours/year			6.352	1.588	1.184	1.184

			Cost Each					
			2015 Bare Costs				Total	Total
Description	Labor-hrs.	Material	Labor	Equip.	Total	In-House	w/O&P
1900	Blower, aerator, annually	1.184	82	57.50		139.50	166	196
1950	Annualized	10.308	154	500		654	830	1,000

PM Components	Labor-hrs.	W	M	Q	S	A
PM System G3025 536 1950						
Clarifier						
1 Check with operating or area personnel for deficiencies.	.035		√	√	√	√
2 Check chain and sprocket for wear, where applicable.	.033		√	√	√	√
3 Grease drive gear fittings.	.150		√	√	√	√
4 Check gearbox oil level; add if necessary.	.046		√	√	√	
5 Drain and refill gearbox oil reservoir.	.681					√
6 Check unit for rust and corrosion; scrape, wire brush, and paint as required.	.105					√
7 Clean clarifier drive and surrounding area.	.098		√	√	√	√
8 Fill out maintenance checklist and report deficiencies.	.022		√	√	√	√
Total labor-hours/period			.384	.384	.384	1.170
Total labor-hours/year			3.072	.768	.384	1.170

			Cost Each					
			2015 Bare Costs				Total	Total
Description	Labor-hrs.	Material	Labor	Equip.	Total	In-House	w/O&P
1900	Clarifier, annually	1.170	29	57		86	107	129
1950	Annualized	5.394	58	262		320	410	500

PM Components		Labor-hrs.	W	M	Q	S	A
PM System G3025 538 1950								
Comminutor								
1	Check with operating or area personnel for deficiencies.	.035			√	√	√
2	Change gearbox oil and lubricate fittings.	.681				√	√
3	Check comminutor and motor for excessive noise, vibration and overheating.	.022			√	√	√
4	Lubricate fittings and check oil in gearbox.	.150			√	√	√
5	Check operation.	.108			√	√	√
6	Fill out maintenance checklist and report deficiencies.	.022			√	√	√
	Total labor-hours/period				.337	1.018	1.018
	Total labor-hours/year				.674	1.018	1.018

				Cost Each				
				2015 Bare Costs			Total	Total
	Description	Labor-hrs.	Material	Labor	Equip.	Total	In-House	w/O&P
1900	Comminutor, annually	1.018	29	49.50		78.50	97	117
1950	Annualized	2.710	77	132		209	258	310

PM Components		Labor-hrs.	W	M	Q	S	A
PM System G3025 540 1950								
Filter, trickling								
1	Lubricate fittings and check oil level in oil reservoir.	.035				√	√
2	Check distribution arms, support cables, clevis pins, and turnbuckles for deterioration.	.325				√	√
3	Check for leaks.	.077				√	√
4	Check operation.	.108				√	√
5	Fill out maintenance checklist and report deficiencies.	.022				√	√
	Total labor-hours/period					.567	.567
	Total labor-hours/year					.567	.567

				Cost Each				
				2015 Bare Costs			Total	Total
	Description	Labor-hrs.	Material	Labor	Equip.	Total	In-House	w/O&P
1900	Filter, trickling, annually	.567	19.30	27.50		46.80	57.50	69
1950	Annualized	1.134	38.50	55		93.50	115	138

For customer support on your Facilities Maintenance & Repair Cost Data, call 877.801.5724.

551

PM Components	Labor-hrs.	W	M	Q	S	A
PM System G3025 542 1950						
Grit drive						
1 Check with operating personnel for obvious deficiencies.	.035				√	√
2 Lubricate fittings and check oil level in gearbox.	.047				√	√
3 Change oil in gearbox.	.681					√
4 Check chain linkage and operation.	.033				√	√
5 Clean area around drive mechanism.	.066				√	√
6 Fill out maintenance checklist and report deficiencies.	.022				√	√
Total labor-hours/period					.203	.884
Total labor-hours/year					.203	.884

				Cost Each				
				2015 Bare Costs			**Total**	**Total**
	Description	Labor-hrs.	Material	Labor	Equip.	Total	In-House	w/O&P
1900	Grit drive, annually	.884	29	43		72	88.50	106
1950	Annualized	1.087	38.50	53		91.50	112	134

PM Components	Labor-hrs.	W	M	Q	S	A
PM System G3025 544 1950						
Mixer, sewage						
1 Check with operating or area personnel for deficiencies.	.035			√	√	√
2 Clean drive motor exterior and check for corrosion on motor exterior and base plate.	.053			√	√	√
3 Check motor operation for vibration, noise, overheating; etc.	.022			√	√	√
4 Check alignment and clearances of shaft and coupler.	.031			√	√	√
5 Tighten or replace loose, missing, or damaged nuts, bolts and screws.	.005			√	√	√
6 Lubricate motor and shaft bearings as required.	.047			√	√	√
7 Clean area around motor.	.066			√	√	√
8 Fill out maintenance checklist and report deficiencies.	.022			√	√	√
Total labor-hours/period				.281	.281	.281
Total labor-hours/year				.562	.281	.281

				Cost Each				
				2015 Bare Costs			**Total**	**Total**
	Description	Labor-hrs.	Material	Labor	Equip.	Total	In-House	w/O&P
1900	Mixer, sewage, annually	.281	9.65	13.65		23.30	28.50	34
1950	Annualized	1.124	19.30	54.50		73.80	93	113

PM Components	Labor-hrs.	W	M	Q	S	A
PM System G3065 310 1950						
Fuel oil storage tank, above ground						
1 Inspect tank for corrosion.	.192					√
2 Inspect floating tank roof, lifter or cone and automatic float gauge, if applicable.	.233					√
3 Inspect fuel tank fire protection devices.	.022					√
4 Inspect berms around fuel storage tank for erosion.	.047					√
5 Inspect non-freeze draw valves.	.060					√
6 Inspect storage tank vacuum and pressure vents.	.179					√
7 Inspect vapor recovery line for deterioration.	.082					√
8 Run pump and check operation.	.159					√
9 Calibrate and adjust fuel oil pressure.	.382					√
10 Inspect fuel strainer and clean as needed.	.415					√
11 Inspect tank electrical grounds for breaks or loose fittings.	1.292					√
12 Inspect liquid level gauge for proper operation.	.191					√
13 Fill out maintenance checklist.	.022					√
Total labor-hours/period						3.276
Total labor-hours/year						3.276

			Cost Each					
			2015 Bare Costs				Total	Total
Description		Labor-hrs.	Material	Labor	Equip.	Total	In-House	w/O&P
1900	Fuel oil storage tank, above ground, annually	3.276	69.50	196		265.50	320	390
1950	Annualized	3.276	69.50	196		265.50	320	390

PM Components	Labor-hrs.	W	M	Q	S	A
PM System G4095 110 1950						
Cathodic protection system						
1 Inspect exterior of rectifier cabinet, all exposed connecting conduit, and fittings for corrosion, deteriorated paint, mechanical integrity and cleanliness, make minor repairs and clean/touch-up paint as required.	.060				√	√
2 Open and inspect interior of rectifier cabinet, inspect wiring for fraying, signs of overheating, deterioration mechanical damage, and loose connections; make adjustments as required.	.138				√	√
3 Check electrical contacts for overheating and pitting; make adjustments.	.075				√	√
4 Clean interior of rectifier cabinet.	.008				√	√
5 Check panel voltmeter and ampmeter with standard test instruments.	.075				√	√
6 Read and record panel meter voltage and current values.	.007				√	√
7 Measure and record rectifier voltage and current with portable instrument.	.059				√	√
8 Check potential at cathode locations; tighten loose connections.	.286				√	√
9 Fill out maintenance checklist and record deficiencies.	.022				√	√
Total labor-hours/period					.730	.730
Total labor-hours/year					.730	.730

	Description	Labor-hrs.	Cost Each				Total In-House	Total w/O&P
			2015 Bare Costs					
			Material	Labor	Equip.	Total		
1900	Cathodic protection system, annually	.730	6.25	40		46.25	56.50	70
1950	Annualized	1.460	12.45	80		92.45	113	140

General Maintenance

Table of Contents

How to Use the General Maintenance Pages

The following is a detailed explanation of a sample entry in the General Maintenance Section. Next to each bold number that follows is the described item with the appropriate component of the sample entry following in parentheses.

1 Division Number/Title (01 93/Facility Maintenance)

Use the General Maintenance Section Table of Contents to locate specific items. This section is classified according to the CSI MasterFormat 2012 system.

2 Line Numbers (01 93 04.40 0030)

Each General Maintenance unit price line item has been assigned a unique 12-digit code based on the CSI MasterFormat classification.

```
                                    ┌──── MasterFormat Level 3
                              ┌───── MasterFormat Level 2 (01 93 00)
                        ┌───── MasterFormat Division (01)
   01    93    04.40         0030
                   └── MasterFormat Level 4
                        └── RSMeans 12-Digit Line Number
```

01 93 Facility Maintenance

01 93 04 – Landscaping Maintenance

01 93 04.15 Edging

		Crew	Daily Output	Labor-Hours	Unit	Material	Labor	Equipment	Total	Total In-House	Total Incl O&P
0010	**EDGING**										
0020	Hand edging, at walks	1 Clam	16	.500	C.L.F.		14.10		14.10	17.55	22
0030	At planting, mulch or stone beds		7	1.143			32		32	40.00	50
0040	Power edging, at walks		88	.091			2.56		2.56	3.19	3.98
0050	At planting, mulch or stone beds		24	.333			9.40		9.40	11.70	14.60

(2015 Bare Costs for Labor/Equipment/Total columns)

01 93 04.40 Lawn Renovation

		Crew	Daily Output	Labor-Hours	Unit	Material	Labor	Equipment	Total	Total In-House	Total Incl O&P
0010	**LAWN RENOVATION**										
0020	Lawn renovations, aerating, 18" walk behind cultivator	1 Clam	95	.084	M.S.F.		2.37		2.37	2.96	3.69
0030	48" tractor drawn cultivator	A-18	750	.011			.30	.47	.77	.84	.98
0040	72" tractor drawn cultivator	"	1100	.007			.21	.32	.53	.58	.67
0050	Fertilizing, dry granular, 4#/M.S.F., drop spreader	1 Clam	24	.333		2.78	9.40		12.18	14.75	18.10
0060	Rotary spreader	"	140	.057		2.78	1.61		4.39	5.05	6

Column markers: 4 (Crew) 5 (Daily Output/Labor-Hours) 6 (Unit) 7 (2015 Bare Costs) 8 (Total In-House) 9 (Total Incl O&P)

3

Description
(Lawn Renovations, etc.)

Each line item is described in detail. Sub-items and additional sizes are indented beneath the appropriate line items. The first line or two after the main item (in boldface) may contain descriptive information that pertains to all line items beneath this boldface listing.

4

Crew (A-18)

The "Crew" column designates the typical trade or crew used for the task. If the task can be accomplished by one trade and requires no power equipment, that trade and the number of workers are listed (for example, "2 Clam"). If the task requires a composite crew, a crew code designation is listed (for example, "A-18"). You'll find full details on all composite crews in the Crew Listings.

- For a complete list of all trades utilized in this book and their abbreviations, see the inside back cover of the book.

Crew A-18	Hr.	Daily	Hr.	Daily	Hr.	Daily	Bare Costs	In House	Incl. O&P
Maintenance Laborer	$28.20	$225.60	$35.10	$280.80	$43.80	$350.40	$28.20	$35.10	$43.80
Farm Tractor w/ Attachment		350.00		350.00		385.00	43.75	43.75	48.13
H.L, Daily Totals		$575.60		$630.80		$735.40	$71.95	$78.85	$91.92

5

Productivity:
Daily Output (750)/ Labor-Hours (.011)

The "Daily Output" represents the typical number of units the designated crew will perform in a normal 8-hour day. To find out the number of days the given crew would require to complete your task, divide your quantity by the daily output. For example:

Quantity	÷	Daily Output	=	Duration
2000 M.S.F.	÷	750 M.S.F./ Crew Day	=	2.7 Crew Days

The "Labor-Hours" figure represents the number of labor-hours required to perform one unit of work. To find out the number of labor-hours required for your particular task, multiply the quantity of the item times the number of labor-hours shown. For example:

Quantity	÷	Productivity Rate	=	Duration
2000 M.S.F.	÷	.011 Labor-Hours/ LF	=	22 Labor-Hours

6

Unit (M.S.F.)

The abbreviated designation indicates the unit of measure upon which the price, production, and crew are based (M.S.F. = Thousand Square Feet). For a complete listing of abbreviations, refer to the Abbreviations Listing in the Reference Section of this book.

7

Bare Costs:

The unit material cost is the "bare" material cost with no overhead and profit included. *Costs shown reflect national average material prices for January of the current year and include delivery to the facility. Small purchases may require an additional delivery charge. No sales taxes are included.*

Labor (.30)

The unit labor cost is derived by multiplying bare labor-hour costs for Crew A-18 by labor-hour units. The bare labor-hour cost is found in the Crew Section under A-18. (If a trade is listed, the hourly labor cost—the wage rate—is found on the inside back cover of the book.)

Labor-Hour CostCrew A-18	X	Labor-Hour Units	=	Labor
$28.20	x	.011	=	$0.30

Equip. (Equipment) (.47)

Equipment costs for each crew are listed in the description of each crew. Tools or equipment whose value justifies purchase or ownership by a contractor are considered overhead as shown on the inside back cover of the book. The unit equipment cost is derived by multiplying the bare equipment hourly cost by the labor-hour units.

Equipment Cost Crew A-18	X	Labor-Hour Units	=	Equip.
$43.75	x	.011	=	$0.47

Total (.77)

The total of the bare costs is the arithmetic total of the three previous columns: mat., labor, and equip

Material	+	Labor	+	Equip.	=	Total
$0.00	+	$0.30	+	$0.47	=	$0.77

8 Total In-House Costs (.84)

"Total In-House Costs" include suggested markups to bare costs for the direct overhead requirements of in-house maintenance staff. The figure in this column is the sum of three components: the bare material cost plus 10% (for purchasing and handling small quantities); the bare labor cost plus workers' compensation and average fixed overhead (per the labor rate table on the inside back cover of the book or, if a crew is listed, from the Crew Listings); and the bare equipment cost.

Material is Bare Material Cost + 10% = $0.00 + $0.00	=	$0.00
Labor for Crew A-18 = Labor-Hour Cost ($35.10) x Labor-Hour Units (0.11)	=	$0.37
Equip. is Bare Equip. Cost	=	$0.47
Total	=	$0.84

9 Total Costs Including O&P (.98)

"Total Costs Including O&P" include suggested markups to bare costs for an outside subcontractor's overhead and profit requirements. The figure in this column is the sum of three components: the bare material cost plus 25% for profit; the bare labor cost plus total overhead and profit (per the labor rate table on the inside back cover of the book or, if a crew is listed, from the Crew Listings); and the bare equipment cost plus 10% for profit.

Material is Bare Material Cost + 25% = $0.00 + $0.00	=	$0.00
Labor for Crew A-18 = Labor-Hour Cost ($43.80) x Labor-Hour Units (0.11)	=	$0.47
Equip. is Bare Equip. Cost + 10% = $0.47 + $0.05	=	$0.51
Total	=	$0.98

01 93 Facility Maintenance

01 93 04 – Landscaping Maintenance

01 93 04.15 Edging

		Crew	Daily Output	Labor-Hours	Unit	Material	2015 Bare Costs Labor	Equipment	Total	Total In-House	Total Incl O&P
0010	**EDGING**										
0020	Hand edging, at walks	1 Clam	16	.500	C.L.F.		14.10		14.10	17.55	22
0030	At planting, mulch or stone beds		7	1.143			32		32	40.00	50
0040	Power edging, at walks		88	.091			2.56		2.56	3.19	3.98
0050	At planting, mulch or stone beds		24	.333			9.40		9.40	11.70	14.60

01 93 04.20 Flower, Shrub and Tree Care

		Crew	Daily Output	Labor-Hours	Unit	Material	2015 Bare Costs Labor	Equipment	Total	Total In-House	Total Incl O&P
0010	**FLOWER, SHRUB & TREE CARE**										
0020	Flower or shrub beds, bark mulch, 3" deep hand spreader	1 Clam	100	.080	S.Y.	3.55	2.26		5.81	6.70	7.95
0030	Peat moss, 1" deep hand spreader		900	.009		2.80	.25		3.05	3.39	3.89
0040	Wood chips, 2" deep hand spreader		220	.036		1.60	1.03		2.63	3.04	3.59
0050	Cleaning		1	8	M.S.F.		226		226	281.00	350
0060	Fertilizing, dry granular, 3 #/C.S.F.		85	.094		8.75	2.65		11.40	12.95	15.05
0070	Weeding, mulched bed		20	.400			11.30		11.30	14.05	17.50
0080	Unmulched bed		8	1			28		28	35.00	44
0090	Trees, pruning from ground, 1-1/2" caliper		84	.095	Ea.		2.69		2.69	3.34	4.17
0100	2" caliper		70	.114			3.22		3.22	4.01	5
0110	2-1/2" caliper		50	.160			4.51		4.51	5.60	7
0120	3" caliper		30	.267			7.50		7.50	9.35	11.70
0130	4" caliper	2 Clam	21	.762			21.50		21.50	26.50	33.50
0140	6" caliper		12	1.333			37.50		37.50	47.00	58.50
0150	9" caliper		7.50	2.133			60		60	75.00	93.50
0160	12" caliper		6.50	2.462			69.50		69.50	86.50	108
0170	Fertilize, slow release tablets	1 Clam	100	.080		.58	2.26		2.84	3.45	4.23
0180	Pest control, spray		24	.333		26.50	9.40		35.90	41.00	47.50
0190	Systemic		48	.167		26.50	4.70		31.20	35.00	40.50
0200	Watering, under 1-1/2" caliper		34	.235			6.65		6.65	8.25	10.30
0210	1-1/2" to 4" caliper		14.50	.552			15.55		15.55	19.35	24
0220	4" caliper and over		10	.800			22.50		22.50	28.00	35
0230	Shrubs, prune, entire bed		7	1.143	M.S.F.		32		32	40.00	50
0240	Per shrub, 3' height		190	.042	Ea.		1.19		1.19	1.48	1.84
0250	4' height		90	.089			2.51		2.51	3.12	3.89
0260	6' height		50	.160			4.51		4.51	5.60	7
0270	Fertilize, dry granular, 3#/M.S.F.		85	.094	M.S.F.	6.60	2.65		9.25	10.55	12.35
0280	Watering, entire bed		7	1.143	"		32		32	40.00	50
0290	Per shrub		32	.250	Ea.		7.05		7.05	8.80	10.95

01 93 04.35 Lawn Care

		Crew	Daily Output	Labor-Hours	Unit	Material	2015 Bare Costs Labor	Equipment	Total	Total In-House	Total Incl O&P
0010	**LAWN CARE**										
0020	Mowing lawns, power mower, 18"-22"	1 Clam	80	.100	M.S.F.		2.82		2.82	3.51	4.38
0030	22"-30"		120	.067			1.88		1.88	2.34	2.92
0040	30"-32"		140	.057			1.61		1.61	2.01	2.50
0050	Self propelled or riding mower, 36"-44"	A-16	300	.027			.75	.23	.98	1.17	1.42
0060	48"-58"		480	.017			.47	.14	.61	.73	.89
0070	Tractor, 3 gang reel, 7' cut		930	.009			.24	.07	.31	.37	.46
0080	5 gang reel, 12' cut		1200	.007			.19	.06	.25	.29	.35
0090	Edge trimming with weed whacker	1 Clam	5760	.001	L.F.		.04		.04	.05	.06

01 93 04.40 Lawn Renovation

		Crew	Daily Output	Labor-Hours	Unit	Material	2015 Bare Costs Labor	Equipment	Total	Total In-House	Total Incl O&P
0010	**LAWN RENOVATION**										
0020	Lawn renovations, aerating, 18" walk behind cultivator	1 Clam	95	.084	M.S.F.		2.37		2.37	2.96	3.69
0030	48" tractor drawn cultivator	A-18	750	.011			.30	.47	.77	.84	.98
0040	72" tractor drawn cultivator	"	1100	.007			.21	.32	.53	.58	.67
0050	Fertilizing, dry granular, 4#/M.S.F., drop spreader	1 Clam	24	.333		2.78	9.40		12.18	14.75	18.10
0060	Rotary spreader	"	140	.057		2.78	1.61		4.39	5.05	6

01 93 04 – Landscaping Maintenance

01 93 04.40 Lawn Renovation

		Crew	Daily Output	Labor-Hours	Unit	Material	2015 Bare Costs Labor	Equipment	Total	Total In-House	Total Incl O&P
0070	Tractor drawn 8' spreader	A-18	500	.016	M.S.F.	2.78	.45	.70	3.93	4.32	4.95
0080	Tractor drawn 12' spreader	"	800	.010		2.78	.28	.44	3.50	3.85	4.40
0090	Overseeding, utility mix, 7#/M.S.F., drop spreader	1 Clam	10	.800		9.40	22.50		31.90	38.50	47
0100	Tractor drawn spreader	A-18	52	.154		9.40	4.34	6.75	20.49	22.50	26
0110	Watering, 1" of water, applied by hand	1 Clam	21	.381			10.75		10.75	13.35	16.70
0120	Soaker hoses	"	82	.098			2.75		2.75	3.42	4.27

01 93 04.65 Raking

		Crew	Daily Output	Labor-Hours	Unit	Material	2015 Bare Costs Labor	Equipment	Total	Total In-House	Total Incl O&P
0010	**RAKING**										
0020	Raking, leaves, by hand	1 Clam	7.50	1.067	M.S.F.		30		30	37.50	46.50
0030	Power blower		45	.178			5		5	6.25	7.80
0040	Grass clippings, by hand		7.50	1.067			30		30	37.50	46.50

01 93 04.85 Walkways and Parking Care

		Crew	Daily Output	Labor-Hours	Unit	Material	2015 Bare Costs Labor	Equipment	Total	Total In-House	Total Incl O&P
0010	**WALKWAYS AND PARKING CARE**										
0020	General cleaning, parking area and walks, sweeping by hand	1 Clam	15	.533	M.S.F.		15.05		15.05	18.70	23.50
0030	Power vacuum		100	.080			2.26		2.26	2.81	3.50
0040	Power blower		100	.080			2.26		2.26	2.81	3.50
0050	Hose down		30	.267			7.50		7.50	9.35	11.70

01 93 04.90 Yard Waste Disposal

		Crew	Daily Output	Labor-Hours	Unit	Material	2015 Bare Costs Labor	Equipment	Total	Total In-House	Total Incl O&P
0010	**YARD WASTE DISPOSAL**										
0030	On site dump or compost heap	2 Clam	24	.667	C.Y.		18.80		18.80	23.50	29
0040	Into 6 C.Y. dump truck, 2 mile haul to compost heap	A-17	125	.064			1.80	1.96	3.76	4.21	4.90
0050	4 mile haul to compost heap, fees not included	"	85	.094			2.65	2.89	5.54	6.20	7.30

01 93 05 – Exterior General Maintenance

01 93 05.95 Window Washing

		Crew	Daily Output	Labor-Hours	Unit	Material	2015 Bare Costs Labor	Equipment	Total	Total In-House	Total Incl O&P
0010	**WINDOW WASHING**										
0030	Minimum productivity	1 Clam	1	8	M.S.F.	4.29	226		230.29	286.00	355
0040	Average productivity		2.50	3.200		4.29	90		94.29	117.00	145
0050	Maximum productivity		4	2		4.29	56.50		60.79	75.00	93

01 93 06 – Interior General Maintenance

01 93 06.20 Carpet Care

		Crew	Daily Output	Labor-Hours	Unit	Material	2015 Bare Costs Labor	Equipment	Total	Total In-House	Total Incl O&P
0010	**CARPET CARE**										
0100	Carpet cleaning, portable extractor with floor wand	1 Clam	20	.400	M.S.F.	8	11.30		19.30	23.00	27.50
0200	Deep carpet cleaning, self-contained extractor										
0210	24" path @ 100 feet per minute	1 Clam	64	.125	M.S.F.	10.05	3.53		13.58	15.45	18.00
0220	20" path		54	.148		10.05	4.18		14.23	16.25	19.00
0230	16" path		43	.186		10.05	5.25		15.30	17.60	20.50
0240	12" path		32	.250		10.05	7.05		17.10	19.85	23.50
0300	Carpet pile lifting, self-contained extractor										
0310	24" path at 90 feet per minute	1 Clam	60	.133	M.S.F.	11.05	3.76		14.81	16.85	19.60
0320	20" path		50	.160		11.05	4.51		15.56	17.80	21
0330	16" path		40	.200		11.05	5.65		16.70	19.20	22.50
0340	12" path		30	.267		11.05	7.50		18.55	21.50	25.50
0400	Carpet restoration cleaning, self-contained extractor										
0410	24" path at 50 feet per minute	1 Clam	32	.250	M.S.F.	12.05	7.05		19.10	22.00	26
0420	20" path		27	.296		12.05	8.35		20.40	23.50	28
0430	16" path		21	.381		12.05	10.75		22.80	26.50	32
0440	12" path		16	.500		12.05	14.10		26.15	31.00	37

01 93 06 – Interior General Maintenance

01 93 06.25 Ceiling Care	Crew	Daily Output	Labor-Hours	Unit	Material	2015 Bare Costs Labor	Equipment	Total	Total In-House	Total Incl O&P
0010 **CEILING CARE**										
0020 Washing hard ceiling from ladder, hand	1 Clam	2.05	3.902	M.S.F.	4.58	110		114.58	142.00	177
0030 Machine	"	2.52	3.175	"	4.58	89.50		94.08	116.00	145
0035 Acoustic tile cleaning, chemical spray, including masking										
0040 Unobstructed floor	4 Clam	5400	.006	S.F.	.06	.17		.23	.28	.34
0045 Obstructed floor	"	4000	.008	"	.06	.23		.29	.35	.43
0048 Chemical spray cleaning and coating, including masking										
0050 Unobstructed floor	4 Clam	4800	.007	S.F.	.13	.19		.32	.37	.45
0055 Obstructed floor	"	3400	.009	"	.13	.27		.40	.47	.57

01 93 06.40 General Cleaning

	Crew	Daily Output	Labor-Hours	Unit	Material	2015 Bare Costs Labor	Equipment	Total	Total In-House	Total Incl O&P
0010 **GENERAL CLEANING**										
0020 Sweep stairs & landings, damp wipe handrails	1 Clam	80	.100	Floor	.09	2.82		2.91	3.61	4.49
0030 Dust mop stairs & landings, damp wipe handrails		62.50	.128		.09	3.61		3.70	4.59	5.70
0040 Damp mop stairs & landings, damp wipe handrails		45	.178		.25	5		5.25	6.50	8.10
0050 Vacuum stairs & landings, damp wipe handrails, canister		41	.195		.09	5.50		5.59	6.95	8.65
0060 Vacuum stairs & landings, damp wipe handrails, backpack		58.22	.137		.09	3.88		3.97	4.92	6.10
0070 Vacuum stairs & landings, damp wipe handrails, upright		31.57	.253		.09	7.15		7.24	9.00	11.20
0080 Clean carpeted elevators		48	.167	Ea.	1.03	4.70		5.73	7.00	8.60
0090 Clean uncarpeted elevators		32	.250		.88	7.05		7.93	9.75	12.05
0100 Clean escalator		8	1		.92	28		28.92	36.00	45
0110 Restrooms, fixture cleaning, toilets		180	.044		.09	1.25		1.34	1.66	2.06
0120 Urinals		192	.042		.08	1.18		1.26	1.55	1.93
0130 Sinks		206	.039		.01	1.10		1.11	1.37	1.71
0140 Utility sink		175	.046		.05	1.29		1.34	1.66	2.06
0150 Bathtub		85	.094		.16	2.65		2.81	3.48	4.32
0160 Shower		85	.094		1.54	2.65		4.19	4.99	6.05
0170 Accessories cleaning, soap dispenser		1280	.006		.01	.18		.19	.23	.28
0180 Sanitary napkin dispenser		640	.013		.01	.35		.36	.45	.56
0190 Paper towel dispenser		1455	.006		.01	.16		.17	.20	.25
0200 Toilet tissue dispenser		1455	.006		.01	.16		.17	.20	.25
0210 Accessories, refilling/restocking, soap dispenser (1 bag)		640	.013		8.85	.35		9.20	10.20	11.60
0220 Sanitary napkin/tampon dispenser (25/50)		384	.021		21	.59		21.59	24.00	27
0230 Paper towel dispenser, folded (2 pkgs of 250)		545	.015		9.35	.41		9.76	10.80	12.35
0240 Paper towel dispenser, rolled (350' roll)		340	.024		4.58	.66		5.24	5.85	6.75
0250 Toilet tissue dispenser (500 2-ply sheet roll)		1440	.006		1.40	.16		1.56	1.74	1.99
0260 Sanitary seat covers (pkg of 250)		960	.008		5.15	.23		5.38	5.95	6.80
0270 Toilet partition cleaning		96	.083		.34	2.35		2.69	3.29	4.08
0280 Ceramic tile wall cleaning		2400	.003	S.F.	.03	.09		.12	.15	.19
0290 General cleaning, empty to larger portable container, wastebaskets		960	.008	Ea.		.23		.23	.29	.36
0300 Ash trays		2880	.003			.08		.08	.10	.12
0310 Cigarette stands		960	.008			.23		.23	.29	.36
0320 Sand urns		640	.013			.35		.35	.44	.55
0330 Pencil sharpeners		2880	.003			.08		.08	.10	.12
0340 Remove trash to central location		290	.028	Lb.		.78		.78	.97	1.21
0347 Bench		480	.017	Ea.	.05	.47		.52	.65	.79
0350 Chairs		960	.008		.02	.23		.25	.31	.39
0360 Chalkboard erasers		960	.008			.23		.23	.29	.36
0370 Chalkboards		16	.500	M.S.F.	4.58	14.10		18.68	22.50	27.50
0380 Desk tops		225	.036	Ea.	.07	1		1.07	1.33	1.65
0390 Glass doors		168	.048	"	.18	1.34		1.52	1.87	2.32
0400 Hard surface audience type seating		40	.200	M.S.F.	5.75	5.65		11.40	13.35	15.95
0409 Locker, metal		320	.025	Ea.	.14	.71		.85	1.03	1.28

For customer support on your Facilities Maintenance & Repair Cost Data, call 877.801.5724.

561

01 93 06.40 General Cleaning		Crew	Daily Output	Labor-Hours	Unit	Material	2015 Bare Costs Labor	Equipment	Total	Total In-House	Total Incl O&P
0410	Microwave ovens	1 Clam	240	.033	Ea.	.03	.94		.97	1.20	1.50
0420	Office partitions, to 48 S.F.		160	.050		.22	1.41		1.63	2.00	2.47
0430	Overhead air vents		175	.046		.02	1.29		1.31	1.62	2.03
0440	Refrigerators		35	.229		.37	6.45		6.82	8.45	10.45
0450	Revolving doors		35	.229		1.07	6.45		7.52	9.20	11.35
0460	Tables		240	.033		.09	.94		1.03	1.27	1.57
0470	Telephone		288	.028		.01	.78		.79	.99	1.23
0480	Telephone booth		190	.042		.56	1.19		1.75	2.10	2.54
0490	Water fountain		480	.017		.03	.47		.50	.62	.77
0500	Wastebaskets		260	.031		.03	.87		.90	1.11	1.39
0510	Dusting, beds		360	.022			.63		.63	.78	.97
0520	Bedside stand		960	.008			.23		.23	.29	.36
0530	Bookcase		960	.008			.23		.23	.29	.36
0540	Calculator		1920	.004			.12		.12	.15	.18
0550	Chair		1340	.006			.17		.17	.21	.26
0560	Coat rack		400	.020			.56		.56	.70	.88
0570	Couch		320	.025			.71		.71	.88	1.10
0580	Desk		550	.015			.41		.41	.51	.64
0590	Desk lamp		1200	.007			.19		.19	.23	.29
0600	Desk trays		1920	.004			.12		.12	.15	.18
0610	Dictaphone		1920	.004			.12		.12	.15	.18
0620	Dresser		480	.017			.47		.47	.59	.73
0630	File cabinets		960	.008			.23		.23	.29	.36
0640	Lockers		2400	.003			.09		.09	.12	.15
0650	Piano		425	.019			.53		.53	.66	.82
0660	Radio		960	.008			.23		.23	.29	.36
0665	Shelf, storage, 1' x 20'		480	.017			.47		.47	.59	.73
0670	Table		670	.012			.34		.34	.42	.52
0680	Table lamp		960	.008			.23		.23	.29	.36
0690	Telephones		960	.008			.23		.23	.29	.36
0700	Television		960	.008			.23		.23	.29	.36
0710	Typewriter		960	.008			.23		.23	.29	.36
0720	Vacuuming, armless chair		825	.010			.27		.27	.34	.42
0730	Armchair		480	.017			.47		.47	.59	.73
0740	Couch		320	.025	↓		.71		.71	.88	1.10
0750	Entry mats		8	1	M.S.F.		28		28	35.00	44
0760	Overhead air vents		480	.017	Ea.		.47		.47	.59	.73
0770	Upholstered auditorium seating	↓	5.27	1.518	M.S.F.		43		43	53.50	66.50

01 93 06.45 Hard Floor Care											
0010	**HARD FLOOR CARE**										
0020	Damp mop w/bucket & wringer, 12 oz. mop head, unobstructed	1 Clam	25	.320	M.S.F.	5	9		14	16.75	20.50
0030	Obstructed		17	.471		5	13.25		18.25	22.00	27
0040	20 oz. mop head, unobstructed		29	.276		5	7.80		12.80	15.20	18.30
0050	Obstructed		20	.400		5	11.30		16.30	19.55	24
0060	24 oz. mop head, unobstructed		35	.229		5	6.45		11.45	13.50	16.25
0070	Obstructed		23	.348		5	9.80		14.80	17.70	21.50
0080	32 oz. mop head, unobstructed		43	.186		5	5.25		10.25	12.05	14.40
0090	Obstructed		29	.276		5	7.80		12.80	15.20	18.30
0100	Dust mop, 12" mop, unobstructed		33	.242			6.85		6.85	8.50	10.60
0110	Obstructed		19	.421			11.85		11.85	14.80	18.40
0120	18" mop, unobstructed		49	.163			4.60		4.60	5.75	7.10
0130	Obstructed	↓	28	.286	↓		8.05		8.05	10.05	12.50

01 93 06.45 Hard Floor Care		Crew	Daily Output	Labor-Hours	Unit	Material	2015 Bare Costs Labor	Equipment	Total	Total In-House	Total Incl O&P
0140	24" mop, unobstructed	1 Clam	61	.131	M.S.F.		3.70		3.70	4.60	5.75
0150	Obstructed		34	.235			6.65		6.65	8.25	10.30
0160	30" mop, unobstructed		74	.108			3.05		3.05	3.79	4.74
0170	Obstructed		42	.190			5.35		5.35	6.70	8.35
0180	36" mop, unobstructed		92	.087			2.45		2.45	3.05	3.81
0190	Obstructed		52	.154			4.34		4.34	5.40	6.75
0200	42" mop, unobstructed		124	.065			1.82		1.82	2.26	2.83
0210	Obstructed		70	.114			3.22		3.22	4.01	5
0220	48" mop, unobstructed		185	.043			1.22		1.22	1.52	1.89
0230	Obstructed		104	.077			2.17		2.17	2.70	3.37
0240	60" mop, unobstructed		246	.033			.92		.92	1.14	1.42
0250	Obstructed		138	.058			1.63		1.63	2.03	2.54
0260	72" mop, unobstructed		369	.022			.61		.61	.76	.95
0270	Obstructed		208	.038			1.08		1.08	1.35	1.68
0280	Sweeping, 8" broom, light soil, unobstructed		19	.421			11.85		11.85	14.80	18.45
0290	Obstructed		14	.571			16.10		16.10	20.00	25
0300	Heavy soil, unobstructed		11	.727			20.50		20.50	25.50	32
0310	Obstructed		9	.889			25		25	31.00	39
0320	12" push broom, light soil, unobstructed		32	.250			7.05		7.05	8.80	10.95
0330	Obstructed		24	.333			9.40		9.40	11.70	14.60
0340	Heavy soil, unobstructed		19	.421			11.85		11.85	14.80	18.45
0350	Obstructed		14	.571			16.10		16.10	20.00	25
0360	16" push broom, light soil, unobstructed		40	.200			5.65		5.65	7.00	8.75
0370	Obstructed		30	.267			7.50		7.50	9.35	11.70
0380	Heavy soil, unobstructed		24	.333			9.40		9.40	11.70	14.60
0390	Obstructed		18	.444			12.55		12.55	15.60	19.45
0400	18" push broom, light soil, unobstructed		44	.182			5.15		5.15	6.40	7.95
0410	Obstructed		33	.242			6.85		6.85	8.50	10.60
0420	Heavy soil, unobstructed		36	.222			6.25		6.25	7.80	9.75
0430	Obstructed		20	.400			11.30		11.30	14.05	17.50
0440	24" push broom, light soil, unobstructed		57	.140			3.96		3.96	4.93	6.15
0450	Obstructed		43	.186			5.25		5.25	6.55	8.15
0460	Heavy soil, unobstructed		34	.235			6.65		6.65	8.25	10.30
0470	Obstructed		26	.308			8.70		8.70	10.80	13.50
0480	30" push broom, light soil, unobstructed		81	.099			2.79		2.79	3.47	4.33
0490	Obstructed		60	.133			3.76		3.76	4.68	5.85
0500	Heavy soil, unobstructed		48	.167			4.70		4.70	5.85	7.30
0510	Obstructed		36	.222			6.25		6.25	7.80	9.75
0520	36" push broom, light soil, unobstructed		100	.080			2.26		2.26	2.81	3.50
0530	Obstructed		75	.107			3.01		3.01	3.74	4.67
0540	Heavy soil, unobstructed		60	.133			3.76		3.76	4.68	5.85
0550	Obstructed		45	.178			5		5	6.25	7.80
0560	42" push broom, light soil, unobstructed		133	.060			1.70		1.70	2.11	2.63
0570	Obstructed		100	.080			2.26		2.26	2.81	3.50
0580	Heavy soil, unobstructed		80	.100			2.82		2.82	3.51	4.38
0590	Obstructed		60	.133			3.76		3.76	4.68	5.85
0600	48" push broom, light soil, unobstructed		200	.040			1.13		1.13	1.40	1.75
0610	Obstructed		150	.053			1.50		1.50	1.87	2.34
0620	Heavy soil, unobstructed		120	.067			1.88		1.88	2.34	2.92
0630	Obstructed		90	.089			2.51		2.51	3.12	3.89
0640	Wet mop/rinse w/bucket & wringer, 12 oz. mop, unobstructed		11	.727		4.58	20.50		25.08	30.50	37.50
0650	Obstructed		8	1		4.58	28		32.58	40.00	49.50
0660	20 oz. mop head, unobstructed		14	.571		4.58	16.10		20.68	25.00	30.50

01 93 06.45 Hard Floor Care		Crew	Daily Output	Labor-Hours	Unit	Material	2015 Bare Costs Labor	Equipment	Total	Total In-House	Total Incl O&P
0670	Obstructed	1 Clam	10	.800	M.S.F.	4.58	22.50		27.08	33.00	40.50
0680	24 oz. mop head, unobstructed		20	.400		4.58	11.30		15.88	19.10	23
0690	Obstructed		15	.533		4.58	15.05		19.63	24.00	29
0700	32 oz. mop head, unobstructed		27	.296		4.58	8.35		12.93	15.45	18.75
0710	Obstructed		20	.400		4.58	11.30		15.88	19.10	23
0720	Dry buffing or polishing, 175 RPM, 12" diameter machine		14.21	.563			15.90		15.90	19.75	24.50
0730	14" diameter machine		16.51	.485			13.65		13.65	17.00	21
0740	17" diameter machine		19.20	.417			11.75		11.75	14.65	18.25
0750	20" diameter machine		22.85	.350			9.85		9.85	12.30	15.35
0760	24" diameter machine		28.99	.276			7.80		7.80	9.70	12.10
0770	350 RPM, 17" diameter machine		30.76	.260			7.35		7.35	9.15	11.40
0780	20" diameter machine		40.60	.197			5.55		5.55	6.90	8.65
0790	24" diameter machine		47.06	.170			4.79		4.79	5.95	7.45
0800	1000 RPM, 17" diameter machine		62.09	.129			3.63		3.63	4.52	5.65
0810	20" diameter machine		67.68	.118			3.33		3.33	4.15	5.20
0820	24" diameter machine		88.79	.090			2.54		2.54	3.16	3.95
0830	27" diameter machine		99.34	.081			2.27		2.27	2.83	3.53
0840	Spray buffing, 175 RPM, 12" diameter		10.66	.750		1.05	21		22.05	27.50	34.50
0850	14" diameter machine		11.95	.669		1.05	18.90		19.95	24.50	31
0860	17" diameter machine		13.80	.580		1.05	16.35		17.40	21.50	27
0870	20" diameter machine		16	.500		1.05	14.10		15.15	18.70	23.50
0880	24" diameter machine		19.05	.420		1.05	11.85		12.90	15.90	19.70
0890	350 RPM, 17" diameter machine		17.39	.460		1.05	12.95		14	17.30	21.50
0900	20" diameter machine		22.13	.362		1.05	10.20		11.25	13.85	17.15
0910	24" diameter machine		29.22	.274		1.05	7.70		8.75	10.75	13.30
0920	1000 RPM, 17" diameter machine		41.02	.195		1.05	5.50		6.55	8.00	9.8
0930	20" diameter machine		44.14	.181		1.05	5.10		6.15	7.50	9.2
0940	24" diameter machine		52.21	.153		1.05	4.32		5.37	6.55	8
0950	27" diameter machine		57.43	.139		1.05	3.93		4.98	6.05	7.4
0960	Burnish, 1000 RPM, 17" diameter machine		60	.133			3.76		3.76	4.68	5.8
0970	2000 RPM, 17" diameter machine		72	.111			3.13		3.13	3.90	4.8
0980	Scrubbing, 12" brush by hand		2	4		5	113		118	146.00	181
0990	175 RPM, 12" diameter machine		10.05	.796		5	22.50		27.50	33.50	41.5
1000	14" diameter machine		11.62	.688		5	19.40		24.40	29.50	36.5
1010	17" diameter machine		14	.571		5	16.10		21.10	25.50	31.5
1020	20" diameter machine		17.61	.454		5	12.80		17.80	21.50	26
1030	24" diameter machine		23.58	.339		5	9.55		14.55	17.40	21
1040	350 RPM, 17" diameter machine		24.12	.332		5	9.35		14.35	17.15	21
1050	20" diameter machine		28.91	.277		5	7.80		12.80	15.20	18.3
1060	24" diameter machine		36.92	.217		5	6.10		11.10	13.10	15.7
1070	Automatic, 21" machine		60.75	.132		5	3.71		8.71	10.10	12
1080	24" machine		83.47	.096		5	2.70		7.70	8.85	10.4
1090	32" machine		128.68	.062		5	1.75		6.75	7.70	8.9
1100	Stripping, separate wet pick-up required, 175 RPM, 17" diameter		6.86	1.166		3.53	33		36.53	45.00	55.5
1110	20" diameter machine		7.68	1.042		3.53	29.50		33.03	40.50	50
1120	350 RPM, 17" diameter machine		8.80	.909		3.53	25.50		29.03	36.00	44.5
1130	20" diameter machine		10.50	.762		3.53	21.50		25.03	30.50	38
1140	Wet pick-up, 12" opening attachment		18	.444			12.55		12.55	15.60	19.4
1150	20" opening attachment		24.12	.332			9.35		9.35	11.65	14.5
1160	Finish application, mop		14	.571		17.05	16.10		33.15	39.00	46.5
1170	Gravity fed applicator		21	.381		17.05	10.75		27.80	32.00	38
1180	Lambswool applicator		17	.471		17.05	13.25		30.30	35.50	42
1190	Sealer application, mop		14	.571		25	16.10		41.10	47.50	56.5

01 93 06 – Interior General Maintenance

01 93 06.45 Hard Floor Care	Crew	Daily Output	Labor-Hours	Unit	Material	2015 Bare Costs Labor	Equipment	Total	Total In-House	Total Incl O&P
1200 Gravity fed applicator	1 Clam	21	.381	M.S.F.	25	10.75		35.75	41.00	48
1210 Lambswool applicator	▼	17	.471	▼	25	13.25		38.25	44.00	52

01 93 06.80 Vacuuming

		Crew	Daily Output	Labor-Hours	Unit	Material	2015 Bare Costs Labor	Equipment	Total	Total In-House	Total Incl O&P
0010	**VACUUMING**										
0020	Vacuum carpet, 12" upright, spot areas	1 Clam	73.68	.109	M.S.F.		3.06		3.06	3.81	4.76
0030	Traffic areas		44.21	.181			5.10		5.10	6.35	7.95
0040	All areas		21.05	.380			10.70		10.70	13.35	16.65
0050	14" upright , spot areas		79.98	.100			2.82		2.82	3.51	4.38
0060	Traffic areas		47.99	.167			4.70		4.70	5.85	7.30
0070	All areas		22.85	.350			9.85		9.85	12.30	15.35
0080	16" upright, spot areas		87.50	.091			2.58		2.58	3.21	4
0090	Traffic areas		52.50	.152			4.30		4.30	5.35	6.65
0100	All areas		25	.320			9		9	11.25	14
0110	18" upright, spot areas		96.53	.083			2.34		2.34	2.91	3.63
0120	Traffic areas		57.92	.138			3.90		3.90	4.85	6.05
0130	All areas		27.58	.290			8.20		8.20	10.20	12.70
0140	20" upright, spot areas		107.66	.074			2.10		2.10	2.61	3.25
0150	Traffic areas		64.60	.124			3.49		3.49	4.35	5.40
0160	All areas		30.76	.260			7.35		7.35	9.15	11.40
0170	22" upright, spot areas		121.73	.066			1.85		1.85	2.31	2.88
0180	Traffic areas		73.04	.110			3.09		3.09	3.84	4.80
0190	All areas		34.78	.230			6.50		6.50	8.05	10.05
0200	24" upright, spot areas		140	.057			1.61		1.61	2.01	2.50
0210	Traffic areas		84	.095			2.69		2.69	3.34	4.17
0220	All areas		40	.200			5.65		5.65	7.00	8.75
0230	14" twin motor upright, spot areas		98.81	.081			2.28		2.28	2.84	3.55
0240	Traffic areas		59.28	.135			3.81		3.81	4.74	5.90
0250	All areas		28.23	.283			8		8	9.95	12.40
0260	18" twin motor upright, spot areas		112	.071			2.01		2.01	2.51	3.13
0270	Traffic areas		67.20	.119			3.36		3.36	4.18	5.20
0280	All areas		32	.250			7.05		7.05	8.80	10.95
0290	Backpack, 12" opening carpet attachment, spot areas		77.77	.103			2.90		2.90	3.61	4.51
0300	Traffic areas		46.66	.171			4.83		4.83	6.00	7.50
0310	All areas		22.22	.360			10.15		10.15	12.65	15.75
0320	14" opening carpet attachment, spot areas		84.84	.094			2.66		2.66	3.31	4.13
0330	Traffic areas		50.90	.157			4.43		4.43	5.50	6.90
0340	All areas		24.24	.330			9.30		9.30	11.60	14.45
0350	16" opening carpet attachment, spot areas		93.35	.086			2.42		2.42	3.01	3.75
0360	Traffic areas		56.01	.143			4.03		4.03	5.00	6.25
0370	All areas		26.67	.300			8.45		8.45	10.55	13.15
0380	18" opening carpet attachment, spot areas		103.67	.077			2.18		2.18	2.71	3.38
0390	Traffic areas		62.20	.129			3.63		3.63	4.51	5.65
0400	All areas		29.62	.270			7.60		7.60	9.50	11.85
0410	20" opening carpet attachment, spot areas		116.66	.069			1.93		1.93	2.41	3
0420	Traffic areas		69.99	.114			3.22		3.22	4.01	5
0430	All areas		33.33	.240			6.75		6.75	8.40	10.50
0440	22" opening carpet attachment, spot areas		133.32	.060			1.69		1.69	2.11	2.63
0450	Traffic areas		79.99	.100			2.82		2.82	3.51	4.38
0460	All areas		38.09	.210			5.90		5.90	7.35	9.20
0470	24" opening carpet attachment, spot areas		155.54	.051			1.45		1.45	1.81	2.25
0480	Traffic areas		93.32	.086			2.42		2.42	3.01	3.76
0490	All areas		44.44	.180			5.10		5.10	6.30	7.90

01 93 Facility Maintenance

01 93 06 – Interior General Maintenance

01 93 06.80 Vacuuming

		Crew	Daily Output	Labor-Hours	Unit	Material	2015 Bare Costs Labor	Equipment	Total	Total In-House	Total Incl O&P
0500	Tank / cannister, 12" opening carpet attachment, spot areas	1 Clam	70	.114	M.S.F.		3.22		3.22	4.01	5
0510	Traffic areas		42	.190			5.35		5.35	6.70	8.35
0520	All areas		20	.400			11.30		11.30	14.05	17.50
0530	14" opening carpet attachment, spot areas		75.67	.106			2.98		2.98	3.71	4.63
0540	Traffic areas		45.40	.176			4.97		4.97	6.20	7.70
0550	All areas		21.62	.370			10.45		10.45	13.00	16.20
0560	16" opening carpet attachment, spot areas		82.32	.097			2.74		2.74	3.41	4.26
0570	Traffic areas		49.39	.162			4.57		4.57	5.70	7.10
0580	All areas		23.52	.340			9.60		9.60	11.95	14.90
0590	18" opening carpet attachment, spot areas		90.30	.089			2.50		2.50	3.11	3.88
0600	Traffic areas		54.18	.148			4.16		4.16	5.20	6.45
0610	All areas		25.80	.310			8.75		8.75	10.90	13.60
0620	20" opening carpet attachment, spot areas		100	.080			2.26		2.26	2.81	3.50
0630	Traffic areas		60	.133			3.76		3.76	4.68	5.85
0640	All areas		28.57	.280			7.90		7.90	9.85	12.25
0650	22" opening carpet attachment, spot areas		112	.071			2.01		2.01	2.51	3.13
0660	Traffic areas		67.20	.119			3.36		3.36	4.18	5.20
0670	All areas		32	.250			7.05		7.05	8.80	10.95
0680	24" opening carpet attachment, spot areas		127.26	.063			1.77		1.77	2.21	2.75
0690	Traffic areas		76.36	.105			2.95		2.95	3.68	4.5
0700	All areas		36.36	.220			6.20		6.20	7.70	9.6
0710	Large area machine, no obstructions, 26" opening		44.44	.180			5.10		5.10	6.30	7.90
0720	28" opening		64	.125			3.53		3.53	4.39	5.5
0730	30" opening		80	.100			2.82		2.82	3.51	4.3
0740	32" opening		120	.067			1.88		1.88	2.34	2.9

01 93 06.90 Wall Care

		Crew	Daily Output	Labor-Hours	Unit	Material	2015 Bare Costs Labor	Equipment	Total	Total In-House	Total Incl O&P
0010	**WALL CARE**										
0020	Dusting walls, hand duster	1 Clam	87	.092	M.S.F.		2.59		2.59	3.23	4.0
0030	Treated cloth		40	.200			5.65		5.65	7.00	8.7
0040	Hand held duster vacuum		53	.151			4.26		4.26	5.30	6.6
0050	Cannister vacuum		32	.250			7.05		7.05	8.80	10.9
0060	Backpack vacuum		45	.178			5		5	6.25	7.8
0070	Washing walls, hand, painted surfaces		2.50	3.200		4.58	90		94.58	117.00	146
0080	Vinyl surfaces		3.25	2.462		4.58	69.50		74.08	91.50	114
0090	Machine, painted surfaces		4.80	1.667		4.58	47		51.58	63.50	78.5
0100	Washing walls from ladder, hand, painted surfaces		1.65	4.848		4.58	137		141.58	175.00	218
0110	Vinyl surfaces		2.15	3.721		4.58	105		109.58	136.00	169
0120	Machine, painted surfaces		3.17	2.524		4.58	71		75.58	93.50	117

01 93 06.95 Window and Accessory Care

		Crew	Daily Output	Labor-Hours	Unit	Material	2015 Bare Costs Labor	Equipment	Total	Total In-House	Total Incl O&P
0010	**WINDOW AND ACCESSORY CARE**										
0025	Wash windows, trigger sprayer & wipe cloth, 12 S.F.	1 Clam	4.50	1.778	M.S.F.	4.29	50		54.29	67.00	83.5
0030	Over 12 S.F.		12.75	.627		4.29	17.70		21.99	26.50	33
0040	With squeegee & bucket, 12 S.F. and under		5.50	1.455		4.29	41		45.29	56.00	69
0050	Over 12 S.F.		15.50	.516		4.29	14.55		18.84	23.00	28
0060	Dust window sill		8640	.001	L.F.		.03		.03	.04	.0
0070	Damp wipe, venetian blinds		160	.050	Ea.	.09	1.41		1.50	1.86	2.3
0080	Vacuum, venetian blinds, horizontal		87.27	.092			2.59		2.59	3.22	4.0
0090	Draperies, in place, to 64 S.F.		160	.050			1.41		1.41	1.76	2.1

01 93 08 – Facility Maintenance Equipment

01 93 08.50 Equipment	Crew	Daily Output	Labor-Hours	Unit	Material	2015 Bare Costs Labor	Equipment	Total	Total In-House	Total Incl O&P
0010 **EQUIPMENT**, Purchase										
0090 Carpet care equipment										
0110 Dual motor vac, 1 H.P., 16" brush				Ea.	1,175			1,175	1,300.00	1,475
0120 Upright vacuum, 12" brush					221			221	243.00	276
0130 14" brush					390			390	430.00	490
0140 Soil extractor, hot water 5' wand, 12" head					3,075			3,075	3,375.00	3,850
0150 Dry foam, 13" brush					1,700			1,700	1,875.00	2,125
0160 24" brush					3,075			3,075	3,375.00	3,850
0240 Floor care equipment										
0260 Polishing, buffing, waxing machine, 175 RPM										
0270 .33 H.P., 20" diam. brush				Ea.	720			720	790.00	900
0280 1 H.P., 16" diam. brush					1,150			1,150	1,275.00	1,450
0290 17" diam. brush					1,275			1,275	1,400.00	1,600
0300 1.5 H.P., 20" diam. brush					1,350			1,350	1,475.00	1,700
0310 20" diam. brush, 1500 RPM					1,100			1,100	1,200.00	1,375
0330 Scrubber, automatic, 2 stage 1 H.P. vacuum motor										
0340 20" diam. brush				Ea.	12,100			12,100	13,300.00	15,100
0350 28" diam. brush				"	17,500			17,500	19,300.00	21,900
1200 Plumbing maintenance equipment										
1220 Kinetic water ram				Ea.	385			385	425.00	480
1240 Cable pipe snake, 104 ft., self feed, electric, .5 H.P.				"	2,625			2,625	2,900.00	3,275
1500 Specialty equipment										
1550 Litterbuggy, 9 H.P. gasoline				Ea.	5,075			5,075	5,575.00	6,350
1560 Propane					3,550			3,550	3,900.00	4,450
1570 Pressure cleaner, hot water, 1000 psi					3,550			3,550	3,900.00	4,450
1580 3500 psi					7,025			7,025	7,725.00	8,775
1590 Vacuum cleaners, steel canister										
1600 Dry only, two stage, 1 H.P.				Ea.	330			330	365.00	415
1610 Wet/dry, two stage, 2 H.P.					655			655	720.00	820
1620 4 H.P.					1,025			1,025	1,125.00	1,275
1630 Wet/dry, three stage, 1.5 H.P.					655			655	720.00	820
1670 Squeegee wet only, 2 stage, 1 H.P.					1,175			1,175	1,300.00	1,475
1680 2 H.P.					1,700			1,700	1,875.00	2,125
1690 Asbestos and hazardous waste dry vacs, 1 H.P.					925			925	1,025.00	1,150
1700 2 H.P.					1,550			1,550	1,700.00	1,950
1710 Three stage, 1 H.P.					545			545	600.00	680
1800 Upholstery soil extractor, dry-foam					3,400			3,400	3,750.00	4,250
1900 Snow removal equipment										
1920 Thrower, 3 H.P., 20", single stage, gas				Ea.	625			625	690.00	780
1930 Electric start					725			725	800.00	905
1940 5 H.P., 24", 2 stage, gas					1,200			1,200	1,325.00	1,500
1950 11 H.P., 36"					2,700			2,700	2,975.00	3,375
2000 Tube cleaning equip., for heat exchangers, condensers, evap.										
2010 Tubes/pipes 1/4"-1", 115 volt, .5 H.P. drive				Ea.	1,650			1,650	1,825.00	2,075
2020 Air powered					1,700			1,700	1,875.00	2,125
2040 1" and up, 115 volt, 1 H.P. drive					1,800			1,800	1,975.00	2,250

For customer support on your Facilities Maintenance & Repair Cost Data, call 877.801.5724.

567

Facilities Audits

Auditing the Facility

Introduction

Capital budget planning begins with compiling a complete inventory of the facility, including equipment, that identifies existing deficiencies. The Facilities Audit is a vehicle for producing such an inventory; it is a system for thoroughly assessing the existing physical condition and functional performance of buildings, grounds, utilities, and equipment. The results of this audit are used to address major and minor, urgent and long-term needs for corrective action, for short- and long-term financial planning. The basic principles presented here can be applied to any scale operation, from a single structure to a facility consisting of multiple building complexes in dispersed locations.

The first step is determining the scope of the audit. Next, the audit team is selected, and a set of forms is designed. A survey team then records field observations on the forms for individual building components. Finally, the information from these forms is summarized, prioritized, presented, and used to:

compare a building's condition and functional performance to that of other buildings within a facility

define regular maintenance requirements

define capital repair and replacement projects in order to eliminate deferred maintenance

develop cost estimates for capital repair and replacement projects

restore functionally obsolete facilities to a usable condition

eliminate conditions that are either potentially damaging to property or present life safety hazards

identify energy conservation measures

The audit ultimately produces a database on the condition of a facility's buildings, identifying the physical adequacy of construction, material, and equipment, and assessing the functional adequacy of the facility to perform its designated purpose. This comprehensive inventory can be used to guide property management decisions and strategic planning of capital assets, such as determining those expenditures necessary to extend the life of a facility.

The audit procedures and forms included here are for the purpose of identifying the condition of buildings and assessing the cost and priority of their repair. While this is a comprehensive stand-alone process, it is for the limited purposes of assessing costs and establishing priorities for repairs. It represents only a portion of the requirements of a full facilities audit. For further information and guidance on conducting a full audit, contact the RSMeans Engineering staff.

The Rating Forms

The audit approach involves conducting a thorough inspection of a facility, then employing a systematic process of completing individual forms for each component, following the order in which building components are assembled. This is a self-evaluation process designed to fit all types of structures. The audit procedure described is a "generic" approach that can be used as a starting point; the facilities manager should incorporate the special characteristics of an organization and its facilities into an individualized facilities audit.

Audits are best performed by the in-house personnel familiar with the building being audited. However, outside consultants may be required when staff is not available or when special expertise is required to further examine suspected deficiencies.

Four types of forms that can be used to conduct the building audit are described here. A complete set of the blank forms ready for photocopying follows.

1. *Building Data Summary*
   This form can be used to summarize all buildings to be covered by the audit.

2. *Facilities Inventory Form*
   This form is designed to collect the basic data about a specific building's characteristics (name, identification number, location, area), including land data. One form is prepared for each building to be audited as summarized on the *Building Data Summary* form.

3. *Standard Inspection Form*
   The standard form for recording observed deficiencies, it provides a uniform way to comment on, prioritize, and estimate the cost to correct those deficiencies. The *Standard Inspection Form* is designed to make use of cost data from the Maintenance & Repair and Preventive Maintenance sections of this publication. Cost data from publications such as *RSMeans Facilities Construction Cost Data* may be utilized as well.

4. *System Checklists*
   For each major building system from foundations through site work, there is a checklist to guide the audit and record the review process. The checklists are intended as general guides and may not be totally comprehensive. Be sure to make any necessary modifications (additions or deletions) required for the specific building being audited. The forms prompt the user on what to check and provide a place to check off or indicate by "Yes" or "No" the presence of deficiencies requiring further detailed analysis on the *Standard Inspection Form*.

Completing the Inspection Forms

The standard inspection forms are designed to be used in conjunction with the system checklists. A system checklist of typical deficiencies for each building system is completed as part of the inspection process. Then for each building deficiency, a standard inspection report is generated listing the identifying information as required on the form. With the deficiency defined and described, the specific work items required to correct the deficiency can be entered, prioritized, and costed in Part 3 of the form. The priority rating is further described below.

Priority Ratings

Priority rating is the inspector's judgment of the priority of an observed deficiency. The rating provides guidance for facilities managers who must review overall deficiencies and develop maintenance schedules and capital budgets. The following priority rating terms, commonly used in maintenance management, are a suggested guideline for priority values. The scale may be expanded as required to fit specific needs.

1. *Emergency*: This designation indicates work that demands immediate attention to repair an essential operating system or building component, or because life safety or property is endangered. A response should occur within two hours of notification.

2. *Urgent*: Work demanding prompt attention to alleviate temporary emergency repairs or a condition requiring correction within one to five days to prevent a subsequent emergency is classified as urgent.

3. *Routine*: A specific completion date can be requested or required for routine work. This includes work that has no short range effect on life safety, security of personnel, or property. Routine work can be planned in detail and incorporated into a trade backlog for scheduling within twelve months.

4. *Deferred*: Projects that can be deferred into the following year's work planning are classified as deferred.

Cost Estimates for Maintenance

A major characteristic of the audit described here is the application of detailed cost estimates. To prepare cost estimates for work required to correct deficiencies, detailed information about the nature and extent of the facility's condition is required at the beginning of the audit. However, when cost estimates are incorporated, the audit process generates highly useful summaries outlining major maintenance costs and priorities. Including cost estimates allows for comparisons to be made in order to prioritize repair and replacement needs on a project-by-project and building-by-building basis.

Other Considerations

A problem which may have to be settled by the audit leader is what to do with mixed-use buildings, those that have recent additions to the original construction, or buildings that have, under emergency circumstances, been forced to house new activities for which they were not designed. Such problems will have to be dealt with individually by the audit team.

The data gathered in a facilities audit must be kept current, gathered consistently, and regularly updated. An annual cycle of inspections is recommended. Each organization initiating a comprehensive audit and continuing with annual cycles of inspections will benefit from organizing data in a systematic format. Conducting annual audits in a consistent format provides the basis for monitoring the backlog of deferred maintenance and preparing effective long-term plans and budgets for maintenance and repair operations.

The forms are designed for ease of field entry of data and manual preparation of summaries and reports. The forms can also be created using basic word processing and spreadsheet software or database programming. Although the self-evaluation process is readily adaptable to data processing programs, not every organization has the capabilities or resources available. For those without computer capabilities, a manual method of storing and updating must be developed and used.

Summarizing the Audit Data

Information collected on the audit forms is compiled and used to determine the condition of the facility's buildings and what should be done to maintain and/or improve them. Summaries of a building audit can serve several purposes, some of which are listed below.

Routine Maintenance Needs: A current property owner may be interested in planning and estimating a program for the cost of routine operations and maintenance.

- *Major Maintenance:* A current or prospective property owner may want to evaluate major maintenance needs to plan and estimate the cost of a corrective program.
- *Deferred Maintenance:* A backlog of routine and major maintenance can be addressed using a survey of conditions to plan and budget a deferred maintenance program.
- *Renovations and/or Additions:* Prior to developing feasibility studies on alternatives to renovations and/or additions, a survey of existing conditions may be necessary.
- *Capital Budgeting and Planning:* A comprehensive audit of facilities conditions will incorporate major maintenance requirements into overall capital needs.

Conclusion

The information and forms for conducting an audit of the facility are included in *RSMeans Facilities Maintenance & Repair Cost Data* to provide a method for applying the cost data to the actual maintenance and repair needs of a specific facility. The Facility Audit presented here is a subset of a more comprehensive audit program. The more comprehensive approach includes assessment of the building's functional performance, as well as additional forms and procedures to incorporate the cost of maintenance and repair into a comprehensive multi-year facility management plan.

BUILDING DATA SUMMARY

Building Number	Building Name	Building Use*	Location / Address	Ownership** Own/Lea/O-L

*   Principal and Special Uses
** Own = Owned
     Lea = Leased
     O-L = Owned/Leased

For customer support on your Facilities Maintenance & Repair Cost Data, call 877.801.5724.

FACILITIES INVENTORY FORM

Building Data

1. Organization Name _____

2. Approved by: Name _____ Title _____ Date _____

3. Building Number _____

4. Property Name _____

5. Property Address _____

6. Property Main Use/Special Use _____

7. Gross Area (Sq. Ft.) _____   8. Net Area (Sq. Ft.) _____

9. Type of Ownership: Owned _____ Leased _____ Owned / Leased _____

10. Book Value $ _____   Year Appraised _____

11. Replacement Value $ _____

12. Year Built _____   Year Acquired _____

13. Year Major Additions _____

Land Data

14. Type of Ownership: Owned _____ Leased _____ Owned / Leased _____

15. Book Value $ _____   Year Appraised _____

16. Year Acquired _____

17. Size in Acres _____

18. Market Value $ _____

Notes _____

_____

_____

_____

STANDARD INSPECTION FORM

1. Facility Inspection Data

Facility #: _____   Name: _____

Component #: _____   Name: _____

Inspector: _____   Date: _____

2. Component Description

_____

_____

_____

3. Component Repair Evaluation

Defi-ciency #	Component Description	Line Number	Pri-ority	Unit	Quant.	Unit Cost	Total In-House	Unit Cost	Total W/O&P

Deficiencies	Causes
_____ Settlement, alignment changes or cracks	_____ Soils - changes in load bearing capacity due to shrinkage, erosion, or compaction. Adjacent construction undermining foundations. Reduced soil cover resulting in frost exposure.
	_____ Design loads - building equipment loads exceeding design loads. Vibration from heavy equipment requiring isolated foundations.
	_____ Structural or occupancy changes - inadequate bearing capacities. Foundation settling.
	_____ Earthquake resistance non-functioning.
_____ Moisture penetration	_____ Water table changes - inadequate drainage.
	_____ Ineffective drains or sump pump/sump pits.
	_____ Roof drainage - storm sewer connections inadequate or defective. Installation of roof drain restrictors, gutters, and downspouts where required.
	_____ Surface drainage - exterior grades should slope away from building and structures.
	_____ Utilities - broken or improperly functioning utility service lines or drains.
	_____ Leakage - wall cracks, opening of construction joints, inadequate or defective waterproofing.
	_____ Condensation - inadequate ventilation, vapor barrier, and/or dehumidification.
_____ Temperature changes	_____ Insulation - improperly selected for insulating value, fire ratings, and vermin resistance.
_____ Surface material deterioration	_____ Concrete, masonry, or stucco - spalling, corrosion of reinforcing, moisture penetration, or chemical reaction between cement and soil.
	_____ Steel or other ferrous metals - corrosion due to moisture or contact with acid-bearing soils.
	_____ Wood - decay due to moisture or insect infestation.
_____ Openings deterioration	_____ Non-functioning of doors, windows, hatchways, and stairways.
	_____ Utilities penetration due to damage, weather, wear, or other cause.

A20 BASEMENT INSPECTION CHECKLIST

Deficiencies	**Causes**

Deficiencies

_____ Floors, concrete - cracking
or arching

_____ Floors, wood - rotting
or arching

_____ Wall deterioration

_____ Crawl space ventilation
and maintenance

Causes

_____ Shrinkage, or settlement of subsoil.
_____ Inadequate drainage.
_____ Movement in exterior walls, or frost heave.
_____ Improper compaction of base. Erosion of base.
_____ Heaving from hydraulic pressure.

_____ Excessive dampness or insect
infestation.
_____ Leak in building exterior.
_____ Lack of ventilation.

_____ Concrete or masonry (see *A10 Foundations*)

_____ Inadequate air circulation due
to blockage of openings in foundation walls.
_____ Moisture barrier ineffective.
_____ Pest control, housekeeping, and proper
drainage.

A20 BASEMENT INSPECTION CHECKLIST

579

The primary materials encountered in the superstructure inspection are concrete, steel, and wood. Typical observations of deficiencies will be observed by: failures in the exterior closure systems of exterior walls, openings, and roofs; cracks; movement of materials; moisture penetration; and discoloration. The exterior visual survey will detect failures of surface materials or at openings that will require further inspection to determine whether the cause was the superstructure system.

Deficiencies	**Causes**

Concrete (columns, walls, and floor and roof slabs)

_____ Overall alignment	_____ Settlement due to design and construction techniques.
	_____ Under designed for loading conditions (see *A10 Foundations*).
_____ Deflection	_____ Expansion and/or contraction due to changes in design loads.
	_____ Original design deficient.
	_____ Original materials deficient.
_____ Surface Conditions	
_____ Cracks	_____ Inadequate design and/or construction due to changes in design loads.
	_____ Stress concentration.
	_____ Extreme temperature changes; secondary effects of freeze-thaw.
_____ Scaling, spalls, and pop-outs	_____ Extreme temperature changes.
	_____ Reinforcement corrosion.
	_____ Environmental conditions.
	_____ Mechanical damage.
	_____ Poor materials.
_____ Stains	_____ Chemical reaction of reinforcing.
	_____ Reaction of materials in concrete mixture.
	_____ Environmental conditions.
_____ Exposed reinforcing	_____ Corrosion of steel.
	_____ Insufficient cover.
	_____ Mechanical damage.

	Deficiencies		Causes

Steel (structural members, stairs, and connections)

_____	Overall alignment	_____	Settlement due to design and construction techniques; improper fabrication.
_____	Deflection or cracking	_____	Expansion and/or contraction.
		_____	Changes in design loads.
		_____	Fatigue due to vibration or impact.
_____	Corrosion	_____	Electrochemical reaction.
		_____	Failure of protective coating.
		_____	Excessive moisture exposure.
_____	Surface deterioration	_____	Excessive wear.

Wood (structural members and connections)

_____	Overall alignment	_____	Settlement; design and construction techniques.
_____	Deflection or cracking	_____	Expansion and/or contraction.
		_____	Changes in design loads.
		_____	Fatigue due to vibration or impact.
		_____	Failure of compression members.
		_____	Poor construction techniques.
		_____	General material failure.
_____	Rot (Decay)	_____	Direct contact with moisture.
		_____	Condensation.
		_____	Omission or deterioration of vapor barrier.
		_____	Poor construction techniques.
		_____	Damage from rodents or insects.

581

For customer support on your Facilities Maintenance & Repair Cost Data, call 877.801.5724.

B20 EXTERIOR CLOSURE INSPECTION CHECKLIST

General Inspection

- Overall appearance
- Displacement
- Paint conditions
- Caulking
- Window & door fit
- Flashing condition

- Material integrity
- Cracks
- Evidence of moisture
- Construction joints
- Hardware condition

Exterior Walls.

_____ Wood (Shingles, weatherboard siding, plywood).

    _____ Paint or surface treatment condition   _____ Moisture penetration

    _____ Rot or decay   _____ Loose, cracked, warped, or broken boards and shingles.

_____ Concrete, Masonry, and Tile (Concrete, brick, concrete masonry units, structural tile, glazed tile, stucco, stone).

    _____ Settlement   _____ Structural frame movement causing cracks

    _____ Construction and expansion joints   _____ Condition of caulking and mortar

    _____ Surface deterioration   _____ Efflorescence and staining

    _____ Parapet movement   _____ Tightness of fasteners

_____ Metal (Corrugated iron or steel, aluminum, enamel coated steel, protected metals).

    _____ Settlement   _____ Structural frame movement

    _____ Condition of bracing   _____ Surface damage due to impact

    _____ Tightness of fasteners   _____ Caulking

    _____ Flashings   _____ Corrosion

_____ Finishes (Mineral products, fiberglass, polyester resins, and plastics).

    _____ Settlement   _____ Structural frame movement

    _____ Surface damage due to impact   _____ Cracks

    _____ Stains   _____ Fasteners

    _____ Adhesion to substrate   _____ Caulking

    _____ Flashings   _____ Lead paint

    _____ Asbestos products

_____ Insulation

    _____ Insulation present   _____ Satisfactory condition

Windows and Doors

    _____ Frame fitting   _____ Frame and molding condition

    _____ Paint or surface finish   _____ Putty and weatherstripping

    _____ Hardware and operating parts   _____ Security

    _____ Cleanliness   _____ Material condition (glass, wood, & metal Panels)

    _____ Rot or corrosion   _____ Screens and storm windows

Shading Devices

    _____ Material conditions   _____ Operations

    _____ Cleanliness

General Appearance

Good_____   Fair_____   Poor_____

_____ **Water tightness**

_____ Evidence of leaks on undersurface	_____ Surface weathering
_____ Faulty material	_____ Faulty design
_____ Faulty application	_____ Standing water
_____ Weather damage	_____ Mechanical damage
_____ Fastening failure	_____ Flashing failure

_____ **Roofing Surface**

_____ Built Up (Felt or bitumen surfacing)

_____ Adhesion	_____ Moisture meter readings
_____ Bare areas	_____ Blisters, wrinkles
_____ Cracks, holes, tears	_____ Fish mouths
_____ Alligatoring	_____ Ballast

_____ Single Ply (Thermosetting, thermoplastic, composites)

_____ Adhesion	_____ Moisture meter readings
_____ Bare areas	_____ Blisters, wrinkles
_____ Cracks	_____ Holes, tears
_____ Seam conditions	_____ Protective coating
_____ Ballast	

_____ Metal Roofing (Preformed, formed)

_____ Corrosion (%)	_____ Protective coating
_____ Scams	_____ Cracks or breaks
_____ Holes	_____ Expansion joints

_____ Shingles & Tiles (Metal, clay, mission, concrete, or others)

_____ Disintegration	_____ Broken or cracked (%)
_____ Missing (%)	_____ Fasteners
_____ Underlayment	

_____ Wood shingles

_____ Cracked	_____ Curled
_____ Missing (%)	

583

For customer support on your Facilities Maintenance & Repair Cost Data, call 877.801.5724.

_____ **Insulation**

    _____ Full coverage (%)   _____ Full thickness (%)

    _____ Satisfactory condition

_____ **Roof Penetrations**

    _____ Weather tightness   _____ Operable

_____ **Flashing**

    _____ Deterioration   _____ Open joints

    _____ Holes or damage   _____ Anchoring

    _____ Protective coating

_____ **Drainage**

    _____ Alignment   _____ Free flowing

    _____ Clamping rings secure   _____ Screens

    _____ Corrosion

General Inspection

    Overall Appearance

            Good _____   Fair _____   Poor _____

    _____ Evidence of moisture   _____ Settlement

    _____ Cracks   _____ Surface condition

    _____ Cleanliness   _____ Stains

    _____ Discoloration

_____ **Paint**

    _____ Repainting necessary   _____ Peeling, cracking, flaking

    _____ Reflectivity   _____ Maintainability

    _____ Lead paint

_____ **Coverings & Coatings**

    _____ Replacement necessary   _____ Peeling

    _____ Rips, tears   _____ Holes

    _____ Adhesion   _____ Seams

_____ **Interior glazing**

    _____ Cracks   _____ Seals

    _____ Frame condition   _____ Missing panes

    _____ Shading devices

General Inspection

- Strength & stability
- Acoustical quality
- Maintainability
- Physical condition
- Adaptability
- Code compliance

Deficiencies   **Causes**

Wall Material

	Deficiencies		Causes
_____	Cracks	_____	Settlement
_____	Holes	_____	Defective material
_____	Looseness	_____	Operational abuse or vibrations
_____	Missing segments	_____	Environmental attack
_____	Water stains	_____	Moisture
_____	Joints	_____	Structural expansion or contraction
_____	Surface appearance	_____	Wind pressure

Hardware

_____	Overall condition	_____	Appearance
_____	Keying system	_____	Operations
_____	Fit	_____	Locksets
_____	Cylinders	_____	Panic devices
_____	Maintainability	_____	Security operations

585

For customer support on your Facilities Maintenance & Repair Cost Data, call 877.801.5724.

General Inspection

Code compliance _____

Maintainability _____

Means of egress _____

Fire ratings _____

Audible & visual device condition _____

Extinguishing systems (see also *08 Mechanical/Plumbing*): _____

Type _____

Lighting system (see also *09 Electrical Lighting/Power*): _____

Type _____

Handicapped accessibility _____

Building user comments _____

_____

Exterior Lighting

Adequacy:

Good _____   Fair _____   Poor _____

Condition _____

Controls (type & location) _____

Fire Alarm Systems

_____	Panel visible	_____	Operational
_____	Pull station condition	_____	Detector conditions

Stairs and Ramps

_____	Exits marked	_____	Hardware operational
_____	Tripping hazards	_____	Surface conditions
_____	Lighting adequate	_____	Handrails

General Inspection

Overall Appearance

Good_____   Fair_____   Poor_____

- Evidence of moisture
- Irregular surface
- Handicapped hazards
- Visible settlement
- Tripping hazards
- Replacement necessary

Carpet (Tufted, tile)

_____ Age
_____ Stains
_____ Holes, tears

_____ Excessive wear
_____ Discoloration
_____ Seam conditions

Resilient (Asphalt tile, cork tile, linoleum, rubber, vinyl)

_____ Broken tiles
_____ Shrinkage
_____ Fading
_____ Porosity

_____ Loose tiles
_____ Lifting, cupping
_____ Cuts, holes
_____ Asbestos present

Masonry (Stone, brick)

_____ Cracks
_____ Joints
_____ Porosity

_____ Deterioration
_____ Stains
_____ Sealing

Monolithic Topping (Concrete, granolithic, terrazzo, magnesite)

_____ Cracks
_____ Joints

_____ Porosity
_____ Sealing

Wood (Plank, strips, block, parquet)

_____ Shrinkage
_____ Excessive wear
_____ Decay

_____ Cupping, warpage
_____ Unevenness
_____ Sealing

587

General Inspection

Overall appearance

Good _____   Fair _____   Poor _____

- Settlement or sagging
- Attachment
- Stains, discoloration
- Suitability
- Code compliance
- Alignment
- Evidence of moisture
- Missing units
- Acoustic quality

Exposed Systems (Unpainted, painted, spray-on, decorative)

_____ Cracks
_____ Surface deterioration
_____ Missing elements
_____ Adhesion

Applied to Structure & Suspended

General Condition:

Good _____   Fair _____   Poor _____

_____ Fasteners
_____ Trim condition

Openings:

_____ Panels
_____ Lighting fixtures
_____ Fire protection

_____ Inserts
_____ Air distribution
_____ Other

588

General Inspection (Passenger Conveying)

Overall appearance (interior)

Good _____   Fair _____   Poor _____

Overall appearance (exterior)

Good _____   Fair _____   Poor _____

_____ Maintenance history available

_____   Regular inspection frequency

_____   Door operations

_____   Control systems

_____   Noise

_____   Code compliance

_____   Handicapped access

_____   Major repairs necessary

_____   Replacement necessary

General Inspection

General appearance

Good ____   Fair ____   Poor ____

Leaks, dripping, running faucets and valves _____

Maintenance history _____

Supply adequacy _____

Sanitation hazards _____

Drain & waste connection _____

Backflow protection _____

Cross connections _____

Fixture quantity _____

Fixture types & conditions _____

Handicapped fixtures _____

Female facilities _____

Metal pipe & fittings corrosion _____

Pipe joints & sealing _____

Pipe insulation _____

Hanger supports & clamps _____

Filters _____

Building user comments _____

____ **Water System**

____ Water pressure adequate	____	Odors, tastes
____ Main cutoff operable	____	Water heating temperature setting
____ Pump condition	____	Insulation condition

____ **Sanitary & Storm System**

____ Flow adequate	____	Cleanouts access
____ Floor drains	____	Chemical resistance
____ Gradient	____	On-site disposal system

____ **Code Requirements**

____ EPA/local permits	____	Other

____ **Fire protection system**

____ Regular inspections	____	Sprinkler heads operable
____ Complies with code	____	Controls operable
____ Hose cabinets functional	____	Water pressure sufficient

General Inspection

General appearance

Good _____   Fair _____   Poor _____

Lubrication: bearings and moving parts _____

Rust and corrosion _____

Motors, fans, drive assemblies, and pumps _____

Wiring and electrical controls _____

Thermostats and automatic temperature controls _____

Thermal insulation and protective coatings _____

Guards, casings, hangers, supports, platforms, and mounting bolts _____

Piping system identification _____

Solenoid valves _____

Burner assemblies _____

Combustion chambers, smokepipes, and breeching _____

Electrical heating units _____

Guards, casings, hangers, supports, platforms, and mounting bolts _____

Steam and hot water heating equipment _____

Accessible steam, water, and fuel piping _____

Traps _____

Humidifier assemblies and controls _____

Strainers _____

Water sprays, weirs, and similar devices _____

Shell-and tube-type condensers _____

Self-contained evaporative condensers _____

Air cooled condensers _____

Compressors _____

Liquid receivers _____

Refrigerant driers, strainers, valves, oil traps, and accessories _____

Building user comments _____

591

Cleaning, maintenance, repair, and replacement

_____ Registers	_____ Grills
_____ Dampers	_____ Draft diverters
_____ Plenum chambers	_____ Supply and return ducts
_____ Louvers	_____ Fire dampers

Air filters

_____ Correct type	_____ Replacement schedule

Heating System Evaluation

_____ Heating capacity	_____ Temperature control
_____ Heating:	
_____ Seasonal	_____ All year
_____ Noise level	_____ Energy consumption
_____ Air circulation & ventilation	_____ Filtration
_____ Humidity control	

Cooling System Evaluation

_____ Cooling capacity	_____ Temperature and humidity control
_____ Cooling all season	_____ Noise level
_____ Energy consumption	_____ Air circulation & ventilation
_____ Filtration	_____ Reliability

Ventilation System Evaluation

_____ Air velocity	_____ Exhaust air systems
_____ Bag collection	_____ Wet collectors
_____ Steam and hot water coils	_____ Electrical heating units
_____ Fire hazards	_____ Fire protective devices

General Inspection

Safety conditions _____

Service capacity, % used, and age _____

Switchgear capacity, % used, and age _____

Feeder capacity, % used, and age _____

Panel capacity _____

Thermo-scanning _____

Maintenance records available _____

Convenience outlets _____

Building user comments _____

Exterior Service

_____ Line drawing
_____ Feed source:
      _____ Utility/owned   _____ Above/below ground
_____ Transformer:
      _____ Transformer tested   _____ Transformer arcing or burning
      _____ Transformer PCB's   _____ Ownership (facility or utility)

Interior Distribution System

_____ Line drawing   _____ Incoming conduit marked
_____ Main circuit breaker marked   _____ Panel boards, junction boxes covered
_____ All wiring in conduit   _____ Conduit properly secured
_____ Panels marked   _____ Panel schedules
_____ Missing breakers

Emergency Circuits

Emergency generator(s):

_____ Condition and age   _____ Auto start and switchover
_____ Testing schedule   _____ Test records available
_____ Service schedule   _____ Service schedule records available
_____ Circuits appropriate   _____ Cooling & exhaust
_____ Fuel storage (capacity)

Emergency lighting/power systems:

_____ Battery operation   _____ Separate power feed
_____ Exit signs   _____ Stairways/corridors
_____ Elevators   _____ Interior areas
_____ HVAC   _____ Exterior

593

For customer support on your Facilities Maintenance & Repair Cost Data, call 877.801.5724.

G20 SITE WORK INSPECTION CHECKLIST

General Inspection

Overall appearance

Good _____   Fair _____   Poor _____

Maintainability _____

Repairs/replacements _____

Code compliance _____

Roads, Walks, and Parking Lots

Surface conditions _____

Subsurface conditions _____

Settling and uplift _____

Cracks, holes _____

Drainage and slope _____

Curbing

Alignment _____

Erosion _____

Repairs/replacements _____

Drainage and Erosion Controls

Surface drainage _____

Manholes, catch basins _____

Vegetation _____

Channels, dikes _____

Retention, detention _____

Drains _____

Parking Lot Controls

Location _____

Operation _____

Repairs/replacements _____

H10 SAFETY INSPECTION CHECKLIST

General Inspection

Code compliance _____

Maintainability _____

Means of egress _____

Fire ratings _____

Audible & visual device condition _____

Extinguishing systems (see also *D20 Mechanical/Plumbing*): _____

      Type _____

Lighting system (see also *D50 Electrical Lighting/Power*): _____

      Type _____

Handicapped accessibility _____

Building user comments _____

_____

Exterior Lighting

Adequacy:

      Good _____   Fair _____   Poor_____

Condition _____

Controls (type & location) _____

Fire Alarm Systems

_____	Panel visible	_____	Operational
_____	Pull station condition	_____	Detector conditions

Stairs and Ramps

_____	Exits marked	_____	Hardware operational
_____	Tripping hazards	_____	Surface conditions
_____	Lighting adequate	_____	Handrails

595

H10 SAFETY INSPECTION CHECKLIST

General Inspection

Code compliance _____

Maintainability _____

Means of egress _____

Fire ratings _____

Audible & visual device condition _____

Extinguishing systems (see also *D20 Mechanical/Plumbing*): _____

      Type _____

Lighting system (see also *D50 Electrical Lighting/Power*): _____

      Type _____

Handicapped accessibility _____

Building user comments _____

_____

Exterior Lighting

Adequacy:

    Good _____   Fair _____   Poor_____

Condition _____

Controls (type & location) _____

Fire Alarm Systems

_____	Panel visible	_____	Operational
_____	Pull station condition	_____	Detector conditions

Stairs and Ramps

_____	Exits marked	_____	Hardware operational
_____	Tripping hazards	_____	Surface conditions
_____	Lighting adequate	_____	Handrails

Reference Section

All the reference information is in one section, making it easy to find what you need to know ...and easy to use the book on a daily basis. This section is visually identified by a vertical black bar on the page edges.

In this Reference Section we've included information on Equipment Rental Costs, Crew Listings, Travel Costs, Reference Tables, Life Cycle Costing, and a listing of Abbreviations.

Table of Contents

Estimating Tips

- This section contains the average costs to rent and operate hundreds of pieces of construction equipment. This is useful information when estimating the time and material requirements of any particular operation in order to establish a unit or total cost. Equipment costs include not only rental, but also operating costs for equipment under normal use.

Rental Costs

- Equipment rental rates are obtained from the following industry sources throughout North America: contractors, suppliers, dealers, manufacturers, and distributors.

- Rental rates vary throughout the country, with larger cities generally having lower rates. Lease plans for new equipment are available for periods in excess of six months, with a percentage of payments applying toward purchase.

- Monthly rental rates vary from 2% to 5% of the purchase price of the equipment depending on the anticipated life of the equipment and its wearing parts.

- Weekly rental rates are about 1/3 the monthly rates, and daily rental rates are about 1/3 the weekly rate.

- Rental rates can also be treated as reimbursement costs for contractor-owned equipment. Owned equipment costs include depreciation, loan payments, interest, taxes, insurance, storage, and major repairs.

Operating Costs

- The operating costs include parts and labor for routine servicing, such as repair and replacement of pumps, filters and worn lines. Normal operating expendables, such as fuel, lubricants, tires and electricity (where applicable), are also included.

- Extraordinary operating expendables with highly variable wear patterns, such as diamond bits and blades, are excluded. These costs can be found as material costs in the Unit Price section.

- The hourly operating costs listed do not include the operator's wages.

Equipment Cost/Day

- Any power equipment required by a crew is shown in the Crew Listings with a daily cost.

- The daily cost of equipment needed by a crew is based on dividing the weekly rental rate by 5 (number of working days in the week), and then adding the hourly operating cost times 8 (the number of hours in a day). This "Equipment Cost/ Day" is shown in the far right column of the Equipment Rental pages.

- If equipment is needed for only one or two days, it is best to develop your own cost by including components for daily rent and hourly operating cost. This is important when the listed Crew for a task does not contain the equipment needed, such as a crane for lifting mechanical heating/ cooling equipment up onto a roof.

- If the quantity of work is less than the crew's Daily Output shown for a Unit Price line item that includes a bare unit equipment cost, it is recommended to estimate one day's rental cost and operating cost for equipment shown in the Crew Listing for that line item.

Mobilization/ Demobilization

- The cost to move construction equipment from an equipment yard or rental company to the job site and back again is not included in equipment rental costs listed in the Reference Section, nor in the bare equipment cost of any unit price line item, nor in any equipment costs shown in the Crew listings.

- Mobilization (to the site) and demobilization (from the site) costs can be found in the Unit Price Section.

- If a piece of equipment is already at the job site, it is not appropriate to utilize mobilization/demobilization costs again in an estimate. ∎

01 54 33 | Equipment Rental

		UNIT	HOURLY OPER. COST	RENT PER DAY	RENT PER WEEK	RENT PER MONTH	EQUIPMENT COST/DAY	
10	**0010**	**CONCRETE EQUIPMENT RENTAL** without operators R015433 -10							**10**
	0200	Bucket, concrete lightweight, 1/2 C.Y.	Ea.	.80	23.50	70	210	20.40	
	0300	1 C.Y.		.90	27.50	83	249	23.80	
	0400	1-1/2 C.Y.		1.10	36.50	110	330	30.80	
	0500	2 C.Y.		1.25	45	135	405	37	
	0580	8 C.Y.		6.10	257	770	2,300	202.80	
	0600	Cart, concrete, self-propelled, operator walking, 10 C.F.		3.25	56.50	170	510	60	
	0700	Operator riding, 18 C.F.		5.35	95	285	855	99.80	
	0800	Conveyer for concrete, portable, gas, 16" wide, 26' long		12.60	123	370	1,100	174.80	
	0900	46' long		13.00	148	445	1,325	193	
	1000	56' long		13.10	157	470	1,400	198.80	
	1100	Core drill, electric, 2-1/2 H.P., 1" to 8" bit diameter		1.77	68.50	205	615	55.15	
	1150	11 H.P., 8" to 18" cores		5.95	113	340	1,025	115.60	
	1200	Finisher, concrete floor, gas, riding trowel, 96" wide		12.75	145	435	1,300	189	
	1300	Gas, walk-behind, 3 blade, 36" trowel		2.15	20	60	180	29.20	
	1400	4 blade, 48" trowel		4.20	27.50	83	249	50.20	
	1500	Float, hand-operated (Bull float) 48" wide		.08	13.35	40	120	8.65	
	1570	Curb builder, 14 H.P., gas, single screw		14.65	248	745	2,225	266.20	
	1590	Double screw		15.35	290	870	2,600	296.80	
	1600	Floor grinder, concrete and terrazzo, electric, 22" path		2.57	158	475	1,425	115.55	
	1700	Edger, concrete, electric, 7" path		1.04	51.50	155	465	39.30	
	1750	Vacuum pick-up system for floor grinders, wet/dry		1.49	81.50	245	735	60.90	
	1800	Mixer, powered, mortar and concrete, gas, 6 C.F., 18 H.P.		8.65	118	355	1,075	140.20	
	1900	10 C.F., 25 H.P.		10.80	143	430	1,300	172.40	
	2000	16 C.F.		11.15	165	495	1,475	188.20	
	2100	Concrete, stationary, tilt drum, 2 C.Y.		6.90	232	695	2,075	194.20	
	2120	Pump, concrete, truck mounted 4" line 80' boom		25.00	865	2,600	7,800	720	
	2140	5" line, 110' boom		32.40	1,150	3,435	10,300	946.20	
	2160	Mud jack, 50 C.F. per hr.		7.30	125	375	1,125	133.40	
	2180	225 C.F. per hr.		9.50	145	435	1,300	163	
	2190	Shotcrete pump rig, 12 C.Y./hr.		14.85	220	660	1,975	250.80	
	2200	35 C.Y./hr.		17.50	235	705	2,125	281	
	2600	Saw, concrete, manual, gas, 18 H.P.		6.75	45	135	405	81	
	2650	Self-propelled, gas, 30 H.P.		13.30	102	305	915	167.40	
	2675	V-groove crack chaser, manual, gas, 6 H.P.		2.35	17.35	52	156	29.20	
	2700	Vibrators, concrete, electric, 60 cycle, 2 H.P.		.46	8.65	26	78	8.90	
	2800	3 H.P.		.60	11.65	35	105	11.80	
	2900	Gas engine, 5 H.P.		2.00	16	48	144	25.60	
	3000	8 H.P.		2.75	15.35	46	138	31.20	
	3050	Vibrating screed, gas engine, 8 H.P.		2.91	71.50	215	645	66.30	
	3120	Concrete transit mixer, 6 x 4, 250 H.P., 8 C.Y., rear discharge		61.55	570	1,715	5,150	835.40	
	3200	Front discharge		72.45	700	2,095	6,275	998.60	
	3300	6 x 6, 285 H.P., 12 C.Y., rear discharge		71.50	660	1,980	5,950	968	
	3400	Front discharge		74.00	705	2,120	6,350	1,016	
20	**0010**	**EARTHWORK EQUIPMENT RENTAL** without operators R015433 -10							**2**
	0040	Aggregate spreader, push type 8' to 12' wide	Ea.	3.05	25.50	76	228	39.60	
	0045	Tailgate type, 8' wide		2.90	32.50	98	294	42.80	
	0055	Earth auger, truck-mounted, for fence & sign posts, utility poles		19.15	440	1,320	3,950	417.20	
	0060	For borings and monitoring wells		46.65	675	2,030	6,100	779.20	
	0070	Portable, trailer mounted		3.30	32.50	98	294	46	
	0075	Truck-mounted, for caissons, water wells		100.15	2,900	8,705	26,100	2,542	
	0080	Horizontal boring machine, 12" to 36" diameter, 45 H.P.		24.20	192	575	1,725	308.60	
	0090	12" to 48" diameter, 65 H.P.		34.35	335	1,000	3,000	474.80	
	0095	Auger, for fence posts, gas engine, hand held		.60	6	18	54	8.40	
	0100	Excavator, diesel hydraulic, crawler mounted, 1/2 C.Y. cap.		24.65	420	1,255	3,775	448.20	
	0120	5/8 C.Y. capacity		32.55	550	1,645	4,925	589.40	
	0140	3/4 C.Y. capacity		35.60	615	1,850	5,550	654.80	
	0150	1 C.Y. capacity		49.70	690	2,075	6,225	812.60	

01 54 33 | Equipment Rental

		UNIT	HOURLY OPER. COST	RENT PER DAY	RENT PER WEEK	RENT PER MONTH	EQUIPMENT COST/DAY	
0200	1-1/2 C.Y. capacity	Ea.	59.80	920	2,760	8,275	1,030	20
0300	2 C.Y. capacity		67.35	1,050	3,180	9,550	1,175	
0320	2-1/2 C.Y. capacity		102.55	1,300	3,925	11,800	1,605	
0325	3-1/2 C.Y. capacity		144.45	2,150	6,420	19,300	2,440	
0330	4-1/2 C.Y. capacity		173.40	2,625	7,880	23,600	2,963	
0335	6 C.Y. capacity		219.80	2,900	8,680	26,000	3,494	
0340	7 C.Y. capacity		222.05	3,025	9,070	27,200	3,590	
0342	Excavator attachments, bucket thumbs		3.20	245	735	2,200	172.60	
0345	Grapples		2.75	193	580	1,750	138	
0346	Hydraulic hammer for boom mounting, 4000 ft lb.		12.40	350	1,045	3,125	308.20	
0347	5000 ft lb.		14.40	425	1,280	3,850	371.20	
0348	8000 ft lb.		21.30	625	1,875	5,625	545.40	
0349	12,000 ft lb.		23.35	750	2,245	6,725	635.80	
0350	Gradall type, truck mounted, 3 ton @ 15' radius, 5/8 C.Y.		43.65	890	2,665	8,000	882.20	
0370	1 C.Y. capacity		47.85	1,025	3,095	9,275	1,002	
0400	Backhoe-loader, 40 to 45 H.P., 5/8 C.Y. capacity		14.90	240	720	2,150	263.20	
0450	45 H.P. to 60 H.P., 3/4 C.Y. capacity		23.40	295	885	2,650	364.20	
0460	80 H.P., 1-1/4 C.Y. capacity		25.60	310	935	2,800	391.80	
0470	112 H.P., 1-1/2 C.Y. capacity		40.95	645	1,930	5,800	713.60	
0482	Backhoe-loader attachment, compactor, 20,000 lb.		5.80	142	425	1,275	131.40	
0485	Hydraulic hammer, 750 ft lb.		3.30	96.50	290	870	84.40	
0486	Hydraulic hammer, 1200 ft lb.		6.30	217	650	1,950	180.40	
0500	Brush chipper, gas engine, 6" cutter head, 35 H.P.		11.10	105	315	945	151.80	
0550	Diesel engine, 12" cutter head, 130 H.P.		27.55	285	855	2,575	391.40	
0600	15" cutter head, 165 H.P.		33.10	335	1,005	3,025	465.80	
0750	Bucket, clamshell, general purpose, 3/8 C.Y.		1.30	38.50	115	345	33.40	
0800	1/2 C.Y.		1.40	45	135	405	38.20	
0850	3/4 C.Y.		1.55	55	165	495	45.40	
0900	1 C.Y.		1.60	58.50	175	525	47.80	
0950	1-1/2 C.Y.		2.55	81.50	245	735	69.40	
1000	2 C.Y.		2.70	90	270	810	75.60	
1010	Bucket, dragline, medium duty, 1/2 C.Y.		.75	23.50	70	210	20	
1020	3/4 C.Y.		.75	24.50	73	219	20.60	
1030	1 C.Y.		.80	26	78	234	22	
1040	1-1/2 C.Y.		1.20	40	120	360	33.60	
1050	2 C.Y.		1.30	45	135	405	37.40	
1070	3 C.Y.		1.95	61.50	185	555	52.60	
1200	Compactor, manually guided 2-drum vibratory smooth roller, 7.5 H.P.		7.25	203	610	1,825	180	
1250	Rammer/tamper, gas, 8"		2.75	46.50	140	420	50	
1260	15"		3.05	53.50	160	480	56.40	
1300	Vibratory plate, gas, 18" plate, 3000 lb. blow		2.70	24.50	73	219	36.20	
1350	21" plate, 5000 lb. blow		3.40	32.50	98	294	46.80	
1370	Curb builder/extruder, 14 H.P., gas, single screw		14.65	248	745	2,225	266.20	
1390	Double screw		15.35	290	870	2,600	296.80	
1500	Disc harrow attachment, for tractor		.44	73.50	221	665	47.70	
1810	Feller buncher, shearing & accumulating trees, 100 H.P.		48.30	755	2,260	6,775	838.40	
1860	Grader, self-propelled, 25,000 lb.		39.50	640	1,925	5,775	701	
1910	30,000 lb.		43.15	650	1,955	5,875	736.20	
1920	40,000 lb.		64.90	1,100	3,275	9,825	1,174	
1930	55,000 lb.		84.15	1,700	5,130	15,400	1,699	
1950	Hammer, pavement breaker, self-propelled, diesel, 1000 to 1250 lb.		29.90	350	1,055	3,175	450.20	
2000	1300 to 1500 lb.		44.85	705	2,110	6,325	780.80	
2050	Pile driving hammer, steam or air, 4150 ft lb. @ 225 bpm		10.40	475	1,420	4,250	367.20	
2100	8750 ft lb. @ 145 bpm		12.40	660	1,975	5,925	494.20	
2150	15,000 ft lb. @ 60 bpm		14.00	790	2,370	7,100	586	
2200	24,450 ft lb. @ 111 bpm		15.00	875	2,630	7,900	646	
2250	Leads, 60' high for pile driving hammers up to 20,000 ft lb.		3.30	80.50	242	725	74.80	
2300	90' high for hammers over 20,000 ft lb.		4.95	141	424	1,275	124.40	

01 54 33 | Equipment Rental

		UNIT	HOURLY OPER. COST	RENT PER DAY	RENT PER WEEK	RENT PER MONTH	EQUIPMENT COST/DAY	
20	2350	Diesel type hammer, 22,400 ft lb.	Ea.	22.55	460	1,380	4,150	456.40	20
	2400	41,300 ft lb.		32.00	540	1,625	4,875	581	
	2450	141,000 ft lb.		51.65	960	2,875	8,625	988.20	
	2500	Vib. elec. hammer/extractor, 200 kW diesel generator, 34 H.P.		54.90	635	1,900	5,700	819.20	
	2550	80 H.P.		100.65	925	2,775	8,325	1,360	
	2600	150 H.P.		190.90	1,775	5,300	15,900	2,587	
	2800	Log chipper, up to 22" diameter, 600 H.P.		63.30	640	1,915	5,750	889.40	
	2850	Logger, for skidding & stacking logs, 150 H.P.		52.75	815	2,445	7,325	911	
	2860	Mulcher, diesel powered, trailer mounted		24.05	212	635	1,900	319.40	
	2900	Rake, spring tooth, with tractor		18.23	345	1,042	3,125	354.25	
	3000	Roller, vibratory, tandem, smooth drum, 20 H.P.		8.80	145	435	1,300	157.40	
	3050	35 H.P.		11.10	247	740	2,225	236.80	
	3100	Towed type vibratory compactor, smooth drum, 50 H.P.		26.00	315	940	2,825	396	
	3150	Sheepsfoot, 50 H.P.		27.35	345	1,035	3,100	425.80	
	3170	Landfill compactor, 220 H.P.		91.00	1,475	4,440	13,300	1,616	
	3200	Pneumatic tire roller, 80 H.P.		16.80	350	1,050	3,150	344.40	
	3250	120 H.P.		25.20	600	1,800	5,400	561.60	
	3300	Sheepsfoot vibratory roller, 240 H.P.		68.95	1,100	3,270	9,800	1,206	
	3320	340 H.P.		100.85	1,575	4,730	14,200	1,753	
	3350	Smooth drum vibratory roller, 75 H.P.		24.85	585	1,755	5,275	549.80	
	3400	125 H.P.		32.55	710	2,135	6,400	687.40	
	3410	Rotary mower, brush, 60", with tractor		22.60	315	950	2,850	370.80	
	3420	Rototiller, walk-behind, gas, 5 H.P.		1.96	51.50	155	465	46.70	
	3422	8 H.P.		3.08	80	240	720	72.65	
	3440	Scrapers, towed type, 7 C.Y. capacity		5.80	113	340	1,025	114.40	
	3450	10 C.Y. capacity		6.60	157	470	1,400	146.80	
	3500	15 C.Y. capacity		7.10	180	540	1,625	164.80	
	3525	Self-propelled, single engine, 14 C.Y. capacity		114.76	1,600	4,830	14,500	1,884	
	3550	Dual engine, 21 C.Y. capacity		173.80	2,175	6,490	19,500	2,688	
	3600	31 C.Y. capacity		231.70	3,075	9,195	27,600	3,693	
	3640	44 C.Y. capacity		287.80	3,950	11,885	35,700	4,679	
	3650	Elevating type, single engine, 11 C.Y. capacity		71.45	985	2,955	8,875	1,163	
	3700	22 C.Y. capacity		139.55	2,350	7,025	21,100	2,521	
	3710	Screening plant 110 H.P. w/5' x 10' screen		25.35	400	1,200	3,600	442.80	
	3720	5' x 16' screen		29.65	505	1,515	4,550	540.20	
	3850	Shovel, crawler-mounted, front-loading, 7 C.Y. capacity		258.85	3,200	9,610	28,800	3,993	
	3855	12 C.Y. capacity		386.45	4,400	13,220	39,700	5,736	
	3860	Shovel/backhoe bucket, 1/2 C.Y.		2.50	66.50	200	600	60	
	3870	3/4 C.Y.		2.55	75	225	675	65.40	
	3880	1 C.Y.		2.65	83.50	250	750	71.20	
	3890	1-1/2 C.Y.		2.75	96.50	290	870	80	
	3910	3 C.Y.		3.15	130	390	1,175	103.20	
	3950	Stump chipper, 18" deep, 30 H.P.		7.82	205	615	1,850	185.55	
	4110	Dozer, crawler, torque converter, diesel 80 H.P.		29.05	400	1,200	3,600	472.40	
	4150	105 H.P.		35.60	530	1,590	4,775	602.80	
	4200	140 H.P.		51.25	795	2,380	7,150	886	
	4260	200 H.P.		77.25	1,275	3,845	11,500	1,387	
	4310	300 H.P.		100.65	1,825	5,460	16,400	1,897	
	4360	410 H.P.		132.60	2,250	6,720	20,200	2,405	
	4370	500 H.P.		171.15	3,225	9,650	29,000	3,299	
	4380	700 H.P.		280.90	4,550	13,660	41,000	4,979	
	4400	Loader, crawler, torque conv., diesel, 1-1/2 C.Y., 80 H.P.		31.25	455	1,360	4,075	522	
	4450	1-1/2 to 1-3/4 C.Y., 95 H.P.		34.60	590	1,765	5,300	629.80	
	4510	1-3/4 to 2-1/4 C.Y., 130 H.P.		53.80	875	2,630	7,900	956.40	
	4530	2-1/2 to 3-1/4 C.Y., 190 H.P.		65.95	1,100	3,300	9,900	1,188	
	4560	3-1/2 to 5 C.Y., 275 H.P.		87.80	1,425	4,295	12,900	1,561	
	4610	Front end loader, 4WD, articulated frame, diesel, 1 to 1-1/4 C.Y., 70 H.P.		19.75	235	705	2,125	299	
	4620	1-1/2 to 1-3/4 C.Y., 95 H.P.		24.80	300	900	2,700	378.40	

01 54 33 | Equipment Rental

		UNIT	HOURLY OPER. COST	RENT PER DAY	RENT PER WEEK	RENT PER MONTH	EQUIPMENT COST/DAY	
20	4650	1-3/4 to 2 C.Y., 130 H.P.	Ea.	29.15	380	1,140	3,425	461.20	20
	4710	2-1/2 to 3-1/2 C.Y., 145 H.P.		33.10	425	1,275	3,825	519.80	
	4730	3 to 4-1/2 C.Y., 185 H.P.		42.40	550	1,650	4,950	669.20	
	4760	5-1/4 to 5-3/4 C.Y., 270 H.P.		67.50	905	2,710	8,125	1,082	
	4810	7 to 9 C.Y., 475 H.P.		114.05	1,700	5,125	15,400	1,937	
	4870	9 - 11 C.Y., 620 H.P.		160.55	2,775	8,290	24,900	2,942	
	4880	Skid steer loader, wheeled, 10 C.F., 30 H.P. gas		10.45	150	450	1,350	173.60	
	4890	1 C.Y., 78 H.P., diesel		20.20	262	785	2,350	318.60	
	4892	Skid-steer attachment, auger		.48	80.50	242	725	52.25	
	4893	Backhoe		.66	111	332	995	71.70	
	4894	Broom		.74	124	372	1,125	80.30	
	4895	Forks		.22	36	108	325	23.35	
	4896	Grapple		.56	92.50	278	835	60.10	
	4897	Concrete hammer		1.06	177	531	1,600	114.70	
	4898	Tree spade		1.03	172	515	1,550	111.25	
	4899	Trencher		.54	90	270	810	58.30	
	4900	Trencher, chain, boom type, gas, operator walking, 12 H.P.		5.00	46.50	140	420	68	
	4910	Operator riding, 40 H.P.		20.15	290	870	2,600	335.20	
	5000	Wheel type, diesel, 4' deep, 12" wide		85.95	845	2,535	7,600	1,195	
	5100	6' deep, 20" wide		92.75	1,925	5,740	17,200	1,890	
	5150	Chain type, diesel, 5' deep, 8" wide		38.35	560	1,675	5,025	641.80	
	5200	Diesel, 8' deep, 16" wide		155.95	3,600	10,770	32,300	3,402	
	5202	Rock trencher, wheel type, 6" wide x 18" deep		21.50	335	1,000	3,000	372	
	5206	Chain type, 18" wide x 7' deep		112.00	2,800	8,395	25,200	2,575	
	5210	Tree spade, self-propelled		14.38	267	800	2,400	275.05	
	5250	Truck, dump, 2-axle, 12 ton, 8 C.Y. payload, 220 H.P.		34.40	227	680	2,050	411.20	
	5300	Three axle dump, 16 ton, 12 C.Y. payload, 400 H.P.		61.00	340	1,015	3,050	691	
	5310	Four axle dump, 25 ton, 18 C.Y. payload, 450 H.P.		71.75	490	1,475	4,425	869	
	5350	Dump trailer only, rear dump, 16-1/2 C.Y.		5.45	138	415	1,250	126.60	
	5400	20 C.Y.		5.90	157	470	1,400	141.20	
	5450	Flatbed, single axle, 1-1/2 ton rating		25.55	68.50	205	615	245.40	
	5500	3 ton rating		30.65	96.50	290	870	303.20	
	5550	Off highway rear dump, 25 ton capacity		73.90	1,275	3,800	11,400	1,351	
	5600	35 ton capacity		82.50	1,425	4,260	12,800	1,512	
	5610	50 ton capacity		102.65	1,700	5,080	15,200	1,837	
	5620	65 ton capacity		105.65	1,700	5,090	15,300	1,863	
	5630	100 ton capacity		151.40	2,850	8,570	25,700	2,925	
	6000	Vibratory plow, 25 H.P., walking		8.45	60	180	540	103.60	
40	0010	**GENERAL EQUIPMENT RENTAL** without operators R015433 -10							40
	0150	Aerial lift, scissor type, to 15' high, 1000 lb. cap., electric	Ea.	3.05	51.50	155	465	55.40	
	0160	To 25' high, 2000 lb. capacity		3.45	66.50	200	600	67.60	
	0170	Telescoping boom to 40' high, 500 lb. capacity, diesel		13.55	320	965	2,900	301.40	
	0180	To 45' high, 500 lb. capacity		14.75	350	1,055	3,175	329	
	0190	To 60' high, 600 lb. capacity		17.20	495	1,490	4,475	435.60	
	0195	Air compressor, portable, 6.5 CFM, electric		.67	12.65	38	114	12.95	
	0196	Gasoline		.81	19	57	171	17.90	
	0200	Towed type, gas engine, 60 CFM		13.55	48.50	145	435	137.40	
	0300	160 CFM		15.80	50	150	450	156.40	
	0400	Diesel engine, rotary screw, 250 CFM		17.05	108	325	975	201.40	
	0500	365 CFM		23.10	132	395	1,175	263.80	
	0550	450 CFM		29.35	165	495	1,475	333.80	
	0600	600 CFM		51.70	228	685	2,050	550.60	
	0700	750 CFM		51.90	237	710	2,125	557.20	
	0800	For silenced models, small sizes, add to rent		3%	5%	5%	5%		
	0900	Large sizes, add to rent		5%	7%	7%	7%		
	0930	Air tools, breaker, pavement, 60 lb.	Ea.	.50	9.65	29	87	9.80	
	0940	80 lb.		.50	10.35	31	93	10.20	
	0950	Drills, hand (jackhammer) 65 lb.		.60	17.35	52	156	15.20	

01 54 33 | Equipment Rental

		UNIT	HOURLY OPER. COST	RENT PER DAY	RENT PER WEEK	RENT PER MONTH	EQUIPMENT COST/DAY	
0960	Track or wagon, swing boom, 4" drifter	Ea.	62.50	880	2,640	7,925	1,028	40
0970	5" drifter		74.45	1,075	3,235	9,700	1,243	
0975	Track mounted quarry drill, 6" diameter drill		128.10	1,600	4,765	14,300	1,978	
0980	Dust control per drill		1.02	23.50	71	213	22.35	
0990	Hammer, chipping, 12 lb.		.55	26	78	234	20	
1000	Hose, air with couplings, 50' long, 3/4" diameter		.03	5	15	45	3.25	
1100	1" diameter		.04	6.35	19	57	4.10	
1200	1-1/2" diameter		.05	9	27	81	5.80	
1300	2" diameter		.07	12	36	108	7.75	
1400	2-1/2" diameter		.11	19	57	171	12.30	
1410	3" diameter		.14	23	69	207	14.90	
1450	Drill, steel, 7/8" x 2'		.05	8.65	26	78	5.60	
1460	7/8" x 6'		.05	9	27	81	5.80	
1520	Moil points		.02	3.33	10	30	2.15	
1525	Pneumatic nailer w/accessories		.58	38.50	115	345	27.65	
1530	Sheeting driver for 60 lb. breaker		.04	6	18	54	3.90	
1540	For 90 lb. breaker		.12	8	24	72	5.75	
1550	Spade, 25 lb.		.45	6.65	20	60	7.60	
1560	Tamper, single, 35 lb.		.55	36.50	109	325	26.20	
1570	Triple, 140 lb.		.82	54.50	164	490	39.35	
1580	Wrenches, impact, air powered, up to 3/4" bolt		.40	12.65	38	114	10.80	
1590	Up to 1-1/4" bolt		.50	23.50	70	210	18	
1600	Barricades, barrels, reflectorized, 1 to 99 barrels		.03	4.60	13.80	41.50	3	
1610	100 to 200 barrels		.02	3.53	10.60	32	2.30	
1620	Barrels with flashers, 1 to 99 barrels		.03	5.25	15.80	47.50	3.40	
1630	100 to 200 barrels		.03	4.20	12.60	38	2.75	
1640	Barrels with steady burn type C lights		.04	7	21	63	4.50	
1650	Illuminated board, trailer mounted, with generator		3.50	130	390	1,175	106	
1670	Portable barricade, stock, with flashers, 1 to 6 units		.03	5.25	15.80	47.50	3.40	
1680	25 to 50 units		.03	4.90	14.70	44	3.20	
1685	Butt fusion machine, wheeled, 1.5 HP electric, 2" - 8" diameter pipe		2.63	167	500	1,500	121.05	
1690	Tracked, 20 HP diesel, 4"-12" diameter pipe		11.14	490	1,465	4,400	382.10	
1695	83 HP diesel, 8" - 24" diameter pipe		30.46	975	2,930	8,800	829.70	
1700	Carts, brick, hand powered, 1000 lb. capacity		.50	83.50	251	755	54.20	
1800	Gas engine, 1500 lb., 7-1/2' lift		4.17	115	345	1,025	102.35	
1822	Dehumidifier, medium, 6 lb./hr., 150 CFM		1.00	62	186	560	45.20	
1824	Large, 18 lb./hr., 600 CFM		2.02	126	378	1,125	91.75	
1830	Distributor, asphalt, trailer mounted, 2000 gal., 38 H.P. diesel		9.95	325	980	2,950	275.60	
1840	3000 gal., 38 H.P. diesel		11.45	355	1,070	3,200	305.60	
1850	Drill, rotary hammer, electric		1.02	26	78	234	23.75	
1860	Carbide bit, 1-1/2" diameter, add to electric rotary hammer		.02	3.61	10.84	32.50	2.35	
1865	Rotary, crawler, 250 H.P.		148.50	2,050	6,185	18,600	2,425	
1870	Emulsion sprayer, 65 gal., 5 H.P. gas engine		2.91	97	291	875	81.50	
1880	200 gal., 5 H.P. engine		7.85	162	485	1,450	159.80	
1900	Floor auto-scrubbing machine, walk-behind, 28" path		4.98	325	970	2,900	233.85	
1930	Floodlight, mercury vapor, or quartz, on tripod, 1000 watt		.44	20.50	62	186	15.90	
1940	2000 watt		.83	41	123	370	31.25	
1950	Floodlights, trailer mounted with generator, 1 - 300 watt light		3.65	71.50	215	645	72.20	
1960	2 - 1000 watt lights		4.85	96.50	290	870	96.80	
2000	4 - 300 watt lights		4.55	93.50	280	840	92.40	
2005	Foam spray rig, incl. box trailer, compressor, generator, proportioner		35.03	490	1,465	4,400	573.25	
2020	Forklift, straight mast, 12' lift, 5000 lb., 2 wheel drive, gas		26.35	202	605	1,825	331.80	
2040	21' lift, 5000 lb., 4 wheel drive, diesel		20.90	240	720	2,150	311.20	
2050	For rough terrain, 42' lift, 35' reach, 9000 lb., 110 H.P.		29.55	485	1,450	4,350	526.40	
2060	For plant, 4 ton capacity, 80 H.P., 2 wheel drive, gas		16.15	93.50	280	840	185.20	
2080	10 ton capacity, 120 H.P., 2 wheel drive, diesel		24.10	162	485	1,450	289.80	
2100	Generator, electric, gas engine, 1.5 kW to 3 kW		3.75	11.35	34	102	36.80	
2200	5 kW		4.80	15.35	46	138	47.60	

		UNIT	HOURLY OPER. COST	RENT PER DAY	RENT PER WEEK	RENT PER MONTH	EQUIPMENT COST/DAY
2300	10 kW	Ea.	9.10	36.50	110	330	94.80
2400	25 kW		10.60	85	255	765	135.80
2500	Diesel engine, 20 kW		12.10	66.50	200	600	136.80
2600	50 kW		23.35	103	310	930	248.80
2700	100 kW		43.30	128	385	1,150	423.40
2800	250 kW		85.15	235	705	2,125	822.20
2850	Hammer, hydraulic, for mounting on boom, to 500 ft lb.		2.55	75	225	675	65.40
2860	1000 ft lb.		4.35	127	380	1,150	110.80
2900	Heaters, space, oil or electric, 50 MBH		2.01	7.65	23	69	20.70
3000	100 MBH		3.76	10.65	32	96	36.50
3100	300 MBH		11.03	38.50	115	345	111.25
3150	500 MBH		18.15	45	135	405	172.20
3200	Hose, water, suction with coupling, 20' long, 2" diameter		.02	3	9	27	1.95
3210	3" diameter		.03	4.33	13	39	2.85
3220	4" diameter		.03	5	15	45	3.25
3230	6" diameter		.11	17.65	53	159	11.50
3240	8" diameter		.27	44.50	133	400	28.75
3250	Discharge hose with coupling, 50' long, 2" diameter		.01	1.33	4	12	.90
3260	3" diameter		.01	2.33	7	21	1.50
3270	4" diameter		.02	3.67	11	33	2.35
3280	6" diameter		.06	9.35	28	84	6.10
3290	8" diameter		.18	30	90	270	19.45
3295	Insulation blower		.78	6	18	54	9.85
3300	Ladders, extension type, 16' to 36' long		.14	22.50	68	204	14.70
3400	40' to 60' long		.19	31	93	279	20.10
3405	Lance for cutting concrete		2.35	71.50	215	645	61.80
3407	Lawn mower, rotary, 22", 5 H.P.		1.80	49	147	440	43.80
3408	48" self propelled		2.98	75	225	675	68.85
3410	Level, electronic, automatic, with tripod and leveling rod		1.20	75.50	226	680	54.80
3430	Laser type, for pipe and sewer line and grade		.73	48.50	145	435	34.85
3440	Rotating beam for interior control		.78	52	156	470	37.45
3460	Builder's optical transit, with tripod and rod		.09	14.65	44	132	9.50
3500	Light towers, towable, with diesel generator, 2000 watt		4.55	93.50	280	840	92.40
3600	4000 watt		4.85	96.50	290	870	96.80
3700	Mixer, powered, plaster and mortar, 6 C.F., 7 H.P.		2.70	20	60	180	33.60
3800	10 C.F., 9 H.P.		2.80	31.50	94	282	41.20
3850	Nailer, pneumatic		.58	38.50	115	345	27.65
3900	Paint sprayers complete, 8 CFM		1.04	69.50	208	625	49.90
4000	17 CFM		1.90	127	380	1,150	91.20
4020	Pavers, bituminous, rubber tires, 8' wide, 50 H.P., diesel		31.15	490	1,465	4,400	542.20
4030	10' wide, 150 H.P.		106.10	1,800	5,370	16,100	1,923
4050	Crawler, 8' wide, 100 H.P., diesel		88.65	1,800	5,385	16,200	1,786
4060	10' wide, 150 H.P.		113.15	2,200	6,610	19,800	2,227
4070	Concrete paver, 12' to 24' wide, 250 H.P.		101.40	1,550	4,675	14,000	1,746
4080	Placer-spreader-trimmer, 24' wide, 300 H.P.		150.25	2,650	7,960	23,900	2,794
4100	Pump, centrifugal gas pump, 1-1/2" diam., 65 GPM		3.90	48.50	145	435	60.20
4200	2" diameter, 130 GPM		5.35	53.50	160	480	74.80
4300	3" diameter, 250 GPM		5.65	55	165	495	78.20
4400	6" diameter, 1500 GPM		30.45	172	515	1,550	346.60
4500	Submersible electric pump, 1-1/4" diameter, 55 GPM		.39	16.35	49	147	12.90
4600	1-1/2" diameter, 83 GPM		.43	18.65	56	168	14.65
4700	2" diameter, 120 GPM		1.50	23.50	70	210	26
4800	3" diameter, 300 GPM		2.69	41.50	125	375	46.50
4900	4" diameter, 560 GPM		12.34	158	475	1,425	193.70
5000	6" diameter, 1590 GPM		18.40	212	635	1,900	274.20
5100	Diaphragm pump, gas, single, 1-1/2" diameter		1.18	49.50	148	445	39.05
5200	2" diameter		4.25	61.50	185	555	71
5300	3" diameter		4.25	61.50	185	555	71

01 54 33 | Equipment Rental

		UNIT	HOURLY OPER. COST	RENT PER DAY	RENT PER WEEK	RENT PER MONTH	EQUIPMENT COST/DAY
5400	Double, 4" diameter	Ea.	6.40	105	315	945	114.20
5450	Pressure washer 5 GPM, 3000 psi		4.90	51.50	155	465	70.20
5460	7 GPM, 3000 psi		6.45	60	180	540	87.60
5500	Trash pump, self-priming, gas, 2" diameter		4.60	21	63	189	49.40
5600	Diesel, 4" diameter		9.15	88.50	265	795	126.20
5650	Diesel, 6" diameter		25.25	147	440	1,325	290
5655	Grout Pump		26.35	268	805	2,425	371.80
5700	Salamanders, L.P. gas fired, 100,000 Btu		3.96	13.65	41	123	39.90
5705	50,000 Btu		2.21	10.35	31	93	23.90
5720	Sandblaster, portable, open top, 3 C.F. capacity		.55	26.50	80	240	20.40
5730	6 C.F. capacity		.95	40	120	360	31.60
5740	Accessories for above		.13	21.50	65	195	14.05
5750	Sander, floor		.70	14.35	43	129	14.20
5760	Edger		.50	14.35	43	129	12.60
5800	Saw, chain, gas engine, 18" long		2.30	21.50	64	192	31.20
5900	Hydraulic powered, 36" long		.75	65	195	585	45
5950	60" long		.75	66.50	200	600	46
6000	Masonry, table mounted, 14" diameter, 5 H.P.		1.32	56.50	170	510	44.55
6050	Portable cut-off, 8 H.P.		2.50	33.50	100	300	40
6100	Circular, hand held, electric, 7-1/4" diameter		.23	4.67	14	42	4.65
6200	12" diameter		.23	8	24	72	6.65
6250	Wall saw, w/hydraulic power, 10 H.P.		9.70	61.50	185	555	114.60
6275	Shot blaster, walk-behind, 20" wide		4.75	293	880	2,650	214
6280	Sidewalk broom, walk-behind		2.52	78.50	235	705	67.15
6300	Steam cleaner, 100 gallons per hour		3.70	76.50	230	690	75.60
6310	200 gallons per hour		5.35	95	285	855	99.80
6340	Tar Kettle/Pot, 400 gallons		15.60	75	225	675	169.80
6350	Torch, cutting, acetylene-oxygen, 150' hose, excludes gases		.30	15	45	135	11.40
6360	Hourly operating cost includes tips and gas		19.00				152
6410	Toilet, portable chemical		.13	21	63	189	13.65
6420	Recycle flush type		.15	25	75	225	16.20
6430	Toilet, fresh water flush, garden hose,		.18	30.50	.91	273	19.65
6440	Hoisted, non-flush, for high rise		.15	24.50	74	222	16
6465	Tractor, farm with attachment		21.50	297	890	2,675	350
6480	Trailers, platform, flush deck, 2 axle, 3 ton capacity		1.45	20	60	180	23.60
6500	25 ton capacity		5.45	117	350	1,050	113.60
6600	40 ton capacity		7.00	163	490	1,475	154
6700	3 axle, 50 ton capacity		7.55	180	540	1,625	168.40
6800	75 ton capacity		9.40	235	705	2,125	216.20
6810	Trailer mounted cable reel for high voltage line work		5.45	260	779	2,325	199.40
6820	Trailer mounted cable tensioning rig		10.85	515	1,550	4,650	396.80
6830	Cable pulling rig		73.05	2,900	8,680	26,000	2,320
6900	Water tank trailer, engine driven discharge, 5000 gallons		7.00	142	425	1,275	141
6925	10,000 gallons		9.50	197	590	1,775	194
6950	Water truck, off highway, 6000 gallons		88.15	770	2,315	6,950	1,168
7010	Tram car for high voltage line work, powered, 2 conductor		6.60	141	423	1,275	137.40
7020	Transit (builder's level) with tripod		.09	14.65	44	132	9.50
7030	Trench box, 3000 lb., 6' x 8'		.56	93.50	280	840	60.50
7040	7200 lb., 6' x 20'		.75	125	375	1,125	81
7050	8000 lb., 8' x 16'		1.08	180	540	1,625	116.65
7060	9500 lb., 8' x 20'		1.21	201	603	1,800	130.30
7065	11,000 lb., 8' x 24'		1.27	211	633	1,900	136.75
7070	12,000 lb., 10' x 20'		1.50	249	748	2,250	161.60
7100	Truck, pickup, 3/4 ton, 2 wheel drive		13.65	58.50	175	525	144.20
7200	4 wheel drive		13.95	73.50	220	660	155.60
7250	Crew carrier, 9 passenger		19.30	86.50	260	780	206.40
7290	Flat bed truck, 20,000 lb. GVW		21.70	125	375	1,125	248.60
7300	Tractor, 4 x 2, 220 H.P.		30.20	197	590	1,775	359.60

01 54 33 | Equipment Rental

		UNIT	HOURLY OPER. COST	RENT PER DAY	RENT PER WEEK	RENT PER MONTH	EQUIPMENT COST/DAY	
40	7410	330 H.P.	Ea.	44.75	270	810	2,425	520	40
	7500	6 x 4, 380 H.P.		51.35	315	945	2,825	599.80	
	7600	450 H.P.		62.30	380	1,145	3,425	727.40	
	7610	Tractor, with A frame, boom and winch, 225 H.P.		33.00	272	815	2,450	427	
	7620	Vacuum truck, hazardous material, 2500 gallons		12.05	290	870	2,600	270.40	
	7625	5,000 gallons		14.26	405	1,220	3,650	358.10	
	7650	Vacuum, HEPA, 16 gallon, wet/dry		.85	18	54	162	17.60	
	7655	55 gallon, wet/dry		.80	27	81	243	22.60	
	7660	Water tank, portable		.16	26.50	80	240	17.30	
	7690	Sewer/catch basin vacuum, 14 C.Y., 1500 gallons		18.53	610	1,830	5,500	514.25	
	7700	Welder, electric, 200 amp		3.88	16.35	49	147	40.85	
	7800	300 amp		5.71	20	60	180	57.70	
	7900	Gas engine, 200 amp		14.65	23.50	70	210	131.20	
	8000	300 amp		16.38	24.50	74	222	145.85	
	8100	Wheelbarrow, any size		.08	13	39	117	8.45	
	8200	Wrecking ball, 4000 lb.		2.35	68.50	205	615	59.80	
50	0010	**HIGHWAY EQUIPMENT RENTAL** without operators	Ea.						50
	0050	Asphalt batch plant, portable drum mixer, 100 ton/hr.	R015433 -10	78.06	1,425	4,270	12,800	1,478	
	0060	200 ton/hr.		89.18	1,500	4,530	13,600	1,619	
	0070	300 ton/hr.		106.34	1,775	5,325	16,000	1,916	
	0100	Backhoe attachment, long stick, up to 185 H.P., 10.5' long		.35	23.50	70	210	16.80	
	0140	Up to 250 H.P., 12' long		.38	25	75	225	18.05	
	0180	Over 250 H.P., 15' long		.53	35	105	315	25.25	
	0200	Special dipper arm, up to 100 H.P., 32' long		1.08	71.50	215	645	51.65	
	0240	Over 100 H.P., 33' long		1.34	89.50	268	805	64.30	
	0280	Catch basin/sewer cleaning truck, 3 ton, 9 C.Y., 1000 gal.		42.70	385	1,160	3,475	573.60	
	0300	Concrete batch plant, portable, electric, 200 C.Y./hr.		22.87	515	1,550	4,650	492.95	
	0520	Grader/dozer attachment, ripper/scarifier, rear mounted, up to 135 H.P.		2.90	56.50	170	510	57.20	
	0540	Up to 180 H.P.		3.90	88.50	265	795	84.20	
	0580	Up to 250 H.P.		5.00	117	350	1,050	110	
	0700	Pvmt. removal bucket, for hyd. excavator, up to 90 H.P.		1.90	53.50	160	480	47.20	
	0740	Up to 200 H.P.		2.15	73.50	220	660	61.20	
	0780	Over 200 H.P.		2.25	85	255	765	69	
	0900	Aggregate spreader, self-propelled, 187 H.P.		53.00	685	2,050	6,150	834	
	1000	Chemical spreader, 3 C.Y.		3.30	43.50	130	390	52.40	
	1900	Hammermill, traveling, 250 H.P.		78.83	2,075	6,240	18,700	1,879	
	2000	Horizontal borer, 3" diameter, 13 H.P. gas driven		6.35	55	165	495	83.80	
	2150	Horizontal directional drill, 20,000 lb. thrust, 78 H.P. diesel		31.15	675	2,025	6,075	654.20	
	2160	30,000 lb. thrust, 115 H.P.		38.90	1,025	3,100	9,300	931.20	
	2170	50,000 lb. thrust, 170 H.P.		55.65	1,325	3,960	11,900	1,237	
	2190	Mud trailer for HDD, 1500 gallons, 175 H.P., gas		31.90	152	455	1,375	346.20	
	2200	Hydromulcher, diesel, 3000 gallon, for truck mounting		23.40	245	735	2,200	334.20	
	2300	Gas, 600 gallon		8.60	98.50	295	885	127.80	
	2400	Joint & crack cleaner, walk behind, 25 H.P.		3.95	50	150	450	61.60	
	2500	Filler, trailer mounted, 400 gallons, 20 H.P.		9.50	210	630	1,900	202	
	3000	Paint striper, self-propelled, 40 gallon, 22 H.P.		7.30	155	465	1,400	151.40	
	3100	120 gallon, 120 H.P.		23.00	395	1,185	3,550	421	
	3200	Post drivers, 6" I-Beam frame, for truck mounting		19.05	380	1,135	3,400	379.40	
	3400	Road sweeper, self-propelled, 8' wide, 90 H.P.		39.80	575	1,720	5,150	662.40	
	3450	Road sweeper, vacuum assisted, 4 C.Y., 220 gallons		75.75	635	1,900	5,700	986	
	4000	Road mixer, self-propelled, 130 H.P.		46.85	765	2,295	6,875	833.80	
	4100	310 H.P.		83.25	2,100	6,315	18,900	1,929	
	4220	Cold mix paver, incl. pug mill and bitumen tank, 165 H.P.		97.00	2,275	6,855	20,600	2,147	
	4240	Pavement brush, towed		3.20	93.50	280	840	81.60	
	4250	Paver, asphalt, wheel or crawler, 130 H.P., diesel		96.65	2,275	6,795	20,400	2,132	
	4300	Paver, road widener, gas 1' to 6', 67 H.P.		48.50	895	2,680	8,050	924	
	4400	Diesel, 2' to 14', 88 H.P.		62.70	1,050	3,135	9,400	1,129	
	4600	Slipform pavers, curb and gutter, 2 track, 75 H.P.		57.20	900	2,700	8,100	997.60	

01 54 33 | Equipment Rental

			UNIT	HOURLY OPER. COST	RENT PER DAY	RENT PER WEEK	RENT PER MONTH	EQUIPMENT COST/DAY	
50	4700	4 track, 165 H.P.	Ea.	41.50	740	2,220	6,650	776	**50**
	4800	Median barrier, 215 H.P.		60.50	1,050	3,150	9,450	1,114	
	4901	Trailer, low bed, 75 ton capacity		10.10	235	705	2,125	221.80	
	5000	Road planer, walk behind, 10" cutting width, 10 H.P.		3.70	33.50	100	300	49.60	
	5100	Self-propelled, 12" cutting width, 64 H.P.		10.40	113	340	1,025	151.20	
	5120	Traffic line remover, metal ball blaster, truck mounted, 115 H.P.		47.95	730	2,190	6,575	821.60	
	5140	Grinder, truck mounted, 115 H.P.		54.60	790	2,375	7,125	911.80	
	5160	Walk-behind, 11 H.P.		4.20	53.50	160	480	65.60	
	5200	Pavement profiler, 4' to 6' wide, 450 H.P.		254.50	3,325	9,960	29,900	4,028	
	5300	8' to 10' wide, 750 H.P.		399.20	4,350	13,055	39,200	5,805	
	5400	Roadway plate, steel, 1" x 8' x 20'		.08	13.35	40	120	8.65	
	5600	Stabilizer, self-propelled, 150 H.P.		49.30	600	1,795	5,375	753.40	
	5700	310 H.P.		99.05	1,675	5,030	15,100	1,798	
	5800	Striper, truck mounted, 120 gallon paint, 460 H.P.		64.20	485	1,455	4,375	804.60	
	5900	Thermal paint heating kettle, 115 gallons		8.20	25.50	77	231	81	
	6000	Tar kettle, 330 gallon, trailer mounted		12.25	56.50	170	510	132	
	7000	Tunnel locomotive, diesel, 8 to 12 ton		31.85	585	1,750	5,250	604.80	
	7005	Electric, 10 ton		26.00	665	1,995	5,975	607	
	7010	Muck cars, 1/2 C.Y. capacity		2.00	24.50	73	219	30.60	
	7020	1 C.Y. capacity		2.25	32.50	97	291	37.40	
	7030	2 C.Y. capacity		2.35	36.50	110	330	40.80	
	7040	Side dump, 2 C.Y. capacity		2.60	45	135	405	47.80	
	7050	3 C.Y. capacity		3.45	50	150	450	57.60	
	7060	5 C.Y. capacity		4.95	63.50	190	570	77.60	
	7100	Ventilating blower for tunnel, 7-1/2 H.P.		2.05	51.50	155	465	47.40	
	7110	10 H.P.		2.28	51.50	155	465	49.25	
	7120	20 H.P.		3.48	67.50	202	605	68.25	
	7140	40 H.P.		5.76	96.50	290	870	104.10	
	7160	60 H.P.		8.77	152	455	1,375	161.15	
	7175	75 H.P.		11.66	207	620	1,850	217.30	
	7180	200 H.P.		23.65	305	910	2,725	371.20	
	7800	Windrow loader, elevating		55.00	1,300	3,935	11,800	1,227	
60	0010	**LIFTING AND HOISTING EQUIPMENT RENTAL** without operators R015433 -10							**6**
	0120	Aerial lift truck, 2 person, to 80'	Ea.	26.30	705	2,120	6,350	634.40	
	0140	Boom work platform, 40' snorkel		12.00	275	825	2,475	261	
	0150	Crane, flatbed mounted, 3 ton capacity R015433 -15		16.35	188	565	1,700	243.80	
	0200	Crane, climbing, 106' jib, 6000 lb. capacity, 410 fpm R312316 -45		39.35	1,650	4,960	14,900	1,307	
	0300	101' jib, 10,250 lb. capacity, 270 fpm		45.95	2,100	6,280	18,800	1,624	
	0500	Tower, static, 130' high, 106' jib, 6200 lb. capacity at 400 fpm		43.20	1,900	5,730	17,200	1,492	
	0600	Crawler mounted, lattice boom, 1/2 C.Y., 15 tons at 12' radius		36.98	625	1,870	5,600	669.85	
	0700	3/4 C.Y., 20 tons at 12' radius		49.31	780	2,340	7,025	862.50	
	0800	1 C.Y., 25 tons at 12' radius		65.75	1,050	3,120	9,350	1,150	
	0900	1-1/2 C.Y., 40 tons at 12' radius		65.75	1,050	3,150	9,450	1,156	
	1000	2 C.Y., 50 tons at 12' radius		69.70	1,225	3,680	11,000	1,294	
	1100	3 C.Y., 75 tons at 12' radius		74.70	1,425	4,305	12,900	1,459	
	1200	100 ton capacity, 60' boom		84.60	1,650	4,950	14,900	1,667	
	1300	165 ton capacity, 60' boom		107.75	1,925	5,785	17,400	2,019	
	1400	200 ton capacity, 70' boom		130.55	2,400	7,210	21,600	2,486	
	1500	350 ton capacity, 80' boom		182.90	3,625	10,845	32,500	3,632	
	1600	Truck mounted, lattice boom, 6 x 4, 20 tons at 10' radius		37.11	1,075	3,260	9,775	948.90	
	1700	25 tons at 10' radius		40.09	1,175	3,550	10,700	1,031	
	1800	8 x 4, 30 tons at 10' radius		43.61	1,250	3,780	11,300	1,105	
	1900	40 tons at 12' radius		46.70	1,325	3,950	11,900	1,164	
	2000	60 tons at 15' radius		53.07	1,400	4,180	12,500	1,261	
	2050	82 tons at 15' radius		59.80	1,475	4,460	13,400	1,370	
	2100	90 tons at 15' radius		67.36	1,625	4,860	14,600	1,511	
	2200	115 tons at 15' radius		76.11	1,800	5,430	16,300	1,695	
	2300	150 tons at 18' radius		83.65	1,900	5,720	17,200	1,813	

		UNIT	HOURLY OPER. COST	RENT PER DAY	RENT PER WEEK	RENT PER MONTH	EQUIPMENT COST/DAY	
60	2350	165 tons at 18' radius	Ea.	90.05	2,025	6,060	18,200	1,932	60
	2400	Truck mounted, hydraulic, 12 ton capacity		42.60	520	1,565	4,700	653.80	
	2500	25 ton capacity		44.95	630	1,885	5,650	736.60	
	2550	33 ton capacity		45.50	645	1,930	5,800	750	
	2560	40 ton capacity		58.85	750	2,250	6,750	920.80	
	2600	55 ton capacity		76.80	855	2,570	7,700	1,128	
	2700	80 ton capacity		100.45	1,375	4,105	12,300	1,625	
	2720	100 ton capacity		94.20	1,425	4,250	12,800	1,604	
	2740	120 ton capacity		100.05	1,525	4,575	13,700	1,715	
	2760	150 ton capacity		127.60	2,000	5,995	18,000	2,220	
	2800	Self-propelled, 4 x 4, with telescoping boom, 5 ton		17.55	225	675	2,025	275.40	
	2900	12-1/2 ton capacity		32.30	360	1,075	3,225	473.40	
	3000	15 ton capacity		33.00	375	1,130	3,400	490	
	3050	20 ton capacity		35.95	445	1,335	4,000	554.60	
	3100	25 ton capacity		37.55	500	1,495	4,475	599.40	
	3150	40 ton capacity		46.00	560	1,675	5,025	703	
	3200	Derricks, guy, 20 ton capacity, 60' boom, 75' mast		27.55	405	1,214	3,650	463.20	
	3300	100' boom, 115' mast		43.21	695	2,090	6,275	763.70	
	3400	Stiffleg, 20 ton capacity, 70' boom, 37' mast		30.11	525	1,580	4,750	556.90	
	3500	100' boom, 47' mast		46.29	845	2,530	7,600	876.30	
	3550	Helicopter, small, lift to 1250 lb. maximum, w/pilot		103.70	3,275	9,800	29,400	2,790	
	3600	Hoists, chain type, overhead, manual, 3/4 ton		.10	.33	1	3	1	
	3900	10 ton		.70	6	18	54	9.20	
	4000	Hoist and tower, 5000 lb. cap., portable electric, 40' high		4.95	233	699	2,100	179.40	
	4100	For each added 10' section, add		.11	18.35	55	165	11.90	
	4200	Hoist and single tubular tower, 5000 lb. electric, 100' high		6.70	325	976	2,925	248.80	
	4300	For each added 6'-6" section, add		.19	31.50	95	285	20.50	
	4400	Hoist and double tubular tower, 5000 lb., 100' high		7.19	360	1,075	3,225	272.50	
	4500	For each added 6'-6" section, add		.21	35	105	315	22.70	
	4550	Hoist and tower, mast type, 6000 lb., 100' high		7.76	370	1,115	3,350	285.10	
	4570	For each added 10' section, add		.13	21.50	65	195	14.05	
	4600	Hoist and tower, personnel, electric, 2000 lb., 100' @ 125 fpm		16.31	990	2,970	8,900	724.50	
	4700	3000 lb., 100' @ 200 fpm		18.62	1,125	3,360	10,100	820.95	
	4800	3000 lb., 150' @ 300 fpm		20.62	1,250	3,760	11,300	916.95	
	4900	4000 lb., 100' @ 300 fpm		21.38	1,275	3,840	11,500	939.05	
	5000	6000 lb., 100' @ 275 fpm		23.06	1,350	4,030	12,100	990.50	
	5100	For added heights up to 500', add	L.F.	.01	1.67	5	15	1.10	
	5200	Jacks, hydraulic, 20 ton	Ea.	.05	2	6	18	1.60	
	5500	100 ton		.40	11.65	35	105	10.20	
	6100	Jacks, hydraulic, climbing w/50' jackrods, control console, 30 ton cap.		2.01	134	402	1,200	96.50	
	6150	For each added 10' jackrod section, add		.05	3.33	10	30	2.40	
	6300	50 ton capacity		3.23	215	646	1,950	155.05	
	6350	For each added 10' jackrod section, add		.06	4	12	36	2.90	
	6500	125 ton capacity		8.45	565	1,690	5,075	405.60	
	6550	For each added 10' jackrod section, add		.58	38.50	115	345	27.65	
	6600	Cable jack, 10 ton capacity with 200' cable		1.68	112	336	1,000	80.65	
	6650	For each added 50' of cable, add		.20	13.35	40	120	9.60	
70	0010	**WELLPOINT EQUIPMENT RENTAL** without operators　R015433 -10							70
	0020	Based on 2 months rental							
	0100	Combination jetting & wellpoint pump, 60 H.P. diesel	Ea.	18.40	330	996	3,000	346.40	
	0200	High pressure gas jet pump, 200 H.P., 300 psi	"	44.06	284	851	2,550	522.70	
	0300	Discharge pipe, 8" diameter	L.F.	.01	.54	1.62	4.86	.40	
	0350	12" diameter		.01	.79	2.38	7.15	.55	
	0400	Header pipe, flows up to 150 GPM, 4" diameter		.01	.49	1.47	4.41	.35	
	0500	400 GPM, 6" diameter		.01	.57	1.72	5.15	.40	
	0600	800 GPM, 8" diameter		.01	.79	2.38	7.15	.55	
	0700	1500 GPM, 10" diameter		.01	.84	2.51	7.55	.60	

01 54 33 | Equipment Rental

			UNIT	HOURLY OPER. COST	RENT PER DAY	RENT PER WEEK	RENT PER MONTH	EQUIPMENT COST/DAY	
70	0800	2500 GPM, 12" diameter	L.F.	.02	1.58	4.74	14.20	1.10	70
	0900	4500 GPM, 16" diameter		.03	2.02	6.07	18.20	1.45	
	0950	For quick coupling aluminum and plastic pipe, add	↓	.03	2.09	6.28	18.85	1.50	
	1100	Wellpoint, 25' long, with fittings & riser pipe, 1-1/2" or 2" diameter	Ea.	.06	4.18	12.54	37.50	3	
	1200	Wellpoint pump, diesel powered, 4" suction, 20 H.P.		7.83	191	574	1,725	177.45	
	1300	6" suction, 30 H.P.		10.70	237	712	2,125	228	
	1400	8" suction, 40 H.P.		14.45	325	976	2,925	310.80	
	1500	10" suction, 75 H.P.		22.27	380	1,141	3,425	406.35	
	1600	12" suction, 100 H.P.		31.72	605	1,810	5,425	615.75	
	1700	12" suction, 175 H.P.	↓	47.41	670	2,010	6,025	781.30	
80	0010	**MARINE EQUIPMENT RENTAL** without operators							80
	0200	Barge, 400 Ton, 30' wide x 90' long R015433-10	Ea.	17.50	1,050	3,180	9,550	776	
	0240	800 Ton, 45' wide x 90' long		21.25	1,300	3,870	11,600	944	
	2000	Tugboat, diesel, 100 H.P.		38.20	217	650	1,950	435.60	
	2040	250 H.P.		79.25	395	1,180	3,550	870	
	2080	380 H.P.		156.80	1,175	3,535	10,600	1,961	
	3000	Small work boat, gas, 16-foot, 50 H.P.		17.15	60	180	540	173.20	
	4000	Large, diesel, 48-foot, 200 H.P.	↓	89.40	1,250	3,740	11,200	1,463	

Crews

Crew No.	Bare Costs		In-House Costs		Incl. Subs O&P		Cost Per Labor-Hour		
							Bare Costs	In House	Incl. O&P
Crew A-3H	Hr.	Daily	Hr.	Daily	Hr.	Daily	Bare Costs	In House	Incl. O&P
1 Equip. Oper. (crane)	$51.70	$ 413.60	$66.00	$ 528.00	$82.05	$ 656.40	$ 51.70	$ 66.00	$ 82.05
1 Hyd. Crane, 12 Ton (Daily)		860.80		860.80		946.88	107.60	107.60	118.36
8 L.H., Daily Totals		$1274.40		$1388.80		$1603.28	$159.30	$173.60	$200.41
Crew A-3I	Hr.	Daily	Hr.	Daily	Hr.	Daily	Bare Costs	In House	Incl. O&P
1 Equip. Oper. (crane)	$51.70	$ 413.60	$66.00	$ 528.00	$82.05	$ 656.40	$ 51.70	$ 66.00	$ 82.05
1 Hyd. Crane, 25 Ton (Daily)		989.60		989.60		1088.56	123.70	123.70	136.07
8 L.H., Daily Totals		$1403.20		$1517.60		$1744.96	$175.40	$189.70	$218.12
Crew A-3J	Hr.	Daily	Hr.	Daily	Hr.	Daily	Bare Costs	In House	Incl. O&P
1 Equip. Oper. (crane)	$51.70	$ 413.60	$66.00	$ 528.00	$82.05	$ 656.40	$ 51.70	$ 66.00	$ 82.05
1 Hyd. Crane, 40 Ton (Daily)		1221.00		1221.00		1343.10	152.63	152.63	167.89
8 L.H., Daily Totals		$1634.60		$1749.00		$1999.50	$204.32	$218.63	$249.94
Crew A-3K	Hr.	Daily	Hr.	Daily	Hr.	Daily	Bare Costs	In House	Incl. O&P
1 Equip. Oper. (crane)	$51.70	$ 413.60	$66.00	$ 528.00	$82.05	$ 656.40	$ 48.45	$ 61.85	$ 76.90
1 Equip. Oper. (oiler)	45.20	361.60	57.70	461.60	71.75	574.00			
1 Hyd. Crane, 55 Ton (Daily)		1469.00		1469.00		1615.90			
1 P/U Truck, 3/4 Ton (Daily)		167.20		167.20		183.92	102.26	102.26	112.49
16 L.H., Daily Totals		$2411.40		$2625.80		$3030.22	$150.71	$164.11	$189.39
Crew A-3L	Hr.	Daily	Hr.	Daily	Hr.	Daily	Bare Costs	In House	Incl. O&P
1 Equip. Oper. (crane)	$51.70	$ 413.60	$66.00	$ 528.00	$82.05	$ 656.40	$ 48.45	$ 61.85	$ 76.90
1 Equip. Oper. (oiler)	45.20	361.60	57.70	461.60	71.75	574.00			
1 Hyd. Crane, 80 Ton (Daily)		2174.00		2174.00		2391.40			
1 P/U Truck, 3/4 Ton (Daily)		167.20		167.20		183.92	146.32	146.32	160.96
16 L.H., Daily Totals		$3116.40		$3330.80		$3805.72	$194.78	$208.18	$237.86
Crew A-3M	Hr.	Daily	Hr.	Daily	Hr.	Daily	Bare Costs	In House	Incl. O&P
1 Equip. Oper. (crane)	$51.70	$ 413.60	$66.00	$ 528.00	$82.05	$ 656.40	$ 48.45	$ 61.85	$ 76.90
1 Equip. Oper. (oiler)	45.20	361.60	57.70	461.60	71.75	574.00			
1 Hyd. Crane, 100 Ton (Daily)		2169.00		2169.00		2385.90			
1 P/U Truck, 3/4 Ton (Daily)		167.20		167.20		183.92	146.01	146.01	160.61
16 L.H., Daily Totals		$3111.40		$3325.80		$3800.22	$194.46	$207.86	$237.51
Crew A-16	Hr.	Daily	Hr.	Daily	Hr.	Daily	Bare Costs	In House	Incl. O&P
1 Maintenance Laborer	$28.20	$225.60	$35.10	$280.80	$43.80	$350.40	$28.20	$35.10	$43.80
1 Lawn Mower, Riding		68.85		68.85		75.73	8.61	8.61	9.47
8 L.H., Daily Totals		$294.45		$349.65		$426.13	$36.81	$43.71	$53.27
Crew A-17	Hr.	Daily	Hr.	Daily	Hr.	Daily	Bare Costs	In House	Incl. O&P
1 Maintenance Laborer	$28.20	$225.60	$35.10	$280.80	$43.80	$350.40	$28.20	$35.10	$43.80
1 Flatbed Truck, Gas, 1.5 Ton		245.40		245.40		269.94	30.68	30.68	33.74
8 L.H., Daily Totals		$471.00		$526.20		$620.34	$58.88	$65.78	$77.54
Crew A-18	Hr.	Daily	Hr.	Daily	Hr.	Daily	Bare Costs	In House	Incl. O&P
1 Maintenance Laborer	$28.20	$225.60	$35.10	$280.80	$43.80	$350.40	$28.20	$35.10	$43.80
1 Farm Tractor w/ Attachment		350.00		350.00		385.00	43.75	43.75	48.13
8 L.H., Daily Totals		$575.60		$630.80		$735.40	$71.95	$78.85	$91.92
Crew B-1	Hr.	Daily	Hr.	Daily	Hr.	Daily	Bare Costs	In House	Incl. O&P
1 Labor Foreman (outside)	$39.60	$316.80	$52.65	$ 421.20	$64.90	$ 519.20	$38.27	$50.85	$62.73
2 Laborers	37.60	601.60	49.95	799.20	61.65	986.40			
24 L.H., Daily Totals		$918.40		$1220.40		$1505.60	$38.27	$50.85	$62.73

For customer support on your Facilities Maintenance & Repair Cost Data, call 877.801.5724.

Crew No.	Bare Costs		In-House Costs		Incl. Subs O&P		Cost Per Labor-Hour		
Crew B-17	Hr.	Daily	Hr.	Daily	Hr.	Daily	Bare Costs	In House	Incl. O&P
2 Laborers	$37.60	$ 601.60	$49.95	$ 799.20	$61.65	$ 986.40	$40.96	$53.52	$66.25
1 Equip. Oper. (light)	48.60	388.80	62.05	496.40	77.15	617.20			
1 Truck Driver (heavy)	40.05	320.40	52.15	417.20	64.55	516.40			
1 Backhoe Loader, 48 H.P.		364.20		364.20		400.62			
1 Dump Truck, 8 C.Y., 220 H.P.		411.20		411.20		452.32	24.23	24.23	26.65
32 L.H., Daily Totals		$2086.20		$2488.20		$2972.94	$65.19	$77.76	$92.90
Crew B-18	Hr.	Daily	Hr.	Daily	Hr.	Daily	Bare Costs	In House	Incl. O&P
1 Labor Foreman (outside)	$39.60	$316.80	$52.65	$ 421.20	$64.90	$ 519.20	$38.27	$50.85	$62.73
2 Laborers	37.60	601.60	49.95	799.20	61.65	986.40			
1 Vibrating Plate, Gas, 21"		46.80		46.80		51.48	1.95	1.95	2.15
24 L.H., Daily Totals		$965.20		$1267.20		$1557.08	$40.22	$52.80	$64.88
Crew B-21	Hr.	Daily	Hr.	Daily	Hr.	Daily	Bare Costs	In House	Incl. O&P
1 Labor Foreman (outside)	$39.60	$ 316.80	$52.65	$ 421.20	$64.90	$ 519.20	$43.34	$57.01	$70.46
1 Skilled Worker	48.65	389.20	63.95	511.60	79.05	632.40			
1 Laborer	37.60	300.80	49.95	399.60	61.65	493.20			
.5 Equip. Oper. (crane)	51.70	206.80	66.00	264.00	82.05	328.20			
.5 S.P. Crane, 4x4, 5 Ton		137.70		137.70		151.47	4.92	4.92	5.41
28 L.H., Daily Totals		$1351.30		$1734.10		$2124.47	$48.26	$61.93	$75.87
Crew B-25B	Hr.	Daily	Hr.	Daily	Hr.	Daily	Bare Costs	In House	Incl. O&P
1 Labor Foreman (outside)	$39.60	$ 316.80	$52.65	$ 421.20	$64.90	$ 519.20	$42.10	$55.06	$ 68.14
7 Laborers	37.60	2105.60	49.95	2797.20	61.65	3452.40			
4 Equip. Oper. (medium)	50.60	1619.20	64.60	2067.20	80.30	2569.60			
1 Asphalt Paver, 130 H.P.		2132.00		2132.00		2345.20			
2 Tandem Rollers, 10 Ton		473.60		473.60		520.96			
1 Roller, Pneum. Whl., 12 Ton		344.40		344.40		378.84	30.73	30.73	33.80
96 L.H., Daily Totals		$6991.60		$8235.60		$9786.20	$72.83	$85.79	$101.94
Crew B-30	Hr.	Daily	Hr.	Daily	Hr.	Daily	Bare Costs	In House	Incl. O&P
1 Equip. Oper. (medium)	$50.60	$ 404.80	$64.60	$ 516.80	$80.30	$ 642.40	$ 43.57	$ 56.30	$ 69.80
2 Truck Drivers (heavy)	40.05	640.80	52.15	834.40	64.55	1032.80			
1 Hyd. Excavator, 1.5 C.Y.		1030.00		1030.00		1133.00			
2 Dump Trucks, 12 C.Y., 400 H.P.		1382.00		1382.00		1520.20	100.50	100.50	110.55
24 L.H., Daily Totals		$3457.60		$3763.20		$4328.40	$144.07	$156.80	$180.35
Crew B-34P	Hr.	Daily	Hr.	Daily	Hr.	Daily	Bare Costs	In House	Incl. O&P
1 Pipe Fitter	$59.75	$ 478.00	$74.70	$ 597.60	$93.20	$ 745.60	$49.82	$63.40	$ 78.85
1 Truck Driver (light)	39.10	312.80	50.90	407.20	63.05	504.40			
1 Equip. Oper. (medium)	50.60	404.80	64.60	516.80	80.30	642.40			
1 Flatbed Truck, Gas, 3 Ton		303.20		303.20		333.52			
1 Backhoe Loader, 48 H.P.		364.20		364.20		400.62	27.81	27.81	30.59
24 L.H., Daily Totals		$1863.00		$2189.00		$2626.54	$77.63	$91.21	$109.44
Crew B-34Q	Hr.	Daily	Hr.	Daily	Hr.	Daily	Bare Costs	In House	Incl. O&P
1 Pipe Fitter	$59.75	$ 478.00	$74.70	$ 597.60	$93.20	$ 745.60	$ 50.18	$ 63.87	$ 79.43
1 Truck Driver (light)	39.10	312.80	50.90	407.20	63.05	504.40			
1 Equip. Oper. (crane)	51.70	413.60	66.00	528.00	82.05	656.40			
1 Flatbed Trailer, 25 Ton		113.60		113.60		124.96			
1 Dump Truck, 8 C.Y., 220 H.P.		411.20		411.20		452.32			
1 Hyd. Crane, 25 Ton		736.60		736.60		810.26	52.56	52.56	57.81
24 L.H., Daily Totals		$2465.80		$2794.20		$3293.94	$102.74	$116.43	$137.25

Crew No.	Bare Costs		In-House Costs		Incl. Subs O&P		Cost Per Labor-Hour		
Crew B-34R	Hr.	Daily	Hr.	Daily	Hr.	Daily	Bare Costs	In House	Incl. O&P
1 Pipe Fitter	$59.75	$ 478.00	$74.70	$ 597.60	$93.20	$ 745.60	$ 50.18	$ 63.87	$ 79.43
1 Truck Driver (light)	39.10	312.80	50.90	407.20	63.05	504.40			
1 Equip. Oper. (crane)	51.70	413.60	66.00	528.00	82.05	656.40			
1 Flatbed Trailer, 25 Ton		113.60		113.60		124.96			
1 Dump Truck, 8 C.Y., 220 H.P.		411.20		411.20		452.32			
1 Hyd. Crane, 25 Ton		736.60		736.60		810.26			
1 Hyd. Excavator, 1 C.Y.		812.60		812.60		893.86	86.42	86.42	95.06
24 L.H., Daily Totals		$3278.40		$3606.80		$4187.80	$136.60	$150.28	$174.49
Crew B-34S	Hr.	Daily	Hr.	Daily	Hr.	Daily	Bare Costs	In House	Incl. O&P
2 Pipe Fitters	$59.75	$ 956.00	$74.70	$1195.20	$93.20	$1491.20	$ 52.81	$ 66.89	$ 83.25
1 Truck Driver (heavy)	40.05	320.40	52.15	417.20	64.55	516.40			
1 Equip. Oper. (crane)	51.70	413.60	66.00	528.00	82.05	656.40			
1 Flatbed Trailer, 40 Ton		154.00		154.00		169.40			
1 Truck Tractor, 6x4, 380 H.P.		599.80		599.80		659.78			
1 Hyd. Crane, 80 Ton		1625.00		1625.00		1787.50			
1 Hyd. Excavator, 2 C.Y.		1175.00		1175.00		1292.50	111.06	111.06	122.16
32 L.H., Daily Totals		$5243.80		$5694.20		$6573.18	$163.87	$177.94	$205.41
Crew B-34T	Hr.	Daily	Hr.	Daily	Hr.	Daily	Bare Costs	In House	Incl. O&P
2 Pipe Fitters	$59.75	$ 956.00	$74.70	$1195.20	$93.20	$1491.20	$ 52.81	$ 66.89	$ 83.25
1 Truck Driver (heavy)	40.05	320.40	52.15	417.20	64.55	516.40			
1 Equip. Oper. (crane)	51.70	413.60	66.00	528.00	82.05	656.40			
1 Flatbed Trailer, 40 Ton		154.00		154.00		169.40			
1 Truck Tractor, 6x4, 380 H.P.		599.80		599.80		659.78			
1 Hyd. Crane, 80 Ton		1625.00		1625.00		1787.50	74.34	74.34	81.77
32 L.H., Daily Totals		$4068.80		$4519.20		$5280.68	$127.15	$141.22	$165.02
Crew B-35	Hr.	Daily	Hr.	Daily	Hr.	Daily	Bare Costs	In House	Incl. O&P
1 Labor Foreman (outside)	$39.60	$ 316.80	$52.65	$ 421.20	$64.90	$ 519.20	$46.91	$60.61	$75.16
1 Skilled Worker	48.65	389.20	63.95	511.60	79.05	632.40			
1 Welder (plumber)	58.70	469.60	73.40	587.20	91.55	732.40			
1 Laborer	37.60	300.80	49.95	399.60	61.65	493.20			
1 Equip. Oper. (crane)	51.70	413.60	66.00	528.00	82.05	656.40			
1 Equip. Oper. (oiler)	45.20	361.60	57.70	461.60	71.75	574.00			
1 Welder, Electric, 300 amp		57.70		57.70		63.47			
1 Hyd. Excavator, .75 C.Y.		654.80		654.80		720.28	14.84	14.84	16.33
48 L.H., Daily Totals		$2964.10		$3621.70		$4391.35	$61.75	$75.45	$91.49
Crew B-37	Hr.	Daily	Hr.	Daily	Hr.	Daily	Bare Costs	In House	Incl. O&P
1 Labor Foreman (outside)	$39.60	$ 316.80	$52.65	$ 421.20	$64.90	$ 519.20	$39.77	$52.42	$64.78
4 Laborers	37.60	1203.20	49.95	1598.40	61.65	1972.80			
1 Equip. Oper. (light)	48.60	388.80	62.05	496.40	77.15	617.20			
1 Tandem Roller, 5 Ton		157.40		157.40		173.14	3.28	3.28	3.61
48 L.H., Daily Totals		$2066.20		$2673.40		$3282.34	$43.05	$55.70	$68.38
Crew B-47H	Hr.	Daily	Hr.	Daily	Hr.	Daily	Bare Costs	In House	Incl. O&P
1 Skilled Worker Foreman (out)	$50.65	$ 405.20	$66.60	$ 532.80	$82.30	$ 658.40	$49.15	$64.61	$79.86
3 Skilled Workers	48.65	1167.60	63.95	1534.80	79.05	1897.20			
1 Flatbed Truck, Gas, 3 Ton		303.20		303.20		333.52	9.47	9.47	10.42
32 L.H., Daily Totals		$1876.00		$2370.80		$2889.12	$58.63	$74.09	$90.28
Crew B-55	Hr.	Daily	Hr.	Daily	Hr.	Daily	Bare Costs	In House	Incl. O&P
2 Laborers	$37.60	$ 601.60	$49.95	$ 799.20	$61.65	$ 986.40	$38.10	$50.27	$ 62.12
1 Truck Driver (light)	39.10	312.80	50.90	407.20	63.05	504.40			
1 Truck-Mounted Earth Auger		779.20		779.20		857.12			
1 Flatbed Truck, Gas, 3 Ton		303.20		303.20		333.52	45.10	45.10	49.61
24 L.H., Daily Totals		$1996.80		$2288.80		$2681.44	$83.20	$95.37	$111.73

For customer support on your Facilities Maintenance & Repair Cost Data, call 877.801.5724.

613

Crew No.	Bare Costs		In-House Costs		Incl. Subs O&P		Cost Per Labor-Hour		
Crew B-80	Hr.	Daily	Hr.	Daily	Hr.	Daily	Bare Costs	In House	Incl. O&P
1 Labor Foreman (outside)	$39.60	$ 316.80	$52.65	$ 421.20	$64.90	$ 519.20	$41.23	$53.89	$66.69
1 Laborer	37.60	300.80	49.95	399.60	61.65	493.20			
1 Truck Driver (light)	39.10	312.80	50.90	407.20	63.05	504.40			
1 Equip. Oper. (light)	48.60	388.80	62.05	496.40	77.15	617.20			
1 Flatbed Truck, Gas, 3 Ton		303.20		303.20		333.52			
1 Earth Auger, Truck-Mtd.		417.20		417.20		458.92	22.51	22.51	24.76
32 L.H., Daily Totals		$2039.60		$2444.80		$2926.44	$63.74	$76.40	$91.45
Crew B-80C	Hr.	Daily	Hr.	Daily	Hr.	Daily	Bare Costs	In House	Incl. O&P
2 Laborers	$37.60	$ 601.60	$49.95	$ 799.20	$61.65	$ 986.40	$38.10	$50.27	$62.12
1 Truck Driver (light)	39.10	312.80	50.90	407.20	63.05	504.40			
1 Flatbed Truck, Gas, 1.5 Ton		245.40		245.40		269.94			
1 Manual Fence Post Auger, Gas		8.40		8.40		9.24	10.57	10.57	11.63
24 L.H., Daily Totals		$1168.20		$1460.20		$1769.98	$48.67	$60.84	$73.75
Crew D-1	Hr.	Daily	Hr.	Daily	Hr.	Daily	Bare Costs	In House	Incl. O&P
1 Bricklayer	$46.15	$369.20	$60.80	$486.40	$75.10	$ 600.80	$42.10	$55.45	$68.50
1 Bricklayer Helper	38.05	304.40	50.10	400.80	61.90	495.20			
16 L.H., Daily Totals		$673.60		$887.20		$1096.00	$42.10	$55.45	$68.50
Crew D-7	Hr.	Daily	Hr.	Daily	Hr.	Daily	Bare Costs	In House	Incl. O&P
1 Tile Layer	$42.80	$342.40	$54.30	$434.40	$67.60	$540.80	$38.17	$48.42	$60.30
1 Tile Layer Helper	33.55	268.40	42.55	340.40	53.00	424.00			
16 L.H., Daily Totals		$610.80		$774.80		$964.80	$38.17	$48.42	$60.30
Crew D-8	Hr.	Daily	Hr.	Daily	Hr.	Daily	Bare Costs	In House	Incl. O&P
3 Bricklayers	$46.15	$1107.60	$60.80	$1459.20	$75.10	$1802.40	$42.91	$56.52	$69.82
2 Bricklayer Helpers	38.05	608.80	50.10	801.60	61.90	990.40			
40 L.H., Daily Totals		$1716.40		$2260.80		$2792.80	$42.91	$56.52	$69.82
Crew G-1	Hr.	Daily	Hr.	Daily	Hr.	Daily	Bare Costs	In House	Incl. O&P
1 Roofer Foreman (outside)	$42.10	$ 336.80	$63.00	$ 504.00	$76.05	$ 608.40	$37.51	$56.17	$67.78
4 Roofers Composition	40.10	1283.20	60.05	1921.60	72.45	2318.40			
2 Roofer Helpers	30.05	480.80	45.00	720.00	54.30	868.80			
1 Application Equipment		193.00		193.00		212.30			
1 Tar Kettle/Pot		169.80		169.80		186.78			
1 Crew Truck		206.40		206.40		227.04	10.16	10.16	11.18
56 L.H., Daily Totals		$2670.00		$3714.80		$4421.72	$47.68	$66.34	$78.96
Crew G-5	Hr.	Daily	Hr.	Daily	Hr.	Daily	Bare Costs	In House	Incl. O&P
1 Roofer Foreman (outside)	$42.10	$ 336.80	$63.00	$ 504.00	$76.05	$ 608.40	$36.48	$54.62	$65.91
2 Roofers Composition	40.10	641.60	60.05	960.80	72.45	1159.20			
2 Roofer Helpers	30.05	480.80	45.00	720.00	54.30	868.80			
1 Application Equipment		193.00		193.00		212.30	4.83	4.83	5.31
40 L.H., Daily Totals		$1652.20		$2377.80		$2848.70	$41.31	$59.45	$71.22
Crew G-8	Hr.	Daily	Hr.	Daily	Hr.	Daily	Bare Costs	In House	Incl. O&P
1 Roofer Composition	$40.10	$320.80	$60.05	$ 480.40	$72.45	$ 579.60	$ 40.10	$ 60.05	$ 72.45
1 Aerial Lift Truck, 80'		634.40		634.40		697.84	79.30	79.30	87.23
8 L.H., Daily Totals		$955.20		$1114.80		$1277.44	$119.40	$139.35	$159.68
Crew L-2	Hr.	Daily	Hr.	Daily	Hr.	Daily	Bare Costs	In House	Incl. O&P
1 Carpenter	$46.95	$375.60	$62.40	$499.20	$76.95	$ 615.60	$41.20	$54.95	$67.72
1 Carpenter Helper	35.45	283.60	47.50	380.00	58.50	468.00			
16 L.H., Daily Totals		$659.20		$879.20		$1083.60	$41.20	$54.95	$67.72

For customer support on your Facilities Maintenance & Repair Cost Data, call 877.801.5724.

Crew No.	Bare Costs		In-House Costs		Incl. Subs O&P		Cost Per Labor-Hour		
Crew Q-1	Hr.	Daily	Hr.	Daily	Hr.	Daily	Bare Costs	In House	Incl. O&P
1 Plumber	$58.70	$469.60	$73.40	$ 587.20	$91.55	$ 732.40	$52.83	$66.05	$82.40
1 Plumber Apprentice	46.95	375.60	58.70	469.60	73.25	586.00			
16 L.H., Daily Totals		$845.20		$1056.80		$1318.40	$52.83	$66.05	$82.40
Crew Q-2	Hr.	Daily	Hr.	Daily	Hr.	Daily	Bare Costs	In House	Incl. O&P
2 Plumbers	$58.70	$ 939.20	$73.40	$1174.40	$91.55	$1464.80	$54.78	$68.50	$85.45
1 Plumber Apprentice	46.95	375.60	58.70	469.60	73.25	586.00	54.78	68.50	85.45
24 L.H., Daily Totals		$1314.80		$1644.00		$2050.80	$54.78	$68.50	$85.45
Crew Q-4	Hr.	Daily	Hr.	Daily	Hr.	Daily	Bare Costs	In House	Incl. O&P
1 Plumber Foreman (inside)	$59.20	$ 473.60	$74.00	$ 592.00	$92.35	$ 738.80	$55.89	$69.88	$87.17
1 Plumber	58.70	469.60	73.40	587.20	91.55	732.40			
1 Welder (plumber)	58.70	469.60	73.40	587.20	91.55	732.40			
1 Plumber Apprentice	46.95	375.60	58.70	469.60	73.25	586.00			
1 Welder, Electric, 300 amp		57.70		57.70		63.47	1.80	1.80	1.98
32 L.H., Daily Totals		$1846.10		$2293.70		$2853.07	$57.69	$71.68	$89.16
Crew Q-5	Hr.	Daily	Hr.	Daily	Hr.	Daily	Bare Costs	In House	Incl. O&P
1 Steamfitter	$59.75	$478.00	$74.70	$ 597.60	$93.20	$ 745.60	$53.77	$67.22	$83.88
1 Steamfitter Apprentice	47.80	382.40	59.75	478.00	74.55	596.40			
16 L.H., Daily Totals		$860.40		$1075.60		$1342.00	$53.77	$67.22	$83.88
Crew Q-6	Hr.	Daily	Hr.	Daily	Hr.	Daily	Bare Costs	In House	Incl. O&P
2 Steamfitters	$59.75	$ 956.00	$74.70	$1195.20	$93.20	$1491.20	$55.77	$69.72	$86.98
1 Steamfitter Apprentice	47.80	382.40	59.75	478.00	74.55	596.40	55.77	69.72	86.98
24 L.H., Daily Totals		$1338.40		$1673.20		$2087.60	$55.77	$69.72	$86.98
Crew Q-7	Hr.	Daily	Hr.	Daily	Hr.	Daily	Bare Costs	In House	Incl. O&P
1 Steamfitter Foreman (inside)	$60.25	$ 482.00	$75.30	$ 602.40	$94.00	$ 752.00	$56.89	$71.11	$88.74
2 Steamfitters	59.75	956.00	74.70	1195.20	93.20	1491.20			
1 Steamfitter Apprentice	47.80	382.40	59.75	478.00	74.55	596.40			
32 L.H., Daily Totals		$1820.40		$2275.60		$2839.60	$56.89	$71.11	$88.74
Crew Q-8	Hr.	Daily	Hr.	Daily	Hr.	Daily	Bare Costs	In House	Incl. O&P
1 Steamfitter Foreman (inside)	$60.25	$ 482.00	$75.30	$ 602.40	$94.00	$ 752.00	$56.89	$71.11	$88.74
1 Steamfitter	59.75	478.00	74.70	597.60	93.20	745.60			
1 Welder (steamfitter)	59.75	478.00	74.70	597.60	93.20	745.60			
1 Steamfitter Apprentice	47.80	382.40	59.75	478.00	74.55	596.40			
1 Welder, Electric, 300 amp		57.70		57.70		63.47	1.80	1.80	1.98
32 L.H., Daily Totals		$1878.10		$2333.30		$2903.07	$58.69	$72.92	$90.72
Crew Q-9	Hr.	Daily	Hr.	Daily	Hr.	Daily	Bare Costs	In House	Incl. O&P
1 Sheet Metal Worker	$55.95	$447.60	$70.95	$ 567.60	$88.30	$ 706.40	$50.35	$63.85	$79.45
1 Sheet Metal Apprentice	44.75	358.00	56.75	454.00	70.60	564.80			
16 L.H., Daily Totals		$805.60		$1021.60		$1271.20	$50.35	$63.85	$79.45
Crew Q-10	Hr.	Daily	Hr.	Daily	Hr.	Daily	Bare Costs	In House	Incl. O&P
2 Sheet Metal Workers	$55.95	$ 895.20	$70.95	$1135.20	$88.30	$1412.80	$52.22	$66.22	$82.40
1 Sheet Metal Apprentice	44.75	358.00	56.75	454.00	70.60	564.80			
24 L.H., Daily Totals		$1253.20		$1589.20		$1977.60	$52.22	$66.22	$82.40
Crew Q-14	Hr.	Daily	Hr.	Daily	Hr.	Daily	Bare Costs	In House	Incl. O&P
1 Asbestos Worker	$52.35	$418.80	$67.90	$543.20	$84.15	$ 673.20	$47.13	$61.13	$75.75
1 Asbestos Apprentice	41.90	335.20	54.35	434.80	67.35	538.80			
16 L.H., Daily Totals		$754.00		$978.00		$1212.00	$47.13	$61.13	$75.75

Crews

Crew No.	Bare Costs		In-House Costs		Incl. Subs O&P		Cost Per Labor-Hour		
Crew Q-15	Hr.	Daily	Hr.	Daily	Hr.	Daily	Bare Costs	In House	Incl. O&P
1 Plumber	$58.70	$469.60	$73.40	$ 587.20	$91.55	$ 732.40	$52.83	$66.05	$82.40
1 Plumber Apprentice	46.95	375.60	58.70	469.60	73.25	586.00			
1 Welder, Electric, 300 amp		57.70		57.70		63.47	3.61	3.61	3.97
16 L.H., Daily Totals		$902.90		$1114.50		$1381.87	$56.43	$69.66	$86.37
Crew Q-16	Hr.	Daily	Hr.	Daily	Hr.	Daily	Bare Costs	In House	Incl. O&P
2 Plumbers	$58.70	$ 939.20	$73.40	$1174.40	$91.55	$1464.80	$54.78	$68.50	$85.45
1 Plumber Apprentice	46.95	375.60	58.70	469.60	73.25	586.00			
1 Welder, Electric, 300 amp		57.70		57.70		63.47	2.40	2.40	2.64
24 L.H., Daily Totals		$1372.50		$1701.70		$2114.27	$57.19	$70.90	$88.09
Crew Q-20	Hr.	Daily	Hr.	Daily	Hr.	Daily	Bare Costs	In House	Incl. O&P
1 Sheet Metal Worker	$55.95	$ 447.60	$70.95	$ 567.60	$88.30	$ 706.40	$51.22	$64.62	$80.50
1 Sheet Metal Apprentice	44.75	358.00	56.75	454.00	70.60	564.80			
.5 Electrician	54.70	218.80	67.70	270.80	84.70	338.80			
20 L.H., Daily Totals		$1024.40		$1292.40		$1610.00	$51.22	$64.62	$80.50
Crew R-3	Hr.	Daily	Hr.	Daily	Hr.	Daily	Bare Costs	In House	Incl. O&P
1 Electrician Foreman	$55.20	$ 441.60	$68.35	$ 546.80	$85.45	$ 683.60	$54.30	$67.62	$84.47
1 Electrician	54.70	437.60	67.70	541.60	84.70	677.60			
.5 Equip. Oper. (crane)	51.70	206.80	66.00	264.00	82.05	328.20			
.5 S.P. Crane, 4x4, 5 Ton		137.70		137.70		151.47	6.88	6.88	7.57
20 L.H., Daily Totals		$1223.70		$1490.10		$1840.87	$61.19	$74.50	$92.04
Crew R-26	Hr.	Daily	Hr.	Daily	Hr.	Daily	Bare Costs	In House	Incl. O&P
2 Electricians	$54.70	$ 875.20	$67.70	$1083.20	$84.70	$1355.20	$54.70	$67.70	$ 84.70
1 Boom Work Pltfm, 40'		261.00		261.00		287.10	16.31	16.31	17.94
16 L.H., Daily Totals		$1136.20		$1344.20		$1642.30	$71.01	$84.01	$102.64

The following table is used to estimate the cost of in-house staff or contractor's personnel to travel to and from the job site. The amount incurred must be added to the in-house total labor cost or contractor's billing application to calculate the total cost with travel.

In-House Staff–Travel Times versus Costs

Travel Time Versus Costs	Crew Abbr. Rate with Mark-ups	Skwk $63.95	Clab $49.95	Clam $35.10	Asbe $67.90	Bric $60.80	Carp $62.40	Elec $67.70	Pord $52.35	Plum $73.40	Rofc $60.05	Shee $70.95	Stpi $74.70
Round Trip Travel Time (hours)	Crew Size												
0.50	1	31.98	24.98	17.55	33.95	30.40	31.20	33.85	26.18	36.70	30.03	35.48	37.35
	2	63.95	49.95	35.10	67.90	60.80	62.40	67.70	52.35	73.40	60.05	70.95	74.70
	3	95.93	74.93	52.65	101.85	91.20	93.60	101.55	78.53	110.10	90.08	106.43	112.05
0.75	1	47.96	37.46	26.33	50.93	45.60	46.80	50.78	39.26	55.05	45.04	53.21	56.03
	2	95.93	74.93	52.65	101.85	91.20	93.60	101.55	78.53	110.10	90.08	106.43	112.05
	3	143.89	112.39	78.98	152.78	136.80	140.40	152.33	117.79	165.15	135.11	159.64	168.08
1.00	1	63.95	49.95	35.10	67.90	60.80	62.40	67.70	52.35	73.40	60.05	70.95	74.70
	2	127.90	99.90	70.20	135.80	121.60	124.80	135.40	104.70	146.80	120.10	141.90	149.40
	3	191.85	149.85	105.30	203.70	182.40	187.20	203.10	157.05	220.20	180.15	212.85	224.10
1.50	1	95.93	74.93	52.65	101.85	91.20	93.60	101.55	78.53	110.10	90.08	106.43	112.05
	2	191.85	149.85	105.30	203.70	182.40	187.20	203.10	157.05	220.20	180.15	212.85	224.10
	3	287.78	224.78	157.95	305.55	273.60	280.80	304.65	235.58	330.30	270.23	319.28	336.15
2.00	1	127.90	99.90	70.20	135.80	121.60	124.80	135.40	104.70	146.80	120.10	141.90	149.40
	2	255.80	199.80	140.40	271.60	243.20	249.60	270.80	209.40	293.60	240.20	283.80	298.80
	3	383.70	299.70	210.60	407.40	364.80	374.40	406.20	314.10	440.40	360.30	425.70	448.20

Installing Contractors–Travel Times versus Costs

Travel Time Versus Costs	Crew Abbr. Rate with O & P	Skwk $79.05	Clab $61.65	Clam $43.80	Asbe $84.15	Bric $75.10	Carp $76.95	Elec $84.70	Pord $64.85	Plum $91.55	Rofc $72.45	Shee $88.30	Stpi $93.20
Round Trip Travel Time (hours)	Crew Size												
0.50	1	39.53	30.83	21.90	42.08	37.55	38.48	42.35	32.43	45.78	36.23	44.15	46.60
	2	79.05	61.65	43.80	84.15	75.10	76.95	84.70	64.85	91.55	72.45	88.30	93.20
	3	118.58	92.48	65.70	126.23	112.65	115.43	127.05	97.28	137.33	108.68	132.45	139.80
0.75	1	59.29	46.24	32.85	63.11	56.33	57.71	63.53	48.64	68.66	54.34	66.23	69.90
	2	118.58	92.48	65.70	126.23	112.65	115.43	127.05	97.28	137.33	108.68	132.45	139.80
	3	177.86	138.71	98.55	189.34	168.98	173.14	190.58	145.91	205.99	163.01	198.68	209.70
1.00	1	79.05	61.65	43.80	84.15	75.10	76.95	84.70	64.85	91.55	72.45	88.30	93.20
	2	158.10	123.30	87.60	168.30	150.20	153.90	169.40	129.70	183.10	144.90	176.60	186.40
	3	237.15	184.95	131.40	252.45	225.30	230.85	254.10	194.55	274.65	217.35	264.90	279.60
1.50	1	118.58	92.48	65.70	126.23	112.65	115.43	127.05	97.28	137.33	108.68	132.45	139.80
	2	237.15	184.95	131.40	252.45	225.30	230.85	254.10	194.55	274.65	217.35	264.90	279.60
	3	355.73	277.43	197.10	378.68	337.95	346.28	381.15	291.83	411.98	326.03	397.35	419.40
2.00	1	158.10	123.30	87.60	168.30	150.20	153.90	169.40	129.70	183.10	144.90	176.60	186.40
	2	316.20	246.60	175.20	336.60	300.40	307.80	338.80	259.40	366.20	289.80	353.20	372.80
	3	474.30	369.90	262.80	504.90	450.60	461.70	508.20	389.10	549.30	434.70	529.80	559.20

For customer support on your Facilities Maintenance & Repair Cost Data, call 877.801.5724.

Table H1010-101   General Contractor's Overhead

There are two distinct types of overhead on a construction project: Project Overhead and Main Office Overhead. Project Overhead includes those costs at a construction site not directly associated with the installation of construction materials. Examples of Project Overhead costs include the following:

1. Superintendent
2. Construction office and storage trailers
3. Temporary sanitary facilities
4. Temporary utilities
5. Security fencing
6. Photographs
7. Clean up
8. Performance and payment bonds

The above Project Overhead items are also referred to as General Requirements and therefore are estimated in Division 1. Division 1 is the first division listed in the CSI MasterFormat but it is usually the last division estimated. The sum of the costs in Divisions 1 through 49 is referred to as the sum of the direct costs.

All construction projects also include indirect costs. The primary components of indirect costs are the contractor's Main Office Overhead and profit. The amount of the Main Office Overhead expense varies depending on the the the following:

1. Owner's compensation
2. Project managers and estimator's wages
3. Clerical support wages
4. Office rent and utilities
5. Corporate legal and accounting costs
6. Advertising
7. Automobile expenses
8. Association dues
9. Travel and entertainment expenses

These costs are usually calculated as a percentage of annual sales volume. This percentage can range from 35% for a small contractor doing less than $500,000 to 5% for a large contractor with sales in excess of $100 million.

Table H1010-102   Main Office Expense

A General Contractor's main office expense consists of many items not detailed in the front portion of the book. The percentage of main office expense declines with increased annual volume of the contractor. Typical main office expense ranges from 2% to 20% with the median about 7.2%

of total volume. This equals about 7.7% of direct costs. The following are approximate percentages of total overhead for different items usually included in a General Contractor's main office overhead. With different accounting procedures, these percentages may vary.

Item	Typical Range			Average
Managers', clerical and estimators' salaries	40%	to	55%	48%
Profit sharing, pension and bonus plans	2	to	20	12
Insurance	5	to	8	6
Estimating and project management (not including salaries)	5	to	9	7
Legal, accounting and data processing	0.5	to	5	3
Automobile and light truck expense	2	to	8	5
Depreciation of overhead capital expenditures	2	to	6	4
Maintenance of office equipment	0.1	to	1.5	1
Office rental	3	to	5	4
Utilities including phone and light	1	to	3	2
Miscellaneous	5	to	15	8
Total				100%

Table H1010-201 Architectural Fees

Tabulated below are typical percentage fees by project size, for good professional architectural service. Fees may vary from those listed depending upon degree of design difficulty and economic conditions in any particular area.

Rates can be interpolated horizontally and vertically. Various portions of the same project requiring different rates should be adjusted proportionately. For alterations, add 50% to the fee for the first $500,000 of project cost and add 25% to the fee for project cost over $500,000.

Architectural fees tabulated below include Structural, Mechanical and Electrical Engineering Fees. They do not include the fees for special consultants such as kitchen planning, security, acoustical, interior design, etc.

Building Types	Total Project Size in Thousands of Dollars						
	100	250	500	1,000	5,000	10,000	50,000
Factories, garages, warehouses, repetitive housing	9.0%	8.0%	7.0%	6.2%	5.3%	4.9%	4.5%
Apartments, banks, schools, libraries, offices, municipal buildings	12.2	12.3	9.2	8.0	7.0	6.6	6.2
Churches, hospitals, homes, laboratories, museums, research	15.0	13.6	12.7	11.9	9.5	8.8	8.0
Memorials, monumental work, decorative furnishings	—	16.0	14.5	13.1	10.0	9.0	8.3

Table H1010-202 Engineering Fees

Typical **Structural Engineering Fees** based on type of construction and total project size. These fees are included in Architectural Fees.

Type of Construction	Total Project Size (in thousands of dollars)			
	$500	$500-1,000	$1,000-5,000	Over $5000
Industrial buildings, factories & warehouses	Technical payroll times 2.0 to 2.5	1.60%	1.25%	1.00%
Hotels, apartments, offices, dormitories, hospitals, public buildings, food stores		2.00%	1.70%	1.20%
Museums, banks, churches and cathedrals		2.00%	1.75%	1.25%
Thin shells, prestressed concrete, earthquake resistive		2.00%	1.75%	1.50%
Parking ramps, auditoriums, stadiums, convention halls, hangars & boiler houses		2.50%	2.00%	1.75%
Special buildings, major alterations, underpinning & future expansion		Add to above 0.5%	Add to above 0.5%	Add to above 0.5%

For complex reinforced concrete or unusually complicated structures, add 20% to 50%.

Table H1010-203 Mechanical and Electrical Fees

Typical **Mechanical and Electrical Engineering Fees** based on the size of the subcontract. The fee structure for both are shown below. These fees are included in Architectural Fees.

Type of Construction	Subcontract Size							
	$25,000	$50,000	$100,000	$225,000	$350,000	$500,000	$750,000	$1,000,000
Simple structures	6.4%	5.7%	4.8%	4.5%	4.4%	4.3%	4.2%	4.1%
Intermediate structures	8.0	7.3	6.5	5.6	5.1	5.0	4.9	4.8
Complex structures	10.1	9.0	9.0	8.0	7.5	7.5	7.0	7.0

For renovations, add 15% to 25% to applicable fee.

619

For customer support on your Facilities Maintenance & Repair Cost Data, call 877.801.5724.

Table H1010-301   Builder's Risk Insurance

Builder's Risk Insurance is insurance on a building during construction. Premiums are paid by the owner or the contractor. Blasting, collapse and underground insurance would raise total insurance costs above those listed. Floater policy for materials delivered to the job runs $.75 to $1.25 per $100 value. Contractor equipment insurance runs $.50 to $1.50 per $100 value. Insurance for miscellaneous tools to $1,500 value runs from $3.00 to $7.50 per $100 value.

Tabulated below are New England Builder's Risk insurance rates in dollars per $100 value for $1,000 deductible. For $25,000 deductible, rates can be reduced 13% to 34%. On contracts over $1,000,000, rates may be lower than those tabulated. Policies are written annually for the total completed value in place. For "all risk" insurance (excluding flood, earthquake and certain other perils) add $.025 to total rates below.

Coverage	Frame Construction (Class 1)			Brick Construction (Class 4)			Fire Resistive (Class 6)		
	Range		Average	Range		Average	Range		Average
Fire Insurance	$.350 to	$.850	$.600	$.158 to	$.189	$.174	$.052 to	$.080	$.070
Extended Coverage	.115 to	.200	.158	.080 to	.105	.101	.081 to	.105	.100
Vandalism	.012 to	.016	.014	.008 to	.011	.011	.008 to	.011	.010
Total Annual Rate	$.477 to	$1.066	$.772	$.246 to	$.305	$.286	$.141 to	$.196	$.180

Table H1010-302   Performance Bond

This table shows the cost of a Performance Bond for a construction job scheduled to be completed in 12 months. Add 1% of the premium cost per month for jobs requiring more than 12 months to complete. The rates are "standard" rates offered to contractors that the bonding company considers financially sound and capable of doing the work. Preferred rates

are offered by some bonding companies based upon financial strength of the contractor. Actual rates vary from contractor to contractor and from bonding company to bonding company. Contractors should prequalify through a bonding agency before submitting a bid on a contract that requires a bond.

Contract Amount	Building Construction Class B Projects			Highways & Bridges					
				Class A New Construction			Class A-1 Highway Resurfacing		
First $ 100,000 bid	$25.00 per M			$15.00 per M			$9.40 per M		
Next 400,000 bid	$ 2,500 plus	$15.00	per M	$ 1,500 plus	$10.00	per M	$ 940 plus	$7.20	per M
Next 2,000,000 bid	8,500 plus	10.00	per M	5,500 plus	7.00	per M	3,820 plus	5.00	per M
Next 2,500,000 bid	28,500 plus	7.50	per M	19,500 plus	5.50	per M	15,820 plus	4.50	per M
Next 2,500,000 bid	47,250 plus	7.00	per M	33,250 plus	5.00	per M	28,320 plus	4.50	per M
Over 7,500,000 bid	64,750 plus	6.00	per M	45,750 plus	4.50	per M	39,570 plus	4.00	per M

Table H1010-401   Workers' Compensation Insurance Rates by Trade

The table below tabulates the national averages for workers' compensation insurance rates by trade and type of building. The average "Insurance Rate" is multiplied by the "% of Building Cost" for each trade. This produces

the "Workers' Compensation" cost by % of total labor cost, to be added for each trade by building type to determine the weighted average workers' compensation rate for the building types analyzed.

Trade	Insurance Rate (% Labor Cost)			% of Building Cost			Workers' Compensation		
	Range		Average	Office Bldgs.	Schools & Apts.	Mfg.	Office Bldgs.	Schools & Apts.	Mfg.
Excavation, Grading, etc.	3.4 % to	21.2%	9.7%	4.8%	4.9%	4.5%	0.47%	0.48%	0.44%
Piles & Foundations	6.1 to	28.3	14.3	7.1	5.2	8.7	1.02	0.74	1.24
Concrete	4.2 to	35.8	12.9	5.0	14.8	3.7	0.65	1.91	0.48
Masonry	4.9 to	36.4	13.7	6.9	7.5	1.9	0.95	1.03	0.26
Structural Steel	6.1 to	101.1	31.7	10.7	3.9	17.6	3.39	1.24	5.58
Miscellaneous & Ornamental Metals	4.3 to	31.5	12.2	2.8	4.0	3.6	0.34	0.49	0.44
Carpentry & Millwork	5.0 to	36.8	14.9	3.7	4.0	0.5	0.55	0.60	0.07
Metal or Composition Siding	6.1 to	69.2	18.4	2.3	0.3	4.3	0.42	0.06	0.79
Roofing	6.1 to	100.3	31.7	2.3	2.6	3.1	0.73	0.82	0.98
Doors & Hardware	4.1 to	36.8	11.3	0.9	1.4	0.4	0.10	0.16	0.05
Sash & Glazing	4.8 to	30.7	13.3	3.5	4.0	1.0	0.47	0.53	0.13
Lath & Plaster	3.1 to	36.2	10.9	3.3	6.9	0.8	0.36	0.75	0.09
Tile, Marble & Floors	3 to	24.8	8.9	2.6	3.0	0.5	0.23	0.27	0.04
Acoustical Ceilings	3.8 to	33.8	8.4	2.4	0.2	0.3	0.20	0.02	0.03
Painting	4.5 to	36.1	11.7	1.5	1.6	1.6	0.18	0.19	0.19
Interior Partitions	5.0 to	36.8	14.9	3.9	4.3	4.4	0.58	0.64	0.66
Miscellaneous Items	2.5 to	132.3	13.4	5.2	3.7	9.7	0.70	0.50	1.30
Elevators	1.3 to	12.8	5.3	2.1	1.1	2.2	0.11	0.06	0.12
Sprinklers	2.9 to	18.5	7.3	0.5	—	2.0	0.04	—	0.15
Plumbing	2.2 to	19.3	7.0	4.9	7.2	5.2	0.34	0.50	0.36
Heat., Vent., Air Conditioning	3.8 to	18.1	8.8	13.5	11.0	12.9	1.19	0.97	1.14
Electrical	2.4 to	13.5	5.8	10.1	8.4	11.1	0.59	0.49	0.64
Total	1.3 % to	132.3%	—	100.0%	100.0%	100.0%	13.61%	12.45%	15.18%
				Overall Weighted Average	13.75%				

Workers' Compensation Insurance Rates by States

The table below lists the weighted average workers' compensation base rate for each state with a factor comparing this with the national average of 14%.

State	Weighted Average	Factor	State	Weighted Average	Factor	State	Weighted Average	Factor
Alabama	20.6%	153	Kentucky	12.5%	93	North Dakota	8.1%	60
Alaska	12.5	93	Louisiana	19.6	145	Ohio	8.1	60
Arizona	13.8	102	Maine	11.4	84	Oklahoma	10.3	76
Arkansas	7.7	57	Maryland	13.5	100	Oregon	11.3	84
California	27.1	201	Massachusetts	11.5	85	Pennsylvania	18.3	136
Colorado	7.3	54	Michigan	13.5	100	Rhode Island	14.7	109
Connecticut	25.9	192	Minnesota	22.8	169	South Carolina	16.9	125
Delaware	14.0	104	Mississippi	14.0	104	South Dakota	14.1	104
District of Columbia	10.4	77	Missouri	13.9	103	Tennessee	12.2	90
Florida	10.8	80	Montana	8.3	61	Texas	9.3	69
Georgia	29.5	219	Nebraska	17.0	126	Utah	8.8	65
Hawaii	8.7	64	Nevada	9.2	68	Vermont	13.3	99
Idaho	10.5	78	New Hampshire	18.8	139	Virginia	8.7	64
Illinois	26.9	199	New Jersey	15.7	116	Washington	9.5	70
Indiana	5.0	37	New Mexico	15.3	113	West Virginia	7.4	55
Iowa	15.5	115	New York	19.0	141	Wisconsin	13.2	98
Kansas	9.2	68	North Carolina	17.3	128	Wyoming	6.5	48
			Weighted Average for U.S. is	13.7% of payroll = 100%				

The weighted average skilled worker rate for 35 trades is 13.5%. For bidding purposes, apply the full value of workers' compensation directly to total labor costs, or if labor is 38%, materials 42% and overhead and profit 20% of total cost, carry 38/80 × 13.5% =6.4% of cost (before overhead and profit)

into overhead. Rates vary not only from state to state but also with the experience rating of the contractor.

Rates are the most current available at the time of publication.

Unemployment Taxes and Social Security Taxes

State unemployment tax rates vary not only from state to state, but also with the experience rating of the contractor. The federal unemployment tax rate is 6.0% of the first $7,000 of wages. This is reduced by a credit of up to 5.4% for timely payment to the state. The minimum federal unemployment tax is 0.6% after all credits.

Social security (FICA) for 2015 is estimated at time of publication to be 7.65% of wages up to $117,000.

Overtime

One way to improve the completion date of a project or eliminate negative float from a schedule is to compress activity duration times. This can be achieved by increasing the crew size or working overtime with the proposed crew.

To determine the costs of working overtime to compress activity duration times, consider the following examples. Below is an overtime efficiency and cost chart based on a five, six, or seven day week with an eight through twelve hour day. Payroll percentage increases for time and one half and double time are shown for the various working days.

Days per Week	Hours per Day	Production Efficiency					Payroll Cost Factors	
		1st Week	2nd Week	3rd Week	4th Week	Average 4 Weeks	@ 1-1/2 Times	@ 2 Times
5	8	100%	100%	100%	100%	100 %	1.000	1.000
	9	100	100	95	90	96	1.056	1.111
	10	100	95	90	85	93	1.100	1.200
	11	95	90	75	65	81	1.136	1.273
	12	90	85	70	60	76	1.167	1.333
6	8	100	100	95	90	96	1.083	1.167
	9	100	95	90	85	93	1.130	1.259
	10	95	90	85	80	88	1.167	1.333
	11	95	85	70	65	79	1.197	1.394
	12	90	80	65	60	74	1.222	1.444
7	8	100	95	85	75	89	1.143	1.286
	9	95	90	80	70	84	1.183	1.365
	10	90	85	75	65	79	1.214	1.429
	11	85	80	65	60	73	1.240	1.481
	12	85	75	60	55	69	1.262	1.524

Sales Tax by State

State sales tax on materials is tabulated below (5 states have no sales tax). Many states allow local jurisdictions, such as a county or city, to levy additional sales tax.

Some projects may be sales tax exempt, particularly those constructed with public funds.

State	Tax (%)	State	Tax (%)	State	Tax (%)	State	Tax (%)
Alabama	4	Illinois	6.25	Montana	0	Rhode Island	7
Alaska	0	Indiana	7	Nebraska	5.5	South Carolina	6
Arizona	5.6	Iowa	6	Nevada	6.85	South Dakota	4
Arkansas	6	Kansas	6.3	New Hampshire	0	Tennessee	7
California	7.5	Kentucky	6	New Jersey	7	Texas	6.25
Colorado	2.9	Louisiana	4	New Mexico	5.13	Utah	4.7
Connecticut	6.35	Maine	5	New York	4	Vermont	6
Delaware	0	Maryland	6	North Carolina	4.75	Virginia	5
District of Columbia	6	Massachusetts	6.25	North Dakota	5	Washington	6.5
Florida	6	Michigan	6	Ohio	5.5	West Virginia	6
Georgia	4	Minnesota	6.88	Oklahoma	4.5	Wisconsin	5
Hawaii	4	Mississippi	7	Oregon	0	Wyoming	4
Idaho	6	Missouri	4.23	Pennsylvania	6	Average	5.04 %

Table H1040-101   Steel Tubular Scaffolding

On new construction, tubular scaffolding is efficient up to 60' high or five stories. Above this it is usually better to use a hung scaffolding if construction permits. Swing scaffolding operations may interfere with tenants. In this case, the tubular is more practical at all heights.

In repairing or cleaning the front of an existing building, the cost of tubular scaffolding per S.F. of building front increases as the height increases above the first tier. The first tier cost is relatively high due to leveling and alignment.

The minimum efficient crew for erection is three workers. For heights over 50', a crew of four is more efficient. Use two or more on top and two at the bottom for handing up or hoisting. Four workers can erect and dismantle about nine frames per hour up to five stories. From five to eight stories, they will average six frames per hour. With 7' horizontal spacing, this will run about 400 S.F. and 265 S.F. of wall surface, respectively. Time for placing

planks must be added to the above. On heights above 50', five planks can be placed per labor-hour.

The cost per 1,000 S.F. of building front in the table below was developed by pricing the materials required for a typical tubular scaffolding system eleven frames long and two frames high. Planks were figured five wide for standing plus two wide for materials.

Frames are 5' wide and usually spaced 7' O.C. horizontally. Sidewalk frames are 6' wide. Rental rates will be lower for jobs over three months' duration.

For jobs under twenty-five frames, add 50% to rental cost. These figures do not include accessories which are listed separately below. Large quantities for long periods can reduce rental rates by 20%.

Item	Unit	Monthly Rent	Per 1,000 S. F. of Building Front	
			No. of Pieces	Rental per Month
5' Wide Standard Frame, 6'-4" High	Ea.	$ 5.50	24	$132.00
Leveling Jack & Plate		2.00	24	48.00
Cross Brace		1.00	44	44.00
Side Arm Bracket, 21"		2.00	12	24.00
Guardrail Post		1.00	12	12.00
Guardrail, 7' section		1.00	22	22.00
Stairway Section		35.00	2	70.00
Walk-Thru Frame Guardrail	▼	2.50	2	5.00
			Total	$357.00
			Per C.S.F., 1 Use/Mo.	$ 35.70

Scaffolding is often used as falsework over 15' high during construction of cast-in-place concrete beams and slabs. Two-foot wide scaffolding is generally used for heavy beam construction. The span between frames depends upon the load to be carried, with a maximum span of 5'.

Heavy duty shoring frames with a capacity of 10,000#/leg can be spaced up to 10' O. C., depending upon form support design and loading.

Scaffolding used as horizontal shoring requires less than half the material required with conventional shoring.

On new construction, erection is done by carpenters.

Rolling towers supporting horizontal shores can reduce labor and speed the job. For maintenance work, catwalks with spans up to 70' can be supported by the rolling towers.

General: The following information on current city cost indexes is calculated for over 930 zip code locations in the United States and Canada. Index figures for both material and installation are based upon the 30 major city average of 100 and represent the cost relationship on July 1, 2013.

In addition to index adjustment, the user should consider:
1. productivity
2. management efficiency
3. competitive conditions
4. automation

5. restrictive union practices
6. unique local requirements
7. regional variations due to specific building codes

The weighted-average index is calculated from about 66 materials and equipment types and 21 building trades. The component contribution of these in a model building is tabulated in Table J1010-011 below.

If the systems component distribution of a building is unknown, the weighted-average index can be used to adjust the cost for any city.

Table J1010-011 Labor, Material, and Equipment Cost Distribution for Weighted Average Listed by System Division

Division No.	Building System	Percentage	Division No.	Building System	Percentage
A	Substructure	6.1%	D10	Services: Conveying	4.3%
B10	Shell: Superstructure	18.7	D20-40	Mechanical	23.1
B20	Exterior Closure	11.4	D50	Electrical	11.4
B30	Roofing	2.8	E	Equipment & Furnishings	2.3
C	Interior Construction	15.9	G	Site Work	4.0
				Total weighted average (Div. A-G)	100.0%

How to Use the Component Indexes: Table J1010-012 below shows how the average costs obtained from this book for each division should be adjusted for your particular building in your city. The example is a building adjusted for New York, NY. Indexes for other cities are tabulated in Division RJ1030. These indexes should also be used when compiling data on the form in Division K1010.

Table J1010-012 Adjustment of "Book" Costs to a Particular City

Systems Division Number	System Description	New York, NY							
		City Cost Index	Cost Factor	Book Cost	Adjusted Cost	City Cost Index	Cost Factor	Book Cost	Adjusted Cost
A	Substructure	146.7	1.467	$ 101,500	$ 148,900				
B10	Shell: Superstructure	133.7	1.337	280,300	374,800				
B20	Exterior Closure	136.4	1.364	190,000	259,200				
B30	Roofing	129.3	1.293	41,800	54,000				
C	Interior Construction	133.5	1.325	270,500	358,400				
D10	Services: Conveying	108.1	1.081	54,200	58,600				
D20-40	Mechanical	126.5	1.265	335,700	424,700				
D50	Electrical	145.5	1.455	187,900	273,400				
E	Equipment	104.4	1.044	28,400	29,600				
G	Site Work	120.2	1.202	79,800	95,900				
Total Cost, (Div. A-G)				$1,570,100	$2,077,500				

Alternate Method: Use the weighted-average index for your city rather than the component indexes. **For example:**

(Total weighted-average index A-G)/100 x (total cost) = Adjusted city cost
New York, NY 1.318 x $1,570,100 = $2,069,400

City to City: To convert known or estimated costs from one city to another, cost indexes can be used as follows:

$$\text{Unknown City Cost} = \text{Known Cost} \times \frac{\text{Unknown City Index}}{\text{Known City Index}}$$

For example: If the building cost in Boston, MA, is $2,000,000, how much would a duplicated building cost in Los Angeles, CA?

$$\text{L.A. Cost} = \$2,000,000 \times \frac{\text{(Los Angeles) } 107.2}{\text{(Boston) } 118.1} = \$1,815,500$$

The table below lists both the Means City Cost Index based on January 1, 1993 = 100 as well as the computed value of an index based on January 1, 2015 costs. Since the January 1, 2015 figure is estimated, space is left to write in the actual index figures as they become available through either the quarterly *Means Construction Cost Indexes* or as printed in the *Engineering News-Record*. To compute the actual index based on January 1, 2015 = 100, divide the Quarterly City Cost Index for a particular year by the actual January 1, 2015 Quarterly City Cost Index. Space has been left to advance the index figures as the year progresses.

Table J1020-011 Historical Cost Indexes

Year	Historical Cost Index Jan. 1, 1993 = 100 Est.	Actual	Current Index Based on Jan. 1, 2015 = 100 Est.	Actual	Year	Historical Cost Index Jan. 1, 1993 = 100 Actual	Current Index Based on Jan. 1, 2015 = 100 Est.	Actual	Year	Historical Cost Index Jan. 1, 1993 = 100 Actual	Current Index Based on Jan. 1, 2015 = 100 Est.	Actual
Oct 2015*					July 2000	120.9	58.5		July 1982	76.1	36.8	
July 2015*					1999	117.6	56.9		1981	70.0	33.9	
April 2015*					1998	115.1	55.7		1980	62.9	30.4	
Jan 2015*	206.7		100.0	100.0	1997	112.8	54.6		1979	57.8	28.0	
July 2014		204.9	99.1		1996	110.2	53.3		1978	53.5	25.9	
2013		201.2	97.3		1995	107.6	52.1		1977	49.5	23.9	
2012		194.6	94.1		1994	104.4	50.5		1976	46.9	22.7	
2011		191.2	92.5		1993	101.7	49.2		1975	44.8	21.7	
2010		183.5	88.8		1992	99.4	48.1		1974	41.4	20.0	
2009		180.1	87.1		1991	96.8	46.8		1973	37.7	18.2	
2008		180.4	87.3		1990	94.3	45.6		1972	34.8	16.8	
2007		169.4	82.0		1989	92.1	44.6		1971	32.1	15.5	
2006		162.0	78.4		1988	89.9	43.5		1970	28.7	13.9	
2005		151.6	73.3		1987	87.7	42.4		1969	26.9	13.0	
2004		143.7	69.5		1986	84.2	40.7		1968	24.9	12.0	
2003		132.0	63.9		1985	82.6	40.0		1967	23.5	11.4	
2002		128.7	62.3		1984	82.0	39.7		1966	22.7	11.0	
2001		125.1	60.5		1983	80.2	38.8		1965	21.7	10.5	

To find the **current cost** from a project built previously in either the same city or a different city, the following formula is used:

$$\text{Present Cost (City X)} = \frac{\text{Current HCI} \times \text{CCI (City X)}}{\text{Previous HCI} \times \text{CCI (City Y)}} \times \text{Former Cost (City Y)}$$

For example: Find the construction cost of a building to be built in San Francisco, CA, as of January 1, 2015 when the identical building cost $500,000 in Boston, MA, on July 1, 1968.

$$\text{Jan. 1, 2015 (San Francisco)} = \frac{\text{(San Francisco) } 206.7 \times 122.7}{\text{(Boston) } 24.9 \times 118.1} \times \$500,000 = \$4,312,500$$

Note: The City Cost Indexes for Canada can be used to convert U.S. national averages to local costs in Canadian dollars.

To Project Future Construction Costs: Using the results of the last five years average percentage increase as a basis, an average increase of 2.6% could be used.

The historical index figures above are compiled from the Means Construction Index Service.

Example:

To estimate and compare the cost of a building in Toronto, ON in 2015 with the known cost of $600,000 (US$) in New York, NY in 2015:

INDEX Toronto = 110.9

INDEX New York = 131.8

$$\frac{\text{INDEX Toronto}}{\text{INDEX New York}} \times \text{Cost New York} = \text{Cost Toronto}$$

$$\frac{110.9}{131.8} \times \$600,000 = .841 \times \$600,000 = \$504,600$$

The construction cost of the building in Toronto is $504,600 (CN$).

*Historical Cost Index updates and other resources are provided on the following website.
http://www.reedconstructiondata.com/rsmeans/chgnotice/456321

625

For customer support on your Facilities Maintenance & Repair Cost Data, call 877.801.5724.

How to Use the City Cost Indexes

What you should know before you begin

RSMeans City Cost Indexes (CCI) are an extremely useful tool to use when you want to compare costs from city to city and region to region.

This publication contains average construction cost indexes for 317 U.S. and Canadian cities covering over 930 three-digit zip code locations, as listed directly under each city.

Keep in mind that a City Cost Index number is a percentage ratio of a specific city's cost to the national average cost of the same item at a stated time period.

In other words, these index figures represent relative construction factors (or, if you prefer, multipliers) for Material and Installation costs, as well as the weighted average for Total In Place costs for each CSI MasterFormat division. Installation costs include both labor and equipment rental costs. When estimating equipment rental rates only, for a specific location, use 01 54 33 EQUIPMENT RENTAL COSTS in the Reference Section.

The 30 City Average Index is the average of 30 major U.S. cities and serves as a National Average.

Index figures for both material and installation are based on the 30 major city average of 100 and represent the cost relationship as of July 1, 2014. The index for each division is computed from representative material and labor quantities for that division. The weighted average for each city is a weighted total of the components listed above it, but does not include relative productivity between trades or cities.

As changes occur in local material prices, labor rates, and equipment rental rates, (including fuel costs) the impact of these changes should be accurately measured by the change in the City Cost Index for each particular city (as compared to the 30 City Average).

Therefore, if you know (or have estimated) building costs in one city today, you can easily convert those costs to expected building costs in another city.

In addition, by using the Historical Cost Index, you can easily convert National Average building costs at a particular time to the approximate building costs for some other time. The City Cost Indexes can then be applied to calculate the costs for a particular city.

Quick Calculations

Location Adjustment Using the City Cost Indexes:

$$\frac{\text{Index for City A}}{\text{Index for City B}} \times \text{Cost in City B} = \text{Cost in City A}$$

Time Adjustment for the National Average Using the Historical Cost Index:

$$\frac{\text{Index for Year A}}{\text{Index for Year B}} \times \text{Cost in Year B} = \text{Cost in Year A}$$

Adjustment from the National Average:

$$\frac{\text{Index for City A}}{100} \times \text{National Average Cost} = \text{Cost in City A}$$

Since each of the other RSMeans publications contains many different items, any *one* item multiplied by the particular city index may give incorrect results. However, the larger the number of items compiled, the closer the results should be to actual costs for that particular city.

The City Cost Indexes for Canadian cities are calculated using Canadian material and equipment prices and labor rates, in Canadian dollars. Therefore, indexes for Canadian cities can be used to convert U.S. National Average prices to local costs in Canadian dollars.

How to use this section

1. Compare costs from city to city.

In using the RSMeans Indexes, remember that an index number is not a fixed number but a ratio: It's a percentage ratio of a building component's cost at any stated time to the National Average cost of that same component at the same time period. Put in the form of an equation:

$$\frac{\text{Specific City Cost}}{\text{National Average Cost}} \times 100 = \text{City Index Number}$$

Therefore, when making cost comparisons between cities, do not subtract one city's index number from the index number of another city and read the result as a percentage difference. Instead, divide one city's index number by that of the other city. The resulting number may then be used as a multiplier to calculate cost differences from city to city.

The formula used to find cost differences between cities for the purpose of comparison is as follows:

$$\frac{\text{City A Index}}{\text{City B Index}} \times \text{City B Cost (Known)} = \text{City A Cost (Unknown)}$$

In addition, you can use RSMeans CCI to calculate and compare costs division by division between cities using the same basic formula. (Just be sure that you're comparing similar divisions.)

2. Compare a specific city's construction costs with the National Average.

When you're studying construction location feasibility, it's advisable to compare a prospective project's cost index with an index of the National Average cost.

For example, divide the weighted average index of construction costs of a specific city by that of the 30 City Average, which = 100.

$$\frac{\text{City Index}}{100} = \text{\% of National Average}$$

As a result, you get a ratio that indicates the relative cost of construction in that city in comparison with the National Average.

3. Convert U.S. National Average to actual costs in Canadian City.

$$\frac{\text{Index for Canadian City}}{100} \times \text{National Average Cost} =$$
$$\text{Cost in Canadian City in \$ CAN}$$

4. Adjust construction cost data based on a National Average.
When you use a source of construction cost data which is based on a National Average (such as RSMeans cost data publications), it is necessary to adjust those costs to a specific location.

$$\frac{\text{City Index}}{100} \quad \text{x} \quad \frac{\text{"Book" Cost Based on}}{\text{National Average Costs}} \quad = \quad \frac{\text{City Cost}}{\text{(Unknown)}}$$

5. When applying the City Cost Indexes to demolition projects, use the appropriate division installation index. For example, for removal of existing doors and windows, use Division 8 (Openings) index.

What you might like to know about how we developed the Indexes

The information presented in the CCI is organized according to the Construction Specifications Institute (CSI) MasterFormat 2014.

To create a reliable index, RSMeans researched the building type most often constructed in the United States and Canada. Because it was concluded that no one type of building completely represented the building construction industry, nine different types of buildings were combined to create a composite model.

The exact material, labor, and equipment quantities are based on detailed analyses of these nine building types, then each quantity is weighted in proportion to expected usage. These various material items, labor hours, and equipment rental rates are thus combined to form a composite building representing as closely as possible the actual usage of materials, labor and equipment used in the North American Building Construction Industry.

The following structures were chosen to make up that composite model:

1. Factory, 1 story
2. Office, 2–4 story
3. Store, Retail
4. Town Hall, 2–3 story
5. High School, 2–3 story
6. Hospital, 4–8 story
7. Garage, Parking
8. Apartment, 1–3 story
9. Hotel/Motel, 2–3 story

For the purposes of ensuring the timeliness of the data, the components of the index for the composite model have been streamlined. They currently consist of:

- specific quantities of 66 commonly used construction materials;
- specific labor-hours for 21 building construction trades; and
- specific days of equipment rental for 6 types of construction equipment (normally used to install the 66 material items by the 21 trades.) Fuel costs and routine maintenance costs are included in the equipment cost.

A sophisticated computer program handles the updating of all costs for each city on a quarterly basis. Material and equipment price quotations are gathered quarterly from cities in the United States and Canada. These prices and the latest negotiated labor wage rates for 21 different building trades are used to compile the quarterly update of the City Cost Index.

The 30 major U.S. cities used to calculate the National Average are:

Atlanta, GA	Memphis, TN
Baltimore, MD	Milwaukee, WI
Boston, MA	Minneapolis, MN
Buffalo, NY	Nashville, TN
Chicago, IL	New Orleans, LA
Cincinnati, OH	New York, NY
Cleveland, OH	Philadelphia, PA
Columbus, OH	Phoenix, AZ
Dallas, TX	Pittsburgh, PA
Denver, CO	St. Louis, MO
Detroit, MI	San Antonio, TX
Houston, TX	San Diego, CA
Indianapolis, IN	San Francisco, CA
Kansas City, MO	Seattle, WA
Los Angeles, CA	Washington, DC

What the CCI does not indicate

The weighted average for each city is a total of the divisional components weighted to reflect typical usage, but it does not include the productivity variations between trades or cities.

In addition, the CCI does not take into consideration factors such as the following:

- managerial efficiency
- competitive conditions
- automation
- restrictive union practices
- unique local requirements
- regional variations due to specific building codes

ALABAMA

DIV. NO.	BUILDING SYSTEMS	BIRMINGHAM MAT.	INST.	TOTAL	HUNTSVILLE MAT.	INST.	TOTAL	MOBILE MAT.	INST.	TOTAL	MONTGOMERY MAT.	INST.	TOTAL	TUSCALOOSA MAT.	INST.	TOTAL
A	Substructure	104.4	81.3	90.3	96.3	76.8	84.4	100.1	69.6	81.5	101.8	63.7	78.5	98.9	69.4	80.9
B10	Shell: Superstructure	103.9	86.4	97.3	104.2	82.1	95.5	103.9	78.1	93.8	103.4	74.5	92.0	104.1	77.7	93.7
B20	Exterior Closure	98.8	77.1	88.4	99.1	68.2	84.3	101.1	56.9	80.0	99.7	52.9	77.3	98.6	67.7	84.3
B30	Roofing	99.6	82.1	93.0	96.6	77.6	89.4	97.6	71.7	87.8	97.8	67.7	86.4	96.7	73.0	87.7
C	Interior Construction	99.6	75.2	89.4	100.2	71.2	88.0	97.4	54.6	79.4	98.1	47.3	76.8	100.2	51.7	79.9
D10	Services: Conveying	100.0	90.1	97.3	100.0	87.1	96.5	100.0	87.1	96.5	100.0	84.5	95.7	100.0	86.2	96.2
D20-40	Mechanical	100.0	70.4	88.1	100.0	61.6	84.5	99.9	60.7	84.1	100.0	34.2	73.4	100.0	34.0	73.4
D50	Electrical	99.0	61.6	79.3	95.3	64.9	79.3	95.7	59.1	76.3	96.3	60.9	77.6	94.8	61.6	77.3
E	Equipment & Furnishings	100.0	75.1	98.7	100.0	71.6	98.5	100.0	52.6	97.6	100.0	42.7	97.1	100.0	44.9	97.2
G	Site Work	97.2	93.6	94.7	89.1	92.5	91.5	95.8	87.8	90.3	95.9	87.8	90.4	89.7	92.7	91.8
A-G	WEIGHTED AVERAGE	100.6	76.7	90.2	99.8	72.1	87.8	99.9	65.4	84.9	99.9	57.1	81.2	99.8	61.2	83.0

ALASKA / ARIZONA

DIV. NO.	BUILDING SYSTEMS	ANCHORAGE MAT.	INST.	TOTAL	FAIRBANKS MAT.	INST.	TOTAL	JUNEAU MAT.	INST.	TOTAL	FLAGSTAFF MAT.	INST.	TOTAL	MESA/TEMPE MAT.	INST.	TOTAL
A	Substructure	134.0	119.9	125.4	131.6	120.1	124.6	132.2	119.9	124.7	93.4	77.4	83.7	97.7	78.5	85.9
B10	Shell: Superstructure	117.7	109.9	114.7	121.0	110.0	116.7	119.5	109.9	115.7	96.5	73.6	87.5	94.7	74.9	86.9
B20	Exterior Closure	160.5	123.0	142.6	147.6	123.0	135.9	155.2	123.0	139.8	117.4	64.1	91.9	107.2	64.2	86.6
B30	Roofing	160.0	117.2	143.8	167.3	118.5	148.8	168.8	117.2	149.2	96.2	69.2	85.9	101.3	65.7	87.8
C	Interior Construction	133.4	120.9	128.1	130.4	121.3	126.6	131.7	120.9	127.2	102.8	60.4	85.0	98.6	65.4	84.7
D10	Services: Conveying	100.0	112.0	103.3	100.0	112.0	103.3	100.0	112.0	103.3	100.0	85.8	96.1	100.0	86.0	96.1
D20-40	Mechanical	100.3	105.0	102.2	100.2	108.0	103.4	100.3	105.0	102.2	100.2	79.0	91.6	100.0	79.0	91.5
D50	Electrical	117.7	117.8	117.7	130.0	117.8	123.5	119.9	117.8	118.8	103.4	61.3	81.1	94.2	61.3	76.8
E	Equipment & Furnishings	100.0	119.0	101.0	100.0	119.0	101.0	100.0	119.0	101.0	100.0	64.1	98.2	100.0	69.0	98.4
G	Site Work	127.9	130.1	129.4	120.7	130.1	127.2	138.4	130.1	132.7	86.5	96.1	93.1	89.5	95.8	93.8
A-G	WEIGHTED AVERAGE	121.1	115.6	118.7	121.0	116.4	119.0	121.3	115.6	118.8	101.3	71.8	88.4	98.6	72.8	87.4

ARIZONA / ARKANSAS

DIV. NO.	BUILDING SYSTEMS	PHOENIX MAT.	INST.	TOTAL	PRESCOTT MAT.	INST.	TOTAL	TUCSON MAT.	INST.	TOTAL	FORT SMITH MAT.	INST.	TOTAL	JONESBORO MAT.	INST.	TOTAL
A	Substructure	97.5	78.2	85.7	91.6	72.8	80.2	95.2	78.0	84.7	88.2	69.8	77.0	86.2	65.5	73.6
B10	Shell: Superstructure	95.8	75.1	87.7	96.3	69.6	85.8	94.9	74.5	86.9	99.5	67.5	86.9	93.4	67.9	83.4
B20	Exterior Closure	101.3	65.7	84.3	103.3	63.5	84.3	100.0	64.1	82.8	97.4	51.8	75.6	95.5	45.9	71.8
B30	Roofing	100.9	67.4	88.2	95.0	64.7	83.5	102.3	64.7	88.0	99.4	57.9	83.6	103.4	50.4	83.3
C	Interior Construction	100.9	65.4	86.0	101.6	50.1	80.0	95.5	63.6	82.1	94.2	58.2	79.1	92.6	49.3	74.4
D10	Services: Conveying	100.0	86.0	96.1	100.0	85.8	96.1	100.0	86.0	96.1	100.0	81.4	94.9	100.0	44.8	84.8
D20-40	Mechanical	99.9	79.0	91.5	100.2	78.9	91.6	100.0	79.0	91.5	100.1	50.6	80.1	100.3	52.5	81.0
D50	Electrical	101.0	66.7	82.9	103.0	61.2	80.9	96.5	61.3	77.9	93.7	64.5	78.3	100.2	63.5	80.8
E	Equipment & Furnishings	100.0	67.9	98.4	100.0	48.4	97.4	100.0	67.6	98.3	100.0	58.8	97.9	100.0	48.3	97.4
G	Site Work	89.8	96.0	94.1	75.9	95.9	89.6	85.8	95.8	92.7	83.4	86.7	85.6	103.8	100.8	101.7
A-G	WEIGHTED AVERAGE	99.2	73.8	88.1	99.2	69.8	86.0	97.4	72.4	86.5	97.2	61.6	81.7	96.7	59.1	80.3

ARKANSAS / CALIFORNIA

DIV. NO.	BUILDING SYSTEMS	LITTLE ROCK MAT.	INST.	TOTAL	PINE BLUFF MAT.	INST.	TOTAL	TEXARKANA MAT.	INST.	TOTAL	ANAHEIM MAT.	INST.	TOTAL	BAKERSFIELD MAT.	INST.	TOTAL
A	Substructure	89.6	70.3	77.8	84.8	70.2	75.9	91.6	55.7	69.7	94.2	118.6	109.1	97.3	117.9	109.9
B10	Shell: Superstructure	99.5	67.3	86.9	100.3	67.2	87.3	95.5	57.1	80.4	101.3	110.6	104.9	103.6	110.5	106.3
B20	Exterior Closure	99.2	52.1	76.7	113.2	52.1	84.0	98.1	36.6	68.7	98.5	118.3	107.9	100.0	116.3	107.8
B30	Roofing	95.6	58.3	81.5	94.4	58.3	80.7	95.1	45.3	76.2	99.5	124.6	109.0	98.0	117.9	105.6
C	Interior Construction	96.1	60.3	81.1	95.9	60.3	80.9	99.5	41.7	75.3	101.9	119.6	109.3	98.5	119.9	107.5
D10	Services: Conveying	100.0	81.4	94.9	100.0	81.4	94.9	100.0	31.5	81.2	100.0	110.5	102.9	100.0	110.5	102.9
D20-40	Mechanical	99.9	56.2	82.3	100.1	52.9	81.1	100.1	53.5	81.3	100.0	118.6	107.5	100.1	117.1	107.0
D50	Electrical	100.7	70.2	84.6	97.0	70.2	82.8	98.9	36.0	65.7	92.2	105.7	99.3	103.6	102.5	103.1
E	Equipment & Furnishings	100.0	62.3	98.1	100.0	62.3	98.1	100.0	41.0	97.0	100.0	122.8	101.2	100.0	122.4	101.1
G	Site Work	93.5	86.7	88.8	85.9	86.7	86.4	95.9	87.0	89.8	100.6	108.4	106.0	100.6	106.2	104.4
A-G	WEIGHTED AVERAGE	98.5	64.0	83.4	99.4	63.3	83.6	98.2	49.2	76.8	99.4	114.9	106.2	100.7	113.6	106.3

CALIFORNIA

DIV. NO.	BUILDING SYSTEMS	FRESNO MAT.	INST.	TOTAL	LOS ANGELES MAT.	INST.	TOTAL	OAKLAND MAT.	INST.	TOTAL	OXNARD MAT.	INST.	TOTAL	REDDING MAT.	INST.	TOTAL
A	Substructure	96.1	120.4	110.9	95.5	118.1	109.3	111.8	128.3	121.9	99.9	117.5	110.6	117.1	123.4	121.0
B10	Shell: Superstructure	103.9	112.7	107.4	100.2	111.5	104.6	103.2	117.5	108.8	100.7	110.4	104.5	104.5	115.8	108.9
B20	Exterior Closure	103.7	120.2	111.6	93.7	121.7	107.1	106.5	133.1	119.2	101.3	116.3	108.5	127.0	121.7	124.5
B30	Roofing	88.6	115.5	98.8	97.6	122.7	107.1	103.2	137.9	116.4	97.3	122.7	107.0	121.8	122.3	122.0
C	Interior Construction	99.5	132.4	113.3	99.7	120.0	108.2	98.4	146.0	118.3	96.9	119.2	106.2	112.5	138.0	123.2
D10	Services: Conveying	100.0	120.2	105.5	100.0	110.5	102.9	100.0	122.7	106.2	100.0	110.8	103.0	100.0	120.2	105.5
D20-40	Mechanical	100.2	111.3	104.7	99.9	118.6	107.5	100.1	147.8	119.4	100.1	118.6	107.6	100.2	109.0	103.7
D50	Electrical	93.8	100.8	97.5	102.0	121.3	112.2	108.5	139.3	124.8	97.7	107.3	102.8	99.2	107.4	103.5
E	Equipment & Furnishings	100.0	136.8	101.9	100.0	123.2	101.2	100.0	154.2	102.8	100.0	122.6	101.2	100.0	147.7	102.4
G	Site Work	103.5	105.5	104.9	97.1	107.3	104.1	126.0	107.7	113.4	103.7	104.2	104.0	126.6	105.2	111.9
A-G	WEIGHTED AVERAGE	100.1	115.3	106.7	99.2	117.5	107.2	103.1	134.3	116.7	99.6	114.4	106.0	107.8	117.7	112.1

CALIFORNIA

DIV. NO.	BUILDING SYSTEMS	RIVERSIDE MAT.	INST.	TOTAL	SACRAMENTO MAT.	INST.	TOTAL	SAN DIEGO MAT.	INST.	TOTAL	SAN FRANCISCO MAT.	INST.	TOTAL	SAN JOSE MAT.	INST.	TOTAL
A	Substructure	98.3	118.2	110.4	96.4	123.5	112.9	100.9	109.5	106.2	117.1	130.6	125.3	113.5	126.5	121.5
B10	Shell: Superstructure	102.3	110.6	105.5	97.1	111.3	102.7	101.0	105.8	102.9	109.1	123.6	114.8	100.1	121.2	108.4
B20	Exterior Closure	94.9	117.7	105.8	97.7	121.0	108.8	99.6	114.7	106.9	107.5	139.8	123.0	117.4	133.6	125.2
B30	Roofing	99.7	123.7	108.8	111.0	121.4	115.0	99.4	107.4	102.4	105.1	141.5	118.9	98.9	140.3	114.6
C	Interior Construction	102.1	119.6	109.4	104.7	133.1	116.6	101.1	112.0	105.7	101.7	146.9	120.7	95.9	145.8	116.8
D10	Services: Conveying	100.0	110.5	102.9	100.0	120.7	105.7	100.0	109.6	102.6	100.0	123.1	106.4	100.0	122.1	106.1
D20-40	Mechanical	100.0	118.6	107.5	100.0	120.8	108.4	99.9	116.9	106.8	100.1	175.0	130.3	100.0	147.3	119.1
D50	Electrical	91.9	105.4	99.1	103.5	108.7	106.2	101.2	98.5	99.8	108.6	157.8	134.6	101.1	155.9	130.1
E	Equipment & Furnishings	100.0	122.7	101.2	100.0	140.0	101.2	100.0	110.3	100.5	100.0	155.2	102.8	100.0	155.0	102.8
G	Site Work	99.3	106.5	104.2	102.9	113.7	110.3	101.5	104.1	103.3	127.3	113.7	118.0	132.3	100.3	110.3
A-G	WEIGHTED AVERAGE	99.4	114.6	106.0	100.5	119.3	108.7	100.5	109.6	104.5	105.2	145.4	122.7	102.5	136.6	117.4

| DIV. NO. | BUILDING SYSTEMS | CALIFORNIA | | | | | | | | | COLORADO | | | | | |
| | | SANTA BARBARA | | | STOCKTON | | | VALLEJO | | | COLORADO SPRINGS | | | DENVER | | |
		MAT.	INST.	TOTAL	MAT.	INST.	TOTAL	MAT.	INST.	TOTAL	MAT.	INST.	TOTAL	MAT.	INST.	TOTAL
A	Substructure	99.8	117.9	110.8	100.7	121.3	113.2	102.6	127.7	117.9	101.5	85.5	91.8	98.5	84.7	90.1
B10	Shell: Superstructure	101.0	110.4	104.7	99.6	114.1	105.3	101.1	115.4	106.7	99.3	84.0	93.3	101.1	82.5	93.8
B20	Exterior Closure	100.3	118.5	109.0	105.1	121.5	112.9	85.9	132.0	107.9	101.7	81.9	92.2	99.1	74.1	87.1
B30	Roofing	94.5	120.9	104.5	102.6	119.8	109.1	105.2	135.0	116.5	103.9	86.7	97.4	102.4	75.5	92.2
C	Interior Construction	97.6	118.7	106.5	102.6	132.4	115.1	105.6	144.8	122.0	100.7	74.1	89.6	102.7	77.8	92.3
D10	Services: Conveying	100.0	110.8	103.0	100.0	120.2	105.5	100.0	120.5	105.6	100.0	94.1	98.4	100.0	90.3	97.3
D20 - 40	Mechanical	100.1	118.6	107.6	100.0	111.3	104.6	100.1	127.5	111.1	100.2	88.2	95.4	100.0	79.9	91.9
D50	Electrical	91.0	110.5	101.3	98.2	110.6	104.8	100.9	122.4	112.3	100.8	82.1	90.9	102.6	83.2	92.4
E	Equipment & Furnishings	100.0	122.7	101.2	100.0	140.1	102.0	100.0	153.1	102.7	100.0	77.8	98.9	100.0	77.5	98.8
G	Site Work	103.7	106.2	105.4	102.2	105.3	104.3	103.9	113.5	110.5	97.2	94.6	95.4	96.4	102.5	100.6
A-G	WEIGHTED AVERAGE	98.9	115.1	106.0	100.9	117.2	108.0	100.1	127.2	111.9	100.4	84.0	93.2	100.8	81.7	92.5

| DIV. NO. | BUILDING SYSTEMS | COLORADO | | | | | | | | | | | | CONNECTICUT | | |
| | | FORT COLLINS | | | GRAND JUNCTION | | | GREELEY | | | PUEBLO | | | BRIDGEPORT | | |
		MAT.	INST.	TOTAL	MAT.	INST.	TOTAL	MAT.	INST.	TOTAL	MAT.	INST.	TOTAL	MAT.	INST.	TOTAL
A	Substructure	109.1	81.2	92.1	110.7	80.2	92.1	96.2	76.7	84.3	102.2	84.4	91.4	107.7	120.9	115.7
B10	Shell: Superstructure	99.7	78.7	91.4	101.7	77.7	92.3	97.1	75.4	88.6	101.1	84.2	94.4	101.5	125.5	111.0
B20	Exterior Closure	110.2	76.8	94.2	122.0	73.0	98.6	101.2	54.2	78.7	100.4	81.8	91.5	107.7	132.8	119.7
B30	Roofing	103.5	71.9	91.5	104.1	67.2	90.1	102.9	70.5	90.6	103.6	84.4	96.3	98.5	127.7	109.6
C	Interior Construction	99.3	69.5	86.8	106.5	71.9	92.0	98.7	71.7	87.4	101.6	76.4	91.1	99.3	126.6	110.8
D10	Services: Conveying	100.0	90.3	97.3	100.0	90.5	97.4	100.0	90.3	97.3	100.0	94.7	98.5	100.0	111.1	103.0
D20 - 40	Mechanical	100.1	77.2	90.9	99.9	82.5	92.9	100.1	77.2	90.8	99.9	76.8	90.6	100.0	116.4	106.6
D50	Electrical	97.4	84.5	90.6	98.2	53.8	74.7	97.4	84.5	90.6	99.1	73.0	85.3	98.5	112.2	105.7
E	Equipment & Furnishings	100.0	77.2	98.8	100.0	75.4	98.7	100.0	82.5	99.1	100.0	77.1	98.8	100.0	122.6	101.2
G	Site Work	108.4	95.9	99.8	130.3	94.7	105.9	95.0	95.3	95.2	121.8	91.2	100.8	107.9	105.3	106.1
A-G	WEIGHTED AVERAGE	101.3	79.0	91.6	104.8	75.3	91.9	98.9	75.5	88.7	101.1	80.3	92.0	101.3	120.8	109.8

| DIV. NO. | BUILDING SYSTEMS | CONNECTICUT | | | | | | | | | | | | | | |
| | | BRISTOL | | | HARTFORD | | | NEW BRITAIN | | | NEW HAVEN | | | NORWALK | | |
		MAT.	INST.	TOTAL	MAT.	INST.	TOTAL	MAT.	INST.	TOTAL	MAT.	INST.	TOTAL	MAT.	INST.	TOTAL
A	Substructure	103.5	120.8	114.1	103.3	120.8	114.0	104.5	120.8	114.5	105.5	121.0	115.0	106.6	121.5	115.7
B10	Shell: Superstructure	100.7	125.4	110.4	104.7	125.4	112.9	97.8	125.4	108.6	98.2	125.4	108.9	101.3	126.1	111.0
B20	Exterior Closure	103.8	132.8	117.6	105.1	132.8	118.3	104.8	132.8	118.2	118.1	132.8	125.1	103.7	134.0	118.2
B30	Roofing	98.7	124.1	108.3	102.7	124.1	110.9	98.7	124.1	108.3	98.7	124.3	108.4	98.7	128.0	109.8
C	Interior Construction	99.4	126.6	110.8	98.6	126.6	110.3	99.4	126.6	110.8	99.4	126.6	110.8	99.4	126.6	110.8
D10	Services: Conveying	100.0	111.1	103.0	100.0	111.1	103.0	100.0	111.1	103.0	100.0	111.1	103.0	100.0	111.1	103.0
D20 - 40	Mechanical	100.0	116.4	106.6	100.0	116.4	106.6	100.0	116.4	106.6	100.0	116.4	106.6	100.0	116.4	106.6
D50	Electrical	98.5	111.8	105.6	97.9	112.8	105.8	98.6	111.8	105.6	98.4	111.8	105.5	98.5	167.2	134.8
E	Equipment & Furnishings	100.0	122.6	101.2	100.0	122.7	101.2	100.0	122.6	101.2	100.0	122.6	101.2	100.0	125.3	101.3
G	Site Work	106.9	105.3	105.8	103.6	105.3	104.7	107.2	105.3	105.9	107.0	106.0	106.3	107.7	105.3	106.0
A-G	WEIGHTED AVERAGE	100.5	120.6	109.3	101.4	120.8	109.8	100.1	120.6	109.1	101.6	120.7	109.9	100.8	128.7	113.0

| DIV. NO. | BUILDING SYSTEMS | CONNECTICUT | | | | | | D.C. | | | DELAWARE | | | FLORIDA | | |
| | | STAMFORD | | | WATERBURY | | | WASHINGTON | | | WILMINGTON | | | DAYTONA BEACH | | |
		MAT.	INST.	TOTAL	MAT.	INST.	TOTAL	MAT.	INST.	TOTAL	MAT.	INST.	TOTAL	MAT.	INST.	TOTAL
A	Substructure	107.7	121.5	116.1	107.6	120.8	115.7	110.5	87.1	96.2	98.0	105.6	102.6	93.6	75.1	82.3
B10	Shell: Superstructure	101.5	126.1	111.2	101.5	125.4	110.9	102.0	97.0	100.0	101.1	110.4	104.7	102.0	81.6	94.0
B20	Exterior Closure	104.0	134.0	118.3	104.0	132.8	117.8	98.1	83.4	91.0	104.1	102.5	103.3	97.8	67.1	83.1
B30	Roofing	98.7	128.0	109.8	98.7	124.3	108.4	101.6	86.5	95.9	98.0	111.8	103.2	96.1	73.9	87.6
C	Interior Construction	99.4	126.6	110.8	99.3	126.6	110.7	100.8	81.3	92.6	93.0	103.5	97.4	97.2	69.5	85.6
D10	Services: Conveying	100.0	111.1	103.0	100.0	111.1	103.0	100.0	101.4	100.4	100.0	86.2	96.2	100.0	86.5	96.3
D20 - 40	Mechanical	100.0	116.4	106.6	100.0	116.4	106.6	100.1	92.1	96.9	100.1	117.4	107.1	99.9	76.1	90.3
D50	Electrical	98.5	167.2	134.8	98.0	112.2	105.5	101.5	105.8	103.8	99.8	111.9	106.2	95.8	54.9	74.2
E	Equipment & Furnishings	100.0	125.3	101.3	100.0	122.6	101.2	100.0	77.2	98.8	100.0	101.9	100.1	100.0	68.1	98.4
G	Site Work	108.3	105.3	106.2	107.7	105.3	106.0	104.1	94.2	97.3	99.4	113.7	109.2	103.9	88.9	93.6
A-G	WEIGHTED AVERAGE	100.9	128.7	113.0	100.9	120.7	109.5	101.1	91.9	97.1	99.3	109.2	103.6	99.0	72.9	87.6

| DIV. NO. | BUILDING SYSTEMS | FLORIDA | | | | | | | | | | | | | | |
| | | FORT LAUDERDALE | | | JACKSONVILLE | | | MELBOURNE | | | MIAMI | | | ORLANDO | | |
		MAT.	INST.	TOTAL	MAT.	INST.	TOTAL	MAT.	INST.	TOTAL	MAT.	INST.	TOTAL	MAT.	INST.	TOTAL
A	Substructure	94.0	73.6	81.6	94.1	67.9	78.1	105.3	76.2	87.5	96.5	74.0	82.8	108.3	75.9	88.5
B10	Shell: Superstructure	100.0	81.3	92.6	100.9	73.8	90.2	111.0	82.3	99.8	100.9	80.9	93.1	103.3	81.7	94.9
B20	Exterior Closure	100.5	68.9	85.4	97.8	62.5	80.9	99.6	71.5	86.2	100.7	71.8	86.9	98.4	66.9	83.3
B30	Roofing	100.4	78.6	92.1	96.3	62.2	83.4	96.5	76.6	89.0	101.8	71.6	90.3	94.8	74.6	87.1
C	Interior Construction	95.5	68.0	84.0	97.2	56.3	80.1	96.3	71.9	86.1	98.0	68.9	85.8	99.7	72.7	88.4
D10	Services: Conveying	100.0	90.1	97.3	100.0	84.7	95.8	100.0	88.0	96.7	100.0	90.7	97.4	100.0	86.5	96.3
D20 - 40	Mechanical	100.0	66.5	86.5	99.9	64.2	85.5	99.9	78.2	91.2	100.0	66.4	86.4	99.9	56.6	82.4
D50	Electrical	96.0	72.7	83.7	95.5	62.2	77.9	97.0	67.8	81.6	100.1	74.9	86.8	97.6	60.0	77.7
E	Equipment & Furnishings	100.0	67.1	98.3	100.0	52.0	97.5	100.0	68.7	98.4	100.0	67.3	98.3	100.0	73.7	98.7
G	Site Work	95.6	76.6	82.6	104.0	88.3	93.2	111.1	88.2	95.4	98.6	76.5	83.4	106.3	88.1	93.8
A-G	WEIGHTED AVERAGE	98.6	72.6	87.2	98.8	66.4	84.7	101.7	76.3	90.6	99.8	73.1	88.2	100.5	70.0	87.2

| DIV. NO. | BUILDING SYSTEMS | FLORIDA | | | | | | | | | | | | | | |
| | | PANAMA CITY | | | PENSACOLA | | | ST. PETERSBURG | | | TALLAHASSEE | | | TAMPA | | |
		MAT.	INST.	TOTAL	MAT.	INST.	TOTAL	MAT.	INST.	TOTAL	MAT.	INST.	TOTAL	MAT.	INST.	TOTAL
A	Substructure	99.0	61.6	76.2	112.8	67.2	85.0	101.2	68.3	81.1	99.3	60.6	75.7	99.6	79.6	87.4
B10	Shell: Superstructure	102.6	70.6	90.0	106.4	74.8	94.0	103.5	77.9	93.5	101.1	69.0	88.5	102.5	88.4	96.9
B20	Exterior Closure	105.5	48.9	78.5	113.7	57.1	86.6	126.0	53.4	91.3	101.0	55.2	79.1	102.5	77.5	90.5
B30	Roofing	96.6	57.6	81.8	96.5	62.4	83.6	100.2	58.6	84.4	102.7	73.1	91.5	100.5	84.8	94.5
C	Interior Construction	96.5	45.7	75.2	95.9	55.1	78.8	95.8	53.1	77.9	100.1	48.2	78.4	97.1	72.8	86.9
D10	Services: Conveying	100.0	46.7	85.4	100.0	45.7	85.1	100.0	56.5	88.1	100.0	69.0	91.5	100.0	86.2	96.2
D20 - 40	Mechanical	99.9	52.3	80.7	99.9	52.5	80.8	100.0	58.4	83.2	100.0	38.9	75.4	100.0	80.8	92.2
D50	Electrical	94.5	59.3	75.9	98.5	55.8	75.9	96.3	61.6	77.9	103.3	60.0	80.4	96.0	61.6	77.8
E	Equipment & Furnishings	100.0	42.0	97.0	100.0	51.1	97.5	100.0	50.1	97.4	100.0	41.7	97.0	100.0	76.1	98.8
G	Site Work	115.2	87.1	95.9	116.0	87.6	96.5	108.8	87.8	94.4	103.4	87.4	92.4	108.9	88.3	94.8
A-G	WEIGHTED AVERAGE	100.2	57.8	81.7	102.7	61.2	84.6	102.6	63.3	85.5	100.8	56.9	81.7	100.1	78.4	90.6

629

GEORGIA

DIV. NO.	BUILDING SYSTEMS	ALBANY MAT.	INST.	TOTAL	ATLANTA MAT.	INST.	TOTAL	AUGUSTA MAT.	INST.	TOTAL	COLUMBUS MAT.	INST.	TOTAL	MACON MAT.	INST.	TOTAL
A	Substructure	93.2	60.0	72.9	103.5	78.7	88.4	99.5	68.5	80.6	93.0	63.7	75.1	92.9	69.4	78.5
B10	Shell: Superstructure	102.2	69.9	89.5	95.0	75.7	87.4	93.2	67.5	83.1	101.7	73.3	90.5	98.0	74.1	88.6
B20	Exterior Closure	99.1	50.3	75.8	95.8	67.5	82.2	93.1	46.1	70.7	99.2	58.3	79.7	103.6	47.5	76.8
B30	Roofing	96.0	62.6	83.3	93.8	72.5	85.7	93.5	59.0	80.4	95.9	64.7	84.0	94.3	64.7	83.1
C	Interior Construction	93.4	43.6	72.5	95.5	73.2	86.1	92.3	61.8	79.5	93.4	53.9	76.8	88.6	50.8	72.8
D10	Services: Conveying	100.0	85.7	96.1	100.0	87.0	96.4	100.0	79.6	94.4	100.0	85.7	96.1	100.0	84.0	95.6
D20 - 40	Mechanical	99.9	68.0	87.0	99.9	70.6	88.1	100.0	61.2	84.4	99.9	63.4	85.2	99.9	66.7	86.6
D50	Electrical	96.9	58.7	76.7	99.0	71.9	84.7	100.4	61.8	80.0	97.1	69.6	82.5	96.1	61.2	77.7
E	Equipment & Furnishings	100.0	38.7	96.9	100.0	70.5	98.7	100.0	69.4	98.4	100.0	52.4	97.6	100.0	51.1	97.5
G	Site Work	97.4	78.2	84.3	97.1	94.9	95.6	94.1	93.2	93.5	97.3	78.2	84.2	97.8	93.3	94.7
A-G	WEIGHTED AVERAGE	98.5	61.3	82.3	97.6	74.4	87.5	96.3	63.6	82.1	98.4	65.4	84.0	97.2	64.7	83.0

GEORGIA / HAWAII / IDAHO

DIV. NO.	BUILDING SYSTEMS	SAVANNAH MAT.	INST.	TOTAL	VALDOSTA MAT.	INST.	TOTAL	HONOLULU MAT.	INST.	TOTAL	STATES & POSS., GUAM MAT.	INST.	TOTAL	BOISE MAT.	INST.	TOTAL
A	Substructure	99.8	61.7	76.6	92.6	60.1	72.8	151.6	123.6	134.5	178.4	80.8	118.9	92.6	85.3	88.2
B10	Shell: Superstructure	100.5	69.5	88.4	100.6	69.1	88.2	127.0	117.9	123.4	145.9	74.3	117.8	100.5	80.9	92.8
B20	Exterior Closure	98.1	52.2	76.2	107.5	51.0	80.5	141.3	127.8	134.8	160.5	45.0	105.3	113.0	81.4	97.9
B30	Roofing	95.0	58.5	81.1	96.1	65.2	84.4	138.8	123.9	133.1	140.9	68.5	113.4	94.6	81.5	89.6
C	Interior Construction	96.0	49.0	76.3	90.4	43.7	70.8	125.8	137.4	130.7	152.1	58.6	113.0	97.2	75.0	87.9
D10	Services: Conveying	100.0	82.2	95.1	100.0	82.2	95.1	100.0	113.1	103.6	100.0	77.6	93.8	100.0	88.9	97.0
D20 - 40	Mechanical	100.0	61.8	84.6	99.9	70.0	87.9	100.3	107.7	103.3	102.6	37.4	76.3	100.0	71.8	88.6
D50	Electrical	101.7	57.4	78.3	95.0	56.2	74.5	107.9	121.1	114.9	153.5	41.2	94.1	98.5	72.8	84.9
E	Equipment & Furnishings	100.0	46.1	97.2	100.0	38.4	96.8	100.0	138.4	102.0	100.0	57.4	97.8	100.0	75.9	98.8
G	Site Work	99.3	79.8	85.9	106.1	78.4	87.1	158.1	106.7	122.8	181.3	102.7	127.3	87.0	96.3	93.4
A-G	WEIGHTED AVERAGE	99.2	60.9	82.5	98.5	61.4	82.3	119.5	120.2	119.8	136.2	58.0	102.1	100.1	78.6	90.7

IDAHO / ILLINOIS

DIV. NO.	BUILDING SYSTEMS	LEWISTON MAT.	INST.	TOTAL	POCATELLO MAT.	INST.	TOTAL	TWIN FALLS MAT.	INST.	TOTAL	CHICAGO MAT.	INST.	TOTAL	DECATUR MAT.	INST.	TOTAL
A	Substructure	102.3	87.7	93.4	94.8	85.1	88.9	97.3	70.6	81.0	105.0	140.6	126.7	99.8	108.5	105.1
B10	Shell: Superstructure	97.4	86.8	93.2	107.1	80.5	96.7	107.6	70.3	92.9	96.3	143.5	114.8	99.9	109.7	103.7
B20	Exterior Closure	122.6	82.8	103.6	109.8	77.0	94.1	118.1	75.5	97.7	100.5	155.4	126.7	92.1	113.5	102.3
B30	Roofing	149.3	81.0	123.3	94.6	73.5	86.6	95.3	74.6	87.5	98.0	145.0	115.9	103.4	109.9	105.8
C	Interior Construction	138.1	83.1	115.0	97.8	75.2	88.3	99.4	50.7	79.0	102.4	156.1	124.9	101.1	114.4	106.6
D10	Services: Conveying	100.0	88.5	97.0	100.0	88.9	97.0	100.0	43.7	84.5	100.0	120.4	105.6	100.0	103.1	100.8
D20 - 40	Mechanical	100.7	85.6	94.6	99.9	71.8	88.6	99.9	69.3	87.6	99.8	133.9	113.6	99.9	98.5	99.4
D50	Electrical	88.9	80.7	84.5	96.2	70.4	82.6	92.3	60.1	75.3	98.3	133.9	117.1	99.5	90.4	94.7
E	Equipment & Furnishings	100.0	81.0	99.0	100.0	75.7	98.8	100.0	48.6	97.4	100.0	155.8	102.9	100.0	113.2	100.7
G	Site Work	92.6	92.3	92.4	86.5	96.3	93.2	92.8	97.2	95.8	106.5	95.8	99.2	94.1	99.0	97.4
A-G	WEIGHTED AVERAGE	108.7	84.9	98.3	101.1	77.4	90.8	102.2	67.3	87.0	99.8	139.7	117.2	99.2	104.9	101.7

ILLINOIS

DIV. NO.	BUILDING SYSTEMS	EAST ST. LOUIS MAT.	INST.	TOTAL	JOLIET MAT.	INST.	TOTAL	PEORIA MAT.	INST.	TOTAL	ROCKFORD MAT.	INST.	TOTAL	SPRINGFIELD MAT.	INST.	TOTAL
A	Substructure	96.5	111.5	105.7	105.1	139.6	126.2	99.4	111.2	106.6	100.1	123.3	114.3	98.7	108.6	104.7
B10	Shell: Superstructure	96.7	117.1	104.7	94.7	140.7	112.8	100.1	114.4	105.7	100.3	129.1	111.6	97.4	110.4	102.5
B20	Exterior Closure	85.8	117.6	101.0	101.8	148.2	124.0	110.0	117.6	113.7	98.5	134.6	115.8	96.7	116.2	106.1
B30	Roofing	97.2	111.0	102.5	97.9	142.7	114.9	99.4	112.9	104.6	101.9	129.7	112.5	104.6	112.6	107.6
C	Interior Construction	94.6	113.5	102.5	100.6	157.4	124.4	99.3	118.2	107.2	99.3	130.4	112.3	103.8	113.2	107.8
D10	Services: Conveying	100.0	103.3	100.9	100.0	119.8	105.4	100.0	103.8	101.1	100.0	112.3	103.4	100.0	103.3	100.9
D20 - 40	Mechanical	99.9	98.8	99.4	99.9	131.2	112.5	99.9	104.9	101.9	100.0	116.5	106.7	99.9	103.3	101.3
D50	Electrical	96.4	103.5	100.1	97.5	137.9	118.9	96.1	96.7	96.4	96.4	131.1	114.7	102.4	92.0	96.9
E	Equipment & Furnishings	100.0	112.3	100.6	100.0	160.6	103.1	100.0	115.1	100.8	100.0	129.3	101.5	100.0	113.3	100.7
G	Site Work	101.6	99.5	100.1	106.3	95.1	98.6	100.5	98.3	99.0	100.1	99.7	99.8	98.7	99.0	98.9
A-G	WEIGHTED AVERAGE	96.4	108.7	101.8	99.2	138.4	116.3	100.5	109.3	104.4	99.5	124.8	110.5	100.0	106.5	102.8

INDIANA

DIV. NO.	BUILDING SYSTEMS	ANDERSON MAT.	INST.	TOTAL	BLOOMINGTON MAT.	INST.	TOTAL	EVANSVILLE MAT.	INST.	TOTAL	FORT WAYNE MAT.	INST.	TOTAL	GARY MAT.	INST.	TOTAL
A	Substructure	103.7	84.8	92.2	98.0	83.1	88.9	96.7	92.9	94.4	107.5	82.2	92.1	106.7	111.0	109.3
B10	Shell: Superstructure	95.2	85.7	91.5	96.7	78.7	89.6	90.9	84.3	88.3	96.1	82.1	90.6	95.8	111.4	101.9
B20	Exterior Closure	95.6	80.8	88.5	106.0	76.9	92.1	103.6	82.3	93.4	97.5	77.8	88.1	96.3	116.0	105.7
B30	Roofing	107.8	75.2	95.5	95.0	79.0	88.9	99.4	84.2	93.6	107.6	77.4	96.1	106.4	108.5	107.2
C	Interior Construction	96.3	81.3	90.0	100.2	80.1	91.8	96.2	82.3	90.4	96.1	75.6	87.5	95.8	118.0	105.1
D10	Services: Conveying	100.0	93.1	98.1	100.0	92.0	97.8	100.0	97.7	99.4	100.0	93.9	98.3	100.0	104.8	101.3
D20 - 40	Mechanical	100.0	78.8	91.5	99.7	78.9	91.3	99.9	79.6	91.7	100.0	72.1	88.7	100.0	106.6	102.7
D50	Electrical	87.0	88.4	87.7	98.9	88.0	93.2	95.1	88.3	91.5	87.7	79.2	83.2	98.7	107.6	103.4
E	Equipment & Furnishings	100.0	82.0	99.1	100.0	81.7	99.1	100.0	81.6	99.1	100.0	74.8	98.7	100.0	115.8	100.8
G	Site Work	94.9	96.1	95.7	86.5	94.6	92.0	91.8	124.3	114.1	96.5	96.0	96.1	95.6	99.6	98.4
A-G	WEIGHTED AVERAGE	97.0	83.8	91.3	99.3	81.8	91.6	97.1	86.9	92.6	97.6	79.1	89.5	98.3	103.6	

INDIANA / IOWA

DIV. NO.	BUILDING SYSTEMS	INDIANAPOLIS MAT.	INST.	TOTAL	MUNCIE MAT.	INST.	TOTAL	SOUTH BEND MAT.	INST.	TOTAL	TERRE HAUTE MAT.	INST.	TOTAL	CEDAR RAPIDS MAT.	INST.	TOTAL
A	Substructure	100.1	88.5	93.0	102.1	83.9	91.0	103.2	84.4	91.7	94.9	92.8	93.6	106.0	84.8	93.1
B10	Shell: Superstructure	95.0	82.8	90.2	98.7	85.3	93.4	94.9	91.1	93.4	91.1	85.5	88.9	94.4	87.5	91.7
B20	Exterior Closure	99.8	80.4	90.6	101.8	80.9	91.8	98.9	79.4	89.6	111.8	80.2	96.7	102.2	78.5	90.9
B30	Roofing	100.7	81.3	93.3	97.7	76.3	89.6	102.1	81.1	94.2	99.5	81.0	92.5	104.0	81.6	95.5
C	Interior Construction	101.0	85.5	94.5	94.8	80.9	89.0	94.5	83.5	89.9	96.5	82.8	90.7	103.9	81.5	94.5
D10	Services: Conveying	100.0	94.3	98.4	100.0	92.0	97.8	100.0	94.3	98.4	100.0	95.3	98.7	100.0	93.0	98.1
D20 - 40	Mechanical	99.9	79.4	91.6	99.7	78.7	91.3	99.9	77.1	90.7	99.9	79.6	91.7	100.2	80.5	92.2
D50	Electrical	101.0	88.4	94.4	90.9	79.6	84.9	99.6	86.6	92.7	93.3	86.7	89.8	98.1	80.6	88.8
E	Equipment & Furnishings	100.0	86.2	99.3	100.0	81.7	99.1	100.0	81.6	99.1	100.0	81.6	99.1	100.0	82.1	99.1
G	Site Work	94.0	99.5	97.8	86.7	94.5	92.0	96.8	95.8	96.1	93.9	124.4	114.8	99.5	95.1	96.5
A-G	WEIGHTED AVERAGE	99.1	84.8	92.9	97.9	82.3	91.1	98.0	84.4	92.1	97.9	86.5	92.9	100.0	83.2	92.7

IOWA

DIV. NO.	BUILDING SYSTEMS	COUNCIL BLUFFS MAT.	INST.	TOTAL	DAVENPORT MAT.	INST.	TOTAL	DES MOINES MAT.	INST.	TOTAL	DUBUQUE MAT.	INST.	TOTAL	SIOUX CITY MAT.	INST.	TOTAL
A	Substructure	108.7	77.7	89.8	103.3	93.5	97.3	100.3	88.2	92.9	102.4	86.8	92.9	107.0	71.7	85.5
B10	Shell: Superstructure	97.9	81.9	91.6	93.9	97.7	95.4	99.7	90.9	96.2	91.9	88.0	90.3	94.3	77.1	87.5
B20	Exterior Closure	103.9	76.2	90.7	100.9	87.0	94.3	101.3	83.1	92.6	101.5	75.6	89.1	99.4	56.3	78.8
B30	Roofing	103.4	65.3	88.9	103.5	88.1	97.6	98.5	79.1	91.2	103.7	74.2	92.5	103.5	66.7	89.5
C	Interior Construction	99.1	72.7	88.0	102.0	93.2	98.3	99.6	83.9	93.0	101.2	74.5	90.0	102.7	65.1	87.0
D10	Services: Conveying	100.0	90.9	97.5	100.0	95.0	98.6	100.0	93.6	98.2	100.0	92.4	97.9	100.0	91.1	97.5
D20 - 40	Mechanical	100.2	74.2	89.7	100.2	92.8	97.2	99.9	79.7	91.7	100.2	75.3	90.1	100.2	75.8	90.4
D50	Electrical	103.6	81.5	91.9	95.6	91.0	93.2	105.4	81.3	92.7	102.1	78.2	89.4	98.1	73.0	84.8
E	Equipment & Furnishings	100.0	71.0	98.5	100.0	90.7	99.5	100.0	80.7	99.0	100.0	73.2	98.6	100.0	66.0	98.3
G	Site Work	101.6	91.4	94.6	98.2	98.6	98.4	103.0	99.7	100.8	97.8	92.4	94.1	108.1	94.9	99.0
A-G	WEIGHTED AVERAGE	100.7	78.1	90.9	99.0	93.1	96.4	100.5	85.2	93.8	99.1	80.2	90.9	99.7	72.6	87.9

IOWA / KANSAS

DIV. NO.	BUILDING SYSTEMS	WATERLOO MAT.	INST.	TOTAL	DODGE CITY MAT.	INST.	TOTAL	KANSAS CITY MAT.	INST.	TOTAL	SALINA MAT.	INST.	TOTAL	TOPEKA MAT.	INST.	TOTAL
A	Substructure	108.4	69.5	84.7	117.7	65.8	86.0	94.3	95.3	94.9	105.8	64.7	80.7	99.1	62.0	76.5
B10	Shell: Superstructure	98.5	76.7	89.9	101.0	67.7	87.9	101.6	98.1	100.2	98.5	68.8	86.8	99.3	75.7	90.0
B20	Exterior Closure	101.3	71.7	87.2	118.0	58.2	89.4	105.6	97.3	101.6	119.2	58.7	90.3	101.8	58.8	81.2
B30	Roofing	103.3	71.5	91.2	99.0	61.4	84.7	93.6	97.5	95.1	98.5	61.9	84.6	96.6	67.0	85.4
C	Interior Construction	96.8	58.1	80.6	99.4	53.3	80.1	95.6	91.5	93.9	98.7	53.6	79.8	99.4	47.3	77.6
D10	Services: Conveying	100.0	93.0	98.1	100.0	39.3	83.3	100.0	57.7	88.4	100.0	90.4	97.4	100.0	48.4	85.8
D20 - 40	Mechanical	100.0	79.8	91.9	100.0	69.0	87.5	99.9	96.4	98.5	100.0	69.1	87.5	100.0	69.1	87.5
D50	Electrical	95.8	55.7	74.6	96.8	73.8	84.6	98.7	97.2	97.9	96.6	73.8	84.5	97.5	71.8	83.9
E	Equipment & Furnishings	100.0	48.1	97.3	100.0	57.9	97.8	100.0	94.5	99.7	100.0	52.5	97.6	100.0	37.9	96.8
G	Site Work	106.6	95.2	98.8	115.7	94.7	101.3	95.9	93.0	93.9	106.2	95.0	98.5	97.8	91.3	93.3
A-G	WEIGHTED AVERAGE	99.5	71.8	87.4	102.7	66.0	86.7	99.5	94.8	97.5	101.5	67.6	86.7	99.5	66.0	84.9

KANSAS / KENTUCKY

DIV. NO.	BUILDING SYSTEMS	WICHITA MAT.	INST.	TOTAL	BOWLING GREEN MAT.	INST.	TOTAL	LEXINGTON MAT.	INST.	TOTAL	LOUISVILLE MAT.	INST.	TOTAL	OWENSBORO MAT.	INST.	TOTAL
A	Substructure	102.1	64.9	79.4	87.9	90.7	89.6	94.3	87.4	90.1	95.3	84.6	88.8	90.9	96.6	94.4
B10	Shell: Superstructure	99.6	70.9	88.3	92.5	87.6	90.6	93.4	85.4	90.3	94.4	85.0	90.7	86.8	89.3	87.8
B20	Exterior Closure	103.9	52.8	79.5	104.7	79.2	92.5	97.5	73.6	86.1	95.2	79.8	87.8	110.6	90.9	101.2
B30	Roofing	97.2	56.0	81.5	86.5	80.7	84.3	99.5	87.0	94.7	96.3	77.9	89.3	99.4	92.8	96.9
C	Interior Construction	101.9	53.3	81.5	95.0	83.2	90.1	95.3	72.7	85.9	93.5	82.8	89.1	94.0	85.8	90.6
D10	Services: Conveying	100.0	87.1	96.5	100.0	59.6	88.9	100.0	88.8	96.9	100.0	85.1	95.9	100.0	105.5	101.5
D20 - 40	Mechanical	99.8	64.8	85.7	99.9	82.7	93.0	100.0	79.5	91.7	100.0	80.9	92.3	99.9	78.6	91.4
D50	Electrical	100.9	73.8	86.5	94.0	81.5	87.4	94.2	81.5	87.5	99.4	81.5	89.9	93.5	81.4	87.1
E	Equipment & Furnishings	100.0	51.3	97.5	100.0	82.9	99.1	100.0	68.9	98.4	100.0	82.8	99.1	100.0	82.7	99.1
G	Site Work	101.7	93.5	96.1	78.8	97.3	91.5	91.5	100.7	97.8	85.0	97.5	93.6	92.0	124.8	114.5
A-G	WEIGHTED AVERAGE	100.7	66.0	85.6	96.2	83.9	90.8	95.6	81.4	90.0	96.6	83.3	90.8	96.3	88.9	93.1

LOUISIANA

DIV. NO.	BUILDING SYSTEMS	ALEXANDRIA MAT.	INST.	TOTAL	BATON ROUGE MAT.	INST.	TOTAL	LAKE CHARLES MAT.	INST.	TOTAL	MONROE MAT.	INST.	TOTAL	NEW ORLEANS MAT.	INST.	TOTAL
A	Substructure	95.1	56.9	71.7	99.1	66.1	79.0	100.3	67.0	80.0	94.8	58.7	72.8	99.2	71.8	82.5
B10	Shell: Superstructure	94.8	59.6	80.9	100.1	66.8	87.0	96.0	67.0	84.6	94.6	60.3	81.1	107.7	70.0	92.9
B20	Exterior Closure	110.7	53.2	83.2	98.7	54.9	77.8	97.4	58.4	78.8	108.2	49.5	80.2	99.6	61.2	81.2
B30	Roofing	95.6	63.0	83.2	96.2	65.9	84.7	96.1	66.6	84.9	95.5	61.8	82.7	94.8	69.9	85.3
C	Interior Construction	100.0	48.4	78.4	95.5	60.9	81.0	96.6	58.7	80.7	99.9	45.7	77.2	98.9	65.2	84.8
D10	Services: Conveying	100.0	51.1	86.6	100.0	83.1	95.4	100.0	83.4	95.4	100.0	46.2	85.2	100.0	85.2	95.9
D20 - 40	Mechanical	100.1	55.3	82.1	100.0	59.5	83.7	100.1	62.0	84.7	100.1	54.3	81.6	100.0	65.1	85.9
D50	Electrical	95.7	57.9	75.7	100.8	61.6	80.1	97.4	65.0	80.3	97.7	59.6	77.6	102.1	73.3	86.9
E	Equipment & Furnishings	100.0	41.8	97.0	100.0	62.7	98.1	100.0	57.2	97.8	100.0	42.3	97.0	100.0	66.4	98.3
G	Site Work	101.5	87.1	91.6	103.1	87.2	92.2	103.5	87.0	92.2	101.4	87.0	91.5	102.4	91.1	94.7
A-G	WEIGHTED AVERAGE	99.4	57.2	81.1	99.1	63.7	83.7	98.1	65.0	83.7	99.3	56.5	80.6	101.4	69.5	87.5

LOUISIANA / MAINE

DIV. NO.	BUILDING SYSTEMS	SHREVEPORT MAT.	INST.	TOTAL	AUGUSTA MAT.	INST.	TOTAL	BANGOR MAT.	INST.	TOTAL	LEWISTON MAT.	INST.	TOTAL	PORTLAND MAT.	INST.	TOTAL
A	Substructure	101.0	58.2	74.9	99.2	87.9	92.3	85.5	103.0	96.2	91.3	103.0	98.4	97.3	103.0	100.8
B10	Shell: Superstructure	99.6	60.4	84.2	104.5	87.0	97.6	93.7	95.4	94.3	97.8	95.4	96.8	105.8	95.4	101.7
B20	Exterior Closure	104.9	50.8	79.0	107.8	65.5	87.6	113.9	98.5	106.5	103.8	98.5	101.3	109.8	98.5	104.4
B30	Roofing	95.3	60.2	82.0	100.0	64.7	86.6	97.7	81.3	91.5	97.6	81.3	91.4	100.1	81.3	93.0
C	Interior Construction	101.7	46.5	78.6	103.4	88.2	97.1	99.2	94.4	97.2	102.4	94.4	99.0	101.6	94.4	98.6
D10	Services: Conveying	100.0	82.6	95.2	100.0	100.2	100.1	100.0	111.7	103.2	100.0	111.7	103.2	100.0	111.7	103.2
D20 - 40	Mechanical	99.9	58.6	83.3	99.9	64.2	85.5	100.1	74.3	89.7	100.1	74.3	89.7	100.0	74.3	89.6
D50	Electrical	103.5	66.6	84.0	100.9	80.1	89.9	97.2	75.8	85.9	99.3	80.1	89.1	98.9	80.1	89.0
E	Equipment & Furnishings	100.0	43.2	97.1	100.0	102.1	100.1	100.0	94.4	99.7	100.0	94.6	99.7	100.0	94.6	99.7
G	Site Work	105.0	87.0	92.7	91.2	101.3	98.1	92.9	101.5	98.8	89.7	101.5	97.8	90.2	101.5	98.0
A-G	WEIGHTED AVERAGE	101.0	59.6	83.0	102.1	79.5	92.3	99.0	89.5	94.9	99.6	90.1	95.5	102.0	90.1	96.8

MARYLAND / MASSACHUSETTS

DIV. NO.	BUILDING SYSTEMS	BALTIMORE MAT.	INST.	TOTAL	HAGERSTOWN MAT.	INST.	TOTAL	BOSTON MAT.	INST.	TOTAL	BROCKTON MAT.	INST.	TOTAL	FALL RIVER MAT.	INST.	TOTAL
A	Substructure	105.5	79.9	89.9	87.7	81.9	84.2	104.7	139.2	125.8	99.7	136.6	122.2	97.7	137.4	121.9
B10	Shell: Superstructure	101.8	85.0	95.2	96.2	84.6	91.6	101.1	139.4	116.1	98.0	136.9	113.2	97.5	138.4	113.6
B20	Exterior Closure	100.0	75.2	88.2	97.2	83.5	90.6	106.2	160.2	132.0	104.1	156.5	129.2	104.5	154.8	128.6
B30	Roofing	102.0	78.1	92.9	100.1	75.9	90.9	103.4	148.3	120.4	100.9	145.0	117.6	100.8	142.2	116.5
C	Interior Construction	99.4	72.5	88.1	97.0	76.3	88.3	102.2	151.7	122.9	100.0	148.1	120.1	99.9	148.5	120.3
D10	Services: Conveying	100.0	88.0	96.7	100.0	91.8	97.8	100.0	116.3	104.5	100.0	115.5	104.2	100.0	116.1	104.4
D20 - 40	Mechanical	100.0	80.6	92.2	100.0	86.2	94.4	100.1	132.4	113.1	100.1	111.1	104.6	100.1	111.2	104.6
D50	Electrical	102.0	91.9	96.6	98.1	81.8	89.5	100.8	135.4	119.1	99.3	100.1	99.7	99.2	100.1	99.7
E	Equipment & Furnishings	100.0	72.7	98.6	100.0	72.8	98.6	100.0	143.0	102.2	100.0	136.4	101.9	100.0	136.3	101.9
G	Site Work	102.4	95.2	97.5	93.1	92.0	92.3	98.4	109.2	105.8	94.2	105.9	102.2	93.1	106.0	101.9
A-G	WEIGHTED AVERAGE	100.8	81.9	92.6	97.6	83.3	91.4	101.6	139.5	118.1	99.9	128.1	112.2	99.7	128.3	112.2

631

MASSACHUSETTS

DIV. NO.	BUILDING SYSTEMS	HYANNIS			LAWRENCE			LOWELL			NEW BEDFORD			PITTSFIELD		
		MAT.	INST.	TOTAL	MAT.	INST.	TOTAL	MAT.	INST.	TOTAL	MAT.	INST.	TOTAL	MAT.	INST.	TOTAL
A	Substructure	88.4	137.1	118.1	100.3	133.3	121.7	94.7	135.3	119.5	90.7	137.4	119.2	96.1	116.4	108.5
B10	Shell: Superstructure	92.5	138.0	110.4	98.8	135.3	113.1	97.5	134.0	111.8	96.1	138.8	112.7	97.6	113.7	104.0
B20	Exterior Closure	101.1	154.8	126.8	110.2	156.9	132.5	100.4	156.1	127.1	103.6	154.8	128.1	100.7	129.9	114.7
B30	Roofing	100.3	143.0	116.6	99.8	144.5	116.8	99.6	144.8	116.8	100.7	142.2	116.5	99.6	116.8	106.1
C	Interior Construction	96.0	148.3	117.9	97.8	147.8	118.7	102.4	147.8	121.4	99.8	148.5	120.2	102.4	119.1	109.4
D10	Services: Conveying	100.0	115.5	104.2	100.0	115.5	104.2	100.0	115.5	104.2	100.0	116.1	104.4	100.0	102.9	100.8
D20 - 40	Mechanical	100.1	111.1	104.6	100.1	124.8	110.0	100.1	125.7	110.4	100.1	111.2	104.6	100.1	101.4	100.6
D50	Electrical	96.8	100.1	98.5	99.1	128.9	114.9	99.6	128.9	115.1	100.0	100.1	100.1	99.6	98.9	99.2
E	Equipment & Furnishings	100.0	136.3	101.9	100.0	137.7	101.9	100.0	137.7	101.9	100.0	136.3	101.9	100.0	110.0	100.5
G	Site Work	91.9	105.8	101.4	95.0	105.9	102.5	93.5	105.9	102.0	91.6	106.0	101.5	94.7	104.3	101.3
A-G	WEIGHTED AVERAGE	97.0	128.1	110.6	100.3	134.6	115.3	99.5	134.5	114.8	99.1	128.3	111.8	99.7	111.3	104.8

MASSACHUSETTS / MICHIGAN

DIV. NO.	BUILDING SYSTEMS	SPRINGFIELD			WORCESTER			ANN ARBOR			DEARBORN			DETROIT		
		MAT.	INST.	TOTAL	MAT.	INST.	TOTAL	MAT.	INST.	TOTAL	MAT.	INST.	TOTAL	MAT.	INST.	TOTAL
A	Substructure	97.0	117.7	109.7	96.7	132.8	118.7	89.4	108.8	101.2	88.2	111.0	102.1	95.4	111.1	104.9
B10	Shell: Superstructure	100.1	114.9	105.9	101.1	131.6	112.5	93.5	115.6	102.2	93.3	117.2	102.7	95.1	107.9	100.1
B20	Exterior Closure	100.6	132.9	116.0	100.3	155.0	126.5	102.3	108.7	105.4	102.3	111.4	106.6	99.0	110.3	104.4
B30	Roofing	99.6	118.6	106.8	99.6	136.9	113.7	100.5	108.1	103.4	98.9	116.6	105.6	97.7	116.6	104.9
C	Interior Construction	102.4	120.0	109.8	102.4	141.1	118.6	96.0	111.9	102.7	95.9	114.6	103.8	96.7	114.6	104.2
D10	Services: Conveying	100.0	104.2	101.2	100.0	106.6	101.8	100.0	105.0	101.4	100.0	106.0	101.6	100.0	106.0	101.6
D20 - 40	Mechanical	100.1	103.3	101.4	100.1	113.6	105.5	100.0	98.8	99.5	100.0	109.2	103.7	100.0	110.2	104.1
D50	Electrical	99.7	98.9	99.2	99.7	105.1	102.5	96.3	108.8	102.9	96.3	107.0	101.9	98.2	107.0	102.9
E	Equipment & Furnishings	100.0	110.2	100.5	100.0	127.2	101.4	100.0	111.9	100.6	100.0	115.8	100.8	100.0	115.8	100.8
G	Site Work	94.0	104.6	101.3	94.0	105.8	102.1	83.0	108.5	93.6	82.7	98.7	93.7	93.8	100.8	98.6
A-G	WEIGHTED AVERAGE	100.2	112.6	105.6	100.2	126.4	111.6	97.1	107.5	101.7	97.0	111.0	103.1	97.8	109.6	102.9

MICHIGAN

DIV. NO.	BUILDING SYSTEMS	FLINT			GRAND RAPIDS			KALAMAZOO			LANSING			MUSKEGON		
		MAT.	INST.	TOTAL	MAT.	INST.	TOTAL	MAT.	INST.	TOTAL	MAT.	INST.	TOTAL	MAT.	INST.	TOTAL
A	Substructure	88.4	96.7	93.5	100.8	89.9	94.2	97.6	91.2	93.7	100.4	96.5	98.0	96.3	89.0	91.8
B10	Shell: Superstructure	93.8	106.5	98.8	96.6	88.3	93.3	98.2	88.9	94.5	95.3	106.1	99.5	96.1	87.9	92.9
B20	Exterior Closure	102.3	97.6	100.1	101.4	87.0	94.5	103.6	87.0	95.6	103.8	95.5	99.8	98.3	87.9	93.4
B30	Roofing	98.3	95.8	97.3	97.3	74.7	88.7	94.5	83.8	90.4	98.7	88.5	94.8	93.5	76.6	87.1
C	Interior Construction	95.7	90.8	93.7	102.8	83.9	94.9	97.4	85.6	92.4	100.7	89.1	95.8	95.2	83.2	90.2
D10	Services: Conveying	100.0	96.5	99.0	100.0	98.7	99.6	100.0	98.2	99.5	100.0	102.2	100.6	100.0	98.1	99.5
D20 - 40	Mechanical	100.0	89.6	95.8	100.0	83.0	93.1	100.0	80.8	92.3	99.9	88.6	95.3	99.9	82.3	92.8
D50	Electrical	96.3	96.0	96.1	99.7	79.1	88.8	94.3	81.8	87.7	98.6	93.3	95.8	94.8	79.1	86.5
E	Equipment & Furnishings	100.0	88.0	99.4	100.0	79.3	98.9	100.0	83.1	99.1	100.0	88.3	99.4	100.0	79.4	98.9
G	Site Work	73.7	97.7	90.1	94.0	87.7	89.7	94.5	87.7	89.8	95.7	97.6	97.0	91.7	87.6	88.9
A-G	WEIGHTED AVERAGE	96.8	96.0	96.4	99.7	85.1	93.3	98.7	85.7	93.0	99.3	94.7	97.3	97.2	84.8	91.8

MICHIGAN / MINNESOTA

DIV. NO.	BUILDING SYSTEMS	SAGINAW			DULUTH			MINNEAPOLIS			ROCHESTER			SAINT PAUL		
		MAT.	INST.	TOTAL	MAT.	INST.	TOTAL	MAT.	INST.	TOTAL	MAT.	INST.	TOTAL	MAT.	INST.	TOTAL
A	Substructure	87.7	95.3	92.3	108.7	106.1	107.1	103.9	118.3	112.7	105.4	104.2	104.7	113.4	115.1	114.5
B10	Shell: Superstructure	93.4	105.1	98.0	103.0	115.9	108.1	100.2	126.3	110.5	101.6	117.4	107.8	103.5	124.5	111.7
B20	Exterior Closure	102.5	89.4	96.2	104.3	116.8	110.3	102.8	126.4	114.1	99.6	115.6	107.3	105.8	126.2	115.5
B30	Roofing	99.1	93.0	96.8	101.7	114.3	106.5	102.8	129.2	112.8	105.0	98.8	102.7	104.2	128.1	113.3
C	Interior Construction	94.7	87.2	91.6	100.3	114.0	106.0	98.5	129.4	111.5	96.7	111.3	102.8	98.8	124.8	109.7
D10	Services: Conveying	100.0	95.2	98.7	100.0	96.1	98.9	100.0	117.4	102.0	100.0	102.5	100.7	100.0	107.1	101.9
D20 - 40	Mechanical	100.0	83.1	93.2	99.8	96.5	98.5	99.9	115.6	106.2	99.9	96.2	98.4	99.9	113.8	105.5
D50	Electrical	93.9	90.5	92.1	98.4	102.4	100.6	101.2	111.4	106.6	98.6	89.2	93.6	97.1	111.4	104.7
E	Equipment & Furnishings	100.0	86.9	99.3	100.0	109.6	100.5	100.0	127.5	101.4	100.0	105.0	100.3	100.0	120.6	101.1
G	Site Work	76.4	97.4	90.8	100.3	99.3	99.6	98.6	104.7	102.8	97.8	98.9	98.5	95.6	100.0	98.6
A-G	WEIGHTED AVERAGE	96.4	91.8	94.4	101.3	107.3	103.9	100.4	120.0	108.9	99.9	104.7	102.0	101.4	117.9	108.6

MINNESOTA / MISSISSIPPI

DIV. NO.	BUILDING SYSTEMS	ST. CLOUD			BILOXI			GREENVILLE			JACKSON			MERIDIAN		
		MAT.	INST.	TOTAL	MAT.	INST.	TOTAL	MAT.	INST.	TOTAL	MAT.	INST.	TOTAL	MAT.	INST.	TOTAL
A	Substructure	96.8	113.6	107.0	103.6	62.5	78.5	104.3	67.0	81.6	98.6	64.6	77.9	99.5	67.3	79.9
B10	Shell: Superstructure	94.1	121.9	105.0	96.7	69.3	86.0	97.2	72.9	87.7	101.0	70.1	88.9	93.8	72.0	85.3
B20	Exterior Closure	98.5	120.3	109.0	101.0	48.5	75.9	123.2	51.8	89.1	104.3	51.4	79.0	98.8	56.7	78.7
B30	Roofing	103.5	118.7	109.3	97.4	56.9	82.0	96.0	61.3	82.8	97.6	60.5	83.5	97.1	63.2	84.2
C	Interior Construction	89.6	120.9	102.7	95.8	53.4	78.0	95.5	63.6	82.1	98.2	58.5	81.5	93.3	63.7	80.9
D10	Services: Conveying	100.0	102.8	100.8	100.0	55.4	87.8	100.0	57.3	88.3	100.0	57.3	88.3	100.0	58.8	88.7
D20 - 40	Mechanical	99.6	113.8	105.3	100.0	49.1	79.5	100.0	47.6	78.8	100.1	57.0	82.7	100.0	59.1	83.5
D50	Electrical	101.8	111.4	106.9	103.8	56.9	79.0	97.1	62.7	78.9	105.4	62.7	82.8	102.6	57.8	78.9
E	Equipment & Furnishings	100.0	115.8	100.8	100.0	55.3	97.7	100.0	67.0	98.3	100.0	59.1	97.9	100.0	64.3	98.2
G	Site Work	93.9	104.3	101.1	100.8	88.0	92.0	103.7	88.1	93.0	97.3	88.1	91.0	97.0	88.3	91.1
A-G	WEIGHTED AVERAGE	96.9	116.0	105.2	99.2	58.1	81.3	101.0	61.7	83.9	100.7	62.2	83.9	97.6	64.1	83.0

MISSOURI

DIV. NO.	BUILDING SYSTEMS	CAPE GIRARDEAU			COLUMBIA			JOPLIN			KANSAS CITY			SPRINGFIELD		
		MAT.	INST.	TOTAL	MAT.	INST.	TOTAL	MAT.	INST.	TOTAL	MAT.	INST.	TOTAL	MAT.	INST.	TOTAL
A	Substructure	92.1	88.5	89.9	95.7	92.3	93.6	107.6	85.0	93.8	100.7	106.0	103.9	101.6	88.4	93.5
B10	Shell: Superstructure	94.6	97.4	95.7	97.4	106.3	100.9	100.2	90.8	96.5	105.0	112.2	107.8	102.6	97.9	100.7
B20	Exterior Closure	106.7	81.2	94.5	120.7	93.4	107.7	97.1	85.4	91.5	100.1	108.1	103.9	107.1	89.0	98.4
B30	Roofing	100.2	84.6	94.3	96.1	86.6	92.5	97.8	77.0	89.9	97.8	106.4	101.1	100.8	80.8	93.2
C	Interior Construction	96.6	83.7	91.2	96.2	84.4	91.2	101.6	76.5	91.1	102.5	107.2	104.5	102.6	80.2	93.2
D10	Services: Conveying	100.0	94.5	98.5	100.0	98.5	99.6	100.0	89.7	97.2	100.0	100.1	100.0	100.0	97.2	99.2
D20 - 40	Mechanical	99.9	96.8	98.7	99.8	96.4	98.4	100.1	72.1	88.8	100.0	104.3	101.7	99.8	74.6	89.6
D50	Electrical	96.9	101.5	99.3	94.8	86.5	90.4	90.4	77.5	83.6	100.3	100.1	100.2	99.2	69.7	83.6
E	Equipment & Furnishings	100.0	85.8	99.3	100.0	78.8	98.9	100.0	73.9	98.7	100.0	104.7	100.2	100.0	76.4	98.8
G	Site Work	91.8	95.0	94.0	103.2	96.8	98.8	105.9	99.5	101.5	99.9	98.2	98.7	101.3	94.9	96.9
A-G	WEIGHTED AVERAGE	98.2	92.4	95.7	100.2	93.9	97.5	99.5	81.7	91.8	101.4	105.7	103.3	101.7	83.7	93.8

MISSOURI / MONTANA

DIV. NO.	BUILDING SYSTEMS	ST. JOSEPH MAT.	INST.	TOTAL	ST. LOUIS MAT.	INST.	TOTAL	BILLINGS MAT.	INST.	TOTAL	BUTTE MAT.	INST.	TOTAL	GREAT FALLS MAT.	INST.	TOTAL
A	Substructure	100.5	97.8	98.9	91.9	104.8	99.7	115.4	77.7	92.4	124.8	78.1	96.3	129.9	76.5	97.3
B10	Shell: Superstructure	101.9	105.4	103.3	98.7	113.6	104.6	107.9	81.6	97.6	104.7	82.0	95.7	108.6	81.1	97.8
B20	Exterior Closure	99.9	97.9	99.0	97.9	113.4	105.3	109.7	72.6	91.9	104.7	75.4	90.7	109.7	73.8	92.5
B30	Roofing	98.3	95.1	97.1	100.3	106.1	102.5	106.7	69.4	92.5	106.5	69.9	92.6	107.1	67.3	92.0
C	Interior Construction	101.9	92.5	98.0	98.8	103.7	100.9	96.7	73.1	86.8	96.1	70.8	85.5	99.4	73.7	88.6
D10	Services: Conveying	100.0	98.2	99.5	100.0	102.2	100.6	100.0	97.9	99.4	100.0	97.9	99.4	100.0	97.9	99.4
D20 - 40	Mechanical	100.1	89.0	95.6	100.0	107.4	103.0	100.1	74.7	89.9	100.2	75.6	90.3	100.2	70.7	88.3
D50	Electrical	98.8	78.6	88.2	99.2	108.3	104.0	98.5	72.2	84.6	105.5	71.7	87.7	97.7	71.9	84.1
E	Equipment & Furnishings	100.0	86.7	99.3	100.0	103.4	100.2	100.0	70.2	98.5	100.0	70.5	98.5	100.0	70.2	98.5
G	Site Work	101.5	93.8	96.2	91.6	98.6	96.4	95.2	96.2	95.9	101.6	96.0	97.7	104.9	96.1	98.9
A-G	WEIGHTED AVERAGE	100.6	93.5	97.5	98.7	107.8	102.7	102.7	77.1	91.5	102.6	77.3	91.6	104.1	76.2	91.9

MONTANA / NEBRASKA

DIV. NO.	BUILDING SYSTEMS	HELENA MAT.	INST.	TOTAL	MISSOULA MAT.	INST.	TOTAL	GRAND ISLAND MAT.	INST.	TOTAL	LINCOLN MAT.	INST.	TOTAL	NORTH PLATTE MAT.	INST.	TOTAL
A	Substructure	110.3	74.0	88.1	100.7	77.0	86.2	112.6	76.5	90.6	93.6	75.6	82.6	113.5	75.3	90.2
B10	Shell: Superstructure	103.8	78.6	93.9	97.4	81.7	91.2	100.7	79.5	92.4	98.6	78.2	90.6	99.1	78.5	91.0
B20	Exterior Closure	108.0	74.3	91.9	108.7	74.8	92.5	109.9	75.9	93.6	103.3	66.9	85.9	100.8	85.7	93.6
B30	Roofing	104.1	69.8	91.1	105.7	79.3	95.7	101.2	71.7	90.0	98.1	70.6	87.6	99.0	80.0	91.8
C	Interior Construction	99.1	63.1	84.0	95.9	71.6	85.7	93.1	75.9	85.9	103.0	70.8	89.5	95.1	74.4	86.4
D10	Services: Conveying	100.0	98.1	99.5	100.0	96.8	99.1	100.0	91.0	97.5	100.0	91.0	97.5	100.0	66.2	90.7
D20 - 40	Mechanical	100.1	70.1	88.0	100.2	69.2	87.7	100.0	78.0	91.1	99.9	78.0	91.0	99.9	76.0	90.2
D50	Electrical	104.6	71.3	87.0	103.2	67.7	84.5	92.2	67.0	78.9	105.4	67.0	85.1	92.6	77.1	84.4
E	Equipment & Furnishings	100.0	62.5	98.1	100.0	69.9	98.5	100.0	71.4	98.5	100.0	66.2	98.3	100.0	71.7	98.6
G	Site Work	97.6	95.7	96.3	84.8	95.7	92.3	106.6	92.9	97.2	92.7	92.9	92.8	106.6	92.1	96.6
A-G	WEIGHTED AVERAGE	102.4	73.8	89.9	99.9	75.5	89.3	100.0	77.1	90.0	100.5	74.9	89.4	99.0	78.3	90.0

NEBRASKA / NEVADA / NEW HAMPSHIRE

DIV. NO.	BUILDING SYSTEMS	OMAHA MAT.	INST.	TOTAL	CARSON CITY MAT.	INST.	TOTAL	LAS VEGAS MAT.	INST.	TOTAL	RENO MAT.	INST.	TOTAL	MANCHESTER MAT.	INST.	TOTAL
A	Substructure	99.2	78.4	86.5	99.5	95.8	97.3	95.3	108.5	103.3	100.5	95.9	97.7	106.3	101.5	103.3
B10	Shell: Superstructure	99.8	76.6	90.7	95.3	100.1	97.2	103.0	109.4	105.5	99.2	100.4	99.7	102.3	96.7	100.1
B20	Exterior Closure	104.0	78.5	91.8	108.1	98.1	103.3	104.8	106.0	105.3	107.1	96.1	101.9	103.9	100.6	102.3
B30	Roofing	95.0	77.8	88.4	103.7	90.8	98.8	113.9	103.8	110.1	99.9	90.8	96.4	101.6	92.7	98.2
C	Interior Construction	102.2	72.8	89.9	98.2	95.7	97.1	97.0	110.8	102.8	96.3	95.8	96.1	102.3	96.7	99.9
D10	Services: Conveying	100.0	89.6	97.2	100.0	89.2	97.0	100.0	103.8	101.0	100.0	89.2	97.0	100.0	106.0	101.6
D20 - 40	Mechanical	99.9	76.8	90.6	100.0	80.1	92.0	100.1	104.4	101.8	100.0	80.1	92.0	99.9	84.4	93.7
D50	Electrical	99.1	82.3	90.2	103.2	96.7	99.7	105.4	119.9	113.1	101.4	96.7	98.9	99.8	83.6	91.2
E	Equipment & Furnishings	100.0	72.3	98.6	100.0	94.5	99.7	100.0	109.2	100.5	100.0	94.5	99.7	100.0	91.5	99.6
G	Site Work	90.1	92.1	91.5	85.0	99.0	94.6	77.0	101.0	93.5	73.9	98.9	91.1	94.0	104.1	100.9
A-G	WEIGHTED AVERAGE	100.2	78.6	90.8	99.7	93.5	97.0	100.9	108.7	104.3	99.6	93.3	96.8	101.4	93.8	98.1

NEW HAMPSHIRE / NEW JERSEY

DIV. NO.	BUILDING SYSTEMS	NASHUA MAT.	INST.	TOTAL	PORTSMOUTH MAT.	INST.	TOTAL	CAMDEN MAT.	INST.	TOTAL	ELIZABETH MAT.	INST.	TOTAL	JERSEY CITY MAT.	INST.	TOTAL
A	Substructure	96.5	101.5	99.5	87.2	101.6	96.0	90.3	122.4	109.9	89.4	125.1	111.1	87.6	124.0	109.8
B10	Shell: Superstructure	98.3	96.7	97.7	93.0	97.1	94.6	98.7	116.0	105.5	94.0	122.7	105.3	96.6	120.9	106.1
B20	Exterior Closure	104.0	100.6	102.4	100.1	99.6	99.9	104.0	130.0	116.4	106.7	130.8	118.2	96.6	130.7	112.9
B30	Roofing	99.7	92.7	97.1	99.4	115.7	105.6	106.0	121.5	111.9	100.9	133.1	113.1	100.3	132.5	112.5
C	Interior Construction	103.5	96.7	100.7	101.3	96.7	99.4	99.9	132.3	113.5	101.2	135.2	115.4	98.9	135.2	114.1
D10	Services: Conveying	100.0	106.0	101.6	100.0	106.0	101.6	100.0	109.4	102.6	100.0	116.9	104.6	100.0	116.9	104.6
D20 - 40	Mechanical	100.1	84.4	93.8	100.1	84.4	93.8	100.0	123.6	109.5	100.1	125.5	110.3	100.1	127.4	111.1
D50	Electrical	101.2	83.6	91.9	99.1	83.6	90.9	96.0	138.1	118.2	93.4	137.7	116.8	97.6	140.1	120.1
E	Equipment & Furnishings	100.0	91.5	99.6	100.0	91.5	99.6	100.0	127.4	101.4	100.0	127.6	101.4	100.0	127.7	101.4
G	Site Work	94.5	104.1	101.1	89.1	104.1	99.4	96.6	105.8	102.9	105.2	106.5	106.1	91.8	106.6	101.9
A-G	WEIGHTED AVERAGE	100.5	93.8	97.6	97.9	94.3	96.4	99.5	124.8	110.5	98.8	127.6	111.3	97.9	127.9	111.0

NEW JERSEY / NEW MEXICO

DIV. NO.	BUILDING SYSTEMS	NEWARK MAT.	INST.	TOTAL	PATERSON MAT.	INST.	TOTAL	TRENTON MAT.	INST.	TOTAL	ALBUQUERQUE MAT.	INST.	TOTAL	FARMINGTON MAT.	INST.	TOTAL
A	Substructure	106.4	125.1	117.8	101.6	125.0	115.9	100.6	121.7	113.5	97.8	76.0	84.5	100.2	76.0	85.4
B10	Shell: Superstructure	101.6	122.7	109.9	95.6	122.6	106.2	100.7	115.1	106.3	104.0	77.8	93.7	102.7	77.8	92.9
B20	Exterior Closure	102.0	130.8	115.8	99.5	130.8	114.5	105.0	129.4	116.7	104.2	60.9	83.5	110.5	60.9	86.8
B30	Roofing	101.2	133.1	113.3	100.8	124.2	109.7	105.7	122.7	112.2	94.6	75.4	87.3	94.9	75.4	87.5
C	Interior Construction	100.3	135.2	114.9	103.4	135.2	116.7	101.3	133.8	114.9	98.0	65.7	84.5	98.6	65.7	84.8
D10	Services: Conveying	100.0	116.9	104.6	100.0	116.9	104.6	100.0	109.3	102.5	100.0	85.2	95.9	100.0	85.2	95.9
D20 - 40	Mechanical	100.1	127.4	111.1	100.1	127.4	111.1	100.1	126.9	110.9	100.1	69.0	87.6	100.0	69.0	87.5
D50	Electrical	101.2	140.1	121.8	97.6	137.7	118.8	99.3	135.4	118.4	91.2	71.2	80.6	88.8	71.2	79.5
E	Equipment & Furnishings	100.0	127.7	101.4	100.0	127.6	101.4	100.0	127.4	101.4	100.0	65.4	98.2	100.0	65.4	98.2
G	Site Work	106.8	106.5	106.6	103.8	106.6	105.7	95.6	106.3	102.9	87.9	103.0	98.3	95.6	103.0	100.7
A-G	WEIGHTED AVERAGE	101.2	128.4	113.0	99.6	127.8	111.9	100.9	125.1	111.5	99.6	72.6	87.8	100.1	72.6	88.1

NEW MEXICO / NEW YORK

DIV. NO.	BUILDING SYSTEMS	LAS CRUCES MAT.	INST.	TOTAL	ROSWELL MAT.	INST.	TOTAL	SANTA FE MAT.	INST.	TOTAL	ALBANY MAT.	INST.	TOTAL	BINGHAMTON MAT.	INST.	TOTAL
A	Substructure	93.7	68.6	78.3	97.7	76.0	84.5	103.5	76.0	86.7	94.7	105.4	101.2	100.2	102.5	101.6
B10	Shell: Superstructure	97.1	72.9	87.6	100.0	77.8	91.2	100.1	77.8	91.3	102.3	107.4	104.3	97.3	110.8	102.6
B20	Exterior Closure	87.5	60.3	74.5	114.0	60.9	88.6	103.5	60.9	83.1	102.2	110.2	106.0	97.8	123.4	110.0
B30	Roofing	86.0	69.7	79.8	99.3	75.4	90.2	96.9	75.4	88.7	106.0	104.7	105.5	107.7	96.8	103.5
C	Interior Construction	96.8	64.9	83.4	96.7	65.7	83.7	100.6	65.7	86.0	99.5	100.7	100.0	94.2	89.8	92.3
D10	Services: Conveying	100.0	82.2	95.1	100.0	85.2	95.9	100.0	85.2	95.9	100.0	99.1	99.8	100.0	97.7	99.4
D20 - 40	Mechanical	100.3	68.7	87.6	99.9	69.0	87.5	100.0	69.0	87.5	100.0	102.8	101.1	100.5	90.2	96.3
D50	Electrical	92.1	71.2	81.0	91.7	71.2	80.9	102.9	71.2	86.2	98.7	104.1	101.6	99.9	105.4	102.8
E	Equipment & Furnishings	100.0	64.6	98.2	100.0	65.4	98.2	100.0	65.4	98.2	100.0	98.2	99.9	100.0	85.4	99.3
G	Site Work	94.8	82.8	86.5	98.0	103.0	101.4	99.7	103.0	102.0	83.2	106.2	99.0	95.7	93.3	94.0
A-G	WEIGHTED AVERAGE	96.1	69.4	84.4	100.0	72.6	88.0	100.8	72.6	88.5	100.1	104.7	102.1	98.5	101.4	99.8

633

NEW YORK

DIV. NO.	BUILDING SYSTEMS	BUFFALO MAT.	INST.	TOTAL	HICKSVILLE MAT.	INST.	TOTAL	NEW YORK MAT.	INST.	TOTAL	RIVERHEAD MAT.	INST.	TOTAL	ROCHESTER MAT.	INST.	TOTAL
A	Substructure	103.1	111.1	108.0	96.1	158.0	133.9	109.0	170.7	146.7	97.7	153.7	131.9	95.0	102.3	99.5
B10	Shell: Superstructure	100.3	104.5	101.9	103.3	156.7	124.3	111.9	167.4	133.7	104.2	148.8	121.7	102.3	103.8	102.9
B20	Exterior Closure	101.9	116.5	108.9	108.9	165.9	136.1	108.0	175.4	136.4	110.6	164.4	136.3	106.0	103.7	104.9
B30	Roofing	101.8	109.6	104.7	107.7	153.8	125.2	108.5	163.2	129.3	108.6	153.5	125.6	103.5	100.9	102.5
C	Interior Construction	99.4	116.1	106.4	94.0	159.9	121.6	96.5	182.6	132.5	94.3	151.9	118.4	103.3	99.8	101.8
D10	Services: Conveying	100.0	104.2	101.2	100.0	124.9	106.8	100.0	129.6	108.1	100.0	124.3	106.7	100.0	100.1	100.0
D20 - 40	Mechanical	100.0	96.7	98.7	99.7	154.4	121.8	100.1	165.5	126.5	99.9	151.4	120.7	100.1	90.3	96.1
D50	Electrical	100.2	102.7	101.5	99.0	143.4	122.5	104.7	181.9	145.5	100.6	133.6	118.1	98.7	94.6	96.6
E	Equipment & Furnishings	100.0	115.8	100.8	100.0	153.3	102.7	100.0	186.1	104.4	100.0	152.8	102.7	100.0	97.3	99.9
G	Site Work	98.7	98.2	98.4	115.0	126.9	123.2	116.5	121.8	120.2	116.8	126.5	123.4	85.9	109.8	102.3
A-G	WEIGHTED AVERAGE	100.3	106.2	102.9	100.9	153.3	123.7	103.4	168.7	131.8	101.6	148.2	121.9	101.1	99.1	100.2

NEW YORK

DIV. NO.	BUILDING SYSTEMS	SCHENECTADY MAT.	INST.	TOTAL	SYRACUSE MAT.	INST.	TOTAL	UTICA MAT.	INST.	TOTAL	WATERTOWN MAT.	INST.	TOTAL	WHITE PLAINS MAT.	INST.	TOTAL
A	Substructure	99.8	105.4	103.2	95.8	99.3	97.9	89.2	97.5	94.2	97.5	101.0	99.6	97.7	140.6	123.9
B10	Shell: Superstructure	101.9	107.4	104.0	99.1	101.1	99.9	96.4	99.9	97.8	97.4	102.3	99.3	99.1	140.3	115.3
B20	Exterior Closure	100.2	110.2	105.0	101.9	103.0	102.4	98.9	101.7	100.2	106.1	105.9	106.0	93.9	148.0	119.7
B30	Roofing	100.4	104.7	102.0	102.4	95.7	99.9	90.6	96.1	92.7	90.9	98.5	93.8	109.2	143.4	122.2
C	Interior Construction	96.4	100.7	98.2	94.3	92.2	93.4	96.1	89.7	93.4	94.9	94.5	94.8	90.1	147.9	114.3
D10	Services: Conveying	100.0	99.1	99.8	100.0	97.5	99.3	100.0	97.7	99.4	100.0	98.2	99.5	100.0	123.5	106.5
D20 - 40	Mechanical	100.2	102.8	101.2	100.2	91.9	96.9	100.2	92.2	97.0	100.2	85.9	94.4	100.3	134.1	114.0
D50	Electrical	98.1	104.1	101.3	100.0	102.2	101.1	98.1	102.2	100.3	100.0	90.4	94.9	95.2	155.2	126.9
E	Equipment & Furnishings	100.0	98.2	99.9	100.0	89.1	99.4	100.0	85.2	99.2	100.0	92.0	99.6	100.0	137.9	101.9
G	Site Work	83.7	106.2	99.2	94.5	105.4	102.0	72.7	104.1	94.3	80.0	105.5	97.5	110.1	118.9	116.1
A-G	WEIGHTED AVERAGE	99.3	104.7	101.7	98.9	98.0	98.5	97.1	97.1	97.1	98.5	96.3	97.5	97.6	141.4	116.7

NEW YORK / NORTH CAROLINA

DIV. NO.	BUILDING SYSTEMS	YONKERS MAT.	INST.	TOTAL	ASHEVILLE MAT.	INST.	TOTAL	CHARLOTTE MAT.	INST.	TOTAL	DURHAM MAT.	INST.	TOTAL	FAYETTEVILLE MAT.	INST.	TOTAL
A	Substructure	107.5	142.3	128.8	108.1	55.7	76.1	112.9	55.3	77.7	99.6	56.9	73.5	112.0	63.0	82.1
B10	Shell: Superstructure	108.2	141.7	121.4	103.9	66.4	89.2	105.8	65.5	90.0	116.7	65.0	96.4	121.4	69.7	101.1
B20	Exterior Closure	100.5	148.0	123.2	96.7	45.6	72.3	100.8	52.1	77.6	94.4	41.2	69.0	93.7	42.7	71.2
B30	Roofing	109.5	144.2	122.7	106.6	43.0	82.5	100.6	44.7	79.4	107.0	46.8	84.1	106.2	46.5	83.6
C	Interior Construction	94.0	151.9	118.3	96.4	41.8	73.5	99.0	43.6	75.8	101.1	43.9	77.1	96.9	57.1	80.2
D10	Services: Conveying	100.0	124.0	106.6	100.0	82.6	95.2	100.0	82.8	95.3	100.0	75.8	93.3	100.0	75.6	93.3
D20 - 40	Mechanical	100.3	134.1	114.0	100.4	53.4	81.4	100.0	54.1	81.5	100.5	52.9	81.3	100.2	52.7	81.0
D50	Electrical	102.9	163.8	135.1	102.2	55.7	77.6	101.2	58.5	78.6	96.1	55.6	74.7	101.4	50.7	74.6
E	Equipment & Furnishings	100.0	144.4	102.3	100.0	40.0	96.9	100.0	41.6	97.0	100.0	44.1	97.1	100.0	64.4	98.2
G	Site Work	115.9	118.9	117.9	102.2	76.3	84.4	105.1	76.3	85.4	99.9	85.2	89.8	101.8	85.2	90.4
A-G	WEIGHTED AVERAGE	102.1	143.7	120.2	100.8	55.3	80.9	101.9	56.8	82.2	102.9	55.3	82.1	104.5	58.1	84.2

NORTH CAROLINA / NORTH DAKOTA

DIV. NO.	BUILDING SYSTEMS	GREENSBORO MAT.	INST.	TOTAL	RALEIGH MAT.	INST.	TOTAL	WILMINGTON MAT.	INST.	TOTAL	WINSTON-SALEM MAT.	INST.	TOTAL	BISMARCK MAT.	INST.	TOTAL
A	Substructure	98.9	57.1	73.4	103.6	58.6	76.2	108.0	57.3	77.1	100.6	59.0	75.2	102.8	62.8	78.4
B10	Shell: Superstructure	110.5	65.2	92.7	105.1	65.7	89.6	103.7	66.4	89.0	108.5	66.4	92.0	103.1	70.7	90.3
B20	Exterior Closure	92.7	44.1	69.5	93.9	44.5	70.3	92.0	44.9	69.5	92.8	41.9	68.5	107.7	57.9	83.9
B30	Roofing	106.8	42.6	82.4	100.5	47.4	80.3	106.6	48.2	84.4	106.8	44.2	83.0	112.6	49.7	88.7
C	Interior Construction	101.2	43.2	76.9	100.5	45.7	77.5	96.8	47.7	76.3	101.2	46.9	78.4	105.5	43.8	79.7
D10	Services: Conveying	100.0	80.1	94.5	100.0	76.1	93.4	100.0	77.2	93.7	100.0	82.7	95.2	100.0	88.6	96.9
D20 - 40	Mechanical	100.4	53.1	84.3	100.0	52.5	80.8	100.4	54.9	82.0	100.5	53.3	81.4	100.1	71.9	88.7
D50	Electrical	95.2	55.7	74.3	98.3	40.4	67.7	102.4	50.7	75.1	95.2	55.7	74.3	102.5	71.6	86.2
E	Equipment & Furnishings	100.0	44.2	97.1	100.0	46.0	97.2	100.0	48.0	97.3	100.0	48.5	97.4	100.0	36.4	96.7
G	Site Work	99.7	85.2	89.8	100.8	85.3	90.1	103.7	76.5	85.0	100.1	85.3	89.9	100.2	96.8	97.8
A-G	WEIGHTED AVERAGE	101.3	55.7	81.4	100.5	54.1	80.3	100.3	55.9	81.0	101.0	56.5	81.6	103.1	66.2	87.0

NORTH DAKOTA / OHIO

DIV. NO.	BUILDING SYSTEMS	FARGO MAT.	INST.	TOTAL	GRAND FORKS MAT.	INST.	TOTAL	MINOT MAT.	INST.	TOTAL	AKRON MAT.	INST.	TOTAL	CANTON MAT.	INST.	TOTAL
A	Substructure	108.4	63.5	81.0	109.6	58.6	78.5	109.6	71.9	86.6	95.8	96.5	96.2	96.4	89.6	92.2
B10	Shell: Superstructure	103.9	71.2	91.1	101.3	65.4	87.2	101.4	78.3	92.3	95.5	87.7	92.5	95.7	79.7	89.4
B20	Exterior Closure	106.4	57.9	83.2	107.9	62.5	86.2	105.5	66.5	86.9	98.8	95.6	97.3	98.4	82.7	90.9
B30	Roofing	107.5	50.0	85.7	107.9	44.7	83.9	107.7	54.4	87.4	111.0	96.6	105.5	112.2	92.6	104.7
C	Interior Construction	101.9	47.9	79.3	102.9	35.4	74.6	102.5	68.3	88.2	105.0	94.4	100.6	101.4	82.1	93.3
D10	Services: Conveying	100.0	88.6	96.9	100.0	32.6	81.5	100.0	88.6	96.9	100.0	98.9	99.7	100.0	97.1	99.2
D20 - 40	Mechanical	100.1	77.0	90.8	100.4	33.3	73.3	100.4	60.2	84.2	100.1	94.6	97.9	100.1	81.4	92.5
D50	Electrical	102.4	68.4	84.4	102.1	53.5	76.4	105.5	75.7	89.8	99.3	93.9	96.5	98.5	92.7	95.5
E	Equipment & Furnishings	100.0	36.7	96.8	100.0	32.2	96.5	100.0	69.6	98.4	100.0	93.1	99.6	100.0	82.7	99.1
G	Site Work	101.9	96.8	98.4	108.4	92.8	97.7	106.0	96.8	99.7	98.3	101.9	100.8	98.4	101.4	100.4
A-G	WEIGHTED AVERAGE	102.7	67.6	87.4	102.7	51.6	80.4	102.7	71.3	89.0	99.9	94.2	97.4	99.3	85.6	93.3

OHIO

DIV. NO.	BUILDING SYSTEMS	CINCINNATI MAT.	INST.	TOTAL	CLEVELAND MAT.	INST.	TOTAL	COLUMBUS MAT.	INST.	TOTAL	DAYTON MAT.	INST.	TOTAL	LORAIN MAT.	INST.	TOTAL
A	Substructure	94.4	88.2	90.6	94.9	100.9	98.6	96.2	89.9	92.4	90.2	86.4	87.9	93.0	95.8	94.7
B10	Shell: Superstructure	96.7	83.8	91.6	96.6	91.6	94.6	96.9	83.4	91.6	95.3	79.8	89.2	95.4	87.0	92.1
B20	Exterior Closure	97.7	84.6	91.4	99.4	103.5	101.4	99.4	91.7	95.7	97.5	83.0	90.6	96.6	102.2	99.3
B30	Roofing	99.2	89.3	95.4	109.7	109.4	109.6	101.0	94.7	98.6	104.8	88.2	98.5	112.1	102.9	108.6
C	Interior Construction	100.9	80.0	92.1	99.1	99.5	99.2	98.2	84.8	92.6	100.9	78.7	91.6	101.4	89.2	96.3
D10	Services: Conveying	100.0	90.9	97.5	100.0	103.2	100.9	100.0	94.7	98.5	100.0	90.6	97.4	100.0	102.8	100.8
D20 - 40	Mechanical	100.0	83.3	93.3	100.0	100.5	100.2	100.1	92.0	96.8	100.9	86.6	95.2	100.1	92.1	96.9
D50	Electrical	93.9	79.4	86.3	98.8	105.0	102.1	97.6	87.9	92.8	92.6	83.0	87.5	98.7	88.0	93.0
E	Equipment & Furnishings	100.0	77.1	98.8	100.0	97.5	99.9	100.0	82.7	99.1	100.0	77.3	98.8	100.0	81.4	99.0
G	Site Work	94.0	102.6	99.9	98.2	102.4	101.1	96.2	98.5	97.7	92.9	102.1	99.2	97.6	103.3	101.5
A-G	WEIGHTED AVERAGE	98.3	84.5	92.2	99.0	100.3	99.6	98.6	89.2	94.5	98.0	84.4	92.1	98.9	93.0	96.3

OHIO / OKLAHOMA

DIV. NO.	BUILDING SYSTEMS	SPRINGFIELD MAT.	INST.	TOTAL	TOLEDO MAT.	INST.	TOTAL	YOUNGSTOWN MAT.	INST.	TOTAL	ENID MAT.	INST.	TOTAL	LAWTON MAT.	INST.	TOTAL
A	Substructure	91.6	87.7	89.2	96.1	96.7	96.5	95.3	93.0	93.9	93.6	56.5	70.9	92.0	60.4	72.7
B10	Shell: Superstructure	95.6	80.9	89.8	98.6	90.4	94.3	95.5	83.8	90.9	101.9	58.1	84.7	104.8	60.7	87.6
B20	Exterior Closure	97.1	82.3	90.0	102.9	98.2	100.6	98.2	92.1	95.3	106.0	56.3	82.2	101.9	56.7	80.3
B30	Roofing	104.6	89.0	98.7	102.5	102.9	102.7	112.3	94.0	105.3	109.3	58.0	89.8	109.1	59.7	90.3
C	Interior Construction	100.0	82.7	92.7	96.4	96.8	96.6	101.4	87.8	95.7	97.7	36.4	72.0	99.5	44.9	76.7
D10	Services: Conveying	100.0	90.6	97.4	100.0	98.9	99.7	100.0	97.7	99.4	100.0	82.7	95.3	100.0	82.7	95.3
D20 - 40	Mechanical	100.9	81.3	93.0	100.1	100.4	100.2	100.1	87.1	94.8	100.3	65.8	86.4	100.3	65.9	86.4
D50	Electrical	92.6	87.8	90.1	97.6	104.7	101.4	98.7	86.0	92.0	94.6	70.9	82.1	96.2	70.9	82.8
E	Equipment & Furnishings	100.0	83.3	99.1	100.0	97.3	99.9	100.0	85.8	99.3	100.0	30.8	96.5	100.0	43.7	97.1
G	Site Work	93.2	100.8	98.4	95.4	99.1	97.9	98.2	102.3	101.0	100.3	90.6	93.6	96.3	92.0	93.4
A-G	WEIGHTED AVERAGE	98.0	84.7	92.2	98.7	98.1	98.4	99.2	89.0	94.7	100.2	60.5	82.9	100.7	62.8	84.2

OKLAHOMA / OREGON

DIV. NO.	BUILDING SYSTEMS	MUSKOGEE MAT.	INST.	TOTAL	OKLAHOMA CITY MAT.	INST.	TOTAL	TULSA MAT.	INST.	TOTAL	EUGENE MAT.	INST.	TOTAL	MEDFORD MAT.	INST.	TOTAL
A	Substructure	88.6	55.5	68.4	94.1	65.5	76.7	94.1	58.3	72.3	100.4	100.8	100.6	102.4	100.7	101.4
B10	Shell: Superstructure	94.0	63.6	82.0	99.3	64.8	85.8	97.5	65.9	85.1	94.3	97.9	95.7	94.0	99.7	95.5
B20	Exterior Closure	104.9	46.4	76.9	105.0	56.7	81.9	99.2	59.5	80.2	98.3	103.2	100.6	102.7	103.2	103.0
B30	Roofing	99.9	47.3	79.9	101.1	61.5	86.0	99.8	60.8	85.0	107.3	89.6	100.6	108.0	91.1	101.6
C	Interior Construction	94.9	35.3	69.9	100.4	56.3	82.0	95.6	40.8	72.6	98.8	97.4	98.2	101.3	97.0	99.5
D10	Services: Conveying	100.0	82.0	95.1	100.0	82.5	95.2	100.0	84.3	95.7	100.0	99.7	99.9	100.0	99.7	99.9
D20 - 40	Mechanical	100.3	60.6	84.3	100.1	65.5	86.2	100.3	63.5	85.4	99.9	100.8	100.2	99.9	100.7	100.2
D50	Electrical	94.5	52.4	72.2	101.7	70.9	85.4	96.5	64.6	79.6	100.1	96.1	98.0	103.9	80.8	91.6
E	Equipment & Furnishings	100.0	31.0	96.5	100.0	61.1	98.0	100.0	37.5	96.8	100.0	98.2	99.9	100.0	97.5	99.9
G	Site Work	91.9	87.1	88.6	95.2	92.4	93.3	97.8	87.5	90.8	96.1	103.0	100.8	103.0	103.0	103.0
A-G	WEIGHTED AVERAGE	97.4	55.7	79.2	100.3	65.7	85.2	98.1	61.6	82.2	98.6	99.3	98.9	100.0	97.1	98.8

OREGON / PENNSYLVANIA

DIV. NO.	BUILDING SYSTEMS	PORTLAND MAT.	INST.	TOTAL	SALEM MAT.	INST.	TOTAL	ALLENTOWN MAT.	INST.	TOTAL	ALTOONA MAT.	INST.	TOTAL	ERIE MAT.	INST.	TOTAL
A	Substructure	102.4	100.9	101.5	98.9	100.8	100.1	90.2	108.2	101.2	95.3	90.8	92.5	95.7	92.5	93.8
B10	Shell: Superstructure	95.5	98.1	96.5	98.9	98.0	98.5	98.1	117.2	105.6	93.4	102.7	97.1	94.4	103.9	98.1
B20	Exterior Closure	98.3	103.2	100.7	101.9	103.2	102.5	99.1	103.3	101.1	91.8	69.8	81.3	85.7	93.2	89.3
B30	Roofing	107.2	95.5	102.8	105.0	93.4	100.6	102.4	122.1	109.9	101.4	93.7	98.5	101.9	92.3	98.2
C	Interior Construction	97.6	97.2	97.4	99.3	97.4	98.5	93.3	107.9	99.4	88.5	87.1	87.9	89.6	90.1	89.8
D10	Services: Conveying	100.0	99.7	99.9	100.0	99.7	99.9	100.0	104.4	101.2	100.0	97.0	99.2	100.0	102.7	100.7
D20 - 40	Mechanical	99.9	100.8	100.2	99.9	100.8	100.2	100.2	112.1	105.0	99.7	82.5	92.8	99.7	92.9	97.0
D50	Electrical	100.4	103.2	101.9	108.0	96.1	101.7	99.1	98.4	98.8	89.3	112.2	101.4	91.1	96.7	94.0
E	Equipment & Furnishings	100.0	98.6	99.9	100.0	98.2	99.9	100.0	114.2	100.7	100.0	83.6	99.2	100.0	86.9	99.3
G	Site Work	98.8	103.0	101.7	92.6	103.0	99.8	93.2	104.3	100.8	95.1	104.1	101.3	92.0	105.0	100.9
A-G	WEIGHTED AVERAGE	98.8	100.4	99.5	100.5	99.4	100.0	97.9	108.6	102.5	94.6	91.9	93.4	94.5	95.8	95.1

PENNSYLVANIA

DIV. NO.	BUILDING SYSTEMS	HARRISBURG MAT.	INST.	TOTAL	PHILADELPHIA MAT.	INST.	TOTAL	PITTSBURGH MAT.	INST.	TOTAL	READING MAT.	INST.	TOTAL	SCRANTON MAT.	INST.	TOTAL
A	Substructure	94.9	96.3	95.7	98.9	128.2	116.8	100.7	100.4	100.5	83.9	99.3	93.3	92.5	96.2	94.7
B10	Shell: Superstructure	102.4	108.4	104.8	101.4	133.6	114.3	99.3	110.9	103.9	96.3	108.8	101.2	100.3	107.6	103.1
B20	Exterior Closure	101.8	91.5	96.9	99.8	132.1	114.7	102.0	103.9	102.9	100.0	95.8	98.0	99.2	95.7	97.6
B30	Roofing	102.5	112.5	106.3	102.8	136.2	115.5	103.1	100.6	102.1	101.9	114.5	106.7	102.3	95.1	99.5
C	Interior Construction	98.7	89.5	94.8	98.5	142.7	117.0	99.8	100.9	100.2	90.8	91.9	91.3	94.3	93.6	94.0
D10	Services: Conveying	100.0	98.9	99.7	100.0	118.0	105.0	100.0	103.0	100.8	100.0	101.9	100.5	100.0	102.9	100.8
D20 - 40	Mechanical	100.1	92.1	96.9	100.0	131.8	112.9	99.9	100.4	100.1	100.1	108.8	103.7	100.2	99.2	99.8
D50	Electrical	96.8	88.3	92.3	97.1	148.4	124.2	97.2	112.2	105.1	99.6	95.5	97.5	99.2	97.7	98.4
E	Equipment & Furnishings	100.0	86.2	99.3	100.0	142.3	102.2	100.0	96.4	99.8	100.0	87.9	99.4	100.0	86.4	99.3
G	Site Work	90.2	103.3	99.2	101.7	100.1	100.6	100.8	104.7	103.5	99.6	112.7	108.6	93.7	104.6	101.2
A-G	WEIGHTED AVERAGE	99.8	95.8	98.0	99.7	133.6	114.5	99.9	104.6	102.0	97.1	102.1	99.3	98.6	99.2	98.9

PENNSYLVANIA / PUERTO RICO / RHODE ISLAND / SOUTH CAROLINA

DIV. NO.	BUILDING SYSTEMS	YORK MAT.	INST.	TOTAL	SAN JUAN MAT.	INST.	TOTAL	PROVIDENCE MAT.	INST.	TOTAL	CHARLESTON MAT.	INST.	TOTAL	COLUMBIA MAT.	INST.	TOTAL
A	Substructure	85.3	97.0	92.4	122.4	37.5	70.6	96.9	122.3	112.4	94.5	64.6	76.3	96.5	58.3	73.2
B10	Shell: Superstructure	96.2	108.7	101.1	118.2	28.3	82.9	101.9	126.1	111.4	101.5	70.7	89.4	99.7	65.5	86.3
B20	Exterior Closure	104.9	91.9	98.7	100.8	17.7	61.1	104.5	130.8	117.1	97.7	45.9	73.0	96.2	42.4	70.5
B30	Roofing	98.9	112.8	104.2	128.5	22.5	88.3	100.6	119.6	107.8	101.5	48.1	81.2	97.1	44.4	77.1
C	Interior Construction	92.3	89.7	91.2	179.7	17.2	111.6	100.8	127.6	112.0	100.4	63.6	85.0	100.6	46.7	78.0
D10	Services: Conveying	100.0	99.1	99.8	100.0	17.8	77.4	100.0	107.2	102.0	100.0	69.1	91.5	100.0	69.1	91.5
D20 - 40	Mechanical	100.2	92.4	97.1	103.3	14.0	86.2	99.9	110.6	104.2	100.5	53.7	81.6	100.0	53.0	81.1
D50	Electrical	91.0	88.3	89.5	125.9	13.2	66.3	99.6	99.0	99.2	99.0	88.8	93.6	99.0	59.7	78.3
E	Equipment & Furnishings	100.0	86.2	99.3	100.0	15.9	95.7	100.0	119.5	101.0	100.0	68.0	98.4	100.0	44.6	97.2
G	Site Work	85.7	103.4	97.8	134.4	91.6	105.0	91.5	105.2	100.9	98.0	85.9	89.7	98.8	85.9	89.9
A-G	WEIGHTED AVERAGE	96.7	95.9	96.3	122.7	24.4	79.8	100.6	117.5	108.0	99.9	65.2	84.8	99.3	56.5	80.6

SOUTH CAROLINA / SOUTH DAKOTA

DIV. NO.	BUILDING SYSTEMS	FLORENCE MAT.	INST.	TOTAL	GREENVILLE MAT.	INST.	TOTAL	SPARTANBURG MAT.	INST.	TOTAL	ABERDEEN MAT.	INST.	TOTAL	PIERRE MAT.	INST.	TOTAL
A	Substructure	86.3	58.2	69.1	86.7	56.0	68.0	87.0	57.8	69.2	101.9	52.1	71.5	99.1	56.9	73.3
B10	Shell: Superstructure	98.0	65.6	85.3	98.8	61.6	84.2	99.0	65.0	85.7	97.1	52.3	79.5	98.3	63.2	84.5
B20	Exterior Closure	97.4	45.2	72.5	94.2	44.7	70.5	95.3	45.1	71.3	108.0	54.3	82.3	108.8	54.3	82.7
B30	Roofing	101.8	45.3	80.3	101.7	45.3	80.3	101.8	45.3	80.3	99.2	44.3	78.3	101.8	42.7	79.4
C	Interior Construction	96.9	47.0	76.0	97.8	48.2	77.0	98.1	48.8	77.5	99.8	38.3	74.0	103.7	39.2	76.6
D10	Services: Conveying	100.0	69.1	91.5	100.0	69.1	91.5	100.0	69.1	91.5	100.0	41.2	83.8	100.0	80.9	94.8
D20 - 40	Mechanical	100.5	53.1	81.4	100.5	53.0	81.3	100.5	53.1	81.3	100.1	39.8	75.8	100.0	64.5	85.7
D50	Electrical	96.8	59.7	77.2	99.1	57.4	77.0	99.1	57.4	77.1	101.7	51.1	75.0	104.8	51.8	76.8
E	Equipment & Furnishings	100.0	44.6	97.2	100.0	44.4	97.2	100.0	44.4	97.2	100.0	36.0	96.7	100.0	36.7	96.8
G	Site Work	106.2	85.9	92.2	101.8	85.5	90.6	101.5	85.5	90.5	99.7	93.5	95.4	99.9	93.6	95.6
A-G	WEIGHTED AVERAGE	98.2	56.9	80.2	98.3	55.8	79.8	98.6	56.7	80.3	100.5	49.6	78.3	101.7	58.4	82.8

635

SOUTH DAKOTA / TENNESSEE

DIV. NO.	BUILDING SYSTEMS	RAPID CITY MAT.	INST.	TOTAL	SIOUX FALLS MAT.	INST.	TOTAL	CHATTANOOGA MAT.	INST.	TOTAL	JACKSON MAT.	INST.	TOTAL	JOHNSON CITY MAT.	INST.	TOTAL
A	Substructure	99.0	56.5	73.1	94.4	58.2	72.3	99.7	69.5	81.3	98.8	60.7	75.6	87.0	59.0	69.9
B10	Shell: Superstructure	98.1	63.0	84.4	97.3	63.9	84.2	98.7	75.9	89.7	98.4	69.0	86.8	93.1	68.3	83.3
B20	Exterior Closure	107.2	56.6	83.0	103.3	54.5	80.0	102.0	53.6	78.8	111.7	50.1	82.2	126.1	47.2	88.4
B30	Roofing	99.6	44.4	78.6	102.0	45.3	80.5	99.4	60.7	84.7	96.0	53.8	80.0	94.4	52.8	78.6
C	Interior Construction	101.8	44.1	77.6	102.2	47.0	79.1	98.2	59.1	81.8	94.6	45.4	74.0	98.3	39.5	73.6
D10	Services: Conveying	100.0	81.2	94.8	100.0	80.9	94.8	100.0	39.9	83.5	100.0	42.8	84.3	100.0	80.6	94.7
D20 - 40	Mechanical	100.1	65.0	85.9	100.0	38.6	75.2	100.2	62.6	85.0	100.1	68.0	87.1	99.9	59.1	83.4
D50	Electrical	98.0	51.8	73.6	101.3	75.1	87.5	102.3	69.2	84.8	101.0	58.5	78.5	92.5	46.3	68.1
E	Equipment & Furnishings	100.0	33.6	96.6	100.0	40.2	96.9	100.0	57.3	97.8	100.0	45.1	97.2	100.0	36.7	96.8
G	Site Work	97.7	93.7	94.9	93.8	95.2	94.8	103.4	97.7	99.5	97.8	96.6	96.9	110.0	86.3	93.8
A-G	WEIGHTED AVERAGE	100.4	59.5	82.6	100.0	57.7	81.5	100.0	66.2	85.2	99.9	61.2	83.0	99.8	56.5	80.9

TENNESSEE / TEXAS

DIV. NO.	BUILDING SYSTEMS	KNOXVILLE MAT.	INST.	TOTAL	MEMPHIS MAT.	INST.	TOTAL	NASHVILLE MAT.	INST.	TOTAL	ABILENE MAT.	INST.	TOTAL	AMARILLO MAT.	INST.	TOTAL
A	Substructure	93.9	68.8	78.6	95.6	71.3	80.8	93.1	73.6	81.2	95.9	68.7	79.3	95.4	65.5	77.2
B10	Shell: Superstructure	98.2	75.8	89.4	99.1	78.8	91.1	98.9	79.1	91.1	103.4	66.9	89.1	100.4	64.2	86.2
B20	Exterior Closure	90.3	56.4	74.1	97.5	64.0	81.5	94.8	60.9	78.6	99.7	64.2	82.7	104.5	63.8	85.1
B30	Roofing	92.5	63.1	81.3	95.5	62.7	83.1	95.6	61.9	82.8	100.0	67.1	87.5	98.0	64.3	85.2
C	Interior Construction	93.3	62.3	80.3	98.5	60.7	82.6	99.3	66.7	85.6	95.1	66.4	83.1	99.0	59.3	82.4
D10	Services: Conveying	100.0	84.5	95.7	100.0	84.5	95.7	100.0	85.1	95.9	100.0	84.1	95.6	100.0	70.1	91.8
D20 - 40	Mechanical	99.9	67.9	87.0	100.0	73.9	89.5	100.0	84.0	93.5	100.3	49.0	79.6	100.0	54.9	83.8
D50	Electrical	98.1	56.0	75.8	101.7	65.3	82.4	97.6	63.1	79.4	99.3	48.7	72.5	101.1	64.1	81.6
E	Equipment & Furnishings	100.0	59.8	97.9	100.0	66.1	98.3	100.0	65.9	98.3	100.0	66.9	98.3	100.0	57.0	97.8
G	Site Work	91.0	87.0	88.3	93.3	93.5	93.4	98.7	97.3	97.8	100.2	88.7	92.3	97.8	88.6	91.5
A-G	WEIGHTED AVERAGE	96.6	66.9	83.7	99.0	71.3	86.9	98.4	74.1	87.8	99.7	62.2	83.3	100.2	63.2	84.1

TEXAS

DIV. NO.	BUILDING SYSTEMS	AUSTIN MAT.	INST.	TOTAL	BEAUMONT MAT.	INST.	TOTAL	CORPUS CHRISTI MAT.	INST.	TOTAL	DALLAS MAT.	INST.	TOTAL	EL PASO MAT.	INST.	TOTAL
A	Substructure	95.1	66.5	77.7	91.0	69.7	78.0	109.0	65.0	82.2	99.3	69.5	81.1	92.8	67.6	77.4
B10	Shell: Superstructure	99.3	63.4	85.2	94.8	69.9	85.0	96.6	69.1	85.8	100.2	72.6	89.4	97.2	65.0	84.6
B20	Exterior Closure	103.8	56.9	81.4	105.3	64.5	85.8	96.4	57.8	78.0	101.4	59.1	81.2	98.9	57.7	79.2
B30	Roofing	104.3	65.6	89.6	101.9	66.5	88.5	104.4	67.2	90.3	92.8	67.4	83.2	99.8	65.1	86.6
C	Interior Construction	100.0	58.2	82.5	92.9	63.2	80.5	107.4	58.1	86.7	99.5	64.5	84.8	94.3	62.7	81.1
D10	Services: Conveying	100.0	83.2	95.4	100.0	86.2	96.2	100.0	85.8	96.1	100.0	84.4	95.7	100.0	83.1	95.4
D20 - 40	Mechanical	100.1	60.1	84.0	100.2	64.1	85.6	100.2	58.9	83.5	100.0	62.4	84.9	100.0	50.3	80.2
D50	Electrical	96.8	63.9	79.4	95.2	70.6	82.2	92.8	60.0	75.5	95.2	64.6	79.0	99.1	57.4	77.0
E	Equipment & Furnishings	100.0	59.8	97.9	100.0	62.1	98.1	100.0	58.3	97.9	100.0	67.8	98.4	100.0	64.6	98.2
G	Site Work	102.9	88.3	92.9	91.7	89.9	90.4	136.8	83.9	100.5	104.4	89.5	94.2	101.4	87.6	91.9
A-G	WEIGHTED AVERAGE	100.0	63.6	84.1	97.4	68.6	84.9	100.9	63.5	84.6	99.5	67.4	85.5	98.0	61.7	82.2

TEXAS

DIV. NO.	BUILDING SYSTEMS	FORT WORTH MAT.	INST.	TOTAL	HOUSTON MAT.	INST.	TOTAL	LAREDO MAT.	INST.	TOTAL	LUBBOCK MAT.	INST.	TOTAL	ODESSA MAT.	INST.	TOTAL
A	Substructure	96.8	68.7	79.7	95.0	70.5	80.1	87.7	65.4	74.1	100.2	65.6	79.1	96.1	68.4	79.2
B10	Shell: Superstructure	100.4	66.8	87.2	97.9	77.5	89.9	95.1	63.0	82.4	106.5	71.0	92.6	102.9	66.5	88.6
B20	Exterior Closure	102.3	58.3	81.3	102.2	67.5	85.6	102.5	62.8	83.5	100.2	65.0	83.4	99.7	58.3	79.9
B30	Roofing	96.0	66.8	84.9	92.5	67.5	83.0	100.2	65.8	87.1	89.1	65.3	80.1	100.0	66.8	87.4
C	Interior Construction	98.5	64.4	84.2	101.3	63.9	85.6	99.5	57.6	82.0	102.1	57.5	83.4	95.1	64.4	82.2
D10	Services: Conveying	100.0	84.1	95.6	100.0	87.1	96.4	100.0	83.4	95.4	100.0	84.5	95.7	100.0	82.6	95.2
D20 - 40	Mechanical	100.0	58.4	83.2	100.1	66.7	86.6	100.2	64.1	85.6	99.8	50.7	80.0	100.3	49.1	79.6
D50	Electrical	94.7	64.5	78.8	96.9	67.5	81.4	95.1	60.9	77.0	97.9	64.2	80.1	99.4	64.1	80.7
E	Equipment & Furnishings	100.0	67.7	98.3	100.0	64.7	98.2	100.0	58.2	97.9	100.0	57.0	97.8	100.0	67.5	98.3
G	Site Work	101.6	88.7	92.8	105.1	87.8	93.2	102.8	88.3	92.8	121.0	85.8	96.8	100.2	88.0	91.8
A-G	WEIGHTED AVERAGE	99.3	65.3	84.5	99.4	70.5	86.8	98.3	64.5	83.6	101.6	63.6	85.0	99.6	63.1	83.7

TEXAS / UTAH

DIV. NO.	BUILDING SYSTEMS	SAN ANTONIO MAT.	INST.	TOTAL	WACO MAT.	INST.	TOTAL	WICHITA FALLS MAT.	INST.	TOTAL	LOGAN MAT.	INST.	TOTAL	OGDEN MAT.	INST.	TOTAL
A	Substructure	89.1	66.9	75.5	89.2	69.0	76.8	92.7	68.5	78.0	92.2	74.3	81.3	91.6	74.3	81.0
B10	Shell: Superstructure	95.3	64.2	83.1	99.7	66.0	86.5	100.4	66.8	87.2	100.0	72.9	89.4	100.6	72.9	89.7
B20	Exterior Closure	98.8	64.4	82.3	101.1	58.1	80.6	101.3	64.2	83.6	114.2	62.4	89.4	100.4	62.4	82.3
B30	Roofing	96.3	67.7	85.4	98.5	66.7	86.4	98.5	67.1	86.6	97.3	68.9	86.5	96.3	68.9	85.9
C	Interior Construction	103.0	57.5	83.9	82.9	63.6	74.8	83.1	66.4	76.1	94.3	56.1	78.3	93.4	56.1	77.8
D10	Services: Conveying	100.0	84.3	95.7	100.0	84.1	95.6	100.0	70.1	91.8	100.0	88.9	97.0	100.0	88.9	97.0
D20 - 40	Mechanical	100.0	63.2	85.1	100.2	54.7	81.8	100.2	54.2	81.6	99.9	69.5	87.6	99.9	69.5	87.6
D50	Electrical	96.4	60.9	77.6	97.6	63.6	79.6	99.3	60.1	78.6	97.5	73.0	84.6	97.9	73.0	84.7
E	Equipment & Furnishings	100.0	57.4	97.8	100.0	67.6	98.3	100.0	67.5	98.3	100.0	55.9	97.7	100.0	55.9	97.7
G	Site Work	101.5	91.9	94.9	98.5	88.7	91.8	99.4	88.7	92.1	95.9	95.5	95.6	85.5	95.5	92.4
A-G	WEIGHTED AVERAGE	98.5	65.1	84.0	96.5	64.1	82.4	97.1	64.4	82.8	99.8	70.1	86.9	98.1	70.1	85.9

UTAH / VERMONT / VIRGINIA

DIV. NO.	BUILDING SYSTEMS	PROVO MAT.	INST.	TOTAL	SALT LAKE CITY MAT.	INST.	TOTAL	BURLINGTON MAT.	INST.	TOTAL	RUTLAND MAT.	INST.	TOTAL	ALEXANDRIA MAT.	INST.	TOTAL
A	Substructure	93.4	73.9	81.5	97.2	74.3	83.2	99.6	96.2	97.5	91.1	96.2	94.2	99.6	81.4	88.5
B10	Shell: Superstructure	99.0	72.8	88.7	104.6	72.8	92.1	102.7	90.5	97.9	98.7	90.4	95.5	104.2	88.0	97.9
B20	Exterior Closure	115.9	62.4	90.3	121.9	62.4	93.4	109.0	92.3	101.0	100.0	92.3	96.3	99.3	74.8	87.6
B30	Roofing	99.0	68.9	87.6	103.4	68.9	90.3	103.8	77.4	93.8	97.2	77.4	89.7	102.3	82.9	94.9
C	Interior Construction	96.5	56.3	79.6	95.4	56.3	79.0	104.7	89.9	98.5	103.0	89.9	97.5	100.4	74.5	89.6
D10	Services: Conveying	100.0	88.9	97.0	100.0	88.9	97.0	100.0	99.0	99.7	100.0	99.0	99.7	100.0	90.8	97.5
D20 - 40	Mechanical	99.9	69.4	87.6	100.1	69.4	87.7	99.9	71.8	88.6	100.1	71.8	88.7	100.3	86.9	94.9
D50	Electrical	98.4	59.8	78.0	100.5	59.8	79.0	102.2	61.2	80.5	101.6	61.2	80.2	97.1	102.4	99.9
E	Equipment & Furnishings	100.0	55.3	97.7	100.0	55.3	97.7	100.0	85.6	99.3	100.0	85.5	99.3	100.0	71.8	98.6
G	Site Work	92.7	93.9	93.5	85.2	95.5	92.2	93.8	100.5	98.4	91.2	100.3	97.5	115.6	89.5	97.7
A-G	WEIGHTED AVERAGE	100.3	68.2	86.3	102.2	68.3	87.4	102.4	83.6	94.2	99.7	83.6	92.7	101.1	85.5	94.3

VIRGINIA

DIV. NO.	BUILDING SYSTEMS	ARLINGTON			NEWPORT NEWS			NORFOLK			PORTSMOUTH			RICHMOND		
		MAT.	INST.	TOTAL	MAT.	INST.	TOTAL	MAT.	INST.	TOTAL	MAT.	INST.	TOTAL	MAT.	INST.	TOTAL
A	Substructure	101.4	79.9	88.3	99.2	71.8	82.4	103.2	72.0	84.2	97.5	67.7	79.4	98.6	66.0	78.7
B10	Shell: Superstructure	103.0	87.6	97.0	102.3	79.0	93.5	102.8	79.2	93.5	100.4	75.6	90.7	105.1	76.1	93.7
B20	Exterior Closure	110.1	73.1	92.4	99.4	56.1	78.7	101.5	56.1	79.8	102.0	55.7	79.9	102.1	55.4	79.8
B30	Roofing	104.4	81.8	95.8	104.3	70.2	91.3	100.7	70.2	89.1	104.3	68.4	90.6	102.2	69.2	89.7
C	Interior Construction	99.4	73.8	88.7	99.8	63.0	84.4	99.4	66.1	85.4	98.3	56.6	80.8	101.6	61.2	84.7
D10	Services: Conveying	100.0	89.7	97.2	100.0	84.5	95.7	100.0	84.5	95.7	100.0	74.8	93.1	100.0	82.5	95.2
D20 - 40	Mechanical	100.3	85.4	94.3	100.3	63.6	85.5	100.0	64.3	85.6	100.3	64.3	85.8	99.9	67.5	86.8
D50	Electrical	94.7	99.2	97.1	92.8	63.5	77.3	95.5	58.3	75.8	91.2	58.3	73.8	96.8	71.9	83.6
E	Equipment & Furnishings	100.0	71.5	98.5	100.0	64.2	98.2	100.0	64.0	98.2	100.0	50.2	97.4	100.0	57.3	97.8
G	Site Work	124.9	87.3	99.1	107.3	88.5	94.4	108.0	89.5	95.3	105.6	88.2	93.7	103.4	88.6	93.2
A-G	WEIGHTED AVERAGE	101.9	84.0	94.1	100.0	68.2	86.1	100.5	68.2	86.4	99.4	65.3	84.5	101.3	68.7	87.1

VIRGINIA / WASHINGTON

DIV. NO.	BUILDING SYSTEMS	ROANOKE			EVERETT			RICHLAND			SEATTLE			SPOKANE		
		MAT.	INST.	TOTAL	MAT.	INST.	TOTAL	MAT.	INST.	TOTAL	MAT.	INST.	TOTAL	MAT.	INST.	TOTAL
A	Substructure	108.7	67.2	83.3	103.3	103.4	103.4	99.1	84.3	90.0	102.0	103.4	102.8	102.0	84.1	91.1
B10	Shell: Superstructure	105.9	75.0	93.8	102.6	96.7	100.3	91.0	83.2	87.9	102.5	97.8	100.7	93.7	82.9	89.5
B20	Exterior Closure	99.0	55.1	78.0	100.7	100.8	100.8	113.7	81.5	98.3	101.9	101.0	101.5	114.1	81.7	98.6
B30	Roofing	103.8	68.3	90.3	103.8	101.9	103.1	145.5	77.9	119.8	102.1	101.9	102.1	141.9	78.7	117.9
C	Interior Construction	99.9	61.4	83.8	102.7	97.4	100.5	117.3	76.7	100.3	102.2	99.9	101.2	116.5	77.9	100.3
D10	Services: Conveying	100.0	72.2	92.4	100.0	98.9	99.7	100.0	86.6	96.3	100.0	100.6	100.2	100.0	86.6	96.3
D20 - 40	Mechanical	100.3	64.8	86.0	100.1	99.1	99.7	100.5	107.6	103.3	100.0	115.1	106.1	100.4	84.3	93.9
D50	Electrical	96.8	54.4	74.4	104.9	98.5	101.5	89.7	94.8	92.4	102.6	107.0	104.9	88.0	79.0	83.3
E	Equipment & Furnishings	100.0	61.0	98.0	100.0	99.9	100.0	100.0	76.7	98.8	100.0	100.3	100.0	100.0	75.9	98.8
G	Site Work	106.4	85.9	92.3	91.7	108.7	103.4	104.9	91.0	95.3	95.0	106.9	103.2	103.9	90.9	95.0
A-G	WEIGHTED AVERAGE	101.5	65.1	85.6	101.6	99.6	100.7	103.1	89.3	97.1	101.4	104.7	102.8	103.4	82.3	94.2

WASHINGTON / WEST VIRGINIA

DIV. NO.	BUILDING SYSTEMS	TACOMA			VANCOUVER			YAKIMA			CHARLESTON			HUNTINGTON		
		MAT.	INST.	TOTAL	MAT.	INST.	TOTAL	MAT.	INST.	TOTAL	MAT.	INST.	TOTAL	MAT.	INST.	TOTAL
A	Substructure	104.5	103.4	103.8	113.6	95.1	102.3	109.1	94.0	99.9	98.4	95.3	96.5	103.3	96.4	99.1
B10	Shell: Superstructure	103.9	96.7	101.0	103.5	92.8	99.3	103.3	88.8	97.6	99.2	97.5	98.5	103.0	99.0	101.4
B20	Exterior Closure	100.0	100.9	100.5	105.2	95.8	100.7	98.5	80.3	89.8	99.6	96.0	97.9	99.4	98.6	99.1
B30	Roofing	103.6	98.2	101.5	104.2	90.8	99.1	103.7	82.4	95.6	99.4	89.1	95.5	103.1	89.0	97.7
C	Interior Construction	102.4	97.5	100.4	98.3	90.5	95.0	101.8	88.0	96.0	101.4	97.0	99.6	99.4	99.0	99.2
D10	Services: Conveying	100.0	100.6	100.2	100.0	56.8	88.1	100.0	97.4	99.3	100.0	102.2	100.6	100.0	102.6	100.7
D20 - 40	Mechanical	100.2	99.2	99.8	100.3	87.0	94.9	100.2	107.2	103.0	100.0	93.4	97.4	100.3	90.5	96.4
D50	Electrical	104.7	99.1	101.7	109.7	101.0	105.1	107.8	94.9	101.0	99.1	91.1	94.8	97.8	95.3	96.5
E	Equipment & Furnishings	100.0	99.9	100.0	100.0	91.0	99.5	100.0	99.2	100.0	100.0	93.1	99.6	100.0	95.2	99.8
G	Site Work	94.5	108.7	104.2	103.5	96.0	98.3	97.2	108.1	104.7	100.9	90.3	93.7	104.8	90.5	95.0
A-G	WEIGHTED AVERAGE	101.8	99.7	100.7	102.7	92.1	98.1	102.0	94.1	98.6	99.9	94.8	97.6	100.7	95.7	98.5

WEST VIRGINIA / WISCONSIN

DIV. NO.	BUILDING SYSTEMS	PARKERSBURG			WHEELING			EAU CLAIRE			GREEN BAY			KENOSHA		
		MAT.	INST.	TOTAL	MAT.	INST.	TOTAL	MAT.	INST.	TOTAL	MAT.	INST.	TOTAL	MAT.	INST.	TOTAL
A	Substructure	99.5	91.4	94.6	99.6	93.5	95.8	99.7	100.2	100.0	103.2	98.2	100.1	104.9	103.0	103.7
B10	Shell: Superstructure	102.5	94.0	99.2	107.8	95.5	99.9	96.0	100.5	97.8	99.0	98.2	98.7	96.8	101.4	98.6
B20	Exterior Closure	96.7	90.5	93.7	107.8	92.7	100.6	96.5	100.4	98.4	112.8	100.1	106.7	99.7	112.4	105.8
B30	Roofing	103.9	88.9	98.2	104.2	89.1	98.5	102.4	84.5	95.6	104.4	83.7	96.6	102.3	108.4	104.6
C	Interior Construction	99.0	93.8	96.8	99.5	96.2	98.1	100.6	101.1	100.8	99.2	100.2	99.6	98.1	102.5	100.0
D10	Services: Conveying	100.0	102.0	100.5	100.0	95.7	98.8	100.0	96.5	99.0	100.0	96.2	99.0	100.0	101.6	100.5
D20 - 40	Mechanical	100.3	88.3	95.4	100.1	93.6	97.6	100.1	87.6	95.0	100.3	85.1	94.2	100.3	97.0	98.8
D50	Electrical	97.2	92.7	94.8	94.4	96.1	95.3	103.8	84.9	93.8	98.2	82.0	89.6	101.5	98.7	100.1
E	Equipment & Furnishings	100.0	89.1	99.4	100.0	93.2	99.7	100.0	98.7	99.9	100.0	98.6	99.9	100.0	97.2	99.9
G	Site Work	109.2	90.3	96.2	110.0	90.0	96.3	96.4	102.9	100.9	99.0	98.9	99.0	100.4	105.3	103.8
A-G	WEIGHTED AVERAGE	100.1	91.8	96.5	101.2	94.3	98.2	99.3	95.3	97.5	101.2	93.4	97.8	99.5	102.2	100.7

WISCONSIN / WYOMING

DIV. NO.	BUILDING SYSTEMS	LA CROSSE			MADISON			MILWAUKEE			RACINE			CASPER		
		MAT.	INST.	TOTAL	MAT.	INST.	TOTAL	MAT.	INST.	TOTAL	MAT.	INST.	TOTAL	MAT.	INST.	TOTAL
A	Substructure	91.5	98.1	95.5	93.4	101.0	98.0	98.7	108.3	104.6	96.9	103.9	101.2	103.1	69.3	82.5
B10	Shell: Superstructure	93.8	97.7	95.3	92.4	98.0	94.6	97.9	102.9	99.9	95.3	101.3	97.7	100.1	66.4	86.9
B20	Exterior Closure	93.0	100.3	96.5	102.5	104.0	103.2	102.3	118.0	109.8	100.8	112.4	106.4	104.1	53.3	79.9
B30	Roofing	101.9	87.6	96.5	102.9	105.2	103.8	100.9	113.4	105.6	102.2	107.9	104.4	106.1	54.5	86.5
C	Interior Construction	99.1	99.9	99.4	103.3	101.2	102.4	101.3	117.8	108.2	99.8	102.5	101.0	105.8	50.3	82.6
D10	Services: Conveying	100.0	96.5	99.0	100.0	99.8	99.9	100.0	102.8	100.8	100.0	101.6	100.5	100.0	92.6	98.0
D20 - 40	Mechanical	100.1	87.5	95.0	100.0	95.6	98.2	100.0	103.5	101.4	100.1	97.0	98.8	99.9	64.1	85.4
D50	Electrical	104.2	85.1	94.1	102.1	92.1	96.8	100.8	99.5	100.1	100.8	98.6	99.6	103.3	64.1	82.6
E	Equipment & Furnishings	100.0	98.7	99.9	100.0	99.1	100.0	100.0	118.1	100.9	100.0	97.2	99.9	100.0	49.8	97.4
G	Site Work	90.5	102.8	98.9	92.7	108.3	103.4	94.9	98.7	97.5	94.9	109.1	104.7	98.4	94.3	95.6
A-G	WEIGHTED AVERAGE	97.8	94.5	96.3	99.1	99.0	99.1	100.0	107.2	103.1	99.1	102.4	100.5	102.0	63.9	85.4

WYOMING / CANADA

DIV. NO.	BUILDING SYSTEMS	CHEYENNE			ROCK SPRINGS			CALGARY, ALBERTA			EDMONTON, ALBERTA			HALIFAX, NOVA SCOTIA		
		MAT.	INST.	TOTAL	MAT.	INST.	TOTAL	MAT.	INST.	TOTAL	MAT.	INST.	TOTAL	MAT.	INST.	TOTAL
A	Substructure	96.3	70.6	80.6	96.8	70.2	80.5	165.2	98.6	124.5	164.3	98.6	124.2	139.0	82.1	104.3
B10	Shell: Superstructure	100.7	67.4	87.6	98.7	67.6	86.5	142.7	92.6	123.0	141.9	92.6	122.5	135.6	79.6	113.6
B20	Exterior Closure	104.5	53.5	80.1	129.9	52.2	92.7	151.4	89.4	121.7	147.2	89.4	119.6	142.4	82.7	113.8
B30	Roofing	100.1	55.1	83.0	102.0	60.0	86.1	117.6	96.1	109.5	122.7	96.1	112.6	118.8	78.7	103.6
C	Interior Construction	104.1	53.2	82.7	105.6	61.6	87.2	106.5	97.3	102.6	106.8	97.3	102.8	104.4	80.3	94.3
D10	Services: Conveying	100.0	92.6	98.0	100.0	82.8	95.3	131.1	96.8	121.7	131.1	96.8	121.7	131.1	67.4	113.6
D20 - 40	Mechanical	99.9	64.1	85.4	99.9	63.3	85.1	101.2	92.9	97.8	101.1	92.9	97.8	100.6	77.6	91.3
D50	Electrical	105.8	68.6	86.1	102.4	62.9	81.5	113.5	101.3	107.1	110.6	101.3	105.7	117.0	81.2	98.1
E	Equipment & Furnishings	100.0	54.4	97.7	100.0	65.7	98.2	131.1	97.4	129.4	131.1	97.4	129.4	131.1	77.8	128.4
G	Site Work	92.4	94.3	93.7	91.3	93.8	93.1	126.6	105.3	112.0	145.1	105.3	117.8	113.2	98.2	102.9
A-G	WEIGHTED AVERAGE	101.5	65.2	85.7	103.7	65.3	87.0	123.5	95.7	111.4	123.1	95.7	111.2	119.6	80.9	102.7

637

DIV. NO.	BUILDING SYSTEMS	CANADA														
		HAMILTON, ONTARIO			KITCHENER, ONTARIO			LAVAL, QUEBEC			LONDON, ONTARIO			MONTREAL, QUEBEC		
		MAT.	INST.	TOTAL	MAT.	INST.	TOTAL	MAT.	INST.	TOTAL	MAT.	INST.	TOTAL	MAT.	INST.	TOTAL
A	Substructure	138.1	100.1	114.9	117.7	93.7	103.0	126.7	92.2	105.7	137.2	95.6	111.8	146.2	97.4	116.4
B10	Shell: Superstructure	126.6	94.1	113.8	117.9	91.0	107.3	113.0	88.4	103.3	127.7	91.8	113.6	132.3	94.3	117.4
B20	Exterior Closure	138.6	99.9	120.1	114.3	96.1	105.6	132.1	84.2	109.2	138.9	97.3	119.0	132.4	91.5	112.8
B30	Roofing	118.1	94.5	109.2	108.4	91.5	101.9	107.7	90.4	101.1	119.0	92.2	108.8	116.4	97.0	109.0
C	Interior Construction	105.9	95.8	101.7	96.0	89.1	93.1	102.3	89.6	97.0	101.9	90.7	97.2	104.1	95.8	100.6
D10	Services: Conveying	131.1	99.4	122.4	131.1	98.4	122.1	131.1	84.9	118.4	131.1	98.8	122.3	131.1	86.4	118.8
D20 - 40	Mechanical	101.3	89.2	96.4	100.4	87.7	95.3	100.4	91.0	96.6	101.4	86.6	95.4	101.2	92.4	97.7
D50	Electrical	108.3	100.5	104.2	111.5	98.0	104.4	112.9	72.6	91.6	107.0	97.5	102.0	112.6	86.8	99.0
E	Equipment & Furnishings	131.1	94.3	129.3	131.1	86.1	128.8	131.1	87.1	128.9	131.1	87.2	128.9	131.1	93.1	129.2
G	Site Work	118.3	112.6	114.4	102.4	104.8	104.1	102.8	99.7	100.7	116.3	105.0	108.5	116.1	99.3	104.6
A-G	WEIGHTED AVERAGE	117.0	96.7	108.2	109.6	92.9	102.3	112.1	87.4	101.3	116.4	93.3	106.3	117.8	93.2	107.1

DIV. NO.	BUILDING SYSTEMS	CANADA														
		OSHAWA, ONTARIO			OTTAWA, ONTARIO			QUEBEC, QUEBEC			REGINA, SASKATCHEWAN			SASKATOON, SASKATCHEWAN		
		MAT.	INST.	TOTAL	MAT.	INST.	TOTAL	MAT.	INST.	TOTAL	MAT.	INST.	TOTAL	MAT.	INST.	TOTAL
A	Substructure	143.3	91.6	111.7	136.9	96.9	112.5	134.9	97.6	112.2	153.5	93.1	116.7	124.8	93.2	105.5
B10	Shell: Superstructure	116.3	89.8	105.9	130.1	92.7	115.4	130.0	94.6	116.1	134.2	88.8	116.3	108.7	88.8	100.9
B20	Exterior Closure	123.0	93.9	109.1	132.6	97.5	115.8	130.5	94.1	113.1	144.4	92.7	119.7	128.5	92.7	111.4
B30	Roofing	109.2	86.0	100.4	124.3	92.6	112.2	118.2	97.2	110.2	123.1	84.1	108.3	106.8	84.1	98.2
C	Interior Construction	99.5	92.4	96.5	107.9	92.0	101.2	106.9	95.8	102.2	106.3	87.4	98.4	101.2	87.4	95.4
D10	Services: Conveying	131.1	98.4	122.1	131.1	96.3	121.6	131.1	86.5	118.9	131.1	69.9	114.3	131.1	71.8	114.8
D20 - 40	Mechanical	100.4	100.7	100.5	101.4	87.2	95.7	101.2	92.4	97.7	100.7	84.8	94.3	100.5	84.8	94.2
D50	Electrical	112.6	90.5	100.9	107.3	98.7	102.8	111.4	86.8	98.4	114.4	92.4	102.8	116.2	92.4	103.6
E	Equipment & Furnishings	131.1	88.1	128.9	131.1	91.0	129.1	131.1	93.1	129.2	131.1	91.9	129.1	131.1	91.9	129.1
G	Site Work	114.1	104.0	107.1	111.5	104.7	106.8	114.4	99.4	104.1	128.4	97.9	107.5	115.1	97.9	103.3
A-G	WEIGHTED AVERAGE	112.3	94.3	104.4	117.3	94.0	107.2	117.1	93.6	106.8	120.7	89.0	106.9	111.1	89.1	101.5

DIV. NO.	BUILDING SYSTEMS	CANADA														
		ST CATHARINES, ONTARIO			ST JOHNS, NEWFOUNDLAND			THUNDER BAY, ONTARIO			TORONTO, ONTARIO			VANCOUVER, BRITISH COLUMBIA		
		MAT.	INST.	TOTAL	MAT.	INST.	TOTAL	MAT.	INST.	TOTAL	MAT.	INST.	TOTAL	MAT.	INST.	TOTAL
A	Substructure	113.9	97.4	103.8	155.2	90.3	115.6	120.4	96.2	105.6	136.7	104.0	116.8	136.1	95.2	111.2
B10	Shell: Superstructure	109.4	93.8	103.3	140.5	85.4	118.8	110.9	92.4	103.6	133.1	98.0	119.3	138.4	90.8	119.7
B20	Exterior Closure	113.9	99.1	106.8	148.5	88.9	120.0	117.2	98.6	108.3	138.3	108.7	124.2	138.1	91.2	115.7
B30	Roofing	108.3	95.5	103.5	127.2	91.9	113.8	108.6	92.8	102.6	123.9	101.7	115.4	125.3	86.3	110.5
C	Interior Construction	94.5	96.4	95.3	107.1	80.9	96.1	96.4	93.8	95.3	103.4	101.4	102.5	105.4	91.3	99.5
D10	Services: Conveying	131.1	72.7	115.1	131.1	71.2	114.7	131.1	73.1	115.2	131.1	101.5	123.0	131.1	95.4	121.3
D20 - 40	Mechanical	100.4	87.8	95.3	100.8	82.1	93.3	100.4	88.0	95.4	101.0	97.3	99.5	101.2	80.5	92.8
D50	Electrical	112.9	98.8	105.5	114.4	83.2	97.9	111.5	97.0	103.8	108.8	101.0	104.7	109.0	83.3	95.4
E	Equipment & Furnishings	131.1	94.1	129.2	131.1	82.9	128.7	131.1	91.3	129.1	131.1	99.3	129.5	131.1	90.7	129.1
G	Site Work	103.3	101.4	102.0	128.6	99.8	108.8	108.2	101.3	103.4	135.0	105.6	114.8	120.5	107.5	111.6
A-G	WEIGHTED AVERAGE	107.6	94.5	101.9	122.7	85.2	106.4	108.8	93.4	102.1	118.3	101.3	110.9	119.5	89.1	106.2

DIV. NO.	BUILDING SYSTEMS	CANADA														
		WINDSOR, ONTARIO			WINNIPEG, MANITOBA											
		MAT.	INST.	TOTAL	MAT.	INST.	TOTAL	MAT.	INST.	TOTAL	MAT.	INST.	TOTAL	MAT.	INST.	TOTAL
A	Substructure	115.9	96.4	104.0	148.6	76.8	104.8									
B10	Shell: Superstructure	110.1	92.7	103.3	143.4	72.5	115.5									
B20	Exterior Closure	113.7	98.2	106.3	139.5	69.7	106.2									
B30	Roofing	108.4	92.7	102.4	120.7	71.5	102.0									
C	Interior Construction	95.2	92.2	93.9	102.7	68.7	88.4									
D10	Services: Conveying	131.1	72.8	115.1	131.1	68.7	114.0									
D20 - 40	Mechanical	100.4	88.0	95.4	101.1	66.9	87.3									
D50	Electrical	116.0	99.0	107.0	113.3	68.0	89.4									
E	Equipment & Furnishings	131.1	89.6	129.0	131.1	69.3	128.0									
G	Site Work	98.7	101.3	100.5	121.5	100.8	107.3									
A-G	WEIGHTED AVERAGE	108.2	93.4	101.7	121.0	71.8	99.5									

Cost Indexes

Costs shown in RSMeans cost data publications are based on national averages for materials and installation. To adjust these costs to a specific location, simply multiply the base cost by the factor and divide by 100 for that city. The data is arranged alphabetically by state and postal zip code numbers. For a city not listed, use the factor for a nearby city with similar economic characteristics.

STATE/ZIP	CITY	MAT.	INST.	TOTAL
ALABAMA				
350-352	Birmingham	100.6	76.7	90.2
354	Tuscaloosa	99.8	61.2	83.0
355	Jasper	100.2	57.5	81.6
356	Decatur	99.8	57.9	81.5
357-358	Huntsville	99.8	72.1	87.8
359	Gadsden	99.9	60.0	82.5
360-361	Montgomery	99.9	57.1	81.2
362	Anniston	98.6	61.8	82.6
363	Dothan	99.0	49.4	77.4
364	Evergreen	98.6	50.7	77.7
365-366	Mobile	99.9	65.4	84.9
367	Selma	98.7	49.9	77.4
368	Phenix City	99.5	54.6	79.9
369	Butler	98.9	49.4	77.3
ALASKA				
995-996	Anchorage	121.1	115.6	118.7
997	Fairbanks	121.0	116.4	119.0
998	Juneau	121.3	115.6	118.8
999	Ketchikan	132.0	115.6	124.8
ARIZONA				
850,853	Phoenix	99.2	73.8	88.1
851,852	Mesa/Tempe	98.6	72.8	87.4
855	Globe	98.8	70.0	86.3
856-857	Tucson	97.4	72.4	86.5
859	Show Low	98.9	72.5	87.4
860	Flagstaff	101.3	71.8	88.4
863	Prescott	99.2	68.9	86.0
864	Kingman	97.6	72.0	86.4
865	Chambers	97.6	71.5	86.2
ARKANSAS				
716	Pine Bluff	99.4	63.3	83.6
717	Camden	97.1	50.1	76.6
718	Texarkana	98.2	49.2	76.8
719	Hot Springs	96.4	51.8	76.9
720-722	Little Rock	98.5	64.0	83.4
723	West Memphis	96.1	62.4	81.4
724	Jonesboro	96.7	59.1	80.3
725	Batesville	94.6	54.6	77.1
726	Harrison	95.9	50.4	76.0
727	Fayetteville	93.4	52.0	75.4
728	Russellville	94.6	53.7	76.8
729	Fort Smith	97.2	61.6	81.7
CALIFORNIA				
900-902	Los Angeles	99.2	117.5	107.2
903-905	Inglewood	94.8	114.1	103.2
906-908	Long Beach	96.2	114.1	104.0
910-912	Pasadena	96.4	114.6	104.3
913-916	Van Nuys	99.1	114.6	105.9
917-918	Alhambra	98.3	114.8	105.5
919-921*	San Diego	100.5	109.6	104.5
922	Palm Springs	97.2	112.0	103.6
923-924	San Bernardino	95.1	111.3	102.1
925	Riverside	99.4	114.6	106.0
926-927	Santa Ana	96.9	111.8	103.4
928	Anaheim	99.4	114.9	106.2
930	Oxnard	99.6	114.4	106.0
931	Santa Barbara	98.9	115.1	106.0
932-933	Bakersfield	100.7	113.6	106.3
934	San Luis Obispo	99.4	112.0	104.9
935	Mojave	96.8	109.8	102.5
936-938	Fresno	100.1	115.3	106.7
939	Salinas	100.2	121.6	109.5
940-941	San Francisco	105.2	145.4	122.7
942,956-958	Sacramento	100.5	119.3	108.7
943	Palo Alto	98.9	134.5	114.4
944	San Mateo	100.9	134.6	115.6
945	Vallejo	100.1	127.2	111.9
946	Oakland	103.1	134.3	116.7
947	Berkeley	102.7	135.1	116.8
948	Richmond	101.8	133.1	115.4
949	San Rafael	103.5	136.7	118.0
950	Santa Cruz	104.5	121.8	112.0

STATE/ZIP	CITY	MAT.	INST.	TOTAL
CALIFORNIA (CONT'D)				
951	San Jose	102.5	136.6	117.4
952	Stockton	100.9	117.2	108.0
953	Modesto	100.8	115.7	107.3
954	Santa Rosa	100.8	134.7	115.6
955	Eureka	102.1	116.4	108.4
959	Marysville	101.3	115.9	107.7
960	Redding	107.8	117.7	112.1
961	Susanville	106.8	118.1	111.7
COLORADO				
800-802	Denver	100.8	81.7	92.5
803	Boulder	97.8	81.3	90.6
804	Golden	99.8	78.9	90.7
805	Fort Collins	101.3	79.0	91.6
806	Greeley	98.9	75.5	88.7
807	Fort Morgan	98.3	79.0	89.9
808-809	Colorado Springs	100.4	84.0	93.2
810	Pueblo	101.1	80.3	92.0
811	Alamosa	102.2	73.7	89.8
812	Salida	101.9	75.3	90.3
813	Durango	102.7	75.7	90.9
814	Montrose	101.4	75.4	90.1
815	Grand Junction	104.8	75.3	91.9
816	Glenwood Springs	102.3	78.1	91.8
CONNECTICUT				
060	New Britain	100.1	120.6	109.1
061	Hartford	101.4	120.8	109.8
062	Willimantic	100.8	120.2	109.2
063	New London	97.1	120.7	107.4
064	Meriden	98.9	120.7	108.4
065	New Haven	101.6	120.7	109.9
066	Bridgeport	101.3	120.8	109.8
067	Waterbury	100.9	120.7	109.5
068	Norwalk	100.8	128.7	113.0
069	Stamford	100.9	128.7	113.0
D.C.				
200-205	Washington	101.1	91.9	97.1
DELAWARE				
197	Newark	98.8	109.2	103.3
198	Wilmington	99.3	109.2	103.6
199	Dover	99.3	109.2	103.6
FLORIDA				
320,322	Jacksonville	98.8	66.4	84.7
321	Daytona Beach	99.0	72.9	87.6
323	Tallahassee	100.8	56.9	81.7
324	Panama City	100.2	57.8	81.7
325	Pensacola	102.7	61.2	84.6
326,344	Gainesville	100.3	66.7	85.7
327-328,347	Orlando	100.5	70.0	87.2
329	Melbourne	101.7	76.3	90.6
330-332,340	Miami	99.8	73.1	88.2
333	Fort Lauderdale	98.6	72.6	87.2
334,349	West Palm Beach	97.4	71.1	85.9
335-336,346	Tampa	100.1	78.4	90.6
337	St. Petersburg	102.6	63.3	85.5
338	Lakeland	99.4	77.8	90.0
339,341	Fort Myers	98.6	72.6	87.3
342	Sarasota	100.8	74.3	89.2
GEORGIA				
300-303,399	Atlanta	97.6	74.4	87.5
304	Statesboro	97.3	55.8	79.2
305	Gainesville	96.0	61.6	81.0
306	Athens	95.5	63.9	81.7
307	Dalton	97.2	59.0	80.5
308-309	Augusta	96.3	63.6	82.1
310-312	Macon	97.2	64.7	83.0
313-314	Savannah	99.2	60.9	82.5
315	Waycross	98.3	59.4	81.4
316	Valdosta	98.5	61.4	82.3
317,398	Albany	98.5	61.3	82.3
318-319	Columbus	98.4	65.4	84.0

STATE/ZIP	CITY	MAT.	INST.	TOTAL
HAWAII				
967	Hilo	114.8	120.2	117.2
968	Honolulu	119.5	120.2	119.8
STATES & POSS.				
969	Guam	136.2	58.0	102.1
IDAHO				
832	Pocatello	101.1	77.4	90.8
833	Twin Falls	102.2	67.3	87.0
834	Idaho Falls	99.8	74.7	88.9
835	Lewiston	108.7	84.9	98.3
836-837	Boise	100.1	78.6	90.7
838	Coeur d'Alene	107.9	82.2	96.7
ILLINOIS				
600-603	North Suburban	99.3	135.1	114.9
604	Joliet	99.2	138.4	116.3
605	South Suburban	99.3	135.1	114.9
606-608	Chicago	99.8	139.7	117.2
609	Kankakee	95.6	131.7	111.3
610-611	Rockford	99.5	124.8	110.5
612	Rock Island	97.2	100.5	98.6
613	La Salle	98.4	124.4	109.7
614	Galesburg	98.2	106.7	101.9
615-616	Peoria	100.5	109.3	104.4
617	Bloomington	97.5	109.4	102.7
618-619	Champaign	101.0	107.6	103.9
620-622	East St. Louis	96.4	108.7	101.8
623	Quincy	98.1	101.0	99.3
624	Effingham	97.3	106.0	101.1
625	Decatur	99.2	104.9	101.7
626-627	Springfield	100.0	106.5	102.8
628	Centralia	95.0	108.5	100.9
629	Carbondale	94.7	107.9	100.5
INDIANA				
460	Anderson	97.0	83.8	91.3
461-462	Indianapolis	99.1	84.8	92.9
463-464	Gary	98.3	110.4	103.6
465-466	South Bend	98.0	84.4	92.1
467-468	Fort Wayne	97.6	79.1	89.5
469	Kokomo	94.8	81.5	89.0
470	Lawrenceburg	94.0	77.7	86.9
471	New Albany	95.1	75.4	86.5
472	Columbus	97.2	80.9	90.1
473	Muncie	97.9	82.3	91.1
474	Bloomington	99.3	81.8	91.6
475	Washington	95.8	81.8	89.7
476-477	Evansville	97.1	86.9	92.6
478	Terre Haute	97.9	86.5	92.9
479	Lafayette	96.9	82.7	90.7
IOWA				
500-503,509	Des Moines	100.5	85.2	93.8
504	Mason City	97.6	65.8	83.8
505	Fort Dodge	97.9	58.2	80.6
506-507	Waterloo	99.5	71.8	87.4
508	Creston	98.2	77.3	89.1
510-511	Sioux City	99.7	72.6	87.9
512	Sibley	98.3	56.4	80.0
513	Spencer	99.9	56.6	81.0
514	Carroll	97.0	71.2	85.8
515	Council Bluffs	100.7	78.1	90.9
516	Shenandoah	97.6	71.2	86.1
520	Dubuque	99.1	80.2	90.9
521	Decorah	98.0	64.2	83.3
522-524	Cedar Rapids	100.0	83.2	92.7
525	Ottumwa	97.9	73.7	87.3
526	Burlington	97.2	75.3	87.6
527-528	Davenport	99.0	93.1	96.4
KANSAS				
660-662	Kansas City	99.5	94.8	97.5
664-666	Topeka	99.5	66.0	84.9
667	Fort Scott	97.9	73.0	87.0
668	Emporia	97.9	69.2	85.4
669	Belleville	99.6	64.9	84.4
670-672	Wichita	100.7	66.0	85.6
673	Independence	100.9	69.2	87.1
674	Salina	101.5	67.6	86.7
675	Hutchinson	96.5	65.0	82.8
676	Hays	100.5	65.4	85.2
677	Colby	101.1	65.0	85.4

STATE/ZIP	CITY	MAT.	INST.	TOTAL
KANSAS (CONT'D)				
678	Dodge City	102.7	66.0	86.7
679	Liberal	100.4	65.5	85.2
KENTUCKY				
400-402	Louisville	96.6	83.3	90.8
403-405	Lexington	96.6	81.4	90.0
406	Frankfort	98.1	80.7	90.5
407-409	Corbin	94.0	71.8	84.3
410	Covington	95.9	91.4	93.9
411-412	Ashland	94.6	99.3	96.6
413-414	Campton	95.8	78.6	88.3
415-416	Pikeville	96.9	88.6	93.3
417-418	Hazard	95.2	78.7	88.0
420	Paducah	93.9	84.5	89.8
421-422	Bowling Green	96.2	83.9	90.8
423	Owensboro	96.3	88.9	93.1
424	Henderson	93.7	86.1	90.4
425-426	Somerset	93.2	79.9	87.4
427	Elizabethtown	92.8	82.0	88.1
LOUISIANA				
700-701	New Orleans	101.4	69.5	87.5
703	Thibodaux	98.2	65.4	83.9
704	Hammond	95.8	54.5	77.8
705	Lafayette	97.9	63.0	82.7
706	Lake Charles	98.1	65.0	83.7
707-708	Baton Rouge	99.1	63.7	83.7
710-711	Shreveport	101.0	59.6	83.0
712	Monroe	99.3	56.5	80.6
713-714	Alexandria	99.4	57.2	81.0
MAINE				
039	Kittery	95.2	84.7	90.6
040-041	Portland	102.0	90.1	96.8
042	Lewiston	99.6	90.1	95.5
043	Augusta	102.1	79.5	92.3
044	Bangor	99.0	89.5	94.9
045	Bath	97.5	81.9	90.7
046	Machias	97.1	85.5	92.0
047	Houlton	97.2	85.6	92.1
048	Rockland	96.3	85.5	91.6
049	Waterville	97.5	79.5	89.7
MARYLAND				
206	Waldorf	97.1	81.7	90.4
207-208	College Park	97.2	83.4	91.2
209	Silver Spring	96.5	83.2	90.7
210-212	Baltimore	100.8	81.9	92.6
214	Annapolis	101.5	82.0	93.0
215	Cumberland	96.8	81.8	90.2
216	Easton	98.3	68.6	85.3
217	Hagerstown	97.6	83.3	91.4
218	Salisbury	98.7	62.1	82.7
219	Elkton	95.9	81.9	89.8
MASSACHUSETTS				
010-011	Springfield	100.2	112.6	105.6
012	Pittsfield	99.7	111.3	104.8
013	Greenfield	97.7	112.3	104.1
014	Fitchburg	96.5	126.5	109.6
015-016	Worcester	100.2	126.4	111.6
017	Framingham	95.7	135.1	112.9
018	Lowell	99.5	134.5	114.8
019	Lawrence	100.3	134.6	115.3
020-022, 024	Boston	101.6	139.5	118.1
023	Brockton	99.9	128.1	112.2
025	Buzzards Bay	94.4	128.1	109.1
026	Hyannis	97.0	128.1	110.6
027	New Bedford	99.1	128.3	111.8
MICHIGAN				
480,483	Royal Oak	94.9	106.9	100.1
481	Ann Arbor	97.1	107.5	101.7
482	Detroit	97.8	109.6	102.9
484-485	Flint	96.8	96.0	96.4
486	Saginaw	96.4	91.8	94.4
487	Bay City	96.6	91.7	94.5
488-489	Lansing	99.3	94.7	97.3
490	Battle Creek	98.3	87.6	93.7
491	Kalamazoo	98.7	85.7	93.0
492	Jackson	96.5	95.1	95.9
493,495	Grand Rapids	99.7	85.1	93.3
494	Muskegon	97.2	84.8	91.8

STATE/ZIP	CITY	MAT.	INST.	TOTAL
MICHIGAN (CONT'D)				
496	Traverse City	96.3	80.1	89.3
497	Gaylord	97.2	83.6	91.3
498-499	Iron mountain	99.3	86.2	93.6
MINNESOTA				
550-551	Saint Paul	101.4	117.9	108.6
553-555	Minneapolis	100.4	120.0	108.9
556-558	Duluth	101.3	107.3	103.9
559	Rochester	99.9	104.7	102.0
560	Mankato	97.2	101.3	99.0
561	Windom	95.9	92.3	94.3
562	Willmar	95.5	101.3	98.0
563	St. Cloud	96.9	116.0	105.2
564	Brainerd	96.9	100.5	98.5
565	Detroit Lakes	98.9	95.1	97.3
566	Bemidji	98.2	98.2	98.2
567	Thief River Falls	97.8	92.4	95.4
MISSISSIPPI				
386	Clarksdale	97.3	52.6	77.8
387	Greenville	101.0	61.7	83.9
388	Tupelo	98.8	54.9	79.7
389	Greenwood	98.4	52.2	78.3
390-392	Jackson	100.7	62.2	83.9
393	Meridian	97.6	64.1	83.0
394	Laurel	98.9	55.8	80.1
395	Biloxi	99.2	58.1	81.3
396	Mccomb	97.4	51.7	77.5
397	Columbus	98.8	53.9	79.2
MISSOURI				
630-631	St. Louis	98.7	107.8	102.7
633	Bowling Green	97.3	94.4	96.0
634	Hannibal	96.2	92.4	94.6
635	Kirksville	99.5	89.9	95.3
636	Flat River	98.1	95.0	96.8
637	Cape Girardeau	98.2	92.4	95.7
638	Sikeston	96.7	92.2	94.8
639	Poplar Bluff	96.2	92.0	94.4
640-641	Kansas City	101.4	105.7	103.3
644-645	St. Joseph	100.6	93.5	97.5
646	Chillicothe	97.6	94.5	96.3
647	Harrisonville	97.1	101.5	99.1
648	Joplin	99.5	81.7	91.8
650-651	Jefferson City	99.9	93.7	97.2
652	Columbia	100.2	93.9	97.5
653	Sedalia	100.2	97.3	98.9
654-655	Rolla	97.9	94.6	96.5
656-658	Springfield	101.7	83.7	93.8
MONTANA				
590-591	Billings	102.7	77.1	91.5
592	Wolf Point	102.3	74.2	90.0
593	Miles City	100.1	74.0	88.7
594	Great Falls	104.1	76.2	91.9
595	Havre	101.2	73.1	89.0
596	Helena	102.4	73.8	89.9
597	Butte	102.6	77.3	91.6
598	Missoula	99.9	75.5	89.3
599	Kalispell	99.0	75.3	88.7
NEBRASKA				
680-681	Omaha	100.2	78.6	90.8
683-685	Lincoln	100.5	74.9	89.4
686	Columbus	98.4	77.1	89.1
687	Norfolk	100.0	76.7	89.8
688	Grand Island	100.0	77.1	90.0
689	Hastings	99.4	79.8	90.9
690	Mccook	98.7	71.1	86.7
691	North Platte	99.0	78.3	90.0
692	Valentine	101.2	72.5	88.7
693	Alliance	101.0	70.5	87.7
NEVADA				
889-891	Las Vegas	100.9	108.7	104.3
893	Ely	99.4	102.2	100.6
894-895	Reno	99.6	93.3	96.8
897	Carson City	99.7	93.5	97.0
898	Elko	98.3	84.9	92.4
NEW HAMPSHIRE				
030	Nashua	100.5	93.8	97.6
031	Manchester	101.4	93.8	98.1

STATE/ZIP	CITY	MAT.	INST.	TOTAL
NEW HAMPSHIRE (CONT'D)				
032-033	Concord	101.1	92.5	97.4
034	Keene	97.2	73.3	86.7
035	Littleton	97.1	76.3	88.0
036	Charleston	96.7	69.5	84.8
037	Claremont	95.9	69.5	84.4
038	Portsmouth	97.9	94.3	96.4
NEW JERSEY				
070-071	Newark	101.2	128.4	113.0
072	Elizabeth	98.8	127.6	111.3
073	Jersey City	97.9	127.9	111.0
074-075	Paterson	99.6	127.8	111.9
076	Hackensack	97.6	128.1	110.9
077	Long Branch	97.1	126.3	109.8
078	Dover	97.9	127.8	110.9
079	Summit	97.9	127.6	110.8
080,083	Vineland	97.5	124.7	109.4
081	Camden	99.5	124.8	110.5
082,084	Atlantic City	98.2	124.8	109.8
085-086	Trenton	100.9	125.1	111.5
087	Point Pleasant	99.2	125.9	110.8
088-089	New Brunswick	99.7	127.6	111.9
NEW MEXICO				
870-872	Albuquerque	99.6	72.6	87.8
873	Gallup	99.5	72.6	87.8
874	Farmington	100.1	72.6	88.1
875	Santa Fe	100.8	72.6	88.5
877	Las Vegas	98.0	72.6	86.9
878	Socorro	97.6	72.6	86.7
879	Truth/Consequences	97.5	69.4	85.2
880	Las Cruces	96.1	69.4	84.4
881	Clovis	98.3	72.5	87.1
882	Roswell	100.0	72.6	88.0
883	Carrizozo	100.3	72.6	88.2
884	Tucumcari	99.0	72.5	87.4
NEW YORK				
100-102	New York	103.4	168.7	131.8
103	Staten Island	99.2	168.4	129.4
104	Bronx	97.7	166.1	127.5
105	Mount Vernon	97.5	141.4	116.6
106	White Plains	97.6	141.4	116.7
107	Yonkers	102.1	143.7	120.2
108	New Rochelle	97.9	141.3	116.8
109	Suffern	97.6	129.0	111.3
110	Queens	100.9	166.8	129.6
111	Long Island City	102.4	166.8	130.5
112	Brooklyn	102.8	168.5	131.4
113	Flushing	102.5	166.8	130.5
114	Jamaica	100.8	166.8	129.6
115,117,118	Hicksville	100.9	153.3	123.7
116	Far Rockaway	102.6	166.8	130.6
119	Riverhead	101.6	148.2	121.9
120-122	Albany	100.1	104.7	102.1
123	Schenectady	99.3	104.7	101.7
124	Kingston	102.6	118.5	109.6
125-126	Poughkeepsie	101.8	136.7	117.0
127	Monticello	101.1	115.1	107.2
128	Glens Falls	94.0	98.6	96.0
129	Plattsburgh	99.1	96.7	98.0
130-132	Syracuse	98.9	98.0	98.5
133-135	Utica	97.1	97.1	97.1
136	Watertown	98.5	96.3	97.5
137-139	Binghamton	98.5	101.4	99.8
140-142	Buffalo	100.3	106.2	102.9
143	Niagara Falls	97.8	106.4	101.5
144-146	Rochester	101.1	99.1	100.2
147	Jamestown	96.8	92.5	94.9
148-149	Elmira	96.7	97.4	97.0
NORTH CAROLINA				
270,272-274	Greensboro	101.3	55.7	81.4
271	Winston-Salem	101.0	56.5	81.6
275-276	Raleigh	100.5	54.1	80.3
277	Durham	102.9	55.3	82.1
278	Rocky Mount	98.5	52.3	78.3
279	Elizabeth City	99.3	51.6	78.5
280	Gastonia	101.6	55.1	81.3
281-282	Charlotte	101.9	56.8	82.2
283	Fayetteville	104.5	58.1	84.2
284	Wilmington	100.3	55.9	81.0
285	Kinston	98.4	52.7	78.5

STATE/ZIP	CITY	MAT.	INST.	TOTAL
NORTH CAROLINA (CONT'D)				
286	Hickory	98.8	53.9	79.2
287-288	Asheville	100.8	55.3	80.9
289	Murphy	99.6	48.3	77.2
NORTH DAKOTA				
580-581	Fargo	102.7	67.6	87.4
582	Grand Forks	102.7	51.6	80.4
583	Devils Lake	102.0	58.6	83.1
584	Jamestown	102.1	45.2	77.3
585	Bismarck	103.1	66.2	87.0
586	Dickinson	102.8	60.9	84.6
587	Minot	102.7	71.3	89.0
588	Williston	101.2	60.9	83.7
OHIO				
430-432	Columbus	98.6	89.2	94.5
433	Marion	95.1	85.3	90.8
434-436	Toledo	98.7	98.1	98.4
437-438	Zanesville	95.6	86.5	91.6
439	Steubenville	96.8	94.9	96.0
440	Lorain	98.9	93.0	96.3
441	Cleveland	99.0	100.3	99.6
442-443	Akron	99.9	94.2	97.4
444-445	Youngstown	99.2	89.0	94.7
446-447	Canton	99.3	85.6	93.3
448-449	Mansfield	96.7	87.3	92.6
450	Hamilton	97.9	85.1	92.3
451-452	Cincinnati	98.3	84.5	92.2
453-454	Dayton	98.0	84.4	92.1
455	Springfield	98.0	84.7	92.2
456	Chillicothe	96.8	93.0	95.1
457	Athens	99.6	80.4	91.2
458	Lima	100.0	86.8	94.3
OKLAHOMA				
730-731	Oklahoma City	100.3	65.7	85.2
734	Ardmore	98.3	61.9	82.4
735	Lawton	100.7	62.8	84.2
736	Clinton	99.6	63.1	83.7
737	Enid	100.2	60.5	82.9
738	Woodward	98.3	63.4	83.1
739	Guymon	99.4	59.7	82.1
740-741	Tulsa	98.1	61.6	82.2
743	Miami	94.8	67.5	82.9
744	Muskogee	97.4	55.7	79.2
745	Mcalester	94.5	60.4	79.7
746	Ponca City	95.1	62.1	80.7
747	Durant	95.1	63.5	81.3
748	Shawnee	96.5	61.8	81.4
749	Poteau	94.3	61.4	80.0
OREGON				
970-972	Portland	98.8	100.4	99.5
973	Salem	100.5	99.4	100.0
974	Eugene	98.6	99.3	98.9
975	Medford	100.0	97.1	98.8
976	Klamath Falls	99.8	97.1	98.6
977	Bend	98.9	99.4	99.1
978	Pendleton	94.8	101.4	97.7
979	Vale	92.9	98.1	95.2
PENNSYLVANIA				
150-152	Pittsburgh	99.9	104.6	102.0
153	Washington	97.0	103.2	99.7
154	Uniontown	97.3	101.7	99.2
155	Bedford	98.2	91.3	95.2
156	Greensburg	98.2	100.9	99.4
157	Indiana	97.1	99.3	98.0
158	Dubois	98.5	96.8	97.8
159	Johnstown	98.1	97.2	97.7
160	Butler	92.5	102.0	96.6
161	New Castle	92.5	97.9	94.8
162	Kittanning	92.9	103.3	97.4
163	Oil City	92.4	96.9	94.4
164-165	Erie	94.5	95.8	95.1
166	Altoona	94.6	91.9	93.4
167	Bradford	95.7	96.4	96.0
168	State College	95.3	92.8	94.2
169	Wellsboro	96.3	92.6	94.7
170-171	Harrisburg	99.8	95.6	98.0
172	Chambersburg	96.2	88.0	92.6
173-174	York	96.7	95.9	96.3
175-176	Lancaster	95.1	89.0	92.4

STATE/ZIP	CITY	MAT.	INST.	TOTAL
PENNSYLVANIA (CONT'D)				
177	Williamsport	93.8	83.0	89.1
178	Sunbury	95.8	95.8	95.8
179	Pottsville	94.9	98.5	96.5
180	Lehigh Valley	95.8	114.0	103.7
181	Allentown	97.9	108.6	102.5
182	Hazleton	95.4	98.3	96.6
183	Stroudsburg	95.3	101.4	97.9
184-185	Scranton	98.6	99.2	98.9
186-187	Wilkes-Barre	95.1	98.7	96.6
188	Montrose	94.7	96.9	95.7
189	Doylestown	94.9	120.6	106.1
190-191	Philadelphia	99.7	133.6	114.5
193	Westchester	95.9	121.5	107.1
194	Norristown	95.0	129.5	110.0
195-196	Reading	97.1	102.1	99.3
PUERTO RICO				
009	San Juan	122.7	24.4	79.8
RHODE ISLAND				
028	Newport	98.4	117.5	106.7
029	Providence	100.6	117.5	108.0
SOUTH CAROLINA				
290-292	Columbia	99.3	56.5	80.6
293	Spartanburg	98.6	56.7	80.3
294	Charleston	99.9	65.2	84.8
295	Florence	98.2	56.9	80.2
296	Greenville	98.3	55.8	79.8
297	Rock Hill	97.8	50.3	77.1
298	Aiken	98.6	69.1	85.7
299	Beaufort	99.3	44.9	75.6
SOUTH DAKOTA				
570-571	Sioux Falls	100.0	57.7	81.5
572	Watertown	99.0	49.1	77.2
573	Mitchell	97.8	48.4	76.3
574	Aberdeen	100.5	49.6	78.3
575	Pierre	101.7	58.4	82.8
576	Mobridge	98.5	48.8	76.8
577	Rapid City	100.4	59.5	82.6
TENNESSEE				
370-372	Nashville	98.4	74.1	87.8
373-374	Chattanooga	100.0	66.2	85.2
375,380-381	Memphis	99.0	71.3	86.9
376	Johnson City	99.8	56.5	80.9
377-379	Knoxville	96.6	66.9	83.7
382	Mckenzie	98.2	59.2	81.2
383	Jackson	99.9	61.2	83.0
384	Columbia	96.9	66.1	83.5
385	Cookeville	98.1	59.1	81.1
TEXAS				
750	Mckinney	98.7	63.1	83.2
751	Waxahackie	98.6	65.9	84.4
752-753	Dallas	99.5	67.4	85.5
754	Greenville	98.8	64.0	83.6
755	Texarkana	98.0	58.4	80.7
756	Longview	98.5	58.7	81.2
757	Tyler	98.5	65.0	83.9
758	Palestine	95.0	64.4	81.7
759	Lufkin	95.6	67.8	83.5
760-761	Fort Worth	99.3	65.3	84.5
762	Denton	98.9	62.5	83.0
763	Wichita Falls	97.1	64.4	82.8
764	Eastland	95.9	63.2	81.6
765	Temple	94.7	61.7	80.3
766-767	Waco	96.5	64.1	82.4
768	Brownwood	98.9	59.4	81.7
769	San Angelo	98.5	59.6	81.5
770-772	Houston	99.4	70.5	86.8
773	Huntsville	98.0	66.7	84.4
774	Wharton	98.9	67.7	85.3
775	Galveston	97.0	70.0	85.2
776-777	Beaumont	97.4	68.6	84.9
778	Bryan	94.9	68.2	83.3
779	Victoria	99.0	65.6	84.4
780	Laredo	98.3	64.5	83.6
781-782	San Antonio	98.5	65.1	84.0
783-784	Corpus Christi	100.9	63.5	84.6
785	Mc Allen	101.0	59.8	83.0
786-787	Austin	100.0	63.6	84.1

STATE/ZIP	CITY	MAT.	INST.	TOTAL
TEXAS (CONT'D)				
788	Del Rio	100.8	63.2	84.4
789	Giddings	97.7	64.0	83.0
790-791	Amarillo	100.2	63.2	84.1
792	Childress	99.3	64.3	84.0
793-794	Lubbock	101.6	63.6	85.0
795-796	Abilene	99.7	62.2	83.3
797	Midland	101.5	64.2	85.2
798-799,885	El Paso	98.0	61.7	82.2
UTAH				
840-841	Salt Lake City	102.2	68.3	87.4
842,844	Ogden	98.1	70.1	85.9
843	Logan	99.8	70.1	86.9
845	Price	100.2	64.6	84.7
846-847	Provo	100.3	68.2	86.3
VERMONT				
050	White River Jct.	98.5	78.3	89.7
051	Bellows Falls	97.2	93.0	95.3
052	Bennington	97.5	93.5	95.8
053	Brattleboro	97.8	92.8	95.6
054	Burlington	102.4	83.6	94.2
056	Montpelier	99.5	82.8	92.2
057	Rutland	99.7	83.6	92.7
058	St. Johnsbury	98.6	77.8	89.6
059	Guildhall	97.4	77.8	88.8
VIRGINIA				
220-221	Fairfax	100.7	84.0	93.4
222	Arlington	101.9	84.0	94.1
223	Alexandria	101.1	85.5	94.3
224-225	Fredericksburg	99.5	82.7	92.1
226	Winchester	100.0	82.2	92.2
227	Culpeper	99.9	80.5	91.4
228	Harrisonburg	100.2	67.2	85.8
229	Charlottesville	100.5	67.1	86.0
230-232	Richmond	101.3	68.7	87.1
233-235	Norfolk	100.5	68.2	86.4
236	Newport News	100.0	68.2	86.1
237	Portsmouth	99.4	65.3	84.5
238	Petersburg	99.4	68.7	86.0
239	Farmville	98.7	52.4	78.5
240-241	Roanoke	101.5	65.1	85.6
242	Bristol	99.4	54.9	80.0
243	Pulaski	99.0	57.3	80.9
244	Staunton	99.7	61.8	83.2
245	Lynchburg	99.9	65.6	85.0
246	Grundy	99.3	52.5	78.9
WASHINGTON				
980-981,987	Seattle	101.4	104.7	102.8
982	Everett	101.6	99.6	100.7
983-984	Tacoma	101.8	99.7	100.9
985	Olympia	101.2	99.4	100.4
986	Vancouver	102.7	92.1	98.1
988	Wenatchee	101.6	86.6	95.0
989	Yakima	102.0	94.1	98.6
990-992	Spokane	103.4	82.3	94.2
993	Richland	103.1	89.3	97.1
994	Clarkston	101.5	80.2	92.2
WEST VIRGINIA				
247-248	Bluefield	98.3	92.9	96.0
249	Lewisburg	99.8	92.4	96.6
250-253	Charleston	99.9	94.8	97.6
254	Martinsburg	99.3	86.4	93.7
255-257	Huntington	100.7	95.7	98.5
258-259	Beckley	97.9	93.3	95.9
260	Wheeling	101.2	94.3	98.2
261	Parkersburg	100.1	91.8	96.5
262	Buckhannon	99.6	95.5	97.8
263-264	Clarksburg	100.1	95.4	98.1
265	Morgantown	100.2	94.4	97.6
266	Gassaway	99.5	93.9	97.1
267	Romney	99.4	92.2	96.3
268	Petersburg	99.3	91.3	95.8
WISCONSIN				
530,532	Milwaukee	100.0	107.2	103.1
531	Kenosha	99.5	102.2	100.7
534	Racine	99.1	102.4	100.5
535	Beloit	98.9	100.7	99.7
537	Madison	99.1	99.0	99.1

STATE/ZIP	CITY	MAT.	INST.	TOTAL
WISCONSIN (CONT'D)				
538	Lancaster	96.7	92.5	94.8
539	Portage	95.3	96.6	95.9
540	New Richmond	96.8	94.6	95.9
541-543	Green Bay	102.1	93.4	97.8
544	Wausau	96.2	93.5	95.0
545	Rhinelander	99.3	91.2	95.8
546	La Crosse	97.8	94.5	96.3
547	Eau Claire	99.3	95.3	97.5
548	Superior	96.6	96.3	96.5
549	Oshkosh	96.9	91.9	94.7
WYOMING				
820	Cheyenne	101.5	65.2	85.7
821	Yellowstone Nat'l Park	99.3	66.3	84.9
822	Wheatland	100.4	64.3	84.7
823	Rawlins	101.8	66.9	86.6
824	Worland	100.0	65.3	84.9
825	Riverton	101.0	65.4	85.5
826	Casper	102.0	63.9	85.4
827	Newcastle	99.9	66.9	85.5
828	Sheridan	102.6	64.3	85.9
829-831	Rock Springs	103.7	65.3	87.0
CANADIAN FACTORS (reflect Canadian currency)				
ALBERTA				
	Calgary	123.5	95.7	111.4
	Edmonton	123.1	95.7	111.2
	Fort McMurray	123.6	92.8	110.2
	Lethbridge	118.4	92.2	107.0
	Lloydminster	113.4	88.9	102.8
	Medicine Hat	113.6	88.2	102.5
	Red Deer	114.1	88.2	102.8
BRITISH COLUMBIA				
	Kamloops	114.5	90.6	104.1
	Prince George	115.4	89.9	104.3
	Vancouver	119.5	89.1	106.2
	Victoria	115.7	87.2	103.3
MANITOBA				
	Brandon	125.2	75.8	103.7
	Portage la Prairie	113.4	74.2	96.3
	Winnipeg	121.0	71.8	99.5
NEW BRUNSWICK				
	Bathurst	111.5	67.3	92.2
	Dalhousie	111.9	67.5	92.5
	Fredericton	117.4	72.5	97.9
	Moncton	111.8	70.1	93.6
	Newcastle	111.6	67.9	92.5
	St. John	114.0	77.9	98.3
NEWFOUNDLAND				
	Corner Brook	129.6	67.6	102.6
	St Johns	122.7	85.2	106.4
NORTHWEST TERRITORIES				
	Yellowknife	135.3	84.2	113.0
NOVA SCOTIA				
	Bridgewater	112.9	74.6	96.2
	Dartmouth	126.1	74.6	103.7
	Halifax	119.6	80.9	102.7
	New Glasgow	124.1	74.6	102.5
	Sydney	122.1	74.6	101.4
	Truro	112.5	74.6	96.0
	Yarmouth	124.0	74.6	102.5
ONTARIO				
	Barrie	117.7	93.0	106.9
	Brantford	114.9	96.9	107.1
	Cornwall	115.0	93.3	105.5
	Hamilton	117.0	96.7	108.2
	Kingston	116.0	93.4	106.2
	Kitchener	109.6	92.9	102.3
	London	116.4	93.3	106.3
	North Bay	127.3	91.1	111.5
	Oshawa	112.3	94.3	104.4
	Ottawa	117.3	94.0	107.2
	Owen Sound	117.8	91.2	106.2
	Peterborough	114.9	93.1	105.4
	Sarnia	114.8	97.5	107.3
	Sault Ste Marie	110.3	92.1	102.3

643

STATE/ZIP	CITY	MAT.	INST.	TOTAL
ONTARIO (CONT'D)				
	St. Catharines	107.6	94.5	101.9
	Sudbury	107.7	93.0	101.3
	Thunder Bay	108.8	93.4	102.1
	Timmins	115.1	91.1	104.6
	Toronto	118.3	101.3	110.9
	Windsor	108.2	93.4	101.7
PRINCE EDWARD ISLAND				
	Charlottetown	120.0	62.8	95.1
	Summerside	126.7	62.8	98.8
QUEBEC				
	Cap-de-la-Madeleine	112.4	87.6	101.6
	Charlesbourg	112.4	87.6	101.6
	Chicoutimi	111.4	92.9	103.3
	Gatineau	112.1	87.4	101.3
	Granby	112.3	87.4	101.4
	Hull	112.2	87.4	101.4
	Joliette	112.5	87.6	101.7
	Laval	112.1	87.4	101.3
	Montreal	117.8	93.2	107.1
	Quebec	117.1	93.6	106.8
	Rimouski	111.8	92.9	103.6
	Rouyn-Noranda	112.0	87.4	101.3
	Saint Hyacinthe	111.6	87.4	101.1
	Sherbrooke	112.4	87.4	101.5
	Sorel	112.5	87.6	101.7
	St Jerome	112.1	87.4	101.3
	Trois Rivieres	124.8	87.6	108.6
SASKATCHEWAN				
	Moose Jaw	110.7	69.5	92.7
	Prince Albert	109.9	67.7	91.5
	Regina	120.7	89.0	106.9
	Saskatoon	111.1	89.1	101.5
YUKON				
	Whitehorse	135.3	68.5	106.2

L1010-101   Minimum Design Live Loads in Pounds per S.F. for Various Building Codes

Occupancy or Use	Uniform (psf)	Concentrated (lbs)
1. Access floor systems		
Office use	50	2000
Computer use	100	2000
2. Armories and drill rooms	100	–
3. Assembly areas		
Fixed seats (fastened to floor)	60	
Follow spot, projections and control rooms	50	
Lobbies	100	
Movable seats	100	–
Stage floors	150	
Platforms (assembly)	100	
Other assembly areas	100	
4. Balconies and decks	Same as occupancy served	–
5. Catwalks	40	300
6. Comices	60	–
7. Corridors		
First floor	100	–
Other floors	Same as occupancy served except as indicated	
8. Dining rooms and restaurants	100	–
9. Elevator machine room grating (on an area of 2 in by 2 in)	–	300
10. Finish light floorplate construction (on area of 1 in by 1 in)	–	200
11. Fire escapes	100	
On single-family dwellings only		–
12. Garages (passenger vehicles only)	40	Note a
13. Hospitals		
Corridors above first floor	80	1000
Operating rooms, laboratories	60	1000
Patient rooms	40	1000
14. Libraries		
Corridors above first floor	80	1000
Reading rooms	60	1000
Stacks	150, Note b	1000
15. Manufacturing		
Heavy	250	3000
Light	125	2000
16. Marquees	75	–
17. Office buildings		
Corridors above first floor	80	2000
Lobbies and first-floor corridors	100	2000
Offices	50	2000
18. Penal institutions		
Cell blocks	40	
Corridors	100	
19. Recreational uses:		
Bowling alleys, poolrooms, and similar uses	75	
Dance halls and ballrooms	100	
Gymnasiums	100	
Reviewing stands, grandstands and bleachers	100	
Stadiums and arenas with fixed seats (fastened to floor)	60	
20. Residential		
One- and two- family dwellings		
Uninhabitable attics without storage, Note c	10	
Uninhabitable attics with storage, Note c, d, e	20	
Habitable attics and sleeping areas, Note e	30	
All other areas	0	
Hotels and multifamily dwellings		
Private rooms and corridors serving them	40	
Public rooms and corridors serving them	100	

645

L1010-101 Minimum Design Live Loads in Pounds per S.F. for Various Building Codes (cont.)

Occupancy or Use	Uniform (psf)	Concentrated (lbs)
21. Roofs		
All roof surfaces subject to maintenance workers		300
Awnings and canopies:		
Fabric construction supported by a skeleton structure	5	-
All other construction	20	
Ordinary flat, pitched and curved roofs (not occupiable)	20	
Where primary roof members are exposed to a work floor, at single panel point of lower cord of roof trusses or any point along primary structural members supporting roofs:		
Over manufacturing, storage warehouses, and repair garages		2000
All other primary roof members		300
Occupiable roofs:		
Roof gardens	100	
Assembly areas	100	
All other similar areas	Note F	Note F
22. Schools		
Classrooms	40	1000
Corridors	80	1000
First-floor corridors	100	1000
23. Scuttles, skylight ribs and accessible ceilings	-	200
24. Sidewalks, vehicular driveways and yards, subject to trucking	250	8000, Note g
25. Stairs and exits		
One- and two- family dwellings	40	300, Note h
All others	100	3000, Note h
26. Storage warehouses		
Heavy	250	
Light	125	-
27. Stores		
Retail		
First floor	100	1000
Upper floors	75	1000
Wholesale, all floors	125	1000
28. Walkways and elevated platforms (other than exitways)	60	-
29. Yards and terraces, pedestrians	100	-

Notes:

a. Floors in garages or portions of buildings used for the storage of motor vehicles shall be designed for the uniformly distributed live loads of Table 1607.1 or the following concentrated loads: (1) for garages restricted to passenger vehicles accomodating not more than nine passengers, 3,000 pounds acting on an area of 4.5 in. by 4.5 in.; (2) for mechanical parking structures without slab or deck that are used for storing passenger vehicles only, 2,250 pounds per wheel.

b. The loading applies to stack room floors that support nonmobile, double-faced library books stacks, subject to the following limitations:
   1. The nominal bookstack unit height shall not exceed 90 inches;
   2. The nominal shelf depth shall not exceed 12 inches for each face; and
   3. Parallel rows of double-faced book sacks shall be separated by aisles not less than 36 inches wide.

c. Uninhabitable attics without storage are those where the maximum clear height between the joists and rafters is less than 42 inches, or where there are not two or more adjacent trusses with web configurations capable of accommmodating an assumed rectangle 42 inches in height by 24 inches in width, or greater, within the plane of the trusses. This live load need not be assumed to act concurrently with any other live load requirements.

d. Uninhabitable attics with storage are those where the maximum clear height between the joists and rafters is 42 inches or greater, or where there are two or more adjacent trusses with web configurations capable of accomodating an assumed rectangle 42 inches in height by 24 inches in width, or greater, within the plane of the trusses.

   The live load need only be applied to those portions of the joists or truss bottom chords where both of the following conditions are met:
     i. The attic area is accessible from an opening not less than 20 inches in width by 30 inches in length that is located where the clear height in the attic is a minimum of 30 inches; and
     ii. The slopes of the joists or truss bottom chords are no greater than two units vertical in 12 units horizontal.

   The remaining portions of the joists or truss bottom chords shall be designed for a uniformly distributed concurrent live load of not less than 10 lb/ft^2.

e. Attic spaces served by stairways other than the pull-down type shall be designed to support the minimum live load specified for habitable attics and sleeping rooms.

f. Areas of occupiable roofs, other than roof gardens and assembly areas, shall be designed for approporate loads as approved by the building official.

g. The concentrated wheel load shall be applied on an area of 4.5 inches by 4.5 inches.

h. The minimum concentrated load on stair treads shall be applied on an area of 2 in by 2 in. This load need not be assumed to act concurrently with the uniform load.

Table L1010-201 Design Weight Per S.F. for Walls and Partitions

Type	Wall Thickness	Description	Weight Per S.F.	Type	Wall Thickness	Description	Weight Per S.F.
Brick	4"	Clay brick, high absorption	34 lb.	Clay tile	2"	Split terra cotta furring	10 lb.
		Clay brick, medium absorption	39			Non load bearing clay tile	11
		Clay brick, low absorption	46		3"	Split terra cotta furring	12
		Sand-lime brick	38			Non load bearing clay tile	18
		Concrete brick, heavy aggregate	46		4"	Non load bearing clay tile	20
		Concrete brick, light aggregate	33			Load bearing clay tile	24
	8"	Clay brick, high absorption	69		6"	Non load bearing clay tile	30
		Clay brick, medium absorption	79			Load bearing clay tile	36
		Clay brick, low absorption	89		8"	Non load bearing clay tile	36
		Sand-lime brick	74			Load bearing clay tile	42
		Concrete brick, heavy aggregate	89		12"	Non load bearing clay tile	46
		Concrete brick, light aggregate	68			Load bearing clay tile	58
	12"	Common brick	120	Gypsum block	2"	Hollow gypsum block	9.5
		Pressed brick	130			Solid gypsum block	12
		Sand-lime brick	105		3"	Hollow gypsum block	10
		Concrete brick, heavy aggregate	130			Solid gypsum block	18
		Concrete brick, light aggregate	98		4"	Hollow gypsum block	15
	16"	Clay brick, high absorption	134			Solid gypsum block	24
		Clay brick, medium absorption	155		5"	Hollow gypsum block	18
		Clay brick, low absorption	173		6"	Hollow gypsum block	24
		Sand-lime brick	138	Structural facing tile	2"	Facing tile	15
		Concrete brick, heavy aggregate	174		4"	Facing tile	25
		Concrete brick, light aggregate	130		6"	Facing tile	38
Concrete block	4"	Solid conc. block, stone aggregate	45	Glass	4"	Glass block	18
		Solid conc. block, lightweight	34		1"	Structural glass	15
		Hollow conc. block, stone aggregate	30	Plaster	1"	Gypsum plaster (1 side)	5
		Hollow conc. block, lightweight	20			Cement plaster (1 side)	10
	6"	Solid conc. block, stone aggregate	50			Gypsum plaster on lath	8
		Solid conc. block, lightweight	37			Cement plaster on lath	13
		Hollow conc. block, stone aggregate	42	Plaster partition (2 finished faces)	2"	Solid gypsum on metal lath	18
		Hollow conc. block, lightweight	30			Solid cement on metal lath	25
	8"	Solid conc. block, stone aggregate	67			Solid gypsum on gypsum lath	18
		Solid conc. block, lightweight	48			Gypsum on lath & metal studs	18
		Hollow conc. block, stone aggregate	55		3"	Gypsum on lath & metal studs	19
		Hollow conc. block, lightweight	38		4"	Gypsum on lath & metal studs	20
	10"	Solid conc. block, stone aggregate	84		6"	Gypsum on lath & wood studs	18
		Solid conc. block, lightweight	62	Concrete	6"	Reinf concrete, stone aggregate	75
		Hollow conc. block, stone aggregate	55			Reinf. concrete, lightweight	36-60
		Hollow conc. block, lightweight	38		8"	Reinf. concrete, stone aggregate	100
	12"	Solid conc. block, stone aggregate	108			Reinf. concrete, lightweight	48-80
		Solid conc. block, lightweight	72		10"	Reinf. concrete, stone aggregate	125
		Hollow conc. block, stone aggregate	85			Reinf. concrete, lightweight	60-100
		Hollow conc. block, lightweight	55		12"	Reinf concrete stone aggregate	150
Drywall	6"	Drywall on wood studs	10			Reinf concrete, lightweight	72-120

Table L1010-202 Design Weight per S.F. for Roof Coverings

Type	Description		Weight lb. Per S.F.	Type	Wall Thickness	Description	Weight lb. Per S.F.
Sheathing	Gypsum	1" thick	4	Metal	Aluminum	Corr. & ribbed, .024" to .040"	.4-.8
	Wood	¾" thick	3		Copper	or tin	1.5-2.5
Insulation	per 1"	Loose	.5		Steel	Corrugated, 29 ga. to 12 ga.	.6-5.0
		Poured in place	2	Shingles	Asphalt	Strip shingles	1.7-2.8
		Rigid	1.5		Clay	Tile	8-16
Built-up	Tar & gravel	3 ply felt	5.5		Slate	¼" thick	9.5
		5 ply felt	6.5		Wood		2-3

647

Table L1010-203 Design Weight per Square Foot for Floor Fills and Finishes

Type		Description	Weight lb. per S.F.	Type		Description	Weight lb. per S.F.
Floor fill	per 1″	Cinder fill	5	Wood	Single 7/8″	On sleepers, light concrete fill	16
		Cinder concrete	9			On sleepers, stone concrete fill	25
		Lightweight concrete	3-9		Double 7/8″	On sleepers, light concrete fill	19
		Stone concrete	12			On sleepers, stone concrete fill	28
		Sand	8		3″	Wood block on mastic, no fill	15
		Gypsum	6			Wood block on ½″ mortar	16
Terrazzo	1″	Terrazzo, 2″ stone concrete	25		per 1″	Hardwood flooring (25/32″)	4
Marble	and mortar	on stone concrete fill	33			Underlayment (Plywood per 1″)	3
Resilient	1/16″-1/4″	Linoleum, asphalt, vinyl tile	2	Asphalt	1-1/2″	Mastic flooring	18
Tile	3/4″	Ceramic or quarry	10		2″	Block on ½″ mortar	30

Table L1010-204 Design Weight Per Cubic Foot for Miscellaneous Materials

Type		Description	Weight lb. per C.F.	Type		Description	Weight lb. per C.F.
Bituminous	Coal, piled	Anthracite	47-58	Masonry	Ashlar	Granite	165
		Bituminous	40-54			Limestone, crystalline	165
		Peat, turf, dry	47			Limestone, oolitic	135
		Coke	75			Marble	173
	Petroleum	Unrefined	54			Sandstone	144
		Refined	50		Rubble, in mortar	Granite	155
		Gasoline	42			Limestone, crystalline	147
	Pitch		69			Limestone, oolitic	138
	Tar	Bituminous	75			Marble	156
Concrete	Plain	Stone aggregate	144			Sandstone & Bluestone	130
		Slag aggregate	132		Brick	Pressed	140
		Expanded slag aggregate	100			Common	120
		Haydite (burned clay agg.)	90			Soft	100
		Vermiculite & perlite, load bearing	70-105		Cement	Portland, loose	90
		Vermiculite & perlite, non load bear	35-50			Portland set	183
	Reinforced	Stone aggregate	150		Lime	Gypsum, loose	53-64
		Slag aggregate	138		Mortar	Set	103
		Lightweight aggregates	30-120	Metals	Aluminum	Cast, hammered	165
Earth	Clay	Dry	63		Brass	Cast, rolled	534
		Damp, plastic	110		Bronze	7.9 to 14% Sn	509
		and gravel, dry	100		Copper	Cast, rolled	556
	Dry	Loose	76		Iron	Cast, pig	450
		Packed	95			Wrought	480
	Moist	Loose	78		Lead		710
		Packed	96		Monel		556
	Mud	Flowing	108		Steel	Rolled	490
		Packed	115		Tin	Cast, hammered	459
	Riprap	Limestone	80-85		Zinc	Cast rolled	440
		Sandstone	90	Timber	Cedar	White or red	24.2
		Shale	105		Fir	Douglas	23.7
	Sand & gravel	Dry, loose	90-105			Eastern	25
		Dry, packed	100-120		Maple	Hard	44.5
		Wet	118-120			White	33
Gases	Air	0C., 760 mm.	.0807		Oak	Red or Black	47.3
	Gas	Natural	.0385			White	47.3
Liquids	Alcohol	100%	49		Pine	White	26
	Water	4°C., maximum density	62.5			Yellow, long leaf	44
		Ice	56			Yellow, short leaf	38
		Snow, fresh fallen	8		Redwood	California	26
		Sea water	64		Spruce	White or black	27

Table L1010-225 Design Weight Per S.F. for Structural Floor and Roof Systems

Type Slab		Description	Weight in Pounds per S.F. — Slab Depth in Inches											
			1"		2"		3"		4"		5"		6"	
Concrete Slab	Reinforced	Stone aggregate	↓	12.5	↓	25	↓	37.5	↓	50	↓	62.5	↓	75
		Lightweight sand aggregate		9.5		19		28.5		38		47.5		57
		All lightweight aggregate		9.0		18		27.0		36		45.0		54
	Plain, nonreinforced	Stone aggregate		12.0		24		36.0		48		60.0		72
		Lightweight sand aggregate		9.0		18		27.0		36		45.0		54
		All lightweight aggregate	↓	8.5		17		25.5		34		42.5		51
Concrete Waffle	19" x 19"	5" wide ribs @ 24" O.C.	6+3	77	8+3	92	10+3	100	12+3	118				
			6+4½	96	8+4½	110	10+4½	119	12+4½	136				
	30" x 30"	6" wide ribs @ 36" O.C.	8+3	83	10+3	95	12+3	109	14+3	118	16+3	130	20+3	155
			8+4½	101	10+4½	113	12+4½	126	14+4½	137	16+4½	149	20+4½	173
Concrete Joist	20" wide form	5" wide rib	8+3	60	10+3	67	12+3	74	14+3	81				
		6" wide rib	↓	63	↓	70	↓	78	↓	86	16+3	94	20+3	111
		7" wide rib									↓	99	↓	118
		5" wide rib	8+4½	79	10+4½	85	12+4½	92	14+4½	99				
		6" wide rib	↓	82	↓	89	↓	97	↓	104	16+4½	113	20+4½	130
		7" wide rib									↓	118	↓	136
	30" wide form	5" wide rib	8+3	54	10+3	58	12+3	63	14+3	68				
		6" wide rib	↓	56	↓	61	↓	67	↓	72	16+3	78	20+3	91
		7" wide rib									↓	83	↓	96
		5" wide rib	8+4½	72	10+4½	77	12+4½	82	14+4½	87				
		6" wide rib	↓	75	↓	80	↓	85	↓	91	16+4½	97	20+4½	109
		7" wide rib									↓	101	↓	115
Wood Joists	Incl. Subfloor	12" O.C.	2x6	6	2x8	6	2x10	7	2x12	8	3x8	8	3x12	11
		16" O.C.	↓	5	↓	6	↓	6	↓	7	↓	7	↓	9

Table L1010-226 Superimposed Dead Load Ranges

Component	Load Range (PSF)
Ceiling	5-10
Partitions	20-30
Mechanical	4-8

Table L1010-301   Design Loads for Structures for Wind Load

Wind Loads: Structures are designed to resist the wind force from any direction. Usually 2/3 is assumed to act on the windward side, 1/3 on the leeward side.

For more than 1/3 openings, add 10 psf for internal wind pressure or 5 psf for suction, whichever is critical.

For buildings and structures, use psf values from Table L1010-301.

For glass over 4 S.F. use values in Table L1010-302 after determining 30' wind velocity from Table L1010-303.

Type Structure	Height Above Grade	Horizontal Load in Lb. per S.F.
Buildings	Up to 50 ft.	15
	50 to 100 ft.	20
	Over 100 ft.	20 + .025 per ft.
Ground signs & towers	Up to 50 ft.	15
	Over 50 ft.	20
Roof structures		30
Glass	See Table below	

Table L1010-302   Design Wind Load in PSF for Glass at Various Elevations (BOCA Code)

Height From Grade	Velocity in Miles per Hour and Design Load in Pounds per S.F.																			
	Vel.	PSF	Vel.	PSF	Vel.	PSF	Vel.	PSF	Vel.	PSF	Vel.	PSF	Vel.	PSF	Vel.	PSF	Vel.	PSF	Vel.	PSF
To 10 ft.	42	6	46	7	49	8	52	9	55	10	59	11	62	12	66	14	69	15	76	19	83	22
10-20	52	9	58	11	61	11	65	14	70	16	74	18	79	20	83	22	87	24	96	30	105	35
20-30*	60	12	67	14	70	16	75	18	80	20	85	23	90	26	95	29	100	32	110	39	120	46
30-60	66	14	74	18	77	19	83	22	88	25	94	28	99	31	104	35	110	39	121	47	132	56
60-120	73	17	82	21	85	12	92	27	98	31	104	35	110	39	116	43	122	48	134	57	146	68
120-140	81	21	91	26	95	29	101	33	108	37	115	42	122	48	128	52	135	48	149	71	162	84
240-480	90	26	100	32	104	35	112	40	119	45	127	51	134	57	142	65	149	71	164	86	179	102
480-960	98	31	110	39	115	42	123	49	131	55	139	62	148	70	156	78	164	86	180	104	197	124
Over 960	98	31	110	39	115	42	123	49	131	55	139	62	148	70	156	78	164	86	180	104	197	124

*Determine appropriate wind at 30' elevation Fig. L1010-303 below.

Table L1010-303   Design Wind Velocity at 30 Ft. Above Ground

SPEEDS ARE FOR NORMAL EXPOSURE WHERE SURFACE FRICTION IS RELATIVELY UNIFORM FOR A FETCH OF ABOUT 25 MILES. IF THE EXPOSURE IS ELEVATED, SUBJECT TO CHANNELING, OR OTHER SPECIAL CONDITIONS AFFECTING THE EXTREME WIND SPEEDS, ADJUSTMENTS MUST BE MADE TO THE MAP VALUES.

650

For customer support on your Facilities Maintenance & Repair Cost Data, call 877.801.5724.

Table L1010-401   Snow Load in Pounds Per Square Foot on the Ground

Based on 50 year
storm for INDUSTRIAL
BUSINESS, MERCANTILE,
& RESIDENTIAL

Table L1010-402   Snow Load in Pounds Per Square Foot on the Ground

Based on 100 year
storm for ASSEMBLY
INSTITUTIONAL, HIGH HAZARD
MOTELS

Table L1010-403 Snow Loads

To convert the ground snow loads on the previous page to roof snow loads, the ground snow loads should be multiplied by the following factors depending upon the roof characteristics.

Note, in all cases $\dfrac{\alpha - 30}{50}$ is valid only for > 30 degrees.

Description	Sketch	Formula	Angle	Conversion Factor = C_f	
				Sheltered	Exposed
Simple flat and shed roofs	Load Diagram	For $\alpha > 30°$ $C_f = 0.8 - \dfrac{\alpha - 30}{50}$	0° to 30°	0.8	0.6
			40°	0.6	0.45
			50°	0.4	0.3
			60°	0.2	0.15
			70° to 90°	0	0
				Case I	Case II
Simple gable and hip roofs	Case I Load Diagram / Case II Load Diagram	$C_f = 0.8 - \dfrac{\alpha - 30}{50}$ $C_f = 1.25\left(0.8 - \left(\dfrac{\alpha - 30}{50}\right)\right)$	10°	0.8	—
			20°	0.8	—
			30°	0.8	1.0
			40°	0.6	0.75
			50°	0.4	0.5
			60°	0.2	0.25
Valley areas of Two span roofs	Case I / Case II	$\beta = \dfrac{\alpha_1 + \alpha_2}{2}$ $C_f = 0.8 - \dfrac{\alpha - 30}{50}$	$\beta \le 10°$ use Case I only $\beta > 10°$ $\beta < 20°$ use Case I & II $\beta \ge 20°$ use Case I, II & III		
Lower level of multi level roofs (or on an adjacent building not more then 15 ft. away)	$W = 2h$	$C_f = 15\dfrac{h}{g}$ h = difference in roof height in feet g = ground snow load in psf w = width of drift For h < 5, w = 10 h >15, w = 30	When $15\dfrac{h}{g} < .8$, use 0.8 When $15\dfrac{h}{g} > 3.0$, use 3.0		

Summary of above:

1. For flat roofs or roofs up to 30°, use 0.8 x ground snow load for the roof snow load.
2. For roof pitches in excess of 30°, conversion factor becomes lower than 0.8.
3. For exposed roofs there is a further 25% reduction of conversion factor.
4. For steep roofs a more highly loaded half span must be considered.
5. For shallow roof valleys conversion factor is 0.8.
6. For moderate roof valleys, conversion factor is 1.0 for half the span.
7. For steep roof valleys, conversion factor is 1.5 for one quarter of the span.
8. For roofs adjoining vertical surfaces the conversion factor is up to 3.0 for part of the span.
9. If snow load is less than 30 psf, use water load on roof for clogged drain condition.

Table L1020-101 Floor Area Ratios

Table below lists commonly used gross to net area and net to gross area ratios expressed in % for various building types.

Building Type	Gross to Net Ratio	Net to Gross Ratio	Building Type	Gross to Net Ratio	Net to Gross Ratio
Apartment	156	64	School Buildings (campus type)		
Bank	140	72	Administrative	150	67
Church	142	70	Auditorium	142	70
Courthouse	162	61	Biology	161	62
Department Store	123	81	Chemistry	170	59
Garage	118	85	Classroom	152	66
Hospital	183	55	Dining Hall	138	72
Hotel	158	63	Dormitory	154	65
Laboratory	171	58	Engineering	164	61
Library	132	76	Fraternity	160	63
Office	135	75	Gymnasium	142	70
Restaurant	141	70	Science	167	60
Warehouse	108	93	Service	120	83
			Student Union	172	59

The gross area of a building is the total floor area based on outside dimensions.

The net area of a building is the usable floor area for the function intended and excludes such items as stairways, corridors, and mechanical rooms.

In the case of a commercial building, it might be considered as the "leasable area."

Table L1020-201 Partition/Door Density

Building Type		Stories	Partition/Density	Doors	Description of Partition
Apartments		1 story	9 SF/LF	90 SF/door	Plaster, wood doors & trim
		2 story	8 SF/LF	80 SF/door	Drywall, wood studs, wood doors & trim
		3 story	9 SF/LF	90 SF/door	Plaster, wood studs, wood doors & trim
		5 story	9 SF/LF	90 SF/door	Plaster, metal studs, wood doors & trim
		6-15 story	8 SF/LF	80 SF/door	Drywall, metal studs, wood doors & trim
Bakery		1 story	50 SF/LF	500 SF/door	Conc. block, paint, door & drywall, wood studs
		2 story	50 SF/LF	500 SF/door	Conc. block, paint, door & drywall, wood studs
Bank		1 story	20 SF/LF	200 SF/door	Plaster, wood studs, wood doors & trim
		2-4 story	15 SF/LF	150 SF/door	Plaster, metal studs, wood doors & trim
Bottling Plant		1 story	50 SF/LF	500 SF/door	Conc. block, drywall, metal studs, wood trim
Bowling Alley		1 story	50 SF/LF	500 SF/door	Conc. block, wood & metal doors, wood trim
Bus Terminal		1 story	15 SF/LF	150 SF/door	Conc. block, ceramic tile, wood trim
Cannery		1 story	100 SF/LF	1000 SF/door	Drywall on metal studs
Car Wash		1 story	18 SF/LF	180 SF/door	Concrete block, painted & hollow metal door
Dairy Plant		1 story	30 SF/LF	300 SF/door	Concrete block, glazed tile, insulated cooler doors
Department Store		1 story	60 SF/LF	600 SF/door	Drywall, metal studs, wood doors & trim
		2-5 story	60 SF/LF	600 SF/door	30% concrete block, 70% drywall, wood studs
Dormitory		2 story	9 SF/LF	90 SF/door	Plaster, concrete block, wood doors & trim
		3-5 story	9 SF/LF	90 SF/door	Plaster, concrete block, wood doors & trim
		6-15 story	9 SF/LF	90 SF/door	Plaster, concrete block, wood doors & trim
Funeral Home		1 story	15 SF/LF	150 SF/door	Plaster on concrete block & wood studs, paneling
		2 story	14 SF/LF	140 SF/door	Plaster, wood studs, paneling & wood doors
Garage Sales & Service		1 story	30 SF/LF	300 SF/door	50% conc. block, 50% drywall, wood studs
Hotel		3-8 story	9 SF/LF	90 SF/door	Plaster, conc. block, wood doors & trim
		9-15 story	9 SF/LF	90 SF/door	Plaster, conc. block, wood doors & trim
Laundromat		1 story	25 SF/LF	250 SF/door	Drywall, wood studs, wood doors & trim
Medical Clinic		1 story	6 SF/LF	60 SF/door	Drywall, wood studs, wood doors & trim
		2-4 story	6 SF/LF	60 SF/door	Drywall, metal studs, wood doors & trim
Motel		1 story	7 SF/LF	70 SF/door	Drywall, wood studs, wood doors & trim
		2-3 story	7 SF/LF	70 SF/door	Concrete block, drywall on metal studs, wood paneling
Movie Theater	200-600 seats	1 story	18 SF/LF	180 SF/door	Concrete block, wood, metal, vinyl trim
	601-1400 seats		20 SF/LF	200 SF/door	Concrete block, wood, metal, vinyl trim
	1401-22000 seats		25 SF/LF	250 SF/door	Concrete block, wood, metal, vinyl trim
Nursing Home		1 story	8 SF/LF	80 SF/door	Drywall, metal studs, wood doors & trim
		2-4 story	8 SF/LF	80 SF/door	Drywall, metal studs, wood doors & trim
Office		1 story	20 SF/LF	200-500 SF/door	30% concrete block, 70% drywall on wood studs
		2 story	20 SF/LF	200-500 SF/door	30% concrete block, 70% drywall on metal studs
		3-5 story	20 SF/LF	200-500 SF/door	30% concrete block, 70% movable partitions
		6-10 story	20 SF/LF	200-500 SF/door	30% concrete block, 70% movable partitions
		11-20 story	20 SF/LF	200-500 SF/door	30% concrete block, 70% movable partitions
Parking Ramp (Open)		2-8 story	60 SF/LF	600 SF/door	Stair and elevator enclosures only
Parking garage		2-8 story	60 SF/LF	600 SF/door	Stair and elevator enclosures only
Pre-Engineered	Steel	1 story	0		
	Store	1 story	60 SF/LF	600 SF/door	Drywall on metal studs, wood doors & trim
	Office	1 story	15 SF/LF	150 SF/door	Concrete block, movable wood partitions
	Shop	1 story	15 SF/LF	150 SF/door	Movable wood partitions
	Warehouse	1 story	0		
Radio & TV Broadcasting		1 story	25 SF/LF	250 SF/door	Concrete block, metal and wood doors
& TV Transmitter		1 story	40 SF/LF	400 SF/door	Concrete block, metal and wood doors
Self Service Restaurant		1 story	15 SF/LF	150 SF/door	Concrete block, wood and aluminum trim
Cafe & Drive-in Restaurant		1 story	18 SF/LF	180 SF/door	Drywall, metal studs, ceramic & plastic trim
Restaurant with seating		1 story	25 SF/LF	250 SF/door	Concrete block, paneling, wood studs & trim
Supper Club		1 story	25 SF/LF	250 SF/door	Concrete block, paneling, wood studs & trim
Bar or Lounge		1 story	24 SF/LF	240 SF/door	Plaster or gypsum lath, wooded studs
Retail Store or Shop		1 story	60 SF/LF	600 SF/door	Drywall metal studs, wood doors & trim
Service Station	Masonry	1 story	15 SF/LF	150 SF/door	Concrete block, paint, door & drywall, wood studs
	Metal panel	1 story	15 SF/LF	150 SF/door	Concrete block, paint, door & drywall, wood studs
	Frame	1 story	15 SF/LF	150 SF/door	Drywall, wood studs, wood doors & trim
Shopping Center	(strip)	1 story	30 SF/LF	300 SF/door	Drywall, metal studs, wood doors & trim
	(group)	1 story	40 SF/LF	400 SF/door	50% concrete block, 50% drywall, wood studs
		2 story	40 SF/LF	400 SF/door	50% concrete block, 50% drywall, wood studs
Small Food Store		1 story	30 SF/LF	300 SF/door	Concrete block drywall, wood studs, wood trim
Store/Apt. above	Masonry	2 story	10 SF/LF	100 SF/door	Plaster, metal studs, wood doors & trim
	Frame	2 story	10 SF/LF	100 SF/door	Plaster, metal studs, wood doors & trim
	Frame	3 story	10 SF/LF	100 SF/door	Plaster, metal studs, wood doors & trim
Supermarkets		1 story	40 SF/LF	400 SF/door	Concrete block, paint, drywall & porcelain panel
Truck Terminal		1 story	0		
Warehouse		1 story	0		

Table L1020-301   Occupancy Determinations

Function of Space	SF/Person Required
Accessory storage areas, mechanical equipment rooms	300
Agriculture Building	300
Aircraft Hangars	500
Airport Terminal	
Baggage claim	20
Baggage handling	300
Concourse	100
Waiting areas	15
Assembly	
Gaming floors (keno, slots, etc.)	11
Exhibit Gallery and Museum	30
Assembly w/ fixed seats	load determined by seat number
Assembly w/o fixed seats	
Concentrated (chairs only-not fixed)	7
Standing space	5
Unconcentrated (tables and chairs)	15
Bowling centers, allow 5 persons for each lane including 15 feet of runway, and for additional areas	7
Business areas	100
Courtrooms-other than fixed seating areas	40
Day care	35
Dormitories	50
Educational	
Classroom areas	20
Shops and other vocational room areas	50
Exercise rooms	50
Fabrication and Manufacturing areas where hazardous materials are used	200
Industrial areas	100
Institutional areas	
Inpatient treatment areas	240
Outpatient areas	100
Sleeping areas	120
Kitchens commercial	200
Library	
Reading rooms	50
Stack area	100
Mercantile	
Areas on other floors	60
Basement and grade floor areas	30
Storage, stock, shipping areas	300
Parking garages	200
Residential	200
Skating rinks, swimming pools	
Rink and pool	50
Decks	15
Stages and platforms	15
Warehouses	500

Excerpted from the 2012 *International Building Code*, Copyright 2011. Washington, D.C.: International Code Council.
Reproduced with permission. All rights reserved. www.ICCSAFE.org

Table L1020-302 Length of Exitway Access Travel (ft.)

Occupancy Type	Without Sprinkler System (feet)	With Sprinkler System (feet)
A, E, F-1, M, R, S-1	200	250
I-1	Not Permitted	250
B	200	300
F-2, S-2, U	300	400
H-1	Not Permitted	75
H-2	Not Permitted	100
H-3	Not Permitted	150
H-4	Not Permitted	175
H-5	Not Permitted	200
I-2, I-3, I-4	Not Permitted	200

Note:

Refer to the 2012 *International Building Code* Section 1016 Exit Access Travel Distance for any exceptions or additions to the information above

Excerpted from the 2012 *International Building Code*, Copyright 2011. Washington, D.C.: International Code Council. Reproduced with permission. All rights reserved. www.ICCSAFE.org

Table L1020-303 Capacity Per Unit Egress Width*

Use Group	Without Fire Suppression System (Inches per Person)*		With Fire Suppression System (Inches per Person)*	
	Stairways	Doors, Ramps and Corridors	Stairways	Doors, Ramps and Corridors
Assembly, Business, Educational, Factory Industrial, Mercantile, Residential, Storage	0.3	0.2	0.2	0.15
Institutional—1	0.3	0.2	0.2	0.15
Institutional—2	—	0.7	0.3	0.2
Institutional—3	0.3	0.2	0.2	0.15
High Hazard	0.7	0.4	0.3	0.2

* 1″ = 25.4 mm

Table L1030-401 Resistances ("R") of Building and Insulating Materials

Material	Wt./Lbs. per C.F.	R per Inch	R Listed Size
Air Spaces and Surfaces			
Enclosed non-reflective spaces, E=0.82			
50° F mean temp., 30°/10° F diff.			
.5"			.90/.91
.75"			.94/1.01
1.50"			.90/1.02
3.50"			.91/1.01
Inside vert. surface (still air)			0.68
Outside vert. surface (15 mph wind)			0.17
Building Boards			
Asbestos cement, 0.25" thick	120		0.06
Gypsum or plaster, 0.5" thick	50		0.45
Hardboard regular	50	1.37	
Tempered	63	1.00	
Laminated paper	30	2.00	
Particle board	37	1.85	
	50	1.06	
	63	0.85	
Plywood (Douglas Fir), 0.5" thick	34		0.62
Shingle backer, .375" thick	18		0.94
Sound deadening board, 0.5" thick	15		1.35
Tile and lay-in panels, plain or acoustical, 0.5" thick	18		1.25
Vegetable fiber, 0.5" thick	18		1.32
	25		1.14
Wood, hardwoods	48	0.91	
Softwoods	32	1.25	
Flooring Carpet with fibrous pad			2.08
With rubber pad			1.23
Cork tile, 1/8" thick			0.28
Terrazzo			0.08
Tile, resilient			0.05
Wood, hardwood, 0.75" thick			0.68
Subfloor, 0.75" thick			0.94
Glass			
Insulation, 0.50" air space			2.04
Single glass			0.91
Insulation Blanket or Batt, mineral, glass or rock fiber, approximate thickness			
3.0" to 3.5" thick			11
3.5" to 4.0" thick			13
6.0" to 6.5" thick			19
6.5" to 7.0" thick			22
8.5" to 9.0" thick			30
Boards			
Cellular glass	8.5	2.63	
Fiberboard, wet felted			
Acoustical tile	21	2.70	
Roof insulation	17	2.94	
Fiberboard, wet molded			
Acoustical tile	23	2.38	
Mineral fiber with resin binder	15	3.45	
Polystyrene, extruded,			
cut cell surface	1.8	4.00	
smooth skin surface	2.2	5.00	
	3.5	5.26	
Bead boards	1.0	3.57	
Polyurethane	1.5	6.25	
Wood or cane fiberboard, 0.5" thick			1.25
Insulation Loose Fill			
Cellulose	2.3	3.13	
	3.2	3.70	
Mineral fiber, 3.75" to 5" thick	2-5		11
6.5" to 8.75" thick			19
7.5" to 10" thick			22
10.25" to 13.75" thick			30
Perlite	5-8	2.70	
Vermiculite	4-6	2.27	
Wood fiber	2-3.5	3.33	
Masonry Brick, Common	120	0.20	
Face	130	0.11	
Cement mortar	116	0.20	
Clay tile, hollow			
1 cell wide, 3" width			0.80
4" width			1.11
2 cells wide, 6" width			1.52
8" width			1.85
10" width			2.22
3 cells wide, 12" width			2.50
Concrete, gypsum fiber	51	0.60	
Lightweight	120	0.19	
	80	0.40	
	40	0.86	
Perlite	40	1.08	
Sand and gravel or stone	140	0.08	
Concrete block, lightweight			
3 cell units, 4"-15 lbs. ea.			1.68
6"-23 lbs. ea.			1.83
8"-28 lbs. ea.			2.12
12"-40 lbs. ea.			2.62
Sand and gravel aggregates,			
4"-20 lbs. ea.			1.17
6"-33 lbs. ea.			1.29
8"-38 lbs. ea.			1.46
12"-56 lbs. ea.			1.81
Plastering Cement Plaster,			
Sand aggregate	116	0.20	
Gypsum plaster, Perlite aggregate	45	0.67	
Sand aggregate	105	0.18	
Vermiculite aggregate	45	0.59	
Roofing			
Asphalt, felt, 15 lb.			0.06
Rolled roofing	70		0.15
Shingles	70		0.44
Built-up roofing .375" thick	70		0.33
Cement shingles	120		0.21
Vapor-permeable felt			0.06
Vapor seal, 2 layers of mopped 15 lb. felt			0.12
Wood, shingles 16"-7.5" exposure			0.87
Siding			
Aluminum or steel (hollow backed) oversheathing			0.61
With .375" insulating backer board			1.82
Foil backed			2.96
Wood siding, beveled, ½" x 8"			0.81

Table L1030-502   Maximum Depth of Frost Penetration in Inches

THIS MAP IS REASONABLY ACCURATE FOR MOST PARTS
OF THE UNITED STATES BUT IS NECESSARILY HIGHLY
GENERALIZED, AND CONSEQUENTLY NOT TOO ACCURATE IN
MOUNTAINOUS REGIONS, PARTICULARLY IN THE ROCKIES.

Table L1040-101 Fire-Resisting Ratings of Structural Elements (in hours)

Description of the Structural Element	No. 1 Fireproof		No. 2 Non Combustible			No. 3 Exterior Masonry Wall			No. 4 Frame	
			Protected		Unprotected	Heavy Timber	Ordinary		Protected	Unprotected
							Protected	Unprotected		
	1A	1B	2A	2B	2C	3A	3B	3C	4A	4B
Exterior, Bearing Walls	4	3	2	1½	1	2	2	2	1	1
Nonbearing Walls	2	2	1½	1	1	2	2	2	1	1
Interior Bearing Walls and Partitions	4	3	2	1	0	2	1	0	1	0
Fire Walls and Party Walls	4	3	2	2	2	2	2	2	2	2
Fire Enclosure of Exitways, Exit Hallways and Stairways	2	2	2	2	2	2	2	2	1	1
Shafts other than Exitways, Hallways and Stairways	2	2	2	2	2	2	2	2	1	1
Exitway access corridors and Vertical separation of tenant space	1	1	1	1	0	1	1	0	1	0
Columns, girders, trusses (other than roof trusses) and framing: Supporting more than one floor	4	3	2	1	0	—	1	0	1	0
Supporting one floor only	3	2	1½	1	0	—	1	0	1	0
Structural members supporting wall	3	2	1½	1	0	1	1	0	1	0
Floor construction including beams	3	2	1½	1	0	—	1	0	1	0
Roof construction including beams, trusses and framing arches and roof deck 15' or less in height to lowest member	2	1½	1	1	0	—	1	0	1	0

Note: a. Codes include special requirements and exceptions that are not included in the table above.

b. Each type of construction has been divided into sub-types which vary according to the degree of fire resistance required. Sub-types (A) requirements are more severe than those for sub-types (B).

c. Protected construction means all structural members are chemically treated, covered or protected so that the unit has the required fire resistance.

Type No. 1, Fireproof Construction — Buildings and structures of fireproof construction are those in which the walls, partitions, structural elements, floors, ceilings, and roofs, and the exitways are protected with approved noncombustible materials to afford the fire-resistance rating specified in Table L1040-101; except as otherwise specifically regulated. Fire-resistant treated wood may be used as specified.

Type No. 2, Noncombustible Construction — Buildings and structures of noncombustible construction are those in which the walls, partitions, structural elements, floors, ceilings, roofs and the exitways are approved noncombustible materials meeting the fire-resistance rating requirements specified in Table L1040-101; except as modified by the fire limit restrictions. Fire-retardant treated wood may be used as specified.

Type No. 3, Exterior Masonry Wall Construction — Buildings and structures of exterior masonry wall construction are those in which the exterior, fire and party walls are masonry or other approved noncombustible materials of the required fire-resistance rating and structural properties. The floors, roofs, and interior framing are wholly or partly wood or metal or other approved construction. The fire and party walls are ground-supported; except that girders and their supports, carrying walls of masonry shall be protected to afford the same degree of fire-resistance rating of the supported walls. All structural elements have the required fire-resistance rating specified in Table L1040-101.

Type No. 4, Frame Construction — Buildings and structures of frame construction are those in which the exterior walls, bearing walls, partitions, floor and roof construction are wholly or partly of wood stud and joist assemblies with a minimum nominal dimension of two inches or of other approved combustible materials. Fire stops are required at all vertical and horizontal draft openings in which the structural elements have required fire-resistance ratings specified in Table L1040-101.

Table L1040-201   Fire Resistance Ratings

Fire Hazard for Fire Walls

The degree of fire hazard of buildings relating to their intended use is defined by "Fire Rating" the occupancy type. Such a rating system is listed in the Table L1040-201 below. This type of rating determines the requirements for fire walls and fire separation walls (exterior fire exposure). For mixed use occupancy, use the higher Fire Rating requirement of the components.

Group	Fire-Resistance Rating (hours)
A, B, E, H-4[1], I, R-1, R-2, U	3
F-1[2], H-3[1], H-5[1], M, S-1	3
H-1[1], H-2[1]	4
F-2[3], S-2, R-3, R-4	2

Note: *The difference in "Fire Hazards" is determined by their occupancy and use.

1. High Hazard: Industrial and storage buildings in which the combustible contents might cause fires to be unusually intense or where explosives, combustible gases or flammable liquids are manufactured or stored.

2. Moderate Hazard: Mercantile buildings, industrial and storage buildings in which combustible contents might cause fires of moderate intensity.

3. Low Hazard: Business buildings that ordinarily do not burn rapidly.

Note:
In Type II or V construction, walls shall be permitted to have a 2-hour fire-resistance rating.

Table 706.4

Excerpted from the 2012 *International Building Code*, Copyright 2011. Washington, D.C.: International Code Council. Reproduced with permission. All rights reserved. www.ICCSAFE.org

Table L1040-202   Interior Finish Classification

Flame Spread for Interior Finishes

The flame spreadability of a material is the burning characteristic of the material relative to the fuel contributed by its combustion and the density of smoke developed. The flame spread classification of a material is based on a ten minute test on a scale of 0 to 100. Cement asbestos board is assigned a rating of 0 and select red oak flooring a rating of 100.

The three classes are listed in Table L1040-202.

The flame spread ratings for interior finish walls and ceilings shall not be greater than the Class listed in the Table L1040-203.

Finish Class	Flame Spread Index	Smoke Developed Index
A	0-25	0-450
B	26-75	0-450
C	76-200	0-450

Section 803.1.1 Interior wall and ceiling finish materials

Excerpted from the 2012 *International Building Code*, Copyright 2011. Washington, D.C.: International Code Council. Reproduced with permission. All rights reserved. www.ICCSAFE.org

For customer support on your Facilities Maintenance & Repair Cost Data, call 877.801.5724.

Table L1040-203 Interior Finish Requirements by Class

Group	Sprinklered			Nonsprinklered		
	Interior exit stairways, interior exit ramps and exit passageways Note a, b	Corridors and enclosure for exit access stairways and exit access ramps	Rooms and enclosed spaces Note c	Interior exit stairways, interior exit ramps and exit passageways Note a, b	Corridors and enclosure for exit access stairways and exit access ramps	Rooms and enclosed spaces Note c
A-1 & A-2	B	B	C	A	A (d)	B (e)
A-3 (f), A-4, A-5	B	B	C	A	A (d)	C
B, E, M, R-1	B	C	C	A	B	C
R-4	B	C	C	A	B	B
F	C	C	C	B	C	C
H	B	B	C	A	A	B
I-1	B	C	C	A	B	B
I-2	B	B	B (h, i)	A	A	B
I-3	A	A (j)	C	A	A	B
I-4	B	B	B (h, i)	A	A	B
R-2	C	C	C	B	B	C
R-3	C	C	C	C	C	C
S	C	C	C	B	B	C
U	No Restrictions			No Restrictions		

Notes:

a. Class C interior finish materials shall be permitted for wainscotting or paneling of not more than 1,000 square feet of applied surface area in the grade lobby where applied directly to a noncombustible base or over furring strips applied to a noncombustible base and fireblocked as required by IBC Section 803.11.1.

b. In other Group I-2 occupancies in buildings less than three stories above grade plane of other than Group I-3, Class B interior finish for nonsprinklered buildings and Class C interior finish for sprinklered buildings shall be permitted in interior exit stairways and ramps.

c. Requirements for rooms and enclosed spaces shall be based upon spaces enclosed by partitions. Where a fire-resistance rating is required for structural elements, the enclosing partitions shall extend from the floor to the ceiling. Partitions that do not comply with this shall be considered enclosed spaces and the rooms or spaces on both sides shall be considered one. In determining the applicable requirements for rooms and enclosed spaces, the specific occupancy thereof shall be the governing factor regardless of the group classification of the building or structure.

d. Lobby areas in Group A-1, A-2, and A-3 occupancies shall not be less than Class B materials.

e. Class C interior finish materials shall be permitted in places of assembly with an occupant load of 300 persons or less.

f. For places of religious worship, wood used for ornamental purposes, trusses, paneling or chancel furnishing shall be permitted.

g. Class B material is required where the building exceeds two stories.

h. Class C interior finish materials shall be permitted in administrative spaces.

i. Class C interior finish materials shall be permitted in rooms with a capacity of four persons or less.

j. Class B materials shall be permitted as wainscotting extending not more than 48 inches above the finished floor in corridors and exit access stairways and ramps.

Section 803.9

Description: This table is primarily for converting customary U.S. units in the left hand column to SI metric units in the right hand column.

In addition, conversion factors for some commonly encountered Canadian and non-SI metric units are included.

Table L1090-101   Metric Conversion Factors

	If You Know		Multiply By		To Find
Length	Inches	x	25.4[a]	=	Millimeters
	Feet	x	0.3048[a]	=	Meters
	Yards	x	0.9144[a]	=	Meters
	Miles (statute)	x	1.609	=	Kilometers
Area	Square inches	x	645.2	=	Square millimeters
	Square feet	x	0.0929	=	Square meters
	Square yards	x	0.8361	=	Square meters
Volume (Capacity)	Cubic inches	x	16,387	=	Cubic millimeters
	Cubic feet	x	0.02832	=	Cubic meters
	Cubic yards	x	0.7646	=	Cubic meters
	Gallons (U.S. liquids)[b]	x	0.003785	=	Cubic meters[c]
	Gallons (Canadian liquid)[b]	x	0.004546	=	Cubic meters[c]
	Ounces (U.S. liquid)[b]	x	29.57	=	Milliliters[c, d]
	Quarts (U.S. liquid)[b]	x	0.9464	=	Liters[c, d]
	Gallons (U.S. liquid)[b]	x	3.785	=	Liters[c, d]
Force	Kilograms force[d]	x	9.807	=	Newtons
	Pounds force	x	4.448	=	Newtons
	Pounds force	x	0.4536	=	Kilograms force[d]
	Kips	x	4448	=	Newtons
	Kips	x	453.6	=	Kilograms force[d]
Pressure, Stress, Strength (Force per unit area)	Kilograms force per square centimeter[d]	x	0.09807	=	Megapascals
	Pounds force per square inch (psi)	x	0.006895	=	Megapascals
	Kips per square inch	x	6.895	=	Megapascals
	Pounds force per square inch (psi)	x	0.07031	=	Kilograms force per square centimeter[d]
	Pounds force per square foot	x	47.88	=	Pascals
	Pounds force per square foot	x	4.882	=	Kilograms force per square meter[d]
Bending Moment Or Torque	Inch-pounds force	x	0.01152	=	Meter-kilograms force[d]
	Inch-pounds force	x	0.1130	=	Newton-meters
	Foot-pounds force	x	0.1383	=	Meter-kilograms force[d]
	Foot-pounds force	x	1.356	=	Newton-meters
	Meter-kilograms force[d]	x	9.807	=	Newton-meters
Mass	Ounces (avoirdupois)	x	28.35	=	Grams
	Pounds (avoirdupois)	x	0.4536	=	Kilograms
	Tons (metric)	x	1000	=	Kilograms
	Tons, short (2000 pounds)	x	907.2	=	Kilograms
	Tons, short (2000 pounds)	x	0.9072	=	Megagrams[e]
Mass per Unit Volume	Pounds mass per cubic foot	x	16.02	=	Kilograms per cubic meter
	Pounds mass per cubic yard	x	0.5933	=	Kilograms per cubic meter
	Pounds mass per gallon (U.S. liquid)[b]	x	119.8	=	Kilograms per cubic meter
	Pounds mass per gallon (Canadian liquid)[b]	x	99.78	=	Kilograms per cubic meter
Temperature	Degrees Fahrenheit	(F-32)/1.8		=	Degrees Celsius
	Degrees Fahrenheit	(F+459.67)/1.8		=	Degrees Kelvin
	Degrees Celsius	C+273.15		=	Degrees Kelvin

[a]The factor given is exact
[b]One U.S. gallon = 0.8327 Canadian gallon
[c]1 liter = 1000 milliliters = 1000 cubic centimeters
 1 cubic decimeter = 0.001 cubic meter

[d]Metric but not SI unit
[e]Called "tonne" in England and "metric ton" in other metric countries

Life Cycle Costing (LCC)
for Evaluating Economic Performance of Building Investments

Introduction

A complete analysis of maintenance costs requires consideration of life cycle costs, or LCC, as it is often called. Under current conditions, facilities managers have to wisely spend their limited operating and maintenance budgets. When deciding how to allocate limited budgets, maintenance professionals are constantly faced with the economic decision of when to stop maintaining and start replacing. Once the decision to replace has been made, LCC analysis can enhance the choice of alternatives as well.

The American Society of Testing and Materials (ASTM) Subcommittee E06.81 on Building Economics has developed a series of standards for evaluating the economic performance of building investments. These standards, which have been endorsed by the American Association of Cost Engineers (AACE), are intended to provide uniformity in both the terminology and calculation methods utilized for Life Cycle Costing. Without these standards as a basis for calculations, life cycle cost comparisons could be misleading or incomplete. This chapter is based on the current ASTM standards which are listed in the references at the end of this section.

Definition of Life Cycle Costing

(ASTM Standard Terminology of Building Economics E833-89)

A technique of economic evaluation that sums over a given study period the cost of initial investment less resale value, replacements, operations, energy use, and maintenance and repair of an investment decision. The costs are expressed in either lump sum present value terms or in equivalent uniform annual values.

This technique for life cycle costing is important to the facilities manager, as it is a valuable tool for evaluating investment alternatives and supports his or her decision-making.

To understand Life Cycle Costing, one should be familiar with the notions of compounding, discounting, present value, and equivalent uniform annual value.

Compounding

Definition of Compounding *(Webster's New Collegiate Dictionary)*

The process of computing the value of an original principal sum based on interest calculated on the sum of the original principal and accrued interest.

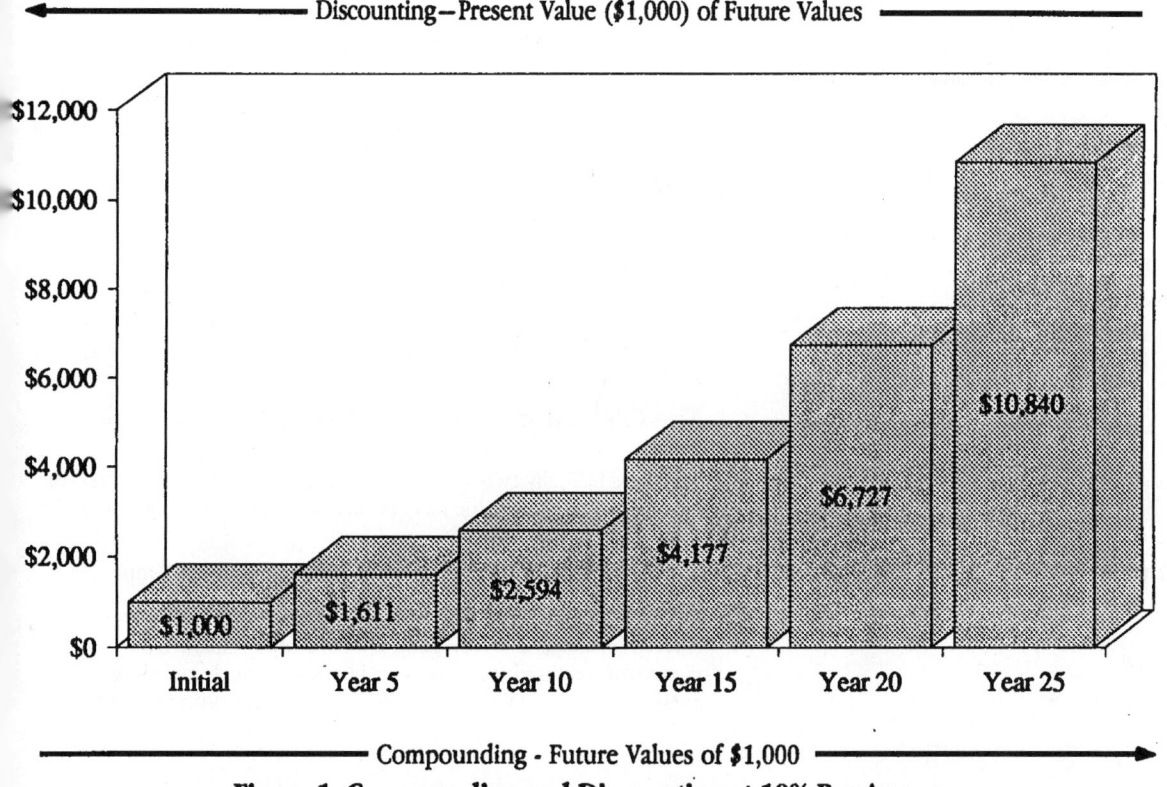

Discounting—Present Value ($1,000) of Future Values

Initial	$1,000	
Year 5	$1,611	
Year 10	$2,594	
Year 15	$4,177	
Year 20	$6,727	
Year 25	$10,840	

Compounding - Future Values of $1,000

Figure 1: Compounding and Discounting at 10% Per Annum

For customer support on your Facilities Maintenance & Repair Cost Data, call 877.801.5724.

663

Figure 1 illustrates compounding. As shown, a principal sum of $1,000 at a compound interest rate of 10% per annum increases to $1,611 in 5 years, $2,594 in 10 years, $4,177 in 15 years, $6,727 in 20 years, and $10,840 in 25 years; this process (compounding) is usually well understood in that it represents the amount accumulated in a bank account for a specific amount at a given interest rate and calculated on the sum of the original principal and accrued interest.

Discounting

ASTM Definition of Discounting

A technique for converting cash flows that occur over time to equivalent amounts at a common time (base time).

It is also evident from Figure 1 that if one had to pay out or meet an obligation of either $1,611 in 5 years, $2,594 in 10 years, $4,177 in 15 years, $6,727 in 20 years, or $10,840 in 25 years, the amount to be invested initially at 10% interest would be $1,000. This amount is referred to as the present value (PV) of a future amount and it is calculated by "discounting" a future value (F) in a specific year at a given rate of interest.

Present Value

ASTM Definition of Present Value

The value of a benefit or cost found by discounting future cash flows to the base time. (Syn. present worth)

The basic mathematics for calculating present values are relatively simple:

Given:

P = Present sum of money
F = Future sum of money
i = Interest or discount rate (expressed as a decimal and not a percentage)
n = Number of years

The future value (F) of a present sum of money (P) may be computed as shown in Table 1, a relatively simple process.

Year	Future Value (F)
1	$P(1+i)$
2	$P(1+i)(1+i) = P(1+i)^2$
3	$P(1+i)^2(1+i) = P(1+i)^3$
n	$P(1+i)^n$

Table 1: The Future Value of a Present Sum of Money

Since $\qquad F = P(1+i)^n$

Therefore $\qquad P = \dfrac{F}{(1+1)^n} = F \times \dfrac{1}{(1+i)^n}$

Example

What is the present value (P) of an anticipated maintenance expense of $300 in year 3 (F) if the interest rate is 10%?

$$P = F \times \frac{1}{(1+i)^n} = 300 \times \frac{1}{(1+0.10)^n} = 300 \times \frac{1}{1.33}$$

$$= 300 \times 0.75$$
$$= \$225$$

From an interest standpoint, this tells us that $225 is the amount that would have to be deposited today into an account paying 10% interest per annum in order to provide $300 at the end of year 3 to meet the anticipated expense. From the economist's point of view, discounting using the investor's minimum acceptable rate of return (10% in this case) makes the economic outlook the same, whether it is the future amount of $300 or a present value of $225.

Note that $(1+i)^n$ is commonly referred to as the *Single Compound Amount (SCA)* factor, and $1/(1+i)^n$ as the *Single Present Value Factor (SPV)*. Both factors may be readily found in standard discount factor tables for a wide range of interest rates and time periods, thus simplifying calculations. The use of these factors in calculating life cycle costs by the formula method is discussed later in this section.

A life cycle cost analysis can be performed in *constant dollars* (dollars of constant purchasing power, net of general inflation) or *current dollars* (actual prices inclusive of general inflation). If you use constant dollars, you must use a "real" discount rate (i.e., a rate which reflects the time value of money, net of general inflation) and differential escalation rates (also net of general inflation) for future cost items. Constant dollars are generally tied to a specific year. If you use current dollars, the discount rates and escalation rate must include general inflation (i.e., they must be "nominal" or "market" rates). In this chapter, all costs are expressed in current dollars and rates in nominal terms. When taxes are included in the analysis, the current dollar approach is generally preferred.

The Cash Flow Method for Calculating Life Cycle Costs

There are basically two approaches to calculating life cycle costs—the *cash flow* method and the *formula* method which, when applicable, is somewhat more simple.

To illustrate the cash flow method, a relatively simple example follows.

Example

A facility manager is considering the purchase of maintenance equipment required for a four-year period for which initial costs, energy costs and maintenance costs of alternative proposals vary.

The financial criteria on which the economic evaluation will be based are the following:

Interest/Discount Rate	–	10%
Energy Escalation Rate	–	8% per annum
Labor and Material Escalation Rate	–	4% per annum
Period of Study	–	4 years

The following data is provided in one of the alternative proposals to be analyzed:

- Initial Capital Cost – $1,000
- Maintenance Costs – A fixed annual cost of $100 per year quoted by the supplier (no escalation to be considered)
- Annual Energy Costs – Initially $100 per year and subject to 8% annual escalation – i.e., $108 for year one.
- Salvage Value – At the end of the four-year period, the equipment has no further useful life and the supplier agrees to purchase it as scrap for $50.

What is the present value of this alternative proposal?

The steps to resolve this problem are the following:

Step 1 – Prepare a Cash Flow Diagram.

Step 2 – Establish a Time Schedule of Costs.

Step 3 – Calculate Annual Net Cash Flows.

Step 4 – Calculate Present Value Factors.

Step 5 – Calculate Present Values of Annual Cash Flows and Life Cycle Costs.

Step 1: Prepare a Cash Flow Diagram

To better understand the problem, cash flow diagrams are often prepared. Revenues are noted as vertical lines above a horizontal time axis, and disbursements are noted as vertical lines below it.

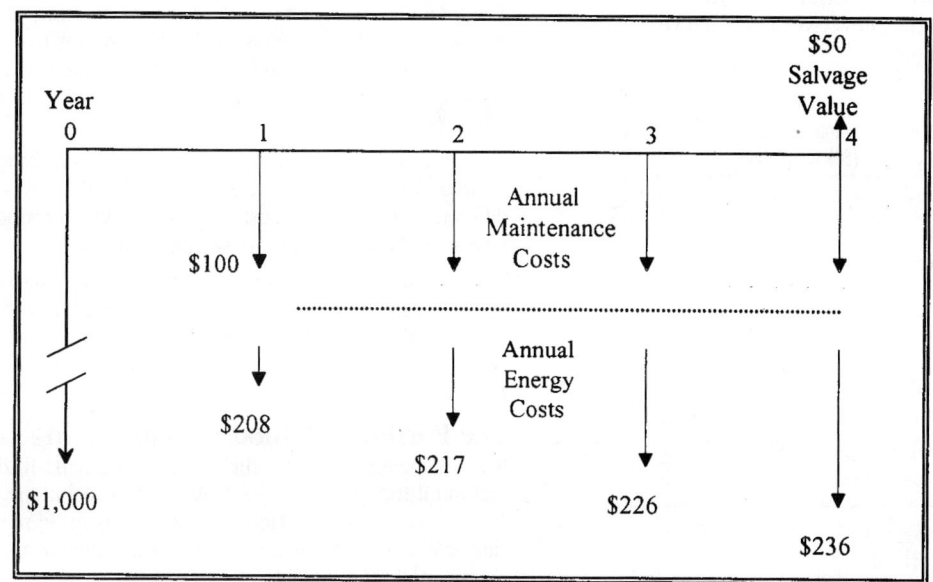

Figure 2: Cash Flow Diagram

For customer support on your Facilities Maintenance & Repair Cost Data, call 877.801.5724.

665

Step 2: Establish a Time Schedule of Costs

Table 2 below presents revenues and disbursements for each year of the study:

A Year	B Capital Costs	C Maintenance Costs	D Energy Cost (8% escalation)
0	1,000	–	–
1	–	100	108
2	–	100	117
3	–	100	126
4	(50)	100	136

Table 2

Note that the disbursements in this example are considered positive, and revenues for the salvage value are negative.

Step 3: Calculate Annual Net Cash Flows

Table 3 below provides the net cash flow (NCF) for each year, which could be a disbursement or a revenue.

A Year	B Capital Costs	C Maintenance Costs	D Energy Cost (8% escalation)	E NCF (B + C + D)
0	1,000	–	–	1,000
1	–	100	108	208
2	–	100	117	217
3	–	100	126	226
4	(50)	100	136	186

Table 3

Step 4: Calculate Present Value Factors

As in a previous example, present value factors based on 10% interest will be calculated for each year to convert the annual net cash flows to present values.

Year n	PV Factor	$\dfrac{1}{(1 + i)^n}$
0		1
1		$\dfrac{1}{(1+ 0.1)^1}$ = 0.91
2		$\dfrac{1}{(1+ 0.1)^2}$ = 0.83
3		$\dfrac{1}{(1+ 0.1)^3}$ = 0.75
4		$\dfrac{1}{(1+ 0.1)^4}$ = 0.68

Table 4

Step 5: Calculate Present Values of Annual Cash Flows and Life Cycle Costs

Annual net cash flows are converted to present values of the base year. Their sum (Total Present Value) will represent the life cycle costs of the alternative.

Year n	NCF	PV Factor	PV $
0	1,000	1	1,000
1	208	0.91	189
2	217	0.83	180
3	226	0.75	170
4	186	0.68	126
PV Life Cycle Cost	–	–	1,665

Table 5

The total life cycle cost of this alternative in terms of today's dollar (if this year is the base year) is $1,665. This means that if a present sum of $1,665 were deposited today at an interest rate of 10%, all expenses could be paid over a four-year period, at which time the bank balance would be zero.

Once the costs of the other proposals have been calculated in present value dollars, the facilities manager will be in a position to quantify the differences in life cycle costs and decide which proposal represents the lowest cost of ownership in the long run.

The approach outlined is a practical one when considering the purchase of equipment such as elevators, controls, etc.; quotations should not only include capital costs, but also the costs of long-term maintenance contracts which could make a higher initial cost proposal lower over the life cycle of the equipment.

The Formula Method for Calculating Life Cycle Cost

It is also possible to solve the previous problem utilizing the formula method if the problem is adaptable to this technique. Business calculators or discount factor tables such as those published by ASTM can be employed to generate the present value factors, and results may be obtained more quickly.

In some cases, the complexity of the calculations (such as when taxes on net revenues are considered) will necessitate the "cash flow" method, with calculations done manually or using a spreadsheet program.

With the formula method approach, the calculations for the life cycle cost of the maintenance equipment alternative analyzed would be as follows:

1. PV of Initial Investment $1,000
2. PV of Maintenance Costs
   Annual Cost (A) x UPV factor
   100 x 3.17 $ 317
3. PV of Energy Costs
   Base Year Cost (A) x UPV*
   100 x 3.82 $ 382
4. Salvage (Scrap) Value
   Future Value (F) x SPV
   (50) x 0.68   (34)
5. Life Cycle Cost / Total Present Value $1,665

Note: The above factors can be found in Discount Factor Tables.

UPV–Uniform Present Value
SPV–Single Present Value
UPV*–Uniform Present Value Modified (*)

The calculated life cycle cost of $1,665 is identical to that obtained with the cash flow method, but is arrived at with less effort.

Equivalent Uniform Annual Value

ASTM Definition of Equivalent Uniform Annual Value

A uniform annual amount equivalent to the project costs or benefits, taking into account the time value of money throughout the study period. (Syn. annual value)

As noted previously in the definition of Life Cycle Costs, these may be expressed as a lump sum present value with a base time reference, or in equivalent uniform annual values (EUAV) over the study period; decision-makers may prefer one or the other, or require both for the analysis of investment alternatives.

To express life cycle costs in annual values, it is necessary to first determine the lump sum present value of an alternative and multiply it by the Uniform Capital Recovery (UCR) factor given the interest rate and time period. This process is identical to calculating annual payments for a mortgage.

Example

The present value life cycle cost for one of the maintenance equipment alternatives analyzed amounted to $1,665; what is the equivalent uniform annual value "A"?

Given: P   (a principal sum or present value life cycle cost) = $1,665

     i = 10%

     n = 4 years

     UCR Factor = 0.32 (from standard factor tables for i)
     = 10%, n = 4 years

     A = equivalent uniform annual value – to be determined

Solution: A = P x UCR
     = $1,665 x 0.32
     = $530

The life cycle cost of the alternative may be expressed as a lump sum, $1,665, or an annual value of $530 based on a four-year time period.

Other Methods of Evaluating Economic Performance

While calculating life cycle costs is a helpful tool for the facilities manager evaluating alternatives, there are other approaches for evaluating economic performance of building investments. Among these other methods are *payback, net benefits,* and *internal rate of return;* these are described in detail in an ASTM compilation of *Building Economics Standards* which are endorsed by the American Association of Cost Engineers (AACE). These standards, as well as adjuncts available such as discount factor tables and a life cycle cost analysis computer program, are included in the References at the end of this section.

ASTM has also developed a Standard Classification of Building Elements and Related Site Work entitled UNIFORMAT II (Designation E-1557-93). One of the many applications for this classification for facilities managers is its use to structure operating and maintenance cost data by building elements or systems, thus facilitating life cycle costing studies. This classification can also be the basis of life cycle cost plans for new facilities that should be developed during the design period with the assistance of the facility manager.

References (Annual Book of ASTM Standards Vol. 04.07)

ASTM Standards

E 833   Terminology of Building Economics

E 917   Practice for Measuring Life Cycle Costs of Buildings and Building Systems

E 964   Practice for Measuring Benefit-to-Cost and Savings-to-Investment Ratios for Buildings and Building Systems

E 1057   Practice for Measuring Internal Rates of Return and Adjusted Internal Rates of Return for Investments in Buildings and Building Systems

E 1074   Practice for Measuring Net Benefits for Investments in Buildings and Building Systems

E 1121   Practice for Measuring Payback for Investments in Buildings and Building Systems

E 1185   Guide for Selecting Economic Methods for Evaluating Investments in Buildings and Building Systems

E 1369   Guide for Selecting Techniques for Treating Uncertainty and Risk in the Economic Evaluation of Buildings and Building Systems

E 1557   Standard Classification for Building Elements and Related Site Work–UNIFORMAT II

ASTM Adjuncts

Discount Factor Tables, Adjunct to Practices E 917, E 964, E 1057, and E 1074.

Computer Program and User's Guide to Building Maintenance, Repair, and Replacement Database for Life Cycle Cost Analysis, Adjunct to Practices E 917, E 964, E 1057, and E 1121.

Note: The Standards noted above, other than E 1557, are available in one publication entitled *ASTM Standards on Building Economics, Second Edition.*

Abbreviations

Abbr.	Meaning
A	Area Square Feet; Ampere
AAFES	Army and Air Force Exchange Service
ABS	Acrylonitrile Butadiene Stryrene; Asbestos Bonded Steel
A.C., AC	Alternating Current; Air-Conditioning; Asbestos Cement; Plywood Grade A & C
ACI	American Concrete Institute
ACR	Air Conditioning Refrigeration
ADA	Americans with Disabilities Act
AD	Plywood, Grade A & D
Addit.	Additional
Adj.	Adjustable
af	Audio-frequency
AFUE	Annual Fuel Utilization Efficiency
AGA	American Gas Association
Agg.	Aggregate
A.H., Ah	Ampere Hours
A hr.	Ampere-hour
A.H.U., AHU	Air Handling Unit
A.I.A.	American Institute of Architects
AIC	Ampere Interrupting Capacity
Allow.	Allowance
alt., alt	Alternate
Alum.	Aluminum
a.m.	Ante Meridiem
Amp.	Ampere
Anod.	Anodized
ANSI	American National Standards Institute
APA	American Plywood Association
Approx.	Approximate
Apt.	Apartment
Asb.	Asbestos
A.S.B.C.	American Standard Building Code
Asbe.	Asbestos Worker
ASCE.	American Society of Civil Engineers
A.S.H.R.A.E.	American Society of Heating, Refrig. & AC Engineers
ASME	American Society of Mechanical Engineers
ASTM	American Society for Testing and Materials
Attchmt.	Attachment
Avg., Ave.	Average
AWG	American Wire Gauge
AWWA	American Water Works Assoc.
Bbl.	Barrel
B&B, BB	Grade B and Better; Balled & Burlapped
B&S	Bell and Spigot
B.&W.	Black and White
b.c.c.	Body-centered Cubic
B.C.Y.	Bank Cubic Yards
BE	Bevel End
B.F.	Board Feet
Bg. cem.	Bag of Cement
BHP	Boiler Horsepower; Brake Horsepower
B.I.	Black Iron
bidir.	bidirectional
Bit., Bitum.	Bituminous
Bit., Conc.	Bituminous Concrete
Bk.	Backed
Bkrs.	Breakers
Bldg., bldg	Building
Blk.	Block
Bm.	Beam
Boil.	Boilermaker
bpm	Blows per Minute
BR	Bedroom
Brg.	Bearing
Brhe.	Bricklayer Helper
Bric.	Bricklayer
Brk., brk	Brick
brkt	Bracket
Brng.	Bearing
Brs.	Brass
Brz.	Bronze
Bsn.	Basin
Btr.	Better
Btu	British Thermal Unit
BTUH	BTU per Hour
Bu.	bushels
BUR	Built-up Roofing
BX	Interlocked Armored Cable
°C	degree centegrade
c	Conductivity, Copper Sweat
C	Hundred; Centigrade
C/C	Center to Center, Cedar on Cedar
C-C	Center to Center
Cab	Cabinet
Cair.	Air Tool Laborer
Cal.	caliper
Calc	Calculated
Cap.	Capacity
Carp.	Carpenter
C.B.	Circuit Breaker
C.C.A.	Chromate Copper Arsenate
C.C.F.	Hundred Cubic Feet
cd	Candela
cd/sf	Candela per Square Foot
CD	Grade of Plywood Face & Back
CDX	Plywood, Grade C & D, exterior glue
Cefi.	Cement Finisher
Cem.	Cement
CF	Hundred Feet
C.F.	Cubic Feet
CFM	Cubic Feet per Minute
CFRP	Carbon Fiber Reinforced Plastic
c.g.	Center of Gravity
CHW	Chilled Water; Commercial Hot Water
C.I., CI	Cast Iron
C.I.P., CIP	Cast in Place
Circ.	Circuit
C.L.	Carload Lot
CL	Chain Link
Clab.	Common Laborer
Clam	Common maintenance laborer
C.L.F.	Hundred Linear Feet
CLF	Current Limiting Fuse
CLP	Cross Linked Polyethylene
cm	Centimeter
CMP	Corr. Metal Pipe
CMU	Concrete Masonry Unit
CN	Change Notice
Col.	Column
CO2	Carbon Dioxide
Comb.	Combination
comm.	Commercial, Communication
Compr.	Compressor
Conc.	Concrete
Cont., cont	Continuous; Continued, Container
Corr.	Corrugated
Cos	Cosine
Cot	Cotangent
Cov.	Cover
C/P	Cedar on Paneling
CPA	Control Point Adjustment
Cplg.	Coupling
CPM	Critical Path Method
CPVC	Chlorinated Polyvinyl Chloride
C.Pr.	Hundred Pair
CRC	Cold Rolled Channel
Creos.	Creosote
Crpt.	Carpet & Linoleum Layer
CRT	Cathode-ray Tube
CS	Carbon Steel, Constant Shear Bar Joist
Csc	Cosecant
C.S.F.	Hundred Square Feet
CSI	Construction Specifications Institute
CT	Current Transformer
CTS	Copper Tube Size
Cu	Copper, Cubic
Cu. Ft.	Cubic Foot
cw	Continuous Wave
C.W.	Cool White; Cold Water
Cwt.	100 Pounds
C.W.X.	Cool White Deluxe
C.Y.	Cubic Yard (27 cubic feet)
C.Y./Hr.	Cubic Yard per Hour
Cyl.	Cylinder
d	Penny (nail size)
D	Deep; Depth; Discharge
Dis., Disch.	Discharge
Db	Decibel
Dbl.	Double
DC	Direct Current
DDC	Direct Digital Control
Demob.	Demobilization
d.f.t.	Dry Film Thickness
d.f.u.	Drainage Fixture Units
D.H.	Double Hung
DHW	Domestic Hot Water
DI	Ductile Iron
Diag.	Diagonal
Diam., Dia	Diameter
Distrib.	Distribution
Div.	Division
Dk.	Deck
D.L.	Dead Load; Diesel
DLH	Deep Long Span Bar Joist
dlx	Deluxe
Do.	Ditto
DOP	Dioctyl Phthalate Penetration Test (Air Filters)
Dp., dp	Depth
D.P.S.T.	Double Pole, Single Throw
Dr.	Drive
DR	Dimension Ratio
Drink.	Drinking
D.S.	Double Strength
D.S.A.	Double Strength A Grade
D.S.B.	Double Strength B Grade
Dty.	Duty
DWV	Drain Waste Vent
DX	Deluxe White, Direct Expansion
dyn	Dyne
e	Eccentricity
E	Equipment Only; East; emissivity
Ea.	Each
EB	Encased Burial
Econ.	Economy
E.C.Y	Embankment Cubic Yards
EDP	Electronic Data Processing
EIFS	Exterior Insulation Finish System
E.D.R.	Equiv. Direct Radiation
Eq.	Equation
EL	elevation
Elec.	Electrician; Electrical
Elev.	Elevator; Elevating
EMT	Electrical Metallic Conduit; Thin Wall Conduit
Eng.	Engine, Engineered
EPDM	Ethylene Propylene Diene Monomer
EPS	Expanded Polystyrene
Eqhv.	Equip. Oper., Heavy
Eqlt.	Equip. Oper., Light
Eqmd.	Equip. Oper., Medium
Eqmm.	Equip. Oper., Master Mechanic
Eqol.	Equip. Oper., Oilers
Equip.	Equipment
ERW	Electric Resistance Welded

668

E.S.	Energy Saver
Est.	Estimated
esu	Electrostatic Units
E.W.	Each Way
EWT	Entering Water Temperature
Excav.	Excavation
excl	Excluding
Exp., exp	Expansion, Exposure
Ext., ext	Exterior; Extension
Extru.	Extrusion
f.	Fiber stress
F	Fahrenheit; Female; Fill
Fab., fab	Fabricated; fabric
FBGS	Fiberglass
F.C.	Footcandles
f.c.c.	Face-centered Cubic
f'c.	Compressive Stress in Concrete; Extreme Compressive Stress
F.E.	Front End
FEP	Fluorinated Ethylene Propylene (Teflon)
F.G.	Flat Grain
F.H.A.	Federal Housing Administration
Fig.	Figure
Fin.	Finished
FIPS	Female Iron Pipe Size
Fixt.	Fixture
FJP	Finger jointed and primed
Fl. Oz.	Fluid Ounces
Flr.	Floor
FM	Frequency Modulation; Factory Mutual
Fmg.	Framing
FM/UL	Factory Mutual/Underwriters Labs
Fdn.	Foundation
FNPT	Female National Pipe Thread
Fori.	Foreman, Inside
Foro.	Foreman, Outside
Fount.	Fountain
fpm	Feet per Minute
FPT	Female Pipe Thread
Fr	Frame
F.R.	Fire Rating
FRK	Foil Reinforced Kraft
FSK	Foil/scrim/kraft
FRP	Fiberglass Reinforced Plastic
FS	Forged Steel
FSC	Cast Body; Cast Switch Box
Ft., ft	Foot; Feet
Ftng.	Fitting
Ftg.	Footing
Ft lb.	Foot Pound
Furn.	Furniture
FVNR	Full Voltage Non-Reversing
FVR	Full Voltage Reversing
FXM	Female by Male
Fy.	Minimum Yield Stress of Steel
g	Gram
G	Gauss
Ga.	Gauge
Gal., gal.	Gallon
gpm, GPM	Gallon per Minute
Galv., galv	Galvanized
GC/MS	Gas Chromatograph/Mass Spectrometer
Gen.	General
GFI	Ground Fault Interrupter
GFRC	Glass Fiber Reinforced Concrete
Glaz.	Glazier
GPD	Gallons per Day
gpf	Gallon per flush
GPH	Gallons per Hour
GPM	Gallons per Minute
GR	Grade
Gran.	Granular
Grnd.	Ground
GVW	Gross Vehicle Weight
GWB	Gypsum wall board
H	High Henry
HC	High Capacity
H.D., HD	Heavy Duty; High Density
H.D.O.	High Density Overlaid
HDPE	High density polyethelene plastic
Hdr.	Header
Hdwe.	Hardware
H.I.D., HID	High Intensity Discharge
Help.	Helper Average
HEPA	High Efficiency Particulate Air Filter
Hg	Mercury
HIC	High Interrupting Capacity
HM	Hollow Metal
HMWPE	high molecular weight polyethylene
HO	High Output
Horiz.	Horizontal
H.P., HP	Horsepower; High Pressure
H.P.F.	High Power Factor
Hr.	Hour
Hrs./Day	Hours per Day
HSC	High Short Circuit
Ht.	Height
Htg.	Heating
Htrs.	Heaters
HVAC	Heating, Ventilation & Air-Conditioning
Hvy.	Heavy
HW	Hot Water
Hyd.; Hydr.	Hydraulic
Hz	Hertz (cycles)
I.	Moment of Inertia
IBC	International Building Code
I.C.	Interrupting Capacity
ID	Inside Diameter
I.D.	Inside Dimension; Identification
I.F.	Inside Frosted
I.M.C.	Intermediate Metal Conduit
In.	Inch
Incan.	Incandescent
Incl.	Included; Including
Int.	Interior
Inst.	Installation
Insul., insul	Insulation/Insulated
I.P.	Iron Pipe
I.P.S., IPS	Iron Pipe Size
IPT	Iron Pipe Threaded
I.W.	Indirect Waste
J	Joule
J.I.C.	Joint Industrial Council
K	Thousand; Thousand Pounds; Heavy Wall Copper Tubing, Kelvin
K.A.H.	Thousand Amp. Hours
kcmil	Thousand Circular Mils
KD	Knock Down
K.D.A.T.	Kiln Dried After Treatment
kg	Kilogram
kG	Kilogauss
kgf	Kilogram Force
kHz	Kilohertz
Kip	1000 Pounds
KJ	Kiljoule
K.L.	Effective Length Factor
K.L.F.	Kips per Linear Foot
Km	Kilometer
KO	Knock Out
K.S.F.	Kips per Square Foot
K.S.I.	Kips per Square Inch
kV	Kilovolt
kVA	Kilovolt Ampere
kVAR	Kilovar (Reactance)
KW	Kilowatt
KWh	Kilowatt-hour
L	Labor Only; Length; Long; Medium Wall Copper Tubing
Lab.	Labor
lat	Latitude
Lath.	Lather
Lav.	Lavatory
lb.; #	Pound
L.B., LB	Load Bearing; L Conduit Body
L. & E.	Labor & Equipment
lb./hr.	Pounds per Hour
lb./L.F.	Pounds per Linear Foot
lbf/sq.in.	Pound-force per Square Inch
L.C.L.	Less than Carload Lot
L.C.Y.	Loose Cubic Yard
Ld.	Load
LE	Lead Equivalent
LED	Light Emitting Diode
L.F.	Linear Foot
L.F. Nose	Linear Foot of Stair Nosing
L.F. Rsr	Linear Foot of Stair Riser
Lg.	Long; Length; Large
L & H	Light and Heat
LH	Long Span Bar Joist
L.H.	Labor Hours
L.L., LL	Live Load
L.L.D.	Lamp Lumen Depreciation
lm	Lumen
lm/sf	Lumen per Square Foot
lm/W	Lumen per Watt
LOA	Length Over All
log	Logarithm
L-O-L	Lateralolet
long.	longitude
L.P., LP	Liquefied Petroleum; Low Pressure
L.P.F.	Low Power Factor
LR	Long Radius
L.S.	Lump Sum
Lt.	Light
Lt. Ga.	Light Gauge
L.T.L.	Less than Truckload Lot
Lt. Wt.	Lightweight
L.V.	Low Voltage
M	Thousand; Material; Male; Light Wall Copper Tubing
M²CA	Meters Squared Contact Area
m/hr.; M.H.	Man-hour
mA	Milliampere
Mach.	Machine
Mag. Str.	Magnetic Starter
Maint.	Maintenance
Marb.	Marble Setter
Mat; Mat'l.	Material
Max.	Maximum
MBF	Thousand Board Feet
MBH	Thousand BTU's per hr.
MC	Metal Clad Cable
MCC	Motor Control Center
M.C.F.	Thousand Cubic Feet
MCFM	Thousand Cubic Feet per Minute
M.C.M.	Thousand Circular Mils
MCP	Motor Circuit Protector
MD	Medium Duty
MDF	Medium-density fibreboard
M.D.O.	Medium Density Overlaid
Med.	Medium
MF	Thousand Feet
M.F.B.M.	Thousand Feet Board Measure
Mfg.	Manufacturing
Mfrs.	Manufacturers
mg	Milligram
MGD	Million Gallons per Day
MGPH	Thousand Gallons per Hour
MH, M.H.	Manhole; Metal Halide; Man-Hour
MHz	Megahertz
Mi.	Mile
MI	Malleable Iron; Mineral Insulated
MIPS	Male Iron Pipe Size
mj	Mechanical Joint
m	Meter
mm	Millimeter
Mill.	Millwright
Min., min.	Minimum, minute

669

Abbreviation	Meaning	Abbreviation	Meaning	Abbreviation	Meaning
Misc.	Miscellaneous	PDCA	Painting and Decorating Contractors of America	SC	Screw Cover
ml	Milliliter, Mainline	P.E., PE	Professional Engineer; Porcelain Enamel; Polyethylene; Plain End	SCFM	Standard Cubic Feet per Minute
M.L.F.	Thousand Linear Feet			Scaf.	Scaffold
Mo.	Month			Sch., Sched.	Schedule
Mobil.	Mobilization	P.E.C.I.	Porcelain Enamel on Cast Iron	S.C.R.	Modular Brick
Mog.	Mogul Base	Perf.	Perforated	S.D.	Sound Deadening
MPH	Miles per Hour	PEX	Cross linked polyethylene	SDR	Standard Dimension Ratio
MPT	Male Pipe Thread	Ph.	Phase	S.E.	Surfaced Edge
MRGWB	Moisture Resistant Gypsum Wallboard	P.I.	Pressure Injected	Sel.	Select
		Pile.	Pile Driver	SER, SEU	Service Entrance Cable
MRT	Mile Round Trip	Pkg.	Package	S.F.	Square Foot
ms	Millisecond	Pl.	Plate	S.F.C.A.	Square Foot Contact Area
M.S.F.	Thousand Square Feet	Plah.	Plasterer Helper	S.F. Flr.	Square Foot of Floor
Mstz.	Mosaic & Terrazzo Worker	Plas.	Plasterer	S.F.G.	Square Foot of Ground
M.S.Y.	Thousand Square Yards	plf	Pounds Per Linear Foot	S.F. Hor.	Square Foot Horizontal
Mtd., mtd., mtd	Mounted	Pluh.	Plumbers Helper	SFR	Square Feet of Radiation
Mthe.	Mosaic & Terrazzo Helper	Plum.	Plumber	S.F. Shlf.	Square Foot of Shelf
Mtng.	Mounting	Ply.	Plywood	S4S	Surface 4 Sides
Mult.	Multi; Multiply	p.m.	Post Meridiem	Shee.	Sheet Metal Worker
M.V.A.	Million Volt Amperes	Pntd.	Painted	Sin.	Sine
M.V.A.R.	Million Volt Amperes Reactance	Pord.	Painter, Ordinary	Skwk.	Skilled Worker
MV	Megavolt	pp	Pages	SL	Saran Lined
MW	Megawatt	PP, PPL	Polypropylene	S.L.	Slimline
MXM	Male by Male	P.P.M.	Parts per Million	Sldr.	Solder
MYD	Thousand Yards	Pr.	Pair	SLH	Super Long Span Bar Joist
N	Natural; North	P.E.S.B.	Pre-engineered Steel Building	S.N.	Solid Neutral
nA	Nanoampere	Prefab.	Prefabricated	SO	Stranded with oil resistant inside insulation
NA	Not Available; Not Applicable	Prefin.	Prefinished		
N.B.C.	National Building Code	Prop.	Propelled	S-O-L	Socketolet
NC	Normally Closed	PSF, psf	Pounds per Square Foot	sp	Standpipe
NEMA	National Electrical Manufacturers Assoc.	PSI, psi	Pounds per Square Inch	S.P.	Static Pressure; Single Pole; Self-Propelled
NEHB	Bolted Circuit Breaker to 600V.	PSIG	Pounds per Square Inch Gauge		
NFPA	National Fire Protection Association	PSP	Plastic Sewer Pipe	Spri.	Sprinkler Installer
NLB	Non-Load-Bearing	Pspr.	Painter, Spray	spwg	Static Pressure Water Gauge
NM	Non-Metallic Cable	Psst.	Painter, Structural Steel	S.P.D.T.	Single Pole, Double Throw
nm	Nanometer	P.T.	Potential Transformer	SPF	Spruce Pine Fir; Sprayed Polyurethane Foam
No.	Number	P. & T.	Pressure & Temperature		
NO	Normally Open	Ptd.	Painted	S.P.S.T.	Single Pole, Single Throw
N.O.C.	Not Otherwise Classified	Ptns.	Partitions	SPT	Standard Pipe Thread
Nose.	Nosing	Pu	Ultimate Load	Sq.	Square; 100 Square Feet
NPT	National Pipe Thread	PVC	Polyvinyl Chloride	Sq. Hd.	Square Head
NQOD	Combination Plug-on/Bolt on Circuit Breaker to 240V.	Pvmt.	Pavement	Sq. In.	Square Inch
		PRV	Pressure Relief Valve	S.S.	Single Strength; Stainless Steel
N.R.C., NRC	Noise Reduction Coefficient/ Nuclear Regulator Commission	Pwr.	Power	S.S.B.	Single Strength B Grade
		Q	Quantity Heat Flow	sst, ss	Stainless Steel
N.R.S.	Non Rising Stem	Qt.	Quart	Sswk.	Structural Steel Worker
ns	Nanosecond	Quan., Qty.	Quantity	Sswl.	Structural Steel Welder
nW	Nanowatt	Q.C.	Quick Coupling	St.; Stl.	Steel
OB	Opposing Blade	r	Radius of Gyration	STC	Sound Transmission Coefficient
OC	On Center	R	Resistance	Std.	Standard
OD	Outside Diameter	R.C.P.	Reinforced Concrete Pipe	Stg.	Staging
O.D.	Outside Dimension	Rect.	Rectangle	STK	Select Tight Knot
ODS	Overhead Distribution System	recpt.	receptacle	STP	Standard Temperature & Pressure
O.G.	Ogee	Reg.	Regular	Stpi.	Steamfitter, Pipefitter
O.H.	Overhead	Reinf.	Reinforced	Str.	Strength; Starter; Straight
O&P	Overhead and Profit	Req'd.	Required	Strd.	Stranded
Oper.	Operator	Res.	Resistant	Struct.	Structural
Opng.	Opening	Resi.	Residential	Sty.	Story
Orna.	Ornamental	RF	Radio Frequency	Subj.	Subject
OSB	Oriented Strand Board	RFID	Radio-frequency identification	Subs.	Subcontractors
OS&Y	Outside Screw and Yoke	Rgh.	Rough	Surf.	Surface
OSHA	Occupational Safety and Health Act	RGS	Rigid Galvanized Steel	Sw.	Switch
		RHW	Rubber, Heat & Water Resistant; Residential Hot Water	Swbd.	Switchboard
Ovhd.	Overhead			S.Y.	Square Yard
OWG	Oil, Water or Gas	rms	Root Mean Square	Syn.	Synthetic
Oz.	Ounce	Rnd.	Round	S.Y.P.	Southern Yellow Pine
P.	Pole; Applied Load; Projection	Rodm.	Rodman	Sys.	System
p.	Page	Rofc.	Roofer, Composition	t.	Thickness
Pape.	Paperhanger	Rofp.	Roofer, Precast	T	Temperature; Ton
P.A.P.R.	Powered Air Purifying Respirator	Rohe.	Roofer Helpers (Composition)	Tan	Tangent
PAR	Parabolic Reflector	Rots.	Roofer, Tile & Slate	T.C.	Terra Cotta
P.B., PB	Push Button	R.O.W.	Right of Way	T & C	Threaded and Coupled
Pc., Pcs.	Piece, Pieces	RPM	Revolutions per Minute	T.D.	Temperature Difference
P.C.	Portland Cement; Power Connector	R.S.	Rapid Start	Tdd	Telecommunications Device for the Deaf
P.C.F.	Pounds per Cubic Foot	Rsr	Riser		
PCM	Phase Contrast Microscopy	RT	Round Trip	T.E.M.	Transmission Electron Microscope
		S.	Suction; Single Entrance; South	temp	Temperature, Tempered, Temporary
		SBS	Styrene Butadiere Styrene	TFFN	Nylon Jacketed Wire

Abbreviations

TFE	Tetrafluoroethylene (Teflon)	U.L., UL	Underwriters Laboratory	w/	With
T. & G.	Tongue & Groove;	Uld.	unloading	W.C., WC	Water Column; Water Closet
	Tar & Gravel	Unfin.	Unfinished	W.F.	Wide Flange
Th., Thk.	Thick	UPS	Uninterruptible Power Supply	W.G.	Water Gauge
Thn.	Thin	URD	Underground Residential	Wldg.	Welding
Thrded	Threaded		Distribution	W. Mile	Wire Mile
Tilf.	Tile Layer, Floor	US	United States	W-O-L	Weldolet
Tilh.	Tile Layer, Helper	USGBC	U.S. Green Building Council	W.R.	Water Resistant
THHN	Nylon Jacketed Wire	USP	United States Primed	Wrck.	Wrecker
THW.	Insulated Strand Wire	UTMCD	Uniform Traffic Manual For Control	W.S.P.	Water, Steam, Petroleum
THWN	Nylon Jacketed Wire		Devices	WT., Wt.	Weight
T.L., TL	Truckload	UTP	Unshielded Twisted Pair	WWF	Welded Wire Fabric
T.M.	Track Mounted	V	Volt	XFER	Transfer
Tot.	Total	VA	Volt Amperes	XFMR	Transformer
T-O-L	Threadolet	VAT	Vinyl Asbestos Tile	XHD	Extra Heavy Duty
tmpd	Tempered	V.C.T.	Vinyl Composition Tile	XHHW,	Cross-Linked Polyethylene Wire
TPO	Thermoplastic Polyolefin	VAV	Variable Air Volume	XLPE	Insulation
T.S.	Trigger Start	VC	Veneer Core	XLP	Cross-linked Polyethylene
Tr.	Trade	VDC	Volts Direct Current	Xport	Transport
Transf.	Transformer	Vent.	Ventilation	Y	Wye
Trhv.	Truck Driver, Heavy	Vert.	Vertical	yd	Yard
Trlr	Trailer	V.F.	Vinyl Faced	yr	Year
Trlt.	Truck Driver, Light	V.G.	Vertical Grain	Δ	Delta
TTY	Teletypewriter	VHF	Very High Frequency	%	Percent
TV	Television	VHO	Very High Output	~	Approximately
T.W.	Thermoplastic Water Resistant	Vib.	Vibrating	Ø	Phase; diameter
	Wire	VLF	Vertical Linear Foot	@	At
UCI	Uniform Construction Index	VOC	Volitile Organic Compound	#	Pound; Number
UF	Underground Feeder	Vol.	Volume	<	Less Than
UGND	Underground Feeder	VRP	Vinyl Reinforced Polyester	>	Greater Than
UHF	Ultra High Frequency	W	Wire; Watt; Wide; West	Z	zone
U.I.	United Inch				

Other RSMeans Products & Services

RSMeans—
a tradition of excellence in construction cost information and services since 1942

For more information visit the RSMeans website at **www.rsmeans.com**

Unit prices according to the latest MasterFormat

Book Selection Guide The following table provides definitive information on the content of each cost data publication. The number of lines of data provided in each unit price or assemblies division, as well as the number of crews, is listed for each book. The presence of other elements such as reference tables, square foot models, equipment rental costs, historical cost indexes, and city cost indexes, is also indicated. You can use the table to help select the RSMeans book that has the quantity and type of information you most need in your work.

Unit Cost Divisions	Building Construction	Mechanical	Electrical	Commercial Renovation	Square Foot	Site Work Landsc.	Green Building	Interior	Concrete Masonry	Open Shop	Heavy Construction	Light Commercial	Facilities Construction	Plumbing	Residential
1	573	393	410	506		509	191	312	456	572	501	246	1042	403	178
2	777	279	85	733		993	178	399	213	776	734	481	1221	286	257
3	1690	340	230	1085		1471	987	355	2032	1689	1685	481	1788	316	388
4	957	21	0	922		720	179	612	1155	925	612	529	1172	0	441
5	1893	158	155	1090		841	1799	1095	720	1893	1037	979	1909	204	721
6	2452	18	18	2110		110	589	1528	281	2448	123	2140	2124	22	2660
7	1584	215	128	1623		581	757	532	524	1581	26	1317	1685	227	1037
8	2082	81	45	2609		265	1123	1753	105	2077	0	2193	2879	0	1526
9	1971	72	26	1767		313	449	2056	390	1911	15	1659	2220	54	1435
10	1026	17	10	642		215	27	842	158	1026	29	511	1118	235	224
11	1087	208	165	546		124	54	932	28	1071	0	230	1107	169	110
12	544	0	2	319		217	137	1654	14	532	0	352	1704	23	310
13	719	140	137	242		343	111	248	66	695	244	83	746	94	79
14	273	36	0	221		0	0	257	0	273	0	12	293	16	6
21	90	0	16	37		0	0	250	0	87	0	68	426	431	220
22	1157	7544	150	1190		1568	1067	838	20	1121	1677	857	7440	9347	712
23	1177	6971	580	928		158	902	776	38	1101	111	878	5187	1918	469
26	1354	454	10081	1021		793	611	1136	55	1349	562	1298	10030	399	617
27	72	0	297	34		13	0	71	0	72	39	52	279	0	22
28	96	58	136	71		0	21	78	0	99	0	40	151	44	25
31	1498	735	610	803		3217	289	7	1208	1443	3256	600	1560	660	611
32	822	53	8	886		4426	356	405	291	793	1841	420	1701	165	468
33	530	1079	538	252		2178	41	0	237	522	2159	128	1699	1286	154
34	107	0	47	4		190	0	0	31	62	193	0	128	0	0
35	18	0	0	0		327	0	0	0	18	442	0	84	0	0
41	60	0	0	32		7	0	22	0	60	30	0	67	14	0
44	75	79	0	0		0	0	0	0	0	0	0	75	75	0
46	23	16	0	0		274	261	0	0	23	264	0	33	33	0
48	12	0	25	0		0	25	0	0	12	17	12	12	0	12
Totals	24719	18967	13899	19673		19853	10154	16158	8022	24231	15597	15566	49880	16421	12682

Assem Div	Building Construction	Mechanical	Electrical	Commercial Renovation	Square Foot	Site Work Landscape	Assemblies	Green Building	Interior	Concrete Masonry	Heavy Construction	Light Commercial	Facilities Construction	Plumbing	Asm Div	Residentia
A		15	0	188	150	577	598	0	0	536	571	154	24	0	1	376
B		0	0	848	2498	0	5658	56	329	1975	368	2089	174	0	2	211
C		0	0	647	926	0	1304	0	1629	146	0	816	250	0	3	588
D		1067	941	712	1859	72	2538	330	825	0	0	1345	1105	1088	4	851
E		0	0	86	260	0	300	0	5	0	0	257	5	0	5	392
F		0	0	0	114	0	114	0	0	0	0	114	3	0	6	357
G		527	447	318	249	3365	729	0	0	535	1350	205	293	677	7	307
															8	760
															9	80
															10	0
															11	0
															12	0
Totals		1609	1388	2799	6056	4014	11241	386	2788	3192	2289	4980	1854	1765		3922

Reference Section	Building Construction Costs	Mechanical	Electrical	Commercial Renovation	Square Foot	Site Work Landscape	Assem.	Green Building	Interior	Concrete Masonry	Open Shop	Heavy Construction	Light Commercial	Facilities Construction	Plumbing	Resi
Reference Tables	yes	yes	yes	yes	no	yes	yes	yes	yes	yes	yes	yes	yes	yes	yes	yes
Models					111			25					50			28
Crews	571	571	571	550		571		571	571	571	548	571	548	550	571	548
Equipment Rental Costs	yes	yes	yes	yes		yes	yes	yes	yes	yes	yes	yes	yes	yes	yes	yes
Historical Cost Indexes	yes	yes	yes	yes	yes	yes	yes	yes	yes	yes	yes	yes	yes	yes	yes	no
City Cost Indexes	yes	yes	yes	yes	yes	yes	yes	yes	yes	yes	yes	yes	yes	yes	yes	yes

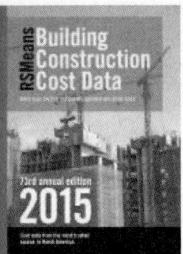

RSMeans Building Construction Cost Data 2015

Offers you unchallenged unit price reliability in an easy-to-use format. Whether used for verifying complete, finished estimates or for periodic checks, it supplies more cost facts better and faster than any comparable source. More than 24,700 unit prices have been updated for 2015. The City Cost Indexes and Location Factors cover more than 930 areas, for indexing to any project location in North America. Order and get *RSMeans Quarterly Update Service* FREE.

$206.95 | Available Sept. 2014 | Catalog no. 60015

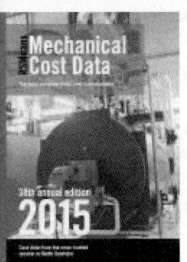

RSMeans Mechanical Cost Data 2015

Total unit and systems price guidance for mechanical construction . . . materials, parts, fittings, and complete labor cost information. Includes prices for piping, heating, air conditioning, ventilation, and all related construction.

Plus new 2015 unit costs for:

- Thousands of installed HVAC/controls, sub-assemblies and assemblies
- "On-site" Location Factors for more than 930 cities and towns in the U.S. and Canada
- Crews, labor, and equipment

$203.95 | Available Oct. 2014 | Catalog no. 60025

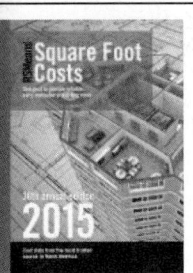

RSMeans Square Foot Costs 2015
Accurate and Easy-to-Use

- **Updated price information** based on nationwide figures from suppliers, estimators, labor experts, and contractors.
- Green building models
- Realistic graphics, offering true-to-life illustrations of building projects
- Extensive information on using square foot cost data, including sample estimates and alternate pricing methods

$218.95 | Available Oct. 2014 | Catalog no. 60055

RSMeans Electrical Cost Data 2015

Pricing information for every part of electrical cost planning. More than 13,800 unit and systems costs with design tables; clear specifications and drawings; engineering guides; illustrated estimating procedures; complete labor-hour and materials costs for better scheduling and procurement; and the latest electrical products and construction methods.

- A variety of special electrical systems, including cathodic protection
- Costs for maintenance, demolition, HVAC/mechanical, specialties, equipment, and more

$208.95 | Available Oct. 2014 | Catalog no. 60035

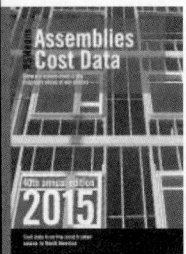

RSMeans Assemblies Cost Data 2015

RSMeans Assemblies Cost Data takes the guesswork out of preliminary or conceptual estimates. Now you don't have to try to calculate the assembled cost by working up individual component costs. We've done all the work for you.

Presents detailed illustrations, descriptions, specifications, and costs for every conceivable building assembly—over 350 types in all—arranged in the easy-to-use UNIFORMAT II system. Each illustrated "assembled" cost includes a complete grouping of materials and associated installation costs, including the installing contractor's overhead and profit.

$334.95 | Available Sept. 2014 | Catalog no. 60065

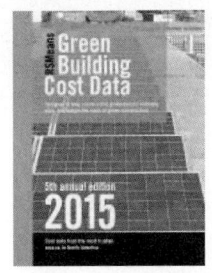

RSMeans Green Building Cost Data 2015

Estimate, plan, and budget the costs of green building for both new commercial construction and renovation work with this fifth edition of *RSMeans Green Building Cost Data*. More than 10,000 unit costs for a wide array of green building products plus assemblies costs. Easily identified cross references to LEED and Green Globes building rating systems criteria.

$170.95 | Available Nov. 2014 | Catalog no. 60555

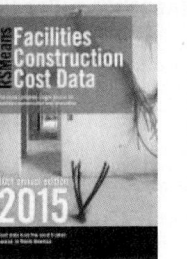

RSMeans Facilities Construction Cost Data 2015

For the maintenance and construction of commercial, industrial, municipal, and institutional properties. Costs are shown for new and remodeling construction and are broken down into materials, labor, equipment, and overhead and profit. Special emphasis is given to sections on mechanical, electrical, furnishings, site work, building maintenance, finish work, and demolition.

More than 49,800 unit costs, plus assemblies costs and a comprehensive Reference Section are included.

$524.95 | Available Nov. 2014 | Catalog no. 60205

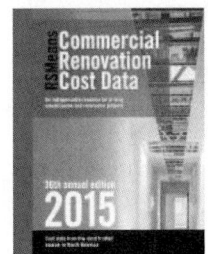

RSMeans Commercial Renovation Cost Data 2015
Commercial/Multi-family Residential

Use this valuable tool to estimate commercial and multi-family residential renovation and remodeling.

Includes: Updated costs for hundreds of unique methods, materials, and conditions that only come up in repair and remodeling, PLUS:

- Unit costs for more than 19,600 construction components
- Installed costs for more than 2,700 assemblies
- More than 930 "on-site" localization factors for the U.S. and Canada

$166.95 | Available Oct. 2014 | Catalog no. 60045

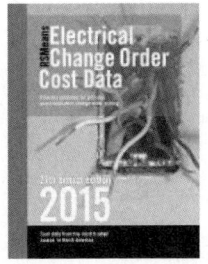

RSMeans Electrical Change Order Cost Data 2015

RSMeans Electrical Change Order Cost Data provides you with electrical unit prices exclusively for pricing change orders—based on the recent, direct experience of contractors and suppliers. Analyze and check your own change order estimates against the experience others have had doing the same work. It also covers productivity analysis, and change order cost justifications. With useful information for calculating the effects of change orders and dealing with their administration.

$199.95 | Available Dec. 2014 | Catalog no. 60235

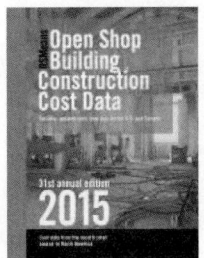

RSMeans Open Shop Building Construction Cost Data 2015

The latest costs for accurate budgeting and estimating of new commercial and residential construction . . . renovation work . . . change orders . . . cost engineering.

RSMeans Open Shop "BCCD" will assist you to:

- Develop benchmark prices for change orders.
- Plug gaps in preliminary estimates and budgets.
- Estimate complex projects.
- Substantiate invoices on contracts.
- Price ADA-related renovations.

$175.95 | Available Dec. 2014 | Catalog no. 60155

Annual Cost Guides

Unit prices according to the latest MasterFormat!

For more information visit the RSMeans website at **www.rsmeans.com**

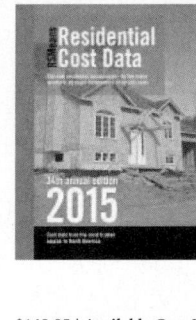

RSMeans Residential Cost Data 2015

Contains square foot costs for 28 basic home models with the look of today, plus hundreds of custom additions and modifications you can quote right off the page. Includes more than 3,900 costs for 89 residential systems. Complete with blank estimating forms, sample estimates, and step-by-step instructions.

Contains line items for cultured stone and brick, PVC trim, lumber, and TPO roofing.

$148.95 | Available Oct. 2014 | Catalog no. 60175

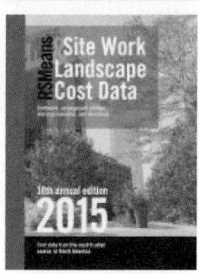

RSMeans Site Work & Landscape Cost Data 2015

Includes unit and assemblies costs for earthwork, sewerage, piped utilities, site improvements, drainage, paving, trees and shrubs, street openings/repairs, underground tanks, and more. Contains more than 60 types of assemblies costs for accurate conceptual estimates.

Includes:

- Estimating for infrastructure improvements
- Environmentally-oriented construction
- ADA-mandated handicapped access
- Hazardous waste line items

$199.95 | Available Dec. 2014 | Catalog no. 60285

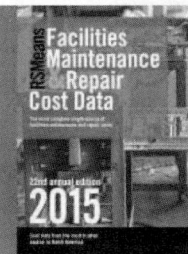

RSMeans Facilities Maintenance & Repair Cost Data 2015

RSMeans Facilities Maintenance & Repair Cost Data gives you a complete system to manage and plan your facility repair and maintenance costs and budget efficiently. Guidelines for auditing a facility and developing an annual maintenance plan. Budgeting is included, along with reference tables on cost and management, and information on frequency and productivity of maintenance operations.

The only nationally recognized source of maintenance and repair costs. Developed in cooperation with the Civil Engineering Research Laboratory (CERL) of the Army Corps of Engineers.

$461.95 | Available Nov. 2014 | Catalog no. 60305

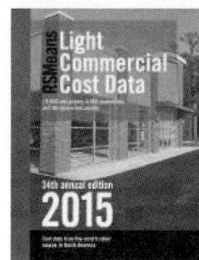

RSMeans Light Commercial Cost Data 2015

Specifically addresses the light commercial market, which is a specialized niche in the construction industry. Aids you, the owner/designer/contractor, in preparing all types of estimates—from budgets to detailed bids. Includes new advances in methods and materials.

Assemblies Section allows you to evaluate alternatives in the early stages of design/planning.

More than 15,500 unit costs ensure that you have the prices you need, when you need them.

$153.95 | Available Nov. 2014 | Catalog no. 60185

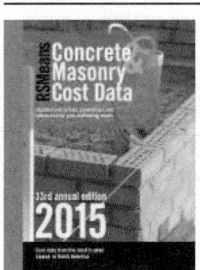

RSMeans Concrete & Masonry Cost Data 2015

Provides you with cost facts for virtually all concrete/masonry estimating needs, from complicated formwork to various sizes and face finishes of brick and block—all in great detail. The comprehensive Unit Price Section contains more than 8,000 selected entries. Also contains an Assemblies (Cost) Section, and a detailed Reference Section that supplements the cost data.

$188.95 | Available Dec. 2014 | Catalog no. 60115

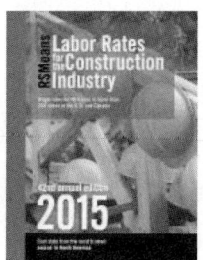

RSMeans Labor Rates for the Construction Industry 2015

Complete information for estimating labor costs, making comparisons, and negotiating wage rates by trade for more than 300 U.S. and Canadian cities. With 46 construction trades in each city, and historical wage rates included for comparison. Each city chart lists the county and is alphabetically arranged with handy visual flip tabs for quick reference.

$450.95 | Available Dec. 2014 | Catalog no. 60125

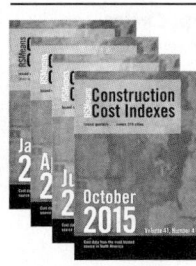

RSMeans Construction Cost Indexes 2015

What materials and labor costs will change unexpectedly this year? By how much?

- Breakdowns for 318 major cities
- National averages for 30 key cities
- Expanded five major city indexes
- Historical construction cost indexes

$359.95 per year (subscription) | Catalog no. 50145
$94.95 individual quarters | Catalog no. 60145 A,B,C,D

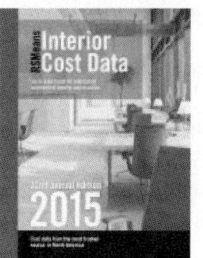

RSMeans Interior Cost Data 2015

Provides you with prices and guidance needed to make accurate interior work estimates. Contains costs on materials, equipment, hardware, custom installations, furnishings, and labor for new and remodel commercial and industrial interior construction, including updated information on office furnishings, and reference information.

$208.95 | Available Nov. 2014 | Catalog no. 60095

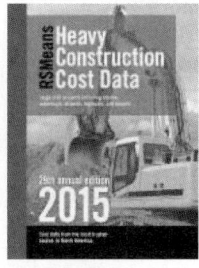

RSMeans Heavy Construction Cost Data 2015

A comprehensive guide to heavy construction costs. Includes costs for highly specialized projects such as tunnels, dams, highways, airports, and waterways. Information on labor rates, equipment, and materials costs is included. Features unit price costs, systems costs, and numerous reference tables for costs and design.

$204.95 | Available Dec. 2014 | Catalog no. 60165

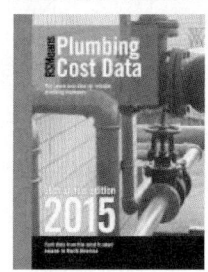

RSMeans Plumbing Cost Data 2015

Comprehensive unit prices and assemblies for plumbing, irrigation systems, commercial and residential fire protection, point-of-use water heaters, and the latest approved materials. This publication and its companion, *RSMeans Mechanical Cost Data*, provide full-range cost estimating coverage for all the mechanical trades.

Contains updated costs for potable water, radiant heat systems, and high efficiency fixtures.

$206.95 | Available Oct. 2014 | Catalog no. 60215

Annual Cost Guides — Unit prices according to the latest MasterFormat!

For more information visit the RSMeans website at **www.rsmeans.com**

RSMeans Online and RSMeans CostWorks CD Comparison Matrix

Both the RSMeans CostWorks CD package and the RSMeans Online estimating tool provide the same comprehensive, up-to-date, and localized cost information to improve planning, estimating, and budgeting with RSMeans—the industry's most-trusted source of construction cost data.

In addition to key differences noted in the comparison matrix below, both products let you do the following:

• Quickly locate costs in the searchable RSMeans database.

• Instantly adjust the data to your specific location for accurate costs anywhere in the U.S. and Canada.

• Create reliable, updated cost estimates in minutes.

• Manage, share, and export your estimates to MS Excel® with ease.

• Access a wide range of reference materials, including crew rates and national average labor rates.

Both the CD and Online options are available in a convenient one-year subscription!

Use the handy comparison matrix below to help you decide the best product for your cost estimating needs!

Why us?

For more than 70 years, RSMeans has provided the quality cost data construction professionals everywhere depend on to estimate with confidence, save time, improve decision-making, and increase profits.

Improve planning and decision-making

Back your estimates with complete, accurate, and up-to-date cost data for informed business decisions.
• Verify construction costs from third parties.
• Check validity of subcontractor proposals.
• Evaluate material and assembly alternatives.

Increase profits

Use RSMeans to estimate projects quickly and accurately, so you can gain an edge over your competition.
• Create accurate and competitive bids.
• Minimize the risk of cost overruns.
• Reduce variability.
• Gain control over costs.

The following matrix summarizes the key comparison points between RSMeans' two electronic cost data products.

	Question	CostWorks CD	RSMeans Online.com
1	How is the data presented?	• The data is organized by book title and sold at comparable book cost. • Several packaged bundles of five titles are available at a cost savings. • Available in a convenient one-year subscription	• The data is organized by book title and sold at comparable book cost. • Several packaged bundles of five titles are available at a cost savings. • RSMeans Online also offers a Complete Library title with all online data at a cost savings. • Available in a convenient one-year subscription
2	Where is the data stored?	With a typical install, the data is read off of your hard drive or off of the CD.	All data is stored on our secure server.
3	How are updates delivered?	Updates are available on a quarterly basis for download from the Costworks HomePort website.	Data is updated automatically on a quarterly basis, and users are notified when updates occur.
4	Where are my projects and estimates stored?	Upon installation, a "My CostWorks Projects" folder is created within the "My Documents" directory. All projects are defaulted to save here unless otherwise specified.	Estimates are stored on our secure server.
5	Does my IT team need to assist me in installing and running the program?	Yes, depending on the level of your permission. See #11 for system requirements.	While the product is entirely Web-based, you may need IT support with permissions for certain tasks. See #11 for system requirements.
6	Can users collaborate on projects/estimates?	Yes, you can copy project folders between computers.	You can create shared folders that can be viewed, copied, and edited by anyone within your account workgroup.

RSMeans Online

Unit prices according to the latest MasterFormat!

685

Seminar Schedule and Professional Development

Comparison matrix continued

7	Can I create and use my own data with the RSMeans database?	Custom lines cannot be added to the CD database. With the Estimator tool, however, you can create and add custom cost lines to any estimate. You can also save custom lines as an Estimator template for use in creating new estimates.	You can create • a custom line within an estimate • a custom cost book consisting of custom unit (not assembly) cost lines, no more than once per quarter **Note** that custom data is searchable using the Include My Custom Data feature.
8	Can I enter my own trade/labor rates and markups?	Although you cannot adjust the trade labor rates, you can add custom markups for each report template using the Estimator tool.	For each quarterly update, you can create one custom unit cost book. You can edit the labor rates by individual trade and the crew equipment costs, and also edit the associated material and equipment markups.
9	What are my reporting options?	The Estimator tool lets you create editable custom reports and report templates. Reports can also be exported to MS Excel®.	You can generate pre-formatted estimate summaries and cost lists in both PDF and Excel formats.
10	How do I find specific cost data?	You can search by CSI line number and by keyword within a division.	You can search by CSI line number and by keyword within a division. There are several additional search options, including keyword and advanced search using special characters.
11	What are the system requirements?	• Operating system – Windows XP or later • PC requirements - 256 MB RAM (recommended), 1024 x 768 High Color Monitor, 800MHz Computer (recommended), 24X CD-ROM, 250 MB Hard Disk Space	• Resolution – Recommended resolution is 1024 x 768 or higher • Compatibility - Internet Explorer version 8.x, 9.x and Mozilla Firefox version 10 and higher • iPad - iOS 5 and higher

2015 RSMeans Seminar Schedule

Note: call for exact dates and details.

Location	Dates	Location	Dates
Seattle, WA	January and August	Jacksonville FL	September
Dallas/Ft. Worth, TX	January	Dallas, TX	September
Austin, TX	February	Charleston, SC	October
Anchorage, AK	March and September	Houston, TX	October
Las Vegas, NV	March and October	Atlanta, GA	November
Washington, DC	April and September	Baltimore, MD	November
Phoenix, AZ	April	Orlando, FL	November
Kansas City, MO	April	San Diego, CA	December
Toronto	May	San Antonio, TX	December
Denver, CO	May	Raleigh, NC	December
San Francisco, CA	June		
Bethesda, MD	June		
Columbus, GA	June		
El Segundo, CA	August		

1-800-334-3509, Press 1

Professional Development

eLearning Training Sessions

Learn how to use *RSMeans Online*® or *RSMeans CostWorks*® CD from the convenience of your home or office. Our eLearning training sessions let you join a training conference call and share the instructors' desktops, so you can view the presentation and step-by-step instructions on your own computer screen. The live webinars are held from 9 a.m. to 4 p.m. eastern standard time, with a one-hour break for lunch.

For these sessions, you must have a computer with high speed Internet access and a compatible Web browser. Learn more at www.rsmeansonline.com or call for a schedule: 1-800-334-3509 and press 1. Webinars are generally held on selected Wednesdays each month.

RSMeans Online® $299 per person

RSMeans CostWorks® CD $299 per person

RSMeans data titles are also available in online format! Go to RSMeansOnline.com for more details.

For more information visit the RSMeans website at **www.rsmeans.com**

RSMeans Online™ Training

Construction estimating is vital to the decision-making process at each state of every project. RSMeansOnline works the way you do. It's systematic, flexible and intuitive. In this one day class you will see how you can estimate any phase of any project faster and better.

Some of what you'll learn:
- Customizing RSMeansOnline
- Making the most of RSMeans "Circle Reference" numbers
- How to integrate your cost data
- Generate reports, exporting estimates to MS Excel, sharing, collaborating and more

Also available: RSMeans Online® training webinar

Facilities Construction Estimating

In this *two-day* course, professionals working in facilities management can get help with their daily challenges to establish budgets for all phases of a project.

Some of what you'll learn:
- Determining the full scope of a project
- Identifying the scope of risks and opportunities
- Creative solutions to estimating issues
- Organizing estimates for presentation and discussion
- Special techniques for repair/remodel and maintenance projects
- Negotiating project change orders

Who should attend: facility managers, engineers, contractors, facility tradespeople, planners, and project managers.

Construction Cost Estimating: Concepts and Practice

This *one-day* introductory course to improve estimating skills and effectiveness starts with the details of interpreting bid documents, ending with the summary of the estimate and bid submission.

Some of what you'll learn:
- Using the plans and specifications for creating estimates
- The takeoff process—deriving all tasks with correct quantities
- Developing pricing, using various sources; how subcontractor pricing fits in
- Summarizing the estimate to arrive at the final number
- Formulas for area and cubic measure, adding waste and adjusting productivity to specific projects
- Evaluating subcontractors' proposals and prices
- Adding Insurance and bonds
- Understanding how labor costs are calculated
- Submitting bids and proposals

Who should attend: project managers, architects, engineers, owner's representatives, contractors, and anyone who's responsible for budgeting or estimating construction projects.

Maintenance & Repair Estimating for Facilities

This *two-day* course teaches attendees how to plan, budget, and estimate the cost of ongoing and preventive maintenance and repair for existing buildings and grounds.

Some of what you'll learn:
- The most financially favorable maintenance, repair, and replacement scheduling and estimating
- Auditing and value engineering facilities
- Preventive planning and facilities upgrading
- Determining both in-house and contract-out service costs
- Annual, asset-protecting M&R plan

Who should attend: facility managers, maintenance supervisors, buildings and grounds superintendents, plant managers, planners, estimators, and others involved in facilities planning and budgeting.

Practical Project Management for Construction Professionals

In this *two-day* course, acquire the essential knowledge and develop the skills to effectively and efficiently execute the day-to-day responsibilities of the construction project manager.

Covers:
- General conditions of the construction contract
- Contract modifications: change orders and construction change directives
- Negotiations with subcontractors and vendors
- Effective writing: notification and communications
- Dispute resolution: claims and liens

Who should attend: architects, engineers, owner's representatives, project managers.

Mechanical & Electrical Estimating

This *two-day* course teaches attendees how to prepare more accurate and complete mechanical/electrical estimates, avoiding the pitfalls of omission and double-counting, while understanding the composition and rationale within the RSMeans mechanical/electrical database.

Some of what you'll learn:
- The unique way mechanical and electrical systems are interrelated
- M&E estimates—conceptual, planning, budgeting, and bidding stages
- Order of magnitude, square foot, assemblies, and unit price estimating
- Comparative cost analysis of equipment and design alternatives

Who should attend: architects, engineers, facilities managers, mechanical and electrical contractors, and others who need a highly reliable method for developing, understanding, and evaluating mechanical and electrical contracts.

Professional Development

RSMeans data titles are also available in online format! Go to RSMeansOnline.com for more details.

For more information visit the RSMeans website at **www.rsmeans.com**

Unit Price Estimating

This interactive *two-day* seminar teaches attendees how to interpret project information and process it into final, detailed estimates with the greatest accuracy level.

The most important credential an estimator can take to the job is the ability to visualize construction and estimate accurately.

Some of what you'll learn:
- Interpreting the design in terms of cost
- The most detailed, time-tested methodology for accurate pricing
- Key cost drivers—material, labor, equipment, staging, and subcontracts
- Understanding direct and indirect costs for accurate job cost accounting and change order management

Who should attend: corporate and government estimators and purchasers, architects, engineers, and others who need to produce accurate project estimates.

RSMeans CostWorks® CD Training

This one-day course helps users become more familiar with the functionality of *RSMeans CostWork*s program. Each menu, icon, screen, and function found in the program is explained in depth. Time is devoted to hands-on estimating exercises.

Some of what you'll learn:
- Searching the database using all navigation methods
- Exporting RSMeans data to your preferred spreadsheet format
- Viewing crews, assembly components, and much more
- Automatically regionalizing the database

This training session requires you to bring a laptop computer to class.

When you register for this course you will receive an outline for your laptop requirements.

Also offering web training for RSMeans CostWorks CD!

Facilities Estimating Using RSMeans CostWorks® CD

Combines hands-on skill building with best estimating practices and real-life problems. Brings you up-to-date with key concepts, and provides tips, pointers, and guidelines to save time and avoid cost oversights and errors.

Some of what you'll learn:
- Estimating process concepts
- Customizing and adapting RSMeans cost data
- Establishing scope of work to account for all known variables
- Budget estimating: when, why, and how
- Site visits: what to look for—what you can't afford to overlook
- How to estimate repair and remodeling variables

This training session requires you to bring a laptop computer to class.

Who should attend: facility managers, architects, engineers, contractors, facility tradespeople, planners, project managers and anyone involved with JOC, SABRE, or IDIQ.

Conceptual Estimating Using RSMeans CostWorks® CD

This *two-day* class uses the leading industry data and a powerful software package to develop highly accurate conceptual estimates for your construction projects. All attendees must bring a laptop computer loaded with the current year *Square Foot Costs* and the *Assemblies Cost Data* CostWorks titles.

Some of what you'll learn:
- Introduction to conceptual estimating
- Types of conceptual estimates
- Helpful hints
- Order of magnitude estimating
- Square foot estimating
- Assemblies estimating

Who should attend: architects, engineers, contractors, construction estimators, owner's representatives, and anyone looking for an electronic method for performing square foot estimating.

Assessing Scope of Work for Facility Construction Estimating

This *two-day* practical training program addresses the vital importance of understanding the SCOPE of projects in order to produce accurate cost estimates for facility repair and remodeling.

Some of what you'll learn:
- Discussions of site visits, plans/specs, record drawings of facilities, and site-specific lists
- Review of CSI divisions, including means, methods, materials, and the challenges of scoping each topic
- Exercises in SCOPE identification and SCOPE writing for accurate estimating of projects
- Hands-on exercises that require SCOPE, take-off, and pricing

Who should attend: corporate and government estimators, planners, facility managers, and others who need to produce accurate project estimates.

Unit Price Estimating Using RSMeans CostWorks® CD

Step-by-step instruction and practice problems to identify and track key cost drivers—material, labor, equipment, staging, and subcontractors—for each specific task. Learn the most detailed, time-tested methodology for accurately "pricing" these variables, their impact on each other and on total cost.

Some of what you'll learn:
- Unit price cost estimating
- Order of magnitude, square foot, and assemblies estimating
- Quantity takeoff
- Direct and indirect construction costs
- Development of contractor's bill rates
- How to use *RSMeans Building Construction Cost Data*

This training session requires you to bring a laptop computer to class.

Who should attend: architects, engineers, corporate and government estimators, facility managers, and government procurement staff.

For more information visit the RSMeans website at **www.rsmeans.com**

Registration Information

Register early . . . Save up to $100! Register 30 days before the start date of a seminar and save $100 off your total fee. *Note: This discount can be applied only once per order. It cannot be applied to team discount registrations or any other special offer.*

How to register Register by phone today! The RSMeans toll-free number for making reservations is **1-800-334-3509, Press 1.**

Two-day seminar registration fee - $935. One-day *RSMeans CostWorks*® training registration fee - $375. To register by mail, complete the registration form and return, with your full fee, to: RSMeans Seminars, 700 Longwater Drive, Norwell, MA 02061.

Government pricing All federal government employees save off the regular seminar price. Other promotional discounts cannot be combined with the government discount.

Team discount program for two to four seminar registrations. Call for pricing: 1-800-334-3509, Press 1

Multiple course discounts When signing up for two or more courses, call for pricing.

Refund policy Cancellations will be accepted up to ten business days prior to the seminar start. There are no refunds for cancellations received later than ten working days prior to the first day of the seminar. A $150 processing fee will be applied for all cancellations. Written notice of cancellation is required. Substitutions can be made at any time before the session starts. **No-shows are subject to the full seminar fee.**

AACE approved courses Many seminars described and offered here have been approved for 14 hours (1.4 recertification credits) of credit by the AACE International Certification Board toward meeting the continuing education requirements for recertification as a Certified Cost Engineer/ Certified Cost Consultant.

AIA Continuing Education We are registered with the AIA Continuing Education System (AIA/CES) and are committed to developing quality learning activities in accordance with the CES criteria. Many seminars meet the AIA/CES criteria for Quality Level 2. AIA members may receive (14) learning units (LUs) for each two-day RSMeans course.

Daily course schedule The first day of each seminar session begins at 8:30 a.m. and ends at 4:30 p.m. The second day begins at 8:00 a.m. and ends at 4:00 p.m. Participants are urged to bring a hand-held calculator, since many actual problems will be worked out in each session.

Continental breakfast Your registration includes the cost of a continental breakfast, and a morning and afternoon refreshment break. These informal segments allow you to discuss topics of mutual interest with other seminar attendees. (You are free to make your own lunch and dinner arrangements.)

Hotel/transportation arrangements RSMeans arranges to hold a block of rooms at most host hotels. To take advantage of special group rates when making your reservation, be sure to mention that you are attending the RSMeans seminar. You are, of course, free to stay at the lodging place of your choice. (**Hotel reservations and transportation arrangements should be made directly by seminar attendees.**)

Important Class sizes are limited, so please register as soon as possible.

Note: Pricing subject to change.

Registration Form

ADDS-1000

Call 1-800-334-3509, Press 1 to register or FAX this form 1-800-632-6732. Visit our website: www.rsmeans.com

Please register the following people for the RSMeans construction seminars as shown here. We understand that we must make our own hotel reservations if overnight stays are necessary.

Full payment of $_____enclosed.

Bill me.

Please print name of registrant(s).

(To appear on certificate of completion)

_____

_____

_____

P.O. #:_____
GOVERNMENT AGENCIES MUST SUPPLY PURCHASE ORDER NUMBER OR TRAINING FORM.

Please mail check to: RSMeans Seminars, 700 Longwater Drive, Norwell, MA 02061 USA

Firm name_____

Address_____

City/State/Zip_____

Telephone no._____ Fax no._____

E-mail address_____

Charge registration(s) to: ☐ MasterCard ☐ VISA ☐ American Express

Account no._____ Exp. date_____

Cardholder's signature_____

Seminar name_____

Seminar City _____

Professional Development

689

RSMeans data titles are also available in online format: Go to RSMeansOnline.com for more details.

For more information visit the RSMeans website at www.rsmeans.com

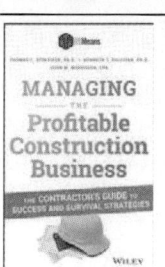

Managing the Profitable Construction Business: The Contractor's Guide to Success and Survival Strategies, 2nd Ed.

by Thomas C. Scheifler, Ph.D.

Learn from a team of construction business veterans led by Thomas C. Schleifer, who is commonly referred to as a construction business "turnaround" expert due to the number of construction companies he has rescued from financial distress. His financial acumen, combined with his practical, hands-on experience, has made him a sought-after private consultant.

$60.00 | 288 pages, hardover | Catalog no. 67370

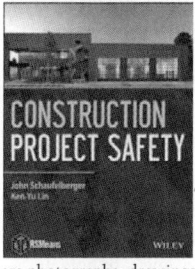

Construction Project Safety

by John Schaufelberger, Ken-Yu Lin

This essential introduction to construction safety for construction management personnel takes a project-based approach to present potential hazards in construction and their mitigation or prevention.

Beginning with an introduction to Accident Prevention Programs and OSHA compliance requirements, the book integrates safety instruction into the building process by following a building project from site construction through interior finish. Reinforcing this applied approach are photographs, drawings, contract documentation, and an online 3D BIM model to help visualize the on-site scenarios.

$85.00 | 320 pages, hardover | Catalog no. 67368

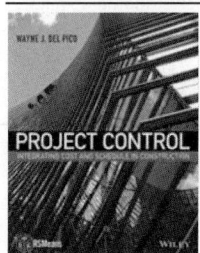

Project Control: Integrating Cost & Schedule in Construction

by Wayne J. Del Pico

Written by a seasoned professional in the field, *Project Control: Integrating Cost and Schedule in Construction* fills a void in the area of project control as applied in the construction industry today. It demonstrates how productivity models for an individual project are created, monitored, and controlled, and how corrective actions are implemented as deviations from the baseline occur.

$65.00 | 240 pages, softcover | Catalog no. 67366

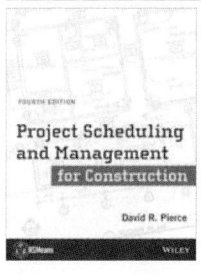

Project Scheduling and Management for Construction, 4th Edition

by David R. Pierce

First published in 1988 by RSMeans, the new edition of *Project Scheduling and Management for Construction* has been substantially revised for both professionals and students enrolled in construction management and civil engineering programs. While retaining its emphasis on developing practical, professional-level scheduling skills, the new edition is a relatable, real world case study.

$95.00 | 272 pages, softcover | Catalog no. 67367

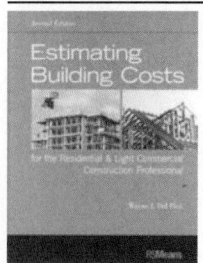

Estimating Building Costs, 2nd Edition
For the Residential & Light Commercial Construction Professional

by Wayne J. Del Pico

Explains, in detail, how to put together a reliable estimate that can be used not only for budgeting, but also for developing a schedule, managing a project, dealing with contingencies, and, ultimately, making a profit.

Completely revised and updated to reflect the CSI MasterFormat 2010 system.

$75.00 | 528 pages, softcover | Catalog no. 67343A

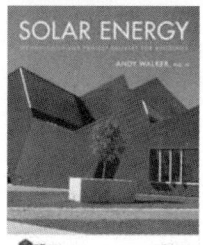

The Gypsum Construction Handbook, 7th Edition

by USG

The tried-and-true Gypsum Construction Handbook is a systematic guide to selecting and using gypsum drywall, veneer plaster, tile backers, ceilings, and conventional plaster building materials. A widely respected training text for aspiring architects and engineers, the book provides detailed product information and efficient installation methodology.

The 7th edition features updates in gypsum products, including ultralight panels, glass-mat panels, paper-faced plastic bead, and ultralightweight joint compound, and modern specialty acoustical and ceiling product guidelines. This comprehensive reference also incorporates the latest in sustainable products.

$40.00 | 576 pages, softcover | Catalog no. 67357B

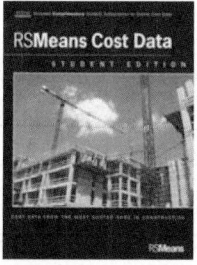

Solar Energy: Technologies & Project Delivery for Buildings

by Andy Walker

An authoritative reference on the design of solar energy systems in building projects, with applications, operating principles, and simple tools for the construction, engineering, and design professionals, the book simplifies the solar design and engineering process and provides sample documentation for the complete design of a solar energy system for buildings.

$85.00 | 320 pages, hardover | Catalog no. 67365

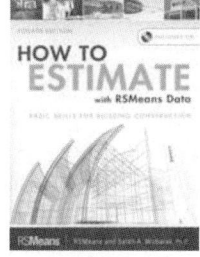

RSMeans Cost Data, Student Edition

Provides a thorough introduction to cost estimating in a print and online package. Clear explanations and example-driven. The ideal reference for students and new professionals. Features include:

- Commercial and residential construction cost data in print and online formats
- Complete how-to guidance on the essentials of cost estimating
- A supplemental website with plans, problem sets, and a full sample estimate

$110.00 | 512 pages, softcover | Catalog no. 67363

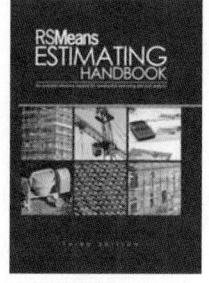

How to Estimate with Means Data & CostWorks, 4th Edition

by RSMeans and Saleh A. Mubarak, Ph.D.

This step-by-step guide takes you through all the major construction items with extensive coverage of site work, concrete and masonry, wood and metal framing, doors and windows, and other divisions. The only construction cost estimating handbook that uses the most popular source of construction data, RSMeans, this indispensible guide features access to the instructional version of *CostWorks* in electronic form, enabling you to practice techniques to solve real-world estimating problems.

$75.00 | 320 pages, softcover | Includes CostWorks CD | Catalog no. 67324C

RSMeans Estimating Handbook, 3rd Edition

Widely used in the industry for tasks ranging from routine estimates to special cost analysis projects. This handbook will help construction professionals:

- Evaluate architectural plans and specifications
- Prepare accurate quantity takeoffs
- Compare design alternatives and costs
- Perform value engineering
- Double-check estimates and quotes
- Estimate change orders

$110.00 | Over 976 pages, hardcover | Catalog No. 67276B

Reference Books

690

For more information visit the RSMeans website at **www.rsmeans.com**

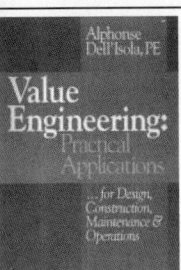

Value Engineering: Practical Applications
For Design, Construction, Maintenance & Operations
by Alphonse Dell'Isola, PE

A tool for immediate application—for engineers, architects, facility managers, owners, and contractors. Includes making the case for VE—the management briefing; integrating VE into planning, budgeting, and design; conducting life cycle costing; using VE methodology in design review and consultant selection; and case studies.

$90.00 | Over 427 pages, illustrated, softcover | Catalog no. 67319A

Risk Management for Design and Construction

Introduces risk as a central pillar of project management and shows how a project manager can prepare to deal with uncertainty. Includes:

- Integrated cost and schedule risk analysis
- An introduction to a ready-to-use system of analyzing a project's risks and tools to proactively manage risks
- A methodology that was developed and used by the Washington State Department of Transportation
- Case studies and examples on the proper application of principles
- Combining value analysis with risk analysis

$125.00 | Over 288 pages, softcover | Catalog no. 67359

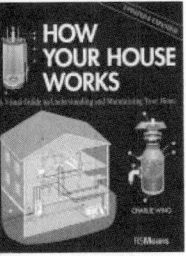

How Your House Works, 2nd Edition
by Charlie Wing

Knowledge of your home's systems helps you control repair and construction costs and makes sure the correct elements are being installed or replaced. This book uncovers the mysteries behind just about every major appliance and building element in your house. See-through, cross-section drawings in full color show you exactly how these things should be put together and how they function, including what to check if they don't work. It just might save you having to call in a professional.

$22.95 | 208 pages, softcover | Catalog no. 67351A

Construction Business Management
by Nick Ganaway

Only 43% of construction firms stay in business after four years. Make sure your company thrives with valuable guidance from a pro with 25 years of success as a commercial contractor. Find out what it takes to build all aspects of a business that is profitable, enjoyable, and enduring. With a bonus chapter on retail construction.

$60.00 | 201 pages, softcover | Catalog no. 67352

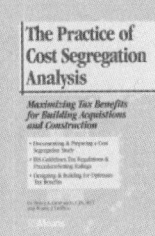

The Practice of Cost Segregation Analysis
by Bruce A. Desrosiers and Wayne J. Del Pico

This expert guide walks you through the practice of cost segregation analysis, which enables property owners to defer taxes and benefit from "accelerated cost recovery" through depreciation deductions on assets that are properly identified and classified.

With a glossary of terms, sample cost segregation estimates for various building types, key information resources, and updates via a dedicated website, this book is a critical resource for anyone involved in cost segregation analysis.

$105.00 | Over 240 pages | Catalog no. 67345

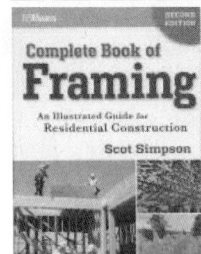

Complete Book of Framing, 2nd Edition
by Scot Simpson

This updated, easy-to-learn guide to rough carpentry and framing is written by an expert with more than thirty years of framing experience. Starting with the basics, this book begins with types of lumber, nails, and what tools are needed, followed by detailed, fully illustrated steps for framing each building element. Framer-Friendly Tips throughout the book show how to get a task done right—and more easily.

$29.95 | 368 pages, softcover | Catalog no. 67353A

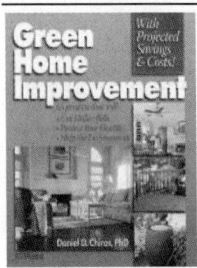

Green Home Improvement
by Daniel D. Chiras, PhD

With energy costs rising and environmental awareness increasing, people are looking to make their homes greener. This book, with 65 projects and actual costs and projected savings, helps homeowners prioritize their green improvements.

Projects range from simple water savers that cost only a few dollars, to bigger-ticket items such as HVAC systems. With color photos and cost estimates, each project compares options and describes the work involved, the benefits, and the savings.

$34.95 | 320 pages, illustrated, softcover | Catalog no. 67355

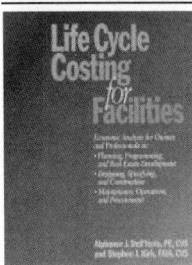

Life Cycle Costing for Facilities
by Alphonse Dell'Isola and Dr. Stephen Kirk

Guidance for achieving higher quality design and construction projects at lower costs! Cost-cutting efforts often sacrifice quality to yield the cheapest product. Life cycle costing enables building designers and owners to achieve both. The authors of this book show how LCC can work for a variety of projects—from roads to HVAC upgrades to different types of buildings.

$110.00 | 396 pages, hardcover | Catalog no. 67341

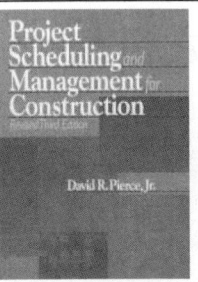

Project Scheduling and Management for Construction, 3rd Edition
by David R. Pierce, Jr.

A comprehensive, yet easy-to-follow guide to construction project scheduling and control—from vital project management principles through the latest scheduling, tracking, and controlling techniques. The author is a leading authority on scheduling, with years of field and teaching experience at leading academic institutions. Spend a few hours with this book and come away with a solid understanding of this essential management topic.

$64.95 | Over 286 pages, illustrated, hardcover | Catalog no. 67247B

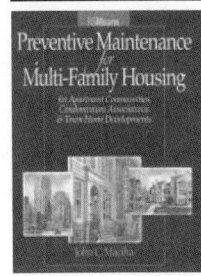

Preventive Maintenance for Multi-Family Housing
by John C. Maciha

Prepared by one of the nation's leading experts on multi-family housing.

This complete PM system for apartment and condominium communities features expert guidance, checklists for buildings and grounds maintenance tasks and their frequencies, and a dedicated website featuring customizable electronic forms. A must-have for anyone involved with multi-family housing maintenance and upkeep.

$95.00 | 290 pages | Catalog no. 67346

Reference Books

691

RSMeans data titles are also available in online format! Go to RSMeansOnline.com for more details.

For more information visit the RSMeans website at **www.rsmeans.com**

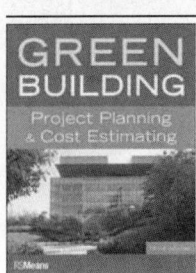

Green Building: Project Planning & Cost Estimating, 3rd Edition

Since the widely read first edition of this book, green building has gone from a growing trend to a major force in design and construction.

This new edition has been updated with the latest in green building technologies, design concepts, standards, and costs. Full-color with all new case studies—plus a new chapter on commercial real estate.

$110.00 | 480 pages, softcover | Catalog no. 67338B

Construction Supervision

Inspires excellence with proven tactics and techniques applied by thousands of construction supervisors. Leadership guidelines carve out a practical blueprint for motivating work performance and increasing productivity through effective communication. Features:

• A unique focus on field supervision and crew management

• Coverage of supervision from the foreman to the superintendent level

• An overview of technical skills whose mastery will build confidence and success for the supervisor

• A detailed view of "soft" management and communication skills

$95.00 | 464 pages, softcover | Catalog no. 67358

The Homeowner's Guide to Mold
By Michael Pugliese

Mold, whether caused by leaks, humidity or flooding, is a real health and financial issue—for homeowners and contractors. This full-color book explains:

• Construction and maintenance practices to prevent mold

• How to inspect for and remove mold

• Mold remediation procedures and costs

• What to do after a flood

• How to deal with insurance companies

$21.95 | 144 pages, softcover | Catalog no. 67344

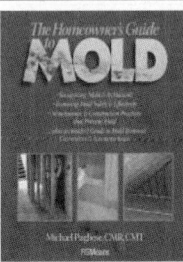

Landscape Estimating Methods, 5th Edition

Answers questions about preparing competitive landscape construction estimates, with up-to-date cost estimates and the new MasterFormat classification system. Expanded and revised to address the latest materials and methods, including new coverage on approaches to green building. Includes:

• Step-by-step explanation of the estimating process

• Sample forms and worksheets that save time and prevent errors

$75.00 | 336 pages, softcover | Catalog no. 67295C

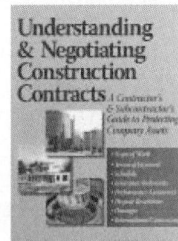

Understanding & Negotiating Construction Contracts
by Kit Werremeyer

Take advantage of the author's 30 years' experience in small-to-large (including international) construction projects. Learn how to identify, understand, and evaluate high risk terms and conditions typically found in all construction contracts—then, negotiate to lower or eliminate the risk, improve terms of payment, and reduce exposure to claims and disputes. The author avoids "legalese" and gives real-life examples from actual projects.

$80.00 | 320 pages, softcover | Catalog no. 67350

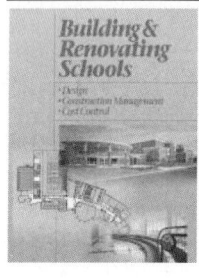

RSMeans Illustrated Construction Dictionary Unabridged 4th Edition, with CD-ROM

Long regarded as the industry's finest, *Means Illustrated Construction Dictionary* is now even better. With nearly 20,000 terms and more than 1,400 illustrations and photos, it is the clear choice for the most comprehensive and current information. The companion CD-ROM that comes with this new edition adds extra features, such as larger graphics and expanded definitions.

$110.00 | Over 880 pages, illust., hardcover | Catalog no. 67292B

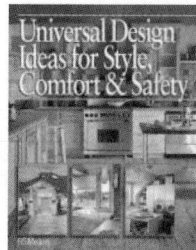

Building & Renovating Schools

This all-inclusive guide covers every step of the school construction process—from initial planning, needs assessment, and design, right through moving into the new facility. A must-have resource for anyone concerned with new school construction or renovation. With square foot cost models for elementary, middle, and high school facilities, and real-life case studies of recently completed school projects.

The contributors to this book—architects, construction project managers, contractors, and estimators who specialize in school construction—provide start-to-finish, expert guidance on the process.

$110.00 | Over 412 pages, hardcover | Catalog no. 67342

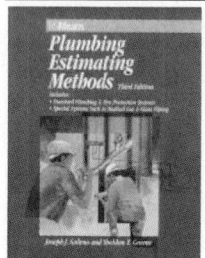

Universal Design Ideas for Style, Comfort & Safety
by RSMeans and Lexicon Consulting, Inc.

Incorporating universal design when building or remodeling helps people of any age and physical ability more fully and safely enjoy their living spaces. This book shows how universal design can be artfully blended into the most attractive homes. It discusses specialized products like adjustable countertops and chair lifts, as well as simple ways to enhance a home's safety and comfort. With color photos and expert guidance, every area of the home is covered. Includes budget estimates that give an idea how much projects will cost.

$21.95 | 160 pages, illustrated, softcover | Catalog no. 67354

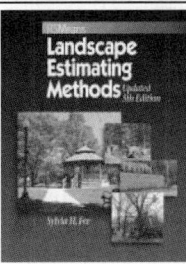

Plumbing Estimating Methods, 3rd Edition
by Joseph J. Galeno and Sheldon T. Greene

Updated and revised! This practical guide walks you through a plumbing estimate, from basic materials and installation methods through change order analysis. *Plumbing Estimating Methods* covers residential, commercial, industrial, and medical systems, and features sample takeoff and estimate forms and detailed illustrations of systems and components.

$65.00 | 381+ pages, softcover | Catalog no. 67283B

Reference Books

2015 Order Form
ORDER TOLL FREE 1-800-334-3509 OR FAX 1-800-632-6732

For more information visit the RSMeans website at www.rsmeans.com

Qty.	Book no.	COST ESTIMATING BOOKS	Unit Price	Total
	60065	Assemblies Cost Data 2015	$334.95	
	60015	Building Construction Cost Data 2015	206.95	
	60045	Commercial Renovaion Cost Data 2015	166.95	
	60115	Concrete & Masonry Cost Data 2015	188.95	
	50145	Construction Cost Indexes 2015 (subscription)	359.95	
	60145A	Construction Cost Index–January 2015	94.95	
	60145B	Construction Cost Index–April 2015	94.95	
	60145C	Construction Cost Index–July 2015	94.95	
	60145D	Construction Cost Index–October 2015	94.95	
	60345	Contr. Pricing Guide: Resid. R & R Costs 2015	39.95	
	60235	Electrical Change Order Cost Data 2015	199.95	
	60035	Electrical Cost Data 2015	208.95	
	60205	Facilities Construction Cost Data 2015	524.95	
	60305	Facilities Maintenance & Repair Cost Data 2015	461.95	
	60555	Green Building Cost Data 2015	170.95	
	60165	Heavy Construction Cost Data 2015	204.95	
	60095	Interior Cost Data 2015	208.95	
	60125	Labor Rates for the Const. Industry 2015	450.95	
	60185	Light Commercial Cost Data 2015	153.95	
	60025	Mechanical Cost Data 2015	203.95	
	60155	Open Shop Building Const. Cost Data 2015	175.95	
	60215	Plumbing Cost Data 2015	206.95	
	60175	Residential Cost Data 2015	148.95	
	60285	Site Work & Landscape Cost Data 2015	199.95	
	60055	Square Foot Costs 2015	218.95	
	62014	Yardsticks for Costing (2014)	189.95	
	62015	Yardsticks for Costing (2015)	201.95	
		REFERENCE BOOKS		
	67329	Bldrs Essentials: Best Bus. Practices for Bldrs	45.00	
	67307	Bldrs Essentials: Plan Reading & Takeoff	50.00	
	67342	Building & Renovating Schools	110.00	
	67261A	The Building Prof. Guide to Contract Documents	75.00	
	67353A	Complete Book of Framing, 2nd Ed.	29.95	
	67146	Concrete Repair & Maintenance Illustrated	75.00	
	67352	Construction Business Management	60.00	
	67364	Construction Law	115.00	
	67358	Construction Supervision	95.00	
	67363	Cost Data Student Edition	110.00	
	67314	Cost Planning & Est. for Facil. Maint.	95.00	
	67230B	Electrical Estimating Methods, 3rd Ed.	75.00	
	67343A	Est. Bldg. Costs for Resi. & Lt. Comm., 2nd Ed.	75.00	
	67276B	Estimating Handbook, 3rd Ed.	110.00	
	67318	Facilities Operations & Engineering Reference	115.00	
	67338B	Green Building: Proj. Planning & Cost Est., 3rd Ed.	110.00	
	67355	Green Home Improvement	34.95	
	67357B	The Gypsum Construction Handbook	40.00	
	67148	Heavy Construction Handbook	110.00	
	67308E	Home Improvement Costs–Int. Projects, 9th Ed.	24.95	
	67309E	Home Improvement Costs–Ext. Projects, 9th Ed.	24.95	
	67344	Homeowner's Guide to Mold	21.95	
	67324C	How to Est.w/Means Data & CostWorks, 4th Ed.	75.00	

Qty.	Book no.	REFERENCE BOOKS (Cont.)	Unit Price	Total
	67351A	How Your House Works, 2nd Ed.	22.95	
	67282A	Illustrated Const. Dictionary, Condensed, 2nd Ed.	65.00	
	67292B	Illustrated Const. Dictionary, w/CD-ROM, 4th Ed.	110.00	
	67362	Illustrated Const. Dictionary, Student Ed.	60.00	
	67348	Job Order Contracting	130.00	
	67295C	Landscape Estimating Methods, 5th Ed.	75.00	
	67341	Life Cycle Costing for Facilities	110.00	
	67370	Managing the Profitable Construction Business	60.00	
	67294B	Mechanical Estimating Methods, 4th Ed.	75.00	
	67345	Practice of Cost Segregation Analysis	105.00	
	67366	Project Control: Integrating Cost & Sched.	95.00	
	67346	Preventive Maint. for Multi-Family Housing	95.00	
	67367	Project Sched. and Manag. for Constr., 4th Ed.	95.00	
	67322B	Resi. & Light Commercial Const. Stds., 3rd Ed.	65.00	
	67359	Risk Management for Design and Construction	125.00	
	67365	Solar Energy: Technology & Project Delivery	85.00	
	67145B	Sq. Ft. & Assem. Estimating Methods, 3rd Ed.	75.00	
	67321	Total Productive Facilities Management	85.00	
	67350	Understanding and Negotiating Const. Contracts	80.00	
	67303B	Unit Price Estimating Methods, 4th Ed.	75.00	
	67354	Universal Design	21.95	
	67319A	Value Engineering: Practical Applications	90.00	

MA residents add 6.25% state sales tax [____]

Shipping & Handling** [____]

Total (U.S. Funds)* [____]

Prices are subject to change and are for U.S. delivery only. *Canadian customers may call for current prices. **Shipping & handling charges: Add 8% of total order for check and credit card payments. Add 8% of total order for invoiced orders.

Send order to: ADDV-1000

Name (please print) _____

Company _____

☐ Company

☐ Home Address _____

_____

City/State/Zip _____

Phone # _____ P.O. # _____

(Must accompany all orders being billed)

Mail to: RSMeans, 700 Longwater Drive, Norwell, MA 02061

693

RSMeans Project Cost Report

By filling out this report, your project data will contribute to the database that supports the RSMeans® Project Cost Square Foot Data. When you fill out this form, RSMeans will provide a $50 discount off one of the RSMeans products advertised in the preceding pages. Please complete the form including all items where you have cost data, and all the items marked (☑).

$50.00 Discount per product for each report you submit

Project Description (NEW construction only, please)

☑ Building Use (Office School . . .) _____

☑ Address (City, State) _____

_____

☑ Total Building Area (SF) _____

☑ Ground Floor (SF) _____

☑ Frame (Wood, Steel . . .) _____

☑ Exterior Wall (Brick, Tilt-up . . .) _____

☑ Basement: (check one)   ☐ Full   ☐ Partial   ☐ None

☑ Number of Stories _____

☑ Floor-to-Floor Height _____

☑ Volume (C.F.) _____

% Air Conditioned _____ Tons _____

Total Project Cost $ _____

Owner _____

Architect _____

General Contractor _____

☑ Bid Date _____

☑ Typical Bay Size _____

☑ Occupant Capacity _____

☑ Labor Force: _____ % Union _____ % Non-Union

☑ Project Description (Circle one number in each line.)

1. Economy   2. Average   3. Custom   4. Luxury

1. Square   2. Rectangular   3. Irregular   4. Very Irregular

Comments _____

_____

A	☑	General Conditions	$
B	☑	Site Work	$
C	☑	Concrete	$
D	☑	Masonry	$
E	☑	Metals	$
F	☑	Wood & Plastics	$
G	☑	Thermal & Moisture Protection	$
GR		Roofing & Flashing	$
H	☑	Doors and Windows	$
J	☑	Finishes	$
JP		Painting & Wall Covering	$

K	☑	Specialties	$
L	☑	Equipment	$
M	☑	Furnishings	$
N	☑	Special Construction	$
P	☑	Conveying Systems	$
Q	☑	Mechanical	$
QP		Plumbing	$
QB		HVAC	$
R	☑	Electrical	$
S	☑	Mech./Elec. Combined	$

Please specify the RSMeans product you wish to receive. Complete the address information.

Product Name _____

Product Number _____

Your Name _____

Title _____

Company _____

☐ Company
☐ Home   Street Address _____

City, State, Zip _____

Email Address _____

Return by mail or fax 888-492-6770.

Method of Payment:

Credit Card # _____

Expiration Date _____

Check _____

Purchase Order _____

Phone Number _____
(if you would prefer a sales call)

RSMeans
Square Foot Costs Department
700 Longwater Drive
Norwell, MA 02061
www.RSMeans.com